Computational Fluid and Solid Mechanics

Series Editor:

Klaus-Jürgen Bathe
Massachusetts Institute of Technology
Cambridge, MA, USA

Dimitris Drikakis · William Rider

High-Resolution Methods for Incompressible and Low-Speed Flows

With 480 Figures and 32 Tables

Springer

Authors:

Prof. Dr. Dimitris Drikakis
Cranfield University
School of Engineering
Dept. of Aerospace Science
MK43 0AL Cranfield
United Kingdom
E-mail: d.drikakis@cranfield.ac.uk

Dr. William Rider
Los Alamos National Laboratory
Computer and Computational
Sciences Division
Mail Stop D413
87545 Los Alamos
U.S.A
E-mail: wjr@lanl.gov

ISBN 3-540-22136-0 Springer Berlin Heidelberg New York

Library of Congress Control Number: 2004107327

Springer is a part of Springer Science + Business Media

springeronline.com

Printed in Germany

Typesetting: Camera ready by author
Cover-Design: Künkel & Lopka, Heidelberg / Design & Production GmbH, Heidelberg
Printed on acid free paper 62/3020/kk - 5 4 3 2 1 0

To my Children

Maria, Anastasia and Sophia

To my Family

Felicia, Rachel and Jackson

Preface by Frank Harlow

The flow of fluids is an important part of numerous scientific and engineering processes, ranging in scale from liquid-drop models of atomic nuclei to cosmic dimensions in the explosion of a star. A breath of air, a drink of water, a swim in the lake, the pulsing of blood in our veins, all involve fluid-flow processes with which we are intimately familiar.

Fluid flows can be classified in two basic categories, depending on the ratio of the material speed (relative to some structural boundary) to the speed of sound in the fluid; this ratio is called the Mach number. In mathematical descriptions of fluid dynamics there are important differences between the properties at very low Mach numbers (often called incompressible flow) and those that take place at Mach numbers above about 0.1. In this book, the emphasis is on flows that take place at very low Mach numbers, with a strong focus on the numerical techniques appropriate for solving problems with accuracy and efficiency on high-speed computers.

Many low-Mach-number fluid flows are essentially incompressible; water waves, hydraulic pumping systems, and air flow around fan blades are all good examples. Low-Mach-number flows can also be highly compressible, as, for example, in pumping air into a tire. The material can have negligible viscosity or such high viscosity as to creep slowly (a chunk of warm tar settling onto a table). The flow may be completely confined by adjacent solid surfaces, or it may have moving interfaces, for example, the dynamics of air around a falling rain drop. The interface may even be a diffusive surface across which the materials pass in both directions, as for example, with salt water adjacent to fresh water.

The earliest of the numerical techniques for solving fluid-flow problems on a big computer were directed to the investigation of relatively high Mach number flows. The challenge was to describe the shocks that can develop in the vicinity of intense energy sources. The idea was to approximate the continuous field of flow by a discrete subdivision of space into computational cells, resembling the approximation of a photograph by dots that can then be printed in a newspaper. In addition, the advancement of the solution through time was divided into finite intervals like the frames of a motion picture. The beauty of high-Mach-number fluid flows is that the results for each time cycle depend only on the data from adjacent cells at the previous time cycle.

For flows at low Mach numbers (the subject of this book), the situation is far more complicated in that the advancement of the solution each time step depends on data from everywhere in the overall domain of interest. Although the first successful techniques for accomplishing the solutions were quite primitive, they nevertheless solved for the first few real problems of interest, and they also served as a foundation for the evolution over the succeeding four decades of remarkably powerful capabilities that are beautifully demonstrated by the discussions in this book.

For the representation of real physical circumstances it is necessary to recognize that fluid flows are very often turbulent. This adds a major challenge to their mathematical and numerical description. The presence of turbulence in a fluid is usually described by either of two calculation techniques. One of these is direct numerical simulation, in which the finite-difference mesh of computation cells is fully three-dimensional and also fine enough to resolve all of the turbulence scales larger than a dissipation scale. The other is to develop turbulence transport equations designed to capture only the essence of what is going on, in close analogy to the way we represent the principal manifestations of molecular fluctuations by introducing the concepts of heat energy, temperature, density, pressure, and any other collective descriptors relevant to the problem at hand. In both of these turbulent-flow techniques, the numerical challenges are severe, as also discussed this book.

With the development of numerical techniques there has been a continuously widening scope of scientific and engineering problems that can be solved. The basic fluid-flow equations are based on the principles of mass, momentum, and energy conservation, but these must be supplemented with descriptions of the fluid stress-strain relations, typically including an equation of state, viscous or non-Newtonian stresses, the effects of plastic deformations, and non-equilibrium processes at high strain rates. In addition, many applications require the coupling of auxiliary physical processes, including phase transitions, energy sources from chemical or nuclear reactions, radiation transport, diffusion of species (for example salt in water), ionization, and the effects of non-inertial reference frames (producing, for example, centrifugal and Coriolis accelerations). Boundary effects such as those of variable surface tension (Marangoni flows) can be highly significant for some applications. A very important class of fluid flows is called multi-field, for which examples include dust or rain in the atmosphere and bubbles in boiling water.

A remarkable feature of this book is that discusses the current state-of-the-art numerical techniques for solving very hard problems with both accuracy and efficiency and presents a class of methods that can address many of the departures from "simple" fluid flows.

Frank Harlow,
Los Alamos, New Mexico, USA
July 2003.

Acknowledgements

Dimitris Drikakis and William J. Rider are particularly grateful to Frank Harlow (Los Alamos) for prefacing the book.

DD would like to acknowledge interaction and discussions with many individuals who contributed either directly or indirectly in developing ideas presented in this book. He would like to thank Tito Toro, Oleg Iliev, Daniela Vassileva and Piotr Smolarkiewicz for fruitful discussions and collaboration on various topics pertinent to the content of the book. DD is grateful to the useful feedback on the manuscript from Sergei Godunov and the various discussions they had on the subject of computational physics. Further, he would like to acknowledge Sergei Utyuzhnikov, Evgeniy Shapiro and George Papadakis for their comments on various chapters, as well as Marco Kalweit's and Jun Ma's help in the preparation of several illustrations.

The writing was completed whilst DD was a member of academic staff in various institutions. These include his tenure at the University of Manchester Institute of Science and Technology (UMIST); Queen Mary College, University of London; and Cranfield University. Parts of the book were also written while DD was a visiting professor at the University of Marseille (Université de la Méditerranée, France) and a visiting scholar at the Isaac Newton Institute (INI) for Mathematical Sciences, University of Cambridge. He is grateful to Tito Toro and Phillipe Le Floch for arranging his stay at INI, as well as to the staff of INI for their kind hospitality. He is also grateful to Daniel Favier and Eric Berton for kindly arranging his visit to the Laboratory of Aerodynamics and Biomechanics in Marseille. Stimulating discussions with many colleagues and collaborators at UMIST, University of London, as well as in the early stages of his academic career at the University of Erlangen-Nuremberg, Germany, and the National Technical University of Athens, have contributed to promote ideas and techniques presented in this book. Particularly, DD would like to thank his former teachers at NTUA for providing him with solid knowledge in several areas of fluid mechanics. DD's research efforts to implement certain methods and techniques presented in this book were partially funded by the Engineering and Physical Sciences Research Council (UK) and European Union.

Most importantly, DD would like to thank his wife Stavroula for her continuous encouragement during the writing of this book and for her enduring, unconditional support throughout his academic career. Without her encouragement and support the completion of this project would not be possible.

WJR would like to acknowledge the efforts and support of many individuals who each enabled to writing of this book. Among the many who he has worked with are Phil Colella and John Bell on approximate projection methods (especially for teaching him the fundamentals of the methods). During that period of time, WJR also found fruitful collaborations with Gerry Puckett and Doug Kothe which led to their joint work on volume of fluid methods and variable density flows. In the years following this, WJR collaborated effectively with Len Margolin on a number of subjects and profited from his support and advice. Under his guidance WJR's work with Dana Knoll on nonlinear methods was conducted.

This work is the culmination of many separate research projects spanning many years. First and foremost among these is the work on interface tracking, incompressible flows with Doug Kothe. Doug's support and encouragement have been fantastic. In his role as supervisor, Doug has provided him with the time and resources to accomplish the task of writing this book. Other individuals have provided assistance in the course of writing including John Turner, Ed Dendy, and James Kamm.

Financial support for WJR's efforts has been provided by Los Alamos National Laboratory under the auspices of the U.S. Department of Energy under Contract W-7405-ENG-36.

Lastly, but most importantly, WJR's family has been essential to his efforts. His wife Felicia's unwavering support of his work has provided him with a constant well of inspiration and the ability to complete this project. His children, Rachel and Jack, provided WJR with the portrait of youth and energy seemingly transferring their boundless energy and joy into his activities.

Dimitris Drikakis,
Essex, United Kingdom

and

William J. Rider
Los Alamos, New Mexico, USA

February 2004

Contents

1. **Introduction** 1

Part I. Fundamental Physical and Model Equations

2. **The Fluid Flow Equations** 7
   2.1 Mathematical Preliminaries 7
   2.2 Kinematic Considerations 9
   2.3 The Equations for Variable Density Flows 10
      2.3.1 The Continuity Equation 10
      2.3.2 The Momentum Equations 11
      2.3.3 The Energy Equation 14
   2.4 Compressible Euler Equations 16
   2.5 Low-Mach Number Scaling 20
   2.6 Boussinesq Approximation 23
   2.7 Variable Density Flow 23
   2.8 Zero Mach Number Combustion 24
   2.9 Initial and Boundary Conditions 25

3. **The Viscous Fluid Flow Equations** 27
   3.1 The Stress and Strain Tensors for a Newtonian Fluid 27
   3.2 The Navier-Stokes Equations for Constant Density Flows 31
   3.3 Non-Newtonian Constitutive Equations for the Shear-Stress Tensor 33
      3.3.1 Generalized Newtonian Fluids 33
      3.3.2 Viscoelastic Fluids 34
      3.3.3 Other Viscoelastic Models 37
   3.4 Alternative Forms of the Advective and Viscous Terms 38
   3.5 Nondimensionalization of the Governing Equations 39
   3.6 General Remarks on Turbulent Flow Simulations 42
   3.7 Reynolds-Averaged Navier-Stokes Equations (RANS) 43
   3.8 Large Eddy Simulation (LES) 47
   3.9 Closing Remarks 49

4. Curvilinear Coordinates and Transformed Equations 51
4.1 Generalized Curvilinear Coordinates 51
4.2 Calculation of Metrics 55
4.3 Transformation of the Fluid Flow Equations 57
4.4 Viscous Terms 60
4.5 Geometric Conservation Law 63

5. Overview of Various Formulations and Model Equations .. 67
5.1 Overview of Various Formulations of the Incompressible Flow Equations 67
5.1.1 Vorticity/Stream-Function Formulation 67
5.1.2 The Vorticity/Vector-Potential Formulation 69
5.1.3 Vorticity-Velocity Formulation 70
5.1.4 Pressure-Poisson Formulation 70
5.1.5 Projection Formulation 71
5.1.6 Artificial-Compressibility Formulation 71
5.1.7 Penalty Formulation 72
5.1.8 Hybrid Formulations 73
5.2 Model Equations 75
5.2.1 Advection-Diffusion Equation 75
5.2.2 Burgers' Equation 76

6. Basic Principles in Numerical Analysis 79
6.1 Stability, Consistency and Accuracy 79
6.2 Fourier Analysis 83
6.2.1 Fourier Analysis of First-Order Upwind 85
6.2.2 Fourier Analysis of Second-Order Upwind 86
6.3 Modified Equation Analysis 90
6.4 Verification via Sample Calculations 94

7. Time Integration Methods 99
7.1 Time Integration of the Flow Equations 99
7.2 Lax-Wendroff-Type Methods 100
7.3 Other Approaches to Time-Centering 102
7.4 Runge-Kutta Methods 103
7.4.1 Second-Order Runge-Kutta 104
7.4.2 Third-Order Runge-Kutta 106
7.4.3 Fourth-Order Runge-Kutta 107
7.4.4 TVD Runge-Kutta Methods Applied to Hyperbolic Conservation Laws 109
7.5 Linear Multi-step Methods 113
7.5.1 Adams-Bashforth Method 113
7.5.2 Adams-Moulton Method 116
7.5.3 Backward Differentiation Formulas 119

8. Numerical Linear Algebra 121
8.1 Basic Numerical Linear Algebra 121
8.2 Basic Relaxation Methods 123
8.3 Conjugate Gradient and Krylov Subspace Methods 126
8.4 Multigrid Algorithm for Elliptic Equations 130
8.5 Multigrid Algorithm as a Preconditioner for Krylov Subspace Methods 138
8.6 Newton's and Newton-Krylov Method 139
8.7 A Multigrid Newton-Krylov Algorithm 140

Part II. Solution Approaches

9. Compressible and Preconditioned-Compressible Solvers 147
9.1 Reconstructing the Dependent Variables 147
9.1.1 Riemann Solvers 148
9.1.2 Basic Predictor-Corrector 152
9.1.3 Characteristic Direct Eulerian 153
9.1.4 Lagrange-Remap Approach 155
9.2 Reconstructing the Fluxes 156
9.2.1 Flux Splitting 157
9.2.2 Flux Splitting Time Integration 158
9.3 Preconditioning for Low Speed Flows 160
9.3.1 Overview of Preconditioning Techniques 160
9.3.2 Preconditioning Choices for Compressible Flows 161
9.3.3 Preconditioning of Numerical Dissipation 167
9.3.4 Differential Preconditioners 169

10. The Artificial Compressibility Method 173
10.1 Basic Formulation 173
10.2 Convergence to the Incompressible Limit 174
10.3 Preconditioning and the Artificial Compressibility Method 176
10.4 Eigenstructure of the Incompressible Equations 177
10.5 Estimation of the Artificial Compressibility Parameter 180
10.6 Explicit Solvers for Artificial Compressibility 183
10.7 Implicit Solvers for Artificial Compressibility 184
10.7.1 Time-Linearized (Euler) Implicit Scheme 184
10.7.2 Implicit Approximate Factorization Method 185
10.7.3 Implicit Unfactored Method 186
10.8 Extension of the Artificial Compressibility to Unsteady Flows 188
10.9 Boundary Conditions 190
10.10 Local Time Step 191
10.11 Multigrid for the Artificial-Compressibility Formulation 192
10.11.1 Rationale for Three-Grid Multigrid 192
10.11.2 FMG-FAS Algorithm 193

10.11.3 Remarks on the Full Approximation Storage (FAS) Procedure . . . 196
10.11.4 Effects of Pre- and Post-Relaxation on the Efficiency of FMG–FAS . . . 197
10.11.5 Transfer Operators . . . 198
10.11.6 Adaptive Multigrid . . . 201

11. Projection Methods: The Basic Theory and the Exact Projection Method . . . 209
11.1 Grids – Variable Positioning . . . 210
11.2 Continuous Projections for Incompressible Flow . . . 211
11.2.1 Continuous Projections for Constant Density Incompressible Flow . . . 212
11.2.2 Continuous Projections for Variable Density Incompressible Flow . . . 213
11.3 Exact Discrete Projections . . . 213
11.3.1 Cell-Centered Exact Projections . . . 214
11.3.2 Vertex-Centered Exact Projections . . . 217
11.3.3 The MAC Projection . . . 219
11.3.4 The MAC Projection Used with Godunov-Type Methods . . . 220
11.3.5 Other Exact Projections . . . 223
11.4 Second-Order Projection Algorithms for Incompressible Flow 223
11.5 Boundary Conditions . . . 225
11.5.1 Solvability . . . 225
11.5.2 Solid Wall Boundary Conditions . . . 227

12. Approximate Projection Methods . . . 237
12.1 Numerical Issues with Approximate Projection Methods . . . 237
12.2 Projection Algorithms for Incompressible Flow . . . 243
12.3 Analysis of Projection Algorithms . . . 244
12.3.1 Basic Definitions for Analysis . . . 244
12.3.2 Analysis of Approximate Projection Algorithms . . . 245
12.3.3 Incremental Velocity Difference Projection . . . 247
12.3.4 Pressure Velocity Difference Projection . . . 248
12.3.5 Incremental Velocity Projection . . . 248
12.3.6 Pressure Velocity Projection . . . 249
12.3.7 Discussion of Analysis Results . . . 249
12.4 Pressure Poisson Equation Methods . . . 250
12.4.1 SIMPLE-Type Methods . . . 251
12.4.2 Implicit High-Resolution Advection . . . 254
12.4.3 Implicit Direct Methods . . . 255
12.5 Filters . . . 256
12.5.1 Classification of Error Modes . . . 256
12.5.2 Projection Filters . . . 258

12.5.3 Velocity Filters 263
12.6 Method Demonstration and Verification 271
12.6.1 Vortex-in-a-Box 271
12.6.2 Inflow with Shear 272
12.6.3 Doubling Periodic Shear Layer 273
12.6.4 Long Time Integration 274
12.6.5 Circular Drop Problem 279
12.6.6 Results Using Various Filters 285

Part III. Modern High-Resolution Methods

13. Introduction to Modern High-Resolution Methods 295
13.1 General Remarks about High-Resolution Methods 295
13.2 The Concept of Nonoscillatory Methods and Total Variation . 301
13.3 Monotonicity 303
13.4 General Remarks on Riemann Solvers 305

14. High-Resolution Godunov-Type Methods for Projection Methods 309
14.1 First-Order Algorithm 309
14.2 High-Resolution Algorithms 316
14.2.1 Piecewise Linear Methods (PLM) 316
14.2.2 Piecewise Parabolic Methods (PPM) 320
14.2.3 Algorithm Verification Tests 323
14.3 Staggered Grid Spatial Differencing 325
14.4 Unsplit Spatial Differencing 327
14.4.1 Least Squares Reconstruction 329
14.4.2 Monotone Limiters and Extensions 333
14.4.3 Monotonic Constrained Minimization 334
14.4.4 Divergence-Free Reconstructions 336
14.4.5 Extending Classical TVD Limiters 336
14.5 Multidimensional Results 340
14.6 Viscous Terms 342
14.7 Stability 343

15. Centered High-Resolution Methods 347
15.1 Lax-Friedrichs Scheme 348
15.2 Lax-Wendroff Scheme 353
15.3 First-Order Centered Scheme 358
15.3.1 Random Choice Method 359
15.3.2 FORCE 361
15.3.3 Variants of the FORCE Scheme 363
15.4 Second- and Third-Order Centered Schemes 364
15.4.1 Nessyahu-Tadmor Second-Order Scheme 364

15.4.2 Two-Dimensional Formulation 367
15.4.3 Third-Order Centered Scheme 369

16. Riemann Solvers and TVD Methods in Strict Conservation Form 373
16.1 The Flux Limiter Approach 373
16.2 Construction of Flux Limiters 374
16.2.1 Flux Limiter for the Godunov/Lax-Wendroff TVD Scheme 375
16.2.2 Flux Limiter for the Characteristics-Based/Lax-Friedrichs Scheme 376
16.3 Other Approaches for Constructing Advective Schemes 382
16.3.1 Positive Schemes 382
16.3.2 Universal Limiter 384
16.4 The Characteristics-Based Scheme 384
16.4.1 Introductory Remarks and Basic Formulation 384
16.4.2 Dimensional Splitting 386
16.4.3 Characteristics-Based Reconstruction in Three Dimensions 389
16.4.4 Reconstructed Characteristics-Based Variables in Two Dimensions 392
16.4.5 High-Order Interpolation 393
16.4.6 Advective Flux Calculation 396
16.4.7 Results 397
16.5 Flux Limiting Version of the CB Scheme 404
16.6 Implementation of the Characteristics-Based Method in Unstructured Grids 404
16.7 The Weight Average Flux Method 406
16.7.1 Basic Formulation 406
16.7.2 TVD Version of the WAF Schemes 408
16.8 Roe's Method 409
16.9 Osher's Method 412
16.10 Chakravarthy-Osher TVD Scheme 414
16.11 Harten, Lax and van Leer (HLL) Scheme 416
16.12 HLLC Scheme 419
16.13 Estimation of the Wave Speeds for the HLL and HLLC Riemann Solvers 420
16.14 HLLE Scheme 421
16.15 Comparison of CB and HLLE Schemes 421
16.16 "Viscous" TVD Limiters 424

17. Beyond Second-Order Methods 429
17.1 General Remarks on High-Order Methods 430
17.2 Essentially Nonoscillatory Schemes (ENO) 433
17.3 ENO Schemes Using Fluxes 436

17.4 Weighted ENO Schemes ... 439
17.4.1 Third-Order WENO Reconstruction ... 441
17.4.2 Fourth-Order WENO Reconstruction ... 442
17.5 A Flux-Based Version of the WENO Scheme ... 444
17.6 Artificial Compression Method for ENO and WENO ... 447
17.7 The ADER Approach ... 448
17.7.1 Linear Scalar Case ... 449
17.7.2 Multiple Dimensions: Scalar Case ... 451
17.7.3 Extension to Nonlinear Hyperbolic Systems ... 453
17.8 Extending and Relaxing Monotonicity in Godunov-Type Methods ... 455
17.8.1 Accuracy and Monotonicity Preserving Limiters ... 455
17.8.2 Extrema and Monotonicity Preserving Methods ... 460
17.8.3 Steepened Transport Methods ... 465
17.9 Discontinuous Galerkin Methods ... 467
17.10 Uniformly High-Order Scheme for Godunov-Type Fluxes ... 469
17.11 Flux-Corrected Transport ... 472
17.12 MPDATA ... 475

Part IV. Applications

18. Variable Density Flows and Volume Tracking Methods ... 479
18.1 Multimaterial Mixing Flows ... 479
18.1.1 Shear Flows ... 480
18.1.2 Rising Bubbles ... 482
18.1.3 Rayleigh-Taylor Instability ... 483
18.2 Volume Tracking ... 490
18.2.1 Fluid Volume Evolution Equations ... 492
18.2.2 Basic Features of Volume Tracking Methods ... 493
18.3 The History of Volume Tracking ... 495
18.4 A Geometrically Based Method of Solution ... 499
18.4.1 A Geometric Toolbox ... 500
18.4.2 Reconstructing the Interface ... 502
18.4.3 Material Volume Fluxes ... 510
18.4.4 Time Integration ... 513
18.4.5 Translation and Rotation Tests ... 515
18.5 Results For Vortical Flows ... 519
18.5.1 Single Vortex ... 521
18.5.2 Deformation Field ... 525

19. High-Resolution Methods and Turbulent Flow Computation ... 529
19.1 Physical Considerations ... 529
19.2 Survey of Theory and Models ... 533
19.3 Relation of High-Resolution Methods and Flow Physics ... 536
19.3.1 Numerical Considerations ... 537
19.3.2 Relation of High-Resolution Methods to Weak Solutions and Turbulence ... 538
19.4 Large Eddy Simulation: Standard and Implicit ... 539
19.5 Numerical Analysis of Subgrid Models ... 543
19.6 ILES Analysis ... 544
19.6.1 Explicit Modeling ... 544
19.6.2 Implicit Modeling ... 546
19.6.3 Limiters ... 547
19.6.4 Energy Analysis ... 549
19.7 Computational Examples ... 552
19.7.1 Burgers' Turbulence (Burgulence) ... 552
19.7.2 Convective Planetary Boundary Layer ... 553

A. MATHEMATICA Commands for Numerical Analysis ... 557
A.1 Fourier Analysis for First-Order Upwind Methods ... 557
A.2 Fourier Analysis for Second-Order Upwind Methods ... 558
A.3 Modified Equation Analysis for First-Order Upwind ... 559

B. Example Computer Implementations ... 563
B.1 Appendix: Fortran Subroutine for the Characteristics-Based Flux ... 563
B.2 Fifth-Order Weighted ENO Method ... 568
B.2.1 Subroutine for Fifth-Order WENO ... 568
B.2.2 Subroutine for Fifth-Order WENO's Third-Order Based Fluxes ... 570
B.2.3 Subroutine Fifth-Order WENO Smoothness Sensors ... 571
B.2.4 Subroutine Fifth-Order WENO Weights ... 572

C. Acknowledgements: Illustrations Reproduced with Permission ... 575

References ... 577

Index ... 615

1. Introduction

The study of incompressible flows is vital to many areas of science and technology. This includes most of the fluid dynamics that one finds in everyday life from the flow of air in a room to most weather phenomena.

In undertaking the simulation of incompressible fluid flows, one often takes many issues for granted. As these flows become more realistic, the problems encountered become more vexing from a computational point-of-view. These range from the benign to the profound. At once, one must contend with the basic character of incompressible flows where sound waves have been analytically removed from the flow. As a consequence vortical flows have been analytically "preconditioned," but the flow has a certain non-physical character (sound waves of infinite velocity).

At low speeds the flow will be deterministic and ordered, i.e., *laminar*. Laminar flows are governed by a balance between the inertial and viscous forces in the flow that provides the stability. Flows are often characterized by a dimensionless number known as the *Reynolds number*, which is the ratio of inertial to viscous forces in a flow. Laminar flows correspond to smaller Reynolds numbers. Even though laminar flows are organized in an orderly manner, the flows may exhibit instabilities and bifurcation phenomena which may eventually lead to transition and turbulence. Numerical modelling of such phenomena requires high accuracy and most importantly to gain greater insight into the relationship of the numerical methods with the flow physics. Practical applications in which laminar flow appears include biological fluid flow, non-Newtonian flows in chemical and mechanical engineering industries, as well as any industrial processes involving heat, fluid flow and mass transport at low Reynolds numbers.

When the flow is dominated by inertial forces it fundamentally changes character. In this case the flow corresponds to higher Reynolds numbers, compared to laminar flows, and in most cases of scientific and engineering interest the flow is in a disordered or *turbulent* state. Mixing of materials in a flow is greatly enhanced by the action of turbulence. This plays a key role in applications ranging from energy production and combustion to the transport of pollutants in various systems (for example, ground water, atmosphere). Other common applications of incompressible turbulent flows involve the flow around vehicles and low speed flows in aeronautics where the fuel efficiency is greatly impacted by the details of the flow.

The high-resolution methods described in this book aim at circumventing the fundamental Godunov's theorem that states that monotone methods are at most first order accurate. A broad class of numerical methods has been developed since the late 1970's. High-resolution methods try to achieve high-order of accuracy and avoid spurious oscillations. They can be broadly defined as nonoscillatory methods. These methods are characterized by their self-adaptive nature where the discrete stencil changes as a function of the solution itself[1]. This is what makes high-resolution methods different than other second- or higher-order (linear) finite difference schemes. The use of high-resolution methods is particularly warranted by the need to simulate flows containing discontinuous phenomena, such as, fluid interfaces and steep shear layers. High-resolution methods have been developed to handle such flow structures without introducing undue oscillations and grossly non-physical values, i.e., positive definite quantities such as density or mass fractions remain positive. For complex flows these are desirable to essential characteristics of the flow solver.

This book will detail the development and implementation of high-resolution numerical methods for solving challenging fluid flow problems where the approximation of incompressibility holds. Many everyday flows are well approximated by incompressible flows. This includes a large fraction of fluid flows of industrial interest.

Our voyage will begin with a description of the governing equations for a variety of flows that can be approximated with the methods we will describe. We provide the basic equations for fluid flow in Chaps. 2 and 3, and by necessity many of the techniques discussed throughout the book can be used to effectively solve all forms of the governing equations. In Chap. 4, we present the transformation of the governing equations in curvilinear coordinates. This approach is commonly used in conjunction with block-structured grids. Further, an overview of various formulations and model equations for incompressible flows is presented in Chap. 5 followed by an introduction to basic numerical analysis in Chap. 6 and discussion on time integration methods in Chap. 7.

A number of basic approaches exist for computing solutions to incompressible flows. These approaches can *all* benefit from the use of high-resolution methods when the flow regime and numerical resolution dictate an emphasis on the (nonlinear) inertial terms. Broadly speaking, these approaches can be divided into two camps:

- direct methods working on the incompressible flow equations,
- and indirect methods using compressible (like) equations and relaxing the solutions to the incompressible flow regime.

[1] Thus these methods result in nonlinear discretizations even when they are implemented to solve linear partial differential equations (PDEs).

We cover the indirect methods in Chap. 9. These approaches heavily use compressible flow technology. Among these are preconditioned flow solvers that aggressively attack the issue of the "stiffness" of the compressible flow equations as the flow speed becomes much smaller than the speed of sound.

The direct approaches may be based on an *artificial compressibility* coupling of the incompressible continuity and momentum equations (covered in depth in Chap. 10) or may produce an elliptic equation for pressure that must be solved. This usually involves great emphasis on numerical linear algebra (covered in Chap. 8) in that the elliptic equation generates (very) large linear systems of equations when approximated. These techniques are called *projection*, *pressure Poisson*, and *pressure correction methods* most commonly. These approaches are covered in depth in Chaps. 11 and 12.

The last portion of the book exclusively covers high-resolution methods. We begin with a general discussion on high-resolution methods (Chap. 13) and further present how *Godunov-type* methods can be used in conjunction with projection methods (Chap. 14). Godunov-type methods depend upon spatial and temporally adaptive interpolation and *Riemann solvers* to effectively solve the inertial terms. Chaps. 15 and 16 describe the commonly used *total variation diminishing* (TVD) methods in their centered and upwind forms. These techniques are founded upon the mathematical theory that arose in the wake of the introduction of Godunov-type methods. In Chap. 17 we cover the plethora of techniques by which high accuracy has been achieved using either Godunov-type or algebraic constructions (TVD-like).

Finally, we close with an overview of two important, challenging flows that can take advantage of the methods we introduce: interfacial flows (Chap. 18) and turbulent flows (Chap. 19). Because high-resolution methods described in this book effectively deal with the inertial terms particularly when the physical viscosity becomes less important, they have great utility in turbulent flows. In Chap. 19 we explore the exciting topic of the turbulence model property encompassed by high-resolution methods. This property has been observed for more than a decade, but only recently gained some theoretical support and a structural explanation for the observed results.

The development of computers and associated computational sciences has been staggeringly quick. The advent of parallel supercomputers as being distinct from the supercomputers of yesterday presents a continual challenge to the working scientist and engineer and has provided them with phenomenal computing power at their fingertips. The supercomputer of a decade ago now rests comfortably in the lap of the traveller sitting on an airplane or at the corner coffee shop. With this power, the advances in numerical methods allows a tremendous degree of simulation capability to be easily accessed. Ultimately, the methods described in this book provide the numerical tool for advancing the simulation capability in fundamental flow studies as well as in a number of practical applications.

Part I

Fundamental Physical and Model Equations

2. The Fluid Flow Equations

The starting point for simulating a fluid flow are the equations of that flow. This chapter will provide the basic description of the fluid flow equations and their mathematical properties.

2.1 Mathematical Preliminaries

In the derivation of the fluid flow equations, as well as throughout the book, we make use of the following notation. The *dot product* of two vectors $\mathbf{a} = (a_1, a_2, a_3)$ and $\mathbf{b} = (b_1, b_2, b_3)$ is a scalar quantity given by

$$\mathbf{a} \cdot \mathbf{b} = a_1 b_1 + a_2 b_2 + a_3 b_3 \; . \tag{2.1}$$

The vector or cross product $\mathbf{a} \times \mathbf{b}$ (it is read "**a** cross **b**") is the vector normal to the plane of **a** and **b**, and is defined by the determinant

$$\mathbf{a} \times \mathbf{b} = \begin{vmatrix} \mathbf{i} & \mathbf{j} & \mathbf{k} \\ a_1 & a_2 & a_3 \\ b_1 & b_2 & b_3 \end{vmatrix} \; . \tag{2.2}$$

The tensor product $\mathbf{a} \otimes \mathbf{b}$ is defined by

$$\mathbf{a} \otimes \mathbf{b} = \begin{pmatrix} a_1 b_1 & a_1 b_2 & a_1 b_3 \\ a_2 b_1 & a_2 b_2 & a_2 b_3 \\ a_3 b_1 & a_3 b_2 & a_3 b_3 \end{pmatrix} \; . \tag{2.3}$$

The gradient (nabla) operator ∇ (in Cartesian coordinates) is defined by

$$\nabla = \mathbf{i}\frac{\partial}{\partial x} + \mathbf{j}\frac{\partial}{\partial y} + \mathbf{k}\frac{\partial}{\partial z} \; . \tag{2.4}$$

The gradient operator as applied to a scalar quantity $\phi(x, y, z)$, where x, y, z are the spatial variables, is a vector defined by

$$\nabla \phi = \left(\frac{\partial \phi}{\partial x}, \frac{\partial \phi}{\partial y}, \frac{\partial \phi}{\partial z} \right)^T \; . \tag{2.5}$$

The notation "grad" for ∇ is also known as *the gradient operator*: grad $\phi \equiv \nabla\phi$. The gradient of the vector **a** is the *second-order tensor* (or simply defined as *tensor*[1])

$$\text{grad}\,\mathbf{a} = \nabla\mathbf{a} = \begin{pmatrix} \frac{\partial a_1}{\partial x} & \frac{\partial a_2}{\partial x} & \frac{\partial a_3}{\partial x} \\ \frac{\partial a_1}{\partial y} & \frac{\partial a_2}{\partial y} & \frac{\partial a_3}{\partial y} \\ \frac{\partial a_1}{\partial z} & \frac{\partial a_2}{\partial z} & \frac{\partial a_3}{\partial z} \end{pmatrix} . \tag{2.6}$$

The dot product of a vector **a** and the operator ∇ is called the *divergence of the vector field*; this is a *scalar* defined by

$$\text{div}\,\mathbf{a} = \nabla\cdot\mathbf{a} = \frac{\partial a_1}{\partial x} + \frac{\partial a_2}{\partial y} + \frac{\partial a_3}{\partial z} . \tag{2.7}$$

The rot operator is defined by

$$\text{rot}\,\mathbf{a} = \left(\frac{\partial a_3}{\partial y} - \frac{\partial a_2}{\partial z}, \frac{\partial a_1}{\partial z} - \frac{\partial a_3}{\partial x}, \frac{\partial a_2}{\partial x} - \frac{\partial a_1}{\partial y}\right)^T , \tag{2.8}$$

and we note that the notation curl **a** is often used instead of rot **a**.
The magnitude of **a** is defined as $|\mathbf{a}| = (\mathbf{a}\cdot\mathbf{a})^{1/2}$.
Let **A** be a tensor defined as

$$\mathbf{A} = \begin{pmatrix} a_{11} & a_{12} & a_{13} \\ a_{21} & a_{22} & a_{23} \\ a_{31} & a_{32} & a_{33} \end{pmatrix} . \tag{2.9}$$

The divergence of the tensor field **A** is a *vector* indicated by the notation div **A** or $\nabla\cdot\mathbf{A}$, and is defined by

$$\text{div}\,\mathbf{A} = \begin{pmatrix} \frac{\partial a_{11}}{\partial x} + \frac{\partial a_{12}}{\partial y} + \frac{\partial a_{13}}{\partial z} \\ \frac{\partial a_{21}}{\partial x} + \frac{\partial a_{22}}{\partial y} + \frac{\partial a_{23}}{\partial z} \\ \frac{\partial a_{31}}{\partial x} + \frac{\partial a_{32}}{\partial y} + \frac{\partial a_{33}}{\partial z} \end{pmatrix} . \tag{2.10}$$

A general tensor can be split into two parts one symmetric and the other anti-symmetric. A symmetric second-order tensor is defined as

$$\boldsymbol{\mathcal{S}} = \begin{pmatrix} s_{11} & s_{12} & s_{13} \\ s_{12} & s_{22} & s_{23} \\ s_{13} & s_{23} & s_{33} \end{pmatrix} . \tag{2.11}$$

[1] A first-order tensor is often referred to as a vector.

Here, the diagonal elements are $s_{ii} = a_{ii}$ and the off diagonal elements are $s_{ij} = (a_{ij} + a_{ji})/2$. The anti-symmetric tensor is defined element by element as $b_{ij} = (a_{ij} - a_{ji})/2$. For a vector $\mathbf{b} = (b_1, b_2, b_3)$, the anti-symmetric second-order tensor is defined as

$$\mathbf{B} = \begin{pmatrix} 0 & b_3 & b_2 \\ -b_3 & 0 & b_1 \\ -b_2 & -b_1 & 0 \end{pmatrix} . \tag{2.12}$$

The Jacobi matrix of the vector $\mathbf{a}$ is given by

$$\begin{pmatrix} \frac{\partial a_1}{\partial x} & \frac{\partial a_1}{\partial y} & \frac{\partial a_1}{\partial z} \\ \frac{\partial a_2}{\partial x} & \frac{\partial a_2}{\partial y} & \frac{\partial a_2}{\partial z} \\ \frac{\partial a_3}{\partial x} & \frac{\partial a_3}{\partial y} & \frac{\partial a_3}{\partial z} \end{pmatrix} . \tag{2.13}$$

Finally, we mention that throughout the book PDE and ODE stand as abbreviations for Partial and Ordinary Differential Equations, respectively.

2.2 Kinematic Considerations

The fluid flow motion can be studied using either the *Lagrangian* or *Eulerian* description. In the *Lagrangian description* one follows the trajectories of the individual fluid particles. This approach is common in certain areas (solid and particle mechanics), but it leads to more complicated analysis and in the case of fluid flow does not give directly the gradients of the fluid velocity. In the *Eulerian description* we do not try to follow any specific fluid particle paths. The velocity field $\mathbf{u}$ is defined as a function of position $\mathbf{r} = (x, y, z)$ and time t and in Cartesian form is given by

$$\mathbf{u}(\mathbf{r}, t) = \mathbf{i}\, u(x, y, z, t) + \mathbf{j}\, v(x, y, z, t) + \mathbf{k}\, w(x, y, z, t) , \tag{2.14}$$

where u, v, and w are the velocity components which are functions of the position in space (x, y, z) and time t.

If $Q(x, y, z, t)$ represents any property of the fluid and $\mathrm{d}x, \mathrm{d}y, \mathrm{d}z, \mathrm{d}t$ represent arbitrary changes in the four independent variables, the total differential change in Q is given by

$$\mathrm{d}Q = \frac{\partial Q}{\partial t}\, \mathrm{d}t + \frac{\partial Q}{\partial x}\, \mathrm{d}x + \frac{\partial Q}{\partial y}\, dy + \frac{\partial Q}{\partial z}\, \mathrm{d}z . \tag{2.15}$$

The spatial increments $\mathrm{d}x, \mathrm{d}y$ and $\mathrm{d}z$ are defined by

$$\mathrm{d}x = u\, \mathrm{d}t, \quad \mathrm{d}y = v\, \mathrm{d}t, \quad \mathrm{d}z = w\, \mathrm{d}t . \tag{2.16}$$

Eq. (2.15) then becomes

$$\frac{\mathrm{d}Q}{\mathrm{d}t} = \frac{\partial Q}{\partial t} + u\frac{\partial Q}{\partial x} + v\frac{\partial Q}{\partial y} + w\frac{\partial Q}{\partial z} \ . \tag{2.17}$$

The term $\mathrm{d}Q/\mathrm{d}t$ is often referred to as substantial derivative, or material derivative, traditionally defined by the symbol $\mathrm{D}Q/\mathrm{D}t$. The term $\partial Q/\partial t$ is the local derivative. In vector form the substantial derivative can also be written as

$$\frac{\mathrm{D}Q}{\mathrm{D}t} = \frac{\partial Q}{\partial t} + (\mathbf{u}\cdot\nabla)Q \ . \tag{2.18}$$

The substantial derivative can be applied to any property of the fluid such as angular velocity, vorticity acceleration and strain rate, in order to find the time rate of change of the property of a fixed element.[2]

If we apply the definition of the substantial derivative for the velocity $\mathbf{u}$, we obtain the particle acceleration vector

$$\frac{\mathrm{D}\mathbf{u}}{\mathrm{D}t} = \frac{\partial \mathbf{u}}{\partial t} + (\mathbf{u}\cdot\nabla)\mathbf{u} = \frac{\partial \mathbf{u}}{\partial t} + \mathbf{u}\cdot \mathrm{grad}\,\mathbf{u} \ . \tag{2.19}$$

Similar to the solid mechanics, a fluid element can perform four different types of motion or deformation: translation (constant velocity), rotation (anti-symmetric part of the velocity gradient tensor), extensional strain (or dilation, the diagonal of the velocity gradient tensor) and shear strain (the off-diagonal parts of the symmetric portion of the velocity gradient tensor).

2.3 The Equations for Variable Density Flows

2.3.1 The Continuity Equation

The continuity equation is obtained by applying the physical principle of *mass conservation*. Consider a flow "tube" of volume V as shown in Fig. 2.1. The total mass inside the control volume is $\int\int\int_V \rho \,\mathrm{d}V$, where ρ is the density of the fluid. The net rate at which mass is flowing out of the control volume through the surface S is $\int\int_S \rho\mathbf{u}\cdot\mathbf{n}\,\mathrm{d}S$, where $\mathbf{n}$ is the outward pointing unit vector (Fig. 2.1). The mass conservation principle implies that

$$\frac{\mathrm{d}}{\mathrm{d}t}\int\int\int_V \rho \,\mathrm{d}V = -\int\int_S \rho\mathbf{u}\cdot\mathbf{n}\,\mathrm{d}S \ . \tag{2.20}$$

The term on left-hand-side of the above equation denotes the time rate of decrease inside the control volume V. Eq. (2.20) is also written as

$$\frac{\mathrm{d}}{\mathrm{d}t}\int\int\int_V \rho \,\mathrm{d}V + \int\int_S \rho\mathbf{u}\cdot\mathbf{n}\,\mathrm{d}S = 0 \ , \tag{2.21}$$

which is the *integral form of the continuity equation.* The surface integral in (2.21) can be transformed to a volume integral using the Gauss's theorem

---

[2] The operator $\mathbf{u}\cdot\nabla$ in Cartesian coordinates is defined as $\mathbf{u}\cdot\nabla = u\partial/\partial x + v\partial/\partial y + w\partial/\partial z$.

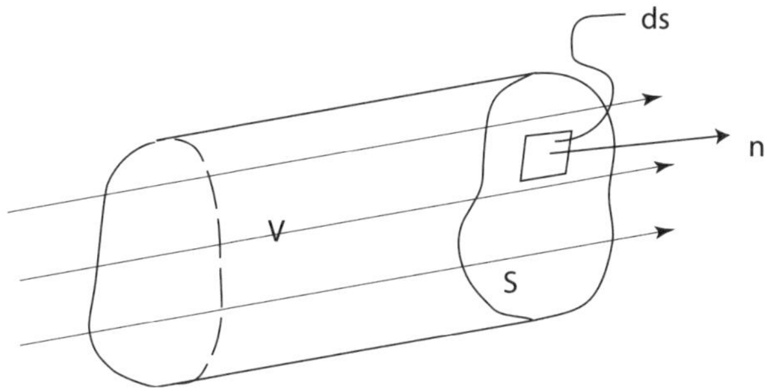

Fig. 2.1. Flow "tube" and elementary surface.

$$\int\int_S \rho\mathbf{u}\cdot\mathbf{n}\,\mathrm{d}S = \int\int\int_V \nabla\cdot(\rho\mathbf{u})\,\mathrm{d}V\,. \tag{2.22}$$

Subsequently, (2.21) is written as

$$\int\int\int_V \left[\frac{\partial\rho}{\partial t} + \nabla\cdot(\rho\mathbf{u})\,\mathrm{d}V\right] = 0\,. \tag{2.23}$$

The above relation is valid for any choice of the control volume V in the fluid and, therefore, the integrand must be equal to zero at all points in the fluid.

$$\frac{\partial\rho}{\partial t} + \nabla\cdot(\rho\mathbf{u}) = 0\,, \tag{2.24}$$

This equation is valid in the limit as the volume goes to zero. Eq. (2.24) is the *differential form of the continuity equation*, also known as *continuity equation.*

2.3.2 The Momentum Equations

The time rate of the change of the fluid momentum of volume V surrounded by a surface S is due to the total force acting on the volume V, as well as to the net transfer of momentum across the surface S. The momentum of fluid in V and the net momentum across the surface S are given by $\int\int\int_V \rho\mathbf{u}\,\mathrm{d}V$ and $-\int\int_S \mathbf{u}(\rho\mathbf{u}\cdot\mathbf{n})\,\mathrm{d}S$, respectively. Thus the conservation law for the momentum can be written as

$$\frac{\mathrm{d}}{\mathrm{d}t}\int\int\int_V \rho\mathbf{u}\,\mathrm{d}V = -\int\int_S \mathbf{u}(\rho\mathbf{u}\cdot\mathbf{n})\,\mathrm{d}S + \int\int\int_V \mathbf{F}_V\,\mathrm{d}V + \int\int_S \mathbf{F}_S\,\mathrm{d}S\,. \tag{2.25}$$

where $\mathbf{F}_V$ and $\mathbf{F}_S$ are the vector resultants of the volume and surface forces, respectively. The volume forces may account for inertial forces, gravitational forces or electromagnetic forces. The vector of the surface forces $\mathbf{F}_S$ is given as $\mathbf{F}_S = \mathbf{n} \cdot \mathbf{S}$, where $\mathbf{S} \equiv \sigma_{ij}$ $(i, j = x, y, z)$ is the stress tensor. The surface forces account for the fluid stresses. The latter are due to the effects of the thermodynamic pressure,[3] p, and the viscous stresses. Thus the stress tensor $\mathbf{S}$ can be written as

$$\mathbf{S} = -p\,\mathbf{I} + \mathbf{T}\,, \tag{2.26}$$

where $p\,\mathbf{I}$ is a spherically symmetric tensor, $\mathbf{I}$ is the unit matrix

$$\mathbf{I} = \begin{pmatrix} 1 & 0 & 0 \\ 0 & 1 & 0 \\ 0 & 0 & 1 \end{pmatrix}, \tag{2.27}$$

and $\mathbf{T} \equiv \tau_{ij}$ is the viscous stress tensor

$$\mathbf{T} = \begin{pmatrix} \tau_{xx} & \tau_{xy} & \tau_{xz} \\ \tau_{yx} & \tau_{yy} & \tau_{yz} \\ \tau_{zx} & \tau_{yz} & \tau_{zz} \end{pmatrix}. \tag{2.28}$$

Note that $\sigma_{ij} = \tau_{ij}$ for $i \neq j$.
Eq. (2.25) is further written as

$$\frac{\mathrm{d}}{\mathrm{d}t}\int\!\!\int\!\!\int_V \rho\mathbf{u}\,\mathrm{d}V = -\int\!\!\int_S \mathbf{u}(\rho\mathbf{u}\cdot\mathbf{n})\,\mathrm{d}S$$
$$+\int\!\!\int_S (-p\,\mathbf{I}\cdot\mathbf{n} + \mathbf{T}\cdot\mathbf{n})\,\mathrm{d}S + \int\!\!\int\!\!\int_V \mathbf{F}_V\,\mathrm{d}V\,. \tag{2.29}$$

Using the Gauss's theorem to convert the surface integrals into volume integrals, the above equation is written as

$$\int\!\!\int\!\!\int_V \frac{\partial(\rho\mathbf{u})}{\partial t}\mathrm{d}V = -\int\!\!\int\!\!\int_V \mathrm{div}\,(\rho\mathbf{u}\otimes\mathbf{u})\,\mathrm{d}V$$
$$-\int\!\!\int\!\!\int_V (\mathrm{grad}p - \mathrm{div}\,\mathbf{T})\mathrm{d}V + \int\!\!\int\!\!\int_V \mathbf{F}_V\,\mathrm{d}V\,. \tag{2.30}$$

The tensor product $\mathbf{u}\otimes\mathbf{u}$ is a second-order tensor given by

$$\mathbf{u}\otimes\mathbf{u} = \begin{pmatrix} u^2 & uv & uw \\ vu & v^2 & vw \\ wu & wv & w^2 \end{pmatrix}. \tag{2.31}$$

[3] The thermodynamic definition of pressure is defined by $p/T = (\partial S/\partial V)_U$, where S is the entropy, U is the internal energy, T is the temperature and V is the volume. The thermodynamic and mechanical pressures are identical in any equilibrium state.

Eq. (2.30) is the integral form of the momentum equation. Since (2.30) is valid for any arbitrary volume V the integrand must vanish

$$\frac{\partial(\rho\mathbf{u})}{\partial t} + \text{div}\ (\rho\mathbf{u} \otimes \mathbf{u}) = -\nabla p + \text{div}\ \mathbf{T} + \mathbf{F}_V\ . \tag{2.32}$$

Eq. (2.32) is the differential form of the momentum equation.[4]

The inviscid (Euler) equations are obtained by neglecting the surface forces thus obtaining

$$\frac{\partial(\rho\mathbf{u})}{\partial t} + \text{div}\ (\rho\mathbf{u} \otimes \mathbf{u}) = -\nabla p + \mathbf{F}_V\ . \tag{2.33}$$

Remark 2.3.1. If the torques within a fluid arise only as the moments of direct forces, we shall call it *non-polar fluid.* For the non-polar fluid we can show that the stress tensor $\mathbf{T}$ is symmetric, i.e., $\tau_{ij} = \tau_{ji}$ $(i, j = x, y, z)$. Let us consider the instantaneous angular acceleration $\dot{\omega} = (\dot{\omega}_x, \dot{\omega}_y, \dot{\omega}_z)$ of a fluid element. For the rotation about the $x-$axis we can write

$$\dot{\omega}_x\ \mathrm{d}I_x = (\tau_{yz}\ \mathrm{d}x\ \mathrm{d}z)\ \mathrm{d}y - (\tau_{zy}\ \mathrm{d}x\ \ \mathrm{d}y)\ \mathrm{d}z = (\tau_{yz} - \tau_{zy})\ \mathrm{d}V\ , \tag{2.34}$$

where $\mathrm{d}I_x$ is the moment of inertia about the $x-$axis. If the angular momentum is conserved, we obtain

$$\tau_{yz} - \tau_{zy} = 0\ . \tag{2.35}$$

The same can also be applied to the other elements of the stress tensor.
The symmetry of the stress tensor can also be confirmed by the following analysis. We write (2.34) as

$$\dot{\omega}_x = (\tau_{yz} - \tau_{zy})\frac{\mathrm{d}V}{\mathrm{d}I_x}\ . \tag{2.36}$$

If we consider a volume of a given shape with a characteristic dimension L then the term $\mathrm{d}I_x$ on the left-hand-side (LHS) of (2.36) is proportional to L^5, whereas $\mathrm{d}V$ on the right-hand-side (RHS) is proportional to L^3. If we let the volume V to shrink on a point (but preserve its shape), then $\tau_{yz} - \tau_{zy} = 0$, if $\dot{\omega}_x$ is not to become infinitely large.
The breaking of the symmetry of the stress tensor would occur if the fluid developed a local moment of momentum which would be proportional to its volume. This can occur in the case that one applies, e.g., an electrostatic field. In the case of a non-symmetric stress tensor, the anti-symmetric part would contribute to the rate of increase of the internal angular momentum.

[4] Note that the *divergence* of a second-order tensor such as $\mathbf{u} \otimes \mathbf{u}$ and $\mathbf{T}$, is a vector.

2.3.3 The Energy Equation

The final equation to introduce for our basic description of fluid flow is the energy equation. This equation is intrinsically linked with thermodynamics, and this link is tied directly to the equilibrium equation of state. One can start directly with the principle of conservation of energy as defined by the first law of thermodynamics,

$$\mathrm{d}e = p\mathrm{d}v + T\mathrm{d}S \, , \tag{2.37}$$

where e is the specific internal energy per unit mass, $v = 1/\rho$ is the specific volume, T is the temperature, and S is the entropy (the amount of order in a system). This expression can be written as a PDE in the Lagrangian frame of reference,

$$\frac{De}{Dt} + p\frac{Dv}{Dt} = T\frac{DS}{Dt} \, , \tag{2.38}$$

In the case of a viscous fluid $p := p +$ viscous stress.
The change in volume can be expressed as $Dv/Dt = -1/\rho\nabla\cdot\mathbf{u}$ (the continuity equation in terms of specific volume). More generally, there can be source terms arising from a gravitational potential, chemical reaction, radiation, etc. Eq. (2.38) holds in the Lagrangian frame of reference and when combined with an expression for the kinetic energy yields the conservation equation for the total energy per unit mass, $E = e + K$, with $K = 1/2|\mathbf{u}|^2$ being the kinetic energy per unit mass. The kinetic energy can be found through vector multiplying the expression for momentum conservation with the velocities. This produces the following equation

$$\rho\frac{DK}{Dt} = -\mathbf{u}\cdot\nabla p + \mathbf{u}\cdot\nabla\cdot\mathbf{T} \, , \tag{2.39}$$

This is then combined with the internal energy, (2.37), to provide an expression for the total energy

$$\rho\frac{DE}{Dt} + \nabla\cdot(p\mathbf{u}) = T\frac{DS}{Dt} + \nabla\cdot(\mathbf{Tu}) \, . \tag{2.40}$$

The term TDS/Dt has the effect of increasing the internal energy as dictated by the second law of thermodynamics.

This equation is appropriate for use in compressible flow especially with shock waves, but for numerical (and analytical) purposes is poorly conditioned for flows that do not exhibit the character of compressibility. For this reason a variety of useful simplifications can be made. These will provide some useful variants of the full conservation law.

We can start by writing (2.40) in full Eulerian form,

$$\frac{\partial\rho E}{\partial t} + \nabla\cdot(\mathbf{u}\rho E + p\mathbf{u}) = \nabla\cdot q + \nabla\cdot(\mathbf{Tu}) \, , \tag{2.41}$$

where the contribution $T\mathrm{d}S$ is written in terms of the heat flux $q = k\nabla T$ (k is the thermal conductivity). If the viscous forces are ignored then the equation simplifies to

$$\frac{\partial \rho e}{\partial t} + \nabla \cdot (\rho \mathbf{u} e) + p\nabla \cdot \mathbf{u} = \nabla \cdot q \,. \tag{2.42}$$

By making assumptions regarding smoothness of the solution, we can then obtain a variety of useful forms for the energy equation using the equation of state.

The first of these forms is cast in terms of temperature and captures the simplest transport of heat by advection or thermal conduction

$$C_v \rho \frac{DT}{Dt} = k\nabla^2 T \,, \tag{2.43}$$

where C_v is the specific heat at constant volume. Another common form of the heat transport equation retains the effects of pressure,

$$C_p \rho \frac{DT}{Dt} + p\nabla \cdot \mathbf{u} = k\nabla^2 T \,. \tag{2.44}$$

Note that (2.43) uses C_v, while (2.44) uses the specific heat at constant pressure, C_p. Both C_v and C_p are assumed to be constants. The first form, (2.43), can be written in a dimensionless form,

$$\frac{D\theta}{Dt} = \kappa \nabla^2 \theta \,, \tag{2.45}$$

by dividing through by $C_v T_0$, where T_0 is a characteristic temperature, i.e., $\theta = T/T_0$ and $\kappa = k/\rho C_v$ (thermal diffusivity).

The equation of state relates the pressure with the internal energy and volume, $p = P(v, e)$. By employing the equation of state for an ideal gas $p = (\gamma - 1)\, e/v$, where $\gamma = v/p \; \partial p/\partial v|_S$ is the adiabatic coefficient which in the case of an ideal gas also signifies the ratio of specific heats (i.e., $\gamma = C_p/C_v$), and making the assumption of adiabatic evolution, a dynamic pressure equation can be found,

$$\frac{Dp}{Dt} + \rho c^2 \nabla \cdot \mathbf{u} = 0 \,, \tag{2.46}$$

where c is the sound speed, $c^2 = \gamma p/\rho$. In the above, we have neglected the heat flux and viscous terms. This form will be quite useful in the development that follows where we will show the close relation of the energy equation to the divergence-free condition for incompressible flow taken as a low-Mach number asymptotic limit of fluid flow. The last form of the energy equation we introduce follows from a combination of (2.46) and (2.44) or an internal energy equation (2.38). This produces an equation for the enthalpy, $h = C_v T + p/\rho = e + p/\rho$,

$$\frac{Dh}{Dt} + \left(c^2 + \frac{p}{\rho}\right) \nabla \cdot \mathbf{u} = 0 \,. \tag{2.47}$$

2.4 Compressible Euler Equations

Eqs. (2.24), (2.32) and (2.41) are the fundamental model of compressible fluid flow and are referred to as the *Navier-Stokes equations.* compressible Euler equations describing the conservation of mass, momentum and energy for an "idealized inviscid fluid". We start by describing these solutions in the Lagrangian frame-of-reference where the frame moves with the material velocity. We choose the Lagrangian frame-of-reference because the equations and their general solutions are significantly simpler than the more common Eulerian frame-of-reference. As such the Lagrangian frame is often employed in the derivation of various numerical methods and directly in some of the original high-resolution methods.

We will also write these equations in a one-dimensional vector form,

$$\frac{D\boldsymbol{U}}{Dt} + \frac{\partial \boldsymbol{E}(\boldsymbol{U})}{\partial m} = 0\,, \tag{2.48}$$

where $\boldsymbol{U} = (v, u, E)^T$ and $\boldsymbol{E} = (-u, p, pu)^T$. The dependent variables are the specific volume $v = 1/\rho$, the fluid velocity, u, and the total energy per unit mass, E. The Lagrangian fluxes include the pressure, p, which depends on the equation of state. The spatial coordinate is the mass, m, because the Lagrangian frame moves with the fluid. This coordinate is defined by the integral of density in space,

$$m = \int_0^x \rho \, \mathrm{d}x\,.$$

Eq. 2.48 is also known generically as a *system of conservation laws* because the vector of unknowns is conserved (excepting boundaries). The changes in the quantities are due to the divergence of fluxes giving rise to the equivalent term, i.e., the *flux form.*

Much of the analytical interest in the solutions to these equations is focused on self similar solutions. Self similar solutions in the compressible Euler equations are functions of space and time, x/t. All of these equations look the same if they are viewed under transformations that maintain the same ratio of x/t.

Continuous solutions are found with the compressible Euler equations. These solutions can be derived with the assistance of rewriting the compressible Euler equations in a "primitive" form,

$$\frac{\partial \boldsymbol{V}}{\partial t} + \mathbf{A}\frac{\partial \boldsymbol{V}}{\partial x} = 0\,, \tag{2.49}$$

where $\boldsymbol{V} = (\rho, u, p)$ and

$$\mathbf{A} = \begin{pmatrix} 0 & \rho & 0 \\ 0 & 0 & 1/\rho \\ 0 & \rho c^2 & 0 \end{pmatrix}.$$

These relations all hold when solutions are smooth. The prototypical smooth self-similar solution is a rarefaction. Rarefactions are defined by the Riemann invariants where the primitive equations are written in terms of x/t and differentials in those terms $V' + \mathbf{A}V' = 0$. One can also write other evolution equations such as entropy, i.e., $S' = 0$. Thus the entropy remains constant in these solutions. This is one of the chief distinguishing characteristics of rarefactions (also known as adiabats for this very reason) from shocks. In a shock entropy is not constant and increases in a compressible fluid as a shock moves through it.

The equations of compressible flow most significantly admit discontinuous solutions such as shock waves. The discontinuous solutions require that the equations be thought of in a more general sense because all notions of smoothness (required for a differential form) are lost. The governing equations are integrated over regions of time and space in order to help define these more general forms. These general forms are known as weak (or integral) solutions because they are less demanding in terms of smoothness than other solutions. One form of these integral equations are known as the Rankine-Hugoniot equations,

$$W[\boldsymbol{U}] = [\boldsymbol{E}(\boldsymbol{U})] \,, \tag{2.50}$$

where W is the speed of the discontinuity, and the notation $[\,]$ denotes a jump in the variables, $\mathbf{U}$, contained within. These equations hold for jumps between piecewise constant states. In the case of the compressible Euler equations the solutions admit two types of discontinuous solutions, shocks and contact discontinuities. In a shock all variables are discontinuous while in a contact the volume (density) and energy are discontinuous and the velocity and pressure are constant.

As mentioned above, shock waves increase the entropy of fluids. This increase in entropy comes about through the residual action of viscosity on the fluid. Although the viscosity has been neglected in the Euler equations, the effects of viscosity are always present in a real fluid. When a discontinuity forms, at a very small scale the flow is actually continuous and characterized by a very large gradient. These gradients become so large that the vanishing size of the viscous coefficients introduces dissipation to the fluid thus increasing the entropy as the second law of thermodynamics demands ($T\mathrm{d}S \geq 0$).

The criterion for shock formation beginning with smooth data is straightforward in one dimension. It depends only on the smallest gradient in the domain and the nonlinearity of the equation of state (convex positive) with respect to density (or volume). One can write the time at which the shock forms (also known as wave breaking) as

$$t_b = \frac{1}{1 - \mathcal{G} \min \nabla \mathbf{u}} \,,$$

if $\nabla \mathbf{u} < 0$ where $\mathcal{G}$ is the non-dimensional curvature of the equation of state. For an ideal gas this is $\mathcal{G} = (\gamma - 1)/2$. More general discussions of this matter can be found in [339, 601].

A good way to understand the solutions of the Euler system is to undertake an analysis of its characteristic structure. This is conducted most simply using the matrix $\mathbf{A}$ from the primitive form of the equations. The starting point is to decompose the matrix into its fundamental modes through an eigen-decomposition. This produces the eigenvalues $\lambda = (-c, 0, c)$ describing the speed of transport in the system (one for every equation). These are the three fundamental modes. The modes $\pm c$ are the acoustic modes describing sound waves when the flow is smooth. The eigenvalue, 0, is the material velocity moving with the frame of reference. The full analysis uses the decomposition $\mathbf{A} = \mathbf{T}\boldsymbol{\Lambda}\mathbf{T}^{-1}$ where $\boldsymbol{\Lambda}$ is the eigenvalue matrix multiplied with the identity matrix,

$$\boldsymbol{\Lambda} = \begin{pmatrix} -c & 0 & 0 \\ 0 & 0 & 0 \\ 0 & 0 & c \end{pmatrix},$$

$\mathbf{T}$ is the right eigenvector matrix

$$\mathbf{T} = \begin{pmatrix} 1 & 1 & 1 \\ -c/\rho & 0 & c/\rho \\ c^2 & 0 & c^2 \end{pmatrix},$$

and $\mathbf{T}^{-1}$ is the left eigenvector matrix (the inverse of the right eigenvectors),

$$\mathbf{T}^{-1} = \begin{pmatrix} 0 & -\rho/(2c) & 1/(2c^2) \\ 1 & 0 & -1/c^2 \\ 0 & \rho/(2c) & 1/(2c^2) \end{pmatrix}.$$

Multiplying the left eigenvectors by the fluctuations in the dependent variables, $(\delta\rho, \delta u, \delta p)^T$, gives the characteristic variables, $-\rho\delta u/(2c) + \delta p/(2c^2)$, $\delta\rho - \delta p/c^2$, $\rho\delta u/(2c) + \delta p/(2c^2)$. These are the modes of information moving at the corresponding characteristic speeds.

These modes are denoted as being either linear or nonlinear depending on whether the characteristic speed changes as a result of the changes in the fluctuating quantity. Obviously the 0 eigenvalue is linear because it is a constant, and the fluid velocity does not depend on the characteristic variables. On the other hand, the sound speed, c, is a function of the pressure thus the acoustic modes are nonlinear. This nonlinearity can be judged by taking the dot product of the right eigenvector with the gradient of the eigenvalue,

$\mathbf{T}_k \cdot \nabla_{\mathbf{V}} \lambda \neq 0$ (k denotes the kth eigenvector). If this product is zero the wave is linear (linearly degenerate). This classification is due to Lax [319].

When dealing with low-speed and incompressible flows, it is far more common to use the Eulerian frame-of-reference. One can derive the Eulerian equations from the Lagrangian through using the conservation of mass and the definition of the substantial derivative. After some manipulation and assumptions of sufficient smoothness a conservation law form can be found,

$$\frac{\partial \mathbf{U}}{\partial t} + \frac{\partial \mathbf{E}(\mathbf{U})}{\partial x} = 0 \,, \tag{2.51}$$

where $\mathbf{U} = (\rho, \rho u, \rho E)^T$ and $\mathbf{E}(\mathbf{U}) = \left(\rho u, \rho u^2 + p, \rho u E + p u\right)^T$. Except for the change of frame, equations (2.48) and (2.51) are equivalent. The primitive form of the equations also follows:

$$\frac{\partial \boldsymbol{V}}{\partial t} + \mathbf{A}\frac{\partial \boldsymbol{V}}{\partial x} = 0 \,, \tag{2.52}$$

where $\boldsymbol{V} = (\rho, u, p)$ and

$$\mathbf{A} = \begin{pmatrix} u & \rho & 0 \\ 0 & u & 1/\rho \\ 0 & \rho c^2 & u \end{pmatrix} .$$

The characteristic analysis of the primitive form now gives the following results. First, the eigenvalues matrix:

$$\boldsymbol{\Lambda} = \begin{pmatrix} u-c & 0 & 0 \\ 0 & u & 0 \\ 0 & 0 & u+c \end{pmatrix} ,$$

next the right eigenvector matrix

$$\mathbf{T} = \begin{pmatrix} 1 & 1 & 1 \\ -c/\rho & 0 & c/\rho \\ c^2 & 0 & c^2 \end{pmatrix} ,$$

and, finally, the left eigenvector matrix

$$\mathbf{T}^{-1} = \begin{pmatrix} 0 & -\rho/(2c) & 1/(2c^2) \\ 1 & 0 & -1/c^2 \\ 0 & \rho/(2c) & 1/(2c^2) \end{pmatrix} .$$

Note that only the eigenvalues have changed in the Eulerian frame. The notions of linear and nonlinear modes also follow the Lagrangian frame results.

A more mathematical and complete description of the Euler equations and their general solutions are given in [339, 341, 383, 543]. Various transformations between the original conservation laws and the primitive form are explained by Huynh [271].

2.5 Low-Mach Number Scaling

Low-Mach number asymptotic analysis provides a distinct range of flow parameters for which the approximation of incompressibility remains valid.

First, we should make some general comments regarding the development of methods applicable to incompressible and low-Mach number flows. We desire equations where the explicit stability is limited by the velocity, $\mathbf{u}$, and not the sound speed, c. In this case, the sound speed is much larger than the velocity, $|\mathbf{u}| \ll c$. This means that the Mach number is much less than one in the following sense $M^2 = (|\mathbf{u}|/c)^2 \ll 1$. This dictates that acoustic phenomena are not of interest, although they can be included into the basic solution framework [119].

When considering the flow of a fluid (gas) in a low-Mach number regime, it is useful to look at asymptotic behavior of the full set of conservation equations in the limit where $M \to 0$. For simplicity, we will give these equations in non-conservation form and neglecting viscous terms,

$$\frac{D\rho}{Dt} + \rho \nabla \cdot \mathbf{u} = 0 \ ,$$

$$\frac{D\mathbf{u}}{Dt} + \frac{1}{\rho} \nabla p = 0 \ ,$$

and

$$\frac{Dp}{Dt} + \rho c^2 \nabla \cdot \mathbf{u} = 0 \ .$$

These equations represent the conservation of mass, momentum (motion), and energy (in the form of the dynamic pressure equation) in the absence of source terms and shock waves. Note that pressure is linked to internal energy through the thermodynamic equation of state.

To scale these equations correctly for the low-speed regime, we introduce the following:[5]

$$\bar{\rho} = \int_\Omega \rho \, \mathrm{d}\Omega \ ,$$

[5] Also assuming we are in a periodic domain, or a closed box, where $\mathbf{u} \cdot \mathbf{n} = 0$ on the boundaries.

$$\bar{p} = \int_{\Omega} p \, \mathrm{d}\Omega \; ,$$

where Ω is the domain of integration (volume or surface in 3-D and 2-D, respectively), and

$$\tilde{p} = \frac{p - \bar{p}}{\gamma \bar{p}} \bar{\rho} U^2 \; ,$$

with U being some characteristic velocity. Defining $c^2 = \gamma\bar{p}/\bar{\rho}$ gives $M^2 = \bar{\rho}U^2/\gamma\bar{p}$, then it is elementary to notice that

$$\tilde{p} = (p - \bar{p})\, M^2 \; .$$

Using the above, the equation of motion is simplified to

$$\bar{\rho}\frac{D\mathbf{u}}{Dt} + \frac{1}{M^2}\nabla\tilde{p} = 0 \; . \tag{2.53}$$

We then expand the variables in M and take the limit as $M \to 0$,

$$\bar{\rho} = \rho_o + M\rho_1 + M^2\rho_2 + O\left(M^3\right) \; ,$$

$$\tilde{p} = p_o + Mp_1 + M^2 p_2 + O\left(M^3\right) \; ,$$

and

$$\mathbf{u} = \mathbf{u}_o + M\mathbf{u}_1 + M^2\mathbf{u}_2 + O\left(M^3\right) \; .$$

Introducing this into our simplified motion equation (2.53) shows that p_o and p_1 are only functions of time (setting the pressure, $\phi = p_2$),[6] and not space, thus we obtain the equation of motion for low-Mach numbers:

$$\bar{\rho}\frac{D\mathbf{u}_o}{Dt} + \nabla\phi = 0 \; ,$$

which is the order of M^2 terms. Now, introducing the above expansions into the energy equation, it can be shown that the divergence of velocity given by

$$\nabla \cdot \mathbf{u}_o = 0 \; ,$$

is a good approximation for $M \ll 1$. Finally, the mass conservation equation can be written as

$$\frac{D\rho}{Dt} = 0 \; .$$

We will now provide a more heuristic derivation of the equations of incompressible flow. The limitation of this approach is that the approximations are not explicitly defined as part of the derivation.

---

[6] In the following sections we will retain the notation ϕ for the pressure.

The continuity (2.24) and momentum (2.33) equations can be simplified if one assumes that the density remains constant throughout the flow field, thus considering that the flow is practically *incompressible.* The continuity equation then becomes

$$\nabla \cdot \mathbf{u} = 0 \,. \tag{2.54}$$

This equation is referred to as *continuity equation for an incompressible fluid*, but also as *incompressibility constraint.* Neglecting the volume forces too, the momentum equations (2.33) can be written as

$$\rho \frac{\partial \mathbf{u}}{\partial t} + \rho \mathrm{div}\, (\mathbf{u} \otimes \mathbf{u}) = -\nabla p \,. \tag{2.55}$$

By dividing (2.55) with ρ we obtain

$$\frac{\partial \mathbf{u}}{\partial t} + \mathrm{div}\, (\mathbf{u} \otimes \mathbf{u}) = -\frac{1}{\rho} \nabla p \,, \tag{2.56}$$

or

$$\frac{\partial \mathbf{u}}{\partial t} + (\mathbf{u} \cdot \nabla)\mathbf{u} = -\frac{1}{\rho} \nabla p \,. \tag{2.57}$$

by making use of the incompressibility condition (2.54). Eqs. (2.54) and (2.57) are *inviscid equations for incompressible flows.* Note, that the pressure used here is an incompressible one and not the thermodynamic pressure. The heuristic notion is that the thermodynamic pressure is constant in the domain where the incompressible flow equations are valid.

Using tensorial indices, the equations can also be written as

$$\frac{\partial u_j}{\partial x_j} = 0 \,, \tag{2.58}$$

$$\frac{\partial u_i}{\partial t} + u_j \frac{\partial u_i}{\partial x_j} = -\frac{1}{\rho} \frac{\partial p}{\partial x_i} \,, \tag{2.59}$$

where $i, j \equiv 1, 2, 3$ for the x, y and z directions, respectively.

Eqs. (2.54) and (2.57) can be expanded for a Cartesian co-ordinates system (x, y, z) as

$$\frac{\partial u}{\partial x} + \frac{\partial v}{\partial y} + \frac{\partial w}{\partial z} = 0 \,, \tag{2.60}$$

$$\frac{\partial u}{\partial t} + \frac{\partial u^2}{\partial x} + \frac{\partial uv}{\partial y} + \frac{\partial uw}{\partial z} = -\frac{1}{\rho} \frac{\partial p}{\partial x} \,, \tag{2.61}$$

$$\frac{\partial v}{\partial t} + \frac{\partial uv}{\partial x} + \frac{\partial v^2}{\partial y} + \frac{\partial vw}{\partial z} = -\frac{1}{\rho} \frac{\partial p}{\partial y} \,, \tag{2.62}$$

$$\frac{\partial w}{\partial t} + \frac{\partial uw}{\partial x} + \frac{\partial vw}{\partial y} + \frac{\partial w^2}{\partial z} = -\frac{1}{\rho} \frac{\partial p}{\partial z} \,. \tag{2.63}$$

A more recent extension of the basic low-Mach number theory has been developed [293, 294, 394, 485]. The newer development includes a more complete connection to fully compressible fluid dynamics and acoustics. This work includes applications as well as numerical methods to solve the equations.

2.6 Boussinesq Approximation

If density variations are small in magnitude,[7] the incompressible flow equations can be augmented to provide useful solutions. The governing equations are the usual mass equation

$$\nabla \cdot \mathbf{u} = \mathbf{0} \,, \tag{2.64}$$

the equation of motion with a source term[8]

$$\frac{\partial \mathbf{u}}{\partial t} + (\mathbf{u} \cdot \nabla)\mathbf{u} + \frac{1}{\rho}\nabla\phi = \nu\nabla^2\mathbf{u} + \mathbf{g} \,, \tag{2.65}$$

with ρ being constant and an energy equation,

$$\frac{D\theta}{Dt} = \kappa\nabla^2\theta. \tag{2.66}$$

The (gravity) source term, $\mathbf{g}$, in (2.65) is $(0, \eta\theta)^T$, where θ is a normalized temperature and η is the thermal expansion coefficient. This represents the effects of gravity acting on small density variations in the flow due to thermal expansion of the fluid.

2.7 Variable Density Flow

For situations where the density variations are not small and are not related to variation in temperature, the following set of equations can be solved. The governing equations are (2.64) and (2.65), plus a density transport equation

$$\frac{D\rho}{Dt} = 0 \,, \tag{2.67}$$

which can also be written as (using $\nabla \cdot \mathbf{u} = 0$)

$$\frac{\partial \rho}{\partial t} + \nabla \cdot (\rho \mathbf{u}) = 0 \,.$$

If in (2.65), $\bar{\rho}$ changes to ρ which is now a function of space and time, we obtain the equation of motion

[7] One could write the density fluctuations in terms of either density or temperature, with the temperature linked to density changes.

[8] ϕ is the pressure according to the low-Mach number scaling.

$$\frac{\partial \mathbf{u}}{\partial t} + (\mathbf{u} \cdot \nabla)\mathbf{u} + \frac{1}{\rho}\nabla\phi = \nu\nabla^2\mathbf{u} + \mathbf{F}_V \ ,$$

which should be used in conjunction with $\nabla \cdot \mathbf{u} = 0$.

Finally, the energy (temperature) equation (2.66) is given by

$$\frac{\partial \theta}{\partial t} + \nabla \cdot (\mathbf{u}\theta) = \kappa\nabla^2\theta \ .$$

2.8 Zero Mach Number Combustion

Another important use of the low-Mach number equations is the modeling of combustion. Modeling of combustion in this manner was first detailed by Sethian and Majda [365]. The main difference with this set of equations is that the flow is not completely incompressible (i.e., $\nabla \cdot \mathbf{u} \neq 0$) although sound waves are still completely absent. The governing equations (in simplified form) are an equation of motion

$$\frac{D\mathbf{u}}{Dt} + \frac{1}{\rho}\nabla\phi = \frac{\nu}{\rho}\nabla^2\mathbf{u} + \mathbf{F}_V \ , \tag{2.68}$$

the energy, or temperature equation,

$$\rho C_p \frac{D\theta}{Dt} = \frac{\mathrm{d}P_o}{\mathrm{d}t} + k\nabla^2\theta + Q_\theta \ , \tag{2.69}$$

with P_o being the thermodynamic mean pressure and Q_θ is a temperature-dependent source term; the species conservation equation,

$$\rho\frac{DZ}{Dt} = \mathcal{D}\nabla \cdot \rho\nabla Z - Q_Z \ , \tag{2.70}$$

where Z is the mass fraction (the ratio of material's mass to total mass), $\mathcal{D}$ is the diffusion of mass concentration and Q_Z denotes a chemical source or sink term; the bulk pressure equation,

$$\frac{\mathrm{d}P_o}{\mathrm{d}t} = \frac{\gamma - 1}{\mathrm{Vol}\,(\Omega)}\left\{\int_\Omega Q_\theta \,\mathrm{d}\Omega + \int_\Gamma \mathbf{n} \cdot (\kappa\nabla\theta - \gamma P_o\mathbf{u})\,\mathrm{d}\Gamma\right\} \ , \tag{2.71}$$

where Ω is the entire computational domain, $\mathrm{Vol}\,(\Omega)$ is its volume, Γ is its boundary and P_o is the thermodynamic pressure and, finally, we have the conservation of mass,

$$\nabla \cdot \mathbf{u} = \frac{\mathrm{d}P_o/\mathrm{d}t + (\gamma - 1)\left(Q_\theta + \kappa\nabla^2\theta\right)}{\gamma P_o} , \tag{2.72}$$

and an equation of state

$$\rho = \frac{P_o}{RT} \ , \tag{2.73}$$

where R is the gas constant.

One should notice that the source term on the conservation of mass equation is a part of the integral source for the bulk pressure equation. In this set of equations, R, C_p, and γ are physical constants that define the thermodynamics of the system.

2.9 Initial and Boundary Conditions

As with any set of differential equations, the initial and boundary conditions play an important role in the solution both analytically and numerically. This is doubly so for incompressible flow. As discussed by Gresho [227, 226, 228], the divergence-free condition permeates every aspect of solving incompressible flow equations. This statement includes the initial and boundary conditions. It is important for solution credibility and stability that several basic principles are followed throughout the solution of these equations.

The omnipotence of $\nabla \cdot \mathbf{u}$ can be seen in the boundary conditions for flow in a box or cavity (all boundary conditions are walls). For inviscid flow, the boundary condition is $(\mathbf{n} \cdot \mathbf{u}) = 0$, and for viscous flows the tangential velocity at the boundary is set to zero as well (assuming non-moving walls). There is also an integral constraint on the velocity that comes from the divergence-free condition

$$\int_{\Omega} \nabla \cdot \mathbf{u} \, \mathrm{d}\Omega \rightarrow \int_{\Gamma} (\mathbf{n} \cdot \mathbf{u}) \, \mathrm{d}\Gamma = 0 \, . \tag{2.74}$$

This equation must be satisfied in both the continuous and discrete case.

Concerning initial conditions, it is required that the initial velocity field be divergence-free in both the domain and the boundary. We will give a numerical procedure to do this later in the book (see Sect. 11.5).

Another tricky situation is posed by outflow boundary conditions. The necessity for outflow boundary conditions arises from the need to truncate a computational domain for the purposes of efficiency. Because the truncation of the domain is artificial, this naturally creates problems. This subject is also an open research question, but we will use a generally accepted practice.

Operationally, the boundary condition will set the pressure on the outflow boundary into hydrostatic equilibrium,

$$(\mathbf{T}\mathbf{n} - p\,\mathbf{n})_{\Gamma} = (\mathbf{F})_{\Gamma} \, ,$$

where $\mathbf{F}$ is force applied by the boundary to the fluid.

Other flow variables will be set so that their derivatives normal to the boundary are zero (homogeneous Neumann conditions)

$$(\mathbf{n} \cdot \nabla \mathcal{V})_{\Gamma} = 0 \, ,$$

with $\mathcal{V}$ being a generic solution variable.

3. The Viscous Fluid Flow Equations

Now that we have introduced the basic flow equations, next the viscous flow equations are given. The viscous equations are complex enough to warrant a full chapter's worth of treatment. This will include the basic structure of the equations and models for the physical properties, which can be quite complex.

3.1 The Stress and Strain Tensors for a Newtonian Fluid

The shear strain rate of a fluid element is defined as the average decrease of the angle between two lines of this element which are considered to be perpendicular under unstrained conditions. The components of the shear-strain rate are defined by

$$\epsilon_{xy} = \epsilon_{yx} = \frac{1}{2}\left(\frac{\partial v}{\partial x} + \frac{\partial u}{\partial y}\right) , \tag{3.1}$$

$$\epsilon_{yz} = \epsilon_{zy} = \frac{1}{2}\left(\frac{\partial w}{\partial y} + \frac{\partial v}{\partial z}\right) , \tag{3.2}$$

$$\epsilon_{xz} = \epsilon_{zx} = \frac{1}{2}\left(\frac{\partial u}{\partial z} + \frac{\partial w}{\partial x}\right) . \tag{3.3}$$

The *dilation*, or extensional strain, is defined as the horizontal length increase of the fluid element in the corresponding direction. Thus, the three extensional-strain rates are given by

$$\epsilon_{xx} = \frac{\partial u}{\partial x}, \quad \epsilon_{yy} = \frac{\partial v}{\partial y}, \quad \epsilon_{zz} = \frac{\partial w}{\partial z}. \tag{3.4}$$

The shear and extensional strains constitute a symmetric[1] second-order tensor

$$\mathbf{E}_{strain} = \epsilon_{ij} = \begin{pmatrix} \epsilon_{xx} & \epsilon_{xy} & \epsilon_{xz} \\ \epsilon_{yx} & \epsilon_{yy} & \epsilon_{yz} \\ \epsilon_{zx} & \epsilon_{zy} & \epsilon_{zz} \end{pmatrix} . \tag{3.5}$$

[1] This is by analogy with solid mechanics.

The strain tensor is associated with three invariants which are independent of direction or choice of axes. The invariants are defined by

$$I_1 = \epsilon_{xx} + \epsilon_{yy} + \epsilon_{zz} \ , \tag{3.6}$$

$$I_2 = \epsilon_{xx}\epsilon_{yy} + \epsilon_{yy}\epsilon_{zz} + \epsilon_{zz}\epsilon_{xx} - \epsilon_{xy}^2 - \epsilon_{yz}^2 - \epsilon_{zx}^2 \ , \tag{3.7}$$

$$I_3 = \begin{vmatrix} \epsilon_{xx} & \epsilon_{xy} & \epsilon_{xz} \\ \epsilon_{yx} & \epsilon_{yy} & \epsilon_{yz} \\ \epsilon_{zx} & \epsilon_{zy} & \epsilon_{zz} \end{vmatrix} \ . \tag{3.8}$$

where (3.8) is a determinant. By substituting the strain rates into the first invariant, one can easily find that $I_1 = \text{div}\ \mathbf{u}$. The strain-rate tensor can also be linked to the velocity gradient tensor $\partial u_i/\partial x_j$ (the velocities u, v, w are obtained for $i = 1, 2, 3$, respectively) and angular velocity of the fluid element. We write the velocity gradient tensor as the sum of symmetric and anti-symmetric parts

$$\frac{\partial u_i}{\partial x_j} = \frac{1}{2}\Big(\frac{\partial u_i}{\partial x_j} + \frac{\partial u_j}{\partial x_i}\Big) + \frac{1}{2}\Big(\frac{\partial u_i}{\partial x_j} - \frac{\partial u_j}{\partial x_i}\Big) = \epsilon_{ij} + \Omega_{ij} \ . \tag{3.9}$$

The term Ω_{ij} is an anti-symmetric tensor that represents the *the rate of rotation* (or *angular velocity*). The rates of rotation in the three axes x, y, z are given by

$$\dot{\omega}_x = \Omega_{yz} = \frac{1}{2}\Big(\frac{\partial w}{\partial y} - \frac{\partial v}{\partial z}\Big) \ , \tag{3.10}$$

$$\dot{\omega}_y = \Omega_{zx} = \frac{1}{2}\Big(\frac{\partial u}{\partial z} - \frac{\partial v}{\partial x}\Big) \ , \tag{3.11}$$

$$\dot{\omega}_z = \Omega_{xy} = \frac{1}{2}\Big(\frac{\partial v}{\partial x} - \frac{\partial u}{\partial y}\Big) \ . \tag{3.12}$$

The vector ω is defined by

$$\omega = 2(\dot{\omega}_x, \dot{\omega}_y, \dot{\omega}_z) \ , \tag{3.13}$$

is called *the vorticity* of the fluid. We can also write

$$\omega = \text{curl}\ \mathbf{u} = \nabla \times \mathbf{u} \ . \tag{3.14}$$

The *principal axes* of the strain rate tensor are those for which the strain rates vanish (Fig. 3.1). The strain-rate tensor then becomes

$$\epsilon_{ij} = \begin{pmatrix} \epsilon_1 & 0 & 0 \\ 0 & \epsilon_2 & 0 \\ 0 & 0 & \epsilon_3 \end{pmatrix} \ , \tag{3.15}$$

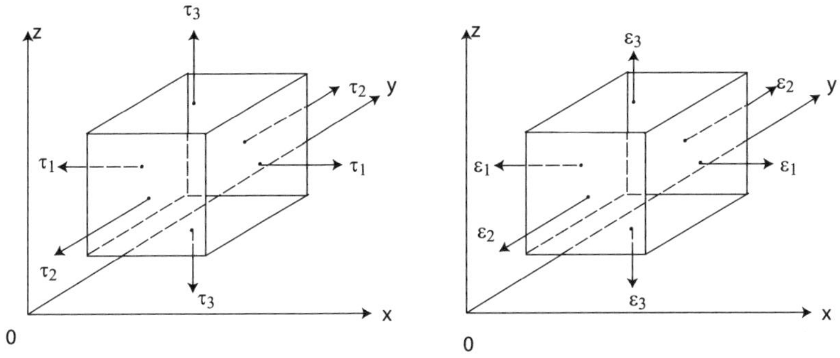

Fig. 3.1. Principal axes for the strain rate tensor and stresses.

where $\epsilon_1, \epsilon_2, \epsilon_3$ are the principal strain rates. As the result of the properties of second-order tensors, the transformation to the principal axes does not affect the sum of the diagonal terms, i.e.,

$$\epsilon_1 + \epsilon_2 + \epsilon_3 = \epsilon_{xx} + \epsilon_{yy} + \epsilon_{zz} \ . \tag{3.16}$$

The simplest relation (beyond the case of an idealized frictionless fluid) for the variation of the shear stress with strain rate is obtained by assuming the following three conditions:

1. The fluid is continuous and its viscous stress tensor $\mathbf{T} = \tau_{ij}$ is a continuous function of the strain rates ϵ_{ij} and local thermodynamic state, but independent of other kinematic quantities. This condition implies that the relation between stress and rate of strain is independent of the rotation of an element, the later being given by the anti-symmetric kinematic tensor Ω_{ij}.
2. The fluid is isotropic, i.e., its properties are independent of the direction. For an isotropic fluid a direct stress acting in it does not produce a shearing deformation.
3. The fluid is homogeneous, i.e., τ_{ij} do not depend explicitly on x, y, z.
4. When the strain rates are zero, i.e., there is no deformation, the only remaining stresses are due to the hydrostatic pressure.

Let us consider the viscous stresses τ_1, τ_2, τ_3 in the principal axes. We define a Newtonian fluid the one for which the stress components depend linearly on the rates of deformation. Since the viscous stress tensor must vanish with vanishing ϵ_i $(i = 1, 2, 3)$, we can write

$$\tau_i = \alpha_{ij}\epsilon_i \ . \tag{3.17}$$

By developing the above relation, we obtain

$$\tau_1 = \alpha_{11}\epsilon_1 + \alpha_{12}\epsilon_2 + \alpha_{13}\epsilon_3 \;, \tag{3.18}$$

$$\tau_2 = \alpha_{21}\epsilon_1 + \alpha_{22}\epsilon_2 + \alpha_{23}\epsilon_3 \;, \tag{3.19}$$

$$\tau_3 = \alpha_{31}\epsilon_1 + \alpha_{32}\epsilon_2 + \alpha_{33}\epsilon_3 \;. \tag{3.20}$$

The assumption of isotropy implies that any permutation of the principal strain rates must effect the same permutation of the principal viscous stresses. Thus, we can permute the $\epsilon_1, \epsilon_2, \epsilon_3$ to $\epsilon_3, \epsilon_1, \epsilon_2$ and obtain

$$\tau_3 = \alpha_{12}\epsilon_1 + \alpha_{13}\epsilon_2 + \alpha_{11}\epsilon_3 \;, \tag{3.21}$$

$$\tau_1 = \alpha_{22}\epsilon_1 + \alpha_{23}\epsilon_2 + \alpha_{21}\epsilon_3 \;, \tag{3.22}$$

$$\tau_2 = \alpha_{32}\epsilon_1 + \alpha_{33}\epsilon_2 + \alpha_{31}\epsilon_3 \;. \tag{3.23}$$

By comparing equations (3.22) and (3.18) we find that $\alpha_{11} = \alpha_{22}$, $\alpha_{12} = \alpha_{23}$ and $\alpha_{13} = \alpha_{21}$. Repeating the same procedure for the permutation $\epsilon_1, \epsilon_2, \epsilon_3$ to $\epsilon_2, \epsilon_3, \epsilon_1$, we obtain

$$\alpha_{11} = \alpha_{22} = \alpha_{33} \;, \tag{3.24}$$

$$\alpha_{12} = \alpha_{21} = \alpha_{23} = \alpha_{32} = \alpha_{13} = \alpha_{31} \;. \tag{3.25}$$

We introduce the coefficients $\lambda + 2\mu$ and λ to denote the elements (3.24) and (3.25), respectively; the precise scope of these coefficients will be shown later. The principal viscous stresses can thus be written as

$$\tau_i = \lambda(\epsilon_1 + \epsilon_2 + \epsilon_3) + 2\mu\epsilon_i \;. \tag{3.26}$$

Due to the assumption of isotropy, (3.26) can be transformed to a general co-ordinate system with the coefficients λ and μ remaining unaffected. Thus, for the viscous stress tensor we obtain

$$\tau_{ij} = \lambda(\epsilon_{xx} + \epsilon_{yy} + \epsilon_{zz})\delta_{ij} + 2\mu\epsilon_{ij} = \lambda I_1 + 2\mu\epsilon_{ij} \;. \tag{3.27}$$

Subsequently, the complete stress tensor σ_{ij} (see (2.26)) is written

$$\sigma_{ij} = (-p + \lambda I_1)\delta_{ij} + 2\mu\epsilon_{ij} \;. \tag{3.28}$$

By analogy to the strain rates, the sum of the three normal stresses $\sigma_{xx} + \sigma_{yy} + \sigma_{zz}$ is a tensor invariant. By defining the mean pressure, $\bar{p}$, as the average compression stress on the fluid element, i.e.,

$$\bar{p} = -\frac{1}{3}\Big(\sigma_{xx} + \sigma_{yy} + \sigma_{zz}\Big) \;, \tag{3.29}$$

and using (3.28), we obtain

$$\bar{p} = p + (\lambda + \frac{2}{3}\mu)I_1 = p + (\lambda + \frac{2}{3}\mu) \;\text{div}\; \mathbf{u} \;. \tag{3.30}$$

According to the above, the thermodynamic pressure p is not equal to the pressure deforming the fluid element. The coefficient $\lambda + 2/3\mu$ is known as *coefficient of bulk viscosity*, where μ is the dynamic viscosity of the fluid.[2] In order to obtain $\bar{p} = p$, Stokes[3] assumed that $\lambda + 2/3\mu = 0$ that gives

$$\lambda = -\frac{2}{3}\mu \, . \tag{3.31}$$

The above assumption is also known as Stokes's hypothesis and is equivalent to the assumption that the thermodynamic pressure is equal to the minus of the one-third of the invariant of the normal stresses, even for the cases where compression or expansion proceed at a finite rate. Using Stokes's hypothesis, we obtain the *constitutive equation* for an isotropic Newtonian fluid

$$\begin{aligned}
\sigma_{xx} &= -p - \frac{2}{3}\,\mu \,\mathrm{div}\,\mathbf{u} + 2\mu\frac{\partial u}{\partial x}\,,\\
\sigma_{yy} &= -p - \frac{2}{3}\,\mu \,\mathrm{div}\,\mathbf{u} + 2\mu\frac{\partial v}{\partial y}\,,\\
\sigma_{zz} &= -p - \frac{2}{3}\,\mu \,\mathrm{div}\,\mathbf{u} + 2\mu\frac{\partial w}{\partial z}\,,\\
\tau_{xy} &= \tau_{yx} = \mu\left(\frac{\partial v}{\partial x} + \frac{\partial u}{\partial y}\right),\\
\tau_{xz} &= \tau_{zx} = \mu\left(\frac{\partial u}{\partial z} + \frac{\partial w}{\partial x}\right),\\
\tau_{yz} &= \tau_{zy} = \mu\left(\frac{\partial w}{\partial y} + \frac{\partial v}{\partial z}\right).
\end{aligned} \tag{3.32}$$

By substitution of the stresses (3.32) into (2.32) we obtain the momentum equations for a Newtonian fluid.

3.2 The Navier-Stokes Equations for Constant Density Flows

Using (2.54), the shear-stress relations for a Newtonian fluid (3.32) can be simplified and, subsequently, the momentum equations (2.32) can be written in a simpler form as

$$\rho\frac{\partial \mathbf{u}}{\partial t} + \rho\,\mathrm{div}\,(\mathbf{u}\otimes\mathbf{u}) = -\nabla p + \mu\nabla^2\mathbf{u}\,. \tag{3.33}$$

Dividing (3.33) with ρ we obtain

$$\frac{\partial \mathbf{u}}{\partial t} + \mathbf{u}\cdot\nabla\mathbf{u} = -\frac{1}{\rho}\nabla p + \nu\nabla^2\mathbf{u}\,, \tag{3.34}$$

by making use of the incompressibility condition (2.54). Equations (2.54) and (3.34) are the *Navier-Stokes equations!incompressible flows.* It has been

[2] μ has units $kg \cdot m^{-1} \cdot sec^{-1}$

[3] G. G. Stokes, Trans. Camb. Phil. Soc., 8, 287-305, 1845.

established in the literature to call the Navier-Stokes equations the system of (2.54) (continuity equation) and (3.67). Using tensorial indices, the equations can also be written

$$\frac{\partial u_j}{\partial x_j} = 0 \,, \tag{3.35}$$

$$\frac{\partial u_i}{\partial t} + \frac{\partial u_i u_j}{\partial x_j} = -\frac{1}{\rho}\frac{\partial p}{\partial x_i} + \nu \frac{\partial^2 u_i}{\partial x_j^2} \,, \tag{3.36}$$

where $i, j \equiv 1, 2, 3$ for the x, y and z directions, respectively.

Equations (2.54) and (3.67) can be expanded for a Cartesian co-ordinates system (x, y, z) as

$$\frac{\partial u}{\partial x} + \frac{\partial v}{\partial y} + \frac{\partial w}{\partial z} = 0 \,, \tag{3.37}$$

$$\frac{\partial u}{\partial t} + \frac{\partial u^2}{\partial x} + \frac{\partial uv}{\partial y} + \frac{\partial uw}{\partial z} = -\frac{1}{\rho}\frac{\partial p}{\partial x} + \nu\left(\frac{\partial^2 u}{\partial x^2} + \frac{\partial^2 u}{\partial y^2} + \frac{\partial^2 u}{\partial z^2}\right) \,, \tag{3.38}$$

$$\frac{\partial v}{\partial t} + \frac{\partial uv}{\partial x} + \frac{\partial v^2}{\partial y} + \frac{\partial vw}{\partial z} = -\frac{1}{\rho}\frac{\partial p}{\partial y} + \nu\left(\frac{\partial^2 v}{\partial x^2} + \frac{\partial^2 v}{\partial y^2} + \frac{\partial^2 v}{\partial z^2}\right) \,, \tag{3.39}$$

$$\frac{\partial w}{\partial t} + \frac{\partial uw}{\partial x} + \frac{\partial vw}{\partial y} + \frac{\partial w^2}{\partial z} = -\frac{1}{\rho}\frac{\partial p}{\partial z} + \nu\left(\frac{\partial^2 w}{\partial x^2} + \frac{\partial^2 w}{\partial y^2} + \frac{\partial^2 w}{\partial z^2}\right) \,. \tag{3.40}$$

The corresponding equations for a general shear-stress tensor $\mathbf{S} \equiv \sigma_{ij}$ (2.26), are written as

$$\frac{\partial u}{\partial t} + \frac{\partial u^2}{\partial x} + \frac{\partial uv}{\partial y} + \frac{\partial uw}{\partial z} = -\frac{1}{\rho}\frac{\partial p}{\partial x} + \frac{\partial \tau_{xx}}{\partial x} + \frac{\partial \tau_{xy}}{\partial y} + \frac{\partial \tau_{xz}}{\partial z} \,, \tag{3.41}$$

$$\frac{\partial v}{\partial t} + \frac{\partial uv}{\partial x} + \frac{\partial v^2}{\partial y} + \frac{\partial vw}{\partial z} = -\frac{1}{\rho}\frac{\partial p}{\partial y} + \frac{\partial \tau_{yx}}{\partial x} + \frac{\partial \tau_{yy}}{\partial y} + \frac{\partial \tau_{yz}}{\partial z} \,, \tag{3.42}$$

$$\frac{\partial w}{\partial t} + \frac{\partial uw}{\partial x} + \frac{\partial vw}{\partial y} + \frac{\partial w^2}{\partial z} = -\frac{1}{\rho}\frac{\partial p}{\partial z} + \frac{\partial \tau_{zx}}{\partial x} + \frac{\partial \tau_{zy}}{\partial y} + \frac{\partial \tau_{zz}}{\partial z} \,. \tag{3.43}$$

Finally, another approximation is obtained when the viscous forces are much larger than the inertia forces. The advective terms can then be neglected thus obtaining the Stokes equations. The Stokes equations (neglecting volume forces) are given by

$$\text{div}\, \mathbf{T} = \nabla p \,, \tag{3.44}$$

or

$$\nabla p = \mu \nabla^2 \mathbf{u} \,, \tag{3.45}$$

for the case of a Newtonian fluid.

3.3 Non-Newtonian Constitutive Equations for the Shear-Stress Tensor

3.3.1 Generalized Newtonian Fluids

There are quite a few fluids of great technical, biological and laboratory importance whose behavior cannot be described by (3.28). These include polymers, biological solutions, soap and cellulose solutions, paints, tars, asphalts and glues, various colloids and crystalline materials. The simplest generalization of the Newtonian fluid is obtained if the dynamic viscosity coefficient is considered to be a function of the rate of strain tensor $\mathbf{E}_{strain}$. In this case, we write

$$\mathbf{T} = 2\,\eta(\mathbf{E}_{strain})\,\mathbf{E}_{strain}\ . \tag{3.46}$$

Eq. (3.46) defines the so-called generalized Newtonian fluids. In regions where η decreases with increasing rate of shear the behavior is termed as *pseudo-plastic*, while in regions where η increases with decreasing rate of shear the behavior is termed as *dilatant*. In order for η to be a scalar function of the tensor $\mathbf{E}_{strain}$, it must depend only on the invariants of $\mathbf{E}_{strain}$. Several forms for η have been proposed in the literature [21, 60] and some of them are presented below.

A form which has been extensively used is the *power-law* formula:

$$\eta = kS^{(n-1)/2}\ , \tag{3.47}$$

where k and n are constants [21, 60] and the parameter S has dimensions of the square of frequency and for constant density fluids is defined as

$$S = 4I_2 = 2\mathrm{tr}(\mathbf{E}^2) = 2\mathbf{E}_{strain} : \mathbf{E}_{strain}\ , \tag{3.48}$$

where tr denotes the trace of a tensor (first tensor invariant I), that is the sum of its diagonal elements.[4] The parameters n and k are constant parameters which are called the power-law index and the consistency, respectively. The former is dimensionless and the latter has units which depend on the value of n, i.e., $Kg/s^{2-n}m$. The Newtonian constitutive equation is obtained when $n = 1$.

In the case where the flow process begins at the instant when the shear stress reaches a certain critical value, the fluid is termed as *Bingham plastic fluid* and η is defined by

[4] The trace of the product of two tensors $\mathbf{A}$ and $\mathbf{B}$, is also indicated by $\mathbf{A} : \mathbf{B} = \mathrm{tr}(\mathbf{A}\cdot\mathbf{B})$. The second invariant, II, of a tensor $\mathbf{A}$ can also be defined using the operation of trace: $II = 1/2[I^2 - \mathrm{tr}(\mathbf{A}^2)]$.

$$\eta = -\left(\mu_0 + \frac{\tau_0}{|\sqrt{\frac{1}{2}(\mathbf{E}_{strain} : \mathbf{E}_{strain})}|}\right) \quad \text{for} \quad 2(\mathbf{T} : \mathbf{T}) > \tau_0^2 \tag{3.49}$$

$$\mathbf{T} = 0 \quad \text{for} \quad \frac{1}{2}(\mathbf{T} : \mathbf{T}) < \tau_0^2 \tag{3.50}$$

Values for the constants τ_0 and μ_0 can be found in [60].

Other forms of the function $\eta(\mathbf{E}_{strain})$ are described by the following models:

- Ostwald - de Waele model

$$\eta = -m|\sqrt{2(\mathbf{E}_{strain} : \mathbf{E}_{strain})}|^{n-1} \; . \tag{3.51}$$

- Reiner - Philippoff model

$$\eta = -\mu_\infty + \frac{\mu_\infty - \mu_0}{1 + (\mathbf{T} : \mathbf{T})/2\tau_0^2} \; . \tag{3.52}$$

- Powell - Eyring model

$$\eta = \eta_\infty + (\eta_0 - \eta_\infty)\frac{\operatorname{arcsinh}(\sqrt{S\lambda})}{\sqrt{S\lambda}} \; . \tag{3.53}$$

A summary of the various constants appearing in the above models can be found in [60].

A more general constitutive relation for the viscous stress tensor is obtained by writing (3.46) in the form

$$\mathbf{T} = \mathbf{g}(\mathbf{E}_{strain}) \; . \tag{3.54}$$

In this case the stresses at a given time will depend solely on the deformation at a given point at that time. Thus, no flow history effects are taken into account and also the dependence on the deformation is assumed to be point-wise.

3.3.2 Viscoelastic Fluids

To describe the behavior of viscoelastic fluids for which the history of the deformation gradient plays an important role in describing the current state of the fluid, more complex constitutive equations need to be employed. Such constitutive relations are obtained by taking into account some important principles of continuum mechanics. These are:

- *Principle of determinism*: the stresses acting in a medium at a certain time depend only on the kinematic history of the medium.
- *Principle of local action*: the value of the stress of a particle is the same for motions which coincide in a small neighborhood of this particle. This principle does not apply in the case of complex media described by theories of elasticity.

- *Principle of material objectivity*: the constitutive equations are independent of the motion of the observer, which may be related to an arbitrary frame of reference. This principle is also referred to as principle of invariance with respect to the frame of reference.
- *Principle of invariance with respect to the system of units.*
- *Principle of invariance with respect to the system of coordinates.*
- *Principle of invariance with respect to the reference configuration*: the constitutive equations are consistent with the properties of homogeneity and inner symmetry of the medium.
- *Principle of fading memory of the medium*: the stresses at certain time are more dependent on the recent history of deformation rather than the distant history of it.

The models of viscoelastic media, can broadly be divided into four types depending on the formulation of the constitutive equation: integral models, differential models, equations of the rate type, and equations of the mixed type. The basic form of the equation for each of the above types is briefly described below.

- *Integral models*: Let assume that $g_k(s_1, \ldots, s_k)$ are isotropic multi-linear tensor functions of k tensor variables $s_1, \ldots, s_k$, which are also symmetric. According to this model, the shear-stress

$$\mathbf{T} = \sum_{i=1}^{m} \int_0^\infty \ldots \int_0^\infty g_k(s_1, \ldots, s_k)[\mathbf{G}^t(s_1), \ldots, \mathbf{G}^t(s_k)] \, \mathrm{d}s_1 \ldots \mathrm{d}s_k \,, \qquad (3.55)$$

  where $\mathbf{G}^t(s)$ involves the deformation history. It has been shown [223] that the function g_k can be uniquely defined. A fluid described by the above constitutive equation is also called an *incompressible fluid of integral type* or *a Green-Rivlin fluid of order* m, where m denotes the order of multiplicity of the integration. The rheological integral equations up to third order have been extensively used for determining experimentally the material properties of a viscoelastic fluid (cf. [359]).
- *Models of differential type*: In this case the properties of the viscoelastic fluid depend on a very short history of the deformation. Constitutive relations can then be obtained by multiple differentiation in time of the deformation gradient. In the case of an isotropic fluid that satisfies the principle of material objectivity we obtain

$$\mathbf{T} = \mathcal{G}(\dot{\mathbf{F}}_t(t), \ldots, \mathbf{F}_t^{(n)}) = \mathcal{M}(\mathbf{A}_1, \ldots, \mathbf{A}_n) \,, \qquad (3.56)$$

  where $\mathbf{F}_t^{(n)}$ denotes high-order derivatives of the deformation gradient.[5] The second equality corresponds to another formulation of the constitutive equation based on the Rivlin-Ericksen tensors $\mathbf{A}_i$ $(i = 1, \ldots, n)$ [459]. The

[5] The deformation gradient tensor (often it is called relative deformation gradient) is defined as $\mathbf{F}_t = \nabla \mathbf{r}$ where $\mathbf{r}(\mathbf{x}, t)$ is the position of a particle.

Rivlin-Ericksen tensors are defined by the gradients of the strain history,[6] $\mathbf{C}(t)$, i.e.,

$$\mathbf{A}_n = \left. \frac{\mathrm{d}^n \mathbf{C}}{\mathrm{d}t'^n} \right|_{t'=t} . \tag{3.57}$$

The function $\mathcal{M}$ is symmetric and anisotropic in its arguments and satisfies the equation

$$\mathbf{Q}\mathcal{M}(\mathbf{A}_1, \ldots, \mathbf{A}_n)\mathbf{Q}^T = \mathcal{M}(\mathbf{Q} \cdot \mathbf{A}_1 \cdot \mathbf{Q}^T, \ldots, \mathbf{Q} \cdot \mathbf{A}_n \cdot \mathbf{Q}^T) , \tag{3.58}$$

for all orthogonal tensors. On the above basis, Rivlin and Ericksen [459] proposed general constitutive equations of the differential type, which can be used for describing the viscoelastic behavior of solids and fluids. A fluid described by such an equation is often referred to as a *Rivlin-Ericksen fluid.* The equation for a Rivlin-Ericksen fluid can be constructed using polynomial tensor functions of n tensor arguments. For example, the equation for an incompressible fluid of the differential type of order 2 is given by

$$\begin{aligned} \mathbf{T} = \alpha_1 \mathbf{A}_1 + \alpha_2 \mathbf{A}_2 + \alpha_3 \mathbf{A}_1^2 + \alpha_4 \mathbf{A}_2^2 + \\ \alpha_5(\mathbf{A}_1 \cdot \mathbf{A}_2 + \mathbf{A}_2 \cdot \mathbf{A}_1) + \alpha_6(\mathbf{A}_1^2 \cdot \mathbf{A}_2 + \mathbf{A}_2 \cdot \mathbf{A}_1^2) \\ + \alpha_7(\mathbf{A}_1 \cdot \mathbf{A}_2^2 + \mathbf{A}_2^2 \cdot \mathbf{A}_1) + \alpha_8(\mathbf{A}_1^2 \cdot \mathbf{A}_2^2 + \mathbf{A}_2^2 \cdot \mathbf{A}_1^2) . \end{aligned} \tag{3.59}$$

The coefficients α_i $(i = 1, \ldots, 8)$ depend on nine joint invariants of $\mathbf{A}_1$ and $\mathbf{A}_2$, where $\mathrm{tr}\mathbf{A}_1 = 0$, and are usually determined experimentally.

- *Models of the rate type*: These models consider that the stress tensor $\mathbf{T}(t)$ and the tensor of the history of the deformation gradient $\mathbf{F}(t)$ satisfy the differential equation

$$\mathbf{T}^{(n)} = \mathcal{S}(\mathbf{T}, \dot{\mathbf{T}}, \ldots, \mathbf{T}^{(n-1)}; \mathbf{F}, \dot{\mathbf{F}}, \ldots, \mathbf{F}^{(n-1)}) , \tag{3.60}$$

where $\mathcal{S}$ is a tensor function. The above equation defines *media of the rate type of order n*, the latter being determined by the order of the derivative of $\mathbf{T}$.

The *Oldroyd constitutive equation* [401, 402] for a viscoelastic fluid falls in this category. According to this equation the stresses (2.26) are given by

$$\begin{aligned} \mathbf{S} + \lambda_1 \widetilde{\mathbf{S}} - \mu_1(\mathbf{S}\,\mathbf{E}_{strain} + \mathbf{E}_{strain}\,\mathbf{S}) = \\ 2\eta_0(\mathbf{E}_{strain} + \lambda_2 \widetilde{\mathbf{E}}_{strain} - \mu_2 \mathbf{E}^2_{strain}) , \end{aligned} \tag{3.61}$$

where $\lambda_1, \lambda_2, \mu_1, \mu_2$ and η_0 are material constants. $\mathbf{S}$ is the stress tensor (2.26), and $\widetilde{\mathbf{S}}$ and $\widetilde{\mathbf{E}}$ are the Oldroyd convected derivatives of the stress and strain-rate tensors

$$\widetilde{\mathbf{S}} = \left. \frac{\mathrm{d}}{\mathrm{d}t} \Big(\mathbf{R}_\tau^T(t) \mathbf{S}(t) \mathbf{R}_\tau(t) \Big) \right|_{\tau=t} , \tag{3.62}$$

$$\widetilde{\mathbf{E}} = \left. \frac{\mathrm{d}}{\mathrm{d}t} \Big(\mathbf{F}_\tau^T(t) \mathbf{E}_{strain}(t) \mathbf{F}_\tau(t) \Big) \right|_{\tau=t} , \tag{3.63}$$

---

[6] The strain tensor can be defined by $\mathbf{C} = \mathbf{F}_t^T \mathbf{F}_t$.

where $\mathbf{R}_\tau$ and $\mathbf{F}_\tau$ are the tensors of rotation and deformation gradient, respectively [401, 402].

- *Mixed-type models*: These models can be obtained by combining ideas emerging from the differential and integral models. For example, the integrand in (3.55) can be assumed to depend on the time derivatives of $\mathbf{G}^t(s)$, thus obtaining an integro-differential type of model. Such models have been proposed by Green and Rivlin [223]. Models of mixed integral-rate type have also been proposed by Oldroyd [401] and Green and Rivlin [223].

3.3.3 Other Viscoelastic Models

In the non-Newtonian fluid mechanics literature one can find various models which have been developed on the basis of the above general theoretical formulations as well as based on experimental observations and data. Some of these models are reported below.

- *Simplified models of differential and rate type*: There are various models which fall in this category; see [510, 600] and references therein. As an example we give the equation for generalized Maxwell fluids [110] as proposed by White and Metzner [600]

$$(\mathbf{I} + \frac{\mu}{G}\frac{\mathrm{d}}{\mathrm{d}t})\mathbf{S} = -2\mu\mathbf{E}_{strain} \;, \tag{3.64}$$

  where $\mathrm{d}/\mathrm{d}t$ denotes the Oldroyd convected derivative, G is the shear modulus of elasticity. The ratio μ/G has dimensions of time and is called *relaxation time.*
- *The Reiner-Rivlin model* [442, 458]: This model is given by

$$\mathbf{S} = -p\mathbf{I} + a_1\mathbf{E}_{strain} + a_2\mathbf{E}^2_{strain} \;, \tag{3.65}$$

  where a_1 and a_2 are functions of the invariants $\mathrm{tr}(\mathbf{E}^2_{strain})$ and $\mathrm{tr}(\mathbf{E}^3_{strain})$.
- *Simplified integral models*: Several simplified integral models have been proposed in the past and a useful summary of such models can be found in various books [21, 527]. One of the most popular models in this category is the viscoelastic model of Bernstein-Kearsley-Zapas [56] which is referred to as *BKZ model.* The BKZ equation is written as

$$\mathbf{T} = \int_{-\infty}^{t} \Big(\frac{\partial \mathcal{U}}{\partial I_C}\mathbf{C}_t^{-1}(\tau) - \frac{\partial \mathcal{U}}{\partial II_C}\mathbf{C}_t(\tau)\Big)\mathrm{d}\tau \;, \tag{3.66}$$

  where $\mathcal{U}$ is the elastic potential defined as function of the time $t-\tau$ and the two principal invariants, $I_{C^{-1}}$ and $II_{C^{-1}}$, of the relative deformation tensor $\mathbf{C}_t^{-1}$. The definition and experimental verification for the potential $\mathcal{U}$ was considered in further studies, [21, 359, 527] and references therein.

The complexity of the modeling of viscoelastic fluids can be further increased if we do not assume that the fluid is isotropic. Anisotropic media may behave like fluids or may be considered as solids being capable to flow. Liquid crystals and suspensions of large oriented molecules fall in the category of anisotropic fluids. Various models have been proposed to account for the anisotropy, for example, [182, 399, 551], among others.

3.4 Alternative Forms of the Advective and Viscous Terms

Alternative forms of (2.32) can be obtained by using the incompressibility constraint. These are discussed below.

- *Divergence form*: Considering that the term $\mathrm{div}(\mathbf{u} \otimes \mathbf{u})$ is often written as $\nabla \cdot (\mathbf{uu})$, we can write (2.32) in their divergence form

$$\frac{\partial \mathbf{u}}{\partial t} + \nabla \cdot (\mathbf{uu}) = -\frac{1}{\rho} \nabla p + \nu \nabla^2 \mathbf{u} \,. \tag{3.67}$$

- *Advective form*: The non-linear term $\mathrm{div}(\mathbf{u} \otimes \mathbf{u})$ (*advective term*) can be written as

$$\mathrm{div}(\mathbf{u} \otimes \mathbf{u}) = \mathbf{u} \cdot \nabla \mathbf{u} + \mathbf{u}(\nabla \cdot \mathbf{u}) = \mathbf{u} \cdot \nabla \mathbf{u} \,, \tag{3.68}$$

  where the last equality is obtained by applying the incompressibility constraint. Using the above we derive the *non-conservative form* of (3.67)

$$\frac{\partial \mathbf{u}}{\partial t} + \mathbf{u} \cdot \nabla \mathbf{u} = -\frac{1}{\rho} \nabla p + \nu \nabla^2 \mathbf{u} \,. \tag{3.69}$$

- *Skew-symmetric form*: This form is obtained by writing the non-linear term as

$$\frac{1}{2}\Big(\mathrm{div}(\mathbf{u} \otimes \mathbf{u}) + \mathbf{u} \cdot \nabla \mathbf{u}\Big) = \mathbf{u} \cdot \nabla \mathbf{u} + \frac{1}{2} \mathbf{u}(\nabla \cdot \mathbf{u}) \,. \tag{3.70}$$

  This formulation was proposed by Temam [529, 530] in order to prove that the computations will be stable. Equation (3.67) is subsequently written

$$\frac{\partial \mathbf{u}}{\partial t} + \mathbf{u} \cdot \nabla \mathbf{u} + \frac{1}{2} \mathbf{u}(\nabla \cdot \mathbf{u}) = -\frac{1}{\rho} \nabla p + \nu \nabla^2 \mathbf{u} \,. \tag{3.71}$$

- *Rotational form*: One can easily show that the Laplacian of the velocity vector, $\nabla^2 \mathbf{u}$, can be written as

$$\nabla^2 \mathbf{u} = \nabla(\nabla \cdot \mathbf{u}) - \nabla \times \nabla \times \mathbf{u} \,, \tag{3.72}$$

  and by applying the incompressibility constraint we obtain

$$\nabla^2 \mathbf{u} = -\nabla \times \nabla \times \mathbf{u} = -\nabla \times \omega \,, \tag{3.73}$$

where $\omega = \nabla \times \mathbf{u}$ is the vorticity.
Furthermore, the non-linear term $\mathbf{u} \cdot \nabla \mathbf{u}$ can be written

$$\mathbf{u} \cdot \nabla \mathbf{u} = \frac{1}{2} \mathrm{div}(\mathbf{u} \otimes \mathbf{u}) - \mathbf{u} \times \nabla \times \mathbf{u} = \frac{1}{\rho} \nabla (\frac{1}{2} \rho \mathbf{u}^2) - \mathbf{u} \times \omega \,. \tag{3.74}$$

The term $\frac{1}{2}\rho\mathbf{u}^2$ is the *dynamic pressure*, and $p_t = p + \frac{1}{2}\rho\mathbf{u}^2$ is *total pressure.* Using the above formula, (3.69) is written as

$$\frac{\partial \mathbf{u}}{\partial t} + \mathbf{u} \times \omega = -\frac{1}{\rho} \nabla p_t + \nu \nabla \times \omega \,. \tag{3.75}$$

In the rotational form (2.54) and (3.75) are solved in conjunction with $\omega = \nabla \times \mathbf{u}$. The resulting system of equations contains only first-order derivatives but more variables need to be computed compared to the advective and skew symmetric form. As Gresho and Sani [231] have pointed out, the term $\omega \times \mathbf{u}$ can become very small when the vorticity and velocity vectors tend to be aligned. The above can occur in certain regions of a turbulent flow field [198]. For irrotational flows $\omega = 0$, the term ∇p_t can be used to define the boundary conditions p_t = constant at the outflow boundary.

- *Quadratically conserving form*: This form of the Navier-Stokes equations was proposed by Heywood et al. [254] and is based on the identity

$$\omega \times \mathbf{u} = \mathbf{u} \cdot \nabla \mathbf{u} - (\nabla \mathbf{u}) \cdot \mathbf{u} = \mathbf{u} \cdot \nabla \mathbf{u} - \mathbf{u} \cdot (\nabla \mathbf{u})^T \,. \tag{3.76}$$

This identity can be used to replace the term $\omega \times \mathbf{u}$ in (3.75). By doing so and also by retaining the Laplacian operator in the calculation of the viscous term, we avoid the curl operation for calculating the vorticity. The momentum equation (3.75) is then written

$$\frac{\partial \mathbf{u}}{\partial t} + \mathbf{u} \cdot \nabla \mathbf{u} - \mathbf{u} \cdot (\nabla \mathbf{u})^T = -\frac{1}{\rho} \nabla p_t - \nu \nabla^2 \mathbf{u} \,. \tag{3.77}$$

The reader can note that the total pressure p_t has been retained in the momentum equation.

The most commonly used forms of the momentum equation are the *divergence* and *advective* forms. These forms will be used throughout the book to present high-resolution schemes for the incompressible Navier-Stokes equations

3.5 Nondimensionalization of the Governing Equations

We define reference variables L, U and ρ for the length, velocity and density, respectively. The relevant dimensionless variables in the Navier-Stokes equations, (3.35) and (3.36), can be defined as

$$u_i^* = \frac{u_i}{U}; \quad x_i^* = \frac{x_i}{L}, \tag{3.78}$$

$$t^* = \frac{tU}{L}; \quad p^* = \frac{p}{\rho U^2}. \tag{3.79}$$

Using the above, we replace the variables u_i, t and p in (3.35) and (3.36) with their dimensionless counterparts and thus obtain

$$\frac{\partial u_j^*}{\partial x_j^*} = 0, \tag{3.80}$$

$$\frac{\partial u_i^*}{\partial t^*} + \frac{\partial u_i^* u_j^*}{\partial x_j^*} = -\frac{\partial p^*}{\partial x_i^*} + \frac{\nu}{UL}\frac{\partial^2 u_i^*}{\partial x_j^* \partial x_j^*}. \tag{3.81}$$

The continuity equation remains invariant, but the momentum equation contains on the right hand side one dimensionless parameter known as the *Reynolds number*

$$Re = \frac{UL}{\nu}. \tag{3.82}$$

Thus the momentum equation is written

$$\frac{\partial u_i^*}{\partial t^*} + \frac{\partial u_i^* u_j^*}{\partial x_j^*} = -\frac{\partial p^*}{\partial x_i^*} + \frac{1}{Re}\frac{\partial^2 u_j^*}{\partial x_j^* \partial x_j^*}. \tag{3.83}$$

To simplify the presentation, the superscript "*" will be omitted throughout the presentation.

The Re number expresses the ratio of inertia to viscous forces. Although the Re number is the most frequently used parameter in the analysis of incompressible flows, there are other dimensionless parameters that can also be defined in order to facilitate the analysis of fluid flow phenomena. These are:

- The *Froude* number:

$$Fr = \frac{U^2}{gL}, \tag{3.84}$$

  where g is the gravitational acceleration. This parameter is obtained if the gravitational force per unit volume (ρg) is included in the equations (see Eq. 2.32). The Froude number is the ratio of inertia to gravity forces and is the dominant parameter in free-surface flows. Examples of flows in which the Froude number can be important include the waves generated by a ship, open channel flows, jet flows.
- The *Weber* number:

$$We = \frac{\rho U^2 L}{\sigma}, \tag{3.85}$$

where σ is the coefficient of surface tension (measured in SI units as Newtons per meter). The Weber number is the ratio of inertia forces to surface tension and is used when surface tension is important. Applications in which the Weber number may be important include droplets, capillary and free surface flows.

- The *Euler* number:

$$\mathrm{Eu} = \frac{\Delta p}{\rho U^2} , \tag{3.86}$$

where Δp is a reference pressure difference. The Euler number is the ratio of inertia forces to pressure forces and is important only if the pressure drops low enough to cause cavitation (vapor formation) in a liquid.

- The *Strouhal* number:

$$St = \frac{fL}{U} , \tag{3.87}$$

where f is the frequency of the unsteady motion. The Strouhal number is an important parameter in time-dependent flows. The frequency f may be associated with the forced unsteady motion which is numerically imposed through the boundary conditions or by the inherent unsteadiness of the flow. Examples of forced unsteady flows are the piston cylinder flow in internal combustion engines, and flows around oscillating aerofoils (for example, helicopter blades); in these cases f is the frequency of the piston and aerofoil oscillation, respectively. An example of inherent unsteadiness is the periodic vortex shedding behind a cylinder, known as a *Kármán vortex street*. If the vortex shedding frequency is near the structural vibration frequency of a body then resonance can occur.

- The *Grashof* number:

$$Gr = \frac{\bar{\beta} \Delta T g L^3 \bar{\rho}^2}{\mu^2} , \tag{3.88}$$

where ΔT denotes temperature difference, $\bar{\rho}$ is a reference density taken at a reference temperature $\bar{T}$, and $\bar{\beta}$ is the coefficient of volume expansion.[7] The Grashof number arises in problems of free convection. Although in this case the fluid density is not constant, the flows may be of very low speed. Such flows fall into the category of the so-called variable density (low-speed) flows (see Chap. 18).

Remark 3.5.1. One can also introduce the *Re*, *St* and Eu numbers simultaneously in the dimensionless form of the incompressible Navier-Stokes equations. This can be done if the dimensionless time and pressure are defined by

$$t^* = tf; \quad p^* = \frac{p}{\Delta p} . \tag{3.89}$$

[7] The coefficient of volume expansion is defined by $\bar{\beta} = -\frac{1}{\rho}\left(\frac{\rho}{T}\right)_p$.

Using (3.78) and (3.89), the momentum equation (3.67) is written in dimensionless form as

$$St\frac{\partial u_i^*}{\partial t^*} + \frac{\partial u_i^* u_j^*}{\partial x_j^*} = -\mathrm{Eu}\frac{\partial p^*}{\partial x_i^*} + \frac{1}{Re}\frac{\partial^2 u_i^*}{\partial x_j^* \partial x_j^*} , \tag{3.90}$$

thus containing the three dimensionless numbers. Even though (3.90) seems to be a more general dimensionless form of the momentum equation, (3.83) is still the commonest utilized form in the analysis and simulation of incompressible flows.

If the time and pressure are nondimensionalized by

$$t^* = \frac{t\nu}{L^2}; \quad p^* = \frac{p\rho L}{\mu U} , \tag{3.91}$$

then (3.36) is written

$$\frac{\partial u_i^*}{\partial t^*} + Re\frac{\partial u_i^* u_j^*}{\partial x_j^*} = -\frac{\partial p^*}{\partial x_i^*} + \frac{\partial^2 u_i^*}{\partial x_j^* \partial x_j^*} . \tag{3.92}$$

Eq. (3.92) is more appropriate for low Reynolds numbers ($Re \to 0$). For $Re = 0$ we obtain the Stokes flow which is dominated solely by viscous effects.

3.6 General Remarks on Turbulent Flow Simulations

The number of grid points (N) that would be required to capture all the length scales in a three dimensional turbulent flow simulation would be

$$N \propto \left(\frac{ul}{\nu}\right)^{9/4} = Re_l^{9/4} , \tag{3.93}$$

where u and l are the characteristic velocity and length scales of the largest eddies in the turbulent flow; Re_l is the corresponding Reynolds number. In engineering and geophysical applications, Re_l is usually of the order of $\mathcal{O}(10^6) - \mathcal{O}(10^8)$ and thus the number of grid points needed to perform simulations would be of the order of $\mathcal{O}(10^{13}) - \mathcal{O}(10^{18})$. Therefore, direct numerical simulation (DNS) of flows at such high Reynolds numbers is beyond the projecting future capacity of parallel computers, unless there is a major breakthrough in computers technology.

There also exist important uncertainties in DNS of turbulent flows arising from:

1. The lack of precise initial and boundary conditions for the smallest scales of motion. This is an unfeasible task if one takes into account that even small perturbations can excite the small scales of motion and trigger flow instabilities.

2. The nonlinear nature of the advective terms in the Navier-Stokes equations results in the lack of unique solutions - in a strict mathematical sense[8] - as well as in the possibility of numerical instabilities and spurious solutions [159, 163].[9]

Additionally, the details of the turbulent motion at the level of the small scales are not required in most of the practical applications. Therefore, all high Reynolds number turbulent flows of practical interest are currently simulated on the basis of averaged forms of the Navier-Stokes equations in conjunction with "closure" assumptions for modeling several correlations arising as a result of the averaging. The numerical approaches that employ averaging forms of the equations are: (i) the "Reynolds-Averaged Navier-Stokes" (RANS),[10] and (ii) the "Large Eddy Simulation" (LES). Below, we present the averaged equations for each of these approaches.

3.7 Reynolds-Averaged Navier-Stokes Equations (RANS)

The starting point for deriving the RANS equations is the Reynolds decomposition[11] of the flow variables $u_i(\mathbf{x},t)$ and $p(\mathbf{x},t)$ into the sum of mean $\bar{u}_i(\mathbf{x}$, $\bar{p}(\mathbf{x})$, and fluctuating components $u_i^{'}(\mathbf{x},t)$ and $p(\mathbf{x},t)$, i.e.,

$$u_i(\mathbf{x},t) = \bar{u}_i(\mathbf{x}) + u_i^{'}(\mathbf{x},t) \ , \tag{3.94}$$

$$p(\mathbf{x},t) = \bar{p}(\mathbf{x}) + p'(\mathbf{x},t) \ , \tag{3.95}$$

where the quantities denoted with a bar are the time-averaged velocities and pressure defined by

$$\bar{u}_i(\mathbf{x}) = \lim_{T\to\infty} \frac{1}{T} \int\limits_t^{t+T} u_i(\mathbf{x},t) \ dt \ , \tag{3.96}$$

$$\bar{p}_i(\mathbf{x}) = \lim_{T\to\infty} \frac{1}{T} \int\limits_t^{t+T} p(\mathbf{x},t) \ dt \ . \tag{3.97}$$

If the mean flow varies slowly in time then (3.94)-(3.97) can be modified as [602]

---

[8] For a comprehensive review on the subject of existence and uniqueness of solutions for the Navier-Stokes equations, we refer the reader to [350].

[9] We will return to the issue of numerical artifacts in subsequent chapters.

[10] This is also referred to as "Statistical Turbulence Modeling" (STM), "Engineering Turbulence Modeling" (ETM), or "Conventional Turbulence Modeling" (CTM).

[11] It was introduced by Osborne Reynolds in 1895 [443].

$$u_i(\mathbf{x},t) = \bar{u}_i(\mathbf{x},t) + u_i'(\mathbf{x},t) \;, \tag{3.98}$$

$$p(\mathbf{x},t) = \bar{p}(\mathbf{x},t) + p'(\mathbf{x},t) \;. \tag{3.99}$$

where

$$\bar{u}_i(\mathbf{x},t) = \frac{1}{T}\int_t^{t+T} u_i(\mathbf{x},t)\,\mathrm{d}t \;, \quad T_1 << t << T_2 \;, \tag{3.100}$$

$$\bar{p}_i(\mathbf{x},t) = \frac{1}{T}\int_t^{t+T} p(\mathbf{x},t)\,\mathrm{d}t \;, \quad T_1 << t << T_2 \;, \tag{3.101}$$

and T_1 and T_2 are the time scales of the fluctuations and slow motion in the flow, respectively. The time scales should differ by several orders of magnitude, but in engineering and geophysical applications very few flows satisfy this condition; this also argues in favor of the LES approach.

We also mention that because of the time averaging the term $\partial\bar{u}_i/\partial t$ should vanish. However, this term may not be zero if, for example, the unsteadiness in the flow is imposed by external factors such as motion of the boundary. In this case the use of time averaged equations to model the unsteady flow is only justified if the unsteadiness due to the imposed boundary conditions is not modulating the inherent unsteadiness due to turbulence. This is equivalent to assuming that there are two dominant time scales in the flow, namely a slow time scale associated with the mean flow and a fast time scale associated with the turbulent fluctuations. This reflects on the time averaging of the time derivative as follows [602]

$$\frac{1}{T}\int_t^{t+T} \frac{\partial}{\partial t}\Big(\bar{u}_i(\mathbf{x},t) + u_i'(\mathbf{x},t)\Big)\,\mathrm{d}t = \frac{\bar{u}_i(\mathbf{x},t+T) - \bar{u}_i(\mathbf{x},t)}{T} + \frac{u_i'(\mathbf{x},t+T) - u_i'(\mathbf{x},t)}{T} \;. \tag{3.102}$$

On the basis of the assumption of two dominant time scales, one can say that the second term in the RHS of (3.102) vanishes because T will be much larger than the time scale of turbulent fluctuations, whereas the first term in the RHS of (3.102) can be approximated by $\partial\bar{u}_i/\partial t$ because T is much smaller than the time scale of the mean flow. The selection of two time scales in the same equation may seem to be controversial but it could be justified by using the two-timing method from perturbation theory [289].

For a physical quantity, A, the following properties also apply

$$\bar{A'} = 0 \,, \tag{3.103}$$

$$\bar{A} = \overline{\bar{A} + A'} = \bar{\bar{A}} \,, \tag{3.104}$$

$$\overline{\frac{\partial A}{\partial x_i}} = \frac{\partial \bar{A}}{\partial x_i} \,. \tag{3.105}$$

For the product of two statistically dependent quantities, A and B, we obtain

$$\begin{aligned}
\overline{AB} &= \overline{(\bar{A} + A')(\bar{B} + B')} \\
&= \overline{\bar{A}\bar{B} + \bar{A}B' + \bar{B}A' + A'B'} \\
&= \overline{\bar{A}\bar{B}} + \overline{\bar{A}B'} + \overline{\bar{B}A'} + \overline{A'B'} \\
&= \bar{A}\bar{B} + \overline{A'B'} \,,
\end{aligned} \tag{3.106}$$

and

$$\overline{\bar{A}B} = \bar{\bar{A}}\bar{B} = \bar{A}\bar{B} \,. \tag{3.107}$$

The quantities A and B are said to be correlated if $\overline{A'B'} \neq 0$; otherwise, they are uncorrelated. Similarly, for the product of three statistically dependent quantities, A, B and C, we obtain

$$\overline{ABC} = \bar{A}\bar{B}\bar{C} + \bar{A}\overline{B'C'} + \bar{B}\overline{A'C'} + \bar{C}\overline{A'B'} + \overline{A'B'C'} \,. \tag{3.108}$$

To derive the RANS equations, we time average (3.35) and (3.36). It can be easily shown that (3.35) reads

$$\frac{\partial \bar{u}_j}{\partial x_j} = 0 \,. \tag{3.109}$$

Using (3.103), (3.106) and (3.107), the time averaged (3.36) gives

$$\frac{\partial \bar{u}_i}{\partial t} + \frac{\partial \bar{u}_i \bar{u}_j}{\partial x_j} = -\frac{1}{\rho}\frac{\partial \bar{p}}{\partial x_i} + \nu \frac{\partial^2 u_i}{\partial x_i^2} - \frac{\partial \overline{u_i' u_j'}}{\partial x_j} \,, \tag{3.110}$$

where the correlation $\overline{u_i' u_j'}$ is known as the *Reynolds-stress tensor* which needs to be defined through a turbulence closure (model) [602, 316, 169]. The latter may be obtained by deriving additional PDEs for the Reynolds stresses. This can be done as follows:

We define the operator

$$\mathcal{N}(u_i) = \frac{\partial u_i}{\partial t} + u_k \frac{\partial u_i}{\partial x_k} + \frac{1}{\rho}\frac{\partial p}{\partial x_i} - \nu \frac{\partial^2 u_i}{\partial x_k^2} = 0 \,, \tag{3.111}$$

and form the following time average

$$\overline{u_i' \mathcal{N}(u_j) + u_j' \mathcal{N}(u_i)} = 0 \,. \tag{3.112}$$

The unsteady term yields

$$
\begin{aligned}
\overline{u_i^{'}\frac{\partial u_j}{\partial t} + u_j^{'}\frac{\partial u_i}{\partial t}} &= \overline{u_i^{'}\frac{\partial}{\partial t}(\bar{u}_j + u_j^{'})} + \overline{u_j^{'}\frac{\partial}{\partial t}(\bar{u}_i + u_i^{'})} \\
&= \overline{u_i^{'}\frac{\partial u_j^{'}}{\partial t}} + \overline{u_j^{'}\frac{\partial u_i^{'}}{\partial t}} \\
&= \frac{\partial \overline{u_i^{'}u_j^{'}}}{\partial t} \ .
\end{aligned} \tag{3.113}
$$

The convective term yields

$$
\begin{aligned}
\overline{u_i^{'}u_k\frac{\partial u_j}{\partial x_k} + u_j^{'}u_k\frac{\partial u_i}{\partial x_k}} &= \overline{u_i^{'}(\bar{u}_k + u_k^{'})\frac{\partial}{\partial x_k}(\bar{u}_j + u_j^{'})} \\
&\quad + \overline{u_j^{'}(\bar{u}_k + u_k^{'})\frac{\partial}{\partial x_k}(\bar{u}_i + u_i^{'})} \\
&= \overline{u_i^{'}\bar{u}_k\frac{\partial u_j^{'}}{\partial x_k}} + \overline{u_i^{'}u_k^{'}\frac{\partial(\bar{u}_j + u_j^{'})}{\partial x_k}} \\
&\quad + \overline{u_j^{'}\bar{u}_k\frac{\partial \bar{u}_i^{'}}{\partial x_k}} + \overline{u_j^{'}u_k^{'}\frac{\partial(\bar{u}_i + u_i^{'})}{\partial x_k}} \\
&= \bar{u}_k\frac{\partial \overline{u_i^{'}u_j^{'}}}{\partial x_k} + \overline{u_i^{'}u_k^{'}}\frac{\partial \bar{u}_j}{\partial x_k} \\
&\quad + \overline{u_j^{'}u_k^{'}}\frac{\partial \bar{u}_i}{\partial x_k} + \overline{u_k^{'}\frac{\partial u_i^{'}u_j^{'}}{\partial x_k}} \\
&= u_k\frac{\partial \overline{u_i^{'}u_j^{'}}}{\partial x_k} + \overline{u_i^{'}u_k^{'}}\frac{\partial \bar{u}_j}{\partial x_k} \\
&\quad + \overline{u_j^{'}u_k^{'}}\frac{\partial \bar{u}_i}{\partial x_k} + \frac{\partial \overline{u_i^{'}u_j^{'}u_k^{'}}}{\partial x_k} \ .
\end{aligned} \tag{3.114}
$$

The viscous terms yield

$$
\begin{aligned}
\overline{\nu\left(u_i^{'}\frac{\partial^2 u_j}{\partial x_k \partial x_k} + u_j^{'}\frac{\partial^2 u_i}{\partial x_k \partial x_k}\right)} &= \overline{\nu u_i^{'}\frac{\partial^2}{\partial x_k \partial x_k}\left(\bar{u}_j + u_j^{'}\right)} \\
&\quad + \overline{\nu u_j^{'}\frac{\partial^2}{\partial x_k \partial x_k}\left(\bar{u}_i + u_i^{'}\right)} \\
&= \overline{\nu u_i^{'}\frac{\partial^2 u_j^{'}}{\partial x_k \partial x_k}} + \overline{\nu u_j^{'}\frac{\partial^2 u_i^{'}}{\partial x_k \partial x_k}} \\
&= \nu\frac{\partial^2 \overline{u_i^{'}u_j^{'}}}{\partial x_k \partial x_k} - 2\nu\overline{\frac{\partial u_i^{'}}{\partial x_k}\frac{\partial u_j^{'}}{\partial x_k}} \ .
\end{aligned} \tag{3.115}
$$

The pressure gradient term gives

$$\frac{1}{\rho}\overline{u_i'\frac{\partial p}{\partial x_j}+u_j'\frac{\partial p}{\partial x_i}} = \frac{1}{\rho}\overline{u_i'\frac{\partial}{\partial x_j}\left(\bar{p}+p'\right)}+\frac{1}{\rho}\overline{u_j'\frac{\partial}{\partial x_i}\left(\bar{p}+p'\right)}$$
$$= \frac{1}{\rho}\overline{u_i'\frac{\partial p'}{\partial x_j}+u_j'\frac{\partial p'}{\partial x_i}} \ . \tag{3.116}$$

The relations (3.113), (3.114), (3.115) and (3.116) give the equations for the Reynolds stress tensor

$$\frac{\partial\overline{u_i'u_j'}}{\partial t}+\bar{u}_k\frac{\partial\overline{u_i'u_j'}}{\partial x_k} = \overline{u_i'u_k'}\frac{\partial \bar{u}_j}{\partial x_k}+\overline{u_j'u_k'}\frac{\partial \bar{u}_i}{\partial x_k}+2\nu\overline{\frac{\partial u_i'}{\partial x_k}\frac{\partial u_j'}{\partial x_k}}$$
$$+\frac{1}{\rho}\overline{u_i'\frac{\partial p'}{\partial x_j}+u_j'\frac{\partial p'}{\partial x_i}}-\nu\frac{\partial^2\overline{u_i'u_j'}}{\partial x_k\partial x_k}+\frac{\partial\overline{u_i'u_j'u_k'}}{\partial x_k} \ . \tag{3.117}$$

Eq. (3.117) contains new unknown correlations on the RHS (the terms under the overbar)[12] which subsequently need to be defined by a turbulence model [169]. The above equation shows exactly the closure problem in the averaged form of the Navier-Stokes equations, i.e. averaging of the equations leads to additional unknown correlations.

3.8 Large Eddy Simulation (LES)

LES is based on the separation of small from the large scales through a filtering operation [498, 324, 346]. Reviews of this approach can be found in [477, 427, 336, 337, 468, 187]. The resolved part of a variable $f(\mathbf{x},t)$ is defined by

$$\overline{f(\mathbf{x},t)} = \int\int\int_{-\infty}^{+\infty}\int G(\mathbf{x}-\mathbf{x}',t-t')f(\mathbf{x}',t')\,\mathrm{d}x'\,\mathrm{d}y'\,\mathrm{d}z'\,\mathrm{d}t' \ , \tag{3.118}$$

where the function G is a weight, averaging or filter function defined as

$$G(\mathbf{x},t) = G_t(t)\prod_{s=1}^{3}G_s(x_s) \ , \tag{3.119}$$

where $G_t(t)$ and $G_s(x_s)$ are the temporal and spatial components of the filter associated with the cutoff scales in space and time, respectively. The spatial and temporal filters remove the fluctuating components in space and time, respectively. In the LES literature, spatial filtering (only) is more often applied [427], i.e.,

[12] There are 22 unknown correlations, in total, accounting for all symmetries.

$$\overline{f(\mathbf{x},t)} = \int\int\limits_{-\infty}^{+\infty}\int G_s(\mathbf{x}-\mathbf{x}')f(\mathbf{x}',t')\,\mathrm{d}x'\,\mathrm{d}y'\,\mathrm{d}z' \,, \tag{3.120}$$

The most commonly used filter functions are

- The "box" or "top-hat" filter

$$G(x-x') = \begin{cases} \dfrac{1}{\Delta} \text{ for } |x-x'| \le \Delta/2 \,, \\ 0 \qquad\qquad \text{otherwise} \,. \end{cases} \tag{3.121}$$

  where Δ is the filter width (the cutoff scale in space).
- The Gaussian filter

$$G(x-x') = \left(\frac{\gamma}{\pi\Delta^2}\right)^{1/2} \exp\left(\frac{-\gamma|x-x'|^2}{\Delta^2}\right) , \tag{3.122}$$

  where γ is a constant usually taken equal to 6.
- The spectral or "sharp" cutoff filter

$$G(x-x') = \frac{\sin(\kappa(x-x'))}{k(x-x')} \,, \quad k = \frac{\pi}{\Delta} \,. \tag{3.123}$$

The fluctuating (unresolved) component, f', is defined by

$$f' = f - \bar{f} \,. \tag{3.124}$$

Also, note that $\bar{f'} \neq 0$. Application of the filtering operation to the Navier-Stokes equations (3.35) and (3.36), yields

$$\frac{\partial \bar{u}_i}{\partial x_i} = 0 \,, \tag{3.125}$$

$$\frac{\partial \bar{u}_i}{\partial t} + \frac{\partial \overline{u_i u_j}}{\partial x_j} = -\frac{1}{\rho}\frac{\partial \bar{p}}{\partial x_i} + \frac{\partial^2 \bar{u}_i}{\partial x_i^2} \,, \tag{3.126}$$

where $\bar{u}_i$ and $\bar{p}$ are the filtered velocities and pressure. Following Leonard [324], (3.125) can also be written

$$\frac{\partial \bar{u}_i}{\partial t} + \frac{\partial \bar{u}_i \bar{u}_j}{\partial x_j} = -\frac{1}{\rho}\frac{\partial \bar{p}}{\partial x_i} + \nu\frac{\partial^2 \bar{u}_i}{\partial x_i^2} - \frac{\partial \tau_{ij}^{sgs}}{\partial x_j} \,, \tag{3.127}$$

where

$$\tau_{ij}^{sgs} = \overline{u_i u_j} - \bar{u}_i \bar{u}_j \,, \tag{3.128}$$

are the *subgrid scale (SGS) stresses*. Using $u_i^{'} = u_i - \bar{u}_i$, the term $\overline{u_i u_j}$ can be written

$$
\begin{aligned}
\overline{u_i u_j} &= \overline{(\bar{u}_i + u'_i)(\bar{u}_j + u'_j)} \\
&= \overline{\bar{u}_i \bar{u}_j} + \overline{\bar{u}_i u'_j} + \overline{u'_i \bar{u}_j} + \overline{u'_i u'_j} \ ,
\end{aligned} \tag{3.129}
$$

By substituting (3.129) into (3.128) we can write the SGS stresses in a triple decomposition form

$$
\tau_{ij}^{sgs} = L_{ij} + C_{ij} + R_{ij} \ , \tag{3.130}
$$

where

$$
L_{ij} = \overline{\bar{u}_i \bar{u}_j} - \bar{u}_i \bar{u}_j \ , \tag{3.131}
$$

$$
C_{ij} = \overline{\bar{u}_i u'_j} + \overline{u'_i \bar{u}_j} \ , \tag{3.132}
$$

$$
R_{ij} = \overline{u'_i u'_j} \ . \tag{3.133}
$$

The term L_{ij} is the Leonard stresses representing interactions between resolved scales that result in large scale contributions. The cross stress tensor C_{ij} represents interactions between resolved and unresolved scales, and the SGS Reynolds stresses, R_{ij}, represent interactions between small, (unresolved) scales. According to the basic principles of modeling in mechanics, the filtered Navier-Stokes equations should be invariant with respect to a Galilean transformation. It has been shown [508] that while τ_{ij}^{sgs}, $L_{ij} + C_{ij}$ and R_{ij} are Galilean invariant, neither L_{ij} nor C_{ij} necessarily are (see also [208]). Because of this, the decomposition (3.131) is not often used and, instead, it is preferred the SGS stresses τ_{ij}^{sgs} to be directly modeled.

3.9 Closing Remarks

- Eqs. (3.125) and (3.127) have been derived considering that all the differentiation operators commute with the filtering operator, i.e., for any function f,

$$
\overline{\frac{\partial f}{\partial x}} = \frac{\partial \bar{f}}{\partial x} \ . \tag{3.134}
$$

  It is verified [211] that (3.134) is valid if the filter width Δ is constant. This is not, however, the case in non-uniform grids where Δ is variable. A systematic analysis of commutation errors can be found in [211, 213].
- The closure problem appearing in the RANS formulation also remains in the LES formulation. However, in the LES formulation the large scales are explicitly resolved and therefore modeling of the unknown stresses τ_{ij}^{sgs} is expected to be more straightforward, though numerical experiments usually prove otherwise.

- There is a direct relationship between existing SGS models and the numerical dissipation mechanism provided by modern high-resolution methods. This issue will be discussed throughout the presentation in the book. We will focus on this matter in Chap. 19.
- The LES approach is more often based on spatial filtering. In geophysical and engineering applications, it is often the case that the scales of interest are large in both space and time. There have been investigations to address the above issue through a space-time filtering (STF) approach [126, 127, 5]. The STF approach aims at eliminating both the high wave number and high frequency components.
- Spatial and temporal discretization schemes also provide a filtering mechanism. Additional filtering effects also arise from the use of multigrid methods as well as iterative algorithms such as multistage Runge-Kutta schemes. Filtering effects due to the numerical discretization will be discussed in Chap. 19.

4. Curvilinear Coordinates and Transformed Equations

The discretization and solution of the Navier-Stokes equations is obtained on an arrangement of points which is called *grid*. There are different ways to generate a grid around a geometry and these are generally classified as curvilinear (body-fitted), unstructured and Cartesian grid methods (Fig. 4.1). A body-fitted grid follows the curvature of the geometry and contains only quadrilateral cells (in two-dimensions) or cubic elements (in three-dimensions). For complex topologies the grid may contain several structured blocks (block-structured grid) (Fig. 4.1). Triangulation of the computational domain results in an unstructured grid in which the connectivity between cells is not based on a logical structure [37, 377, 391, 586] (Fig. 4.1). Hybrid approaches based on prismatic grids have also been proposed [283]. In the case of unstructured grids explicitly tracking of the neighboring cells is required.

If a rectangular grid is used for a non-rectangular geometry then the intersection of the grid lines with the geometry will result in non-quadrilateral cells. The grid generation approach based on rectangular grids even for complex (non-rectangular) geometries is referred to as *Cartesian grid methods.* There have been over the past few years an increasing interest in developing *Cartesian grid methods* as an alternative to *block-structured* body-fitted and *unstructured* grid methods [3, 93, 114, 418].

The body-fitted grids are used to overcome the numerical difficulties arising from the appearance of non-quadrilateral cells near the surface in the case of Cartesian grids. This chapter presents the transformation of the incompressible Navier-Stokes equations for the case of non-uniform, moving and non-moving, *body-fitted grids.*

4.1 Generalized Curvilinear Coordinates

Body-fitted grids follow the shape of the boundary and can be uniform or non-uniform in terms of the grid spacing. Body-fitted grids require the introduction of generalized curvilinear coordinates (ξ, η, ζ). The coordinate lines (ξ, η, ζ) are defined such that the geometrical boundary corresponds to a coordinate line ξ = const, η = const or ζ = const. Since the independent variables in the governing equations include also the time t, in the gener-

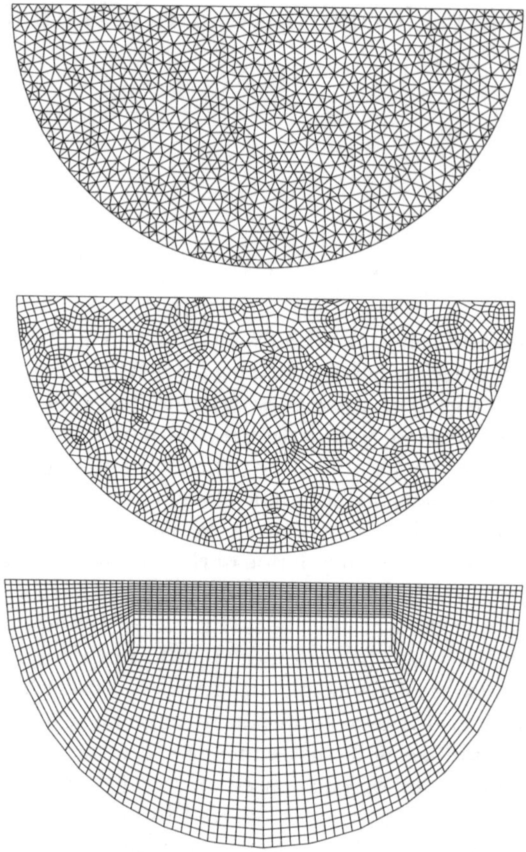

Fig. 4.1. Examples of grid generation on a half circle, triangulation, unstructured quadrilateral and block-structured (body-fitted) grids. The meshes were generated by Jim Warsa of Los Alamos National Laboratory.

alized coordinate system a new time variable τ should be considered. The transformation of the coordinates is then defined as

$$\left.\begin{aligned} \xi &\equiv \xi(x,y,z,t) \\ \eta &\equiv \eta(x,y,z,t) \\ \zeta &\equiv \zeta(x,y,z,t) \\ \tau &\equiv \tau(t) \end{aligned}\right\} . \tag{4.1}$$

In the CFD literature τ is almost always considered to be equal to t and thus the transformation with respect to time is simply defined as $\tau = t$.

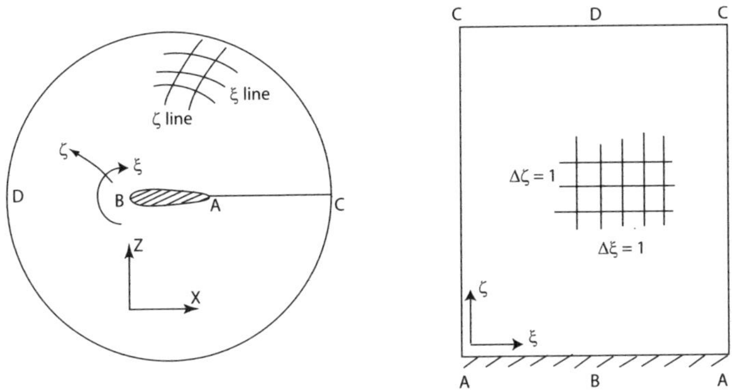

Fig. 4.2. O-type arrangement for the physical and computational domains for a two-dimensional topology.

The body-fitted grid introduces an extra difficulty associated with the differentiation of the spatial derivatives on a non-uniform grid. To overcome this difficulty, (4.1) should transform the curvilinear grid in physical space to a rectangular uniform grid (called *the computational plane*) in terms of (ξ, η, ζ) (Figs. 4.2 and 4.3). The transformation should be such that there is one-to-one correspondence of the grid points from the physical to the computational plane and since the grid is uniform one can consider that the grid spacing in the *computational plane* is $\Delta\xi = \Delta\eta = \Delta\zeta = 1$, without loss of generality. One can choose between different types of arrangement in the transformation from the physical to the computational plane. O- and C-type arrangements are shown in Figs. 4.2 and 4.3, respectively. In the case of an O-type arrangement periodic boundary conditions are applied along the AC boundary line. In the case of a C-type arrangement periodic boundary conditions are applied along AB and CD.

Since the solution of the equations is easier to be performed on the *computational plane*, the spatial derivatives appearing in the Navier-Stokes equations with respect to the Cartesian coordinates need to be transformed in terms of the curvilinear coordinates. Below we present the transformation of the derivatives for the case of a two-dimensional coordinate system.

Using the chain rule of differentiation, the partial derivative $\partial/\partial x$ is written

$$\frac{\partial}{\partial x} = \left(\frac{\partial}{\partial \xi}\right)\frac{\partial \xi}{\partial x} + \left(\frac{\partial}{\partial \eta}\right)\frac{\partial \eta}{\partial x} + \left(\frac{\partial}{\partial \tau}\right)\frac{\partial \tau}{\partial x} \,. \tag{4.2}$$

In (4.2), when we differentiate with respect to a variable in one plane we hold constant the rest independent variables in the same plane. Similarly, we can transform the derivatives $\partial/\partial y$ and $\partial/\partial t$

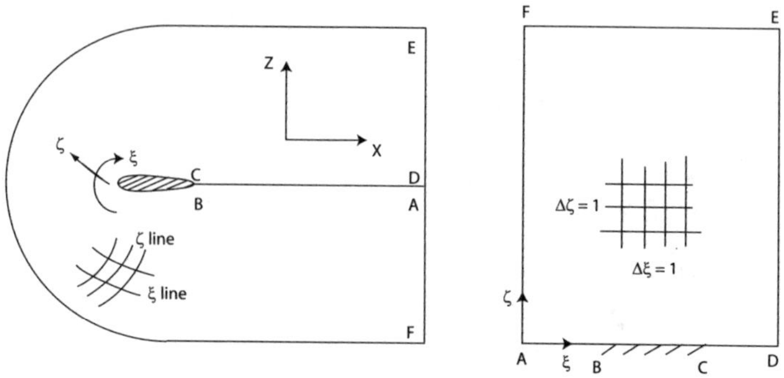

Fig. 4.3. C-type arrangement for the physical and computational domains for a two-dimensional topology.

$$\frac{\partial}{\partial y} = \left(\frac{\partial}{\partial \xi}\right)\frac{\partial \xi}{\partial y} + \left(\frac{\partial}{\partial \eta}\right)\frac{\partial \eta}{\partial y} + \left(\frac{\partial}{\partial \tau}\right)\frac{\partial \tau}{\partial y}, \tag{4.3}$$

$$\frac{\partial}{\partial t} = \left(\frac{\partial}{\partial \xi}\right)\frac{\partial \xi}{\partial t} + \left(\frac{\partial}{\partial \eta}\right)\frac{\partial \eta}{\partial t} + \left(\frac{\partial}{\partial \tau}\right)\frac{\partial \tau}{\partial t}. \tag{4.4}$$

Defining that $\tau = t$ the above differentiations can be further simplified

$$\frac{\partial}{\partial x} = \left(\frac{\partial}{\partial \xi}\right)\frac{\partial \xi}{\partial x} + \left(\frac{\partial}{\partial \eta}\right)\frac{\partial \eta}{\partial x}, \tag{4.5}$$

$$\frac{\partial}{\partial y} = \left(\frac{\partial}{\partial \xi}\right)\frac{\partial \xi}{\partial y} + \left(\frac{\partial}{\partial \eta}\right)\frac{\partial \eta}{\partial y}, \tag{4.6}$$

$$\frac{\partial}{\partial t} = \left(\frac{\partial}{\partial \xi}\right)\frac{\partial \xi}{\partial t} + \left(\frac{\partial}{\partial \eta}\right)\frac{\partial \eta}{\partial t} + \frac{\partial}{\partial \tau}. \tag{4.7}$$

For a non-moving grid (4.7) gives

$$\frac{\partial}{\partial t} = \frac{\partial}{\partial \tau}. \tag{4.8}$$

The viscous terms in the Navier-Stokes equations contain second-order derivatives and these also need to be transformed. This is obtained by differentiating twice the derivatives $\partial/\partial x$ and $\partial/\partial y$ and following a similar program. Let us consider the derivative $\partial/\partial x$ from (4.5) and differentiate with respect to x

$$\frac{\partial^2}{\partial x^2} = \frac{\partial}{\partial x}\left[\left(\frac{\partial}{\partial \xi}\right)\xi_x + \left(\frac{\partial}{\partial \eta}\right)\eta_x\right] =$$

$$\left(\frac{\partial}{\partial \xi}\right)\frac{\partial^2 \xi}{\partial x^2} + \frac{\partial \xi}{\partial x}\left(\frac{\partial^2}{\partial x \partial \xi}\right) + \frac{\partial \eta}{\partial x}\left(\frac{\partial^2}{\partial x \partial \eta}\right) + \left(\frac{\partial}{\partial \eta}\right)\frac{\partial^2 \eta}{\partial x^2}. \tag{4.9}$$

Using (4.5), we can replace the derivative $\partial/\partial x$ in (4.9) thus obtaining

$$\frac{\partial^2}{\partial x^2} = \xi_{xx}\left(\frac{\partial}{\partial \xi}\right) + \eta_{xx}\left(\frac{\partial}{\partial \eta}\right) + \xi_x^2\left(\frac{\partial^2}{\partial \xi^2}\right) + \eta_x^2\left(\frac{\partial^2}{\partial \eta^2}\right) + 2\eta_x\xi_x\left(\frac{\partial^2}{\partial \eta \partial \xi}\right), \tag{4.10}$$

$$\frac{\partial^2}{\partial y^2} = \xi_{yy}\left(\frac{\partial}{\partial \xi}\right) + \eta_{yy}\left(\frac{\partial}{\partial \eta}\right) + \xi_y^2\left(\frac{\partial^2}{\partial \xi^2}\right) + \eta_y^2\left(\frac{\partial^2}{\partial \eta^2}\right) + 2\eta_y\xi_y\left(\frac{\partial^2}{\partial \eta \partial \xi}\right), \tag{4.11}$$

where the indices indicate partial differentiation and have been introduced to simplify the derived formulas. The derivatives $\partial\xi/\partial x, \partial\xi/\partial y, \partial\eta/\partial x$ and $\partial\eta/\partial y$ are called *metrics* and cannot be calculated in a closed form since the transformation (4.1) is not often known in the form of an analytic relation. Their calculation is obtained through the Cartesian coordinates on the physical domain.

4.2 Calculation of Metrics

Let us consider the total differentials $\mathrm{d}x$ and $\mathrm{d}y$ given by

$$\mathrm{d}x = \frac{\partial x}{\partial \xi}\,\mathrm{d}\xi + \frac{\partial x}{\partial \eta}\,\mathrm{d}\eta\,, \tag{4.12}$$

$$\mathrm{d}y = \frac{\partial y}{\partial \xi}\,\mathrm{d}\xi + \frac{\partial y}{\partial \eta}\,\mathrm{d}\eta\,, \tag{4.13}$$

which can be written in a matrix form as

$$\begin{pmatrix} \mathrm{d}x \\ \mathrm{d}y \end{pmatrix} = \begin{pmatrix} \frac{\partial x}{\partial \xi} & \frac{\partial x}{\partial \eta} \\ \frac{\partial y}{\partial \xi} & \frac{\partial y}{\partial \eta} \end{pmatrix} \begin{pmatrix} \mathrm{d}\xi \\ \mathrm{d}\eta \end{pmatrix}. \tag{4.14}$$

The last equation can be inverted with respect to the differentials $\mathrm{d}\xi$ and $\mathrm{d}\eta$ thus yielding

$$\begin{pmatrix} \mathrm{d}\xi \\ \mathrm{d}\eta \end{pmatrix} = \begin{pmatrix} \frac{\partial x}{\partial \xi} & \frac{\partial x}{\partial \eta} \\ \frac{\partial y}{\partial \xi} & \frac{\partial y}{\partial \eta} \end{pmatrix}^{-1} \begin{pmatrix} \mathrm{d}x \\ \mathrm{d}y \end{pmatrix}. \tag{4.15}$$

Let us now consider the total differentials $d\xi$ and $d\eta$ which can be written, in a matrix form, as

$$\begin{pmatrix} d\xi \\ d\eta \end{pmatrix} = \begin{pmatrix} \frac{\partial \xi}{\partial x} & \frac{\partial \xi}{\partial y} \\ \frac{\partial \eta}{\partial x} & \frac{\partial \eta}{\partial y} \end{pmatrix} \begin{pmatrix} dx \\ dy \end{pmatrix} . \tag{4.16}$$

By comparing (4.15) and (4.16) we can establish a relationship for the metrics

$$\begin{pmatrix} \frac{\partial \xi}{\partial x} & \frac{\partial \xi}{\partial y} \\ \frac{\partial \eta}{\partial x} & \frac{\partial \eta}{\partial y} \end{pmatrix} = \begin{pmatrix} \frac{\partial x}{\partial \xi} & \frac{\partial x}{\partial \eta} \\ \frac{\partial y}{\partial \xi} & \frac{\partial y}{\partial \eta} \end{pmatrix}^{-1} . \tag{4.17}$$

Using matrix algebra, the inverse matrix on the right hand side of the last equation can be calculated thus obtaining

$$\begin{pmatrix} \xi_x & \xi_y \\ \eta_x & \eta_y \end{pmatrix} = \begin{pmatrix} y_\eta & -x_\eta \\ -y_\xi & x_\xi \end{pmatrix} \frac{1}{x_\xi y_\eta - x_\eta y_\xi} , \tag{4.18}$$

The term $x_\xi y_\eta - x_\eta y_\xi$ is the Jacobian of the coordinates transformation

$$J \equiv \left| \frac{\partial(x,y)}{\partial(\xi,\eta)} \right| = \begin{vmatrix} x_\xi & y_\xi \\ x_\eta & y_\eta \end{vmatrix} = \begin{vmatrix} x_\xi & x_\eta \\ y_\xi & y_\eta \end{vmatrix} = x_\xi y_\eta - x_\eta y_\xi . \tag{4.19}$$

From (4.18), we obtain

$$\begin{aligned} \xi_x &= \frac{1}{J}\, y_\eta, \quad \eta_x = -\frac{1}{J}\, y_\xi , \\ \xi_y &= -\frac{1}{J}\, x_\eta, \quad \eta_y = \frac{1}{J}\, x_\xi . \end{aligned} \tag{4.20}$$

Using (4.20) we can see that

$$\frac{\partial}{\partial \xi}(J\xi_x) + \frac{\partial}{\partial \eta}(J\eta_x) = \frac{\partial}{\partial \xi}(y_\eta) - \frac{\partial}{\partial \eta}(y_\xi) = \frac{\partial^2 y}{\partial \xi \partial \eta} - \frac{\partial^2 y}{\partial \xi \partial \eta} = 0 , \tag{4.21}$$

and

$$\begin{aligned} \frac{\partial}{\partial \xi}(J\xi_y) + \frac{\partial}{\partial \eta}(J\eta_y) &= \frac{\partial}{\partial \xi}(-x_\eta) - \frac{\partial}{\partial \eta}(x_\xi) = \\ &\qquad -\frac{\partial^2 x}{\partial \xi \partial \eta} + \frac{\partial^2 x}{\partial \xi \partial \eta} = 0 . \end{aligned} \tag{4.22}$$

By substituting (4.20) into (4.5) and (4.6), we can also write

$$\frac{\partial}{\partial x} = \frac{1}{J}\left[\left(\frac{\partial}{\partial \xi}\right)y_\eta - \left(\frac{\partial}{\partial \eta}\right)y_\xi\right], \tag{4.23}$$

$$\frac{\partial}{\partial y} = \frac{1}{J}\left[\left(\frac{\partial}{\partial \eta}\right)x_\xi - \left(\frac{\partial}{\partial \xi}\right)x_\eta\right]. \tag{4.24}$$

The above formulas can be used to transform the first-order spatial derivatives in the Navier-Stokes equations.

4.3 Transformation of the Fluid Flow Equations

We now can rewrite the Navier-Stokes equations using the transformed derivatives. We consider (3.37) and (3.41) to (3.43), and write them in a matrix form as

$$\frac{\partial \mathbf{U}}{\partial t} + \frac{\partial \mathbf{E}}{\partial x} + \frac{\partial \mathbf{F}}{\partial y} = \frac{\partial \mathbf{R}}{\partial x} + \frac{\partial \mathbf{S}}{\partial y}, \tag{4.25}$$

where

$$\mathbf{U} = \begin{pmatrix} 0 \\ u \\ v \end{pmatrix}, \quad \mathbf{E} = \begin{pmatrix} u \\ u^2 + p/\rho \\ uv \end{pmatrix}, \quad \mathbf{F} = \begin{pmatrix} v \\ uv \\ v^2 + p/\rho \end{pmatrix}, \tag{4.26}$$

$$\mathbf{R} = \begin{pmatrix} 0 \\ \tau_{xx} \\ \tau_{xy} \end{pmatrix}, \quad \mathbf{S} = \begin{pmatrix} 0 \\ \tau_{yx} \\ \tau_{yy} \end{pmatrix}. \tag{4.27}$$

The usefulness of the above is that we can handle the system of equations as a single equation thus reducing the algebraic operations. By multiplying (4.25) with J and transforming the Cartesian derivatives using (4.5)-(4.7), (4.25) is written

$$\begin{aligned} J\frac{\partial \mathbf{U}}{\partial \tau} + J\frac{\partial \mathbf{E}}{\partial \xi}\xi_x + J\frac{\partial \mathbf{E}}{\partial \eta}\eta_x + J\frac{\partial \mathbf{F}}{\partial \xi}\xi_y + J\frac{\partial \mathbf{F}}{\partial \eta}\eta_y = \\ J\frac{\partial \mathbf{R}}{\partial \xi}\xi_x + J\frac{\partial \mathbf{R}}{\partial \eta}\eta_x + J\frac{\partial \mathbf{S}}{\partial \xi}\xi_y + J\frac{\partial \mathbf{S}}{\partial \eta}\eta_y . \end{aligned} \tag{4.28}$$

The term $J(\partial \mathbf{E}/\partial \xi)\xi_x$ can be written as

$$J\frac{\partial \mathbf{E}}{\partial \xi}\xi_x = \frac{\partial (J\mathbf{E}\xi_x)}{\partial \xi} - \mathbf{E}\frac{\partial}{\partial \xi}(J\xi_x). \tag{4.29}$$

Similarly, we can write

$$\left.\begin{aligned}
J\frac{\partial \mathbf{E}}{\partial \eta}\eta_x &= \frac{\partial(J\mathbf{E}\eta_x)}{\partial \eta} - \mathbf{E}\frac{\partial}{\partial \eta}(J\eta_x)\\
J\frac{\partial \mathbf{F}}{\partial \xi}\xi_y &= \frac{\partial(J\mathbf{F}\xi_y)}{\partial \xi} - \mathbf{F}\frac{\partial}{\partial \xi}(J\xi_y)\\
J\frac{\partial \mathbf{F}}{\partial \eta}\eta_y &= \frac{\partial(J\mathbf{F}\eta_y)}{\partial \eta} - \mathbf{F}\frac{\partial}{\partial \eta}(J\eta_y)\\
J\frac{\partial \mathbf{R}}{\partial \xi}\xi_x &= \frac{\partial(J\mathbf{R}\xi_x)}{\partial \xi} - \mathbf{R}\frac{\partial}{\partial \xi}(J\xi_x)\\
J\frac{\partial \mathbf{R}}{\partial \eta}\eta_x &= \frac{\partial(J\mathbf{R}\eta_x)}{\partial \eta} - \mathbf{R}\frac{\partial}{\partial \eta}(J\eta_x)\\
J\frac{\partial \mathbf{S}}{\partial \xi}\xi_y &= \frac{\partial(J\mathbf{S}\xi_y)}{\partial \xi} - \mathbf{S}\frac{\partial}{\partial \xi}(J\xi_x)\\
J\frac{\partial \mathbf{S}}{\partial \eta}\eta_y &= \frac{\partial(J\mathbf{S}\eta_y)}{\partial \eta} - \mathbf{S}\frac{\partial}{\partial \eta}(J\eta_y)
\end{aligned}\right\} . \tag{4.30}$$

Using (4.29)-(4.30), (4.28) is written

$$\begin{aligned}
J\frac{\partial \mathbf{U}}{\partial \tau} &+ \frac{\partial}{\partial \xi}\Big(J\mathbf{E}\xi_x + J\mathbf{F}\xi_y - J\mathbf{R}\xi_x - J\mathbf{S}\xi_y\Big)\\
&+ \frac{\partial}{\partial \eta}\Big(J\mathbf{E}\eta_x + J\mathbf{F}\eta_y - J\mathbf{R}\eta_x - J\mathbf{S}\eta_y\Big)\\
-\mathbf{E}&\Big[\frac{\partial}{\partial \xi}(J\xi_x) + \frac{\partial}{\partial \eta}(J\eta_x)\Big] - \mathbf{F}\Big[\frac{\partial}{\partial \xi}(J\xi_y) + \frac{\partial}{\partial \eta}(J\eta_y)\Big]\\
+\mathbf{R}\Big[\frac{\partial}{\partial \xi}(J\xi_x) &+ \frac{\partial}{\partial \eta}(J\eta_x)\Big] + \mathbf{S}\Big[\frac{\partial}{\partial \xi}(J\xi_y) + \frac{\partial}{\partial \eta}(J\eta_y)\Big] = 0 .
\end{aligned} \tag{4.31}$$

Using (4.21) and (4.22), the terms in the brackets of the LHS of (4.31) become zero and (4.31) is simplified as

$$\frac{\partial \bar{\mathbf{U}}}{\partial t} + \frac{\partial \bar{\mathbf{E}}}{\partial \xi} + \frac{\partial \bar{\mathbf{F}}}{\partial \eta} = \frac{\partial \bar{\mathbf{R}}}{\partial \xi} + \frac{\partial \bar{\mathbf{S}}}{\partial \eta} , \tag{4.32}$$

where

$$\left.\begin{aligned}
\bar{\mathbf{U}} &= J\mathbf{U} \\
\bar{\mathbf{E}} &= J(\mathbf{E}\xi_x + \mathbf{F}\xi_y) \\
\bar{\mathbf{F}} &= J(\mathbf{E}\eta_x + \mathbf{F}\eta_y) \\
\bar{\mathbf{R}} &= J(\mathbf{R}\xi_x + \mathbf{S}\xi_y) \\
\bar{\mathbf{S}} &= J(\mathbf{R}\eta_x + \mathbf{S}\eta_y)
\end{aligned}\right\} . \tag{4.33}$$

Eq. (4.32) is the matrix form of the Navier-Stokes equations in curvilinear coordinates and strong conservation form. This form was first derived in [588] and [589]. It can be easily shown that by substituting (4.20) into the bar fluxes (4.33), the Jacobian, J, cancels out, i.e.,

$$\bar{\mathbf{E}} = \mathbf{E}y_\eta - \mathbf{F}x_\eta \, . \tag{4.34}$$

The expression (4.34) is preferred in the development of CFD programs since it reduces the numerical operations.

Using the same approach for three-dimensional flows an equation similar to (4.32) can be obtained

$$\frac{\partial \bar{\mathbf{U}}}{\partial \tau} + \frac{\partial \bar{\mathbf{E}}}{\partial \xi} + \frac{\partial \bar{\mathbf{F}}}{\partial \eta} + \frac{\partial \bar{\mathbf{G}}}{\partial \zeta} = \frac{\partial \bar{\mathbf{R}}}{\partial \xi} + \frac{\partial \bar{\mathbf{S}}}{\partial \eta} + \frac{\partial \bar{\mathbf{L}}}{\partial \zeta} \, , \tag{4.35}$$

where

$$\left.\begin{aligned}
\bar{\mathbf{U}} &= J\mathbf{U} \\
\bar{\mathbf{E}} &= J(\mathbf{E}\xi_x + \mathbf{F}\xi_y + \mathbf{G}\xi_z) \\
\bar{\mathbf{F}} &= J(\mathbf{E}\eta_x + \mathbf{F}\eta_y + \mathbf{G}\zeta_z) \\
\bar{\mathbf{G}} &= J(\mathbf{E}\zeta_x + \mathbf{F}\zeta_y + \mathbf{G}\zeta_z) \\
\bar{\mathbf{R}} &= J(\mathbf{R}\xi_x + \mathbf{S}\xi_y + \mathbf{L}\xi_z) \\
\bar{\mathbf{S}} &= J(\mathbf{R}\eta_x + \mathbf{S}\eta_y + \mathbf{L}\eta_z) \\
\bar{\mathbf{L}} &= J(\mathbf{R}\zeta_x + \mathbf{S}\zeta_y + \mathbf{L}\zeta_z)
\end{aligned}\right\} , \tag{4.36}$$

where $\mathbf{G}$ and $\mathbf{L}$ denote the inviscid and viscous Cartesian fluxes, respectively, in the $z-$direction. The three-dimensional Cartesian fluxes are written in a matrix form as

$$\mathbf{U} = \begin{pmatrix} 0 \\ u \\ v \\ w \end{pmatrix} , \tag{4.37}$$

$$\mathbf{E} = \begin{pmatrix} u \\ u^2 + p/\rho \\ uv \\ uw \end{pmatrix}, \quad \mathbf{F} = \begin{pmatrix} v \\ vu \\ v^2 + p/\rho \\ vw \end{pmatrix}, \quad \mathbf{G} = \begin{pmatrix} w \\ wu \\ wv \\ w^2 + p/\rho \end{pmatrix}, \tag{4.38}$$

$$\mathbf{R} = \begin{pmatrix} 0 \\ \tau_{xx} \\ \tau_{xy} \\ \tau_{xz} \end{pmatrix}, \quad \mathbf{S} = \begin{pmatrix} 0 \\ \tau_{yx} \\ \tau_{yy} \\ \tau_{yz} \end{pmatrix}, \quad \mathbf{L} = \begin{pmatrix} 0 \\ \tau_{zx} \\ \tau_{zy} \\ \tau_{zz} \end{pmatrix}. \tag{4.39}$$

The Jacobian and metrics for a three-dimensional grid are defined by

$$J = x_\xi \left(y_\eta z_\zeta - y_\zeta z_\eta\right) + x_\eta \left(y_\zeta z_\xi - y_\xi z_\zeta\right) + x_\zeta \left(y_\xi z_\eta - y_\eta z_\xi\right), \tag{4.40}$$

$$\left.\begin{aligned} \xi_x &= \frac{y_\eta z_\zeta - y_\zeta z_\eta}{J}, \quad \xi_y = \frac{-x_\eta z_\zeta + x_\zeta z_\eta}{J}, \quad \xi_z = \frac{x_\eta y_\zeta - x_\zeta y_\eta}{J} \\ \eta_x &= \frac{-y_\xi z_\zeta + y_\zeta z_\xi}{J}, \quad \eta_y = \frac{x_\xi z_\zeta - x_\zeta z_\xi}{J}, \quad \eta_z = \frac{-x_\xi y_\zeta + x_\zeta y_\xi}{J} \\ \zeta_x &= \frac{y_\xi z_\eta - y_\eta z_\xi}{J}, \quad \zeta_y = \frac{-x_\xi z_\eta + x_\eta z_\xi}{J}, \quad \zeta_z = \frac{x_\xi y_\eta - x_\eta y_\xi}{J} \end{aligned}\right\}. \tag{4.41}$$

4.4 Viscous Terms

The discretization of the viscous terms of the incompressible flow equations is fairly straightforward since these terms do not encompass any nonlinearities at least with respect to Newtonian fluids. In this case, the viscous matrices in (4.35) are written in curvilinear coordinates as follows

$$\bar{\mathbf{R}} = J(\mathbf{R}\xi_x + \mathbf{S}\xi_y + \mathbf{L}\xi_z), \tag{4.42}$$

$$\bar{\mathbf{S}} = J(\mathbf{R}\eta_x + \mathbf{S}\eta_y + \mathbf{L}\eta_z), \tag{4.43}$$

$$\bar{\mathbf{L}} = J(\mathbf{R}\zeta_x + \mathbf{S}\zeta_y + \mathbf{L}\zeta_z). \tag{4.44}$$

where

$$\mathbf{R} = \begin{pmatrix} 0 \\ DUDX\ \xi_x + DUDY\ \xi_y + DUDX\ \xi_z \\ DVDX\ \xi_x + DVDY\ \xi_y + DVDX\ \xi_z \\ DWDX\ \xi_x + DWDY\ \xi_y + DWDX\ \xi_z \end{pmatrix}, \tag{4.45}$$

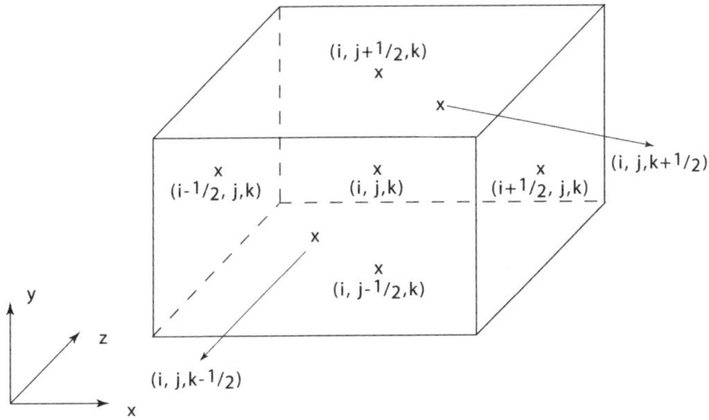

Fig. 4.4. Three-dimensional cell notation used in the discretization of the viscous fluxes.

$$\mathbf{S} = \begin{pmatrix} 0 \\ DUDX\ \eta_x + DUDY\ \eta_y + DUDX\ \eta_z \\ DVDX\ \eta_x + DVDY\ \eta_y + DVDX\ \eta_z \\ DWDX\ \eta_x + DWDY\ \eta_y + DWDX\ \eta_z \end{pmatrix}, \tag{4.46}$$

$$\mathbf{L} = \begin{pmatrix} 0 \\ DUDX\ \zeta_x + DUDY\ \zeta_y + DUDX\ \zeta_z \\ DVDX\ \zeta_x + DVDY\ \zeta_y + DVDX\ \zeta_z \\ DWDX\ \zeta_x + DWDY\ \zeta_y + DWDX\ \zeta_z \end{pmatrix}. \tag{4.47}$$

and

$$DUDX = \frac{\partial u}{\partial \xi}\frac{\partial \xi}{\partial x} + \frac{\partial u}{\partial \eta}\frac{\partial \eta}{\partial x} + \frac{\partial u}{\partial \zeta}\frac{\partial \zeta}{\partial x}, \tag{4.48}$$

$$DVDX = \frac{\partial v}{\partial \xi}\frac{\partial \xi}{\partial x} + \frac{\partial v}{\partial \eta}\frac{\partial \eta}{\partial x} + \frac{\partial v}{\partial \zeta}\frac{\partial \zeta}{\partial x}, \tag{4.49}$$

$$DWDX = \frac{\partial w}{\partial \xi}\frac{\partial \xi}{\partial x} + \frac{\partial w}{\partial \eta}\frac{\partial \eta}{\partial x} + \frac{\partial w}{\partial \zeta}\frac{\partial \zeta}{\partial x}, \tag{4.50}$$

$$DUDY = \frac{\partial u}{\partial \xi}\frac{\partial \xi}{\partial y} + \frac{\partial u}{\partial \eta}\frac{\partial \eta}{\partial y} + \frac{\partial u}{\partial \zeta}\frac{\partial \zeta}{\partial y}, \tag{4.51}$$

$$DVDY = \frac{\partial v}{\partial \xi}\frac{\partial \xi}{\partial y} + \frac{\partial v}{\partial \eta}\frac{\partial \eta}{\partial y} + \frac{\partial v}{\partial \zeta}\frac{\partial \zeta}{\partial y}, \tag{4.52}$$

$$DWDY = \frac{\partial w}{\partial \xi}\frac{\partial \xi}{\partial y} + \frac{\partial w}{\partial \eta}\frac{\partial \eta}{\partial y} + \frac{\partial w}{\partial \zeta}\frac{\partial \zeta}{\partial y}, \tag{4.53}$$

$$DUDZ = \frac{\partial u}{\partial \xi}\frac{\partial \xi}{\partial z} + \frac{\partial u}{\partial \eta}\frac{\partial \eta}{\partial z} + \frac{\partial u}{\partial \zeta}\frac{\partial \zeta}{\partial z}\,, \tag{4.54}$$

$$DVDZ = \frac{\partial v}{\partial \xi}\frac{\partial \xi}{\partial z} + \frac{\partial v}{\partial \eta}\frac{\partial \eta}{\partial z} + \frac{\partial v}{\partial \zeta}\frac{\partial \zeta}{\partial z}\,, \tag{4.55}$$

$$DWDZ = \frac{\partial w}{\partial \xi}\frac{\partial \xi}{\partial z} + \frac{\partial w}{\partial \eta}\frac{\partial \eta}{\partial z} + \frac{\partial w}{\partial \zeta}\frac{\partial \zeta}{\partial z}\,. \tag{4.56}$$

The flux derivative $\partial \mathbf{R}/\partial \xi$ can be discretized using the intercell flux values (see Fig. 4.4)

$$\frac{\partial \mathbf{R}}{\partial \xi} = \mathbf{R}_{i+1/2,j,k} - \mathbf{R}_{i-1/2,j,k}\,. \tag{4.57}$$

For completeness we will show how the discretization of the metrics and velocity derivatives associated with the $\mathbf{R}$ flux in ξ direction can be obtained. The metrics and velocity derivatives associated with the fluxes in η and ζ directions can be similarly obtained. The derivatives (4.41) can be discretized as follows

$$\begin{aligned}
x_\eta|_{i-1/2,j,k} &= \frac{1}{2}(x_{i,j+1,k} + x_{i,j+1,k+1} - x_{i,j,k} - x_{i,j,k+1})\,,\\
y_\eta|_{i-1/2,j,k} &= \frac{1}{2}(y_{i,j+1,k} + y_{i,j+1,k+1} - y_{i,j,k} - y_{i,j,k+1})\,,\\
z_\eta|_{i-1/2,j,k} &= \frac{1}{2}(z_{i,j+1,k} + z_{i,j+1,k+1} - z_{i,j,k} - z_{i,j,k+1})\,,
\end{aligned} \tag{4.58}$$

$$\begin{aligned}
x_\xi|_{i-1/2,j,k} &= 0.125(x_{i+1,j,k} + x_{i+1,j+1,k} + x_{i+1,j+1,k+1} + x_{i+1,j,k}\\
&\quad - x_{i-1,j,k} - x_{i-1,j+1,k+1} - x_{i-1,j+1,k+1} - x_{i-1,j,k+1})\,,\\
y_\xi|_{i-1/2,j,k} &= 0.125(y_{i+1,j,k} + y_{i+1,j+1,k} + y_{i+1,j+1,k+1} + y_{i+1,j,k}\\
&\quad - y_{i-1,j,k} - y_{i-1,j+1,k+1}\\
&\quad - y_{i-1,j+1,k+1} - y_{i-1,j,k+1})\,,\\
z_\xi|_{i-1/2,j,k} &= 0.125(z_{i+1,j,k} + z_{i+1,j+1,k} + z_{i+1,j+1,k+1} + z_{i+1,j,k}\\
&\quad - z_{i-1,j,k} - z_{i-1,j+1,k+1} - z_{i-1,j+1,k+1} - z_{i-1,j,k+1})\,,
\end{aligned} \tag{4.59}$$

$$\begin{aligned}
x_\zeta|_{i-1/2,j,k} &= 0.5(x_{i,j+1,k} + x_{i,j+1,k+1} - x_{i,j,k} - x_{i,j,k+1})\,,\\
y_\zeta|_{i-1/2,j,k} &= 0.5(y_{i,j+1,k} + y_{i,j+1,k+1} - y_{i,j,k} - y_{i,j,k+1})\,,\\
z_\zeta|_{i-1/2,j,k} &= 0.5(z_{i,j+1,k} + z_{i,j+1,k+1} - z_{i,j,k} - z_{i,j,k+1})\,.
\end{aligned} \tag{4.60}$$

The intercell velocity derivatives are calculated[1] as follows

---

[1] We always refer to the computational plane.

$$
\begin{aligned}
u_\xi|_{i-1/2,j,k} &= u_{i+1,j,k} - u_{i,j,k} \, , \\
v_\xi|_{i-1/2,j,k} &= v_{i+1,j,k} - v_{i,j,k} \, , \\
w_\xi|_{i-1/2,j,k} &= w_{i+1,j,k} - w_{i,j,k} \, ,
\end{aligned} \tag{4.61}
$$

$$
\begin{aligned}
u_\eta|_{i-1/2,j,k} &= 0.25(u_{i,j+1,k} + u_{i-1,j+1,k} - u_{i,j-1,k} - u_{i-1,j-1,k}) \, , \\
v_\eta|_{i-1/2,j,k} &= 0.25(v_{i,j+1,k} + v_{i-1,j+1,k} - v_{i,j-1,k} - v_{i-1,j-1,k}) \, , \\
w_\eta|_{i-1/2,j,k} &= 0.25(w_{i,j+1,k} + w_{i-1,j+1,k} - w_{i,j-1,k} - w_{i-1,j-1,k}) \, ,
\end{aligned} \tag{4.62}
$$

$$
\begin{aligned}
u_\zeta|_{i-1/2,j,k} &= 0.25(u_{i,j,k+1} + u_{i-1,j,k+1} - u_{i,j,k-1} - u_{i-1,j,k-1}) \, , \\
v_\zeta|_{i-1/2,j,k} &= 0.25(v_{i,j,k+1} + v_{i-1,j,k+1} - v_{i,j,k-1} - v_{i-1,j,k-1}) \, , \\
w_\zeta|_{i-1/2,j,k} &= 0.25(w_{i,j,k+1} + w_{i-1,j,k+1} - w_{i,j,k-1} - w_{i-1,j,k-1}) \, .
\end{aligned} \tag{4.63}
$$

Similarly, one can obtain the discretization of the flux derivatives $\partial \mathbf{S}/\partial \eta$ and $\partial \mathbf{L}/\partial \zeta$.

4.5 Geometric Conservation Law

The "geometric conservation law" (GCL) was introduced by Thomas and Lombard [531] in order to maintain the global conservation as well as conservation within the local cell (volume) element under time dependent mappings that result from boundary motion.

Let us consider a fluid volume V which is bounded by a close surface S that is moving with a local velocity $\mathbf{v}_s$. The mass conservation principle (2.20) is then written

$$
\frac{\partial}{\partial t} \int\int\int_V \rho \, \mathrm{d}V + \int\int_S \rho(\mathbf{u} - \mathbf{v}_s) \cdot \mathbf{n} \, \mathrm{d}S = 0 \, . \tag{4.64}
$$

In a finite volume, finite difference or finite element method, the flow field is subdivided into a number of cells of volume ΔV, bounded by cell faces S_{ijk} (in 3-D) or S_{ij} (in 2-D). The time variation of the cell volume in terms of the area, orientation and velocity of the cell face is given by

$$
\frac{\mathrm{d}}{\mathrm{d}t} \int_{\Delta V_i} \mathrm{d}V = \sum_j \int_{S_{ij}} \mathbf{v}_s \cdot \mathbf{n} \, \mathrm{d}S \, , \tag{4.65}
$$

where i and j denote cell and cell-face indices. The corresponding relation for the entire flow region is written as

$$
\frac{\mathrm{d}}{\mathrm{d}t} \int_V \mathrm{d}V = \int_S \mathbf{v}_s \cdot \mathbf{n} \, \mathrm{d}S \, . \tag{4.66}
$$

Eq. (4.66) is called the "geometric conservation law" [531].

As shown in the preceding section, the transformed equations in a curvilinear coordinate system contain the Jacobian of the transformation, J. Because J represents a volume element in the transformed coordinates system, its values should be consistent with the effective volume element ΔV_i of each grid cell i. The above is especially important in the case of moving grids, as well as in the context of the finite difference approach where mass conservation is preserved only if conservative difference operators are employed in the numerical discretization of (2.24). A differential equation of the GCL can also be derived.

Let us define the velocity of any point on the surface S in the Cartesian coordinate system in terms of its components x, y, and z

$$\mathbf{v}_s = (x_\tau, y_\tau, z_\tau) , \tag{4.67}$$

where the index τ denotes differentiation in time. Using the divergence theorem to replace the surface integral in (4.66) by a volume integral and transforming the volume integral into curvilinear coordinates, we obtain

$$\frac{\mathrm{d}}{\mathrm{d}t}\int_V J \,\mathrm{d}\xi\,\mathrm{d}\eta\,\mathrm{d}\zeta = \int_V J(\nabla\cdot\mathbf{v}_s)\,\mathrm{d}\xi\,\mathrm{d}\eta\,\mathrm{d}\zeta , \tag{4.68}$$

where $\mathrm{d}V = \mathrm{d}x\ \mathrm{d}y\ \mathrm{d}z = J\ \mathrm{d}\xi\ \mathrm{d}\eta\ \mathrm{d}\zeta$, and

$$\begin{aligned} J &= \frac{\partial(x,y,z)}{\partial(\xi,\eta,\zeta)} \\ &= x_\xi(y_\eta z_\zeta - y_\zeta z_\eta) + x_\eta(y_\zeta z_\xi - y_\xi z_\zeta) + x_\zeta(y_\xi z_\eta - y_\eta z_\xi) . \end{aligned} \tag{4.69}$$

We also note that the curvilinear coordinates (ξ, η, ζ) are constructed as boundary fitted coordinates considering that both S and V are fixed in time. Therefore, the time derivative operator on the left hand side of (4.68) can move inside the integral.

The term $J(\nabla\cdot\mathbf{v}_s)$ can be written

$$\begin{aligned} J(\nabla\cdot\mathbf{v}_s) &= J\Big(\nabla\xi\cdot(\mathbf{v}_s)_\xi + \nabla\eta\cdot(\mathbf{v}_s)_\eta + \nabla\zeta\cdot(\mathbf{v}_s)_\zeta\Big) \\ &= (J\nabla\xi\cdot\mathbf{v}_s)_\xi + (J\nabla\eta\cdot\mathbf{v}_s)_\eta + (J\nabla\zeta\cdot\mathbf{v}_s)_\zeta \\ &\quad -\mathbf{v}_s\cdot\Big[(J\nabla\xi)_\xi + (J\nabla\eta)_\eta + (J\nabla\zeta)_\zeta\Big] . \end{aligned} \tag{4.70}$$

Using the following identities

$$\xi_t = -\nabla\xi\cdot\mathbf{v}_s, \quad \eta_t = -\nabla\eta\cdot\mathbf{v}_s, \quad \zeta_t = -\nabla\zeta\cdot\mathbf{v}_s , \tag{4.71}$$

(4.70) can be written as

$$\int_V \Big[J_\tau + (J\xi_t)_\xi + (J\eta_t)_\eta + (J\zeta_t)_\zeta\Big]\, \mathrm{d}\xi\, \mathrm{d}\eta\, \mathrm{d}\zeta =$$
$$\int_V \Big[x_\tau\Big((J\xi_x)_\xi + (J\eta_x)_\eta + (J\zeta_x)_\zeta\Big)$$
$$+y_\tau\Big((J\xi_y)_\xi + (J\eta_y)_\eta + (J\zeta_y)_\zeta\Big) +$$
$$z_\tau\Big((J\xi_z)_\xi + (J\eta_z)_\eta + (J\zeta_z)_\zeta\Big)\Big]\, \mathrm{d}\xi\, \mathrm{d}\eta\, \mathrm{d}\zeta \ . \tag{4.72}$$

Since the RHS of (4.72) vanishes and V is fixed in time, we obtain

$$J_\tau + (J\xi_t)_\xi + (J\eta_t)_\eta + (J\zeta_t)_\zeta = 0 \, . \tag{4.73}$$

The above equation is the differential form of GCL and has a similar form with the transformed flow conservation equations (4.32). For a moving coordinate system, the transformed fluxes are given by

$$\bar{\mathbf{E}} = J(\mathbf{U}\xi_t + \mathbf{E}\xi_x + \mathbf{F}\xi_y + \mathbf{G}\xi_z) \, , \tag{4.74}$$

$$\bar{\mathbf{F}} = J(\mathbf{U}\eta_t + \mathbf{E}\eta_x + \mathbf{F}\eta_y + \mathbf{G}\eta_z) \, , \tag{4.75}$$

$$\bar{\mathbf{G}} = J(\mathbf{U}\zeta_t + \mathbf{E}\zeta_x + \mathbf{F}\zeta_y + \mathbf{G}\zeta_z) \, , \tag{4.76}$$

whereas the viscous fluxes (4.39) remain intact.

To avoid numerical errors introduced by the motion of the grid, (4.73) should be solved in conjunction with (4.35). Equation (4.73) requires initial values which can be directly deducted by the geometric definition of the Jacobian and these values can be retained on the boundaries. Further, numerical errors can be avoided if the same discretization scheme is used to discretize the metrics in both (4.73) and (4.36).

5. Overview of Various Formulations and Model Equations

In this chapter we summarize the various formulations for the incompressible Navier-Stokes equations. Model equations such as the linear advection, linear advection-diffusion[1] and Burgers' equations, are often used as the starting point for the development of numerical schemes. These model equations are presented at the end of this chapter.

5.1 Overview of Various Formulations of the Incompressible Flow Equations

In the case of incompressible flows the continuity and momentum equations are decoupled due to the absence of the pressure variable from the former. Although it is straightforward to realize some of the advantages and disadvantages of each of the formulations, it would be rather difficult to draw an overall judgment regarding which formulation (method) is superior as the results are problem specific. This is because accuracy and efficiency within each formulation depend on several factors including the choice of discretization of spatial derivatives (advective terms in particular), explicit versus implicit time discretization, efficiency in conjunction with iterative solvers (e.g., multigrid), implementation of boundary conditions, versatility in the implementation in complex geometries, parallelization, and accuracy at low and high Reynolds number flows.

The most widely used formulations, i.e., the artificial compressibility, pressure-Poisson and projection formulations, will be described and discussed in more detail in subsequent chapters.

5.1.1 Vorticity/Stream-Function Formulation

One possible way to avoid the appearance of the pressure in the governing equations is to introduce the vorticity and stream function as dependent variables. The vorticity is defined by (3.14)

$$\omega = \text{curl}\,\mathbf{u} = \nabla \times \mathbf{u}\,. \tag{5.1}$$

[1] It is also called convection-diffusion equation.

In three dimensions the vorticity is a vector, but in two dimensions it is a scalar. Let us consider the two-dimensional version of the momentum equations (3.38) and (3.39). By differentiating (3.38) and (3.39), with respect to y and x, respectively, and subsequently subtracting the latter from the former, we obtain

$$\frac{\partial \omega}{\partial t} + u\frac{\partial \omega}{\partial x} + v\frac{\partial \omega}{\partial y} = \nu(\frac{\partial^2 \omega}{\partial x^2} + \frac{\partial^2 \omega}{\partial y^2}) , \tag{5.2}$$

where

$$\omega = \frac{\partial u}{\partial y} - \frac{\partial v}{\partial x} . \tag{5.3}$$

With the aid of the continuity equation, (5.2) can be written

$$\frac{\partial \omega}{\partial t} + \frac{\partial u\omega}{\partial x} + \frac{\partial v\omega}{\partial y} = \nu(\frac{\partial^2 \omega}{\partial x^2} + \frac{\partial^2 \omega}{\partial y^2}) . \tag{5.4}$$

The velocities u and v can be calculated by introducing the stream function ψ

$$u = \frac{\partial \psi}{\partial y} , \quad v = -\frac{\partial \psi}{\partial x} . \tag{5.5}$$

By substituting (5.5) into (5.3), we obtain

$$\frac{\partial^2 \psi}{\partial x^2} + \frac{\partial^2 \psi}{\partial y^2} = \omega . \tag{5.6}$$

The last equation is a Poisson equation for the stream function subject to given vorticity. Equations (5.4), (5.5) and (5.6) are the three equations that consist the vorticity/stream-function formulation of the incompressible fluid flow equations. Note that the velocities u and v in (5.4) can also be replaced by the stream function (5.5).

Equation (5.4) is a parabolic equation given that the velocities are known. The system (5.4), (5.5) and (5.6) can be used both for steady and unsteady flows. Alternatively, (5.6) can be replaced by a pseudo-transient approach

$$\frac{\partial \psi}{\partial \tau} - \left(\frac{\partial^2 \psi}{\partial x^2} + \frac{\partial^2 \psi}{\partial y^2} - \omega\right) = 0 . \tag{5.7}$$

Eq. (5.7) provides a coupled approach for solving the stream function (5.6) and vorticity (5.4) equations.

The implementation of the vorticity/stream-function formulation requires careful treatment of the boundary conditions. The system is elliptic in space. For the two-dimensional case the stream-function derivatives[2] can be calculated using (5.5) if the velocity components are known, e.g., at inflow and

[2] Note that ψ appears only in the derivative form.

far-field boundaries. No boundary condition is required for ω. At out-flow boundaries $\partial^2\omega/\partial x^2 = 0$ is often very small and can thus be neglected from (5.2); $\partial\omega/\partial x = 0$ and $\partial^2\psi/\partial x^2 = 0$ have also been proposed as alternative boundary conditions [26, 460].

At solid boundaries the following boundary conditions for the stream-function can be implemented

$$\psi = 0 \, , \quad \frac{\partial \psi}{\partial n} = 0 \, , \tag{5.8}$$

where n denotes direction normal to the boundary.

Finally, the pressure can be calculated by

$$\frac{\partial^2 p}{\partial x^2} + \frac{\partial^2 p}{\partial y^2} = 2\left[\frac{\partial^2 \psi}{\partial x^2}\frac{\partial^2 \psi}{\partial y^2} - \left(\frac{\partial^2 \psi}{\partial x \partial y}\right)^2\right] , \tag{5.9}$$

with Neumann boundary conditions for p obtained by the momentum equations.

In two dimensional flows the vorticity/stream-function formulation has an advantage over the primitive variables Navier-Stokes equations because it avoids the explicit solution of the continuity equation. In three-dimensional flows, however, the formulation leads to six variables compared to four dependent variables in the case of primitive variables Navier-Stokes. For three-dimensional flows, two formulations can be employed [190]: (i) the vorticity/vector-potential formulation, and (ii) the vorticity-velocity formulation. These are presented below

5.1.2 The Vorticity/Vector-Potential Formulation

The stream-function is replaced by a three-component vector potential $\Psi = (\Psi_x, \Psi_y, \Psi_z)$ defined by

$$\mathbf{u} = \text{curl}\, \Psi \, , \tag{5.10}$$

where

$$u = \frac{\partial \Psi_z}{\partial y} - \frac{\partial \Psi_y}{\partial z} , \tag{5.11}$$

$$v = \frac{\partial \Psi_x}{\partial z} - \frac{\partial \Psi_z}{\partial x} , \tag{5.12}$$

$$w = \frac{\partial \Psi_y}{\partial x} - \frac{\partial \Psi_x}{\partial y} . \tag{5.13}$$

The three-component vorticity $\Omega = (\Omega_x, \Omega_y, \Omega_z)$ equation is given by

$$\frac{\partial \Omega}{\partial t} + \nabla \cdot (\mathbf{u}\Omega) - (\Omega \cdot \nabla)\mathbf{u} - \nu \nabla^2 \Omega = 0 \, , \tag{5.14}$$

and the three-dimensional equivalent of (5.6) is written [190, 191]

$$\nabla \Psi = -\Omega \,. \tag{5.15}$$

Boundary conditions similar to the two dimensional case can be extended to three dimensions [25, 255].

5.1.3 Vorticity-Velocity Formulation

This formulation is based on (5.14) and three Poisson equations for the velocity components [190]

$$\nabla^2 u = \frac{\partial \Omega_y}{\partial z} - \frac{\partial \Omega_z}{\partial y} \,, \tag{5.16}$$

$$\nabla^2 v = \frac{\partial \Omega_z}{\partial x} - \frac{\partial \Omega_x}{\partial z} \,, \tag{5.17}$$

$$\nabla^2 w = \frac{\partial \Omega_x}{\partial y} - \frac{\partial \Omega_y}{\partial x} \,. \tag{5.18}$$

The last three equations are derived by using

$$\Omega = \operatorname{curl} \mathbf{u} \,. \tag{5.19}$$

and the continuity equation.

5.1.4 Pressure-Poisson Formulation

The pressure-Poisson formulation applies the continuity equation (divergence constraint) to the equation of motion in order to derive an explicit equation for pressure. Typically, the equation is simplified using the explicit knowledge that the divergence of velocity (and its time derivative) is zero. Taking the divergence of (3.69) yields the following,

$$\nabla \cdot \left[\frac{\partial \mathbf{u}}{\partial t} + \mathbf{u} \cdot \nabla \mathbf{u} + \frac{\mathbf{1}}{\rho} \nabla \mathbf{p} - \nu \nabla^{\mathbf{2}} \mathbf{u} \right] = 0 \,. \tag{5.20}$$

Because $\nabla \cdot \mathbf{u} = \mathbf{0}$ holds at all times, $\nabla \cdot \partial \mathbf{u}/\partial t = 0$, and the equation can be simplified into a time-invariant form. This can be rearranged and simplified to give the pressure Poisson equation,

$$\nabla \cdot \frac{1}{\rho} \nabla p = -\nabla \cdot \left[\mathbf{u} \cdot \nabla \mathbf{u} - \nu \nabla^{\mathbf{2}} \mathbf{u} \right] \,. \tag{5.21}$$

This form is valid in the case where ρ is variable, and in the pure incompressible flow it can be rewritten as

$$\nabla^2 p = -\rho \nabla \cdot \left[\mathbf{u} \cdot \nabla \mathbf{u} - \nu \nabla^{\mathbf{2}} \mathbf{u} \right] \,.$$

The LHS is the pressure-Poisson operator, and the RHS is the divergence of the remainder of the equation of motion without the time derivative. If one includes boundary conditions, this is the pressure Poisson equation.

5.1.5 Projection Formulation

The projection formulation encompasses similarities with the pressure Poisson approach. The form presented below ultimately produces a Poisson equation that is solved for the "pressure" in the incompressible flow. This pressure is viewed as a potential field used to enforce a divergence-free velocity.

The basic goal with projection methods is to advance a vector (velocity) field, $\mathbf{V} = (V^x, V^y, V^z)^T$ by some convenient means disregarding the solenoidal nature of $\mathbf{V}$, then recover the desired solenoidal vector field, $\mathbf{V}^d$ ($\nabla \cdot \mathbf{V}^d = 0$). We use the notation $\mathbf{V}$ to denote either the velocity, $\mathbf{u}$ or its time derivative $\partial \mathbf{u}/\partial t$ since either can be projected.

The operation to recover a solenoidal vector field is a projection, $\mathcal{P}$, which has the effect

$$\mathbf{V}^d = \mathcal{P}(\mathbf{V}) \ .$$

The projection accomplishes this through the decomposition of the velocity field into divergence-free and curl-free parts. This is known as a Hodge or Helmholtz decomposition [109]. The curl-free portion will be denoted as the gradient of a potential, $\nabla\varphi$. This decomposition can be written

$$\mathbf{V} = \mathbf{V}^d + \nabla\varphi \ . \tag{5.22}$$

Taking the divergence of (5.22) gives

$$\nabla \cdot \mathbf{V} = \nabla \cdot \mathbf{V}^d + \nabla \cdot \nabla\varphi \rightarrow \nabla \cdot \mathbf{V} = \nabla^2\varphi \ .$$

Once φ is known, then the solenoidal vector field is found through

$$\mathbf{V}^d = \mathbf{V} - \nabla\varphi \ . \tag{5.23}$$

Based on the above the projection operator can be written

$$\mathcal{P} = \mathbf{I} - \nabla (\nabla \cdot \nabla)^{-1} \nabla \cdot \ .$$

After the application of $\mathcal{P}$ to a vector field, $\mathbf{V}$, this field will be divergence-free. An important aspect of projections is that they are idempotent, i.e., $\mathcal{P}^2 = \mathcal{P}$, thus repeated application of the operator will not change the result.

In the context of incompressible flow, the equations are utilized in a manner such that the divergence-free velocity constraint is ignored than imposed via the above Helmholtz decomposition. A detailed description of the approximate and exact projection methods is presented in Chaps. 11 and 12.

5.1.6 Artificial-Compressibility Formulation

Another primitive-variable-based approach is the artificial compressibility (AC) method proposed by Chorin [104]. The continuity and momentum equations are coupled by the system

$$\frac{1}{\beta}\frac{\partial p}{\partial \tau} + \nabla \cdot \mathbf{u} = 0 \ , \tag{5.24}$$

$$\frac{\partial \mathbf{u}}{\partial t} + \mathbf{u} \cdot \nabla \mathbf{u} = -\frac{1}{\rho} \nabla p + \nu \nabla^2 \mathbf{u} , \tag{5.25}$$

where β is the artificial compressibility parameter and τ is a pseudo-time. For steady flows $\tau \equiv t$ and the system can be solved by a time marching scheme to provide steady steady solutions. The artificial compressibility approach results in hyperbolic and hyperbolic-parabolic equations for inviscid and viscous incompressible flows, respectively, which are often less computationally expensive to be numerically solved than elliptic equations.

The classical formulation of Chorin [104], suitable for steady-state problems, is extended to transient flows [384, 505, 469, 470] via an approach referred sometimes to as dual-time stepping. The overall idea of the dual-time stepping can be summarized as follows: At each instant t, the augmented pseudo-compressible system

$$\frac{1}{\beta} \frac{\partial p}{\partial \tau} + \nabla \cdot \mathbf{u} = 0 , \tag{5.26}$$

$$\frac{\partial \mathbf{u}}{\partial \tau} + \mathbf{u} \cdot \nabla \mathbf{u} = -\frac{\partial \mathbf{u}}{\partial t} - \frac{1}{\rho} \nabla p + \nu \nabla^2 \mathbf{u} , \tag{5.27}$$

is integrated in a pseudo time τ to a steady state, assuming an artificial speed of sound $\sqrt{\beta}$. The attenuation forcing on the RHS of the momentum equation damps the flow divergence to zero at the rate $(\Delta t)^{-1}$, where Δt is the solution's time step. In the steady-state at $\tau = t + \Delta t$, all $\partial / \partial \tau$ terms vanish and the damping term on the RHS becomes the $\partial \mathbf{u} / \partial t$ derivative of the original equations.

A detailed presentation of the artificial-compressibility approach both for steady and unsteady flows is given in Chap. 10.

5.1.7 Penalty Formulation

Another possibility to remove the ellipticity of the equations, thus allowing the use of explicit schemes, is the penalty (P) method. Application of the penalty function concept of Courant [122] to the finite element analysis of incompressible flows was proposed by Temam [529] and Zienkiewicz [373, 621, 622]. Subsequently, the method attracted interest in the finite element community and was implemented by several researchers to the solution of two- and three-dimensional flow problems [267, 439, 440]. In addition, the method was also considered in the context of finite difference and finite volume formulations [190, 423, 436, 529]. According to the penalty method, the continuity equation is replaced by

$$\frac{1}{\epsilon} p + \nabla \cdot \mathbf{u} = 0 , \quad \text{or} \quad p = -\epsilon \nabla \cdot \mathbf{u} , \tag{5.28}$$

where ϵ is coefficient with large values. As $1/\epsilon \to \infty$, $\nabla \cdot \mathbf{u}$ will tend to zero thus satisfying the incompressibility condition. Taking the divergence of (5.28) to replace pressure in the momentum equations (3.34), one obtains

$$\frac{\partial \mathbf{u}}{\partial t} + \mathbf{u} \cdot \nabla \mathbf{u} = (\frac{\epsilon}{\rho} + \nu) \nabla^2 \mathbf{u} \, . \tag{5.29}$$

The last equation is a convection-diffusion equation with diffusion coefficient $\epsilon/\rho + \nu$. Since (5.29) contains no pressure, it can be solved directly to provide the velocity field. The characteristic time for the artificial diffusion problem must be much shorter compared to the physical time scales. This requires $\epsilon >> \nu$, However, similarly with the artificial compressibility method, it is not clear how large ϵ should be. Note that large values of ϵ (compared to ν) lead to $\nabla^2 \mathbf{u} \simeq 0$. Usually, ϵ takes finite values which can be determined through numerical experimentation [436].

5.1.8 Hybrid Formulations

Ramshaw and Mesina [436] have proposed hybrid formulations which combine the advantages of the P and AC methods. Both the P and AC methods share certain advantages, namely: (i) they facilitate the implementation of explicit numerical schemes and are therefore well suited for parallel computations; (ii) they are easy to implement compared to elliptic solvers. On the other hand, both methods exhibit some stiffness regarding the numerical convergence, which is particularly associated with large values of the parameters β (5.24) and ϵ (5.28).

Ramshaw and Mesina [436] observed that the P method is very effective at attenuating short wavelength components of the advective terms and that the AC method is very effective in numerical solutions where long wavelength errors predominate. In other words, the mathematical character of the P and AC methods with respect to the errors in velocity divergence is diffusive and dispersive, respectively. The above considerations led them to combine the P and AC methods and thus propose a hybrid formulation (P-AC) given by

$$\frac{\partial p}{\partial t} = -\beta \nabla \cdot \mathbf{u} - \epsilon \frac{\partial}{\partial t} (\nabla \cdot \mathbf{u}) \, , \tag{5.30}$$

where both the coefficients β and ϵ are large. The first term on the RHS of (5.30) introduces artificial sound waves and the second term dissipates them. The second term has the character of an artificial bulk viscosity. When $\beta = 0$ the formulation reduces to the P formulation while if $\epsilon = 0$ the AC formulation is obtained. Eq. (5.30) is solved in conjunction with the momentum equation (3.34) given a set of initial and boundary conditions. The boundary conditions do not pose any particular problem, for example, values for the pressure (or pressure gradient) and velocities shall be given at the open boundaries, and the no-slip condition should be set on rigid walls (or the velocity of the boundary in the case of moving walls). However, the initial condition for pressure poses difficulties similar to the case of the P and AC methods.

In [436] a hybridization ratio was defined

$$\chi = \frac{\beta \Delta t}{2\epsilon} , \tag{5.31}$$

where the values of $\chi = 0$ and $\chi = \infty$ correspond to the original P and AC methods, respectively. By performing Fourier stability analysis of the simplest explicit (two dimensional) version of the hybrid method and using centered spatial differences, the following stability criterion can be obtained [436]

$$\Delta t \, \frac{\beta \Delta t + 2(\epsilon + \nu)}{\Delta^2} < 1 , \tag{5.32}$$

where $\Delta^2 = 1/(\Delta x^2 + \Delta y^2)^{-1}$, with Δx and Δy being the spatial steps in the x and y directions, respectively. To estimate the values of β and ϵ, Ramshaw and Mesina [436] proposed the formulas

$$\beta = f\chi \frac{\Delta^2}{(\chi + 1)\Delta t^2} , \tag{5.33}$$

$$\epsilon = f \frac{\Delta^2}{2(\chi + 1)\Delta t} , \tag{5.34}$$

where f is a coefficient ("safety factor") such that $f < 1$. The last equations provide the largest values of β and ϵ that are consistent with stability.

The hybrid method was developed by Ramshaw and Mesina [436] mainly having in mind time-accurate transient solutions, though the formulation is also applicable to steady state calculations. The hybrid method was further investigated by Dukowicz [167] in contrast to the pressure-Poisson equation in conjunction with a conjugate-gradient (CG) method. Dukowicz [167] selected the conjugate-gradient (CG) method instead of the successive over relaxation (SOR) method because it is equally to the hybrid P-AC method well suited to parallel computations. His computations showed that the P-AC method is less efficient than the pressure-Poisson/CG method.

An alternative hybrid formulation in conjunction with a fully implicit scheme has also been proposed by McHugh and Ramshaw [381]. According to this, the continuity equation is replaced by the AC formulation (5.24) while the momentum equation is written

$$\frac{\partial \mathbf{u}}{\partial \tau} + \mathbf{u} \cdot \nabla \mathbf{u} = -\frac{\partial \mathbf{u}}{\partial t} - \nabla(p - b\nabla \cdot \mathbf{u}) + \nu \nabla^2 \mathbf{u} , \tag{5.35}$$

where τ is a pseudo-time and b is an artificial bulk viscosity. These equations provide a coupling in pseudo-time τ. Eq. (5.35) should be iterated in

τ in order to obtain a converged solution for each real time t. McHugh and Ramshaw [381] also presented a Fourier dispersion and convergence analysis for this hybrid formulation in order to estimate the optimum values of the AC parameter and bulk viscosity.

Another hybrid formulation has also been proposed by Brooks and Hughes [82]. They combine the P and AC methods by letting the velocity divergence to be a linear combination of p and $\partial p/\partial t$. The difference of Ramshaw's and Mesina's formulation with the one proposed by Brooks and Hughes [82] is that the former adds the P and AC contributions to the pressure, whereas the latter adds them to the divergence.

5.2 Model Equations

In the development of numerical methods for the solution of fluid flow equations, model equations are often employed. These are the linear advection and advection-diffusion equations, as well as the Burgers' equation.

5.2.1 Advection-Diffusion Equation

The linear advection-diffusion equation for a scalar transport, c, is written as

$$\frac{\partial c}{\partial t} + u\frac{\partial c}{\partial x} = \kappa\frac{\partial^2 c}{\partial x^2} + \tilde{S}\,, \tag{5.36}$$

where u is a given constant (linearized advection speed), κ is a diffusion coefficient, and $\tilde{S}$ are source terms which, for example, may arise from chemical reactions and/or radiation. The scalar transport may represent temperature, T, or mass concentration (e.g., chemical species, air pollution). In the case of heat conduction problems, $\kappa = k/(\rho C_p)$, where k is the thermal conductivity and C_p is the heat capacity for constant pressure. Equation (5.36) is parabolic in time. For $\kappa = 0$ (and $\tilde{S} = 0$), we obtain the linear advection equation

$$\frac{\partial c}{\partial t} + u\frac{\partial c}{\partial x} = 0\,, \tag{5.37}$$

which is a hyperbolic equation (linear advection an important model equation).

We can make (5.36) dimensionless by defining a characteristic length l_0, a characteristic velocity u_0, and a characteristic value c_0 for the scalar transport c; the time becomes dimensionless by l_0/u_0. Neglecting the source term in order to simplify the analysis, (5.36) yields

$$\frac{\partial c^*}{\partial t^*} + u^*\frac{\partial c^*}{\partial x^*} = \frac{\kappa}{u_0 l_0}\frac{\partial^2 c^*}{\partial x^{*2}}\,. \tag{5.38}$$

By omitting the superscript "*", we write

$$\frac{\partial c}{\partial t} + u\frac{\partial c}{\partial x} = \frac{1}{\mathrm{Pe}}\frac{\partial^2 c}{\partial x^2} , \tag{5.39}$$

where $\mathrm{Pe} = u_0 l_0/\kappa$ is the Peclet number.

The above non-dimensionalization is appropriate when the time scale is set by advection. If the time scale is set by diffusion then it is preferred the time to become dimensionless by l_0^2/κ; this is similar to the derivation of the dimensionless Navier-Stokes equations for low Reynolds number flows (Sec. 3.5). Then the advection-diffusion equation is written

$$\frac{\partial c}{\partial t} + \mathrm{Pe}\, u\frac{\partial c}{\partial x} = \frac{\partial^2 c}{\partial x^2} . \tag{5.40}$$

Similar to the Reynolds number, the Peclet number represents ratio of advective to diffusive processes.

The linear advection equation is obtained if one sets in (5.36) the RHS equal to zero, i.e.,

$$\frac{\partial c}{\partial t} + u\frac{\partial c}{\partial x} = 0 . \tag{5.41}$$

The boundary conditions can be of Dirichlet type

$$c(0) = c_1 , \quad c(1) = c_2 , \tag{5.42}$$

combination of Dirichlet/Neumann, e.g.,

$$c(0) = c_1 , \quad \left.\frac{\mathrm{d}\, c(x)}{\mathrm{d}x}\right|_{x=1} = c_2 , \tag{5.43}$$

or periodic (in space),

$$c(0,t) = c(1,t) , \tag{5.44}$$

where c_1 and c_2 are known constant values of the scalar transport c at the boundaries of the domain.

5.2.2 Burgers' Equation

The Burgers' equation has been characterized as *a model for all reasons* [188]. The one-dimensional Burgers' equation is

$$\frac{\partial U}{\partial t} + U\frac{\partial U}{\partial x} = \nu\frac{\partial^2 U}{\partial x^2} , \tag{5.45}$$

this can also be written in conservation form

$$\frac{\partial U}{\partial t} + \frac{\partial U^2/2}{\partial x} = \nu\frac{\partial^2 U}{\partial x^2} , \tag{5.46}$$

which is important with shocks.
where U is a velocity-like dependent variable and ν is a viscosity-like parameter. In non-dimensional form the viscosity will be the inverse of the Reynolds number, $\nu = 1/Re$. Interestingly, the first steady solution to (5.45) was given by Bateman [40], but the equation has finally taken its name from the extensive work of Burgers [90, 91, 92].

The one-dimensional Burgers' equation can be used for modeling a variety of flow problems that includes: shock propagation, turbulence, acoustic transmission, traffic simulation and wave propagation in a thermoelastic medium. In fact, the equation can be used to model any nonlinear wave propagation problem subject to dissipation arising from viscosity, heat conduction, chemical reaction, mass diffusion and thermal radiation. A thorough review of Burgers' equation and its numerical solution is given in [188] and [189], while about thirty-five exact solutions have been classified in a tabular form in [54]. One of the most straightforward solutions is for the case where $U = -\tilde{U}$ and $U = \tilde{U}$ at $x = -\infty$ and $x = \infty$, respectively.[3] For this case, the exact (steady) solution is given by

$$U(x) = -\tilde{U} \tanh\left(\frac{\tilde{U}x}{2\nu}\right) , \tag{5.47}$$

which describes a single shock of width $l = \nu/\tilde{U}$.

The solutions to Burgers' equation can encompass both steep gradients and statistics which resemble to hydrodynamic turbulence. In the context of high-resolution methods for incompressible flows, we are particularly interested in the analogy of turbulent-like solutions of Burgers' equation (*Burgers' turbulence*) and Navier-Stokes turbulence. An extensive theoretical work on *Burgers' turbulence* (Burgulence) [236] has been presented by Kraichnan [303, 304]. Results on Burgers' turbulence are presented in Chap. 19. The most important differences and similarities between Burgers' and incompressible Navier-Stokes equations are:

1. In both Burgers' and Navier-Stokes equations the advective term conserves $\int U(x,t)\,\mathrm{d}x$ and $\int U^2(x,t)\,\mathrm{d}x$ (a quadratic nonlinearity).

2. In both cases, the advective term tends to produce regions of steep velocity gradients. This implies a transfer of excitation from lower to higher wave-number components of the velocity field [303]. On the other hand Burgers' equation produces shock wave solutions while incompressible flows do not.

3. The most important difference between Burgers' and Navier-Stokes equations is the lack of the pressure term from the former. The pressure

[3] For computational purposes, $x = x_L$ and $x = x_R$, respectively, where x_L and x_R are chosen to be sufficiently large.

gradient term is a continuous source of random motion, whereas the randomness in Burgers' solution is due to the initial data and, as a result, the initial randomness is reduced with time.

4. The Navier-Stokes solutions for turbulence lead to an inertial range of eddy sizes with energy spectrum $E(\kappa) \propto \kappa^{-5/3}$, where κ is the wave number [39]. In the case of Burgers' turbulence, we obtain $E(\kappa) \propto \kappa^{-2}$ [90, 303].

6. Basic Principles in Numerical Analysis

As a general rule it is a good idea for someone using a numerical method to have a grasp on the properties of the solutions it produces, or more properly the errors that one can expect. This includes the knowledge of the conditions under which one can expect a method to be stable, and produce reliable results. Below, we briefly introduce a set of basic tools that can be used to begin to examine these issues. We start with an introduction to the fundamental definition of accuracy, stability and consistency (the last two provide some hope of convergence as more computational resources are used and the first relates to how fast convergence can be observed) as applied in one dimension (we choose time as a prelude to Chap. 7). Next, we introduce the Fourier or von Neumann stability analysis which can provide a window into each of these key issues for PDEs. Next, we discuss the modified equation approach which compliments Fourier analysis and allows the study of the nonlinear properties of equations and methods. Finally, we introduce the basic concepts in verification of computational algorithms as a quality assurance measure.

6.1 Stability, Consistency and Accuracy

Here, we will introduce the basic concepts of accuracy and stability associated with time integration methods. These concepts generalize to spatial integration, but it is useful to examine them in the context of time (single variable). This material is covered in numerous books, articles and web pages, but in order to make our discussion self-contained and consistent, we will provide it here briefly. The methods discussed are standard and largely applicable to the spatial (or coupled time and space) differencing associated with the advection-diffusion equation discussed in Chaps. 13 through 17. We recommend the recent book by Ascher and Petzold [19] for a more complete discussion of issues associated with the integration of ordinary differential equations. For more complete introduction especially suited for computational fluid dynamics we recommend [339, 341, 315, 171, 543].

When designing a method two issues immediately come to mind: to what degree of accuracy is the problem solved? and under what conditions are the results reliable (i.e., stable and accurate)? We might further recommend that the users of a given method are well-served by considering these issues carefully as they provide one with expectations prior to using a method. Moreover these expectations should be carefully confirmed upon implementing a method in a computer code.

A starting point to answer these questions can be stated via the *Lax equivalence theorem* which states *that consistency and stability are necessary and sufficient for convergence.* It is notable that this theorem only truly holds for linear equations although its guidance has proven useful for nonlinear equations as well. By consistency we mean that a method is a viable approximation to the original differential equation we are solving. We will supply a more exacting criteria for consistency shortly. Stability implies that the solutions are well bounded, i.e., that small perturbations do not grow without bound. The satisfaction of both of these conditions implies that as the step size is made smaller the numerical solution will converge to the correct solution of the (linear) differential equation. Note that these conditions are associated with *linear* differential equations and for nonlinear differential equations no such theorem exists[1].

The concept of accuracy involves the estimate of the quality of the approximation as a function of the step size, $h = t^{n+1} - t^n$, the differencing in time from one discrete time step to the next. The standard tool to handle accuracy is the Taylor series expansion in time,

$$U(t+h) = U(t) + hU_t + \frac{1}{2}h^2U_{tt} + \frac{1}{6}h^3U_{ttt} + \text{H.O.T.}\,, \tag{6.1}$$

where H.O.T. stands for the higher order terms neglected in the equation. In this case, the next term in the expansion is $1/24h^4U_{tttt}$. Given the Taylor series expansion, we can design or analyze the accuracy of a given difference method by substituting this expansion for the discrete values of U^k, $k = n+1, n, n-1, n-2, \cdots$ at the mesh points, and expanding around a single point, $U(t)$. This yields linear or nonlinear equations that can be solved for the coefficients of the method.

For example, the forward Euler method (the foundation of all explicit methods), for the ODE, $U_t = f(U,t)$ is

$$\frac{U^{n+1} - U^n}{h} = f(U^n, t^n)\,, \tag{6.2}$$

where $U^{n+1} \equiv U(t+h)$ and $U^n \equiv U(t)$. This method is only first-order accurate. Using the Taylor series (6.1) and retaining only the first two ordered terms in the expansion, the difference formula (6.2) is written as

[1] Generally speaking, numerical analysis is primarily confined to linear equations, and then extended rather successfully in practice to nonlinear equations. While this approach is heuristic and ad hoc in nature, it is guided by the rigorous results that can be achieved for linear equations. One exception is the use of the "modified equation" approach where an effective differential equation is derived for a numerical method as discussed in Sect. 6.3.

$$U_t = f(U,t) + \frac{1}{2} h U_{tt} \, . \tag{6.3}$$

The non-ordered term is simply the original ODE, thus providing a demonstration of consistency. This produces a working definition of consistency: *The Taylor series expansion for a method recovers the original differential equation and a sequence of ordered error terms. The key for consistency is the recovery of the original differential equation.* The last term in (6.3), proportional to h, is the leading order truncation error establishing the order of accuracy of the discretization method, i.e., in the present case as a first-order accurate method. For second and third-order methods the leading order error term would be proportional to h^2 and h^3, respectively, and so on for higher orders. Thus, for a first-order method we would expect errors to decrease linearly with decreasing time step. A second-order method would provide a quadratic reduction in error with decreasing time step. From this we see that high-order methods are more efficient than low-order methods as long as the complexity and cost of the method does not overwhelm the scaling advantages. All of these considerations only hold when the time step is small enough for the asymptotic behavior of the method to exhibit itself. Outside the asymptotic range for a method the results will be generally of lower order, but far less predictable and often dependent upon other factors. Finally, it is absolutely essential to test the method (as implemented) and demonstrate that the solutions behave as expected given the analysis.

Next, we will introduce the concept of 0-stability. This is the most basic requirement for a method in that it is stable for $h \to 0$. Without meeting this criterion a method is utterly hopeless. Generally speaking, methods that involve more than two time levels can potentially violate this requirement. The general class of methods having this form are the linear multistep methods (LMM). On the other hand, Runge-Kutta methods are not susceptible to 0-instability due to their construction directly from Taylor series expansions (as opposed to polynomial interpolation and integration as in LMMs). Stability is found through solving for the eigenvalues of the characteristic polynomial. The derivation of the characteristic polynomial is briefly discussed below and for more detailed description and discussion we refer the reader to previous published numerical analysis textbooks [207, 314, 19, 286, 256].

For the ODE $U_t = f(U,t)$ using a third-order accurate LMM, we obtain

$$\frac{U^{n+1} + 4U^n - 5U^{n-1}}{6h} = \frac{2}{3} f(U^n, t^n) + \frac{1}{3} f\left(U^{n-1}, t^{n-1}\right) \, . \tag{6.4}$$

Defining the operator H by $HU^n = U^{n+1}$ or, for a general k-step method by $H^k U^n = U^{n+k}$, and taking $h \to 0$ (6.4) is written as

$$\mathcal{P}(H) U^n = 0 \, , \tag{6.5}$$

where $\mathcal{P}$ is a polynomial of H. Since the stability of (6.4) depends solely on the transient solution of the homogeneous equation, it is sufficient to study the solution of (6.5). The characteristic equation of (6.4), and subsequently of (6.5), is defined by

$$\mathcal{P}(A) = \sum_{k=1}^{3} \Big(a_k - hf\left(U^n, t^n\right) \Big) A^k = 0 \,, \tag{6.6}$$

where $a_2 = 1$, $a_1 = 4$ and $a_0 = -5$. If one takes $h = 0$ that equation will produce the characteristic polynomial, $A^2 + 4A - 5$, with roots, $A = 1$ and $A = -5$. The second root has $|A| > 1$ indicating instability. Therefore, this method is not 0-stable.

Now we will show how to compute more general conditions for stability. The approach we will take is applied to a linear equation, $u_t = \lambda u$, with $\lambda = a + b\imath$ ($\imath = \sqrt{-1}$) where a and b are constant coefficients. This approach will be extended to PDEs with von Neumann analysis in the next section. The route to finding the stability bounds are straightforward, one substitutes the linear function into the formula for the difference method. The different time levels are denoted by powers of an amplification factor, A, per time step. We can then determine the conditions on the step size that produce $|A| \leq 1$. Again, speaking quite generally accuracy requirements often render the stability bounds to academic interest because of the small time step sizes required to achieve a given level of accuracy.

As shown in the above example of 0-instability, LMMs require one to solve a polynomial equation. In that case one must take the worst case of the roots, i.e., $\max |A_k|$ (for a k-step method) to determine stability. Note that there will be one root fewer than the number of time levels used by the method. One of the roots is associated with the accurate solution, while the others are called *spurious* or *computational* roots. From these roots the accuracy of the method can be shown for linear equations by comparing to roots of A, which are expanded via Taylor series with respect to the exact solution of the model equation, $u_t = \lambda u$.

Expanding the difference method in h shows the accuracy of the method. For example, take the forward Euler method, (6.2), which produces a characteristic polynomial,

$$A = 1 + h\left(a + b\imath\right) \,. \tag{6.7}$$

We can plot $|A|$ against values of ah and bh. This is shown in Fig. 6.1 where the values $|A| > 1$ have been masked out[2]. For the linear model equation, the leading order truncation error is

$$\frac{1}{2} h\lambda \,.$$

This is the same expression as the form derived by the Taylor series (substituting the function for the time derivative). The Taylor series expansion

[2] This analysis is easily conducted with the aid of a modern algebraic manipulation program like Mathematica [603].

as applied to the multiple roots of the LMM method is used to differentiate the accurate root from the spurious root. The accurate root will reproduce one or more correct terms in the Taylor series (corresponding to the correct answer) thus demonstrating its accuracy. The spurious roots will not provide an approximation to any of the terms in the Taylor series of the correct solution.

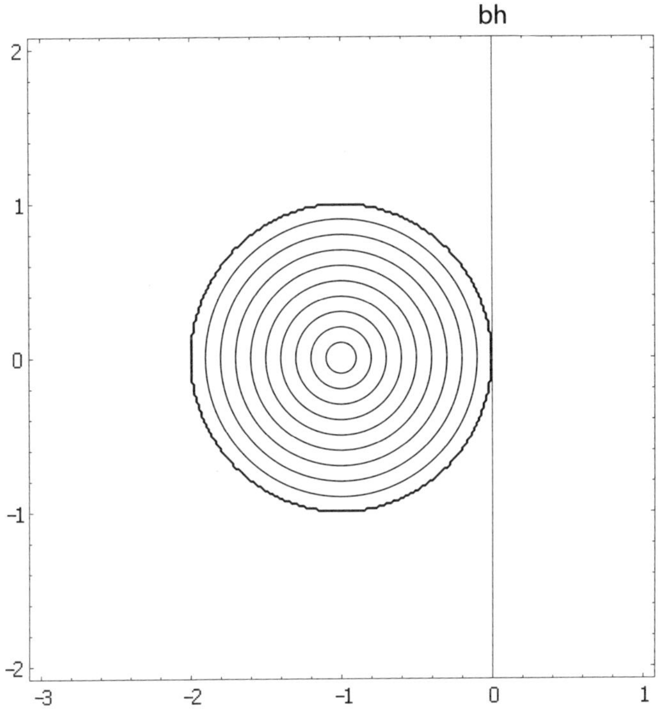

Fig. 6.1. The stability region, $|A| \leq 1$ for the forward Euler method. The axes of the plot are the real, ah, and imaginary, bh, values corresponding to the linear ODE $u_t = (a + b\imath)u$. Contours are only displayed on the interior of the stability region. Note that for stability the method must have $ha < 0$ so that the numerical integrator is dissipative (contractive).

6.2 Fourier Analysis

Fourier analysis is the most commonly used method used to examine the properties of numerical techniques for PDEs. It is also known as von Neumann stability analysis because John von Neumann first used this method [447][3]. This technique is a fundamental step in the analysis of a method for PDEs. It will uncover the basic linear properties of a method. Most important among these properties is the stability of a method as described by its amplification factor. The availability of high quality symbolic algebra software enables simpler more accessible approaches to the analysis of numerical methods.

Errors in the numerical solution of hyperbolic problems are generally classified as being of either a damping or dispersive variety. As will be seen later, a useful numerical scheme must contain some minimal amount of dissipation to remain stable and produce physical solutions. This is exactly what was needed for the stability of the integration of ODEs. This dissipation damps out error which would otherwise grow in an unbounded fashion. Lack of sufficient damping will result in dispersive errors that can cause non-physical maxima and minima to be created in the solution by the numerical scheme. Dispersive (phase) errors produce changes in the velocity of a wave. In numerical solutions, waves will travel at different speeds depending upon the wavelength of the wave compared with the numerical grid size [548].

Phase errors result in information being transported at a numerical velocity below or above the true velocity of this information. Typically, von Neumann stability analysis [447, 403, 256, 190] is used to analyze these errors. The process consists of replacing the dependent variables by Fourier series. For a discrete Fourier series the approximate expression can be written as

$$U_j^n(x_j, n\Delta t) \approx \sum_{l=-N/2}^{N/2} C_l^n e^{2\pi \imath l j/N} , \tag{6.8}$$

where the domain is of length $N\Delta x$ and the positions of the discrete points in the domain are defined by $x_j = j\Delta x$. A general technique that can be used to analyze the stability of any linear method is to use the principle of superposition, i.e., instead of using the entire Fourier series one can use only one term of the series. Defining $\theta_l = 2\pi lj/N$ and writing the solution of the linear equation that is under examination in terms of the ratio $G_l = C_l^n / C_l^{n-1}$, we obtain

$$U_j^n = \frac{C_l^n}{C_l^{n-1}} \frac{C_l^{n-1}}{C_l^{n-2}} \cdots \frac{C_l^1}{C_l^0} C_l^0 \, e^{\imath \theta_l j} . \tag{6.9}$$

However, for linear equations the ratios G_l do not depend on n. In other words, this ratio is the same at every time step. Therefore, (6.9) gives

$$U_j^n = G_l^n C_l^0 e^{\imath \theta_l j} , \tag{6.10}$$

[3] The technique was introduced by von Neumann in February 1947 in a series of lectures in Los Alamos. The technique was first published as Los Alamos Report, LA-657, detailing the numerical solution of parabolic differential equations.

where n in G_l^n is a power. For the sake of simplicity, one can assume that $C_l^0 = 1$ (any other value except 0 could be employed without affecting the results of the stability analysis). By dropping the subscript l, we can write

$$U_j^n = G^n e^{\imath j\theta} . \tag{6.11}$$

The magnitude of G is called the amplification factor. Generally, the expression of G will be a combination of real and imaginary trigonometric terms and is transformed in order to extract useful information. This is accomplished by taking the functional form of G into two pieces: an amplification factor and a phase angle,

$$G(\imath j\theta) = |A| e^{\imath\phi} , \tag{6.12}$$

where $|A|$ is the magnitude of G and $\phi = \tan^{-1}(\mathrm{Im}(G)/\mathrm{Re}(G))$, the ratio of the imaginary and real parts of G, is the phase angle.

For stability, $|A|$ must be less than or equal to one for all θ, but this implies damping. Small values of A imply excessive damping. For the scalar wave equation, the exact phase speed is known[4] so that the ratio of this to the numerical phase speed can be taken. If this quantity is less than one the error is lagging, if it is greater than one the error is leading. These errors have a spectrum of values which can have a large range of values. As far as the magnitude of the amplification factor is concerned, the spectral values must all be less than one (in absolute value). At different wave-numbers the phase error can vary leading at some wave-numbers while lagging at others.

6.2.1 Fourier Analysis of First-Order Upwind

As an example consider a first-order upwind method for the linear advection equation $U_t + cU_x = 0$ (where U is a scalar and c is a constant)

$$U_j^{n+1} = U_j^n - \mathcal{C}\left(U_j^n - U_{j-1}^n\right) ,$$

where $c > 0$ has been assumed ($\mathcal{C} = c\Delta t/\Delta x$ is the Courant or CFL number). Substituting the Fourier series for $U_j^{n+1} = G^{n+1}e^{\imath j\theta}$ and $U_j^n = G^n e^{\imath j\theta}$ in (6.2.1) gives

$$G^{n+1}e^{\imath j\theta} = G^n e^{\imath j\theta} - \mathcal{C}\left(G^n e^{\imath j\theta} - G^n e^{\imath(j-1)\theta}\right) .$$

Dividing by $G^n e^{\imath j\theta}$, we obtain

$$G = 1 - \mathcal{C}\left(1 - e^{\imath-\theta}\right) .$$

We separate this form into its real,$1 - \mathcal{C}(1 - \cos(-\theta))$, and imaginary, $\sin(-\theta)$, parts and form the amplification factor $|A|$,

[4] The exact wave-speed is $\mathcal{C}\theta$ where $\mathcal{C}$ is the CFL number. The CFL number is the ratio of the distance covered by a wave to the size of the numerical grid. It is named for Courant, Friedrichs and Lewy [123] who first introduced this concept.

$$|A| = \sqrt{\left(1 - \mathcal{C}\left(1 - \cos\theta\right)\right)^2 + \left(\mathcal{C}\sin\theta\right)^2} \,.$$

Simply plotting this function for $\theta \in [0, \pi]$ confirms that this method is stable for a value of $0 \le \mathcal{C} \le 1$. This is shown in Fig. 6.2. The amplitude can also be expanded in a Taylor series around $\theta = 0$ to help describe its error. In the case of the amplitude the error is

$$|A| \approx 1 - \frac{1}{2}\mathcal{C}\left(1 - \mathcal{C}\right)\theta^2 + \mathcal{O}\left(\theta^4\right) \,.$$

The stability condition $|A| \le 1$ gives $0 \le \mathcal{C} \le 1$.

The phase error is the second part of the error. This describes how accurately signals move on the grid. One manner of describing this error is an angle formed by the arc tangent of the ratio of the imaginary and real parts of the error and the exact wave-speed,

$$\phi\left(\theta\right) = \tan^{-1}\frac{\frac{-\sin(-\theta)}{1-\mathcal{C}(1-\cos(-\theta))}}{\mathcal{C}\theta} \,.$$

Again, the order of the error can be examined using a Taylor series expansion,

$$\phi \approx 1 - \left(\frac{1}{6} - \frac{\mathcal{C}}{2} + \frac{\mathcal{C}^2}{3}\right)\theta^2 + \mathcal{O}\left(\theta^4\right) \,.$$

Here, the error goes to zero (the second term on the RHS) as $\mathcal{C} = 1$, but grows larger as $\mathcal{C} = 0$. This is a third-order error because the phase error has been divided by one order in its formula $\mathcal{C}\theta$.

This sort of analysis can easily be conducted in Mathematica and Appendix A.1 shows the commands for the method and results above. The upwind method is simple enough to conduct by hand, but as methods become more complex it rapidly exceeds the skill and patience of most people. With computer algebra the analysis can be conducted in a fast and error free manner while allowing method complexity to grow.

6.2.2 Fourier Analysis of Second-Order Upwind

We will also show how this analysis can be extended to second-order methods. The commands for Mathematica are given in Appendix A.2 and builds off of the first-order methodology.

The basic steps are the same as the first-order method's analysis. The single addition is the definition of the edge value, $U_{j+1/2}$. In the upwind method this value (for $\mathcal{C} > 0$) is $U_{j+1/2} = U_j^n$. In the second order method this value changes to an extrapolated value using a linear interpolation,

$$U_{j+1/2} = U_j^n + 1/2\left(1 - \mathcal{C}\right)S_j; \quad S_j = \frac{1}{2}\left(U_{j+1}^n - U_{j-1}^n\right) \,.$$

The term proportional to $(1 - \mathcal{C})$ encodes both the extrapolation and the advancement of the method to second-order in time. This comes from $S_j =$

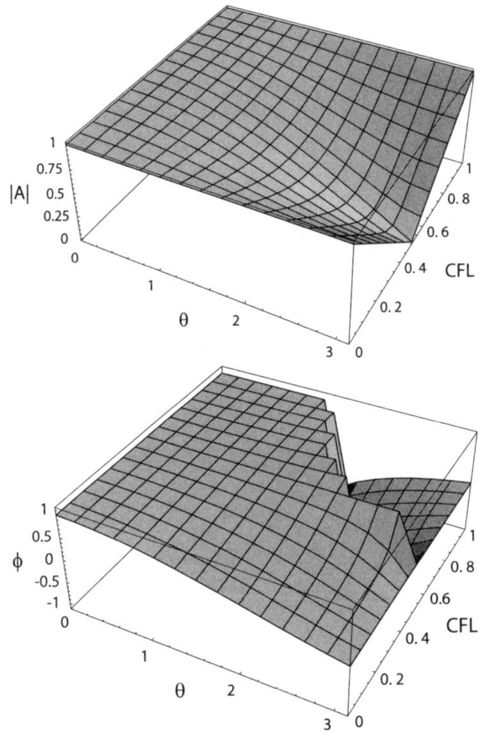

Fig. 6.2. The amplitude $|A|$ and phase ϕ of the upwind method plotted as a function of the CFL number and frequency θ. The axis for the CFL number has a range from zero to one and the axis for the frequency has a range from zero to $\pi \approx 3.14$. Because the amplitude is less than one, the method is stable.

$\partial U/\partial x \; \Delta x$ and $U_t = -c \; U_x$. This is Fromm's method first introduced in 1968 [199].

The overall results are shown for the amplitude and phase of the method in Fig. 6.3. The accuracy can be examined via Taylor series. The amplitude error is

$$|A| \approx 1 - \left(\frac{\mathcal{C}}{8} - \frac{\mathcal{C}^2}{4} + \frac{\mathcal{C}^3}{4} - \frac{\mathcal{C}^4}{8}\right)\theta^4 + \mathcal{O}\left(\theta^6\right) .$$

and the phase error is

$$\phi \approx 1 + \left(\frac{1}{12} - \frac{\mathcal{C}}{4} + \frac{\mathcal{C}^2}{6}\right)\theta^2 + \mathcal{O}\left(\theta^4\right) .$$

The amplitude error is higher order than the first-order method (fourth-order), so that the leading order error is the phase error (same order phase error as first-order upwind, but first-order upwind has a leading amplitude error). While the phase errors are of the same order as the first-order upwind method, the size of that leading order error is smaller with the second-order method. The general properties are otherwise similar to first-order upwind.

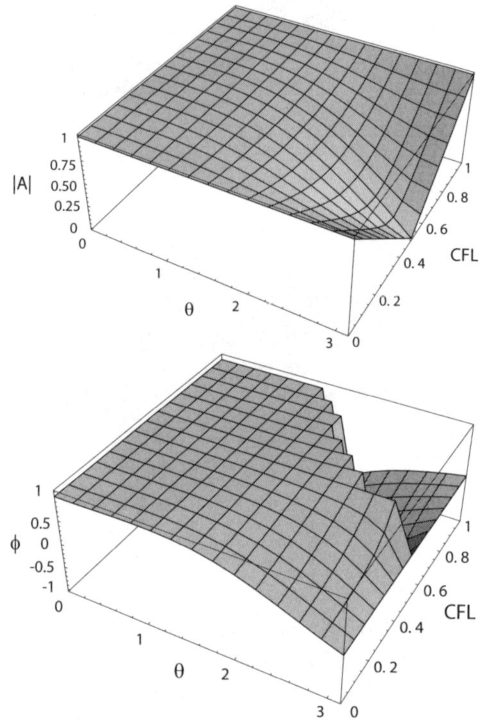

Fig. 6.3. The amplitude $|A|$ and phase ϕ of the Fromm's method plotted as a function of the CFL number and frequency θ. The axis for the CFL number has a range from zero to one and the axis for the frequency has a range from zero to $\pi \approx 3.14$. Because the amplitude is less than one, the method is stable.

The last step is to test the analysis by coding the method and running it on problems. We show such an exercise in Fig. 6.4 by solving the advection equation for a sine wave (initial condition) which is like the long wavelength Fourier analysis. The Fromm's method is much more accurate than the upwind method. Note that we have not used many grid points. In this way the problems in the solution will be magnified and are easier to see. If we had used a large number of points we would have learned less because the errors are smaller.

The next step is to measure the error. Because the waveform used is analytical measuring the error, the difference between the exact and numerical solution, is straightforward. Table 6.1 gives the sum of the absolute value

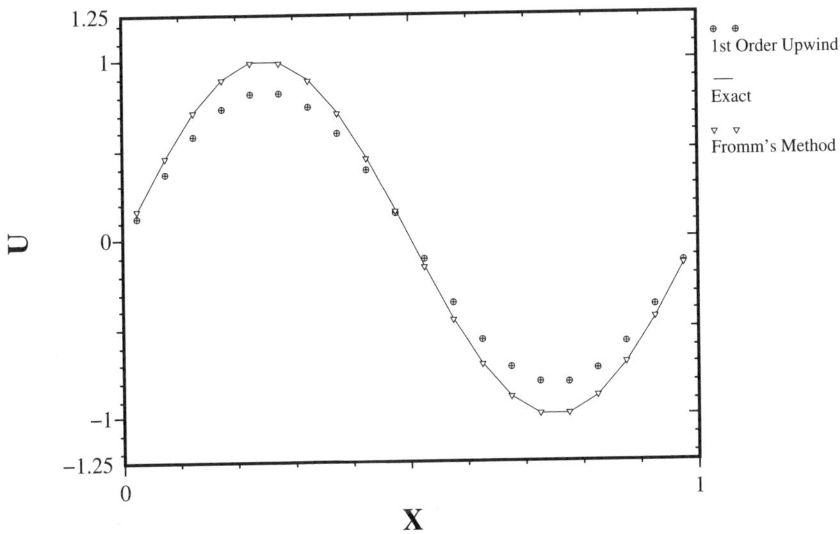

Fig. 6.4. The solution of a single period of a sine wave using the upwind and Fromm's method with $C = 0.8$ for the advection equation.

of the errors for different number of grid points. A first-order method will ideally reduce the error by a factor of two as the mesh is refined while the second-order method reduces the error by a factor of four. We can see that both methods perform as expected.

Table 6.1. The sum of errors for the sine wave advected with first-order upwind and Fromm's method. The error is simple the average of the absolute value of the difference between the exact and numerical solutions.

Grid	Upwind	Fromm's
20 cells	1.15×10^{-1}	4.52×10^{-3}
40 cells	5.99×10^{-2}	1.02×10^{-3}
80 cells	3.07×10^{-2}	2.49×10^{-4}

6.3 Modified Equation Analysis

The approach of using the modified equation has a long history. The most concrete starting point is the paper by Hirt [259] although similar work exists even earlier without the explicit name or broader formalism introduced by Hirt. The method has some of its chief application to the heuristic understanding of errors in finite difference algorithms. In that work and other early papers the notion of even order errors being associated with physical diffusion processes and odd order errors being associated with dispersive errors was forwarded. Indeed these notions were central to the proof of entropy satisfying solutions arising from upstream differencing [249]. This was the consequence of the negative definite dissipation associated with the second-order diffusive truncation error.

The method was refined and popularized by Warming and Hyett [592]. In [592] fully discrete methods were studied by systematically replacing time derivative by fully discrete spatial derivatives. This method has enjoyed broad usage in a variety of applications, although hyperbolic partial differential equations is the most common. A great deal of recent theoretical development has been focused in the numerical solution of ODEs (see [233] in particular). The basic approach of modified equation analysis has also been used to describe the basic properties of implicit turbulence models [369], discussed at length in Chap. 19. In particular, in [369] numerical experiments were performed to verify that the modified equation produces a differential form which when integrated produces results consistent with the original method. More recently, Knoll et al. [295] have utilized the modified equation approach in quantifying the impact of nonlinearly converged methods as opposed to methods that do not converge the nonlinearities in the equations. Their approach shares many quantitative touches with [369] as they verify their analysis via computational experimentation. We will follow this course of action and demonstrate the predictive power of the modified equation analysis as well as where it breaks down.

Here, we will discuss the specific techniques used in our analysis. The general approach is greatly enabled by the availability of high quality symbolic mathematics packages (e.g., Maple, Mathematica). The analysis is characterized by several specific steps:

1. Write the computational algorithm in a pseudo-code form into the symbolic algebra package.
2. Expand the discrete method in a Taylor series, and collect the equation into terms ordered in h, the discrete spacing (time or space). Should the equation be in divergence form, integrate the error in space to evaluate the modified fluxes. There are limits to this technique because of lack smoothness in function such as min( ) and max( ). However many of such functions can "cleverly" be rewritten in an algebraic form. For example, the functions can be rewritten

$$\max(a,b) = \frac{1}{2}(a+b) + \frac{1}{2}|a-b| ,$$

and

$$\min(a,b) = \frac{1}{2}(a+b) - \frac{1}{2}|a-b| .$$

3. Analyze the equation in terms of specific functions that one has interest in studying either as initial data or effective forms which the solution evolves toward.
   a) Compute the error, and verify the modified equation analysis.
   b) Analyze the energy behavior of the method; does it dissipate energy locally?
   c) The rate of convergence expected under these circumstances can also be examined.
4. Confirm the results of the analysis in actual computational implementations with nonlinear equations.

Of course, standard von Neumann (Fourier) stability analysis is essential and is also greatly enhanced and simplified by the same symbolic algebra software (e.g., see [455, 340, 342]). For a more pedagogical approach to using this software see [203]. The linear stability analyzed by that method is essential, but it is intrinsically linear. Far too often the nonlinear behavior is left for fully experimental investigations (i.e., with the actual computer code). Nonlinear analysis is often quite heuristic, experimental, and folklore ridden. The above list is an attempt to lift the capability of nonlinear analysis to a somewhat higher level.

As a concrete example of each of these techniques we will analyze upwind differencing on Burgers' equation, (5.46), $E(U) = 1/2U^2$. This analysis is the example given in Appendix A. The upwind differencing can be written through its flux,

$$E_{j+1/2}(U) = \frac{1}{2}\left[E(U_j) + E(U_{j+1})\right] - \frac{1}{2}\left|E'_{j+1/2}\right|(U_{j+1} - U_j) , \quad (6.13)$$

and conservation form differencing (detailed presentation of various flux forms is given in Chaps. 15 and 16). The notation, E', is the derivative of the flux with respect to U, $\partial E/\partial U$. The spatial truncation error in the flux is obtained by taking the truncation error from the flux difference and integrate with respect to x. This gives

$$\frac{1}{2}h|E'|U_x + \frac{1}{6}h^2\left[E'U_{xx} + E''(U_x)^2\right] + h^3\frac{1}{24}|E'|U_{xxx} , \quad (6.14)$$

where $h = \Delta x$. For Burgers' equation (6.14), $E' = U$ and $E'' = 1$, the spatial truncation error is written as

$$\frac{1}{2}h|U|U_x + \frac{1}{6}h^2\left[UU_{xx} + (U_x)^2\right] + h^3\frac{1}{24}|U|U_{xxx} .$$

Taking as test function $U(x) = \tanh(10x)$ we can evaluate the scheme for various values of h. The full nonlinear examination of this method is displayed in Fig. 6.5. Even though the discretization is coarse, the modified equation produces a good representation of the error (to within less than 10 percent).

Given a difference scheme the modified equation for the method is given by replacing the discrete values with appropriately centered Taylor series expansions. For linear methods and equations this is quite straightforward, but if either the method or the equation being solved is nonlinear, it rapidly becomes complicated. Nonetheless, symbolic algebra makes all of this tractable.

Moreover, we can apply this analysis to time differencing. For example, if we solve the equation $\mathbf{U}_t = -\mathbf{E}_x(\mathbf{U})$ and discretize only in time (rather than space, i.e., semi-discrete form) some interesting results can be obtained. Take a second-order Runge-Kutta method [495], with the first Runge-Kutta step given by $\mathbf{U}^1 = -\mathbf{E}_x(\mathbf{U}^n)$; and the second Runge-Kutta step given by $\mathbf{U}^{n+1} = 1/2\left[\mathbf{U}^n + \mathbf{U}^1 - \mathbf{E}_x\left(\mathbf{U}^1\right)\right]$. Using a Taylor series in time and then systematically replace time derivatives with spatial derivatives, we obtain the following modified equation

$$\mathbf{U}_t = -\mathbf{E}_x(\mathbf{U}) + (\mathbf{E}'\Delta t)^2\left(\frac{1}{4}\mathbf{E}''(\mathbf{U}_x)^2 + \frac{1}{6}\mathbf{E}'\mathbf{U}_{xx}\right)_x .$$

Another issue concerns the fact that linear accuracy of methods does not generally hold for nonlinear hyperbolic equations as they advance in time because of the spontaneous generation of singularities (shocks). A scheme that is high-order accurate for a linear equation may only be limited to second-order accuracy in the nonlinear case. Take the example of a fourth-order estimate for the edge value $\mathbf{U}_{j+1/2}$ on a grid,

$$\mathbf{U}_{j+1/2} \approx \frac{7\left(\mathbf{U}_j + \mathbf{U}_{j+1}\right) - \left(\mathbf{U}_{j-1} + \mathbf{U}_{j+2}\right)}{12} .$$

This will yield a second-order approximation to the nonlinear flux, $\mathbf{E}\left(\mathbf{U}_{j+1/2}\right)$, with error,

$$\frac{h^2}{24}\mathbf{E}''(\mathbf{U}_x)^2 .$$

For full fourth-order accuracy one uses

$$\mathbf{E}_{j+1/2}(\mathbf{U}) = \frac{7\Big[\mathbf{E}(\mathbf{U}_j) + \mathbf{E}(\mathbf{U}_{j+1})\Big] - \Big[\mathbf{E}(\mathbf{U}_{j-1}) + \mathbf{E}(\mathbf{U}_{j+2})\Big]}{12} ,$$

with leading order truncation error in the flux,

$$-\frac{h^4}{30}\Big[\mathbf{E}'\mathbf{U}_{xxxx} + 3\mathbf{E}''(\mathbf{U}_{xx})^2 + 4\mathbf{E}''\mathbf{U}_x\mathbf{U}_{xxx} + 6\mathbf{E}'''\mathbf{U}_{xx}(\mathbf{U}_x)^2 + \mathbf{E}''''(\mathbf{U}_x)^4\Big] .$$

When the error incurred with these two methods is examined using the function $\tanh(10x)$, with $h = 0.05$, we find that these methods give similar errors. This comparison is given in Fig. 6.6. However, in terms of the leading order

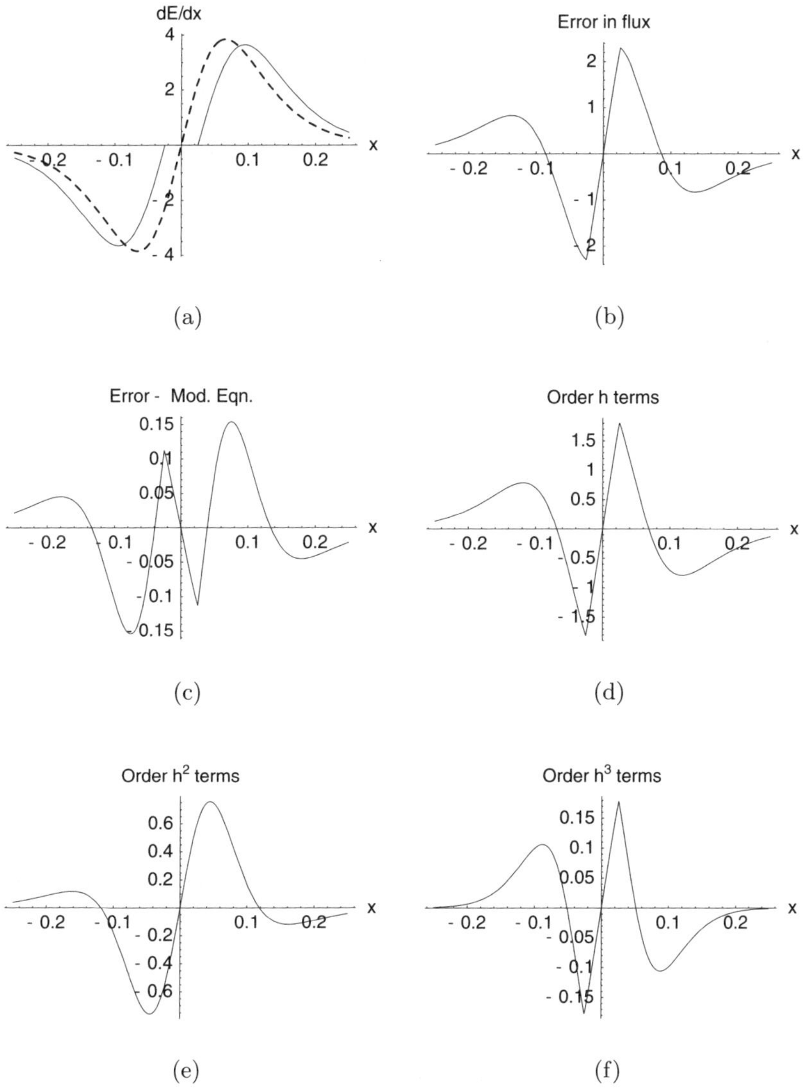

Fig. 6.5. The plots supporting the analysis of first-order upwind for Burgers' equation on the function $\tanh(10x) \in [0, 1]$ using $h = 0.05$: (a) the true (analytically calculated; dashed line) versus approximate (i.e., discrete value of $\mathrm{d}E/\mathrm{d}x$; solid line) flux derivative; (b) the the actual error in the flux calculation; (c) the estimate of error given by the first two terms in the modified equation; (d) the order h term in the modified equation; (e) the order h^2 term in the modified equation; and (f) the order h^3 term in the modified equation.

truncation error the true fourth-order scheme produces a larger error. This is reflected in the actual error in the flux difference as well. To some extent this result demonstrates that the order of error does not always dictate the actual error outside the asymptotic range of convergence for a method. This will be our first encounter with the difference between differencing the dependent variable or the flux of dependent variable. These lead to somewhat different high-resolution numerical methods and these issues will be explored further in Chaps. 13, 16 and 17.

While the leading order truncation error gives a significant amount of information, in many ways it is not conclusive or sufficient. Ultimately, the performance as implemented in code is the arbiter of efficacy of any given method.

6.4 Verification via Sample Calculations

The concept of method verification is essential to modern simulation. Verification is largely a mathematical issue where one goes through a systematic determination that a method has been implemented correctly. Typically this is accomplished via error analysis where the central issue is the rate of convergence of methods as the computational grid is refined. The goal is to check whether a given method converges toward the correct unique solution at the rate consistent with the order of accuracy of the method. Validation is the determination of whether a given method produces a reliable representation of reality. This task necessarily involves comparisons with experimental data.

In recent years, verification and validation has attracted the attention of many professional societies and organizations (e.g., ASME, AIAA, AICHE, IEEE, ERCOFTAC, the US EPA, US DOE, and the UK EPSRC). Various standards have been published and a greater burden has been placed upon computer codes to provide evidence that the methods used are correct. The book by Roache forms an excellent introduction to the finer points of this issue [462]. First and foremost is the understanding that these processes are inherently quantitative in nature.

In the following the semantics makes a real difference and the definitions of the terms below should be kept firmly in mind:

Verification: The determination via convergence analysis of whether a method is implemented correctly. A given method should produce its theoretical rate of convergence on smooth solutions. Colloquially, this can be thought of as "are the equations being solved correctly?"

Validation: The determination that the model represented by the equations is correct. This process involves comparison with experimental data. Colloquially, this can be thought of as "are the correct equations being solved correctly?"

Calculation Verification: The first portion of validation is calculation verification. This is the systematic mesh refinement of a simulation in order

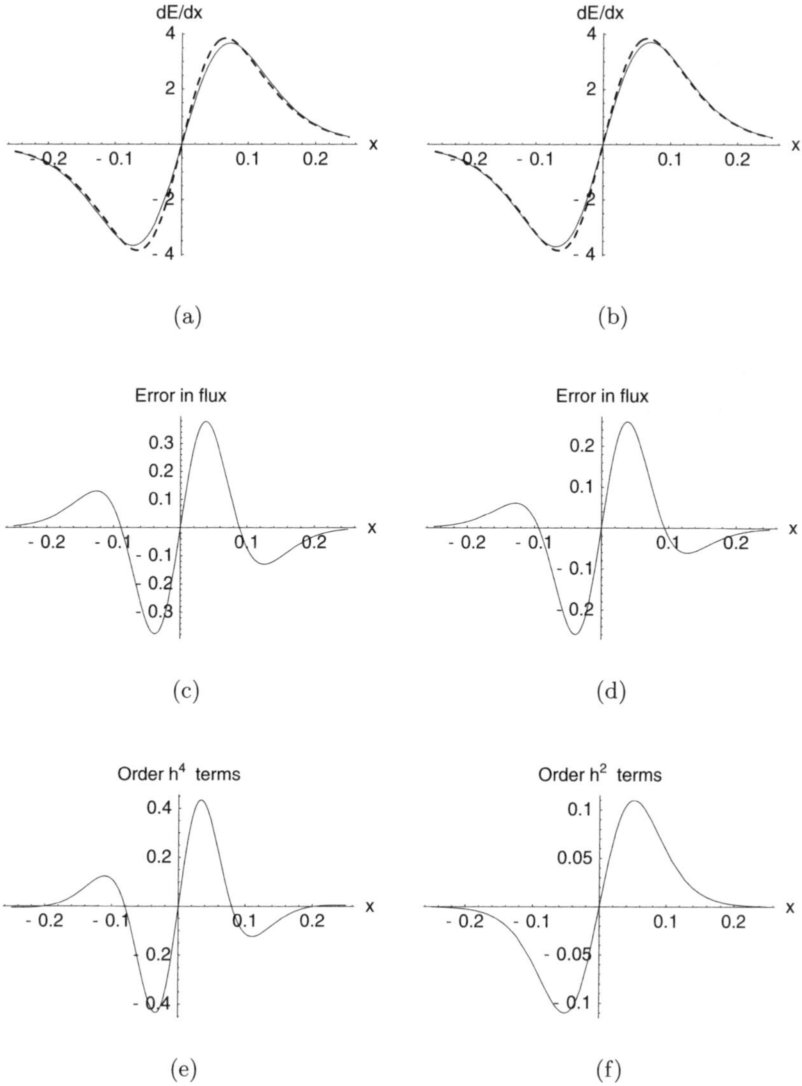

Fig. 6.6. A comparison of a fourth-order flux-based scheme (plots a,c,e) and a second-order scheme with fourth-order interpolation for **U** (plots b,d,f) . The plots (a) and (b) are the analytic (dashed line) and approximate (solid line) flux differences as calculated for the two schemes; (c) and (d) show the error in the flux difference; and (e) and (f) show the fourth-order error in the modified equation.

to determine that solution is converging and provide an estimate of the numerical error in the simulation as well as a greater degree of confidence in the results.

Accuracy: is defined as the quality of deviating slightly from fact. For our purposes, this definition is refined as the measured error for a given solution. There is also a distinction between order of accuracy and numerical accuracy. For reasonable grid resolution, methods with a higher order of accuracy can be accompanied by significantly larger numerical error than the lower order method. This naturally leads to our next definition.

Fidelity: This is defined as exact correspondence with fact. A solution that possesses fidelity is one that is physically meaningful. A method is considered to be of high-fidelity when it produces solutions that are accurate relative to the computational resources (the mesh size) applied to them. For example, interface tracking mechanisms can increase solution fidelity by maintaining interface discontinuities as the interface is advected and/or undergoing topological change.

Robustness: This is the property of being powerfully built or sturdy. A robust method will not fail in a catastrophic manner, but rather "degrade gracefully". Robustness implies that the algorithm can be used with confidence on a difficult problem. The degree to which the degradation is graceful is subject to interpretation. A robust method should produce physically reasonable results beyond the point where accuracy is expected or achieved.

At its simplest, verification is done through mesh doubling. If one has an analytic solution, S_a, than the process works as follows:

1. Compute the solution on a sequence of two grids, $2h$ and h.
2. Compute the errors on each of these grids, $E_{2h} = |S_{2h} - S_a|$, and $E_h = |S_h - S_a|$.
3. These errors can then be converted to a suitable norm and compute the convergence rate from

$$n = \ln\left(\frac{\|E_{2h}\| \,/\, \|E_h\|}{2}\right) .$$

Without an analytic solution the process changes slightly. This is often referred to as "self-convergence":

1. Compute the solution on a sequence of three grids, $4h$, $2h$ and h.
2. Compute the errors under the assumption that the finer grid (smaller mesh) is better. Compute the errors on each of these grids, $E_{2h} = |S_{4h} - S_{2h}|$, and $E_h = |S_{2h} - S_h|$.
3. These errors can then be converted to a suitable norm and compute the convergence rate from

$$n = \ln\left(\frac{\|E_{2h}\| \,/\, \|E_h\|}{2}\right) .$$

Here, the assumption that the finer grid provides a better solution is made. Fortunately, the results of a grid convergence test produce a confirmation of this assumption. If the convergence rate is not positive the assumption is shown to be false. One must note that in the absence of exact solutions and insufficient solution smoothness, self-convergence does not guarantee that a method produces correct solutions.

7. Time Integration Methods

In this chapter, we will discuss various approaches to achieving time accurate solutions to the equations of low-speed or incompressible flow. Of course, these methods will be useful for integrating other systems of equations, but our principle focus will be low-speed or incompressible flows. We will cover several of the commonly used approaches to integration in time. The principal approaches are

- Lax-Wendroff-type (LW).
- Runge-Kutta (R-K), including total variation diminishing (TVD)[1] R-K or Strongly Stability Preserving (SSP) R-K.
- Linear Multi-step Methods (LMM).
- Implicit Methods.

Some of the above methods will be discussed more specifically in Chap. 10 in relation to their implementation in conjunction with the artificial compressibility method. Each of these methods has specific advantages and drawbacks. We will briefly cover each of these and introduce some of the specific methods for each of the above types. First, we will cover some basic principles employed in the construction, implementation, analysis and evaluation of these methods. This discussion will also serve to introduce the basic issues involved when considering a method for use with integrating the low-speed or incompressible flow equations.

7.1 Time Integration of the Flow Equations

For the purposes of this chapter we will consider the spatial discretization as an abstract operator. The fluid flow is described by equations of motion subject to a constraint imposed by the divergence-free condition. As was

[1] The total variation is defined as follows: In a discretized domain, a variable U is a function of the mesh and its total variation at a time instant n, is given by $\mathrm{TV}(U^n) \equiv \mathrm{TV}(U(t)) = \sum_{j=-\infty}^{+\infty} |U_{j+1}^n - U_j^n|$, where U is assumed to be either 0 or constant as the index j approaches the infinity, in order to obtain finite total variation. The total variation, $\mathrm{TV}(U(t))$ is a decreasing function of time, i.e., $\mathrm{TV}(U^{n+1}) \leq \mathrm{TV}(U^n)$ (see Sect. 13.2 for further discussion).

introduced in Chap. 5, there is no explicit equation for the pressure, *this is of extreme importance*. Various approaches to overcome this problem are discussed in Chaps. 10, 11 and 12.

Issues related to time-integration can also be studied considering the differentiation index of partial differential-algebraic equations [19, 374]. The index of a system of equations defines the nature of explicit or hidden constraints and characterizes the difficulty of solving the system. Simply stated a direct algebraic constraint typically produces an index one system. If the constraint on variables is implicit this will produce a higher index problem. Incompressible flow is an index two system[2] implying that it is quite difficult to solve thus many methods exhibit a loss of accuracy. This is indeed observed with respect to the solution of these equations as accurate solution of the pressure is often not achieved. While there are other explanations for this behavior, it is consistent with the index of the equations and the expectations arising from their solution.

Another approach quite often taken for high index problems is the index reduction where a transformation is applied to the system of equations to derive more explicit equations for the unknowns [19]. An example of this practice are pressure Poisson equation-based approaches to the incompressible flow equations. Care must be taken to ensure that the index reduction does not introduce anomalous solutions that satisfy the lower index system, but fail to solve the original high index problem.

Now, we will embark upon our introduction to the various methods for integrating systems of equations. Our starting point will be Lax-Wendroff methods where time and space differencing is interchanged to provide some unique characteristics as compared with the more traditional methods introduced later in the chapter. A detailed discussion of Lax-Wendroff time differencing for incompressible flows is given in Chap. 14.

7.2 Lax-Wendroff-Type Methods

Lax-Wendroff methods [321] are inherited from the tradition of compressible fluid dynamics solvers. These methods are focused on second-order methods. Another manner that is useful to describe these methods is as combined time and space differencing [171]. In this class of methods time derivative are replaced by spatial derivatives as described by the partial differential equations.

It is useful to review the development of Lax-Wendroff and some of its more useful derivative forms. Take the general form of a conservation law

$$\frac{\partial U}{\partial t} + \frac{\partial E\left(U\right)}{\partial x} = 0 \,, \tag{7.1}$$

[2] This is because the constraint is applied to velocity, but the variable effected is pressure.

with some assumptions about the smoothness of derivatives (i.e., the Cauchy-Kovalevskaya theorem[3]) we can derive equations for the derivatives of this equation. As we will be developing a second-order method we need the following:

$$\frac{\partial^2 U}{\partial x \partial t} + \frac{\partial^2 E(U)}{\partial x^2} = 0 , \tag{7.2}$$

$$\frac{\partial^2 U}{\partial t^2} + \frac{\partial^2 E(U)}{\partial t \partial x} = 0 . \tag{7.3}$$

Next, expand the spatial terms in the above equations in terms of $E'(U)$ and $E''(U)$ where $E'(U) = \partial E(U)/\partial U$ (the Jacobian). Its useful to remember that

$$\frac{\partial U}{\partial t} + \frac{\partial E(U)}{\partial x} = 0 \rightarrow \frac{\partial U}{\partial t} + E'(U)\frac{\partial U}{\partial x} = 0 ,$$

for sufficiently smooth functions. We can now introduce the necessary terms to cancel the first-order error in Taylor series expansions in time and space,

$$\frac{\partial}{\partial t}\left(\frac{\partial U}{\partial x}\right) + E'(U)\frac{\partial^2 U}{\partial x^2} + E''(U)\left(\frac{\partial U}{\partial x}\right)^2 = 0 . \tag{7.4}$$

By approximating these derivatives with centered derivatives, the Lax-Wendroff method can be derived which is consistent with the Taylor series to second-order,

$$\begin{aligned} \frac{U_j^{n+1} - U_j^n}{\Delta t} &+ \frac{E\left(U_{j+1}^n\right) - E\left(U_{j-1}^n\right)}{\Delta x} \\ &= \frac{\Delta t\left[E\left(U_{j+1}^n\right) - 2E\left(U_j^n\right) + E\left(U_{j-1}^n\right)\right]}{\Delta x^2} . \end{aligned} \tag{7.5}$$

Richtmyer [446] derived a useful variant in a predictor-corrector format. The predictor formula is

$$U_{j+1/2}^{n+1/2} = \frac{1}{2}\left(U_j^n + U_{j+1}^n\right) - \frac{\Delta t}{2\Delta x}\left[E\left(U_{j+1}^n\right) - E\left(U_j^n\right)\right] , \tag{7.6}$$

and the corrector is

$$\frac{U_j^{n+1} - U_j^n}{\Delta t} + \frac{E\left(U_{j+1/2}^{n+1/2}\right) - E\left(U_{j-1/2}^{n+1/2}\right)}{\Delta x} = 0 . \tag{7.7}$$

Yet another variant is the MacCormack scheme using alternating backward and forward differences [362],

$$\frac{U_j^{n+1,*} - U_j^n}{\Delta t} + \frac{E\left(U_j^n\right) - E\left(U_{j-1}^n\right)}{\Delta x} = 0 , \tag{7.8}$$

---

[3] The Cauchy-Kovalevskaya theorem basically states that an analytic solution of a partial differential equation exists through a Taylor expansion (and ample application of the chain rule).

and the forward differenced corrector,

$$\frac{U_j^{n+1} - U_j^{n+1,*}}{\Delta t} + \frac{E\left(U_{j+1}^{n+1,*}\right) - E\left(U_j^{n+1,*}\right)}{\Delta x} = 0 . \tag{7.9}$$

Still higher order methods can be found by continuing this process replacing third-order (and then fourth-order) time derivatives with spatial derivatives. The drawback of this method is that as the order increases the complexity of the algorithm increases geometrically for nonlinear equations. One of the best examples of this process can be found in [251]. More recently Toro et al. [545] have returned to this method. Additionally, Qiu and Shu [434] have applied Lax-Wendroff techniques for time accuracy for weighted essentially nonoscillatory methods (WENO) methods. For ENO-type differencing, the simplicity of method-of-lines approaches has supplanted the Lax-Wendroff style methods for time accuracy of greater than second-order.

These principles can be generally applied to incompressible flow. The key difference is the necessity of the divergence-free velocity field, at each (sub)step. While not absolutely necessary, the failure to enforce the divergence-free velocity at the mid-time-step has been shown to be prone to a weak nonlinear instability at large CFL numbers [266]. If the CFL number is kept at less than a half there is no evidence of this instability (the authors of the Book have verified this behavior). This experience acts as a general word of caution for integrating the incompressible flow equations with respect to pressure.

7.3 Other Approaches to Time-Centering

The chief mechanism to achieve time accuracy in this class of methods is to exchange spatial derivatives for time derivatives. As the PDE becomes more complicated, it becomes more important to include all of the terms in computing the differential balances. For incompressible flows the most important term is the pressure in its coupling to the divergence-free condition. The pressure gradient can be included explicitly, but (nonlinear) stability is enhanced by solving the pressure at the time-level appropriate for accuracy.

The chief decisions that one has to make is whether to include inter-cell coupling in the method. This can take several distinct forms: upwinding or Riemann solvers applied to the convective fluxes, the pressure solution and the treatment of viscous terms (all these issues are discussed in a much greater detail in the subsequent chapters). With upwinding the time-centered values one has the advantage of having a unique-single-valued velocity at the cell edges. This velocity field is useful in providing for a pressure solution thus providing for appropriate feedback to the velocity field.

Another alternative scheme for completing the time-centering is Hancock's scheme. Given the old time, cell-edge values of $U_{i+1/2,j}^n$ (i, j are grid indices on a 2-D domain and n stands for the old time instant) the time-centering is computed without the use of characteristic extrapolation. Therefore, the time-derivative is completely taken from the underlying differential equation. For instance, if

$$\frac{\partial U}{\partial t} + \nabla \cdot E(U) = 0 \, ,$$

then the time-centered value at $(i+1/2, j)$ would be given by

$$U^{n+1/2}_{i+1/2,j} = U^{n}_{i+1/2,j} - \frac{\Delta t}{2} \nabla \cdot E(U^n) \, ,$$

where the derivatives $\nabla \cdot E(U^n)$ are evaluated cell-by-cell using the spatial derivatives. Another alternative is to evaluate the derivatives by solving for the coupling between zones (a Riemann solution for compressible flow and advection plus pressure solve for incompressible flows). The convection is computed as it is in Chap. 14, but without characteristic extrapolation. This causes the stability condition for the algorithm to change to

$$\sum_{i=1}^{\text{dimensions}} \frac{|U_i| \, \Delta t}{\Delta x_i} \leq 1 \, .$$

Now we will discuss more classical methods for solving systems equations by considering them as systems of ODEs (method-of-lines approach).

7.4 Runge-Kutta Methods

Runge-Kutta methods are commonly used for integrating ODEs where the evaluation of the function $f(u,t)$ is thought of as being inexpensive. The presentation is done in terms of this general function because the time integration methods apply to general (ordinary and partial) differential equations. Accuracy is built up through solving the ODE in a series of steps (or stages). The incompressible flow equations are typically viewed as having expensive function evaluations because of the elliptic equation arising from the pressure solution. Thus, the number of stages in a Runge-Kutta method is equal to the number of pressure-equation solutions. This character is essential to the accuracy and stability of the overall method for incompressible flow. In other words, all intermediate velocity fields employed by the algorithm are made divergence-free. Another key advantage of Runge-Kutta methods is that time step changes are dealt with simplicity. This is because the methods are self-contained within a time step and do not require the storage across more than one time step. For LMMs the method becomes more complicated for uneven time step sizes because it is derived through a sequence of interpolation and integration over several time levels.

7.4.1 Second-Order Runge-Kutta

In Chap. 6 we introduced the first-order explicit Runge-Kutta method: the forward Euler method. There are several second-order Runge-Kutta algorithms that are all essentially identical for linear equations. They are differentiated by their cost in a number of function evaluations as well as by the nonlinear error and stability properties.[4] Perhaps the simplest and best known of these is

$$\left.\begin{aligned} \frac{U^1 - U^n}{h} &= \tfrac{1}{2} f\left(U^n, t^n\right) \\ \frac{U^{n+1} - U^n}{h} &= f\left(U^1, t^{n+1/2}\right) \end{aligned}\right\}, \tag{7.10}$$

where $h = \Delta t$. Its stability is shown in Fig. 7.1. The truncation error for this method is

$$\frac{1}{6} h^2 U_{ttt}.$$

One of the second-order Runge-Kutta variants is also known as a TVD Runge-Kutta method[5] [493], more recently denoted as a strongly stability preserving (SSP) method [220]. It is also known as Heun's method [19, 403],

$$\left.\begin{aligned} \frac{U^1 - U^n}{h} &= f\left(U^n, t^n\right) \\ \frac{U^{n+1} - U^n}{h} &= \tfrac{1}{2}\left[f\left(U^n, t^n\right) + f\left(U^1, t^{n+1}\right)\right] \end{aligned}\right\}. \tag{7.11}$$

This class of methods permits the solution to retain certain favorable characteristics of the nonlinear spatial differencing. As Shu states [494] if one is concerned about the nonlinear stability of the integration procedure, then these methods deserve consideration. The conditions for these methods to have this property are fairly restrictive. At higher order the method becomes less attractive due to more restrictive stability conditions and non-standard spatial operators required. If these properties are not required, the linear stability of these methods is equivalent to the standard Runge-Kutta methods. This highlights the intrinsic difference between linear and nonlinear stability in a numerical method.

For incompressible flows computed using high-resolution methods, Runge-Kutta methods were first employed by Sussman et al. [522] and later by Shu and E [495] using ENO spatial discretizations (ENO schemes are presented in

[4] For simply hyperbolic PDEs the stability is defined by a CFL number, but we are considering more general properties of the differential equations.

[5] TVD Runge-Kutta methods were developed and presented by Shu and Osher [493] for hyperbolic PDEs (see Sect. 7.4.4).

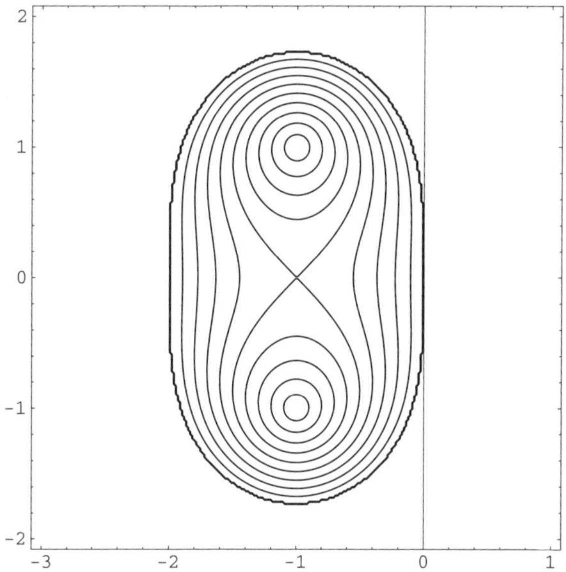

Fig. 7.1. The stability region, $|A| \leq 1$, for the second-order Runge-Kutta method. The horizontal axis describes the real part of the linear operator while the vertical axis describes the imaginary part of the generic operator.

Chap. 17). In both cases the projection operator (pressure solve) was applied once per stage.

As with the second-order method there is a variant of an SSP method producing larger CFL limits with the cost of more function evaluations. For example, a four-stage method is given by

$$\left.\begin{aligned}
\frac{U^1 - U^n}{h} &= \tfrac{1}{2} f\left(U^n, t^n\right) \\
\frac{U^2 - U^1}{h} &= \tfrac{1}{2} f\left(U^1, t^{n+1/2}\right) \\
\frac{U^3 - \frac{1}{3}U^1 - \frac{2}{3}U^n}{h} &= \tfrac{1}{6} f\left(U^1, t^{n+1/2}\right) \\
\frac{U^{n+1} - U^2}{h} &= \tfrac{1}{2} f\left(U^2, t^{n+1/2}\right)
\end{aligned}\right\} . \tag{7.12}$$

This carries a CFL limit of 2 rather than 1 as in (7.11).

For nonlinear problems there is a difference in (7.11) if the final step is evaluated using a midpoint rule

$$\frac{U^{n+1} - U^n}{h} = f\left(\frac{U^n + U^1}{2}, t^{n+1/2}\right). \tag{7.13}$$

The midpoint rule, (7.13), produces smaller coefficients on the truncation for terms proportional to $\partial^2 f/\partial U^2$ and $\partial^3 f/\partial U^3$.

7.4.2 Third-Order Runge-Kutta

Heun's third-order method is [19, 403]

$$\left.\begin{aligned}
\frac{U^1 - U^n}{h} &= \tfrac{1}{3} f\left(U^n, t^n\right) \\
\frac{U^1 - U^n}{h} &= \tfrac{2}{3} f\left(U^n, t^n\right) \\
\frac{U^{n+1} - U^n}{h} &= \tfrac{1}{4}\left[f\left(U^n, t^n\right) + 3f\left(U^2, t^{n+2/3}\right)\right]
\end{aligned}\right\}. \tag{7.14}$$

The truncation error of this method is

$$\frac{1}{24} h^3 U_{tttt}.$$

The stability of the method is displayed in Fig. 7.2.

The TVD third-order method is also quite commonly used because it has the same stability conditions as the second-order methods. The form of the solution is

$$\left.\begin{aligned}
\frac{U^1 - U^n}{h} &= f\left(U^n, t^n\right) \\
\frac{U^2 - U^n}{h} &= \tfrac{1}{4}\left[f\left(U^n, t^n\right) + f\left(U^1, t^{n+1}\right)\right] \\
\frac{U^{n+1} - U^n}{h} &= \\
\frac{1}{6}\Big[f\left(U^n, t^n\right) &+ 4f\left(U^2, t^{n+1/2}\right) + f\left(U^1, t^{n+1}\right) \Big]
\end{aligned}\right\}. \tag{7.15}$$

As in second-order Runge-Kutta, recent work [509] has shown a wider variety of these methods producing larger CFL limits with the cost of more function evaluations. For example, a four-stage method is given by

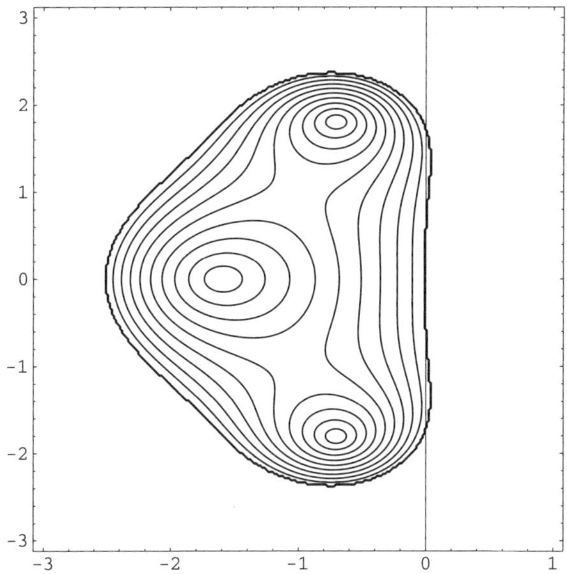

Fig. 7.2. The stability region, $|A| \leq 1$, for the third-order Heun's Runge-Kutta method.

$$\left.\begin{aligned}\frac{U^1 - U^n}{h} &= \tfrac{1}{2} f\left(U^n, t^n\right)\\ \frac{U^2 - U^1}{h} &= \tfrac{1}{2} f\left(U^1, t^{n+1/2}\right)\\ \frac{U^{n+1} - \frac{2}{3}U^2 - \frac{1}{3}U^n}{h} &= \tfrac{1}{3}\left[f\left(U^2, t^{n+1/2}\right) + f\left(U^1, t^{n+1}\right)\right]\end{aligned}\right\} . \qquad (7.16)$$

This carries a CFL limit of 2 rather than 1 as in (7.11).

7.4.3 Fourth-Order Runge-Kutta

We will only give the classical fourth-order scheme. The method consists of the following four stages that build up the solution to design accuracy,

$$\left.\begin{aligned}
\frac{U^1 - U^n}{h} &= \tfrac{1}{2} f\left(U^n, t^n\right) \\
\frac{U^2 - U^n}{h} &= \tfrac{1}{2} f\left(U^1, t^{n+1/2}\right) \\
\frac{U^3 - U^n}{h} &= f\left(U^2, t^{n+1/2}\right) \\
\frac{U^{n+1} - U^n}{h} &= \tfrac{1}{6}\Big[\, f\left(U^n, t^n\right) \\
&\quad + 2f\left(U^1, t^{n+1/2}\right) + 2f\left(U^2, t^{n+1/2}\right) + f\left(U^3, t^{n+1}\right)\Big]
\end{aligned}\right\} . \tag{7.17}$$

This method is one of the workhorse algorithms for the non-ODE-expert to integrate differential equations in time. The truncation error of this method is

$$\frac{1}{120} h^4 U_{ttttt}.$$

The stability of the method is displayed in Fig. 7.3. The reader would observe that Runge-Kutta methods gain larger stability regions with growing order.

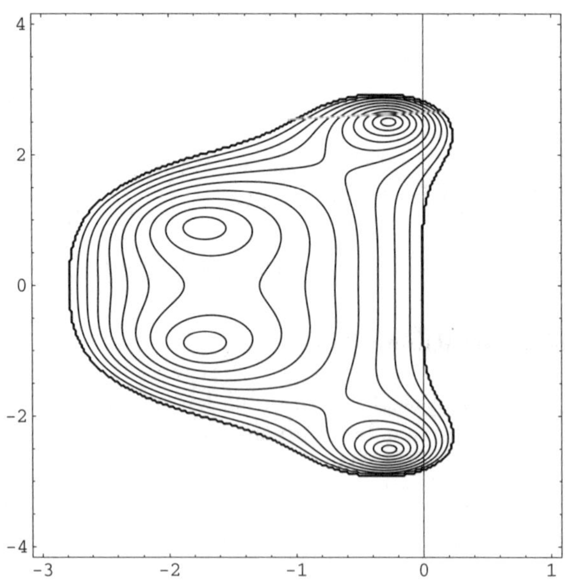

Fig. 7.3. The stability region, $|A| \leq 1$, for the fourth-order Runge-Kutta method.

Note that for fourth-order of accuracy, the TVD Runge-Kutta methods begin to incur a reduced stability (CFL) limit, and/or require non-standard spatial operators (i.e., anti-upwind schemes). As a consequence, they tend to be much less popular than the second- or third-order versions of this class of method. For initial value problems in fluid dynamics it has been found that the relative gain in accuracy and solution-quality reaches the point of diminishing returns at third-order.

7.4.4 TVD Runge-Kutta Methods Applied to Hyperbolic Conservation Laws

The family of explicit Runge-Kutta schemes of various orders of accuracy was introduced by Shu and Osher [493, 495] for hyperbolic conservation laws. Their goal was to develop r-th order approximations of a differential equation

$$U_t = L(U) \,, \tag{7.18}$$

where $L(U) = -\nabla \cdot E(U)$ is a spatial operator. We also define the operator $\mathcal{L}(U) = L(U) + \mathcal{O}(\Delta x^r)$ as an r-th order approximation to L. The general explicit Runge-Kutta method is given by

$$U^i = U^0 + \Delta t \sum_{k=0}^{i-1} c_{ik} \mathcal{L}\big(U^{(k)}\big), \quad i = 1, 2, ..., m \,, \tag{7.19}$$

$$U^{(0)} = U^n, \quad U^{(m)} = U^{n+1} \,, \tag{7.20}$$

where c_{ik} are coefficients the values of which depend on the order of accuracy of the scheme (see discussion below). If the differential equation contains source terms and/or the boundary conditions are time dependent, the operator $\mathcal{L}$ depends explicitly on time and in this case the Runge-Kutta method takes a more complicated form

$$U^i = U^0 + \Delta t \sum_{k=0}^{i-1} c_{ik} \mathcal{L}\big(U^{(k)}, t^{(0)} + d_k \Delta t\big) \,, \tag{7.21}$$

where

$$d_k = \sum_{i=0}^{k-1} c_{kl} \,. \tag{7.22}$$

To produce explicit schemes which satisfy the total variation diminishing (TVD) condition,[6] we can rewrite (7.19) as follows [495]

[6] The TVD condition states that the total variation of the solution (see Sect. 13.2 for its definition) with respect, for example, to spatial dimension x is uniformly bounded with respect to t; see Chap. 13 for further discussion on the total variation.

$$U^{(i)} = \sum_{k=0}^{i-1} \left[a_{ik} U^{(k)} + b_{ik} \Delta t \mathcal{L}(U^{(k)}) \right] , \tag{7.23}$$

where

$$\sum_{k=0}^{i-1} a_{ik} = 1, \quad \text{and} \quad b_{ik} = c_{ik} - \sum_{l=k+1}^{i-1} c_{lk} a_{il} .$$

On the basis of (7.23), different orders of accuracy Runge-Kutta methods can be obtained. This is done by using Taylor series expansions and choose coefficients a_{ik} and b_{ik} that optimize the CFL restriction. We list below the various versions of the Runge-Kutta schemes as proposed by Shu and Osher [495]:

- *Second-order of accuracy*:
  The coefficients are given by

$$\left. \begin{aligned} a_{20} &= 1 - a_{21} \\ b_{20} &= 1 - \frac{1}{2b_{10}} - a_{21} b_{10} \\ b_{21} &= \frac{1}{2b_{10}} \end{aligned} \right\} , \tag{7.24}$$

  where b_{10} and a_{21} are free parameters. The optimal scheme is given by

$$\left. \begin{aligned} U^{(1)} &= U^{(0)} + \Delta t \mathcal{L}\big(U^{(0)}\big) \\ U^{(2)} &= U^{(0)} + \tfrac{1}{2} \Delta t \mathcal{L}\big(U^{(0)}\big) + \tfrac{1}{2} \Delta t \mathcal{L}\big(U^{(1)}\big) \end{aligned} \right\} , \tag{7.25}$$

  with the CFL coefficient being equal to 1. Equation (7.25) is the modified Euler method [207].

- *Third-order of accuracy*:
  The coefficients are given by

$$\left. \begin{aligned} a_{32} &= 1 - a_{31} - a_{30} , \\ b_{32} &= \frac{3b_{10} - 2}{6P(b_{10} - P)} , \\ b_{21} &= \frac{1}{6b_{10} b_{32}} , \end{aligned} \right\}$$

$$\left.\begin{aligned} b_{31} &= \frac{1/2 - a_{32}b_{10}b_{21} - Pb_{32}}{b_{10}} \,, \\ b_{30} &= 1 - a_{31}b_{10}a_{32}P - b_{31} - b_{32} \,, \\ b_{20} &= P - a_{21}b_{10} - b_{21} \,, \end{aligned}\right\} \tag{7.26}$$

where $P = b_{20} + a_{21}b_{10} + b_{21}$, a_{21}, a_{30}, a_{31} and b_{10} are free parameters. The following (optimum) version of the scheme has been proposed [495]

$$\left.\begin{aligned} U^{(1)} &= U^{(0)} + \Delta t \mathcal{L}\big(U^{(0)}\big) \\ U^{(2)} &= \tfrac{3}{4}U^{(0)} + \tfrac{1}{4}U^{(1)} + \tfrac{1}{4}\Delta t \mathcal{L}\big(U^{(1)}\big) \\ U^{(3)} &= \tfrac{1}{3}U^{(0)} + \tfrac{2}{3}U^{(2)} + \tfrac{2}{3}\Delta t \mathcal{L}\big(U^{(2)}\big) \end{aligned}\right\}, \tag{7.27}$$

with the CFL coefficient being equal to 1.

- *Fourth-order of accuracy*:
  The fourth-order version is written as [207]

$$\left.\begin{aligned} U^{(1)} &= U^{(0)} + \tfrac{1}{2}\Delta t \mathcal{L}\big(U^{(0)}\big) \\ U^{(2)} &= \tfrac{1}{2}U^{(0)} - \tfrac{1}{4}\Delta t \widetilde{\mathcal{L}}\big(U^{(0)}\big) + \tfrac{1}{2}U^{(1)} + \tfrac{1}{2}\Delta t \mathcal{L}\big(U^{(1)}\big) \\ U^{(3)} &= \tfrac{1}{9}U^{(0)} - \tfrac{1}{9}\Delta t \widetilde{\mathcal{L}}\big(U^{(0)}\big) + \tfrac{2}{9}U^{(1)} - \tfrac{1}{3}\Delta t \widetilde{\mathcal{L}}\big(U^{(1)}\big) \\ &\quad + \tfrac{2}{3}U^{(2)} + \Delta t \mathcal{L}\big(U^{(2)}\big) \\ U^{(4)} &= \tfrac{1}{3}U^{(0)} - \tfrac{1}{6}\Delta t \mathcal{L}\big(U^{(1)}\big) + \tfrac{1}{3}U^{(2)} + \tfrac{1}{3}\big(U^{(3)}\big) \\ &\quad + \tfrac{1}{6}\Delta t \mathcal{L}\big(U^{(3)}\big) \end{aligned}\right\}, \tag{7.28}$$

with the CFL coefficient being equal to 2/3.
The operator $\widetilde{\mathcal{L}}$ is also a discrete approximation to the spatial operator L, which satisfies the TVD condition under the same CFL restriction.

- *Fifth-order of accuracy*:
  The formula for the fifth-order method is given by [314, 495]

$$
\left.
\begin{aligned}
U^{(1)} &= U^{(0)} + \tfrac{1}{2}\Delta t\mathcal{L}\big(U^{(0)}\big) \\
U^{(2)} &= \tfrac{3}{4}U^{(0)} + \tfrac{1}{4}U^{(1)} + \tfrac{1}{8}\Delta t\mathcal{L}\big(U^{(1)}\big) \\
U^{(3)} &= \tfrac{3}{8}U^{(0)} - \tfrac{1}{8}\Delta t\widetilde{\mathcal{L}}\big(U^{(0)}\big) + \tfrac{1}{8}U^{(1)} - \tfrac{1}{16}\Delta t\widetilde{\mathcal{L}}\big(U^{(1)}\big) \\
&\quad + \tfrac{1}{2}U^{(2)} + \tfrac{1}{2}\Delta t\mathcal{L}\big(U^{(2)}\big) \\
U^{(4)} &= \tfrac{1}{4}U^{(0)} - \tfrac{5}{64}\Delta t\widetilde{\mathcal{L}}\big(U^{(0)}\big) + \tfrac{1}{8}U^{(1)} - \tfrac{13}{64}\Delta t\widetilde{\mathcal{L}}\big(U^{(1)}\big) \\
&\quad + \tfrac{1}{8}U^{(2)} + \tfrac{1}{8}\Delta t\mathcal{L}\big(U^{(2)}\big) + \tfrac{1}{2}U^{(3)} + \tfrac{9}{16}\Delta t\mathcal{L}\big(U^{(1)}\big) \\
U^{(5)} &= \tfrac{89537}{2880000}U^{(0)} + \tfrac{2276219}{40320000}\Delta t\mathcal{L}\big(U^{(0)}\big) + \tfrac{407023}{2880000}U^{(1)} \\
&\quad + \tfrac{407023}{672000}\Delta t\mathcal{L}\big(U^{(1)}\big) + \tfrac{1511}{12000}U^{(2)} + \tfrac{1511}{2800}\Delta t\mathcal{L}\big(U^{(2)}\big) \\
&\quad + \tfrac{87}{200}U^{(3)} - \tfrac{261}{140}\widetilde{\mathcal{L}}\big(U^{(3)}\big) + \tfrac{4}{15}U^{(4)} + \tfrac{8}{7}\mathcal{L}\big(U^{(4)}\big) \\
U^{(6)} &= \tfrac{4}{9}U^{(0)} + \tfrac{1}{15}U^{(1)} - \tfrac{8}{45}\widetilde{\mathcal{L}}\big(U^{(1)}\big) + \tfrac{8}{45}U^{(3)} \\
&\quad + \tfrac{2}{3}\Delta t\mathcal{L}\big(U^{(3)}\big) + \tfrac{14}{45}U^{(5)} + \tfrac{7}{90}\Delta t\mathcal{L}\big(U^{(5)}\big)
\end{aligned}
\right\} , \qquad (7.29)
$$

with the CFL coefficient being equal to 7/30.

Another class of methods introduced recently are the natural continuous extension (NCE) methods [57]. There are also a large class of implicit Runge-Kutta methods based on various quadratures. We will not cover these methods explicitly and the details about these methods can be found in a number of textbooks that exclusively cover numerical methods for ODEs (e.g., Ascher and Petzold [19]).

7.5 Linear Multi-step Methods

Linear multi-step methods are another general class of ODE methods. These are less commonly associated with the solution of hyperbolic PDEs [19]. They have the advantage of only requiring one function evaluation per time step. Thus, they are favored for expensive function evaluations such as the pressure Poisson equation. On the other hand, these methods require the storage of the function for one or more preceding time steps. Moreover, the necessity of using the function evaluations from previous time steps make the start up for these methods problematic. This is usually handled with the use of lower order methods or Runge-Kutta methods until sufficient memory has been built up.

7.5.1 Adams-Bashforth Method

The second-order version of this method is quite commonly used and takes a simple form,

$$\frac{U^{n+1} - U^n}{h} = \frac{1}{2}\left[3f\left(U^n, t^n\right) - f\left(U^{n-1}, t^{n-1}\right)\right] . \tag{7.30}$$

One of the first things to notice is that the truncation error takes a different form

$$-\frac{5h^2}{12} f\left(u, t\right) .$$

On the negative side, the stability is more limiting for this method than second-order Runge-Kutta methods, being roughly half their value. This is shown in Fig. 7.4. Note that multi-step methods have k (k is the step number) amplification factors and one must look at the worst of these to determine stability.

As noted earlier, variable time step sizes also cause some difficulty in relation to that formulas change between different time steps. For variable time step sizes one can use Lagrange interpolation over the time interval t^n to t^{n+1} and then integrate the interpolated function. This yields the following method,

$$\frac{U^{n+1} - U^n}{h^n} = f\left(U^n, t^n\right) + \frac{h^n}{2h^{n-1}}\left[f\left(U^n, t^n\right) + f\left(U^{n-1}, t^{n-1}\right)\right] , \tag{7.31}$$

where $h^n = t^{n+1} - t^n$ and $h^{n-1} = t^n - t^{n-1}$. Another approach stated in a manner similar to Runge-Kutta methods, is given by

$$\left.\begin{aligned} \frac{U^1 - U^n}{h^n} &= \frac{1}{2}\frac{U^n - U^{n-1}}{h^{n-1}} \\ \frac{U^{n+1} - U^n}{h} &= f\left(U^1, t^{n+1/2}\right) \end{aligned}\right\} . \tag{7.32}$$

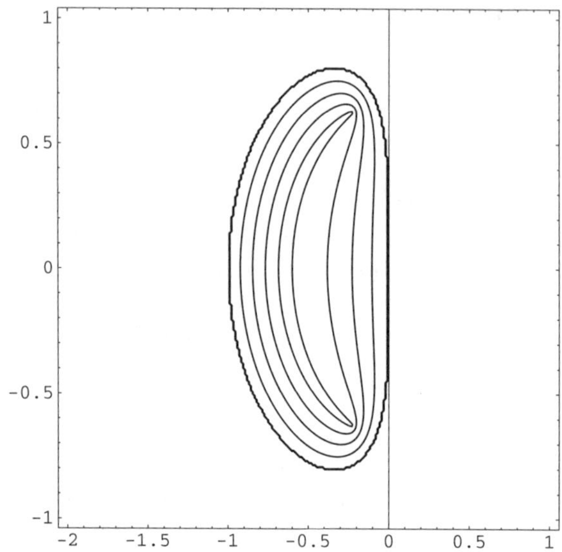

Fig. 7.4. The stability region, $\max_k |A_k| \leq 1$, for the second-order Adams-Bashforth method. For multi-step methods k is the step number.

A back-substitution of the first stage into the second confirms that this method is linearly equivalent to the classical second-order Adams-Bashforth method (7.30). This approach requires the storage of the time derivative from the previous time step. The more classical approach to this problem is to conduct an interpolation over the previous two time steps and integrate thus producing the following,

$$\frac{U^{n+1} - U^n}{h^n} = \left(1 - \frac{h^n}{h^{n-1}}\right) f\left(U^n, t^n\right) - \frac{h^n}{h^{n-1}} f\left(U^{n-1}, t^{n-1}\right) . \quad (7.33)$$

The third-order Adams-Bashforth method is

$$\frac{U^{n+1} - U^n}{h} = \frac{1}{12}\Big[23 f\left(U^n, t^n\right) \\ -16 f\left(U^{n-1}, t^{n-1}\right) + 5 f\left(U^{n-2}, t^{n-2}\right)\Big] . \quad (7.34)$$

The Adams-Bashforth methods are contrasted with Runge-Kutta methods in that the stability region decreases with increasing order. This is easily seen in comparing Figs. 7.4 and 7.5.

There are also linear multi-step methods of a SSP type [220, 494, 509]. These methods are the generalization of TVD time discretizations sharing the same qualities and should be employed if strong nonlinear stability is desired [494]. The second-order method can be written,

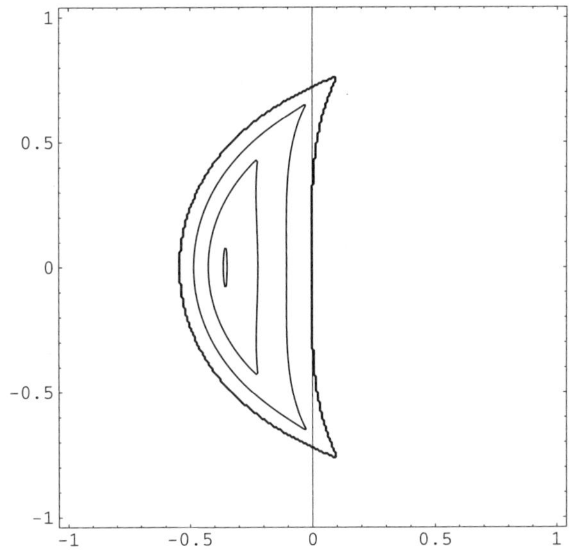

Fig. 7.5. The stability region, $\max_k |A_k| \leq 1$, for the third-order Adams-Bashforth method.

$$\frac{4U^{n+1} - 3U^n - U^{n-1}}{6h} = f\left(U^n, t^n\right) . \qquad (7.35)$$

This method is stable for a relatively larger region than it is SSP, i.e., a time step limit of 1/2.

There is also a third-order SSP-LMM method, which has a reasonable form,

$$\frac{27U^{n+1} - 16U^n - 11U^{n-3}}{60h} = \frac{4}{5} f\left(U^n, t^n\right) + \frac{1}{5} f\left(U^{n-2}, t^{n-2}\right) . \qquad (7.36)$$

This method has a stability limit of 1/2 for retaining the properties of SSP methods.

General Remark:

Each of the various methods for integrating the incompressible flow equations has its distinct advantages and disadvantages. In the case of pressure-Poisson methods, the main issue is the number of pressure Poisson solutions that are required per unit time step. Thus, one must factor in the CFL limit (or the CFL number that is practical for a given accuracy) into this.

Perhaps one of the practical trade offs can be seen by comparing the CFL limit and number of steps between the third-order Runge-Kutta and Adams-Bashforth methods. The three steps of the Runge-Kutta method require a CFL number of less than 1.5 with one pressure solution per step. The Adams-

Bashforth method has a CFL limit of 0.73 with one pressure solution (overall). Thus, by efficiency alone the Adams-Bashforth method is preferable.

7.5.2 Adams-Moulton Method

Implicit methods will entail more complex solution algorithms as the velocity and pressure solution are coupled. In general, this will also require the solution of coupled nonlinear equations. Linearizations can be performed, and indeed are required for effective preconditioners, but must be used with extreme caution without achieving some degree of nonlinear convergence. This can be found through the use of an efficient modern Newton's method like Newton-Krylov [86]. Using explicit methods only the pressure solution necessarily entails numerical linear algebra. As before, stability of the solution depends on having a (approximate or exact) divergence-free velocity field. With the implicit solution this property comes automatically.

Adams-Moulton methods are implicit in that the update formula contains the current time level [19]. These methods are derived like the explicit Adams-Bashforth methods, but the interpolant includes the current time level (and one level back in time for a given order of accuracy). The basic method for the time derivative is the the backward Euler method,

$$\frac{U^{n+1} - U^n}{h} = f\left(U^{n+1}, t^{n+1}\right) . \tag{7.37}$$

Like the forward Euler method a first-order accuracy is obtained. Unlike the explicit Euler scheme, the method is extremely stable and in fact is unstable for a small region only. Fig. 7.6 shows the stability properties of the method.

The next method in a sequence that raises the accuracy of time integration is the Crank-Nicholson method. This is simply a trapezoidal integration rule, but unlike the explicit case, one must implicitly compute the new time values. The update formula is quite simple,

$$\frac{U^{n+1} - U^n}{h} = \frac{1}{2}\left[f\left(U^n, t^n\right) + f\left(U^{n+1}, t^{n+1}\right)\right] . \tag{7.38}$$

One obvious variation of this method is the implicit midpoint rule,

$$\frac{U^{n+1} - U^n}{h} = \frac{1}{2} f\left(\frac{U^n + U^{n+1}}{2}, \frac{t^n + t^{n+1}}{2}\right) . \tag{7.39}$$

Both of the above variants have desirable stability properties. The stability region is shown in Fig. 7.7. The characteristic of being stable for all dissipative operators is known as "A-stability".[7] Crank-Nicolson and implicit midpoint methods are all A-stable methods.

[7] A-stable methods are those which have regions of stability containing the whole of the left-hand half plane in the stability diagram, i.e., Re$G < 0$, where G is defined in Sect. 6.2.

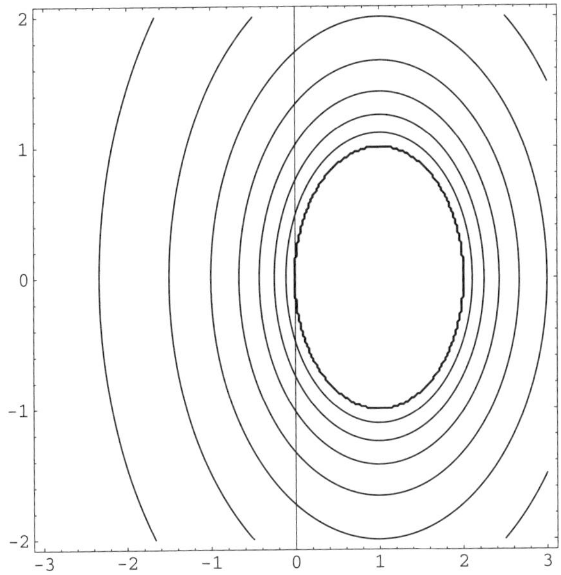

Fig. 7.6. The stability region, $\max_k |A_k| \leq 1$, for the first-order backward Euler method.

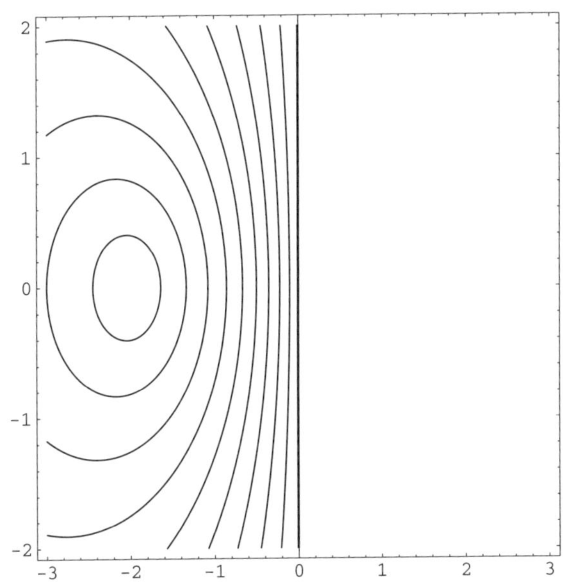

Fig. 7.7. The stability region, $\max_k |A_k| \leq 1$, for the second-order Crank-Nicholson method.

In the Adams-Moulton family of methods, the next method in the sequence is a third-order method,

$$\frac{U^{n+1} - U^n}{h} = \frac{1}{12}\Big[5f\left(U^{n+1}, t^{n+1}\right)$$
$$+8f\left(U^n, t^n\right) - f\left(U^{n-1}, t^{n-1}\right)\Big] . \quad (7.40)$$

In this case the price paid for this extra order of accuracy is quite high in terms of stability. Rather than unconditional stability for negative real functions, there is now an uncomfortably small stability region. This is shown in Fig. 7.8. As the order of Adams-Moulton methods increases, the stability region continues to shrink.

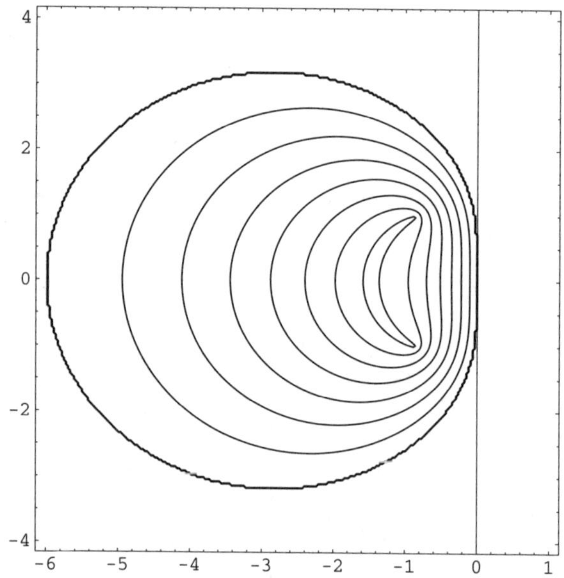

Fig. 7.8. The stability region, $\max_k |A_k| \leq 1$, for the third-order Adams-Moulton method.

An alternative to treating these methods as implicit methods are predictor-corrector methods where an explicit method is used to "predict" the time advanced solution which is substituted into the implicit difference formula in the "corrector" step. Typically, this results in an enhanced stability region over the purely explicit method. As a simple example consider the coupling of forward Euler with the trapezoidal rule, producing Heun's method.

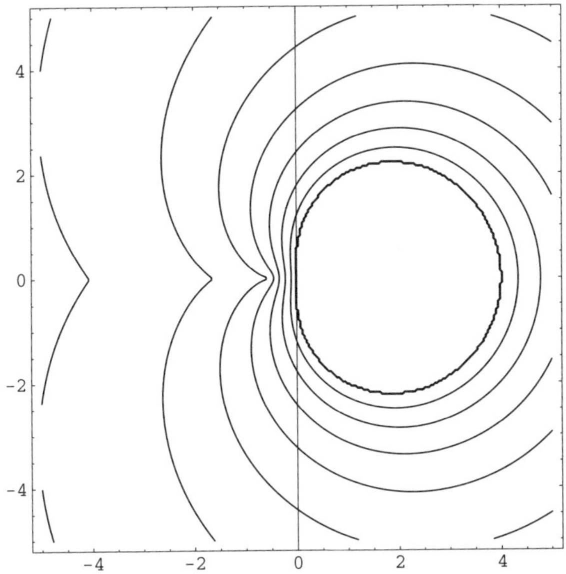

Fig. 7.9. The stability region, $\max_k |A_k| \leq 1$, for the second-order backward differentiation formula.

7.5.3 Backward Differentiation Formulas

The last category of methods we will introduce are backwards differentiation formulas (BDFs) [19]. These methods are quite popular for solving stiff systems of equations. This is because of their simple form and large stability regions. For the same order of accuracy these methods have much larger stability regions than the Adams-Moulton methods. BDF methods are chiefly characterized by only evaluating the function at the advance time, $n+1$. Accuracy is achieved through high-order approximations to U_t.

BDFs and Adams-Moulton methods share their first-order incarnation, the backward Euler method. The second-order BDF update equation is different however, i.e.,

$$\frac{3U^{n+1} - 4U^n + U^{n-1}}{2h} = f\left(U^{n+1}, t^{n+1}\right). \tag{7.41}$$

The large stability region is shown in Fig. 7.9.

The third-order method includes one more time level in the approximation of U_t,

$$\frac{11U^{n+1} - 18U^n + 9U^{n-1} - 2U^{n-2}}{6h} = f\left(U^{n+1}, t^{n+1}\right). \tag{7.42}$$

Its stability region is slightly smaller than the second-order method, but nevertheless is large and significantly larger than the Adams-Moulton method.

8. Numerical Linear Algebra

When incompressible flows are solved efficient solution of linear systems of equation is essential in most cases. The only exception is when the time marching approach is taken with the artificial compressibility method. Even there the utilization of multigrid ideas is useful for improving the efficiency of the solution (see Chap. 10). There are a number of basic approaches each raising the efficiency of the solution as well as its complexity. The chief driving force in this progression is the desire to solve larger systems of equations (i.e., finer and finer meshes). Any approach will incur a cost in terms of number of operators per degree of freedom in the linear system. The simplest methods will incur a cost that scales with the cube of the number of degrees of freedom (N^3). By the end of the chapter, we will discuss methods that have a cost that is linear in the number of degrees of freedom (N). We will discuss each of these approaches briefly because far better and complete references are readily available. Nonetheless, these methods are essential to the toolbox of the practitioner and as such they are important to be introduced.

Below, we cover these basic techniques, a few specific tips for their use and provide pointers to more complete descriptions of,

- basic numerical linear algebra [136, 549];
- basic relaxation methods [475, 136, 550];
- conjugate gradient and Krylov subspace methods [475, 286, 224, 549, 136, 33];
- multigrid for elliptic equations [77, 78, 550, 595];
- multigrid as a preconditioner for Krylov subspace methods;
- and Newton's method implemented via a Newton-Krylov algorithm [86].

8.1 Basic Numerical Linear Algebra

One of the most well developed areas of applied mathematics is numerical linear algebra. In particular, the area of dense linear algebra is in a mature state with most of the current research being done in the area of sparse linear algebra. Because sparse linear systems arise as a consequence of discretizing partial differential equations, the efficiency of the numerical solutions of incompressible flows is dependent upon the best research in numerical linear algebra. Conversely, the maturity of dense linear algebra provides one with access to a number of high quality textbooks and software [136, 219, 549]. Despite this mature state of development of dense linear algebra these techniques require large operation counts (intrinsically scaling as N^3, where N is the number of equations in the system).[1] More recent techniques for sparse systems scale with N or N^2 and are preferred because N becomes large (as a mesh is refined).

Typically, the linear algebra problem is typically stated as

$$\mathbf{A}\mathbf{x} = \mathbf{b} \,, \tag{8.1}$$

where $\mathbf{A}$ is the linear system, $\mathbf{x}$ is the solution vector and $\mathbf{b}$ is the right hand side. Each equation is written as follows:

$$a_{1,1}x_1 + a_{1,2}x_2 + \ldots + a_{1,n-1}x_{n-1} + a_{n,n}x_n = b_1 \,.$$

If one has a guess for the solution $\mathbf{x}_k$, then the residual of the solution can be found

$$\mathbf{r}_k = \mathbf{b} - \mathbf{A}\mathbf{x}_k \,, \tag{8.2}$$

which is enormously useful in constructing iterative algorithms to solve (8.1).

Another notable term is the operation known as a *matrix-vector* multiplication, here being $\mathbf{A}\mathbf{x}_k$. This is often the most expensive portion of any given algorithm. The other important operation is the inner product,

$$\mathbf{y}^T\mathbf{z} = (\mathbf{y}, \mathbf{z}) = \sum_i y_i z_i \,,$$

where y_i and z_i are the individual components of the vectors $\mathbf{y}$ and $\mathbf{z}$.

The most well known method is Gaussian elimination. In this method, the system is reduced equation-by-equation into an upper triangular form, then the unknowns are solved via back-substitution. The back-substitution is quite efficient because the last equation is explicit, and its solution then renders the next to the last equation explicit and so on. The entire sequence of operations scales as N^2. Special techniques known as pivoting have been defined to avoid any issues that might arise during the elimination due to round-off error [549]. In general, the stability of this method is dependent upon the ordering of the equations. As such numerical analysis has grown up around Gaussian elimination and the method is very well developed.[2] This reflects the change in emphasis for dense numerical linear algebra as the focus of research changes to applications other than partial differential equations.

A second very common form is known as LU decomposition, where $\mathbf{L}$ is a lower triangular system and $\mathbf{U}$ is an upper triangular system. This technique

---

[1] There are very complex algorithms can lower the operation count to $N^{2.31}$, but these methods cannot be recommended for general use.

[2] Indeed there have been interesting musings about the consequences of the historical emphasis on Gaussian elimination. Perhaps a better choice would have been QR-decomposition and the least squares problem [136].

is advantageous when the same system of equations must be solved repeatedly with different right hand sides, **b**. This comes at an increased cost for the initial decomposition (a factor of two as N becomes large), but the repeated back-substitution is extremely fast (order N).

If the system A contains a small number of non-zero terms per equation it is called sparse. For systems of equations arising from the discretization of partial differential equations this is typically the case. Moreover, the discrete systems often have an accommodating structure for the efficient solution of the system (without pivoting). There are sparse matrix methods derived for dealing with these systems that are similar to the general dense linear algebra methods, but take into account the structure (or limited bandwidth) of the system. Otherwise, the usual methods for dense system begin to fill in non-zero entries and destroy the special structure of the system. For details and theory on all of these issues the book by Golub and van Loan [219] is highly recommended. In addition, a vast array of basic introductions to numerical analysis will discuss these methods.

The solution of linear systems is the *classical* focus of numerical linear algebra. More recently, the focus has shifted to minimization and least square problems as exemplified by QR decomposition and singular value decomposition (SVD) [549, 136]. Implicit in these methods are orthogonalization ideas which also connect to modern iterative methods in the body of Krylov subspace methods [475]. The emphasis of the "new" numerical linear algebra significantly broadens the horizons of applications. The former emphasis on linear systems of equations is a direct consequence of a focus on partial differential equations, but modern problems are well served by the new focus.

Perhaps one of the most important aspects of dense linear algebra is readily available high quality software. The most common implementations are LINPAC [143] and LAPACK [14]. Both of these collections of software are of high quality and general in their use. These can be found easily at various repositories on the Internet. We also found them as part of the distribution of software with the Linux operating system!

8.2 Basic Relaxation Methods

The first methods developed for sparse systems were relaxation methods. For solving systems of equations, these methods have been almost entirely supplanted by more modern methods because of their poor efficiency (also N^3). The names of common methods belie their age, Jacobi, and Gauss-Seidel are the basic techniques with their inventors being some of the fathers of modern mathematics. However these methods are sometimes preferred over dense methods because they do not require an exact solution of the equations. Thus, if the convergence rate is sufficiently fast, or the accuracy for solving the system is low enough, these methods provide a lower cost alternative to dense methods. Furthermore, the elliptic pressure equation is structured so that this class of methods works very well (the pressure equation is discussed in Chaps. 11 and 12). Convergence is measured by the norm (usually L_2) of the residual $\mathbf{r}_k = \mathbf{A}\mathbf{x}_k - \mathbf{b}$, $\|\mathbf{r}_k\|_2 = (\mathbf{r}_k, \mathbf{r}_k)^{1/2}$.

However, as the system of equations becomes larger these advantages are rapidly offset by the scaling of operations (think about the diffusion of error as the discrete size becomes larger, time is replaced by iteration count, the linear cost of operations per iteration and the quadratic evolution of the error gives the N^3 order of operations). As these methods are important as preconditioners for Krylov methods and an intrinsic part of multigrid method we will spend some effort on introducing them. Their role in multigrid is closely related to their ability to reduce high frequency errors preferentially (frequency in the sense of the Fourier transform, i.e., errors that are close spatially). Just as high frequency errors are reduced effectively, low frequency errors are not impacted. This character is essential in the construction of multigrid methods where the idea is to only iterate on errors that are effectively high frequency and efficiently reduced.

These methods are closely related to simple time marching discretizations for parabolic equations, and work much as the approach to a steady-state solution using a finite time step. In fact, this structure gives the method the name "relaxation" for its physical feel. The first method to take this form is the Jacobi method [77], which acts just like an explicit update of a diffusion equation. The Jacobi method is naturally vector or parallel, but requires the storage of two solution vectors. This is a consequence of the dependence of the new iterate on only the previous iterates data. In its simplest form the Jacobi method works as follows: consider a five-point two-dimensional Laplacian operator (equation i, j) defined abstractly giving[3]

$$a^{i,j}_{i-1,j} x_{i-1,j} + a^{i,j}_{i+1,j} x_{i+1,j} + a^{i,j}_{i,j-1} x_{i,j-1} + a^{i,j}_{i,j+1} x_{i,j+1} + a^{i,j}_{i,j} x_{i,j} = b^{i,j}_{i,j} ,$$

the iteration defining the $(k+1)$th values from the kth values is

$$x^{k+1}_{i,j} = \frac{1}{a^{i,j}_{i,j}} \Big(b^{i,j}_{i,j} - a^{i,j}_{i-1,j} x^{k}_{i-1,j} - a^{i,j}_{i+1,j} x^{k}_{i+1,j} - a^{i,j}_{i,j-1} x^{k}_{i,j-1} - a^{i,j}_{i,j+1} x^{k}_{i,j+1} \Big) . \quad (8.3)$$

The next method is Gauss-Seidel which modifies the Jacobi method by using the most up to date values for the relaxation. The iterations can be ordered so that the method has data dependency that allows parallel or vector operations. This is most commonly associated with the "red-black" or checkerboard ordering. The basic pattern of the ordering is displayed in Fig. 8.1. This iteration is shown abstractly as for the Jacobi iteration,

[3] The superscripts and subscripts (i, j) refer to the (i, j)th equation and the index of the term in that equation, respectively.

Fig. 8.1. Here we show the checkboard pattern used for ordering the red-black Gauss-Seidel iteration (like a checker or chess board).

$$x_{i,j}^{k+1} = \frac{b_{i,j}^{i,j} - a_{i-1,j}^{i,j} x_{i-1,j}^{k+1} - a_{i+1,j}^{i,j} x_{i+1,j}^{k} - a_{i,j-1}^{i,j} x_{i,j-1}^{k+1} - a_{i,j+1}^{i,j} x_{i,j+1}^{k}}{a_{i,j}^{i,j}} .$$

The portions of this iteration depending upon the new iterate will vary according to the ordering of the variables.

The convergence of these methods can be accelerated by selecting an over-relaxation parameter which has the effect of taking a larger "time step". This technique is known as successive over-relaxation (SOR) and is usually applied to the Gauss-Seidel method [430]. In this method, a weighted average of the new and old iterate is taken to maximize the convergence rate,

$$x_{i,j}^{k+1} := \omega x_{i,j}^{k+1} + (1-\omega)\, x_{i,j}^{k} ,$$

where $\omega > 1$.

A more general way to discuss these iterations is to employ the same sort of matrix splitting used in the LU-decomposition. The matrix $\mathbf{A}$ is decomposed into three matrices: $\mathbf{L}$, a lower triangular matrix; $\mathbf{U}$, an upper triangular matrix; and $\mathbf{D}$ a diagonal matrix. The Jacobi iteration can then be written,

$$\mathbf{x}^{k+1} = \mathbf{D}^{-1}\left(\mathbf{L}\mathbf{x}^{k} + \mathbf{U}\mathbf{x}^{k}\right) ,$$

where $\mathbf{D}$ being diagonal is trivial to invert. Gauss-Seidel can be written as

$$\mathbf{x}^{k+1} = \left(\mathbf{L} + \mathbf{D}\right)^{-1}\left(\mathbf{U}\mathbf{x}^{k}\right) ,$$

where the ordering of the variables determines which entries are denoted by $\mathbf{L}$ and $\mathbf{U}$.

Line-by-line relaxation methods can also be used. This uses the tridiagonal matrix algorithm (TDMA) to solve the equations implicitly on a grid line, holding the values in the stencil off the line constant (they are put on the right hand side $\mathbf{b}$). TDMA is well known (see Numerical Recipes for example [430]). This iteration can proceed either after the fashion of Jacobi or Gauss-Seidel depending on whether the off-line values are updated in the sweep ordering. This sort of relaxation is important in multigrid methods as well sequential solution methods such as SIMPLE [415] (see Chap. 12).

8.3 Conjugate Gradient and Krylov Subspace Methods

Conjugate gradient methods now have supplanted both relaxation methods and banded solvers because of their efficiency. While these methods can be temperamental, they tend to be robust and in the case of symmetric positive definite systems, the conjugate gradient method is extremely fool-proof. The conjugate gradient method was invented in 1952, but did not achieve its present day prominence until the late 1970's. Originally, this method was viewed as an esoteric way to solve a linear system exactly. Kershaw [287] realized in 1978 that this method with a preconditioner could greatly improve efficiency if it were used to obtain an approximate solution. This made the method an iterative method that produced an exact solution as the number of iterations approached the system size, but very good solutions could be obtained for far fewer iterations. The scaling of this method (i.e., the number of operations that it takes to solve the system as measured by the number of equations in the system) is $N^{5/4}$ to $N^{3/2}$ depending on the scaling of the number of iterations, $N^{1/4}$ to $N^{1/2}$, to achieve a level of convergence of the residual.

Since that time Krylov solvers for non-symmetric systems and preconditioners for either the symmetric or non-symmetric case have become an active research area. In recent years, these methods have become quite mature with textbooks becoming available [475, 286, 224, 136, 549, 33] and the use of these methods is being extended to many applications outside differential equations. As we will see shortly, the methods are composed of relatively simple elements which are orchestrated for good efficiency. These are dominated by discrete inner products and one or two matrix-vector multiplications.

Ultimately, the efficiency of this broad class of methods is intimately related to the effectiveness of the preconditioning. A variety of preconditioners are typically used including incomplete Cholesky (or incomplete LU decomposition for non-symmetric systems) [219]. Incomplete Cholesky preconditioning involves stopping the elimination at some early stage to avoid the inevitable fill-in of the sparse matrix. The relaxation methods introduced earlier in this chapter are also effective. By increasing the number of iterations of the preconditioner, the number of Krylov iterations can be reduced. While the number of iterations is reduced, the number of iterations grows with the size of the linear system. To avoid this issue multigrid methods ideally offer a number of iterations that is constant with problem size. This type of preconditioning alludes to the use of multigrid as a sort of "ultimate iterative" method with "perfect" scaling.

A simple way to think about how these methods work is by considering the eigenvalue decomposition of a system. Krylov methods can be used to estimate the eigenvalues of a system, $\mathbf{A}$, and are closely related to general eigen-analysis methods. Indeed, iterative eigen-analysis methods form the basis for many schemes (e.g., GMRES's relation to the Arnoldi algorithm) [475]. The eigenvalues can be ordered from largest to smallest. Krylov methods effectively pick off the largest eigenvalues and form a polynomial basis for the solution. This basis is the Krylov subspace. The polynomial is defined by powers of the system matrix and the residual vectors, $\mathbf{Ar}$. As the remaining eigenvalues become close to unity, the residual becomes very small and the successive terms of the polynomial add little to the solution and convergence is at hand. Successful preconditioning will tend to cluster the eigenvalues near unity thus reducing the number of large eigenvalues with which to form the polynomial.

For the symmetric case the basic method is the conjugate gradient method which has the major cost of a single matrix-vector multiplication. It can include the cost of another matrix-vector multiplication (or its equivalent) through the application of a preconditioner. The algorithm proceeds as follows:

Algorithm 1 *[Preconditioned Conjugate Gradient Method]*

1. Start with the linear system $\mathbf{A}$, the right hand side, $\mathbf{b}$, and an initial guess, $\mathbf{x}^0$, $k = 0$.
2. Compute the initial preconditioned vector $\mathbf{z}_0 = \mathbf{M}^{-1}\mathbf{r}_0$, this step will solve some equation $\mathbf{M}\mathbf{z}_0 = \mathbf{r}_0$ where M is an easy to invert approximation to $\mathbf{A}$. The closer $\mathbf{M}$ is to $\mathbf{A}$ the more effective the preconditioning is and the harder it is to invert (in general).
3. Compute the initial residual $\mathbf{r}_0 = \mathbf{b} - \mathbf{A}\mathbf{x}_0$, $\mathbf{p}_0 = \mathbf{r}_0$.
4. $\mathbf{q}_k = \mathbf{A}\mathbf{p}_k$.
5. $\alpha_k = (\mathbf{r}_k, \mathbf{r}_k) / (\mathbf{q}_k, \mathbf{p}_k)$.
6. $\mathbf{x}_{k+1} = \mathbf{x}_k + \alpha_k \mathbf{p}_k$.
7. $\mathbf{r}_{k+1} = \mathbf{r}_k - \alpha_k \mathbf{q}_k$.
8. $\mathbf{z}_{k+1} = \mathbf{M}^{-1}\mathbf{r}_{k+1}$.
9. $\beta_k = (\mathbf{r}_{k+1}, \mathbf{z}_{k+1} / \mathbf{r}_k, \mathbf{z}_k)$.
10. $\mathbf{p}_{k+1} = \mathbf{z}_{k+1} + \beta_k \mathbf{p}_k$.
11. Advance the iteration counter $k := k + 1$.
12. Test for convergence, is $\|\mathbf{r}_{k+1}\|_2 < \epsilon$, $\epsilon \approx 0$, if true, exit iteration, else return to step 4.

The key to this algorithm is a simple short recursion that produces an orthogonal basis of vectors. This property is the direct consequence of the symmetric (semi)positive definite nature of the systems of equations that this algorithm applies to. On occasion it may be necessary to recompute $\mathbf{r}_k$ from its definition because the identity used in steps 2 and 8 of the algorithm only holds for exact arithmetic and can drift if the system $\mathbf{A}$ is sufficiently poorly conditioned and a large number of iterations are taken. The down-side to this step is that this will double the number of matrix-vector multiplications in an iteration. As a consequence a compromise is to compute the residual from the definition on the order of every 100 iterations.

If the system is indefinite, but symmetric, the minimum residual, MINRES algorithm is recommended, but if the system is non-symmetric, the algorithmic doors are wide open, but unclear. Many algorithms attempt to form an approximately symmetric system by effectively multiplying the system $\mathbf{A}$ by its approximate transpose $\mathbf{A}^T$. A large number of methods have been derived along these lines and form a veritable alphabet soup CGS [506], BiCG [194], BiCGStab [563], QMR [196], TFQMR [195] and many others. Each of these methods is classified as a Lanzcos method denoting the approximate nature of the minimization procedure. The unfortunate side-effect of $\mathbf{A}^T\mathbf{A}$ is a squaring of the condition number of the system. With the approximation of this step, the impact is not as dire, but the exactness or actual minimization of the solution is lost as well. The condition number is a measure of how difficult a system of equation is to solve and is defined by the ratio of the largest-to-smallest eigenvalues of $\mathbf{A}$. A large value signifies a system of equations that is difficult to solve (accurately). The second important aspect of these methods is the loss of the exact solution property found with classical CG (or GMRES introduced next). This property while not practically essential, but generally leads to robustness and reliability for the method.

For non-symmetric system the most reliable algorithm is GMRES which has the property of exactly solving the system in N iterations. Unfortunately, this property incurs a steep cost because vectors must be stored for all previous iterates. This algorithm works by producing a set of orthogonal vectors through the application of a Gram-Schmidt algorithm to these vectors. For robust use the GMRES algorithm can be written out as shown below:

Algorithm 2 *[Left Preconditioned Generalized Minimum Residual Algorithm]*

1. Start with the linear system $\mathbf{A}$, the right hand side, $\mathbf{b}$, and an initial guess, $\mathbf{x}^0$, $k = 0$.
2. Compute the initial residual $\mathbf{r}_0 = \mathbf{b} - \mathbf{A}\mathbf{x}_0$.
3. Compute the initial preconditioned vector $\mathbf{r}_0 := \mathbf{M}^{-1}\mathbf{r}_0$, $\beta = \|\mathbf{r}_0\|_2$ and $\mathbf{v}_1 = \mathbf{r}_0/\beta$.
4. $\mathbf{w} = \mathbf{M}^{-1}\mathbf{A}\mathbf{v}_{k+1}$.
5. Now loop over all the iterations, $i = 1 \ldots k$ (this one and all the previous ones) performing a Gram-Schmidt orthogonalization, $h_{i,k} = (\mathbf{w}, \mathbf{v}_{k+1})$, $\mathbf{w} := \mathbf{w} - h_{i,j}\mathbf{v}_{k+1}$.
6. $h_{k+2,k+1} = \|\mathbf{w}\|_2$ and $\mathbf{v}_{k+2} = \mathbf{w}/h_{k+2,k+1}$.

7. $V_k = [\mathbf{v}_1, \ldots, \mathbf{v}_k]$ and $H_k = h_{i,j}$, $i = 1, \ldots, j+1$, $j = 1, \ldots, k$.
8. Solve $y_k = \min \|\beta \mathbf{e}_1 - H_k \mathbf{y}\|_2$, $\mathbf{x}_k = \mathbf{x}_0 + V_k \mathbf{y}_k$ where $\mathbf{e}_1$ is a unit vector providing the connection to the residual.
9. Advance the iteration counter $k := k + 1$.
10. Test for convergence, is $\|\mathbf{r}_{k+1}\|_2 < \epsilon$, $\epsilon \approx 0$, if true, exit iteration, else $\mathbf{x}_0 = \mathbf{x}_m$ and return to step 4.

The step where the solution is constructed involves solution of a least squares problem (QR decomposition is recommended). This is then constructed using all the previous vectors $\mathbf{v}_i$.

Because of the need to apply the Gram-Schmidt process to all the directional vectors, the algorithm is quite costly and the cost grows with each iteration. This places a rather extreme premium on the quality of the preconditioning. To combat this expense often the number of vector iterations is limited and the iteration is restarted. With this process comes a benefit in storage and operation cost, but the exact solution is sacrificed (the minimization property becomes approximate as in the Lanzcos methods). While this step is often pragmatic, it also reduces the robustness of GMRES substantially. Thus, if the preconditioning is effective enough to eliminate the need for restarting the iteration, GMRES becomes extremely attractive. Another alternative is a "flexible" variant of GMRES that allows for varying the preconditioner from step-to-step. This would allow for adaptive behavior in the preconditioner.

Another important use of GMRES is in the Newton-Krylov methods. An important aspect of this method is the scaling of the vectors as they are of a reliable scale. This is important in the computation of the Freshet derivatives used to approximate the action of the Jacobian of Newton's method.[4]

In the interests of time and space we will not introduce further methods such as the broad class of inexact Krylov methods based on some sort of approximate transpose [475]. Generally speaking, these methods are both less expensive, and less reliable than GMRES or its variants.

A classical version of the Lanczos algorithm for bi-orthogonalization can be found in the Bi-conjugate gradient method. It is typical of non-symmetric Krylov methods not requiring large amounts of memory as GMRES. The price for this is efficiency of iteration and assurance of convergence. Nevertheless, this method and others like have found much success in recent years.

A description of the Bi-conjugate Gradient Stabilized Method (BiCGstab) is given below:

Algorithm 3 *[Preconditioned Bi-conjugate Gradient Stabilized Method (BiCGstab)]*

1. Start with the linear system $\mathbf{A}$, the right hand side, $\mathbf{b}$, and an initial guess, $\mathbf{x}^0$, $k = 0$.

[4] A Freshet derivative is formed using a finite difference where the function is evaluated at a state and a small perturbation from that state

2. Compute the initial residual $\mathbf{r}_0 = \mathbf{b} - \mathbf{A}\mathbf{x}_0$, $\mathbf{r}^* = \mathbf{r}_0$ (generally choose $\mathbf{r}^*$ such that $(\mathbf{r}_0, \mathbf{r}^*) \neq 0$); $\mathbf{p}_0 = \mathbf{r}_0$, $k = 0$.
3. Start the iteration:
4. If $k = 0$ then $\mathbf{p}_k = \mathbf{r}_k$.
5. Compute the initial preconditioned vector $\mathbf{p}_k^* = \mathbf{M}^{-1}\mathbf{p}_k$ and this step will solve some equation $\mathbf{M}\mathbf{p}_0 = \mathbf{r}_0$ where $\mathbf{M}$ is an easy to invert approximation to $\mathbf{A}$. The closer $\mathbf{M}$ is to $\mathbf{A}$ the more effective the preconditioning is and the harder it is to invert (in general).
6. $\rho_k = (\mathbf{r}_k, \mathbf{r}^*)$, if $\rho_k = 0$ stop the method has failed.
7. If $k \neq 0$ then $\beta = \rho_k / \rho_{k-1}$,
8. $\mathbf{q}_k = \mathbf{A}\mathbf{p}_k^*$,
9. $\alpha = \rho_k / \left(\mathbf{p}_k^*, \mathbf{q}\right)$,
10. $\mathbf{s} = \mathbf{r}_k - \alpha \mathbf{q}_k$.
11. If $(\mathbf{s}, \mathbf{s}) < \epsilon$, $\mathbf{x} = \mathbf{x}_k + \alpha \mathbf{p}_k$ and exit.
12. Solve $\mathbf{s}^* = \mathbf{M}^{-1}\mathbf{s}$,
13. $\mathbf{t} = \mathbf{A}\mathbf{s}^*$,
14. $\omega = (\mathbf{t}, \mathbf{s}) / (\mathbf{t}, \mathbf{t})$,
15. $\mathbf{x}_{k+1} = \mathbf{x}_k + \alpha \mathbf{p}_k + \omega \mathbf{s}^*$,
16. $\mathbf{r}_{k+1} = \mathbf{s} - \omega \mathbf{t}$, and $\mathbf{r}_{k+1}^* = \mathbf{r}_k^* - \alpha \mathbf{q}_k^*$ $\beta = (\rho_k/\rho_{k-1})(\alpha/\omega)$, $\mathbf{p}_{k+1} = \mathbf{r}_k + \beta\left(\mathbf{p}_k - \omega \mathbf{q}_\mathbf{k}\right)$.
17. Advance the iteration counter $k := k + 1$.
18. Test for convergence, is $\|\mathbf{r}_{k+1}\|_2 < \epsilon$, $\epsilon \approx 0$, if true, exit iteration, else return to step 4.

Note that for a symmetric positive definite system that conjugate gradient solves naturally, BiCGstab costs roughly twice as much because of the standard matrix-vector multiply and the matrix-transpose-vector operation. For non-symmetric systems convergence is not guaranteed.

8.4 Multigrid Algorithm for Elliptic Equations

Alternatively, the aforementioned linear systems can be solved via a multigrid algorithm. Good basic references on the multigrid method are various textbooks [77, 595, 78, 550]. The reason for choosing multigrid is the optimality of the scaling of this algorithm. The multigrid methods covered in Chap. 10 are specialized for hyperbolic partial differential equations and differ in details from the algorithms described below.

The origin of the multigrid method is found in the papers of Fedorenko [186] and Bakhvalov [27], and later on in the work of Brandt [73]. Many of the multigrid developments for partial differential equations have been reviewed in books, for example, [238, 595]. Most of the developments and applications of multigrid for incompressible flows are related to elliptic systems of equations and mainly to SIMPLE-type approaches (covered in more depth in

Chap. 12), while fewer studies have been dealt with the development of multigrid methods in conjunction with the artificial compressibility (AC) method [157, 158, 184, 347, 491]. We will cover this development in Chap. 10.

Why multigrid? The operations count. Multigrid scales linearly and the number of operations needed to achieve a solution of a given accuracy is a (large) constant multiplying N. Thus, as N grows larger, eventually multigrid (where it works) will be the fastest route to a solution. A reasonable fall-back position are Krylov methods. A better position is to combine the two approaches from the outset and use multigrid to precondition an appropriately chosen Krylov method.

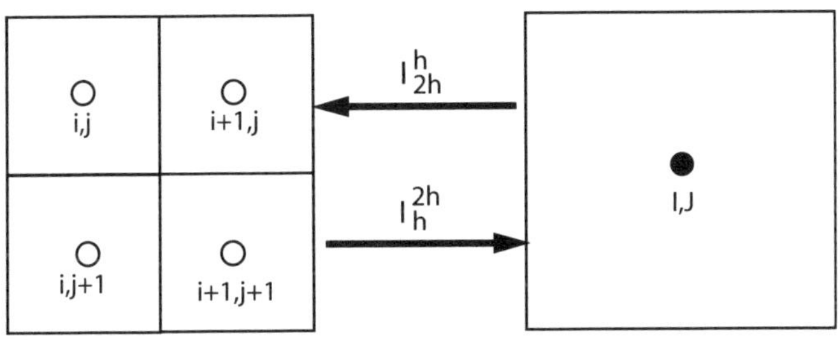

Fig. 8.2. This figures shows the relation of multigrid levels to one another with the inter-grid transfer operators. Indices i, j refer to the finer of two levels, while I, J refer to the coarser grid. The operator I_h^{2h} transfers values from the fine to the coarse grid and I_{2h}^h transfers coarse grid values to the fine grid.

Our focus here is simple cell-centered algorithms. Far better and more general, but complex methods have also been developed. For instance algebraic multigrid methods have achieved greater and greater capability over the past few years. These methods can be applied to quite general problems including unstructured grid, and a variety of discretization of complex physics. For simpler cases having complex or challenging discretizations, there are black box multigrid methods [137, 138]. If the circumstances are far from simple and the solution of a linear system must be done efficiently, these methods should be considered. To solve the pressure (Poisson) equation on simple grids, one obtains a system of equations which is symmetric, positive and definite thus simple approaches work well.

Before describing the algorithm, we will define several of the operators used in the multigrid method. The reader should note that this is a cell-centered multigrid and the formulation is somewhat different than the familiar vertex-centered multigrid. See Wesseling [595] for more details and general theory. The inter-grid transfer operators are defined differently depending on whether the five- or nine-point stencil is being solved. This is because for rigorously defined multigrid algorithms these transfer operators depend directly upon the stencils themselves. The basic nomenclature used in this description is shown by Fig. 8.2. For the five-point stencil they are

$$I_h^{2h} = \begin{pmatrix} 1 & 1 \\ 1 & 1 \end{pmatrix},$$

and

$$I_{2h}^h = \frac{1}{4}\begin{pmatrix} 1 & 1 \\ 1 & 1 \end{pmatrix}.$$

For the nine point operator, we use a bilinear weighting for the coarse to fine operator,

$$I_{2h}^h = \frac{1}{16}\begin{pmatrix} 1 & 3 & 3 & 1 \\ 3 & 9 & 9 & 3 \\ 3 & 9 & 9 & 3 \\ 1 & 3 & 3 & 1 \end{pmatrix}.$$

The methods used to relax the equations on any given grid are quite similar. First, all the equations are scaled by the cell area at that level $\Delta x^k \Delta y^k$ or in axisymmetric coordinates $r^k \Delta r^k \Delta z^k$. Both relaxation steps can be Gauss-Seidel iterations that sweep the grid on alternating grid points to aid vectorization of the algorithm. For the five-point operator, the method is a red-black method which means that the grid points where $i + j$ is odd are relaxed first, followed by the points where $i + j$ is even.

We can see the effect of this iteration on the multigrid framework by looking at its damping. This can be done through Fourier analysis by plotting the symbol (the amplitude in terms of the frequency) of the method [288]. For the red-black Gauss-Seidel, the first pass over the data is effectively a Jacobi iteration because of the order that the equations are updated, with the second step being the same, but using the updated data from the adjacent cells. By replacing the variable with a Fourier transform and taking the symbol,[5] the overall Fourier expression from the equations, from the first step and processing it through the second step, the symbol of iteration (and thus damping via the difference in the amplification from one) can be found. This is computed by examining the amplification factor for the symbol, i.e., its absolute value. This is shown in Fig. 8.3; the desirable character for this plot is the amplification factor going to zero and having this occurred preferentially at high wave numbers $\alpha \rightarrow \pi$, where $\alpha = \pi x/L_x$ or $\pi y/L_y$ for the $x-$ and $y-$ directions, respectively (L_x and L_y are the lengths of the domain in the $x-$ and $y-$ directions). Damping high wave number error is the key to multigrid

[5] The characteristic polynomial derived via the Fourier transform is called the symbol of the PDE.

because the high wave number error will be represented at larger scales as coarser grids are used in a multigrid cycle.

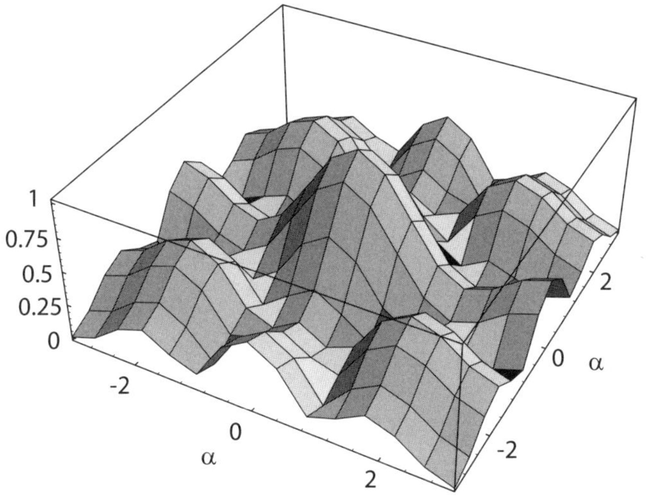

Fig. 8.3. The damping of error for a standard Laplacian operator by a red-black Gauss-Seidel iteration. The perpendicular axis shows the amplification factor for the symbol (see text). The axes (α) represent either the x or y directions and α refers to the wave number for x and y. The iteration is quite effective in damping the error in the high-frequency region.

If we are concerned about preserving the symmetry of a given solution, we cannot use the Gauss-Seidel iteration. This point has been noted by Smolarkiewicz and Margolin [502] as the basis for the development of conjugate gradient solvers. We follow a different course by using an effective symmetric method, a Jacobi iteration. In order to get the necessary performance from the iteration, we use two weighted Jacobi [77, 78] sweeps to define one iteration. The first sweep uses a weight of 1/2 and a second one with a weight of 1. The weighted iteration uses a linear combination of the result from an iteration with the final value,

$$\mathbf{x}^{k+1} := w\mathbf{x}^{k+1} + (1-w)\,\mathbf{x}^k \,,$$

where $\mathbf{x}^{k+1}$ is the result of the Jacobi iteration, (8.3), and w is the weight.

This combination seems to give good damping of high-frequency error and approximately equals the damping of a single red-black Gauss Seidel sweep. It ends up being slightly more expensive than the red-black Gauss-Seidel, but not prohibitively so. The damping is displayed in Fig. 8.4. We denote this combination of sweeps as a "composite Jacobi" sweep.

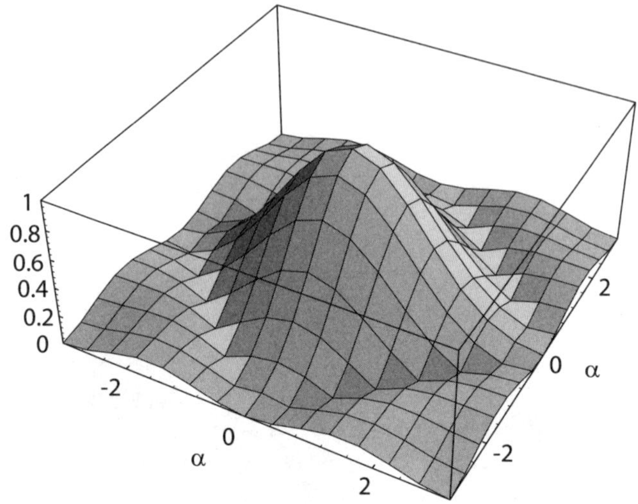

Fig. 8.4. The damping of error for a standard Laplacian operator by a composite Jacobi iteration. The perpendicular axis shows the amplification factor for the symbol (see text). The axes (α) represent either the x or y directions and α refers to the wave number for x and y. The iteration is quite effective in damping the error in the high-frequency region. Overall the effectiveness is less than that of the red-black Gauss-Seidel, but it does preserve symmetries in the solution.

For the nine-point operator a four color method is required to vectorize the Gauss-Seidel iteration [1]. This is an extension of the red-black iteration, but with four colors instead of two (each cell in a two-by-two block is "colored" and the grid is swept by cycling through each of the cells with the same color). Unfortunately, the Gauss-Seidel for the nine-point Laplacian is not efficient in reducing high-frequency error as shown in Fig. 8.5. A point Jacobi iteration, (8.3), is effective as shown in Fig. 8.6.

Each time a multigrid level is visited the equations can be relaxed (or not). We have used the option of relaxing twice at each visit, but this is dependent on details of the problem, discretization and the computer architecture. On the finest level the equations are solved "exactly" with a preconditioned conjugate gradient method. These equations are symmetric-positive-semi-definite so we do not have to worry about using a Krylov-space method to solve the system exactly.

One of the most important tasks in setting up the multigrid algorithm is the process of approximating the linear equations on the coarse grid. One approach would be to use inter-grid transfer functions to define a variational or Galerkin coarse grid operator,

$$L^{2h} = I_h^{2h} L^h I_{2h}^h ,$$

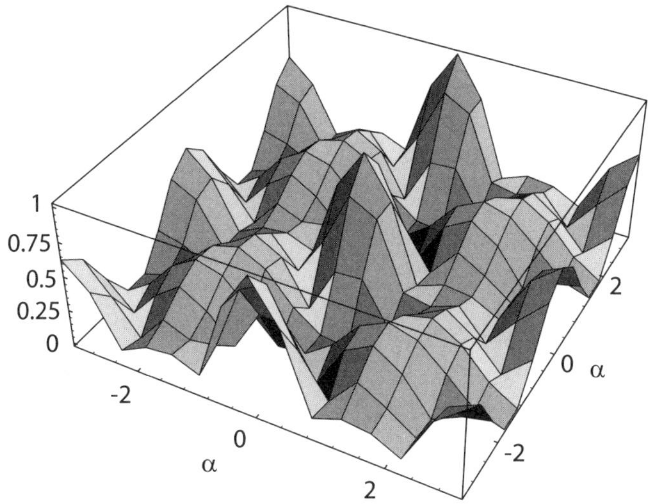

Fig. 8.5. The damping of error for a nine-point Laplacian operator by a four color Gauss-Seidel iteration. The perpendicular axis shows the amplification factor for the symbol (see text). The axes (α) represent either the x or y directions and α refers to the wave number for x and y. The iteration is only somewhat effective in damping the error in the high-frequency region.

where L^h is the Laplacian at mesh spacing h. Because of the complication and expense of this step, we have implemented a simpler, less expensive approach based in part on suggestions in [4].

For many problems the Laplacian has variable coefficients defined by the cell-coupling coefficient, σ, in the operator $\nabla \cdot \sigma \nabla \varphi$ (where φ is a generic function). The operator remains the same, as do the boundary conditions on the coarse grids, but σ must be defined at each level. The basic idea is to construct coarse grid approximations that give the same average value in a cell of the quantity $\sigma \nabla \varphi$ as on the finer levels.

In using the cell-centered multigrid framework, the control volume derivation of the equations comes quite naturally. In two dimensions, coarse grid cells are formed from four fine cells. As such, the flux through one edge of a coarse cell should sum from the two corresponding fine grid cell-edges. Using this heuristic argument, the value of σ on the coarse grid is computed to give the above characteristic. On the fine grid, we compute and store the value of σ on each cell edge. For the coarse grids, these values are used to define edge values of σ by averaging σ over the coarse grid edge. This process can be summarized in a couple of equations; for the x-edges

$$\sigma_{I-1/2,J} = 1/2\left(\sigma_{i-1/2,j} + \sigma_{i-1/2,j+1}\right) ,$$

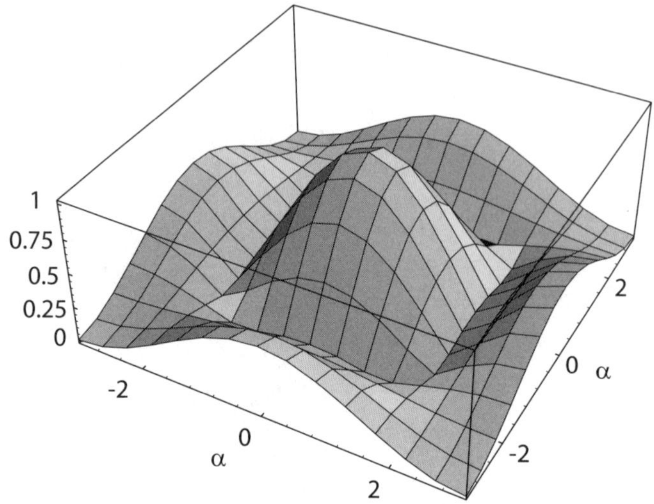

Fig. 8.6. The damping of error for a nine-point Laplacian operator by a Jacobi iteration. The perpendicular axis shows the amplification factor for the symbol (see text). The axes (α) represent either the x or y directions and α refers to the wave number for x and y. The iteration is quite effective in damping the error in the high-frequency region.

and for the y-edges

$$\sigma_{I,J-1/2} = 1/2\left(\sigma_{i,j-1/2} + \sigma_{i+1,j-1/2}\right) .$$

This process is identical to the five-point Laplacian defined in [313, 351].

In the case of the bilinear operator for the pressure, the situation is somewhat more difficult because σ is averaged on a cell vertex. The value of σ on the coarse grid vertices is computed by using a bilinear weighted average of the fine grid values of σ. The expression for this is

$$\begin{aligned}\sigma_{I-1/2,J-1/2} = \frac{1}{16}\Big[&4\sigma_{i-1/2,j-1/2} + 2(\sigma_{i+1/2,j-1/2} + \sigma_{i-1/2,j+1/2}\\ &+\sigma_{i-3/2,j-1/2} + \sigma_{i-1/2,j-3/2}) + \sigma_{i+1/2,j+1/2} +\\ &+\sigma_{i+1/2,j-3/2} + \sigma_{i-3/2,j+1/2} + \sigma_{i-3/2,j-3/2}\Big] .\end{aligned} \tag{8.4}$$

Based on the performance of each multigrid solver, it is likely that the nine-point scheme may be improved through appealing to the same control volume approach used with the five-point operator multigrid.

The convergence criterion that is used in this algorithm is

$$\|\mathbf{r}\|_2 \leq \mathrm{tol}\,\|\mathbf{b}\|_2 .$$

In practical problems we have chosen the error tolerance as $\mathrm{tol} = 1 \times 10^{-8}$.

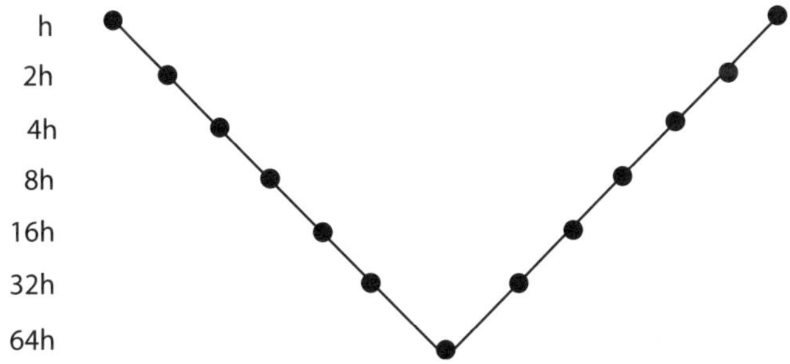

Fig. 8.7. One V-cycle is shown schematically. The multigrid algorithm coarsens the solution through 6 coarser grids solving the system "exactly" at the level $64h$ then moves through the grids in reverse order to complete the cycle on the finest grid.

A "V-cycle" multigrid is shown in Fig. 8.7. Of course there are other cycles to use like the W-cycle (with a myriad of options), and the F-cycle, which starts at the coarse level rather than the fine level. The algorithm using the tools introduced above is given below.

Algorithm 4 *[Cell-Centered Multigrid]*

1. Compute the coefficients for the linear system $\mathbf{Ax} = \mathbf{b}$ for all grids with initial guess $\mathbf{x}^o$.
2. Compute the initial residual, $\mathbf{r} = \mathbf{b} - \mathbf{Ax}^o$.
3. Compute $\|\mathbf{b}\|$ and $\|\mathbf{r}\|$. If $\|\mathbf{r}\| < \varepsilon \|\mathbf{b}\|$ then exit.
4. Begin multigrid V-cycle.
5. For $k = 0$ to $k_{\max -1}$ relax on grid k to get $\mathbf{r}^k = \mathbf{b}^k - \mathbf{A}^k\mathbf{x}^k$. The initial guess if $k \neq 0$ is $\mathbf{x}^k = 0$. Transfer the residual to the $k+1$ grid $\mathbf{b}^{k+1} = I_h^{2h}\mathbf{r}^k$.
6. On the $k_{\max}$ grid solve the equations "exactly".
7. For $k = k_{\max -1}$ to 0, relax on grid k to get $\mathbf{r}^k = \mathbf{b}^k - \mathbf{A}^k\mathbf{x}^k$. Transfer the solution, $\mathbf{x}^k$, to the $k-1$ grid via $\mathbf{x}^{k-1} = \mathbf{x}^{k-1} + I_{2h}^h\mathbf{x}^k$.
8. Compute a new error estimate $\|\mathbf{r}\|^0$. If $\|\mathbf{r}\|^0 < \varepsilon \|\mathbf{b}\|$ exit, otherwise go to step 4.

Further acceleration of the numerical solution can be obtained by combining multigrid methods and parallel computing [486, 487, 161] in spite of the fact that the parallel efficiency[6] of the multigrid algorithm is worse than the single-grid method due to the increase of communication between processors. This further deteriorates when several coarse grids are used in combination

[6] The parallel efficiency represents the time loss in parallel computations due to communication lag between processors during which computations cannot take place.

with a large number of processors. The numerical efficiency that represents the increase (due to algorithmic changes) in the number of iterations necessary to reach convergence is similar both for the multigrid and for the single-grid algorithm [486]. For laminar flows multigrid implementation results in significant reduction, by several order of magnitudes, both of the number of iterations and of the computing time. In complex turbulent flows the multigrid performance is reduced due to the use of highly stretched grids and solution stiffness at high Reynolds numbers. Examples from complex engineering applications show that multigrid can provide four to five times acceleration compared to the single-grid algorithm [15].

8.5 Multigrid Algorithm as a Preconditioner for Krylov Subspace Methods

In recent years it has become more and more evident that the combination of Krylov methods with their relative robustness and multigrid with its optimal scaling provides the "best" linear algebra capability. In many cases multigrid turns out to be temperamental at best and fragile at worst. By using multigrid in the context of a preconditioner for Krylov methods robustness can be recovered. This phenomenon is relatively easy to explain. The failure of multigrid is due to a small number of eigen-modes that are not effectively in the range of any of multigrid operators. These modes dominate the residual error, and are effectively solved by a Krylov method since they form the basis that Krylov constructs. Where multigrid is robustly functioning, the Krylov iteration produces a small overhead, and perhaps a small degree of acceleration of the solution (reducing the number of iterations by one).

The algorithm for multigrid changes little except for the nature of the solution and its right hand side. For preconditioning (see Algorithm 1 for preconditioned conjugate gradient method in Sect. 8.3) the right hand side is a function of the residual, **r**, and a vector that is not the solution that is solved for, i.e., **z**. The matrix **M** is simply the multigrid iteration matrix formed from the sequence of relaxations, coarsenings and interpolation operators. If the system is symmetric then care must be taken to produce a symmetric iteration matrix. This entails the use of the same number of relaxations on both down and upward sweeps (and a symmetric multigrid cycle). If one uses a non-symmetric relaxation (i.e., Gauss-Seidel), the order of evaluation must be inverted from downward to upward strokes of the cycle.

The efficiency gain of multigrid is perhaps most important with the use of GMRES in order to control the number of vectors that must be stored and made orthogonal. In this case one has much more freedom in forming the preconditioner, which does not have to be symmetric. Using a symmetric operator **A** to form **M** for solving a non-symmetric (even indefinite) system [452] is also an attractive preconditioning strategy.

8.6 Newton's and Newton-Krylov Method

In this part of the chapter we will introduce methods for solving nonlinear systems of equations. While there are a variety of methods to achieve this, Newton's method holds out the greatest promise because of its potential for quadratic convergence. Efficiency is also a key issue and concerns about this have prompted the creation of Newton-Krylov methods which combine the convergence of Newton's method with efficiency for sparse systems of equations offered by Krylov methods. If one chooses an implicit discretization for the flow equations (discussed in Chap. 7), these methods offer an attractive alternative.

Newton's method is both powerful and "dangerous". The power lies in its theoretical quadratic convergence rate, the danger lies in its propensity to diverge if the approximation is too far from the solution. As such most of the complexity associated with the method is related to keeping the method from diverging under circumstances that the quadratic convergence is not seen.

Like much of numerical analysis the starting point for Newton's method is a Taylor series expansion. Starting with the general problem stated rather abstractly,

$$F(\mathbf{x}) = 0 \,, \tag{8.5}$$

where F is a general nonlinear function of $\mathbf{x}$ we expand around some state (initial guess, $k = 0$, or the kth iteration),

$$F(\mathbf{x}) = F\left(\mathbf{x}^k\right) + \frac{\partial F\left(\mathbf{x}^k\right)}{\partial(\mathbf{x}^k)}\left(\mathbf{x}^{k+1} - \mathbf{x}^k\right) + \text{H.O.T.} \,, \tag{8.6}$$

truncating the expansion with one term. The term $\left(\mathbf{x}^{k+1} - \mathbf{x}^k\right)$ is usually abbreviated as $\delta\mathbf{x}^k$ and is used to advance the solution through,

$$\mathbf{x}^{k+1} = \mathbf{x}^k + \delta\mathbf{x}^k \,.$$

For systems of equations the derivative $J\left(\mathbf{x}^k\right) = \partial F\left(\mathbf{x}^\mathbf{k}\right)/\partial\mathbf{x}^\mathbf{k}$ is a matrix known as a Jacobian.

In order to advance this method the Jacobian must be inverted. Since we express the problem as (8.5), then (8.6) can be rearranged to yield the following,

$$\delta\mathbf{x}^k = -J\left(\mathbf{x}^k\right)^{-1} F\left(\mathbf{x}^k\right) \,. \tag{8.7}$$

Quite often the update given is too large and must be reduced,

$$\mathbf{x}^{k+1} = \mathbf{x}^k + \alpha\delta\mathbf{x}^k, \tag{8.8}$$

where α is an under-relaxation parameter to avoid divergence. Some of the algorithms are quite advanced and are required to produce robust stable solutions.

Forming the Jacobian is often quite expensive or difficult. Often a good choice is computing it via Fesquet derivatives as

$$J(\mathbf{x}) = \frac{F(\mathbf{x}+\epsilon) - F(\mathbf{x})}{\epsilon} ,$$

where $\epsilon \approx 0$. This idea forms the heart of Newton-Krylov methods where the vector is a Krylov vector which is substituted for $\mathbf{x}$, and the evaluation of J replaces the matrix-vector multiplication. For example, in GMRES we get the following set of operations,

$$\mathbf{A}\mathbf{v} = \frac{F(\mathbf{x}+\epsilon\mathbf{v}) - F(\mathbf{x})}{\epsilon} . \tag{8.9}$$

Here, two evaluations of the nonlinear function have replaced the matrix-vector multiplication.

Finally, we mention that the multigrid algorithm can be used to attack nonlinear problems directly (rather than as a preconditioner or solver with Newton's method or Newton Krylov). This is the FAS or "Full Approximation Scheme". In this method, the multigrid acts on the nonlinear problem directly in the course of a multilevel mulitgrid algorithm (see Sect. 10.11 for further discussion).

8.7 A Multigrid Newton-Krylov Algorithm

The starting point for the development of nonlinearly convergent methods is a standard linearized solution method. It is this typical linearization which, if applied iteratively to the same time step, constitutes a Picard-type (or successive substitution) nonlinear solver. Later, we show that this linearized solver forms the basis of the nonlinear preconditioning.

For nonlinear problems solution may diverge if the step size is too large and as a consequence care must be taken. We employ an under-relaxation factor ξ that is defined by $\xi = \min(1, 1/\|\delta\mathbf{x}/\mathbf{x}\|)$ to robustly deal with convergence difficulties often encountered during the early stages of a nonlinear iteration. Both of the nonlinear iteration methods considered are inexact [134], we use 10^{-2} times the current nonlinear residual to define the linear convergence tolerance. This limits the amount of work which is used to produce solutions that poorly approximate the nonlinear solution. Convergence within a time step is determined by the norm, $\|F(\mathbf{x})\|_2$, dropping below a value like 10^{-6}.

As we will see preconditioning is the heart of the problem, and the Picard solver shown below (Algorithm 5) only differs from the Newton solver (Algorithm 6) in the matrix-vector product (step 2c).

Algorithm 5 *[Multigrid Picard-type Nonlinear Solver]*

1. Start the nonlinear iteration, $k = 0$.
2. Compute the nonlinear residual, $\mathbf{r} = -F(\mathbf{x})$.
   a) Start the Krylov iteration to solve $\mathbf{A}\delta\mathbf{x} = \mathbf{r}$, $n = 0$. Initialize the Krylov vector with $\mathbf{v}_n = \mathbf{r}_n$.

   b) Compute the preconditioned Krylov vector, $\mathbf{A}\tilde{\mathbf{M}}^{-1}\mathbf{v}$, using a multigrid V-cycle to approximate the solution to $\mathbf{A}\mathbf{y}_n = \mathbf{v}_n$. $\tilde{\mathbf{M}}^{-1}$ is the approximate inverse of $\mathbf{A}$.
   c) Perform the matrix-vector multiply through the operation $\mathbf{w}_n = \mathbf{A}\mathbf{y}_n$.
   d) Complete the Krylov iteration (constructing a new Krylov vector, $\mathbf{v}_{n+1}$) and compute the Krylov convergence; if converged, exit, otherwise $n := n + 1$ and go to (b).
3. Compute the (damped) update to the full nonlinear problem.
4. Check for nonlinear convergence, if converged, exit, otherwise, $k := k+1$, go to 2.

This algorithm forms the foundation of a more sophisticated algorithm. In other words, a convergent nonlinear Picard-type iteration preconditions Newton's method. As will be seen shortly, the only difference between the two algorithms is the form of the matrix-vector product used in the Krylov algorithm. Both methods use a multigrid preconditioned Krylov method as an inner iteration with different connections to the full nonlinear problem. In developing our method we build upon our earlier efforts to combine multigrid as a preconditioner for Newton-Krylov methods [297].

First, we define the nonlinear functions that are being solved in abstract form as $F(\mathbf{x})$. Our goal is to execute an inexact Newton iteration. This updates the dependent variables by approximately solving,

$$J\left(\mathbf{x}^k\right)\delta\mathbf{x}^{k+1} = -F\left(\mathbf{x}^k\right) , \tag{8.10}$$

where k is the iteration index and

$$\mathbf{x}^{k+1} = \mathbf{x}^k + \xi\delta\mathbf{x}^{k+1} , \tag{8.11}$$

to solve $F(\mathbf{x}) = 0$. J is the Jacobian of $F(\mathbf{x})$ whose elements are defined by $J_{i,j} = \partial F(x_i)/\partial x_j$. To implement a Krylov method we only need to represent the matrix-vector product rather than explicitly represent the matrix. This allows the definition of matrix-free (Jacobian-free) algorithm [86] with an approximation,

$$J\mathbf{v} \approx \frac{F(\mathbf{x} + \epsilon\mathbf{v}) - F(\mathbf{x})}{\epsilon} , \tag{8.12}$$

where $\mathbf{v}$ is a Krylov vector and $\epsilon = \rho(1 + \|\mathbf{x}\|)$ and $\rho = 10^{-8}$.

In order for this algorithm to be effective, a preconditioner must be employed. In this case we need to approximate $J\tilde{\mathbf{M}}^{-1}\mathbf{v}$ (the approximate inverse used as a preconditioner) which is done in two steps:

1. Approximately solve the linear system $\mathbf{M}\mathbf{y} = \mathbf{v}$, where we choose $\mathbf{M}$ as the linear system $\mathbf{A}$ from the Picard-type iteration with an approximate solution computed with a single multigrid V-cycle.

2. Approximate the Jacobian via

$$J\tilde{\mathbf{M}}^{-1}\mathbf{v} = J\mathbf{y} \approx \frac{F(\mathbf{x} + \epsilon\mathbf{y}) - F(\mathbf{x})}{\epsilon}.$$

Here $\mathbf{y}$ is referred to as a preconditioned Krylov vector. Symbolically, this can be compactly represented as $J\tilde{\mathbf{M}}^{-1}\mathbf{v}$ with $\tilde{\mathbf{M}}^{-1}$ referring to the approximate inversion of $\mathbf{A}$ accomplished with the multigrid V-cycle. The overall Newton-Krylov iteration takes the symbolic form, $\left(J\tilde{\mathbf{M}}^{-1}\right)\left(\tilde{\mathbf{M}}\delta\mathbf{x}\right) = -F(\mathbf{x})$ which is known as right preconditioning. Right preconditioning is sometimes preferable because it yields the true residual of the system of equations directly rather than its preconditioned value.

The chief advantage of this method is that the actual Jacobian is never formed. The necessary element for this approach to be successful is good preconditioning. This process should not be confused with the process of numerically approximating the elements of the Jacobian via numerically evaluated (Frechet) derivatives.

To summarize, one can apply the Picard-type linearization of the governing equation as the preconditioner. This is simultaneously the most important and subtle aspect of the method. Despite the asymmetry and potential indefiniteness of the nonlinear system, the symmetric positive definite preconditioner can be used. In using this approximation, the only presence of the true Jacobian is found in the matrix-free matrix-vector product in the Krylov iteration.

Symbolically, this algorithm can be stated in the following way:

Algorithm 6 *[Newton-Krylov with Picard-type Multigrid Preconditioning]*

1. Start the nonlinear iteration, $k = 0$.
2. Compute the nonlinear residual, $\mathbf{r} = -F(\mathbf{x})$.
   a) Start the Krylov iteration to solve $J\delta\mathbf{x} = \mathbf{r}$, $n = 0$. Initialize the Krylov vector with $\mathbf{v}_n = \mathbf{r}_n$.
   b) Compute the preconditioned Krylov vector, $J\tilde{\mathbf{M}}^{-1}\mathbf{v}$, Using a multigrid V-cycle to approximate the solution to $\mathbf{A}\mathbf{y}_n = \mathbf{v}_n$.
   c) Perform the matrix-vector product through the operation $\mathbf{w}_n = [F(\mathbf{x} + \epsilon\mathbf{y}_n) - F(\mathbf{x})]/\epsilon$.
   d) Complete the Krylov iteration (constructing a new Krylov vector, $\mathbf{v}_{n+1}$) and compute convergence, if converged, exit, otherwise $n := n + 1$ and go to (b). This is the step that distinguishes Algorithm 6 from Algorithm 5 and is the heart of the matrix-free Krylov method.
3. Compute the (damped) update to the full nonlinear problem.
4. Check for nonlinear convergence, if converged, exit, otherwise, $k := k+1$, go to 2.

The only difference between the Picard-type and Newton iteration is the matrix-free implementation of GMRES algorithm in Newton's method.

Viewed in this light, the matrix-free Newton's method can be viewed as accelerating the convergence of the simpler Picard iteration. The convergence tolerance of the linear problem is adaptive on each nonlinear step. We make note that we favor the use of GMRES because of its superior stability properties [296].

Part II

Solution Approaches

9. Compressible and Preconditioned-Compressible Solvers

High-resolution methods were developed specifically for compressible high-speed flows. The most vexing challenge in high speed flows is successfully computing shock waves where all variables are discontinuous. High-resolution methods provided the means to compute shocks without (significant) oscillations while mitigating the dissipative solutions commonly associated with first-order solutions. They also allow higher accuracy of flows away from discontinuities where the flow is "smooth." It is important to understand the algorithmic structure of the foundational methods used in conjunction with high resolution methods. One consideration that is central to these methods is the use of conservation form. This is a consequence of the Lax-Wendroff theorem [321]: *If a difference equation is in conservation form and is consistent with the original conservation law as well as stable it will converge to the correct weak solution of that conservation law.* This powerful theorem is wise to adhere to because it supplies what little assurances can be made in solving flows with shocks.

If one has a compressible flow solver sometimes the simplest thing to do is run the code at a low Mach number. The downside of using this approach is efficiency. Nonetheless this is common practice because it is often faster to run an inefficient calculation than write an efficient code. In other cases the low speed or incompressible solutions are used to more fully test the range of applicability for a compressible flow code. In an explicit algorithm the time step size is determined by the sum of the fluid velocity and sound speed, while the dynamical scale of incompressible flows is set by the fluid velocity. This naturally leads to the issue of preconditioning where this chapter ends.

We begin with the basic Godunov-type method in its various incarnations. This will be followed by a description of the explicit flux-splitting methods. We close with a description of low-speed preconditioners for making compressible flow solvers more efficient for these flows.

9.1 Reconstructing the Dependent Variables

The first high-resolution methods for compressible flows used interpolations of the dependent variables to discretize the equations. This builds directly upon the work of Godunov who described the methods in terms of piecewise constant “slabs” of material. These lines of thinking were extended first by van Leer [571] with piecewise linear interpolants and later by Colella and Woodward [120] with piecewise parabolic interpolants. The more general extension came with the ENO (interpolation) schemes where the interpolants were theoretically extended to arbitrarily high-order [248].

9.1.1 Riemann Solvers

The fundamental first-order method begins with the introduction of piecewise constant functions for the dependent variables in each zone. Thus, $\mathbf{U}(x) \equiv \mathbf{U}_j$, $x \in [x_j - \Delta x/2, x_j + \Delta x/2]$. The flow field is updated using

$$\mathbf{U}_j^{n+1} = \mathbf{U}_j^n - \frac{\Delta t}{\Delta x}\left[\mathbf{E}_{j+1/2}\left(\mathbf{U}_j^n, \mathbf{U}_{j+1}^n\right) - \mathbf{E}_{j-1/2}\left(\mathbf{U}_{j-1}^n, \mathbf{U}_j^n\right)\right]. \quad (9.1)$$

Notice that at the interface between zones the interpolants are double-valued. More generally the values used in determining the fluxes are denoted with subscripts L and R for“left” and “right” meaning the values directly to the left and right of the interface, respectively.

Before examining the algorithm for a system of equations, consider a simpler setting, a linear advection equation, $U_t + aU_x = 0$. The direction of transport is determined by a and if this is greater than zero information travels left-to-right on a grid. The simplest Godunov method is simply upwind differencing

$$U_j^{n+1} = U_j^n - \frac{\Delta t\, a}{\Delta x}\left(U_j^n - U_{j-1}^n\right). \quad (9.2)$$

If $a < 0$ this formula is replaced by a difference biased in the other direction,

$$U_j^{n+1} = U_j^n - \frac{\Delta t\, a}{\Delta x}\left(U_{j+1}^n - U_j^n\right). \quad (9.3)$$

These formulas can be combined to give a simple form using the following recipe,

$$U_j^{n+1} = U_j^n - \frac{\Delta t}{\Delta x}\left(E_{j+1/2} - E_{j-1/2}\right).$$

where

$$E_{j+1/2} = \frac{a}{2}\left(U_j^n + U_{j+1}^n\right) - \frac{|a|}{2}\left(U_{j+1}^n - U_j^n\right). \quad (9.4)$$

The reader can confirm for themselves that the application of (9.4) reproduces either (9.2) or (9.3) in the appropriate case. In the more general case (9.4) is a Riemann solver.

While (9.4) is an exact Riemann solution for the simple linear model equation, its direct extension to systems is an approximation. This leads to the term of *approximate Riemann solver*. Through the use of characteristic variables, the same principle can be applied to systems of equations. Generically, direct extension of (9.4) is

$$\mathbf{E}_{j+1/2} = \frac{1}{2}\left(\mathbf{E}_j^n + \mathbf{E}_{j+1}^n\right) - \mathbf{T}_{j+1/2}\frac{\left|\boldsymbol{\Lambda}_{j+1/2}\right|}{2}\mathbf{T}_{j+1/2}^{-1}\left(\mathbf{U}_{j+1}^n - \mathbf{U}_j^n\right). \quad (9.5)$$

This method is complete once the definition of the eigenstructure, $\mathbf{T}\boldsymbol{\Lambda}\mathbf{T}^{-1}$, of the equations at the cell interface $j+1/2$ is known. For very simple cases without strong shocks, the arithmetic mean suffices. This is sufficient for many if not most low-speed flows. Much greater detail will be given on approximate Riemann solvers in Chap. 16. Below, we will discuss the use of the much more elaborate *exact Riemann solver*. Its complicated form offers a sufficient explanation for the popularity of approximate Riemann solvers.

To compute a single value for the numerical flux the double-valued $\mathbf{U}$ must be converted into single values. This process is computed using the Riemann solution which is the analytical solution to the interaction between two semi-infinite states, i.e., the Riemann problem (also known as the shock tube solution). For an ideal gas the Riemann problem can be solved exactly [221]. Godunov solved the Riemann problem for the Euler equations of gas dynamics using Newton's method. Remember that this solution is self-similar in the ratio x/t. This method is shown in algorithmic form in Fig. 9.1. For the Euler equations in Lagrangian coordinates the information available from this solution is sufficient to advance the equations because $\mathbf{E}\left(\mathbf{U}_L, \mathbf{U}_R\right) = \left(-u_*, p_*, p_* u_*\right)^T$.[1]

In the end, approximate Riemann solvers covered in Chap. 16 have almost entirely supplanted exact Riemann solvers. It is useful to recognize that simply applying a single iteration of the exact Riemann solver in Fig. 9.1 itself constitutes an approximate Riemann solver.

In Eulerian coordinates more must be done before constructing the correct fluxes. The Riemann solution algorithm only produces the pressure and velocity at the contact discontinuity, which conveniently is the cell-edge in Lagrangian coordinates, in Eulerian coordinate this is not generally (or usually) the case. However, using this information the state at the cell interface can be determined. In an Eulerian flow the cell interface is fixed thus does not move, i.e., $x/t = 0$. In finding the correct values the solution must be interrogated to correctly place the interface. Using the wave speeds one must determine whether the interface is in the pre- or post-shock region, in a state to the direct left or right of the contact discontinuity or embedded in a rarefaction. Once the interface has been placed its fluid state must be evaluated. The simplest case is when the flow is completely supersonic. If the flow is supersonic then the Riemann solution is a pure initial left or right state. In the end it is sufficient to approximate the solution in x/t space. The procedure goes as follows: The first step is to classify the left and right facing waves. This is based on the sign of the particle velocity with a positive particle velocity

[1] The star values are the internal states of the Riemann solution where the exact solution has a great deal of structure in x/t space. For example, u_* and p_* are correct between the nonlinear waves that bound the evolution of the Euler system, but if the solution is a single wave they do not exist (i.e., like a shock).

Initial condition, $(\rho_L, u_L, e_L, p_L)^T$, $(\rho_R, u_R, e_R, p_R)^T$

$p_* = \frac{1}{2}(\rho_L + \rho_R) + \frac{1}{2}\sqrt{\frac{\gamma(p_L + p_R)(\rho_L + \rho_R)}{4}}(u_L + u_R)$

begin

Do While not converged

begin

begin

if $p_* > p_L$ **then**

$$M_L = \sqrt{\gamma \rho_L p_L}\sqrt{\frac{\gamma+1}{2\gamma}\frac{p_*}{p_L} + \frac{\gamma-1}{2\gamma}}$$

$$(u_{L*})' = -\frac{(M_L)^2 + \gamma\rho_L p_L}{2(M_L)^3}$$

else

$$M_L = \frac{\gamma-1}{2\gamma}\sqrt{\gamma\rho_L p_L}\left(1 - \frac{p_*}{p_L}\right)$$

$$(u_{L*})' = -\frac{1}{\gamma\rho_L p_L\left(\frac{p_*}{p_L}\right)^{\frac{\gamma-1}{2\gamma}}}$$

endif

if $p_* > p_R$ **then**

$$M_R = \sqrt{\gamma \rho_R p_R}\sqrt{\frac{\gamma+1}{2\gamma}\frac{p_*}{p_R} + \frac{\gamma-1}{2\gamma}}$$

$$(u_{R*})' = -\frac{(M_L)^2 + \gamma\rho_R p_R}{2(M_R)^3}$$

else

$$M_R = \frac{\gamma-1}{2\gamma}\sqrt{\gamma\rho_L p_L}\left(1 - \frac{p_*}{p_R}\right)$$

$$(u_{R*})' = -\frac{1}{\gamma\rho_R p_R\left(\frac{p_*}{p_R}\right)^{\frac{\gamma-1}{2\gamma}}}$$

endif

$$u_{L*} = u_L - \frac{p_* - p_L}{M_L}$$

$$u_{R*} = u_R - \frac{p_* - p_R}{M_R}$$

$$p_* = p_* - \frac{u_{L*} - u_{R*}}{(u_{L*})' - (u_{R*})'}$$

end

check convergence

end

$u_* = \frac{1}{2}(u_{L*} + u_{R*})$

Fig. 9.1. The algorithm for computing the exact Riemann solution using Newton's method; γ is the ratio of specific heat capacities of the gas and e is the specific internal energy.

indicating a shock and a negative particle velocity indicating an expansion. The state 0 is either L or R as appropriate. The specific volume across the shock is given by

$$V_* = V_0\left(1 - \frac{U_p}{W_s}\right),$$

where V_0 is the unshocked/unrarified specific volume, $W_s = \rho c_s$ is the Lagrangian wave speed, with c_s being the shock speed, and $U_p = u_L - u_*$ or $U_p = u_* - u_R$. The use of isentropic conditions across the rarefaction wave giving

$$V_* = V_0\left(1 - \frac{p_* - p_0}{\gamma p_0}\right).$$

For a shock the Rankine-Hugoniot conditions determine the specific internal energy,

$$e_* - e_0 - \frac{1}{2}\left(p_0 + p_*\right)\left(V_* - V_0\right).$$

In the case of a rarefaction, an isentropic relation at constant C_V (specific heat capacity at constant volume V) provides an approximation for the specific internal energy,

$$e_* = e_0\left(\frac{V_*}{V_0}\right)^{-\Gamma_0},$$

where $\Gamma_0 = 1/\rho(\partial p/\partial e) = \gamma - 1$ is the Grüneisen coefficient.

With these quantities computed all that remains is to compute the solution as a function of x/t with the goal in Eulerian coordinates to evaluate the solution at $x/t = 0$ with $M = U_p/c$. A rarefaction is bounded by the wave speeds,

$$u_0 \pm c_0,$$

and using one definition of the fundamental derivative, $\mathcal{G}_0 = 1/c_0(d\left(\rho c\right)/d\rho)$,

$$c_* = \frac{c_0}{1 - \rho_*\left(\mathcal{G} - 1\right)\left(V_0 - V_*\right)},$$

then the bounding velocity is

$$u_* \pm c_*.$$

For $u_* > 0$ if the left state is shocked, and $W_{L,e} > 0$, the left state is present at $x/t = 0$ otherwise, the "$*$" state is present. For rarefied flow the situation is a bit more complicated. Three alternatives are possible, if $u_L - c_L > 0$, the left state is present, and if $u_* - c_{L,*} < 0$, the "$*$" state is present.

The last option is a bit more involved, $x/t = 0$ lies in the fan. The starting point is a linear approximation for the velocity,

$$u\,(x/t) = \xi u_* + (1-\xi)\,u_L\ ,$$

where $\xi = (x/t - u_L + c_L)\,/\,(u_* - c_{L,*} - u_L + c_L)$. The rest follows directly, with the pressure given by

$$p\,(x/t) = \xi p_* + (1-\xi)\,p_L\ .$$

The specific volume is

$$V\,(x/t) = V_L\left(1 - \frac{p\,(x/t) - p_L}{\gamma p_L}\right)\ .$$

If $u_* < 0$ the construction given above is applied to the right state with the appropriate sign conventions changed.

Now we will move to higher order methods in a relatively simple setting.

9.1.2 Basic Predictor-Corrector

For higher than first-order methods a number of things change, but the considerations for the Riemann solver remain fixed. Time accuracy must also rise and the method described in this section produces a simple setting for achieving this. With higher order methods the first principle difference is that the piecewise constant interpolants are replaced by higher order ones. For example, a piecewise linear interpolant is most often used, $\mathbf{U}\,(x)\,, x \in [x_j - \Delta x/2, x_j + \Delta x/2] = \mathbf{U}_j + \partial\mathbf{U}_j/\partial x\,(x - x_j)$. Solutions are still two valued at cell interfaces. A first-order in time method now appears as follows

$$\begin{aligned}\mathbf{U}_j^{n+1} &= \mathbf{U}_j^n \\ &- \frac{\Delta t}{\Delta x}\Big[\mathbf{E}\left(\mathbf{U}^n\,(x_j + \Delta x/2)\,, \mathbf{U}^n\,(x_{j+1} - \Delta x/2)\right) \\ &- \mathbf{E}\left(\mathbf{U}^n\,(x_{j-1} + \Delta x/2)\,, \mathbf{U}^n\,(x_j - \Delta x/2)\right)\Big]\ .\end{aligned}$$

To achieve better accuracy the edge values must be centered in time

$$\begin{aligned}\mathbf{U}_j^{n+1} &= \mathbf{U}_j^n \\ &- \frac{\Delta t}{\Delta x}\Big[\mathbf{E}\left(\mathbf{U}^{n+1/2}\,(x_j + \Delta x/2)\,, \mathbf{U}^{n+1/2}\,(x_{j+1} - \Delta x/2)\right) \\ &- \mathbf{E}\left(\mathbf{U}^{n+1/2}\,(x_{j-1} + \Delta x/2)\,, \mathbf{U}^{n+1/2}\,(x_j - \Delta x/2)\right)\Big]\ .\end{aligned}$$

This is a second-order method.

One may use the original governing equation to do this. For example, $\mathbf{U}_j^{n+1/2} = \mathbf{U}_j^n + \Delta t/2\,(\partial\mathbf{U}/\partial t)_j^n$ and using the Lax-Wendroff technique, $(\partial\mathbf{U}/\partial t)_j^n = -\,(\partial\mathbf{E}/\partial x)_j^n$. Fig. 9.2 shows two manners in which this can viewed either as integrating the differential equation over the time step or characteristic tracing to the time-centered value.

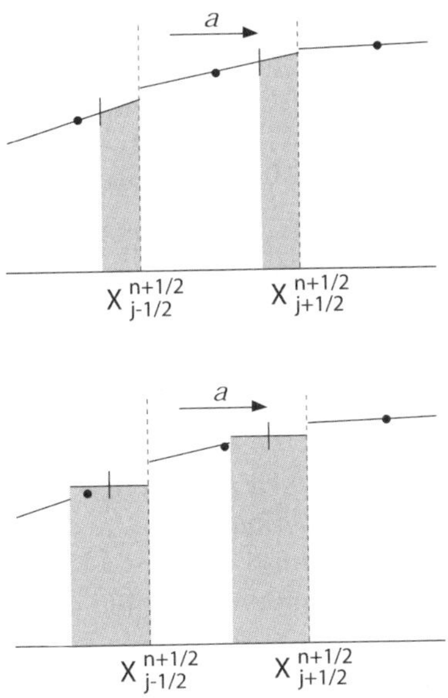

Fig. 9.2. The figure shows two complementary views of time differencing. In the top case the characteristics are traced back half a time step to "time-center" the edge data. In the bottom case the reconstructions are integrated over a time step to produce a time-averaged value. In both cases the procedure will produce a valid second-order approximation to the time derivative, $\partial \mathbf{U}/\partial t$. These are equivalent for second-order methods, but the time-integrated approach generalizes to higher order reconstructions. The shaded regions are those swept by the characteristics.

For the compressible Euler equations this does not always lead to the best results. In this case it is advisable to use the primitive variables for these steps, $(\rho, \rho u, \rho E)^T \to (\rho, u, p)^T$ (where E is the total energy per unit mass) to produce the time centered values. One can then transform back to the conservative variables as needed and use the resulting solution to the Riemann problem to advance the solution in time. The method is also known as Hancock's method.

9.1.3 Characteristic Direct Eulerian

In many cases the best possible approach is to work with characteristic variables. This acts to separate the system of equations into a set of locally decoupled scalar equations. While this decomposition is only valid in a very localized sense in both space and time, it allows the equations to be discretized in a manner that is most appropriate for each wave in the system.

Whether the characteristics are derived directly from the conserved variables, $(\rho, \rho u, \rho E)^T$, or the primitive variables, $(\rho, u, p)^T$, the characteristics are equivalent. Before discussing the method it is useful to digress to the simple linear wave equation, $U_t + aU_x = 0$. For the sake of simplicity we will assume that $a > 0$, and that we are using a piecewise linear local polynomial expansion, $U(x)_j = U_j + S_j(x - x_j)$, where S_j is an approximation to $\partial U/\partial x$ It is useful to write the method down in a sequence of steps, first defining the edge extrapolated values as

$$U_j^n(x_j \pm \Delta x/2) = U_j^n \pm \frac{\Delta x}{2} S_j ,$$

and the time-centered value of

$$U_j^{n+1/2}(x_j \pm \Delta x/2) = U_j^n(x_j \pm \Delta x/2) - \frac{\Delta t\, a}{2} S_j .$$

This has made explicit use of the substitution, $U_t = -aU_x$. For further discussion we make a change in notation using $U_{j+1/2,L} = U_j(x_j + \Delta x/2)$ and $U_{j-1/2,R} = U_j(x_j - \Delta x/2)$. Because we have assumed that $a > 0$, the Riemann solution is simple and the update formula becomes,

$$U_j^{n+1} = U_j^n - \frac{a\,\Delta t}{\Delta x}\left(U_{j+1/2}^{n+1/2} - U_{j-1/2}^{n+1/2}\right) . \tag{9.6}$$

Notice that the ratio, $a\Delta t/\Delta x$ is the Courant number, $\mathcal{C}$.

The characteristic method is based on this approach as applied to the locally decoupled equations. One important caution is that the update must be applied to the fully coupled set of equations where the coupling of the fluxes is accomplished using a Riemann solver (approximate).

For systems of equations the algorithm proceeds along a similar path to the above scalar equation algorithm. First, the characteristic variables are defined for the entire stencil width using the left eigenvectors defined for the cell j,

$$\mathbf{W}_k^n - \mathbf{T}^{-1}\left(\mathbf{U}_j^n\right)\mathbf{U}_k^n, \quad k_{\min} \le k \le k_{\max} ,$$

where $k_{\min}$ and $k_{\max}$ define the stencil width. The reconstruction of the variables is then performed with the characteristic variables (for example piecewise linear)

$$\mathbf{W}_j(x) = \mathbf{W}_j + \mathbf{S}_j(x - x_j) .$$

Next, the edge- and time-centered values of the characteristic variables are given by,

$$\mathbf{W}_j^{n+1/2}(x + j \pm \Delta x/2) = \mathbf{W}_j^n \pm \left(\frac{\Delta x}{2} - \frac{\Delta t\, \boldsymbol{\lambda}_j}{2}\right)\mathbf{S}_j .$$

Here, the vector $\boldsymbol{\lambda}_j$ are the eigenvalues or characteristic speeds ordered appropriately for the characteristic variables. Finally, the right eigenvector is used to transform the variables back into their original form,

$$\mathbf{U}_j^{n+1/2}(x_j \pm \Delta x/2) = \mathbf{T}\left(\mathbf{U}_j^n\right)\mathbf{W}_{j\pm 1/2}^{n+1/2} .$$

The method then proceeds to use these values to solve the Riemann problem and update the cell quantities as before,

$$\begin{aligned}\mathbf{U}_j^{n+1} &= \mathbf{U}_j^n \\ &- \frac{\Delta t}{\Delta x}\left[\mathbf{E}\left(\mathbf{U}^{n+1/2}\left(x_j + \Delta x/2\right), \mathbf{U}^{n+1/2}\left(x_{j+1} - \Delta x/2\right)\right)\right. \\ &- \left.\mathbf{E}\left(\mathbf{U}^{n+1/2}\left(x_{j-1} + \Delta x/2\right), \mathbf{U}^{n+1/2}\left(x_j - \Delta x/2\right)\right)\right] .\end{aligned}$$

As a final bit of subtlety, the entire characteristic procedure can be cast in terms of the primitive variables. The advantage of the primitive variables is that the eigenvectors are generally much simpler while being equivalent to the eigenvectors defined in terms of the conserved variables.

9.1.4 Lagrange-Remap Approach

There is an older algorithm that is worth some mention due to its continued use, the Lagrange-Remap approach. Here, the equations are solved first in the Lagrangian or moving frame-of-reference and then remapped (or interpolated) back onto the original grid. These methods were used in the earliest versions of high-resolution Godunov methods by van Leer [571] and then with the PPM method of Colella and Woodward [120].

In Lagrangian coordinates the Euler equations can be put in conservation form by writing them as a function of the mass coordinate m defined as

$$m = \int \rho \, \mathrm{d}x \ , \quad \text{or } \mathrm{d}m = \rho \, \mathrm{d}m \ .$$

The system of equations is then

$$\frac{\partial V}{\partial t} - \frac{\partial u}{\partial m} = 0 \ , \tag{9.7}$$

$$\frac{\partial u}{\partial t} + \frac{\partial p}{\partial m} = 0 \ , \tag{9.8}$$

and

$$\frac{\partial E}{\partial t} + \frac{\partial pu}{\partial m} = 0 \ . \tag{9.9}$$

This system also has three characteristic speeds, $-C$, 0, and C, where $C^2 = \gamma p V$ is the Lagrangian sound speed for an ideal gas. The ideal gas equation of state in terms of the Lagrangian variables is $p = \left(E - 1/2u^2\right)(\gamma - 1)/V$. In this setting the Riemann solver defined in Fig. 9.1 is sufficient to provide an update to the equations (only p_* and u_* are needed). The remap can be cast directly as a set of advection equations (equivalent since the distance to be interpolated is equal to $u_* \Delta t$),

$$V_j^{n+1} = V_j^n + \frac{\Delta t}{\rho_j \Delta x}\left(u_{*j+1/2}^{n+1/2} - u_{*j-1/2}^{n+1/2}\right) , \tag{9.10}$$

$$u_j^{n+1} = u_j^n - \frac{\Delta t}{\rho_j \Delta x}\left(p_{*j+1/2}^{n+1/2} - p_{*j-1/2}^{n+1/2}\right), \tag{9.11}$$

and

$$E_j^{n+1} = E_j^n - \frac{\Delta t}{\rho_j \Delta x}\left(p_{*j+1/2}^{n+1/2} u_{*j+1/2}^{n+1/2} - p_{*j-1/2}^{n+1/2} u_{*j-1/2}^{n+1/2}\right), \tag{9.12}$$

The schematic of the different configurations possible is shown in Fig. 9.3. Both the Lagrange and remap steps are differenced using the same methods discussed earlier in this chapter. We note that the remap takes place on an intrinsically uneven mesh even when the original mesh is evenly spaced.

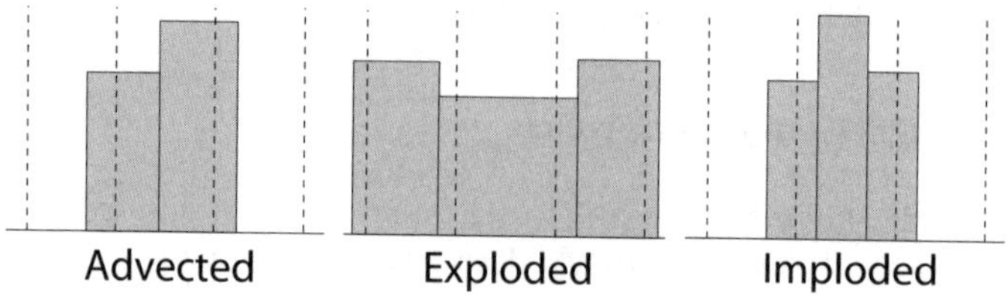

Fig. 9.3. The major configurations for remapping in the Lagrange-Remap approach depending on the divergence of velocity. The dashed lines are the cell face positions.

It is important to recognize that the arbitrary Lagrangian-Eulerian methods [260] use high resolution methods in their remap phase. These methods are highly capable solvers for the compressible flow equations. The history and basic methodology of ALE methods is well covered in Benson's review article [53]. We will not consider these methods further here and one should consult recent articles and reports [417, 12, 13] for further details.

9.2 Reconstructing the Fluxes

Another major approach for high-resolution methods is to spatially difference the fluxes. In this case, the fluxes are first computed and then differenced to produce the fundamental update formula for the conservations laws,

$$\mathbf{U}_j^{n+1} = \mathbf{U}_j^n - \frac{\Delta t}{\Delta x}\left(\mathbf{E}_{j+1/2} - \mathbf{E}_{j-1/2}\right) . \tag{9.13}$$

These fluxes must ultimately incorporate the same type of information as the differencing of the dependent variables as well as the Riemann solvers.

As Merriman has pointed out [385] there is a fundamental difference with a flux reconstruction. Implicit in the presentation of the explicit algorithm is that the cell variables, $\mathbf{U}_j$, are averaged over a cell (control volume), thus

$$\mathbf{U}_j = \int_{x_j-\Delta x/2}^{x_j+\Delta x/2} \mathbf{U}(x)\ \mathrm{d}x = \mathcal{A}\left[\mathbf{U}(x)\right] .$$

In this case one can write the effective partial differential equation as

$$\frac{\partial \mathcal{A}(\mathbf{U})}{\partial t} + \frac{\mathbf{E}\left(\mathbf{U}_{j+1/2}\right) - \mathbf{E}\left(\mathbf{U}_{j-1/2}\right)}{\Delta x} = 0 \,. \tag{9.14}$$

Thus given cell-averages of $\mathbf{U}$ we reconstruct values at the cell boundaries to evaluate fluxes.

Flux reconstruction literally inverts this process. The update equation now reads,

$$\frac{\partial \mathbf{U}}{\partial t} + \frac{\mathcal{A}^{-1}\mathbf{E}\left(\mathbf{U}_{j+1/2}\right) - \mathcal{A}^{-1}\mathbf{E}\left(\mathbf{U}_{j-1/2}\right)}{\Delta x} = 0 \,. \tag{9.15}$$

where $\mathcal{A}^{-1}$ is the (abstract) inverse of the integral averaging operator and $\mathbf{U}$ denotes approximate point values at x_j. The fluxes themselves are reconstructed using the same algorithmic principles as the dependent variables were reconstructed.

9.2.1 Flux Splitting

In its most fundamental form the flux splitting algorithm is an upwind method. The basic idea is to produce fluxes that constrain either entirely left- or right-going information,

$$\mathbf{E}_j = \mathbf{E}_j^- + \mathbf{E}_j^+ \,,$$

where $\partial \mathbf{E}^-/\partial \mathbf{U} < 0$ and $\partial \mathbf{E}^+/\partial \mathbf{U} > 0$. In order to understand how to construct high resolution algorithms it is useful to reduce to our old standby, the linear advection equation. Recall the upwind flux introduced by (9.4) which can be rewritten as

$$E_{j+1/2} = \frac{1}{2}\left(a + |a|\right) U_j + \frac{1}{2}\left(a - |a|\right) U_{j+1} \,. \tag{9.16}$$

Similarly the system case (9.5) can be rewritten,

$$\begin{aligned} \mathbf{E}_{j+1/2} = {} & \frac{1}{2}\left(\mathbf{E}_j + \mathbf{T}_{j+1/2}\left|\boldsymbol{\Lambda}_{j+1/2}\right|\mathbf{T}_{j+1/2}^{-1}\mathbf{U}_j\right) \\ & + \frac{1}{2}\left(\mathbf{E}_{j+1} - \mathbf{T}_{j+1/2}\left|\boldsymbol{\Lambda}_{j+1/2}\right|\mathbf{T}_{j+1/2}^{-1}\mathbf{U}_{j+1}\right) \,. \end{aligned} \tag{9.17}$$

In an overall sense the equation for updating the equations in a first-order manner is

$$\mathbf{U}_j^{n+1} = \mathbf{U}_j^n - \frac{\Delta t}{\Delta x}\left(\mathbf{E}_j^+ + \mathbf{E}_{j+1}^- - \mathbf{E}_{j-1}^+ - \mathbf{E}_j^-\right) \,. \tag{9.18}$$

Making these methods into high-resolution methods requires more care. Again, let us begin with the scalar case. For the given stencil of a method, the fluxes are split into the forward and backward components,

$$E_k = E_k^- + E_k^+ , \quad k_{\min} < k < k_{\max} \,,$$

where k defines the number of the eigenvalues and eigenvectors. One then reconstructs the positive and negative fluxes at the cell boundaries. For numerical stability these reconstructions are biased by the transport directions. For example, one can take our prototypical case of a piecewise linear interpolant,

$$E_{j+1/2} = E_j^+ + \frac{1}{2}\,\Delta x\; S_j^+ + E_{j+1}^- - \frac{1}{2}\,\Delta x\; S_{j+1}^- \,.$$

In the case of a scalar equation this is completely equivalent to reconstructing the dependent variables followed by a Riemann solver. This will not be the case for a system of equations.

For a system of equations, the fluxes are split, but if the system is nonlinear this splitting must be local, and there will be one splitting for each cell interface. This splitting should also be conducted in characteristic variables, where the characteristic decomposition is a function of the dependent variables:

$$\mathbf{e}_k^- = \mathbf{T}^{-1}\left(\mathbf{U}_{j+1/2}\right)\mathbf{E}_k^- \;,\; k_{\min} < k < k_{\max} \;,$$

and

$$\mathbf{e}_k^+ = \mathbf{T}^{-1}\left(\mathbf{U}_{j+1/2}\right)\mathbf{E}_k^+ \;,\; k_{\min} < k < k_{\max} \,.$$

Here, the splitting is conducted on the basis of the eigenvalues of $\partial\mathbf{E}/\partial\mathbf{U}$ evaluated at $j + 1/2$. The reconstructions are then performed in the same biased fashion as for the scalar case,

$$\mathbf{e}_{j+1/2}^+ = \mathbf{e}_j^+ + \frac{1}{2}\,\Delta x\;\mathbf{s}_j^+ \,,$$

and

$$\mathbf{e}_{j+1/2}^- = \mathbf{e}_{j+1}^- + \frac{1}{2}\,\Delta x\;\mathbf{s}_{j+1}^- \,.$$

The flux is then transformed back to physical variables using

$$\mathbf{E}_{j+1/2} = \mathbf{T}\left(\mathbf{U}_{j+1/2}\right)\left(\mathbf{e}_{j+1/2}^- + \mathbf{e}_{j+1/2}^+\right) \,.$$

9.2.2 Flux Splitting Time Integration

We will now describe several common forms of time integration applied to flux-splitting methods. This subject has been introduced in Chap. 7, but the development here is more specific to the flux-splitting approach as described in the literature. One is the method-of-lines approach typified by the TVD Runge-Kutta method [493]. The method is very straightforward and simple, the flux splitting defines a spatial operator,

$$\mathbf{L}\left(\mathbf{U}_j\right) = -\left(\mathbf{E}_{j+1/2} - \mathbf{E}_{j-1/2}\right) \,.$$

The time integration then uses this operator in conjunction with a series of stages to advance the flow time accurately. A starting point for advancing the equations is

$$\frac{\partial \mathbf{U}_j}{\partial t} = -\mathbf{L}_j, \tag{9.19}$$

The prototypical second-order example is Heun's method,

$$\mathbf{U}_j^1 = \mathbf{U}_j^n + \Delta t\, \mathbf{L}\left(\mathbf{U}_j^n\right) , \tag{9.20}$$

and

$$\mathbf{U}_j^{n+1} = \frac{1}{2}\left(\mathbf{U}_j^n + \mathbf{U}_j^1\right) + \frac{\Delta t}{2}\mathbf{L}\left(\mathbf{U}_j^1\right) . \tag{9.21}$$

Perhaps the most common method is the third-order method,

$$\mathbf{U}_j^1 = \mathbf{U}_j^n + \Delta t\mathbf{L}\left(\mathbf{U}_j^n\right) , \tag{9.22}$$

$$\mathbf{U}_j^2 = \frac{1}{3}\mathbf{U}_j^n + \frac{2}{3}\mathbf{U}_j^1 + \frac{2\Delta t}{3}\mathbf{L}\left(\mathbf{U}_j^n\right) , \tag{9.23}$$

and

$$\mathbf{U}_j^{n+1} = \frac{3}{4}\mathbf{U}_j^n + \frac{1}{4}\mathbf{U}_j^2 + \frac{\Delta t}{4}\mathbf{L}\left(\mathbf{U}_j^2\right) . \tag{9.24}$$

The other approach is the Lax-Wendroff differencing [251, 434]. In this case, one more directly approximates a Taylor series for the time accuracy,

$$\mathbf{U}\left(t + \Delta t\right) = \mathbf{U}\left(t\right) + \Delta t\frac{\partial \mathbf{U}}{\partial t} + \frac{\Delta t^2}{2}\frac{\partial^2 \mathbf{U}}{\partial t^2} + \frac{\Delta t^3}{6}\frac{\partial^3 \mathbf{U}}{\partial t^3} + \ldots \tag{9.25}$$

Using a Taylor series expansion and substituting space derivatives the next term in the expansion is found,

$$\frac{\partial^2 \mathbf{U}}{\partial t^2} = -\frac{\partial\left(\mathbf{E}_{\mathbf{U}}\mathbf{U}_t\right)}{\partial x} , \tag{9.26}$$

where $\mathbf{E}_{\mathbf{U}} = \partial \mathbf{E}/\partial \mathbf{U}$. This term can be computed with centered spatial derivatives of an order of one less than that of the flux reconstruction. The next term is similar,

$$\frac{\partial^3 \mathbf{U}}{\partial t^3} = -\frac{\partial\left(\mathbf{E}_{\mathbf{U}}\mathbf{U}_{tt} + \mathbf{E}_{\mathbf{U}\mathbf{U}}\left(\mathbf{U}_t\right)^2\right)}{\partial x} . \tag{9.27}$$

This term can be computed with spatial derivatives of two orders less than that of the flux reconstruction.

9.3 Preconditioning for Low Speed Flows

9.3.1 Overview of Preconditioning Techniques

Preconditioning techniques in CFD aim at overcoming stiffness in the solution of the Euler and Navier-Stokes equations. In respect of preconditioning techniques there are two main streams of research. Firstly, the development of preconditioners for low-Mach number and incompressible flows [102, 553, 554, 576, 258]. The artificial compressibility method by Chorin [104] for incompressible flows can also be viewed as a preconditioning technique. Secondly, the design of preconditioners for alleviating discrete stiffness in the Euler and Navier-Stokes equations. This can be achieved by a variety of approaches such as clustering high frequency eigenvalues away from the origin and provide rapid damping by a multistage scheme [424], directional coarsening multigrid algorithms [393] and implicit ADI (alternating direction implicit) preconditioners [6, 88]. The second category of preconditioners aims mainly at improving the solution efficiency of the compressible Euler and Navier-Stokes equations (see also discussion on multigrid methods in Chaps. 8 and 10).

The idea of preconditioning was originally used to accelerate the solution of elliptic equations. For example, if one considers the algebraic system obtained after discretizing the governing PDEs

$$\mathbf{A}\mathbf{u} = \mathbf{b} \,, \tag{9.28}$$

where $\mathbf{A}$ is a positive definite matrix, the solution can be accelerated by replacing (9.28) with

$$\mathbf{P}\mathbf{A}\mathbf{u} = \mathbf{P}\mathbf{b} \,, \tag{9.29}$$

where $\mathbf{P}$ is the preconditioning matrix (or simply the *preconditioner*) and should be nonsingular in order to obtain the same solutions for (9.29) and (9.28) [554]. The requirement for designing efficient preconditioners is $\kappa(\mathbf{PA}) << \kappa(\mathbf{A})$, where $\kappa(\mathbf{A}) = \max|\lambda_i|/|\lambda_j|$ for all the eigenvalues of $\mathbf{A}$. Also, $\mathbf{P}$ should be easily inverted.

There are three main types of preconditioners [554]:

- Jacobi preconditioning: In this case the preconditioner is a diagonal matrix (block-diagonal for a system of equations) and introduces a Jacobi iteration to the solution.
- Differential preconditioner: $\mathbf{P}$ or $\mathbf{P}^{-1}$ has non-zero elements in the same positions as the non-zero elements of $\mathbf{A}$.
- Global preconditioner which does not satisfy any of the above conditions. In this case the preconditioner can be based on the algebraic properties of the matrix $\mathbf{A}$ without considering the connection of $\mathbf{A}$ and the differential equation.

Turkel [554] has shown that an algorithm developed for the compressible equations does not (necessarily) result in an accurate algorithm for the incompressible equations unless preconditioning is implemented. The difficulty in obtaining accurate and efficient solutions for the compressible flow equations at low Mach numbers is associated with the large disparity in the acoustic wave speed $u + c$ and the waves convected by the flow velocity u. This disparity can be removed if the time derivative of the compressible flow equations is multiplied by the preconditioning matrix to reduce the acoustic wave speed to the flow velocity. In the case of elliptic equations, the preconditioner is implemented to the inversion matrix of the system of equations in order to accelerate the convergence.

9.3.2 Preconditioning Choices for Compressible Flows

We consider the time-dependent Euler equations in terms of the primitive variables with respect to the variables (p, u, v, S) with S being the entropy, $\mathrm{d}S = \mathrm{d}p - c^2 \mathrm{d}\rho$,

$$\frac{\partial p}{\partial t} + \left(u\frac{\partial p}{\partial x} + v\frac{\partial p}{\partial y}\right) + (\rho c^2)\left(\frac{\partial u}{\partial x} + \frac{\partial v}{\partial y}\right) = 0\,, \tag{9.30}$$

$$\frac{\partial u}{\partial t} + u\frac{\partial u}{\partial x} + v\frac{\partial u}{\partial y} + \frac{1}{\rho}\frac{\partial p}{\partial x} = 0\,, \tag{9.31}$$

$$\frac{\partial v}{\partial t} + u\frac{\partial v}{\partial x} + v\frac{\partial v}{\partial y} + \frac{1}{\rho}\frac{\partial p}{\partial y} = 0\,, \tag{9.32}$$

$$\frac{\partial S}{\partial t} + u\frac{\partial S}{\partial x} + v\frac{\partial S}{\partial y} = 0\,. \tag{9.33}$$

The system can be written in two-dimensional matrix preconditioned form

$$\mathbf{P}^{-1}\frac{\partial \mathbf{W}}{\partial t} + \mathbf{A}\frac{\partial \mathbf{W}}{\partial x} + \mathbf{B}\frac{\partial \mathbf{W}}{\partial y} = 0\,, \tag{9.34}$$

where $\mathbf{P}$ is the preconditioning matrix; $\mathbf{A}$ and $\mathbf{B}$ are the Jacobian matrices;[2] and $\mathbf{W}$ is the solution vector defined as $\mathbf{W} = (p, u, v, S)$. The choice of $\mathbf{W}$ is not unique and the Jacobian matrices depend on it. The choice of $\mathbf{W} = (p, u, v, S)$ [556, 552, 553] yields

$$\mathbf{A} = \begin{pmatrix} u & \rho c^2 & 0 & 0 \\ 1/\rho & u & 0 & 0 \\ 0 & 0 & u & 0 \\ 0 & 0 & 0 & u \end{pmatrix}, \quad \mathbf{B} = \begin{pmatrix} v & 0 & \rho c^2 & 0 \\ 0 & v & 0 & 0 \\ 1/\rho & 0 & v & 0 \\ 0 & 0 & 0 & v \end{pmatrix}. \tag{9.35}$$

[2] Strictly speaking, $\mathbf{A}$ and $\mathbf{B}$ are the Jacobi matrices and $|\mathbf{A}|$ and $|\mathbf{B}|$ are their Jacobians. In the CFD literature, it has been established $\mathbf{A}$ and $\mathbf{B}$ to be called Jacobians thus this term will be used throughout the book.

There have been proposed various choices for the preconditioner **P** [552, 553, 556, 102, 103, 576, 103, 386, 577]. To satisfy that the problem is well-posed , **P** must have an inverse as well as be positive definite. Generalization of the preconditioners proposed in [552, 553] and [103] was suggested by Turkel et al. [556] according to which the matrices **P** and $\mathbf{P}^{-1}$ are defined as

$$\mathbf{P} = \begin{pmatrix} \tilde{\beta}\tilde{M}_r^2/c^2 & 0 & 0 & -\tilde{\beta}\tilde{M}_r^2\delta/c^2 \\ -\alpha u/(\rho c^2) & 1 & 0 & \alpha u\delta/(\rho c^2) \\ -\alpha v/(\rho c^2) & 0 & 1 & \alpha v\delta/(\rho c^2) \\ 0 & 0 & 0 & 1 \end{pmatrix} , \tag{9.36}$$

$$\mathbf{P}^{-1} = \begin{pmatrix} c^2/(\tilde{\beta}\tilde{M}_r^2) & 0 & 0 & \delta \\ -\alpha u/(\rho\tilde{\beta}\tilde{M}_r^2) & 1 & 0 & 0 \\ -\alpha v/(\rho\tilde{\beta}\tilde{M}_r^2) & 0 & 1 & 0 \\ 0 & 0 & 0 & 1 \end{pmatrix} . \tag{9.37}$$

The parameters δ and α are free parameters which are defined by numerical experiments.[3] For $\delta = 1$ and $\alpha = 0$, one obtains the preconditioner of [102]. For $\delta = 0$, we obtain Turkel's preconditioner [553]. To reduce the number of parameters involved in the preconditioning, one can combine $\tilde{\beta}$ and $\tilde{M}_r^2$ and define directly $\tilde{\beta}\tilde{M}_r^2$ as one parameter [556]. If the parameter $\tilde{\beta}\tilde{M}_r^2$ is very small then there will be numerical difficulties with inverting the preconditioning matrix since it will essentially become a singular matrix. Further, the values of $\tilde{\beta}\tilde{M}_r^2$ should not become much smaller than the flow velocity [129, 554]; a complete theory to explain this has not yet been proposed. Choi and Merkle [103] suggest M_r^2 to be defined as function of the local Mach number, $M = \sqrt{u^2+v^2}/c$. A direct definition for $\tilde{\beta}\tilde{M}_r^2$ is given in [552, 553]:

$$\tilde{\beta}\tilde{M}_r^2 = \min\Big(\max(\hat{K}_1(u^2+v^2), \hat{K}_2(u_{\mathrm{ref}}^2 + v_{\mathrm{ref}}^2)), c^2\Big) , \tag{9.38}$$

where

$$\hat{K}_1 = K_1\Big[1 + \frac{(1 - K_1 M_{\mathrm{ref}}^2)}{K_1 M_{\mathrm{ref}}^4} M^2\Big] . \tag{9.39}$$

In [553], the free stream velocity and Mach number are used as free parameters.[4] The coefficient K_1 takes values between 1 and 1.1. The coefficient K_2 depends on α, the grid size near the stagnation point [556, 552] and may

[3] Note that different symbols are used in [103, 556]. In relation to [556], here we have replaced β and M_r^2 of [556] with $\tilde{\beta}$ and $\tilde{M}_r^2$, respectively.

[4] The parameter M_0 used in [556] should actually be M_∞ or M_{ref} as in (9.38).

also depend on the local (cell) Reynolds number [103]. Considering simultaneously $\tilde{\beta}\tilde{M}^2 > M_{\text{ref}}$, and $\alpha = 0$, the problem becomes free of preconditioning thus avoiding problems near shock waves. It is recommended to use no preconditioning for transonic and supersonic flows [556]. To avoid numerical difficulties for $M = 0$, the parameter K_2 is utilized to prevent very small values of $\tilde{\beta}\tilde{M}^2$ by assigning to it a value as a percentage of the inflow speed. The values of K_2 are obtained by numerical experimentation and the success of the implementation depends on the user's experience. Note that in [553] the corresponding β^2 (that is equivalent to $\tilde{\beta}\tilde{M}_r^2$ in (9.38)) has been defined as $\beta^2 = \max\Big(\min(K_1 M^2, \beta^2_{\text{min}}, 1)\Big)$, where $\beta^2_{\text{min}} = K_2 M^2_{\text{ref}}$ and $K_1 \sim 1$. At stagnation flow conditions, this would result in $\beta^2 = 1$; the formula should be

$$\beta^2 = \min\Big(\max(K_1 M^2, \beta^2_{\text{min}}, 1)\Big) , \tag{9.40}$$

which leads to the dependence of β^2 on the free stream flow conditions. An alternative definition for β_{min} has been proposed in [130] according to which β_{min} depends on local pressure gradients.

The matrix $\mathbf{PA}$ is written as

$$\mathbf{PA} = \begin{pmatrix} \dfrac{\tilde{\beta}\tilde{M}_r^2 u}{c^2} & \rho\tilde{\beta}\tilde{M}_r^2 & 0 & -\dfrac{\tilde{\beta}\tilde{M}_r^2 u\delta}{c^2} \\ \dfrac{1}{\rho}\Big(1 - \dfrac{\alpha u^2}{\rho c^2}\Big) & (1-\alpha)u & 0 & \dfrac{\alpha u^2 \delta}{\rho c^2} \\ -\dfrac{\alpha u v}{\rho c^2} & -\alpha v & u & \dfrac{\alpha u v \delta}{\rho c^2} \\ 0 & 0 & 0 & u \end{pmatrix} , \tag{9.41}$$

and the eigenvalues are defined by

$$\lambda_0 = u , \tag{9.42}$$

$$\lambda_1 = \frac{1}{2}\Big[fu - \sqrt{d^2u^2 + 4\tilde{\beta}M_r^2(1 - \frac{u^2}{c^2})}\Big] , \tag{9.43}$$

$$\lambda_2 = \frac{1}{2}\Big[fu + \sqrt{d^2u^2 + 4\tilde{\beta}M_r^2(1 - \frac{u^2}{c^2})}\Big] , \tag{9.44}$$

where $f = 1 - \alpha + \tilde{\beta}M_r^2/c^2$. The right eigenvector is given by

$$\mathbf{R} = \begin{pmatrix} \rho\tilde{\beta}\tilde{M}_r^2 & \rho\tilde{\beta}\tilde{M}_r^2 & 0 & 0 \\ \lambda_2 - \dfrac{\tilde{\beta}M_r^2}{c^2}u & \lambda_1 - \dfrac{\tilde{\beta}M_r^2}{c^2}u & 0 & \dfrac{u\delta}{\rho c^2} \\ -\dfrac{\alpha uv\lambda_2}{u-\lambda_2} & \dfrac{\alpha uv\lambda_1}{u-\lambda_1} & 1 & 0 \\ 0 & 0 & 0 & 1 \end{pmatrix} . \tag{9.45}$$

The preconditioner presented above was defined for the set of variables $\mathbf{W} = (p, u, v, S)$, but can also be derived for other sets of variables using the formula

$$\mathbf{P}_n = \frac{\partial \mathbf{W}_n}{\partial \mathbf{W}} \, \mathbf{P} \, \frac{\partial \mathbf{W}}{\partial \mathbf{W}_n} , \tag{9.46}$$

where the variables $\mathbf{W}_n$ can be any of the following

$$\mathbf{W}_1 = (\rho, \rho u, \rho v, E)^T , \tag{9.47}$$
$$\mathbf{W}_2 = (p, \rho u, \rho v, E)^T , \tag{9.48}$$
$$\mathbf{W}_3 = (p, \rho u, \rho v, E - h_\infty \rho)^T , \tag{9.49}$$
$$\mathbf{W}_4 = (p, u, v, T)^T . \tag{9.50}$$

In the above, $E = p/\left[\rho(\gamma-1)\right] + 0.5(u^2+v^2)$, T is the temperature and h_∞ is the far field enthalpy (free stream enthalpy). The derivatives of (9.46) are given by

$$\frac{\partial \mathbf{W}_1}{\partial \mathbf{W}} = \begin{pmatrix} \dfrac{1}{c^2} & 0 & 0 & -\dfrac{1}{c^2} \\ \dfrac{u}{c^2} & \rho & 0 & -\dfrac{u}{c^2} \\ \dfrac{v}{c^2} & 0 & \rho & -\dfrac{v}{c^2} \\ \dfrac{1}{\gamma-1} + \dfrac{M^2}{2} & \rho u & \rho v & -\dfrac{M^2}{2} \end{pmatrix} , \tag{9.51}$$

$$\frac{\partial \mathbf{W}}{\partial \mathbf{W}_1} = \begin{pmatrix} \frac{\gamma-1}{2}(u^2+v^2) & (1-\gamma)u & (1-\gamma)v & \gamma-1 \\ -\frac{u}{\rho} & \frac{1}{\rho} & 0 & 0 \\ -\frac{v}{\rho} & 0 & \frac{1}{\rho} & 0 \\ \frac{\gamma-1}{2}(u^2+v^2)-c^2 & (1-\gamma)u & (1-\gamma)v & \gamma-1 \end{pmatrix}, \quad (9.52)$$

$$\frac{\partial \mathbf{W}_2}{\partial \mathbf{W}} = \begin{pmatrix} 1 & 0 & 0 & 0 \\ \frac{u}{c^2} & \rho & 0 & -\frac{u}{c^2} \\ \frac{v}{c^2} & 0 & \rho & -\frac{v}{c^2} \\ \frac{1}{\gamma-1}+\frac{M^2}{2} & \rho u & \rho v & -\frac{M^2}{2} \end{pmatrix}, \quad (9.53)$$

$$\frac{\partial \mathbf{W}}{\partial \mathbf{W}_2} = \begin{pmatrix} 1 & 0 & 0 & 0 \\ -\frac{u}{(\gamma-1)Q^2} & \frac{v^2-u^2}{2Q^2} & -\frac{uv}{Q^2} & \frac{u}{Q^2} \\ -\frac{v}{(\gamma-1)Q^2} & -\frac{uv}{Q^2} & \frac{u^2-v^2}{2Q^2} & \frac{v}{Q^2} \\ 1-\frac{2}{(\gamma-1)M^2} & -2\frac{u}{M^2} & -2\frac{v}{M^2} & \frac{2}{M^2} \end{pmatrix}, \quad (9.54)$$

where $Q^2 = \rho(u^2+v^2)/2$.

$$\frac{\partial \mathbf{W}_3}{\partial \mathbf{W}} = \begin{pmatrix} 1 & 0 & 0 & 0 \\ \frac{u}{c^2} & \rho & 0 & -\frac{u}{c^2} \\ \frac{v}{c^2} & 0 & \rho & -\frac{v}{c^2} \\ \frac{1}{\gamma} - 1 + \frac{M^2}{2} - \frac{h_\infty}{c^2} & \rho u & \rho v & -\frac{M^2}{2} + \frac{h_\infty}{c^2} \end{pmatrix}, \quad (9.55)$$

$$\frac{\partial \mathbf{W}}{\partial \mathbf{W}_3} = \begin{pmatrix} 1 & 0 & 0 & 0 \\ -\frac{u}{(\gamma-1)\rho G} & \frac{G-u^2}{\rho G} & -\frac{uv}{\rho G} & \frac{u}{\rho G} \\ -\frac{v}{(\gamma-1)\rho G} & -\frac{uv}{\rho G} & \frac{G-v^2}{\rho G} & -\frac{v}{\rho G} \\ 1-\frac{c^2}{(\gamma-1)G} & -\frac{uc^2}{G} & -\frac{vc^2}{G} & -\frac{c^2}{G} \end{pmatrix}, \quad (9.56)$$

where $G = (u^2 + v^2)/2 + h_\infty$.

$$\frac{\partial \mathbf{W}_4}{\partial \mathbf{W}} = \begin{pmatrix} 1 & 0 & 0 & 0 \\ 0 & 1 & 0 & 0 \\ 0 & 0 & 1 & 0 \\ \frac{(\gamma-1)T}{\gamma p} & 0 & 0 & \frac{T}{\gamma p} \end{pmatrix}, \quad (9.57)$$

$$\frac{\partial \mathbf{W}}{\partial \mathbf{W}_4} = \begin{pmatrix} 1 & 0 & 0 & 0 \\ 0 & 1 & 0 & 0 \\ 0 & 0 & 1 & 0 \\ 1-\gamma & 0 & 0 & \frac{\gamma p}{T} \end{pmatrix}. \quad (9.58)$$

The above choices of preconditioner aim at enabling the compressible Euler equations to be solved at low Mach numbers. Van Leer et al. [576] proposed to diagonalize the two-dimensional equations for supersonic flow if a preconditioner is included in the system. In terms of the variables $\mathbf{W} = (p, u, v, S)^T$, Van Leer's et al. preconditioner is written

$$\mathbf{P} = \begin{pmatrix} \frac{\hat{\tau}}{\hat{\beta}^2}M^2 & -\frac{\hat{\tau}}{\hat{\beta}^2}\frac{u}{c} & -\frac{\hat{\tau}}{\hat{\beta}^2}\frac{v}{c} & 0 \\ -\frac{\hat{\tau}}{\hat{\beta}^2}\frac{u}{c} & \left(\frac{\hat{\tau}}{\hat{\beta}^2}+1\right)q_u + \hat{\tau}q_v & \left(\frac{\hat{\tau}}{\hat{\beta}^2}+1\right)q_{uv} & 0 \\ -\frac{\hat{\tau}}{\hat{\beta}^2}\frac{v}{c} & \left(\frac{\hat{\tau}}{\hat{\beta}^2}+1\right)q_{uv} & \left(\frac{\hat{\tau}}{\hat{\beta}^2}+1\right)q_v + \hat{\tau}q_u & 0 \\ 0 & 0 & 0 & 1 \end{pmatrix}, \quad (9.59)$$

where

$$q_u = \frac{u^2}{u^2+v^2}\,, \quad (9.60)$$

$$q_v = \frac{v^2}{u^2+v^2}\,, \quad (9.61)$$

$$q_{uv} = \frac{uv}{u^2+v^2}\,. \quad (9.62)$$

and

$$\hat{\beta} = \begin{cases} \sqrt{1-M^2}\,, & M < 1 \\ \sqrt{M^2-1}\,, & M \geq 1\,, \end{cases} \quad (9.63)$$

$$\hat{\tau} = \begin{cases} \sqrt{1-M^2}\,, & M < 1 \\ \sqrt{1-M^{-2}}\,, & M \geq 1\,, \end{cases} \quad (9.64)$$

The preconditioner (9.59) is symmetric thus leads to a well-posed problem at least for non-zero Mach numbers.

9.3.3 Preconditioning of Numerical Dissipation

The analysis presented in the preceding section concerns preconditioning of the system of PDEs in order to obtain convergence at low Mach numbers. Further investigations [555, 235, 234, 554, 278, 395, 557, 558, 594] have shown that to obtain/improve accuracy at low Mach numbers, the artificial viscosity of centered-based schemes and the numerical dissipation of Riemann solvers should also be preconditioned. Turkel et al. [555] has shown that the elements of the artificial viscosity matrix in x-direction (similarly in the other two directions) should be (for two-dimensional problems in terms of **W** variables)

$$\begin{pmatrix} \mathcal{O}(1/M^2) & \mathcal{O}(1/M) & \mathcal{O}(1/M) & \mathcal{O}(1) \\ \mathcal{O}(1/M) & \mathcal{O}(1) & \mathcal{O}(1) & \mathcal{O}(1) \\ \mathcal{O}(1/M) & \mathcal{O}(1) & \mathcal{O}(1) & \mathcal{O}(1) \\ \mathcal{O}(1) & \mathcal{O}(1) & \mathcal{O}(1) & \mathcal{O}(1) \end{pmatrix} . \tag{9.65}$$

It has been proved [292] that for an initial pressure field that scales with the square of the Mach number and for an initial velocity field that approaches the divergence free (velocity) field, the solutions to compressible flow equations remain uniformly bounded as the Mach number approaches the zero limit and that the limit solutions satisfy the equations for incompressible flows. Direct relevance to this topic also have the studies in [237, 305, 87] which show general treatment of hyperbolic problems featuring different time scales.

Many of the high-resolution methods discussed in this book are extension of the first-order Godunov method [215]. The advective flux derivative $\partial\mathbf{E}/\partial x$ is discretized at the center of the control volume (i, j) using the values of the intercell fluxes, i.e., $\partial\mathbf{E}/\partial x = (\mathbf{E}_{i+1/2,j} - \mathbf{E}_{i-1/2,j})/\Delta x$. The first-order Godunov flux (in x-direction) is written

$$\mathbf{E}_{i+1/2,j} = \frac{1}{2}(\mathbf{E}_{i,j} + \mathbf{E}_{i+1,j}) - \frac{1}{2}|\mathbf{A}|(\mathbf{U}_{i+1,j} - \mathbf{U}_{i,j}) , \tag{9.66}$$

where the second term on the RHS is the wave-speed dependent term ($\mathbf{A}$ is the Jacobian matrix). Guillard and Viozat [235, 234] have shown, using Roe's scheme [463], that low Mach number scaling [292] in numerical solutions can be achieved if the wave-speed dependent term $|\mathbf{A}|\Delta\mathbf{U}$ ($\Delta\mathbf{U} = \mathbf{U}_{i+1,j} - \mathbf{U}_{i,j}$ for a first-order method) is replaced by $\mathbf{P}^{-1}|\mathbf{PA}|\Delta\mathbf{U}$, where $\mathbf{P}$ is the preconditioning matrix. By doing so, the wave-speed dependent term associated with the continuity and momentum equations is augmented by a factor $1/M$. Bearing in mind that the original system of equations (without preconditioning) contains terms of $\mathcal{O}(1/M)$, the preconditioned system will now contain terms of $\mathcal{O}(1/M^2)$ and $\mathcal{O}(1)$ thus leading to pressure solutions which are scaled with M^2. They showed that the above treatment of the upwind scheme does not require any further modification of the first term on the RHS of (9.66) or the time-dependent term of the equations. A general way to combine preconditioning with a TVD approximation is also discussed in [560] and is applied to subsonic and supersonic flows at thermochemical equilibrium.

Preconditioning for both the artificial viscosity term (in centered non-high-resolution schemes) and the PDE can be used and the preconditioning matrices do not have necessarily to be the same. The one-dimensional system (preconditioned using $\mathbf{P}$) including the preconditioned artificial viscosity term (using a preconditioning matrix Q) on the RHS is written

$$\mathbf{P}^{-1}\frac{\partial\mathbf{W}}{\partial t} + \mathbf{A}\frac{\partial\mathbf{W}}{\partial x} = \tilde{c}\frac{\partial}{\partial x}\left(\mathbf{Q}^{-1}|\mathbf{QA}|\frac{\partial\mathbf{W}}{\partial x}\right) , \tag{9.67}$$

where the term on the RHS is the artificial viscosity term ($\tilde{c}$ is a constant). **Q** and **P** may be the same, but other choices are also possible. For example, both matrices can retain the same form using, however, different $\tilde{\beta}$ [554].

9.3.4 Differential Preconditioners

So far we have discussed preconditioners in the form of matrices which are based on the solution variables. These preconditioners cannot address the issue of numerical efficiency in a computational mesh in which different waves co-exist simultaneously. For a scalar equation

$$\frac{\partial u}{\partial t} + a\frac{\partial u}{\partial x} + b\frac{\partial u}{\partial y} = 0 \, , \tag{9.68}$$

where a and b are constants, one defines the aspect ratio as

$$\frac{a/\Delta x}{b/\Delta y} \, . \tag{9.69}$$

This is the time that one wave requires to travel in the x-direction relative to the time in the $y-$direction. This should ideally be equal or close to 1 in order to obtain fast convergence. In this case the time step chosen will be the same appropriate for all directions.

In a system of equations there are many waves, e.g., acoustic, shear and entropy waves, and although the mesh may be appropriate for one family of waves may not be appropriate for the other. One possibility is to define a preconditioner as the absolute value of **A** thus resulting in waves of the same speed. This is possible in one dimension but not in two or three dimensions, especially when the Jacobian matrices do not commute.

Differential preconditioners try to alleviate the above difficulty by defining them to have as elements derivatives of the dependent variables. Let us consider the compressible Euler equations, using the variables $\mathbf{W} = (p, u, v, S)^T$, written in the form

$$\frac{\partial \mathbf{W}}{\partial t} + \mathbf{LW} = 0 \, , \tag{9.70}$$

where the matrix **L** is defined by [553]

$$\mathbf{L} = \begin{pmatrix} \mathcal{Q} & \rho c^2 \partial_x & \rho c^2 \partial_y & 0 \\ (1/\rho)\partial_x & \mathcal{Q} & 0 & 0 \\ (1/\rho)\partial_y & 0 & \mathcal{Q} & 0 \\ 0 & 0 & 0 & \mathcal{Q} \end{pmatrix} . \tag{9.71}$$

The term $\mathcal{Q}$ is given by

$$\mathcal{Q} = u\partial_x + v\partial_y \, , \tag{9.72}$$

where $\partial_x \equiv \partial/\partial x$ and $\partial_y \equiv \partial/\partial y$.

The system (9.70) can be preconditioned as

$$\frac{\partial \mathbf{W}}{\partial t} + \mathbf{P}_d \mathbf{L} \mathbf{W} = 0 \,, \tag{9.73}$$

where $\mathbf{P}_d$ is the differential preconditioner

$$\mathbf{P}_d = \begin{pmatrix} \mathcal{Q}^2 & -\rho c^2 \partial_x \mathcal{Q} & -\rho c^2 \partial_y \mathcal{Q} & 0 \\ -(1/\rho)\partial_x & \mathcal{Q}^2 - c^2 \partial_y^2 & c^2 \partial_x \partial_y & 0 \\ -(1/\rho)\partial_y & c^2 \partial_x \partial_y & \mathcal{Q}^2 - c^2 \partial_x^2 & 0 \\ 0 & 0 & 0 & \mathcal{D} \end{pmatrix} . \tag{9.74}$$

where $\mathcal{D} = \mathcal{Q}^2 - c^2(\partial_x^2 + \partial_y^2)$. The above matrices satisfy the following equalities

$$\mathbf{P}_d \mathbf{L} = \mathcal{Q}\mathcal{D}\mathbf{I} \,, \quad \text{and} \quad \mathbf{P}_d^{-1} = \mathcal{D}^{-1}\mathcal{Q}^{-1}\mathbf{L} \,. \tag{9.75}$$

Alternative differential preconditioners have also been proposed [553, 554]. For example, if one considers the variables $\mathbf{W} = c(dp/\rho c, u, v, S)$ then the matrix $\mathbf{L}$ in (9.70) is written as

$$\mathbf{L} = \begin{pmatrix} \mathcal{Q} & c\,\partial_x & c\,\partial_y & 0 \\ c\,\partial_x & \mathcal{Q} & 0 & 0 \\ c\,\partial_y & 0 & \mathcal{Q} & 0 \\ 0 & 0 & 0 & \mathcal{Q} \end{pmatrix} . \tag{9.76}$$

A differential preconditioner for (9.76) can be defined by

$$\mathbf{P}_e = \begin{pmatrix} \mathcal{Q} & -c\,\partial_x & -c\,\partial_y & 0 \\ -c\,\partial_x & 1 + c^2\partial_x^2 & c^2\partial_x\partial_y & 0 \\ -c\,\partial_y & c^2\partial_x\partial_y & 1 + c^2\partial_y^2 & 0 \\ 0 & 0 & 0 & 1 \end{pmatrix} . \tag{9.77}$$

The matrix (9.77) can be obtained as

$$\mathbf{P}_e = \boldsymbol{\mathcal{P}}^T \boldsymbol{\mathcal{P}} \,, \tag{9.78}$$

where

$$\boldsymbol{\mathcal{P}} = \begin{pmatrix} \mathcal{Q} & -c\,\partial_x & -c\,\partial_y & 0 \\ 0 & 1 & 0 & 0 \\ 0 & 0 & 1 & 0 \\ 0 & 0 & 0 & 1 \end{pmatrix} , \tag{9.79}$$

and

$$\mathcal{P}^T = \begin{pmatrix} \mathcal{Q} & 0 & 0 & 0 \\ -c\,\partial_x & 1 & 0 & 0 \\ -c\,\partial_y & 0 & 1 & 0 \\ 0 & 0 & 0 & 1 \end{pmatrix} . \tag{9.80}$$

The matrices $\mathcal{P}^T$ and $\mathcal{P}$ have the property that can diagonalize $\mathbf{L}$ by a congruent transformation, i.e.,

$$\mathcal{P}\mathbf{L}\mathcal{P}^T = \begin{pmatrix} \mathcal{D}\mathcal{Q} & 0 & 0 & 0 \\ 0 & \mathcal{Q} & 0 & 0 \\ 0 & 0 & \mathcal{Q} & 0 \\ 0 & 0 & 0 & \mathcal{Q} \end{pmatrix} . \tag{9.81}$$

Because $\mathcal{P}\mathbf{L}\mathcal{P}^T = (\mathcal{P}^T)^{-1}(\mathcal{P}^T P\mathbf{L})\mathcal{P}$, the congruent transform is similar to a preconditioning with a positive definite matrix $\mathbf{P}_e = \mathcal{P}^T\mathcal{P}$ that contains fewer derivatives than $\mathbf{P}_d$ [553].

Another form of differential preconditioner is the residual smoothing [276]. Residual smoothing is applied directly to the residuals of the equation by solving

$$\Big(1 - c_1\frac{\partial^2}{\partial x^2}\Big)\Big(1 - c_2\frac{\partial^2}{\partial y^2}\Big)\mathcal{R}_{\text{new}} = \mathcal{R}_{\text{old}} , \tag{9.82}$$

where $\mathcal{R}_{\text{old}}$ and $\mathcal{R}_{\text{new}}$ are the residuals before and after smoothing. The parameters c_1 and c_2 are constants which can be defined on the basis of stability analysis for the time stepping scheme. Residual smoothing can be used in conjunction with explicit schemes and multigrid methods to improve the time step values and smoothing properties of the solution.

10. The Artificial Compressibility Method

The main difficulty with the solution of the incompressible flow equations is the decoupling of the continuity and momentum equations due to the absence of the pressure (or density) term from the former. As was mentioned in previous chapters, one can obtain a Poisson equation for the pressure. This equation can be used for calculating the pressure field, given a velocity field that satisfies the incompressibility condition both inside the computational domain as well as at the boundaries. Chorin [104] proposed an entirely different approach in order to overcome the difficulty of the pressure decoupling. This approach is called *artificial compressibility* (AC). In this chapter, we present the artificial compressibility method, the eigenstructure of the system of fluid flow equations, preconditioning, explicit and implicit solvers, as well as the implementation of multigrid techniques in conjunction with the AC method.

10.1 Basic Formulation

For steady flows, Chorin [104] introduced the auxiliary system of equations

$$\frac{1}{\beta}\frac{\partial p}{\partial \tau} + \frac{\partial u_j}{\partial x_j} = 0 \,, \tag{10.1}$$

$$\frac{\partial u_i}{\partial t} + \frac{\partial u_i u_j}{\partial x_j} = -\frac{1}{\rho}\frac{\partial p}{\partial x_i} + \nu \frac{\partial^2 u_i}{\partial x_j^2} \,, \tag{10.2}$$

where β is the artificial compressibility (AC) parameter; for the case of steady flow $\tau \equiv t$. The parameter β is a disposable parameter, analogous to a relaxation parameter, which enables the system of (10.1)-(10.2) to converge to a solution that satisfies the incompressibility condition.

The above system also has similarities with the equations of motion for a compressible fluid at low Mach numbers. Making this analogy, we can relate the artificial compressibility parameter to an artificial speed of sound

$$c = \sqrt{\beta} \,. \tag{10.3}$$

The artificial compressibility method results in hyperbolic and hyperbolic-parabolic equations for inviscid and viscous incompressible flows, respectively, which are often less computationally expensive to be solved than elliptic equations.

Temam [530] has shown that for fixed β (10.1) and (10.2) can also be used for time-dependent problems. He considered the system

$$\frac{1}{\beta}\frac{\partial p}{\partial \tau} + \frac{\partial \widetilde{u}_j}{\partial x_j} = 0 \,, \tag{10.4}$$

$$\frac{\partial \widetilde{u}_i}{\partial t} + \frac{\partial \widetilde{u}_i \widetilde{u}_j}{\partial x_j} + \frac{1}{2}\widetilde{u}_i \frac{\partial \widetilde{u}_i}{\partial x_i} = -\frac{1}{\rho}\frac{\partial p}{\partial x_i} + \nu \frac{\partial^2 \widetilde{u}_i}{\partial x_j^2} \,. \tag{10.5}$$

and showed that under suitable hypotheses, solutions of $\widetilde{u}_i$ exist and that these solutions converge to a solution of u_i in the limit $1/\beta \to 0$. The term $1/2(\widetilde{u}_i \partial \widetilde{u}_i / \partial x_i)$ is a stabilization term that Temam used in order to carry out his analysis.

The artificial-compressibility method based on (10.1) and (10.2) can be used for steady flows. The extension of the method to unsteady flows is obtained by (5.26) and (5.27).

The discretization schemes and solvers developed for artificial compressibility have many similarities with the methods developed for compressible flows. Therefore, numerical developments gained from compressible flows can be transferred to incompressible flows.

10.2 Convergence to the Incompressible Limit

Chang and Kwak [100] have shown that the solution obtained using the artificial compressibility formulation converges to the incompressible limit as the steady state is approached. Let us assume, for simplicity, the one dimensional pseudo-compressible equations in dimensionless form

$$\frac{1}{\beta}\frac{\partial p}{\partial t} + \frac{\partial u}{\partial x} = 0 \,, \tag{10.6}$$

$$\frac{\partial u}{\partial t} + 2u\frac{\partial u}{\partial x} = -\frac{\partial p}{\partial x} + \frac{1}{Re}\frac{\partial^2 u}{\partial x^2} \,. \tag{10.7}$$

Following [100] the velocity and pressure can be expressed by the sum of a steady-state incompressible solution, u_s and p_s, and the unsteady velocity, $\widetilde{u}$, and pressure, $\widetilde{p}$, introduced as the result of the pseudo-compressible unsteady system (10.6) and (10.7), i.e.,

$$u(x,t) = u_s(x) + \widetilde{u}(x,t) \,, \quad p(x,t) = p_s(x) + \widetilde{p}(x,t) \,. \tag{10.8}$$

Note that

- u_s satisfies the incompressibility condition $\partial u_s / \partial x = 0$.

- u_s satisfies in steady state the momentum equation

$$2u_s \frac{\partial u_s}{\partial x} = -\frac{\partial p_s}{\partial x} + \frac{1}{Re}\frac{\partial^2 u_s}{\partial x^2} \,. \tag{10.9}$$

- The terms $\widetilde{u}$ and $\widetilde{p}$ are small compared to their steady state counterparts and, therefore, linearization with respect to $\widetilde{u}$ can be allowed.

By substituting (10.8) into (10.6) and (10.7) and making use of the above points, we obtain a linearized system of equations for $\widetilde{u}$ and $\widetilde{p}$

$$\frac{1}{\beta}\frac{\partial \widetilde{p}}{\partial t} + \frac{\partial \widetilde{u}}{\partial x} = 0 \,, \tag{10.10}$$

$$\frac{\partial \widetilde{u}}{\partial t} + 2u_s \frac{\partial \widetilde{u}}{\partial x} = -\frac{\partial \widetilde{p}}{\partial x} + \frac{1}{Re}\frac{\partial^2 \widetilde{u}}{\partial x^2} \,. \tag{10.11}$$

Cross-differentiation of (10.10) and (10.11) will eliminate the coupling of velocity and pressure and the following equations can be obtained

$$\frac{\partial^2 p}{\partial t^2} + 2u_s \frac{\partial^2 p}{\partial t \partial x} - \beta \frac{\partial^2 p}{\partial x^2} = \beta \frac{1}{Re}\frac{\partial^3 \widetilde{u}}{\partial x^3} \,, \tag{10.12}$$

$$\frac{\partial^2 u}{\partial t^2} + 2u_s \frac{\partial^2 u}{\partial t \partial x} - \beta \frac{\partial^2 u}{\partial x^2} = -\beta \frac{1}{Re}\frac{\partial^3 \widetilde{u}}{\partial x^2 \partial t} \,. \tag{10.13}$$

The system of (10.12) and (10.13) can also be written as

$$\left[\frac{\partial}{\partial t} + (u_s + c)\frac{\partial}{\partial x}\right]\left[\frac{\partial}{\partial t} + (u_s - c)\frac{\partial}{\partial x}\right]\begin{pmatrix} p \\ u \end{pmatrix} = \begin{pmatrix} \beta \dfrac{1}{Re}\dfrac{\partial^3 \widetilde{u}}{\partial x^3} \\ \\ -\beta \dfrac{1}{Re}\dfrac{\partial^3 \widetilde{u}}{\partial x^2 \partial t} \end{pmatrix} , \tag{10.14}$$

where the pseudo-speed of sound is given by

$$c = \sqrt{u_s^2 + \beta} \,. \tag{10.15}$$

Considering positive and negative waves travelling downstream and upstream with velocities $u_s + c$ and $u_s - c$, respectively, approximate solutions to (10.14) can be formulated as follows [100]

$$p^+ \text{or} u^+ \approx \mathcal{F}\Big\{\alpha[x - (u_s + c)t]\Big\} \exp\Big[-\frac{\alpha^2 t}{2Re}(1 + M^2)\Big] \,, \tag{10.16}$$

$$p^- \text{or} u^- \approx \mathcal{G}\Big\{\alpha[x - (u_s - c)t]\Big\} \exp\Big[-\frac{\alpha^2 t}{2Re}(1 - M^2)\Big] \,, \tag{10.17}$$

where α is the wave number and M is a pseudo Mach number defined by

$$M = \frac{u_s}{\sqrt{u_s^2 + \beta}} < 1 \,. \tag{10.18}$$

The pseudo Mach number is always less than one for all $\beta > 0$ and, therefore, the pseudo waves vanish as time progresses. The rate of decay will depend on the wave number.

The above analysis shows that the artificial compressibility formulation asymptotically converges to the incompressible limit when the steady state is reached.

10.3 Preconditioning and the Artificial Compressibility Method

The artificial compressibility method can be considered as a kind of preconditioning method to march the incompressible equations to a steady state solution. Both the artificial compressibility and other preconditioners that were discussed in Chap. 9.3 refer to symmetric hyperbolic systems, which lead to a well-posed problem for steady state solutions. For such systems a positive-definite preconditioning matrix ensures that the number of boundary conditions remains unchanged as well as that the time direction has not been reversed; in this case all the boundary conditions should change.

A generalization of the system (10.1)-(10.2) has been proposed by Turkel [552] by replacing (10.2) with[1]

$$\frac{\alpha u_i}{\beta}\frac{\partial p}{\partial t} + \frac{\partial u_i}{\partial t} + u_j \frac{\partial u_i}{\partial x_j} = -\frac{1}{\rho}\frac{\partial p}{\partial x_i} + \nu \frac{\partial^2 u_i}{\partial x_j^2} , \tag{10.19}$$

in non-conservative form, or with

$$\frac{(\alpha+1)}{\beta} u_i \frac{\partial p}{\partial t} + \frac{\partial u_i}{\partial t} + \frac{\partial u_i u_j}{\partial x_j} = -\frac{1}{\rho}\frac{\partial p}{\partial x_i} + \nu \frac{\partial^2 u_i}{\partial x_j^2} , \tag{10.20}$$

in "conservative" form[2].

Let us consider the nonconservative form (10.19) in conjunction with the continuity equation (10.1) for inviscid two-dimensional flows and write it in a matrix form as

$$\mathbf{P}^{-1}\frac{\partial \mathbf{U}}{\partial t} + \boldsymbol{\mathcal{A}}\frac{\partial \mathbf{U}}{\partial x} + \boldsymbol{\mathcal{B}}\frac{\partial \mathbf{U}}{\partial y} = 0 , \tag{10.21}$$

where $\mathbf{U} = (p, u, v)^T$ and

$$\mathbf{P}^{-1} = \begin{pmatrix} 1/\beta & 0 & 0 \\ \alpha u/\beta & 1 & 0 \\ \alpha v/\beta & 0 & 1 \end{pmatrix} , \tag{10.22}$$

[1] Turkel's [552] β^2 is the the same with the artificial parameter β used throughout this book.

[2] The system is not truly conservative for time-dependent flows, but it turns to a conservative system when steady flow is obtained.

$$\boldsymbol{\mathcal{A}} = \begin{pmatrix} 0 & 1 & 0 \\ 1 & u & 0 \\ 0 & 0 & u \end{pmatrix} , \tag{10.23}$$

$$\boldsymbol{\mathcal{B}} = \begin{pmatrix} 0 & 0 & 1 \\ 0 & v & 0 \\ 1 & 0 & v \end{pmatrix} . \tag{10.24}$$

After multiplication with $\mathbf{P}$ (from the left), (10.21) gives

$$\frac{\partial \mathbf{U}}{\partial t} + \boldsymbol{\mathcal{A}}\mathbf{P}\frac{\partial \mathbf{U}}{\partial x} + \boldsymbol{\mathcal{B}}\mathbf{P}\frac{\partial \mathbf{U}}{\partial y} = 0 . \tag{10.25}$$

The system (10.25) is hyperbolic since its eigenvalues are real. For example, the eigenvalues of $\boldsymbol{\mathcal{A}}\mathbf{P}$ are

$$\lambda_1 = u , \quad \lambda_{2,3} = \frac{(1-\alpha)u \pm \sqrt{(1-\alpha)^2 u^2 + 4\beta}}{2} . \tag{10.26}$$

These are always real numbers thus the system is hyperbolic. The parameter β can be defined by (9.40) (Chap. 9.3; β^2 of (9.40) is the same with β here) or, alternatively, by [552]

$$\beta = \begin{cases} (u^2+v^2)(2-\alpha) , & \alpha < 1 \\ (u^2+v^2)\alpha , & \alpha \geq 1 , \end{cases} \tag{10.27}$$

In [552], results for the above formulation were not presented. More recently, numerical experiments using the above preconditioning formulation have been performed [416] and showed that this formulation may have strong dissipative effects on the predictions of flows that feature separation and instabilities.

In the next section we will discuss the eigenstructure of the equations for the original artificial compressibility method [104] which can be obtained by (10.19) (for $\alpha = 0$) and (10.20) (for $\alpha = -1$), respectively, in conjunction with (10.1).

10.4 Eigenstructure of the Incompressible Equations

We consider the one dimensional inviscid counterpart of the Navier-Stokes equations in a Cartesian co-ordinate system. Using the artificial compressibility method the system is written in a matrix form as

$$\frac{\partial \mathbf{U}}{\partial t} + \frac{\partial \mathbf{E}}{\partial x} = 0 , \tag{10.28}$$

where

$$\mathbf{U} = \begin{pmatrix} p \\ u \\ v \end{pmatrix} \; ; \quad \mathbf{E} = \begin{pmatrix} \beta u \\ u^2 + p \\ uv \end{pmatrix} . \tag{10.29}$$

The Jacobian matrix is given by

$$\mathbf{A} = \frac{\partial \mathbf{E}}{\partial \mathbf{U}} = \begin{pmatrix} 0 & \beta & 0 \\ 1 & 2u & 0 \\ 0 & v & u \end{pmatrix} , \tag{10.30}$$

and it can be easily found that the eigenvalues are

$$\lambda_0 = u \, , \quad \lambda_1 = u + c \, , \quad \lambda_2 = u - c \tag{10.31}$$

where c is a pseudo-speed of sound

$$c = \sqrt{u^2 + \beta} \, . \tag{10.32}$$

Since the eigenvalues of the system are real and distinct, the system is hyperbolic.

For the most general case of a three-dimensional curvilinear system that includes motion of the grid, the inviscid incompressible equations in conjunction with the artificial compressibility are written as

$$\frac{\partial \bar{\mathbf{U}}}{\partial t} + \frac{\partial \bar{\mathbf{E}}}{\partial \xi} + \frac{\partial \bar{\mathbf{F}}}{\partial \eta} + \frac{\partial \bar{\mathbf{G}}}{\partial \zeta} = 0 \, , \tag{10.33}$$

where

$$\bar{\mathbf{U}} = J\mathbf{U} = J(p, u, v, w)^T \, , \tag{10.34}$$

$$\bar{\mathbf{E}} = J(\mathbf{E}\xi_x + \mathbf{F}\xi_y + \mathbf{G}\xi_z) \, , \tag{10.35}$$

$$\bar{\mathbf{F}} = J(\mathbf{E}\eta_x + \mathbf{F}\eta_y + \mathbf{G}\eta_z) \, , \tag{10.36}$$

$$\bar{\mathbf{G}} = J(\mathbf{E}\zeta_x + \mathbf{F}\zeta_y + \mathbf{G}\zeta_z) \, , \tag{10.37}$$

The matrices $\bar{\mathbf{E}}, \bar{\mathbf{F}}$ and $\bar{\mathbf{G}}$ are given by

$$\boldsymbol{\mathcal{E}} = J \begin{pmatrix} \beta Q \\ uk_t + pk_x + uQ \\ vk_t + pk_y + vQ \\ wk_t + pk_z + wQ \end{pmatrix} , \tag{10.38}$$

where $\boldsymbol{\mathcal{E}} = \bar{\mathbf{E}}$ and $\bar{\mathbf{F}}, \bar{\mathbf{G}}$ for $k = \xi, \eta$ and ζ, respectively, and

$$Q = uk_x + vk_y + wk_z \,. \tag{10.39}$$

The Jacobian $\mathbf{A}$ of the flux vector $\boldsymbol{\mathcal{E}}$ is given by

$$\mathbf{A} = \begin{pmatrix} 0 & \beta k_x & \beta k_y & \beta k_z \\ k_x & Q + uk_x + k_t & uk_y & uk_z \\ k_y & vk_x & Q + vk_y + k_t & vk_z \\ k_z & wk_x & wk_y & Q + wk_z + k_t \end{pmatrix} , \tag{10.40}$$

The matrix (10.40) can also be written, using a similarity transform, as

$$\mathbf{A} = \mathbf{T} \boldsymbol{\Lambda} \mathbf{T}^{-1} \,, \tag{10.41}$$

where $\boldsymbol{\Lambda}$ is a diagonal matrix containing the eigenvalues

$$\boldsymbol{\Lambda} = \begin{pmatrix} \lambda_0 & 0 & 0 & 0 \\ 0 & \lambda_0 & 0 & 0 \\ 0 & 0 & \lambda_1 & 0 \\ 0 & 0 & 0 & \lambda_2 \end{pmatrix} , \tag{10.42}$$

The eigenvalues are given by

$$\lambda_0 = Q + k_t \,, \quad \lambda_1 = Q + \frac{k_t}{2} + c \,, \quad \text{and } \lambda_2 = Q + \frac{k_t}{2} - c \,. \tag{10.43}$$

where

$$c = \sqrt{(Q + k_t/2)^2 + \beta(k_x^2 + k_y^2 + k_z^2)} \,. \tag{10.44}$$

The right $\mathbf{T}$ and left $\mathbf{T}^{-1}$ eigenvector matrices are given by

$$\mathbf{T} = \begin{pmatrix} 0 & 0 & \beta(c - k_t/2) & -\beta(c + k_t/2) \\ x_l & x_n & u\lambda_1 + \beta k_x & u\lambda_2 + \beta k_x \\ y_l & y_n & v\lambda_1 + \beta k_y & v\lambda_2 + \beta k_y \\ z_l & z_n & w\lambda_1 + \beta k_z & w\lambda_2 + \beta k_z \end{pmatrix} , \tag{10.45}$$

$$\mathbf{T}^{-1} = \frac{1}{c^2 - k_t^2/4} \begin{pmatrix} t_{11} & t_{12} & t_{13} & t_{14} \\ t_{21} & t_{22} & t_{23} & t_{24} \\ t_{31} & t_{32} & t_{33} & t_{34} \\ t_{41} & t_{42} & t_{43} & t_{44} \end{pmatrix} , \tag{10.46}$$

where

$$\left.\begin{aligned}
t_{11} &= x_n(vk_z - wk_y) + y_n(wk_x - uk_z) + z_n(uk_y - vk_x) \\
t_{12} &= y_n(\lambda_1 w + \beta k_z) + z_n(\lambda_1 v + \beta k_y) \\
t_{13} &= z_n(\lambda_1 u + \beta k_x) - x_n(\lambda_1 w + \beta k_z) \\
t_{14} &= x_n(\lambda_1 v + \beta k_y) - y_n(\lambda_1 u + \beta k_x) \\
t_{21} &= x_l(wk_y - vk_z) + y_l(uk_z - wk_x) + z_l(vk_x - uk_y) \\
t_{22} &= -y_l(\lambda_1 w + \beta k_z) + z_l(\lambda_1 v + \beta k_y) \\
t_{23} &= -z_l(\lambda_1 u + \beta k_x) + x_l(\lambda_1 w + \beta k_z) \\
t_{24} &= -x_l(\lambda_1 v + \beta k_y) + y_l(\lambda_1 u + \beta k_x) \\
t_{31} &= -\lambda_2(c + k_t/2)/(2\beta c) \\
t_{32} &= k_x(c + k_t/2)/(2c) \\
t_{33} &= k_y(c + k_t/2)/(2c) \\
t_{34} &= k_z(c + k_t/2)/(2c) \\
t_{41} &= -\lambda_1(c - k_t/2)/(2\beta c) \\
t_{42} &= k_x(c - k_t/2)/(2c) \\
t_{43} &= k_y(c - k_t/2)/(2c) \\
t_{44} &= k_z(c - k_t/2)/(2c)
\end{aligned}\right\} . \tag{10.47}$$

The terms x_l and x_n are defined by

$$x_l = \frac{\partial x}{\partial \phi}, \quad x_n = \frac{\partial x}{\partial \psi}, \tag{10.48}$$

where $\phi = \eta, \zeta$ and ξ for the fluxes $\bar{\mathbf{E}}, \bar{\mathbf{F}}$ and $\bar{\mathbf{G}}$, respectively; and $\psi = \zeta, \xi$ and η for the fluxes $\bar{\mathbf{E}}, \bar{\mathbf{F}}$ and $\bar{\mathbf{G}}$, respectively.

10.5 Estimation of the Artificial Compressibility Parameter

The selection of the artificial compressibility parameter can affect the convergence of the computations in both steady and time-dependent problems. One possible way to determine optimum values for β is to derive a relation between pseudo-pressure waves and vorticity spreading [100].

Neglecting viscous effects in (10.14) and taking into account that at steady state $u_s \equiv u$, the characteristic equations are obtained

$$\left[\frac{\partial}{\partial t} + (u + c)\frac{\partial}{\partial x}\right]\begin{pmatrix} p^+ \\ u^+ \end{pmatrix} = 0, \tag{10.49}$$

$$\left[\frac{\partial}{\partial t} + (u-c)\frac{\partial}{\partial x}\right]\begin{pmatrix} p^- \\ u^- \end{pmatrix} = 0 \,. \tag{10.50}$$

Waves denoted by "+" propagate downstream with a velocity $u+c$ and waves denoted by "-" propagate upstream with a velocity $c-u$.

To establish a criterion for an optimum value of β, Chang and Kwak [100] employed the problem of internal flow in a channel of width x_r and length L (normalized by x_r). They considered the interaction of upstream propagating waves with vorticity spreading through the characteristic equation

$$\frac{\partial u^-}{\partial t} + (u-c)\frac{\partial u^-}{\partial x} = -\frac{\partial}{\partial t}(\nu\frac{\partial^2 u}{\partial x^2}) \equiv -\frac{\partial \tau_w}{\partial t} \,, \tag{10.51}$$

where τ_w is a viscous stress.

The rate of spreading for vorticity thickness, δ, in a laminar flow is approximately given by

$$\frac{\mathrm{d}\delta^2}{\mathrm{d}t} \simeq 4\nu = \frac{4}{Re} \,, \quad \text{where} \quad Re = \frac{ux_r}{\nu} \,. \tag{10.52}$$

Using the non-dimensionalization

$$\tilde{x} = \frac{x}{L} \,, \quad \tilde{t} = \frac{(u-c)t}{L} \,, \tag{10.53}$$

and defining $t_\nu = \frac{4}{Re}t$, such that $\frac{\mathrm{d}\delta^2}{\mathrm{d}t_\nu} \simeq 1$, (10.51) can be written

$$\frac{\partial u^-}{\partial \tilde{t}} + \frac{\partial u^-}{\partial \tilde{x}} = -\frac{4L}{(u-c)Re}\frac{\partial \tau_w}{\partial t_\nu} \,, \tag{10.54}$$

In (10.54) the variations of the wave with respect to $\tilde{t}$ and $\tilde{x}$, as well as the variation of the shear stress term with respect to t_ν, are of the order of 1. Therefore, the interactions of the waves and viscous effects can be completely decoupled if

$$\frac{4L}{(c-u)Re} << 1 \,. \tag{10.55}$$

Substituting the artificial speed of sound (10.32) into (10.55) and consider that the dimensionless flow speed is $u=1$, we obtain the following criterion for β

$$\beta >> \left(1 + \frac{4L}{Re}\right)^2 - 1 \,. \tag{10.56}$$

For a turbulent flow one obtains

$$\frac{\mathrm{d}\delta^2}{\mathrm{d}t} \simeq \nu_t = \frac{1}{Re_t} \,, \tag{10.57}$$

where ν_t is the eddy viscosity and $Re_t = u\delta/\nu_t$. Following a similar analysis, one can derive a similar criterion for turbulent flows [100], i.e.,

$$\beta >> \left(1 + \frac{L}{Re_t}\right)^2 - 1 \tag{10.58}$$

Eqs. (10.56) and (10.58) set a possible lower limit for the parameter β. The upper limit will depend on the numerical algorithm.

Using the above criteria, Chang and Kwak [100] estimated a value of $\beta = 5$ and reported calculations for the flow in a turnaround duct similar to the hot gas manifold in the Space Shuttle main engine. They also pointed out that the values of β can affect more the convergence of an internal flow than an external one.

An alternative way to define the artificial compressibility parameter was discussed by McHugh and Ramshaw [381]. They employed a "damped" artificial compressibility scheme which is written in a differential form as

$$\frac{\partial p}{\partial \tau} = -\beta \frac{\partial u_j}{\partial x_j} \,, \tag{10.59}$$

$$\frac{\partial u_i}{\partial \tau} = -\frac{\partial u_i}{\partial t} - \frac{\partial u_i u_j}{\partial x_j} - -\frac{\partial}{\partial x_i}\left(\frac{p}{\rho} - b\frac{\partial u_i}{\partial x_j}\right) + \nu \frac{\partial^2 \widetilde{u}_i}{\partial x_j^2} \,. \tag{10.60}$$

where τ is the artificial time, t is the real time and $b > 0$ is an artificial bulk viscosity. These equations can be used for both steady and unsteady problems. In the unsteady case, one iterates the solution with respect to τ for each real time, until the derivatives $\partial/\partial\tau$ vanish.

McHugh and Ramshaw [381] conducted a linearized (differential) Fourier dispersion analysis for (10.59) and (10.60). The purpose of this analysis was to understand the wave propagation and damping characteristics of the scheme in a simplified setting obtained by neglecting the convective terms and considering an one-dimensional flow in $x-$direction. The solutions can be expressed as Fourier modes of the form

$$p = \hat{p} \exp\Big[\imath(\kappa x - \omega t)\Big] \,, \tag{10.61}$$

$$u = \hat{u} \exp\Big[\imath(\kappa x - \omega t)\Big] \,, \tag{10.62}$$

where κ is the wave number and ω is the angular frequency; the wavelength $\lambda = 2\pi/\kappa$ should be much larger than Δx for applying the continuous analysis. Further, the longest wavelength in a computational domain of length L is $\lambda = 2L$ which corresponds to a wave number $\kappa = \kappa_L = \pi/L$. By substituting (10.61) into (10.59) and (10.60), McHugh and Ramshaw [381] found that non-trivial solutions exist only when ω and κ are related by the dispersion relation

$$\frac{\omega}{\kappa} = -\imath a_0 \pm \sqrt{\beta - a_0} \,, \tag{10.63}$$

where a_0 is a critical value of β given by

$$a_0 = \frac{1}{2}\Big[\frac{1}{\kappa \Delta t} + (b+\nu)\kappa\Big]. \tag{10.64}$$

The imaginary and real parts of ω determine the rate of growth (or decay) of the Fourier mode and wave speed, $c = \sqrt{\beta - a_0}$, respectively. Note that the wave speed should vanish for $\beta < a_0$. Considering small viscosities, the magnitude of a_0 is mainly governed by Δt. When Δt is larger then a_0 becomes small and, therefore, $c \approx \sqrt{\beta}$. However, for a certain combination of β and κ there is a critical value of Δt below which a_0 exceeds β and therefore the artificial sound waves cease to exist [381]. It has, however, been found [436, 437, 438] that combination of artificial waves and numerical diffusion result in better convergence properties. To preserve the existence of artificial sound waves: $\beta > a_0$. The dominant term in (10.64) is inversely proportional to κ and, therefore, the largest value for a_0 occurs when κ takes its smallest value $\kappa = \pi/L$ [381]. Subsequently, the maximum a_0 is defined by

$$a_{0,max} = \frac{1}{2}\Big[\frac{1}{\pi \Delta t/L} + (b+\nu)\frac{\pi}{L}\Big] . \tag{10.65}$$

Although values of β greater than $a_{0,max}$ preserve the artificial waves, the resulting $\Delta\tau$ becomes smaller. However, as pointed out in [381], numerical experiments and heuristic convergence rate analysis indicate that even though $\Delta\tau$ becomes smaller for $\beta > a_{0,max}^2$, the convergence not only does not deteriorate but on the contrary is improved.

Based on the above arguments, McHugh and Ramshaw [381] proceeded further to propose values for β and b. They set $c \sim U$, where U is the maximum flow speed, and suggested that $\beta \sim a_{0,max}^2 + U^2$ or, alternatively

$$\beta = d_1(a_{0,max} + U^2) , \tag{10.66}$$

where d_1 is a constant of the order of unity.

Furthermore, for the bulk viscosity b they proposed

$$b = \max\Big(0, d_2 U \Delta x - \nu - \frac{L^2}{\pi^2 \Delta t}\Big) , \tag{10.67}$$

where d_2 is a constant of the order of unity.

10.6 Explicit Solvers for Artificial Compressibility

The artificial-compressibility formulation facilitates the implementation of both explicit and implicit schemes. Explicit formulations are very popular because they are easy to be programmed on both serial and parallel computer architectures.

A family of explicit Runge-Kutta schemes of various orders of accuracy was introduced by Shu and Osher [495]. These methods were described in detail in Chap. 7.

10.7 Implicit Solvers for Artificial Compressibility

Various implicit implementations in conjunction with the artificial compressibility approach have been proposed [89, 146, 252, 310, 376, 381, 81, 469, 470, 512, 526, 414]. The most widely used are variants of implicit approximate factorization and implicit-unfactored methods. These are presented below.

10.7.1 Time-Linearized (Euler) Implicit Scheme

Eq. (10.33) augmented by the viscous fluxes as have been defined by (4.33), can be written as

$$J\frac{\partial \mathbf{U}}{\partial t} = -\mathrm{RHS}(\mathbf{U}) \,, \tag{10.68}$$

where RHS for a two-dimensional problem is given by

$$\mathrm{RHS}(\mathbf{U}) = \frac{\partial \bar{\mathbf{E}}}{\partial \xi} + \frac{\partial \bar{\mathbf{F}}}{\partial \eta} - \frac{\partial \bar{\mathbf{R}}}{\partial \xi} - \frac{\partial \bar{\mathbf{S}}}{\partial \eta} \,. \tag{10.69}$$

First-order backward implicit discretization yields

$$J\frac{\Delta \mathbf{U}^n}{\Delta t} = -\mathrm{RHS}(\mathbf{U}^{n+1}) \,, \tag{10.70}$$

where $\Delta \mathbf{U}^n = \mathbf{U}^{n+1} - \mathbf{U}^n$, and the steady state is achieved by driving iteratively $\Delta \mathbf{U}^n$ to machine zero. The RHS of the last equation can be linearized around the time level n as [80]

$$\mathrm{RHS}(\mathbf{U}^{n+1}) = \mathrm{RHS}(\mathbf{U}^n) + \boldsymbol{\mathcal{S}}^n + \mathcal{O}(\Delta \mathbf{U}^n)^2 \,, \tag{10.71}$$

where $\boldsymbol{\mathcal{S}}^n(\Delta \mathbf{U}^n)$ contains the Jacobians of the inviscid and viscous flux vectors, i.e.,

$$\boldsymbol{\mathcal{S}}^n = \left(\frac{\partial \bar{\mathbf{E}}}{\partial \mathbf{U}}\right)^n \Delta \mathbf{U}^n_\xi + \left(\frac{\partial \bar{\mathbf{F}}}{\partial \mathbf{U}}\right)^n \Delta \mathbf{U}^n_\eta - \left(\frac{\partial \bar{\mathbf{R}}}{\partial \mathbf{U}}\right)^n \Delta \mathbf{U}^n_\xi - \left(\frac{\partial \bar{\mathbf{S}}}{\partial \mathbf{U}}\right)^n \Delta \mathbf{U}^n_\eta \,.$$

Using (10.71), (10.70) is written

$$\left(\frac{J \cdot \mathbf{I}}{\Delta t} + \boldsymbol{\mathcal{S}}^n\right)\Delta \mathbf{U}^n = -\mathrm{RHS}(\mathbf{U}^n) \,. \tag{10.72}$$

Equation (10.72) is a first-order accurate linearized implicit formulation. Second-order time accuracy can also be obtained if the time derivative in (10.68) is discretized using three-point backward time difference [469, 470]. Third-order discretization of the term $\boldsymbol{\mathcal{S}}^n$ increases the band width of the block-matrix inversion and destroys its diagonal dominance which is important since it facilitates the implementation of approximate matrix inversion methods [89].

10.7.2 Implicit Approximate Factorization Method

The approximate factorization method [42, 79] is an extension of the alternating direction implicit (ADI) method [144] to the system of the Euler and Navier-Stokes equations. Direct solution of (10.72) is expensive, especially for three-dimensional problems, thus an approximate factorization scheme [81] is a viable choice. An approximate factorization procedure for (10.72) yields

$$\Big[\boldsymbol{\mathcal{D}}+\boldsymbol{\mathcal{S}}_1^n(\cdot)\Big]\boldsymbol{\mathcal{D}}^{-1}\Big[\boldsymbol{\mathcal{D}}+\boldsymbol{\mathcal{S}}_2^n(\cdot)\Big]\Delta\mathbf{U}^n=-\mathrm{RHS}(\mathbf{U}^n)\,, \tag{10.73}$$

where the matrix $\boldsymbol{\mathcal{D}}$, and the difference operators $\boldsymbol{\mathcal{S}}_1^n(\cdot)$ and $\boldsymbol{\mathcal{S}}_2^n(\cdot)$ are functions of the flux approximations and flux Jacobians, and should satisfy

$$\boldsymbol{\mathcal{D}}(\cdot)+\boldsymbol{\mathcal{S}}_1^n(\cdot)+\boldsymbol{\mathcal{S}}_2^n(\cdot)=\frac{J\cdot\mathbf{I}}{\Delta t}(\cdot)+\boldsymbol{\mathcal{S}}^n(\cdot)\,. \tag{10.74}$$

With $\boldsymbol{\mathcal{D}},\boldsymbol{\mathcal{S}}_1^n$, and $\boldsymbol{\mathcal{S}}_2^n$ defined, the approximate factorization (10.73) can be solved using approximate factorization in two steps which are associated with lower and upper triangular matrices (labeled LU/AF)

$$\begin{aligned}\Big[\boldsymbol{\mathcal{D}}+\boldsymbol{\mathcal{S}}^n{}_1(\cdot)\Big]\Delta\mathbf{U}^*&=-\mathrm{RHS}(\mathbf{U}^n)\,,\\ \Big[\boldsymbol{\mathcal{D}}+\boldsymbol{\mathcal{S}}_2^n(\cdot)\Big]\Delta\mathbf{U}^n&=\boldsymbol{\mathcal{D}}\Delta\mathbf{U}^*\,.\end{aligned} \tag{10.75}$$

For a two-dimensional problem, Briley et al. [81] have specified $\boldsymbol{\mathcal{D}},\boldsymbol{\mathcal{S}}_1^n$, and $\boldsymbol{\mathcal{S}}_2^n$ as (see Fig. 10.1 for grid notation)

$$\boldsymbol{\mathcal{D}}(\cdot)=\Big[\frac{J\cdot\mathbf{I}}{\Delta t}+\alpha(\tilde{\mathbf{A}}^+_{i,j}-\tilde{\mathbf{A}}^-_{i,j})+\alpha(\tilde{\mathbf{B}}^+_{i,j}-\tilde{\mathbf{B}}^-_{i,j})\Big](\cdot)_{i,j}\,, \tag{10.76}$$

$$\boldsymbol{\mathcal{S}}_1^n(\cdot)=(1-\alpha)(\tilde{\mathbf{A}}^+_{i,j}+\tilde{\mathbf{B}}^+_{i,j})(\cdot)_{i,j}-\tilde{\mathbf{A}}^+_{i-1,j}(\cdot)_{i-1,j} \tag{10.77}$$
$$-\tilde{\mathbf{B}}^+_{i,j-1}(\cdot)_{i,j-1}\,, \tag{10.78}$$

$$\boldsymbol{\mathcal{S}}_2^n(\cdot)=(1-\alpha)(-\tilde{\mathbf{A}}^-_{i,j}-\tilde{\mathbf{B}}^-_{i,j})(\cdot)_{i,j}+\tilde{\mathbf{A}}^-_{i+1,j}(\cdot)_{i+1,j} \tag{10.79}$$
$$+\tilde{\mathbf{B}}^-_{i,j+1}(\cdot)_{i,j+1}\,. \tag{10.80}$$

The Jacobians $\tilde{\mathbf{A}}^\pm$ and $\tilde{\mathbf{B}}^\pm$ are defined by

$$\tilde{\mathbf{A}}^\pm=\frac{\partial\mathbf{e}_{i\pm1/2,j}}{\partial\mathbf{U}_{i,j}}\,, \tag{10.81}$$
$$\tilde{\mathbf{B}}^\pm=\frac{\partial\mathbf{f}_{i,j\pm1/2}}{\partial\mathbf{U}_{i,j}}\,, \tag{10.82}$$

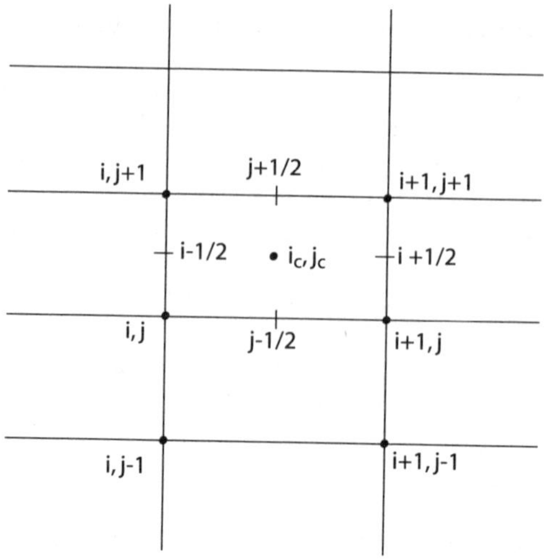

Fig. 10.1. Grid notation used in the implicit discretization. The primitive variables can be stored on the cell center or on the cell vertices. The description of the implicit approximate factorization in the text is based on the definition of the variables on the cell vertices.

where $\mathbf{e} = \bar{\mathbf{E}} + \bar{\mathbf{R}}$ and $\mathbf{f} = \bar{\mathbf{F}} + \bar{\mathbf{S}}$. The inviscid fluxes depend on a two- or a four-point stencil in terms of $\mathbf{U}$, for first and higher order approximations, respectively, while the viscous fluxes depend on a six-point stencil in terms of $\mathbf{U}$. Although the flux Jacobians are required at each point in the grid stencil, it has been found [81] that the linearization can be confined to a stencil depending only on $\mathbf{U}_i$ and $\mathbf{U}_{i+1}$.

The parameter α $(0 \le \alpha \le 1)$ is adjusted to accelerate convergence. For $\alpha = 0$ and $\alpha = 1$ the above scheme is also referred to as *two-pass* and *modified two-pass* method [81].

The Jacobians can be computed either analytically (see Sec. 10.4) or numerically [406]. For example, the numerical approximation of the kth column of $\mathbf{A}$ is given by

$$\tilde{\mathbf{A}}^{\pm}_{i,j} = \frac{\mathbf{e}_{i\pm1/2,j}(\mathbf{U}_{i,j} + \epsilon \mathbf{e}_k) - \mathbf{e}_{i\pm1/2,j}(\mathbf{U}_{i,j})}{\epsilon} , \tag{10.83}$$

where $\mathbf{e}_k$ is the kth unit vector and ϵ is a small number ($\epsilon = 10^{-6} \div 10^{-5}$).

10.7.3 Implicit Unfactored Method

Eq. (10.72) in two dimensions (ξ and η) can be written

$$\frac{\Delta(J\mathbf{U})^n}{\Delta t} + (\mathbf{A}^n_{inv}\Delta\mathbf{U})_\xi + (\mathbf{B}^n_{inv}\Delta\mathbf{U})_\eta - (\mathbf{A}^n_{vis}\Delta\mathbf{U})_\xi - (\mathbf{B}^n_{vis}\Delta\mathbf{U})_\eta = -\mathrm{RHS}(\mathbf{U}^n) \, , \tag{10.84}$$

where the subscripts inv and vis denote inviscid and viscous flux Jacobians, respectively.
A Newton-type method can be obtained if a sequence of approximations $\mathbf{q}^\nu$, such that $\lim_{\nu>1}\mathbf{q}^\nu \to \mathbf{U}^{n+1}$ is defined between two time steps n and $n+1$, respectively.[3] Equation (10.84) is then written

$$\frac{\Delta(J\mathbf{q}^{\nu+1})}{\Delta t} + (\mathbf{A}^n_{inv}\Delta\mathbf{q})_\xi + (\mathbf{B}^n_{inv}\Delta\mathbf{q})_\eta - (\mathbf{A}^n_{vis}\Delta\mathbf{q})_\xi - (\mathbf{B}^n_{vis}\Delta\mathbf{q})_\eta = J\frac{\mathbf{U}^n - \mathbf{q}^\nu}{\Delta t} - \mathrm{RHS}(\mathbf{U}^n) \, , \tag{10.85}$$

where

$$\mathbf{q}^{\nu+1} = \mathbf{q}^\nu + \Delta\mathbf{q}^{\nu+1} \, . \tag{10.86}$$

The superscript ν denotes Newton sub-iterations.

Second-order accurate discretizations for $(\mathbf{A}^n_{inv}\Delta\mathbf{U})_\xi$, $(\mathbf{B}^n_{inv}\Delta\mathbf{U})_\eta$ and $(\mathbf{A}^n_{vis}\Delta\mathbf{U})_\xi$, $(\mathbf{B}^n_{vis}\Delta\mathbf{U})_\eta$ have been proposed in [29, 152, 146, 483] in conjunction with the compressible Navier-Stokes equations. Similar discretizations can be used in the case of the incompressible flow equations. For example, the terms in the ξ-direction can be discretized as,

$$(\mathbf{A}^n_{inv}\Delta\mathbf{q})_\xi - (\mathbf{A}^n_{vis}\Delta\mathbf{q})_\xi = (\mathbf{A}^n_{inv}\Delta\mathbf{q})_{i+1/2,j} - (\mathbf{A}^n_{inv}\Delta\mathbf{q})_{i-1/2,j} - (\mathbf{A}^n_{vis})_{i,j}(\Delta\mathbf{q}_{i-1,j} - 2\Delta\mathbf{q}_{i,j} + \Delta\mathbf{q}_{i+1,j}) \, , \tag{10.87}$$

where the superscript $\nu+1$ has been dropped from $\Delta\mathbf{q}$, for simplicity,[4] and

$$(\mathbf{A}^n_{inv}\Delta\mathbf{q})_{i+1/2,j} = (\mathbf{T}\Lambda^+\mathbf{T}^{-1})_{i+1/2,j}(\Delta\mathbf{q})^+_{i+1/2,j} + (\mathbf{T}\Lambda^-\mathbf{T}^{-1})_{i+1/2,j}(\Delta\mathbf{q})^-_{i+1/2,j} \, , \tag{10.88}$$

$$(\mathbf{A}^n_{inv}\Delta\mathbf{q})_{i-1/2,j} = (\mathbf{T}\Lambda^+\mathbf{T}^{-1})_{i-1/2,j}(\Delta\mathbf{q})^+_{i-1/2,j} + (\mathbf{T}\Lambda^-\mathbf{T}^{-1})_{i-1/2,j}(\Delta\mathbf{q})^-_{i-1/2,j} \, . \tag{10.89}$$

$\lambda^+_k = \max(0, \lambda_k)$ and $\lambda^-_k = \min(0, \lambda_k)$ are the elements of the diagonal matrices of the positive and negative eigenvalues of $\mathbf{A}_{inv}$, respectively. The terms $\Delta\mathbf{q}^+$ and $\Delta\mathbf{q}^-$ can be defined by

[3] $\mathbf{q}$ denotes the conservative solution vector $\mathbf{U}$ at each Newton sub-iteration.
[4] Also note that $\Delta\xi = \Delta\eta = 1$ in the computational plane.

$$\left.\begin{aligned}\Delta\mathbf{q}^{+}_{i+1/2,j} &= \widetilde{\beta}\Delta\mathbf{q}_{i,j} + (1-\widetilde{\beta})(1.5\Delta\mathbf{q}_{i,j} - 0.5\Delta\mathbf{q}_{i-1,j})\\ \Delta\mathbf{q}^{-}_{i+1/2,j} &= \Delta\mathbf{q}_{i+1,j}\\ \Delta\mathbf{q}^{-}_{i-1/2,j} &= \widetilde{\beta}\Delta\mathbf{q}_{i,j} + (1-\widetilde{\beta})(1.5\Delta\mathbf{q}_{i,j} - 0.5\Delta\mathbf{q}_{i+1,j})\\ \Delta\mathbf{q}^{+}_{i-1/2,j} &= \Delta\mathbf{q}_{i-1,j}\end{aligned}\right\}\,, \tag{10.90}$$

where $\widetilde{\beta}$ is a sensor function defined by the maximum of the eigenvalues at the cell faces. Similarly, one can obtain discretization for the term $(B^n_{inv}\Delta\mathbf{q})_\eta - (B^n_{vis}\Delta\mathbf{q})_\eta$. Equation (10.85) can be solved by Gauss-Seidel relaxation techniques. For example, (10.85) can be written as

$$J\,(DIAG)^{\nu}_{i,j}\,\Delta\mathbf{q}^{\mu+1,\nu}_{i,j} = -\omega^*(RHS)^{\nu}_{i,j} + (ODIAG)^{\mu,\nu}_{i,j} + J\frac{\mathbf{U}^n - \mathbf{q}^\nu}{\Delta t}\,, \tag{10.91}$$

where $(DIAG)^n_{i,k}$ contains the diagonal elements of the eigenvalue-split inviscid Jacobians, the viscous Jacobians and the term $\mathbf{I}/\Delta t$; $(ODIAG)_{i,j}$ includes the off-diagonal elements and is function of $\Delta\mathbf{U}_{i+1,j}$, $\Delta\mathbf{U}_{i-1,j}$, $\Delta\mathbf{U}_{i,j+1}$, $\Delta\mathbf{U}_{i,j-1}$. The number of Gauss-Seidel (μ) and Newton steps (ν) depends on the flow problem, grid size and details of the numerical scheme employed for discretizing the advective (inviscid) fluxes. The under-relaxation parameter $\omega^* = 0.1 - 1$ is used to compensate for the different orders of accuracy between left and right hand sides.

10.8 Extension of the Artificial Compressibility to Unsteady Flows

The artificial compressibility approach was originally proposed for steady state problems [104]. The extension of the method to unsteady flows can be obtained by adding a pseudotime derivative to the momentum equation [384, 505, 469, 470, 76, 290, 148, 147] thus obtain the system

$$\frac{1}{\beta}\frac{\partial p}{\partial \tau} + \frac{\partial u_j}{\partial x_j} = 0\,, \tag{10.92}$$

$$\frac{\partial u_i}{\partial \tau} + \frac{\partial u_i u_j}{\partial x_j} = -\frac{\partial u_i}{\partial t} - \frac{1}{\rho}\frac{\partial p}{\partial x_i} + \nu\frac{\partial^2 u_i}{\partial x_j^2}\,. \tag{10.93}$$

The above system can be iterated in the pseudotime τ until the divergence-free flow field is satisfied, i.e., $\partial p/\partial\tau = 0$ and $\partial u_i/\partial\tau = 0$. The solution in pseudotime can be obtained by using explicit or implicit methods. The above procedure is similar to the dual-time stepping procedure that is also used to solve the compressible flow equations for both steady and unsteady flows. In this case a pseudotime derivative of the unknown solution vector is added to both the momentum and continuity equations and the system is solved at each real time step until convergence is obtained at the inner iteration level.

The velocity derivative $\partial u_i/\partial t$ in (10.93) can be discretized by a first order

$$\frac{\partial u_i}{\partial t} = \frac{u_i^{n+1} - u_i^n}{\Delta t} , \tag{10.94}$$

or a second order backward differencing scheme

$$\frac{\partial u_i}{\partial t} = \frac{1.5u_i^{n+1} - 2u_i^n + 0.5u_i^{n-1}}{\Delta t} , \tag{10.95}$$

An alternative to the backward time differencing scheme is to use the Crank-Nicolson approximation. Soh and Goodrich [505] have employed the Crank-Nicolson scheme in conjunction with the artificial compressibility approach to obtain time accurate solutions. The momentum equation is discretized as

$$\frac{u_i^{n+1} - u_i^n}{\Delta t} + \frac{1}{2}\Big[\mathcal{H}(u_i^{n+1}, p^{n+1}) + \mathcal{H}(u_i^n, p^n)\Big] = 0 , \tag{10.96}$$

where

$$\mathcal{H}(u_i, p) = \frac{\partial u_i u_j}{\partial x_j} + \frac{1}{\rho}\frac{\partial p}{\partial x_i} - \nu \frac{\partial^2 u_i}{\partial x_j^2} .$$

After introducing $\hat{u}_i^{n+1} = u_i^{n+1} - u_i^n$ and $\hat{p}_i^{n+1} = p_i^{n+1} - p_i^n$, (10.96) is written

$$\hat{u}_i^{n+1} + \frac{\Delta t}{2}\mathcal{H}(u_i^n + \hat{u}_i^{n+1}, p^n + \hat{p}^{n+1}) = -\frac{\Delta t}{2}\mathcal{H}(u_i^n, p^n) . \tag{10.97}$$

The above system can be preconditioned by adding a pseudotime pressure and velocity derivatives to the continuity and momentum equations, i.e.,

$$\frac{1}{\beta}\frac{\partial \hat{p}}{\partial \tau} + \frac{\partial \hat{u}_j}{\partial x_j} = 0 , \tag{10.98}$$

$$\frac{\partial \hat{u}_i}{\partial \tau} + \hat{u}_i + \frac{\Delta t}{2}\mathcal{H}(u_i^n + \hat{u}_i, p^n + \hat{p}) = -\frac{\Delta t}{2}\mathcal{H}(u_i^n, p^n) . \tag{10.99}$$

where $\hat{u}_i = \hat{u}_i^* - u_i^n$ and $\hat{p} = \hat{p}^* - p^n$. The asterisk denotes transient values in pseudotime [505]; $\hat{u}_i^*$ and $\hat{p}^*$ eventually become $\hat{u}_i^{n+1}$ and $\hat{p}^{n+1}$ when the steady state in pseudotime is achieved. The terms $\mathcal{H}(u_i^n, p^n)$ serve as external forcing terms that drive the flow variables from n to $n+1$.

10.9 Boundary Conditions

Boundary conditions have to specified in all boundaries around the computational domain. Preliminary remarks on the issue of boundary conditions have been made in Sect. 2.9.

The wall boundary conditions are specified in the form of a Dirichlet condition. For a viscous flow (the natural scenario) the boundary condition on the wall is no-penetration and no-slip, i.e., both the normal and tangential velocity components are set equal to zero. If the problem includes a moving wall, then the given wall velocity is specified on the wall boundary in conjunction with the no-penetration condition. The pressure at a wall boundary is obtained by setting the pressure gradient normal to the wall to be zero.

The boundary conditions at the inflow and outflow boundaries can be defined by considering the characteristic waves traveling in and out of the computational domain. For each positive (or negative) eigenvalue there is a wave propagating information in the positive (or negative) direction. Therefore, we can use the eigenvalues to determine the characteristic waves that bring information from the interior of the domain to the boundaries.

An implicit implementation of boundary conditions was proposed in [469]. Let us consider the one-dimensional counterpart of (10.68) and (10.69),

$$J\frac{\partial \mathbf{U}}{\partial t} = -\frac{\partial \bar{\mathbf{E}}}{\partial \xi} = -\mathbf{T}\mathit{\Lambda}\mathbf{T}^{-1}\frac{\partial \bar{\mathbf{U}}}{\partial \xi} , \tag{10.100}$$

where use of the left and right eigenvector matrices has been made on the right hand side. We can also write

$$J\mathbf{T}^{-1}\frac{\partial \mathbf{U}}{\partial t} = -\mathit{\Lambda}\mathbf{T}^{-1}\frac{\partial \mathbf{U}}{\partial \xi} . \tag{10.101}$$

From the above system, one can obtain a system of scalar equations which have the form of wave equations. The direction of the wave is determined by the sign of the eigenvalues. The selection of the characteristic waves can be made by multiplying the last equation with a diagonal matrix $\boldsymbol{\mathcal{T}}$ which has an element of one in the position of the eigenvalue that is to be selected and values zero elsewhere, i.e.,

$$J\boldsymbol{\mathcal{T}}\mathbf{T}^{-1}\frac{\partial \mathbf{U}}{\partial t} = -\boldsymbol{\mathcal{T}}\mathit{\Lambda}\mathbf{T}^{-1}\frac{\partial \mathbf{U}}{\partial \xi} , \tag{10.102}$$

which can be written in an Euler implicit formulation as

$$\left(\frac{J\boldsymbol{\mathcal{T}}\mathbf{T}^{-1}}{\Delta t} + \boldsymbol{\mathcal{T}}\mathit{\Lambda}\mathbf{T}^{-1}\frac{\partial}{\partial \xi}\right)(\mathbf{U}^{n+1} - \mathbf{U}^{n}) = -\boldsymbol{\mathcal{T}}\mathit{\Lambda}\mathbf{T}^{-1}\frac{\partial \mathbf{U}^{n}}{\partial \xi} . \tag{10.103}$$

Rogers and Kwak [469] completed the above set of equations by specifying some variables to be held constant. These are defined through a vector $\mathbf{Q}$ according to

$$\frac{\partial \mathbf{Q}}{\partial t} = \frac{\partial \mathbf{Q}}{\partial \mathbf{U}} \frac{\partial \mathbf{U}}{\partial t} = 0 \,, \tag{10.104}$$

and write (10.103) as

$$\Big(\frac{J\mathcal{T}\mathbf{T}^{-1}}{\Delta t} + \mathcal{T}\Lambda\mathbf{T}^{-1}\frac{\partial}{\partial \xi} + \frac{\partial \mathbf{Q}}{\partial \mathbf{U}}\Big)(\mathbf{U}^{n+1} - \mathbf{U}^n) = -\mathcal{T}\Lambda\mathbf{T}^{-1}\frac{\partial \mathbf{U}^n}{\partial \xi} \,. \tag{10.105}$$

Let us now consider the case of inflow and outflow boundary conditions. Assuming that the fluid is traveling in the positive direction the eigenvalues (10.43) are $\lambda_0 < 0, \lambda_1 > 0$ and $\lambda_2 < 0$ (also assuming a non-moving grid).[5] The negative eigenvalues is the one to be chosen for specifying the definition of the boundary conditions (thus the third diagonal element of the matrix $\mathcal{T}$ will be one). The variables to be specified in the vector $\mathbf{Q}$ can be (for example, in 2-D)

$$\mathbf{Q} = \begin{pmatrix} p + 1/2(u^2 + v^2 + w^2) \\ 0 \\ v \end{pmatrix} \,, \tag{10.106}$$

or

$$\mathbf{Q} = \begin{pmatrix} 0 \\ u \\ v \end{pmatrix} \,. \tag{10.107}$$

The former set of variables is preferred for cases where the inflow velocity is not known whereas the latter one one is preferred when the velocity profile is specified.

At the outflow boundary, assuming that the fluid leaving the domain is traveling in the positive direction, there are two characteristic waves traveling out of the computational domain, i.e., $\lambda_0 > 0, \lambda_1 > 0$ and $\lambda_2 < 0$. The vector $\mathbf{Q}$ can be defined as $\mathbf{Q} = (p, 0, 0)^T$.

Finally, symmetry and periodic boundary conditions may be applied if the flow configuration allows this kind of conditions.

10.10 Local Time Step

In flow problems where a steady state flow is desired, the number of iterations to achieve steady-state convergence can be reduced by using a variable time step. This depends on the local flow changes (fluid velocities) and the grid spacing. For example, the local time step Δt can be defined by

[5] If the incoming fluid is traveling in the negative direction the sign of the eigenvalues will be $\lambda_0 > 0, \lambda_1 > 0$ and $\lambda_2 < 0$.

$$\Delta t = \frac{CFL}{\max\limits_{m}\{\mu_m\}} , \tag{10.108}$$

where $\mu_m = \max\{(|\lambda_1|, |\lambda_2|)\sqrt{\chi_x^2 + \chi_y^2 + \chi_z^2}\}$, $m = 1, 2, \ldots, 6$ (3-D) and m=1,2,...,4 (2-D) are the volume cell-face pointers, λ_1 and λ_2 are the eigenvalues (10.43) at the cell faces, and χ stands for ξ, η and ζ.

10.11 Multigrid for the Artificial-Compressibility Formulation

The principles and properties of multigrid method have been discussed in a number of textbooks in the past, e.g., [238, 595] and review articles [597]. The method was originally developed for accelerating the solution of elliptic PDEs [73], but can also be implemented in conjunction with the AC form of the Navier-Stokes equations to speed up convergence. In this case, the multigrid is implemented similar to the Jameson's multigrid procedure developed for the solution of the compressible Euler equations [273, 275, 274] and later on applied to the solution of the compressible Navier-Stokes equations [355, 352, 353, 307].

Many of the basic concepts introduced in Chap. 8 regarding the multigrid method for elliptic equations are also applicable to the present discussion, but the specific details are different due to the fundamental nature of the equations. Implementation of different variants of multigrid method in conjunction with AC have been presented in the literature for computations of laminar flows [157, 348], free surface flows (based on Euler simulation) [184], turbulent flows [491, 347, 618], incompressible low-Mach number flows [587], unsteady flows featuring instabilities and transition to turbulence [366], as well as in conjunction with adaptive grids [348] and adaptive solvers [158].

Relevant to the multigrid implementation for AC, is also work on multigrid methods in conjunction with the preconditioned Euler/Navier-Stokes equations for low-Mach number two-dimensional, steady and unsteady flows [125, 361, 511].

In the following paragraphs, we discuss the main components of the multigrid method.

10.11.1 Rationale for Three-Grid Multigrid

In principle, multigrid can be implemented in conjunction with several grid levels, i.e., five, six or even more, and this is often the case when multigrid is employed to solve elliptic equations such as the pressure-Poisson equation. However, one can argue in favor of using a smaller number of grid levels[6] based on the following reasons:

[6] In [157] this is called *short-multigrid*.

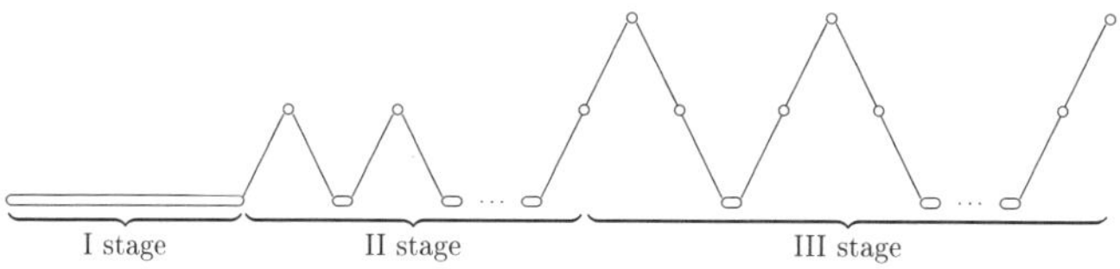

Fig. 10.2. Schematic of the full multigrid (FMG) for three grids. Stage I: The solution is obtained on the coarse grid. Stage II: Two-grid multigrid algorithm solution. Stage III: Three-grid multigrid algorithm.

- If the grid on which the equations are to be solved is not fine enough, then the coarsest grid will not encompass a sufficient number of grid points to provide a good correction for the fine grid. Numerical experiments have shown that in the case of (very) coarse grids the efficiency of the multigrid is significantly reduced. This has been observed in theoretical investigations of multilevel algorithms for non-symmetric (e.g., [584, 605]) and nonlinear problems (e.g., [606]).
- A smaller number of grids (*short-multigrid*) improves the efficiency of parallel computations, as has been demonstrated in previous studies by Ålund et al. [11], Axelsson and Neytcheva [23, 24] and Drikakis et al. [154, 161, 162].
- The use of several grid levels increases the complexity of the computer code and memory requirements.

10.11.2 FMG-FAS Algorithm

For the case of steady flows, the solution of the equations is initially obtained on a sequence of coarser grids and this solution can be used as an initial guess for the multigrid procedure. Computations are performed on the coarse grid in order to provide a good initial guess for the intermediate grid and the same procedure is repeated on the intermediate grid in order to provide a good initial guess for the finest grid. Using this initial guess the three-grid multigrid procedure can be initialized This is the full multigrid – full approximation storage algorithm (FMG-FAS), schematically shown in Fig. 10.2.

The so-called V-cycle implementation of the multigrid algorithm for three grids (fine, intermediate and coarse grids) is shown in Fig. 10.3. The main steps in the solution procedure for the FMG-FAS algorithm of Fig. 10.2 are listed below (also in [157]) where: P stands for the prolongation operator; N_{cg} denotes the Navier-Stokes solution on the coarse grid and 0_{cg} the initial

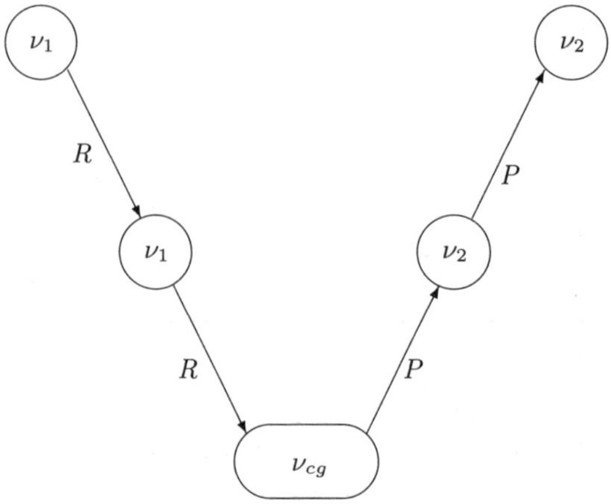

Fig. 10.3. Schematic of the V-cycle for a three-grid multigrid algorithm.

condition used for the solution; S_{ig} stands for the relaxation procedure on the intermediate grid; R stands for the restriction operator; $\bar{V}$ stands for the coarse grid function (see discussion later regarding full approximation storage – FAS):

auxiliary stage I – single grid solution

$U_{cg} := N_{cg}^{-1} 0_{cg}$	compute coarsest grid solution
$U_{ig}^0 := P\ U_{cg}$	prolongation - initial guess on the intermediate grid

auxiliary stage II – multigrid sweeps on two grids

repeat

$U_{ig} := S_{ig}(U_{ig}, 0_{ig}, \nu_1)$	ν_1 = pre-smoothings
$d_{ig} := N_{ig} U_{ig}$	compute intermediate grid defect
$d_{cg} := R\ d_{ig}$	restriction of the defect to the coarsest grid

repeat

$f_{cg} := -d_{cg} + N_{cg}\bar{V}_{cg}$ compute right hand side on the coarsest grid

$V_{cg} := N_{cg}^{-1} f_{cg}$ compute coarsest grid approximate solution

$c_{cg} := V_{cg} - \bar{V}_{cg}$ compute correction on the coarsest grid

$c_{ig} := P\ c_{cg}$ prolongation of the correction to the intermediate grid

$U_{ig} := U_{ig} + c_{ig}$ correct solution on the intermediate grid

$U_{ig} := S_{ig}(U_{ig}, 0_{ig}, \nu_2)$ ν_2 post-smoothing iterations

until the steady state solution on the intermediate grid is achieved

$U_{fg}^0 := P\ U_{ig}$ prolongation - initial guess on the finest grid

stage III – multigrid sweeps on three grids (V-cycles)

repeat

$U_{fg} := S_{fg}(U_{fg}, 0_{fg}, \nu_1)$ ν_1 pre-smoothings

$d_{fg} := N_{fg} U_{fg}$ compute finest grid defect

$d_{ig} := R\ d_{fg}$ restriction of the defect to the intermediate grid

$f_{ig} := -d_{ig} + N_{ig}\bar{V}_{ig}$ compute right hand side on the intermediate grid

$V_{ig} := S_{ig}(V_{ig}, f_{ig}, \nu_1)$ ν_1 pre-smoothings

$d_{ig} := -f_{ig} + N_{ig} V_{ig}$ compute intermediate grid defect

$d_{cg} := R\ d_{ig}$ restriction of the defect to the coarse grid

$f_{cg} := -d_{cg} + N_{cg}\bar{V}_{cg}$ compute right hand side on the coarse grid

$V_{cg} := N_{cg}^{-1} f_{cg}$ compute coarse grid approximate solution *repeat*

$c_{cg} := V_{cg} - \bar{V}_{cg}$	compute correction on the coarse grid
$c_{ig} := P\, c_{cg}$	prolongation of the correction to the intermediate grid
$V_{ig} := V_{ig} + c_{ig}$	correct solution on the intermediate grid
$V_{ig} := S_{ig}(V_{ig}, f_{ig}, \nu_2)$	ν_2 post-smoothing iterations
$c_{ig} := V_{ig} - \bar{V}_{ig}$	compute correction on the intermediate grid
$c_{fg} := P\, c_{ig}$	prolongation of the correction to the finest grid
$U_{fg} := U_{fg} + c_{fg}$	correct solution on the finest grid
$U_{fg} := S_{fg}(U_{fg}, 0_{fg}, \nu_2)$	ν_2 post-smoothing iterations

until the steady state solution on the finest grid is achieved

The solution algorithm utilized for single grid calculations is also used as a relaxation procedure (S) on the fine (S_{fg}) *and intermediate grids* (S_{ig}), and as a solver on the coarse grid. The Navier-Stokes solver used on the coarse and intermediate grids is, however, slightly different than the original single-grid solver. This is due to the fact that the RHS of the Navier–Stokes equations is identically zero inside the domain only in the case of the single-grid algorithm. In the case of the multigrid method the right-hand-side of the equations on the coarse and intermediate grids is not zero due to the additional terms (e.g., $N_{ig}V_{ig}$) arising from the FAS linearization procedure.

The multigrid method can be used in time-dependent flow problems, as well. The implementation is schematically shown in Fig. 10.4. In this case, V-cycles are performed at each time step but without utilizing the FMG procedure (auxiliary stages I and II). Larger time steps can be used on the coarse and intermediate grids to further reduce the number of iterations on these grid levels.

10.11.3 Remarks on the Full Approximation Storage (FAS) Procedure

The FAS algorithm was proposed by Brandt [73] and is also discussed in detail in [238, 595]. When linear problems are to be solved, the solution on the fine grid can be directly computed on coarser grids using the same solution matrix,

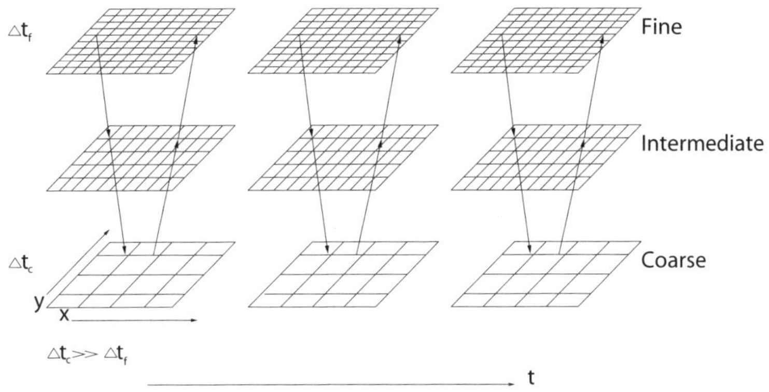

Fig. 10.4. Schematic representation of V-cycle multigrid for time-dependent flow problems using larger time steps on the coarse and intermediate grids.

but having in the RHS of the equations the *restricted defect.* However, this will not lead to efficient solutions in the case of nonlinear problems. In this case the multigrid corrections are formed as differences between some basic - reference solution - and the currently computed approximation of this solution [157]. That is why the three-grid FAS algorithm requires the calculation of the so-called *coarse-grid* functions. In the case of the three-grid multigrid these functions need to be defined for the coarse, $\bar{V}_{cg}$, and intermediate grid, $\bar{V}_{ig}$, respectively [157]. In the original Brandt's algorithm (henceforth labeled FAS-1) these functions are computed as projections of the current intermediate and finest grid solutions onto the coarse and intermediate grids, respectively:

$$\bar{V}_{cg} = R\ V_{ig}, \quad \bar{V}_{ig} = R\ U_{fg}.$$

where R is the restriction operator. Another approach (henceforth labeled FAS-2) has been proposed in [157] according to which the computed, through FMG, solutions on the coarse, U_{cg}, and intermediate grids, U_{ig}, are utilized as coarse and intermediate grid functions in the FAS algorithm, i.e., $\bar{V}_{cg} = U_{cg}$ and ,$\bar{V}_{ig} = U_{ig}$. It has been found [157] that the FAS-2 implementation improves the performance of multigrid, especially in the case of fine grids.

10.11.4 Effects of Pre- and Post-Relaxation on the Efficiency of FMG–FAS

The efficiency of the multigrid solution depends on the relaxation steps at different grid levels, i.e., pre- (ν_1) and post-relaxation (ν_2) iterations, as well as iterations on the coarse grid (ν_{cg}) (Fig. 10.3). Tables 10.1 and 10.2 show the effects of these parameters on the number of multigrid V-cycles and work units for a three-dimensional laminar flow ($Re = 100$) in a rectangular channel [157]. In the case of "unsteady-type" multigrid, we do not have to solve the equations on the coarse grid down to the convergence limit in order to obtain better multigrid efficiency. This is demonstrated in Table 10.1 where the better results are obtained for $\nu_{cg} = 100$ rather than $\nu_{cg} = 400$.

As expected, the number of the pre- (ν_1)and post-relaxations (ν_2) also affects multigrid efficiency. The results in Table 10.2 (for a 3-D straight channel flow) [157] indicate that multigrid is more efficient when only post-relaxations are performed. Similar conclusions have been drawn for the case of linear PDEs by de Zeeuw [132] for the case of linear problems.

Table 10.1. Effects of the coarse-grid iterations on the multigrid performance.

ν_{cg}	MG sweeps	Work units
100	105	2429
400	86	2447
30	210	4332

Table 10.2. Effects of pre- and post-relaxation iterations on the multigrid performance for three-dimensional laminar flow computations in a rectangular channel (see also Sect. 16.4.7).

ν_1	ν_2	MG sweeps	Work units
0	15	105	2429
10	50	80	5521
5	10	188	2900

10.11.5 Transfer Operators

The implementation of multigrid requires to define the restriction and prolongation operators. A simple way to define the restriction operator can be simply defined by considering that any coarse-grid control volume (CV) consists of eight fine grid CVs (in three dimensions). In simple geometries, this can be achieved by covering the computational domain with a coarse grid and further refine it in such a way that any coarse-grid volume is split into eight fine-grid volumes. For complex geometries it is better to first generate the finest grid, and then to construct the coarser grids by eliminating lines of the fine grid [157]. Then, the restriction operator is defined by the weighted summation of all the values over the fine-grid CVs.

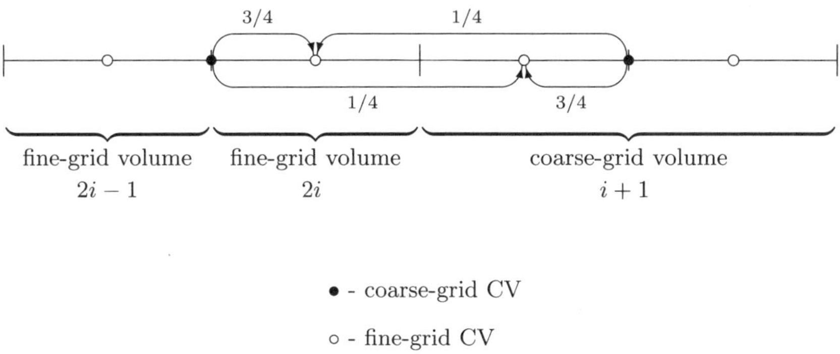

Fig. 10.5. Schematic of the linear prolongation in 1-D case.

Multigrid algorithms can be implemented using different prolongation operators. The simplest definition of the prolongation operator is the linear interpolation (Fig. 10.5). If U^f and U^c are the values of the variable U on the fine and coarse grids, respectively, and let us assume that in one dimension (1D) the fine-grid cells with indices $(2i-1)$ and $(2i)$ will form a coarse-grid cell (i), then the fine-grid values are obtained by the coarse-grid ones using

$$U^f_{2i} = \frac{1}{4}U^c_{i+1} + \frac{3}{4}U^c_i \,, \quad U^f_{2i+1} = \frac{3}{4}U^c_{i+1} + \frac{1}{4}U^c_i \,. \tag{10.109}$$

For two- and three-dimensional cases, bilinear or trilinear prolongation formulas can be obtained by combining one-dimensional linear interpolation. If $(2i, 2j, 2k)$ are the indices of the fine-grid cell in three dimensions, the trilinear prolongation operator is then defined by

$$\begin{aligned} U^f_{2i,2j,2k} = {} & \frac{1}{64}U^c_{i+1,j+1,k+1} + \frac{3}{64}U^c_{i+1,j+1,k} + \\ & \frac{3}{64}U^c_{i+1,j,k+1} + \frac{3}{64}U^c_{i,j+1,k+1} + \frac{9}{64}U^c_{i+1,j,k} + \\ & \frac{9}{64}U^c_{i,j+1,k} + \frac{9}{64}U^c_{i,j,k+1} + \frac{27}{64}U^c_{i,j,k} \,. \end{aligned} \tag{10.110}$$

Another prolongation operator is the piecewise constant prolongation. This operator is defined by

$$U^f_{2i-1} = U^c_i \ , \quad U^f_{2i} = U^c_i \ , \tag{10.111}$$

and

$$\begin{aligned} U^f_{2i-1,2j-1,2k-1} &= U^f_{2i-1,2j-1,2k} = \\ U^f_{2i-1,2j,2k-1} &= U^f_{2i-1,2j,2k} = \\ U^f_{2i,2j-1,2k-1} &= U^f_{2i,2j-1,2k} = \\ U^f_{2i,2j,2k-1} &= U^f_{2i,2j,2k} = U^c_{ijk} \ . \end{aligned} \tag{10.112}$$

for one- and three-dimensional problems, respectively.[7]

Inspired by the idea of upwind discretization, we can make prolongation to depend on the sign of the advective velocity; this is labeled as *upwind piecewise constant prolongation* [157]. In 1-D, this operator is given by

$$U^f_{2i} = \begin{cases} U^f_{2i+1} = U^c_i & \text{for } u_i > 0 \ , \\ U^f_{2i+1} = U^c_{i+1} & \text{for } u_i < 0 \ . \end{cases} \tag{10.113}$$

where u_i is the advective velocity on the cell center. The extension to 3-D is straightforward.

A combination of upwind prolongation in the streamwise direction and bilinear in the cross-stream plane (*mixed-prolongation*) has also been proposed [157].

In the case of the bilinear and trilinear prolongation on non-uniform grids, geometrical factors (grid-weighted averages) such as the distances between grid nodes, can be taken into account in order to account for grid non-uniformities.

Most of the multigrid studies in literature employ the trilinear interpolation as a prolongation operator. It is recommended to use the trilinear operator for second-order derivatives (viscous terms) and the piecewise operator for first-order derivatives (convective terms). The optimum choice also depends on the numerical scheme employed for the discretization of the equations (the advective scheme in particular), grid size and stretching (near the wall boundaries), as well as physical scales in space and time. Thus there is no golden rule for the optimum parameters and sometimes the optimum choices emerge as a result of the user's experience.

Results using different prolongation operators for computations of the three-dimensional laminar flow in a straight channel [157] are given in Table 10.3. The following operators or combinations of them have been considered:

1. *Mixed-prolongation* for u, v and w after the auxiliary stage, trilinear prolongation for p and corrections;

[7] Note that the piecewise constant prolongation does not encompass any grid-weighted averaging.

2. trilinear prolongation at all stages of the FMG–FAS procedure;
3. piecewise constant prolongation at all stages of the FMG–FAS procedure;
4. upwind piecewise constant prolongation for u, v and w after the auxiliary stage, and trilinear prolongation for p and corrections.

In this table the number of MG sweeps on the fine-grid, required for steady state solution with accuracy of 10^{-6} for the L_2-norm of the u–residual, is shown. The work units are calculated by taking into account that one relaxation step at the grid level l is equivalent to $1/8^{l-1}$ work units of the finest grid ($l = 1$). The pre- and post-relaxation sweeps have also been included in the total work units. The acceleration that is obtained by using the first prolongation choice (see list above) accelerates the solution by a factor of (about) 27.

Table 10.3. Effects of prolongation operators on the multigrid performance for three-dimensional computations in a rectangular channel. The work units for single grid computation are 67,200. The type of prolongation operators is explained in the text.

Type of prolongation	MG sweeps	Work units
(1)	105	2429
(2)	137	2875
(3)	231	4988
(4)	226	4886

10.11.6 Adaptive Multigrid

Adaptivity in scientific computations can be exploited in many different ways, e.g., in the context of grid adaptation, see e.g. [74, 378], in the context of minimization (reduction) of the computational domain (known as local-solution method) [412, 164, 165], or in the form of other numerical techniques such as the sparse-grid approach [616, 232]. The aim of all the above is to improve numerical efficiency in terms of memory and computing time.

Adaptivity in conjunction with multilevel techniques was first proposed by Brandt [73, 75] and Bai and Brandt [74] for solving elliptic problems. In the above papers to perform additional smoothing operations near the known singularity of the solution and/or near the boundary, as well as to exclude some subdomains from the relaxation sweeps were proposed as possible solution strategies. In the context of nonlinear problems, the idea of obtaining local solution of the equations in selective parts of the computational domain during the iterations was initially explored in [412] and was more systematically implemented in [164, 165] in conjunction with the compressible Euler/Navier-Stokes equations. A pointwise adaptive-smoothing algorithm was also developed and theoretically investigated in [473, 474] in connection with multigrid solutions of linear elliptic equations. The algorithm in [473, 474] is based on the Southwell method [507] for hand-solving (!) systems of linear algebraic equations. The Southwell method is a variant of the Gauss-Seidel method, exploiting adaptive ordering of unknowns, based on the range of residuals.

In [158] an adaptive multigrid method, called adaptive-smoothing MG (or AS-MG), was developed to solve the incompressible Navier-Stokes equations in their artificial compressibility formulation. The method of [158] uses the Navier-Stokes solver (called *smoother*) to solve the equations on subsets (Ω_s) of the grid Ω. The subset domain Ω_s is formed adaptively during the solution and is defined as the area of the computational domain in which the solution has not yet converged; convergence is monitored by a prescribed threshold criterion. In Ω_s the flow experiences significant variations compared to the rest of the computational domain. The multigrid method is utilized to accelerate the solution of the equations in Ω_s. Note that the adaptive multigrid procedure does not involve movement, addition or elimination of grid nodes from the initial computational domain, but simply selection of areas of Ω for locally performing the solution of the equations using multigrid.

The selection of Ω_s is based on adaptivity criteria which can be associated with fixed parameters (*static-adaptivity*)[8] or with dynamically-defined parameters (*dynamic-adaptivity*).[9] In the *static-adaptivity* case information about the residuals at the current iteration (or current time step) is used to reconstruct the subset Ω_s. In the *dynamic-adaptivity* information about the residuals at consecutive iterations is utilized [158].

In addition to the multigrid method and Navier-Stokes solver, the development of the adaptive-smoothing algorithm requires: (i) an algorithm, i.e., the adaptivity criterion, to reconstruct Ω_s, based on the local convergence behavior of the iterative solver; and (ii) optimization of the *global exchange* of information between different subdomains[10] in order to avoid *stagnation* (constant residuals values) or divergence of the iterative solver. The above issues are discussed below.

Adaptivity Criterion. The objective is to form subsets Ω_s of the computational domain that contain those CVs $P \in \Omega$ for which the residuals are relatively large. One option is to put in Ω_s only the CV corresponding to the maximum residual value and repeat this procedure at every iteration. Such a pointwise adaptive-smoothing algorithm has been proposed for elliptic linear

---

[8] The adaptivity parameters remain the same throughout the solution.

[9] In this case the adaptivity parameters are calculated during the solution.

[10] Ω_s may consist of a number of computational cells scattered in different locations of the computational domain.

problems in [473, 474]. However, this requires reconstruction of the Ω_s every time where the residual reaches its maximum value and this may result in significant additional computational work. A more efficient alternative is to work with larger subsets Ω_s.

The identification of *large residuals* can be done either with respect to the convergence criterion or with respect to the current norm of the residuals. The AC approach solves a steady state problem by performing pseudo-time steps. Therefore, the norm of the current residuals of the steady state problem is equal to the norm of the time derivative term, i.e., $\| J\mathbf{U}_t \|$. Let us denote by $res(P)$ the value of the maximum component of the discrete analogue of $J\mathbf{U}_t$ on a CV P, and let ε be the required accuracy of the iterative solution of the steady state problem. In other words, the convergence criterion on the finest grid should be $\| res \|_{C(\Omega)} < \varepsilon$.
To reconstruct the subset Ω_s in the least expensive way, one can "freeze" the residuals for several time steps in those CVs where the residuals have relatively small values. Let us denote by $\widetilde{res}(P)$ the last computed residual on CV P to distinguish it from the "true" residual $res(P)$. To understand the difference between the above two residuals, consider two neighboring cells, Q and P, where Q belongs to Ω_s, but not P. At the next adaptive-smoothing step, the solution is updated only for the cell Q, and the residual is also computed there. Because the residual corresponding to the cell P depends on the solution in cell Q (if the entire domain was solved), the "true" residual in P has to be recomputed after the solution in Q has been updated. However, this would be computationally expensive. Therefore, the recommended approach is to use the last computed residual in P, instead of the "true" one. Three adaptivity criteria for reconstructing Ω_s have been proposed in [158]:

- *Absolute criterion:* $\Omega_s = \{P : |\widetilde{res}(P)| \geq \gamma\varepsilon,\ P \in \Omega\}$,
- *Relative C criterion:* $\Omega_s = \{P : |\widetilde{res}(P)| \geq \gamma \| \widetilde{res} \|_{C(\Omega)},\ P \in \Omega\}$,
- *Relative L_2 criterion:* $\Omega_s = \{P : |\widetilde{res}(P)| \geq \gamma \| \widetilde{res} \|_{L_2(\Omega)},\ P \in \Omega\}$,

where γ ($\gamma \geq 0$) is a parameter that controls the size of Ω_s. For $\gamma = 0$ we obtain the full domain Ω ($\Omega_s \equiv \Omega$).

For static-adaptivity γ remains constant during the computation, whereas for dynamic-adaptivity γ changes values as the solution evolves. The value of γ determines the number of CVs involved in Ω_s. The variable γ prevents the cases where the number of CVs reduces very quickly (then the solution may diverge) or the number of CVs reduces slowly (then convergence speedup may not be significant). The adaptive-smoothing procedure utilizes information about the convergence behavior of the solver to decide how fast the number of CVs in Ω_s will be reduced. For variable γ the following formula has been proposed [158]

$$\gamma = \begin{cases} \gamma_{max}, & q \leq 1 \\ \gamma_{max} + \dfrac{q-1}{q_{max}-1}(\gamma_{min} - \gamma_{max}), & 1 < q < q_{max} \\ \gamma_{min}, & q \geq q_{max} \end{cases} \tag{10.114}$$

where γ_{min}, γ_{max} and $q_{max} \geq 1$ are given parameters. The rate of convergence q is defined by

$$q = \frac{\widetilde{res}^{n}_{max}}{\widetilde{res}^{*}_{max}}, \tag{10.115}$$

$$\widetilde{res}^{n}_{max} = \max_{\Omega}\{\widetilde{res}(P)\}, \tag{10.116}$$

with n, $1 \leq n \leq \nu$, being the current iteration on the corresponding grid in the current MG sweep, and

$$\widetilde{res}^{*}_{max} = \begin{cases} \widetilde{res}^{n-1}_{max}, & \widetilde{res}^{n}_{max} < \widetilde{res}^{n-1}_{max} \\ \min\limits_{i<n}\{\widetilde{res}^{i}_{max}\}, & \widetilde{res}^{n}_{max} \geq \widetilde{res}^{n-1}_{max} \end{cases} \tag{10.117}$$

In (10.114) $q_{max} = 1$ corresponds to change of γ from γ_{max} to γ_{min} (as long as the norm of the residual increases). In this case only two values of γ are used. However, for $q_{max} > 1$, we allow γ to take some intermediate values, as well. As a result, γ decreases gradually if a moderate increase of the residuals occurs. In the case of the artificial compressibility approach the residuals are computed during the pseudotime subiteration. Thus, no additional operations for implementing the adaptive-smoothing are required.

Table 10.4. Acceleration of the convergence using MG, and AS-MG in conjunction with static (rows 3 and 4 corresponding to $\gamma_{min} = \gamma_{max}$ and dynamic (rows 4 and 5) adaptivity for the flow around a NACA 0012 airfoil flow case at $a = 10^o$ and $Re = 1000$.

Method	γ_{min}	γ_{max}	MG sweeps	Acceleration
SG				1.00
MG			228	6.18
AS-MG	0.1	0.1	358	15.32
AS-MG	0.2	0.2	316	23.05
AS-MG	0	0.2	369	20.0
AS-MG	0	0.4	562	14.77

Global Exchange. Global exchange of information between different parts of the domain can be achieved either through the solution of the equations on the entire coarse grid or by performing a complete smoothing after every successive $(n_s - 1)$ adaptive smoothings. Complete smoothing means that the equations are solved in all CVs of Ω.

The implementation of the adaptive multigrid procedure is shown below:

$\Omega_s = \Omega$

$compute\ res^0_{max}$

$for\ n = 1\ to\ \nu\ do$

  $for\ all\ P \in \Omega_s\ perform\ a\ smoothing\ with\ a\ time\ stepping\ procedure$

  $compute\ \widetilde{res}^n_{max}$

  $if\ (\Omega_s \not\equiv \Omega)\ then$

    $i_s = i_s + 1$

  $else$

    $if\ (res^n_{max} < \varepsilon)\ exit$

    $i_s = 0$

  $end\ if$

  $if\ (i_s = n_s - 1\ or\ n = \nu - 1)\ then$

    $\Omega_s = \Omega$

  $else$

    $reconstruct\ \Omega_s$

    $if\ (\Omega_s = \emptyset)\ \Omega_s = \Omega$

  $end\ if$

$end\ do$

where i_s is the current adaptive-smoothing iteration.

Computational Example Using Adaptive Multigrid. We demonstrate the performance of the AS-MG for the flow around the NACA 0012 at ten degrees $\alpha = 10°$ angle of incidence and $Re = 1,000$. The results remain the same regardless the adaptivity criterion that is implemented (Fig. 10.6). The distribution of the residuals after the end of a certain number of AS-MG sweeps is shown in Fig. 10.7. In the AS-MG algorithm the number of CVs

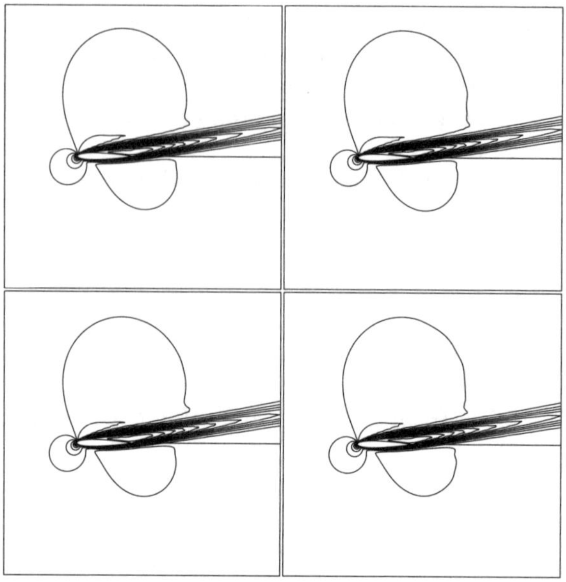

Fig. 10.6. Isovelocity (u velocity) contours for the flow around the NACA 0012 airfoil at $Re = 1000, \alpha = 10°$, using different adaptivity criteria: absolute (top right), relative C (bottom left) and relative L_2 criterion (bottom right); the top left plot shows the solution as obtained by the multigrid method without adaptive smoothing.

involved in the computation varies during the iterations. Fig. 10.8 shows the variation of the number of CVs, on the finest grid only, during iterations, for the absolute (constant $\gamma = 1$), relative C (constant $\gamma = 0.2$) and relative L_2 (constant $\gamma = 1$) criteria, respectively. The dots in these figures correspond to the number of CVs during the post-smoothing iterations of each multigrid sweep. The x-axis in Fig. 10.8 includes both the number of complete (in Ω) and adaptive smoothings (in Ω_s). The horizontal line Fig. 10.8 corresponds to the number of CVs when no adaptation of the multigrid solution is utilized, i.e., the number of CVs does not change. Although the absolute criterion gradually leads to a continuous reduction of the CVs involved in the computation, the acceleration is less than the one obtained by the relative C as well as by the relative L_2 criterion. This is due to the fact that the absolute criterion requires more MG sweeps than the other two criteria. Various numerical experiments suggest that the best performance is achieved through a balance of the size of active set (i.e., number of CVs) and number of MG sweeps [158].

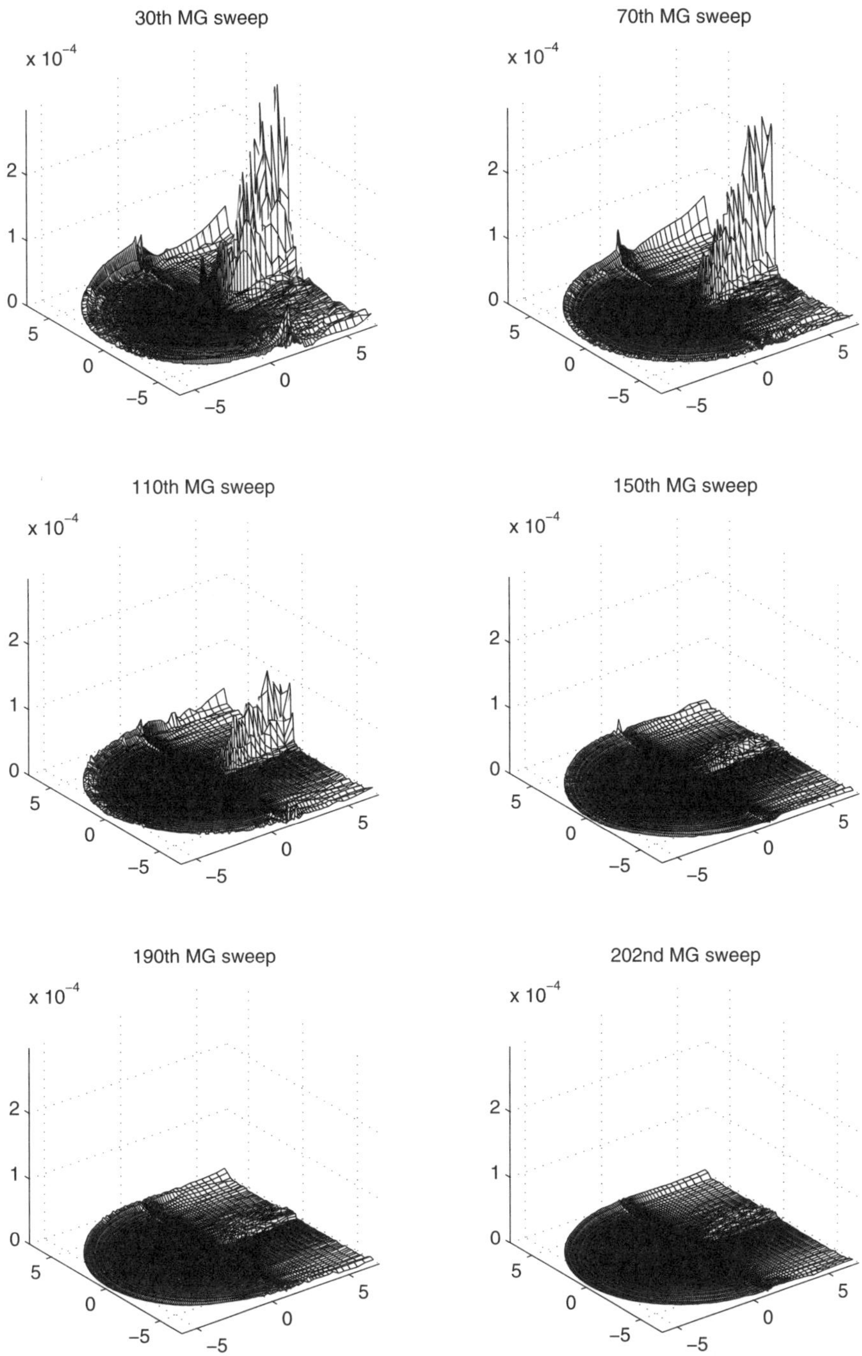

Fig. 10.7. Distribution of residuals at the end of certain MG sweeps during the AS-MG solution.

Table 10.4 shows the acceleration achieved using AS-MG with static and dynamic adaptivity against the multigrid and single-grid performance. Although greater acceleration is achieved for constant γ (static adaptivity), the acceleration factor is close to the one achieved by using variable γ, but the variable γ has been found, in general, to provide more stable solutions.

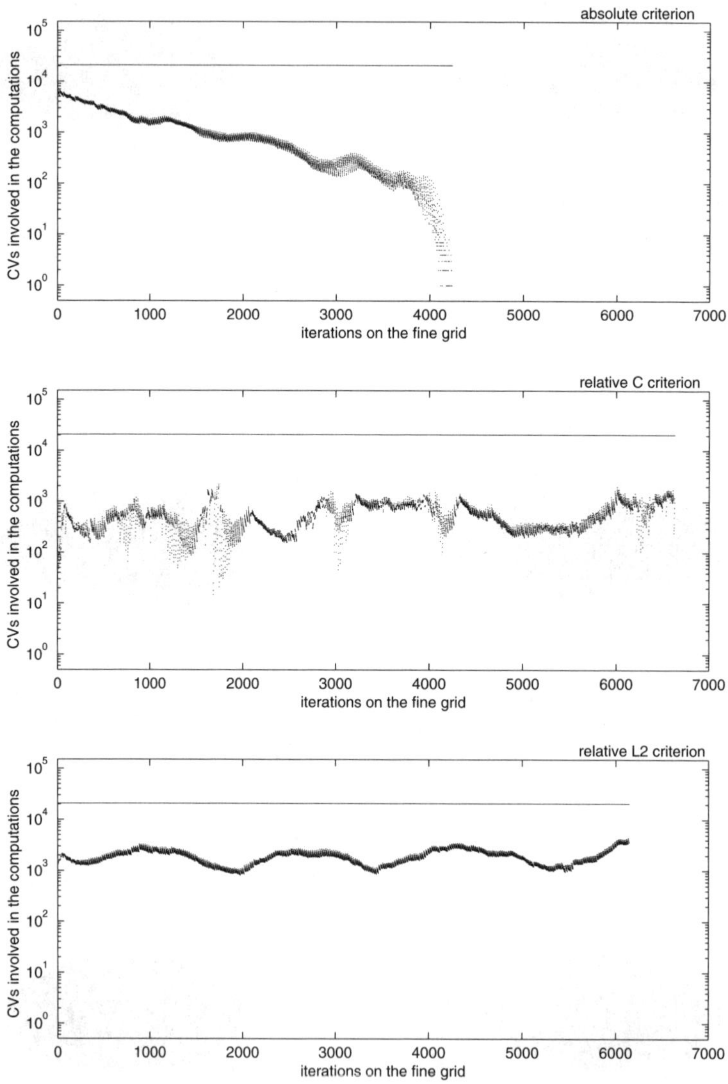

Fig. 10.8. Variation of the number of CVs involved in the AS-MG computation on the finest grid using different adaptivity criteria.

11. Projection Methods: The Basic Theory and the Exact Projection Method

This chapter introduces the basic aspects of a projection method and discuss the numerical method known as the "exact" projection. This will serve to motivate the discussion in the next chapter regarding approximate projection methods. Projection methods as a numerical method for computing incompressible flows were introduced by Chorin [105, 106]. This method used the discrete version of the continuous projection described below and will be the first numerical algorithm we will describe. The marker-and-cell (MAC) method is perhaps the first projection method [244] as well as being important as a step in approximate projections in that it can be used to construct a divergence-free velocity field for advection (the MAC projection covered in Sect. 11.3.3). Yet another method was developed by Strikwerda [516] using third-order differences. The basic method will be covered in more detail in Sect. 11.3.5.

Recent advances in the use of projection methods are based on the work of Bell, Colella, and Glaz [45] and began the connection of projection methods to high-resolution methods. Their work was built upon the form of the exact projection defined by van Kan [580]. E and Shu describe an extension of the basic projection methodology of Bell, Colella, and Glaz [176]. This method uses spectral techniques for the pressure solution and ENO methods for the advection. This work includes other high resolution methods than originally considered and method-of-lines time integration. This provides a window into the extension of the basic method to a wider variety of time integration method such as those covered in Chap. 7 (and Chap. 12).

The term "exact" projection refers to the property that the discrete divergence of a vector field is intended to be identically "zero" after applying the projection to a vector field.[1] This is differentiated from an "approximate" projection where the discrete divergence is a function of the truncation error of the scheme covered in Chap. 12. One should always keep in mind that the numerical value of the divergence is only zero for the precise numerical operator defined in projection, but will be the value of the truncation error for other divergence operators.

[1] The actual level of discrete divergence of velocity will be proportional to the tolerance of the solution of the linear system of equations if this is done via an iterative method.

In this chapter we will cover several topics: the formulation of projection algorithms starting with continuous mathematical operators, then describe their incarnation as discrete numerical operators. Finally, we will discuss boundary conditions (a particularly controversial topic) and conditions of solvability of the linear systems that result from discretizing projection methods. Other perspectives on recent developments with respect to projection methods can be found in the works such as E and Liu [175], and the more numerically focused work of Brown et al. [83]. For a discussion of projection methods from the perspective of finite element method, the book by Gresho and Sani [230] is recommended.

Our emphasis in this chapter and the following one are the practical numerical aspects of computing with projection methods. Because of our focus on high-resolution methods, mathematical rigor is more difficult to apply. We comment on this because there has been substantial activity in the area of mathematical analysis of these methods in recent years. Interested readers should consult these works for details as well as applicability of the results to their problems of interest. Recommended starting points for examining these matters can be found in [84, 173, 174, 175, 265, 389, 598, 599].

11.1 Grids – Variable Positioning

Variants of projection methods are most simply differentiated by the positioning of variables on the grid. Most often the velocity and pressure variables are placed in different locations (*staggered grid* arrangement) in order to suppress the decoupling of differencing stencils. This theme will be expanded upon further in this chapter as this decoupling complicates the implementation of projection methods. Furthermore, the residual impact of this decoupling will necessitate the development of "grid coupling" mechanisms in the form of filters described in detail in the next chapter. These issues are present in both simple and complex grid topologies, but are most acute on simple structured grids where algebraic simplifications amplify the decoupling.

The issue of where to place variables is as old as computational fluid dynamics. This issue first arose in the classic paper by von Neumann and Richtmyer in 1950 [590]. There, the compressible flow equations are solved using a staggered grid to avoid pressure-velocity decoupling. Since that paper, the centering of variables has preoccupied computational physicists. One of the first papers to address this issue systematically was written by Arakawa [17] who coined the terms "A", "B" and "C" grids.

Fig. 11.1 shows the differences between grid types and the appropriate labeling of the grids that is used throughout this chapter and the next. The "A" grids come in two flavors: a control volume form where the variables are often viewed as integral average values over a cell, and the finite difference form. The control volume form (*collocated grid*) is shown in Fig. 11.1. On the "B" grid (also known as *vertex* or *node grid*) the velocities are cell centered and the pressure is defined on the vertices, or vice-versa (Fig. 11.1). On the "C" grid (also known as *MAC grid*) the pressure is cell centered and the velocities are defined at the cell faces (Fig. 11.1). From the above, we understand that in more than one dimension both the vertex and MAC grid encompass a form of staggering.

The variable positioning has far reaching consequences for nearly every aspect of the algorithm. A staggered grid will generally improve the coupling of the velocity and pressure fields, while somewhat complicating the advection algorithm and general indexing in the implementation of the method. Another consideration is the ease and accuracy of boundary conditions that must invariably be applied. Most successors to the von Neumann-Richtmyer method typically use the vertex grid [497, 95]. Of course, even more complex labeling ideas are possible, but these three options form the majority of choices made in the literature.

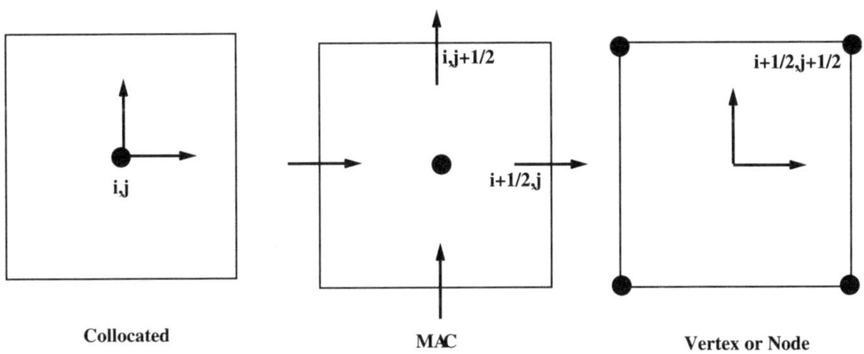

Fig. 11.1. The major types of grids with positioning of variables is shown. We will be concentrating on collocated grids, but the other two grids play an important role in the methodology. The "A" grid is cell-centered, the "B" grid is vertex-centered and the "C" grid is the MAC grid.

11.2 Continuous Projections for Incompressible Flow

This section will cover the basic mathematics of projections first in the context of standard incompressible flow where the density of the fluid is constant. Next, we will generalize this development to flows where the density is variable. This development will provide the foundation for the numerical methods discussed in both this and next chapter.

11.2.1 Continuous Projections for Constant Density Incompressible Flow

Our basic goal with projection methods is to advance a velocity field, $\mathbf{V} = (V^x, V^y, V^z)^T$ by some convenient means disregarding the solenoidal nature of $\mathbf{V}$, then recover the desired solenoidal vector field, $\mathbf{V}^d$ ($\nabla \cdot \mathbf{V}^d = 0$).[2] The means to this end is a projection, $\mathcal{P}$, which has the effect

$$\mathbf{V}^d = \mathcal{P}(\mathbf{V}) \; .$$

The projection accomplishes this through the decomposition of the velocity field into parts that are divergence-free and curl-free. This is known as a Hodge or Helmholtz decomposition [109]. The curl-free portion will be denoted by the gradient of a potential, $\nabla\varphi$. This decomposition can be written

$$\mathbf{V} = \mathbf{V}^d + \nabla\varphi \; . \tag{11.1}$$

This equation holds the key for computing solutions to the incompressible flow equations using projections.

Taking the divergence of (11.1) gives

$$\nabla \cdot \mathbf{V} = \nabla \cdot \mathbf{V}^d + \nabla \cdot \nabla\varphi \rightarrow \nabla \cdot \mathbf{V} = \nabla \cdot \nabla\varphi \; .$$

Once φ has been computed, then the solution can be found through

$$\mathbf{V}^d = \mathbf{V} - \nabla\varphi \; . \tag{11.2}$$

We can then write the projection operator based on these practical steps toward the solution given above

$$\mathcal{P} = \mathbf{I} - \nabla \left(\nabla \cdot \nabla\right)^{-1} \nabla \cdot \; .$$

After the application of $\mathcal{P}$ to a vector field, $\mathbf{V}$, this field will be divergence-free. We can also write $\nabla\varphi = \mathcal{Q}(\mathbf{V})$ with $\mathcal{Q} = \mathbf{I} - \mathcal{P}$. One of the most important aspects of projections is that they are idempotent, i.e., $\mathcal{P}^2 = \mathcal{P}$, or repeated application of the operator will not change the result. Also, it can be shown that the norm of the operator is less than or equal to one, $\|\mathcal{P}(\mathbf{V})\|_2 \leq \|\mathbf{V}\|_2$. Thus, it is easy to prove that the operator is stable [106].

A more general and complete introduction to this theory can be found in the book by Chorin and Marsden [109].

This development can be applied to produce a pressure Poisson equation. Operationally, the discrete versions of the pressure Poisson equation fall into either exact or approximate projection categories largely dependent upon the chosen mesh staggering and the accompanying discrete divergence and gradient operators.

---

[2] We use the notation $\mathbf{V}$ to denote either the velocity, $\mathbf{u}$ or its time derivative $\partial_t \mathbf{u}$ since either can be projected in the methods described in this chapter.

11.2.2 Continuous Projections for Variable Density Incompressible Flow

In many physical cases the fluids varying in density while still being essentially incompressible. In these cases the high resolution methods focused upon in this book are especially useful. Often these situations are characterized by sharp interfaces and flow gradients. For example, the Rayleigh-Taylor instability can be described by this case. Below, we show that the basic projection formalism can be extended without difficulty to cover this circumstance.

For variable density flows (11.1) is written

$$\rho \mathbf{V} = \rho \mathbf{V}^d + \nabla \varphi \,,$$

or

$$\mathbf{V} = \mathbf{V}^d + \sigma \nabla \varphi \,,$$

where $\sigma = 1/\rho$. In a straightforward fashion, the elliptic equation for φ is

$$\nabla \cdot \mathbf{V} = \nabla \cdot \sigma \nabla \varphi \,. \tag{11.3}$$

The correction equation is

$$\mathbf{V}^d = \mathbf{V} - \sigma \nabla \varphi \,.$$

The projection operators now become

$$\mathcal{P}_\sigma = \mathbf{I} - \sigma \nabla \left(\nabla \cdot \sigma \nabla \right)^{-1} \nabla \cdot \,,$$

and

$$\mathcal{Q}_\sigma = \mathbf{I} - \mathcal{P}_\sigma \,.$$

The constant density projection operators can be denoted by $\mathcal{P}_0$ and $\mathcal{Q}_0$, but are the same as the variable density operators for a constant σ. Unless otherwise noted, we will discuss variable density projections for the remainder of this chapter.

Note, the variable density situations also represent many challenging situations for numerical methods and produce many specialized techniques. By and large, the extra techniques and numerical difficulties are not present for constant density flows with relatively smooth gradients (i.e., well-resolved).

11.3 Exact Discrete Projections

In order to make the previous discussion useful for computing flows, we must define a discrete analog to the aforementioned operators. The discrete case is fraught with difficulties associated with placing the continuous framework discussed above onto a discrete grid. The process of dealing with the discrete nature of the approximations will generate most of the details in the following pages.

Let D be the discrete divergence, and $\mathbf{G}$, the discrete gradient. Choosing to let the continuous case be our guide, we require that the operators be skew adjoint, $D = -\mathbf{G}^T$, or as an inner product,

$$(D\mathbf{V}, \varphi)_s = -(\sigma\mathbf{G}\varphi, \mathbf{V})_v ,$$

where $(,)$ is a discrete inner product (defined below) over grid cells with s and v referring to scalar and vector products, respectively.

11.3.1 Cell-Centered Exact Projections

The vector inner product is density weighted

$$(\mathbf{V}_1, \mathbf{V}_2)_v = \sum_i \sum_j V_{1,i,j} V_{2,i,j} \rho_{i,j} .$$

With this said, the discrete operators on a collocated grid are

$$D_{i,j,k}\mathbf{V} = \frac{V^x_{i+1,j,k} - V^x_{i-1,j,k}}{2\Delta x} + \frac{V^y_{i,j+1,k} - V^y_{i,j-1,k}}{2\Delta y} + \frac{V^z_{i,j,k+1} - V^z_{i,j,k-1}}{2\Delta z} , \tag{11.4}$$

or in axisymmetric coordinates

$$D_{i,j}\mathbf{V} = \frac{r_{i+1/2}\left(V^r_{i+1,j} + V^r_{i,j}\right) - r_{i-1/2}\left(V^r_{i-1,j} + V^r_{i,j}\right)}{2r_i\Delta r} + \frac{V^z_{i,j+1} - V^z_{i,j-1}}{2\Delta z} , \tag{11.5}$$

and

$$\sigma\mathbf{G}_{i,j,k}\varphi = \begin{pmatrix} \dfrac{\varphi_{i+1,j,k} - \varphi_{i-1,j,k}}{2\rho_{i,j,k}\Delta x} \\[2ex] \dfrac{\varphi_{i,j+1,k} - \varphi_{i,j-1,k}}{2\rho_{i,j,k}\Delta y} \\[2ex] \dfrac{\varphi_{i,j,k+1} - \varphi_{i,j,k-1}}{2\rho_{i,j,k}\Delta z} \end{pmatrix} .$$

The discrete projection operators form a non-standard discrete Laplacian through $L = D\sigma\mathbf{G}$ that decouples a two-dimensional grid into four distinct grids as shown in Fig. 11.2 (eight in three dimensions). Aspects of this decoupling persist with approximate projection method on cell-centered grids and are associated with the form of the divergence operator. These will become manifestly obvious through the necessity and the design of filters for the deviations from being divergence-free associated with approximate projections.

The four distinct grids are coupled through boundary conditions, but the decoupling is worrisome and creates problems for solution techniques [266, 312]. This has shown itself particularly when the equations contain highly localized source terms such as those that exist in combustion applications.

This discrete Laplacian taken from (11.3) is

$$\begin{aligned} L_{i,j,k}\varphi = & \frac{\sigma_{i+1,j,k}\left(\varphi_{i+2,j,k} - \varphi_{i,j,k}\right) - \sigma_{i-1,j,k}\left(\varphi_{i,j,k} - \varphi_{i-2,j,k}\right)}{4\Delta x^2} \\ & + \frac{\sigma_{i,j+1,k}\left(\varphi_{i,j+2,k} - \varphi_{i,j,k}\right) + \sigma_{i,j-1,k}\left(\varphi_{i,j,k} - \varphi_{i,j-2,k}\right)}{4\Delta y^2} \\ & + \frac{\sigma_{i,j,k+1}\left(\varphi_{i,j,k+2} - \varphi_{i,j,k}\right) + \sigma_{i,j-1,k}\left(\varphi_{i,j,k} - \varphi_{i,j,k-2}\right)}{4\Delta z^2} \end{aligned} \quad (11.6a)$$

or in axisymmetric coordinates

$$\begin{aligned} L_{i,j}\varphi = & \frac{r_{i+1/2}\sigma_{i+1,j}\left(\varphi_{i+2,j} - \varphi_{i,j}\right) - r_{i-1/2}\sigma_{i-1,j}\left(\varphi_{i,j} - \varphi_{i-2,j}\right)}{4r_i\Delta r^2} \\ & + \frac{\left(r_{i+1/2} - r_{i-1/2}\right)\sigma_{i,j}\left(\varphi_{i+1,j} - \varphi_{i-1,j}\right)}{4r_i\Delta r^2} \\ & + \frac{\sigma_{i,j+1}\left(\varphi_{i,j+2} - \varphi_{i,j}\right) + \sigma_{i,j-1}\left(\varphi_{i,j} - \varphi_{i,j-2}\right)}{4\Delta z^2} \,. \end{aligned} \quad (11.6b)$$

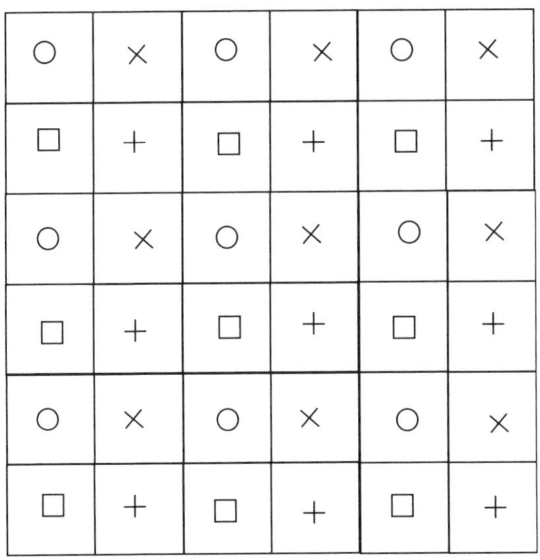

Fig. 11.2. This grid shows the four decoupled grids for a two-dimensional cell-centered exact projection. The sub-stencils are denoted by $\circ$, $+$, $\times$ and $\Box$.

The discrete divergence will be zero only for the divergence operator used to define the projection. The divergence will be nonzero for other divergence-difference stencils that can be defined for the same velocity data (see Chap. 12 for a discussion of other divergence operators as associated with filters to remove unwanted concentrations of divergence errors).

We can analyze a number of properties of the method through Fourier analysis. The Fourier transform and its inverse are

$$\hat{\psi}(\alpha_x, \alpha_y) = \frac{1}{2\pi} \int_{-\infty}^{\infty} \psi(x, y)\, e^{-\imath(\alpha_x x + \alpha_y y)}\, dx\, dy ,$$

and

$$\psi(x, y) = \int_{-\infty}^{\infty} \hat{\psi}(\alpha_x, \alpha_y)\, e^{\imath(\alpha_x x + \alpha_y y)}\, d\alpha_x\, d\alpha_y ,$$

where $\imath = \sqrt{-1}$ and α_x and α_y are the wave numbers in the x- and y-directions, respectively. These will be introduced into the finite-difference equations to produce their symbol, $\Lambda(\cdot)$. This can be analyzed to determine the properties of the operator. The null-space of the operator can be seen with the symbol. The number of zeros of the symbol in the range $\alpha \in [0, \pi]$ show the dimension of the null-space. For instance, $\Lambda(D\mathbf{G})$ has four zeros in two dimensions leading to a null-space that gives the decoupled grids described above (as seen in Fig. 11.2).

We can find the truncation error through expanding the Fourier transform of the discrete equations in a Taylor series and comparing them with the exact transform. The continuous first-order operators $\nabla\cdot$ and ∇ have the symbols

$$\Lambda(\nabla\cdot) = (\imath\alpha_x, \imath\alpha_y, \imath\alpha_z) ,$$

and

$$\Lambda(\nabla) = (\imath\alpha_x, \imath\alpha_y, \imath\alpha_z)^T ,$$

respectively. The continuous projection Laplacian $\nabla \cdot \nabla$ has the symbol

$$\Lambda(\nabla \cdot \nabla) = -\alpha_x^2 - \alpha_y^2 - \alpha_z^2 .$$

Expanding the Fourier transforms of the discrete forms of each will give their truncation error. For the D it is

$$\left(-\frac{1}{6}\alpha_x^2 + \mathcal{O}\left(\alpha_x^4\right), -\frac{1}{6}\alpha_y^2 + \mathcal{O}\left(\alpha_y^4\right), -\frac{1}{6}\alpha_z^2 + \mathcal{O}\left(\alpha_z^4\right)\right) .$$

The truncation error for $\mathbf{G}$ is

$$\left(-\frac{1}{6}\alpha_x^2 + \mathcal{O}\left(\alpha_x^4\right), -\frac{1}{6}\alpha_y^2 + \mathcal{O}\left(\alpha_y^4\right) - \frac{1}{6}\alpha_z^2 + \mathcal{O}\left(\alpha_z^4\right)\right)^T .$$

The Laplacian, $D\mathbf{G}$, has an error

$$-\frac{1}{3}\alpha_x^2 - \frac{1}{3}\alpha_y^2 - \frac{1}{3}\alpha_z^2 + \mathcal{O}\left(\alpha_x^4, \alpha_y^4, \alpha_z^4\right) .$$

This also demonstrates that the projection is second-order accurate in space. Proper centering of the variables will give second-order temporal accuracy.

The projection operators are now defined

$$\mathbf{P}_\sigma = \mathbf{I} - \sigma\mathbf{G}\left(D\sigma\mathbf{G}\right)^{-1} D\ ,$$

and

$$\mathbf{Q}_\sigma = \sigma\mathbf{G}\left(D\sigma\mathbf{G}\right)^{-1} D\ .$$

These are the discrete analogs of $\mathcal{P}$ and $\mathcal{Q}$. They are also designed to have the desirable properties of the continuous projections carry over as well (thus $\mathbf{P} = \mathbf{P}^2$ and $\|\mathbf{P}(\mathbf{V})\|_2 \leq \|\mathbf{V}\|_2$). For this reason they are also known as "exact projections."

It is a simple matter to show that the norm of this discrete projection operator is like that of the continuous case. We will work with the constant density projection. Recognizing that the D and $\mathbf{G}$ are defined as central difference operators, i.e.,

$$D = \left(\delta_x, \delta_y, \delta_z\right)\ ,$$

and

$$\mathbf{G} = \left(\delta_x, \delta_y, \delta_z\right)^T\ ,$$

allows the projection operator to be written in a convenient form. This matrix form is

$$\mathbf{P} = \mathbf{I} - \frac{1}{\delta_x^2 + \delta_y^2 + \delta_z^2}\begin{bmatrix} \delta_x\delta_x & \delta_x\delta_y & \delta_x\delta_z \\ \delta_y\delta_x & \delta_y\delta_y & \delta_y\delta_z \\ \delta_z\delta_x & \delta_z\delta_y & \delta_z\delta_z \end{bmatrix}\ .$$

The eigenvalues of this matrix are zero and one, thus showing that $\|\mathbf{P}\| \leq 1$.

One down side to this projection is the requisite linear algebra problem that must be solved to implement the method. Typically, a Krylov method (e.g., conjugate gradient) can be used. Preconditioning can be done with a standard method. More efficient multigrid schemes are a challenge because of the decoupling, but have been successful [266].

11.3.2 Vertex-Centered Exact Projections

Moving to a grid where the pressures are centered at the vertices of the mesh cells produces a different method. The decoupling is still present, but less severe, producing a true "checkerboard" mode in it appearance. This is related to the improved velocity-pressure coupling on this grid and the smaller dimension of the null space of the Laplacian, $D\sigma\mathbf{G}$. Again, aspects of this decoupling will present themselves in association with approximate projections on the same grid staggering and filtering algorithms. The filtering methods based on this discretization will be described in Chap. 12. This method retains the advantage of centering other variables (velocities and density) the cell-centers.

With this said, the discrete operators on a vertex-staggered grid are

$$D_{i+1/2,j+1/2,k+1/2}\mathbf{V} = \frac{\sum_{j'=0,1}\sum_{k'=0,1} V^x_{i+1,j+j',k+k'} - \sum_{j'=0,1}\sum_{k'=0,1} V^x_{i,j+j',k+k'}}{8\Delta x} + \frac{\sum_{i'=0,1}\sum_{k'=0,1} V^y_{i+i',j+1,k+k'} - \sum_{i'=0,1}\sum_{k'=0,1} V^y_{i+i',j,k+k'}}{8\Delta y} + \frac{\sum_{i'=0,1}\sum_{j'=0,1} V^z_{i+i',j+j',k+1} - \sum_{i'=0,1}\sum_{j'=0,1} V^z_{i+i',j+j',k}}{8\Delta z},$$

and

$$\sigma\mathbf{G}_{i,j,k}\varphi = \begin{pmatrix} \frac{\sum_{j'=-1/2,1/2}\sum_{k'=-1/2,1/2}\varphi_{i+1/2,j+j',k+k'}}{8\rho_{i,j,k}\Delta x} - \frac{\sum_{j'=-1/2,1/2}\sum_{k'=-1/2,1/2}\varphi_{i-1/2,j+j',k+k'}}{8\rho_{i,j,k}\Delta x} \\ \hdashline \frac{\sum_{i'=-1/2,1/2}\sum_{k'=-1/2,1/2}\varphi_{i+i'j+1/2,k+k'}}{8\rho_{i,j,k}\Delta y} - \frac{\sum_{i'=-1/2,1/2}\sum_{k'=-1/2,1/2}\varphi_{i+i',j-1/2,k+k'}}{8\rho_{i,j,k}\Delta y} \\ \hdashline \frac{\sum_{i'=-1/2,1/2}\sum_{j'=-1/2,1/2}\varphi_{i+i',j+j',k+1/2}}{8\rho_{i,j,k}\Delta z} - \frac{\sum_{j'=-1/2,1/2}\sum_{k'=-1/2,1/2}\varphi_{i+i',j+j',k-1/2}}{8\rho_{i,j,k}\Delta z} \end{pmatrix}.$$

It is important to recognize that the divergence and gradient operators are retained in approximate projection methods. Despite the complexity of these operators, the constant density Laplacian is surprisingly simple (setting $h = \Delta x = \Delta y = \Delta z$),

$$L_{i,j,k}\varphi = \frac{\sum_{i'=-1/2,3/2}\sum_{j'=-1/2,3/2}\sum_{-1/2,3/2}\varphi_{i+i',j+j',k+k'}}{h^2} - \frac{8\varphi_{i+1/2,j+1/2,k+1/2}}{h^2}, \tag{11.7}$$

We will dispense with writing the variable density Laplacian because important cancellation does not occur. The form of this Laplacian gives an insight into the null space of this operator as the stencil is rotated to the vertices of the zone rather than along coordinate lines as with the classical Laplacian.

Linear algebra is much the same as the cell-centered projection with Krylov methods being simple and effective [45], and more efficient schemes like multigrid producing a distinct challenge. Unlike cell-centered exact projections, we are not aware of any successful multigrid method, although nothing necessarily precludes success in this endeavor.

Remark 11.3.1. Some of the earlier work with exact projections used a rather different formulation of the "pressure" equation [45, 50, 522]. Rather than define the equation in terms of some potential field, a stream function was used where

$$\mathbf{V} = -\nabla \times \Psi .$$

Because $\nabla \times \mathbf{V} = \nabla \times \mathbf{V}^d$ and $\nabla \times \nabla\varphi = 0$, we can take the curl of (11.1) to get the equation

$$\nabla \times \nabla \times \Psi = \nabla \times \mathbf{V} .$$

In two dimensions, this simplifies to

$$\nabla \cdot \nabla\Psi = -\nabla \times \mathbf{V} ,$$

with boundary data on solid walls of $\Psi = 0$. This Dirichlet data makes for easier linear algebra than the Neumann conditions on a pressure equation. The problem with this is that it makes things much worse in three dimensions because the equation does not simplify to an elliptic equation, but to three elliptic equations. For variable density, we get

$$\nabla \cdot \rho\nabla\Psi = -\nabla \times \rho\mathbf{V} .$$

11.3.3 The MAC Projection

Harlow and Welch's MAC method developed at Los Alamos was in a sense the first projection method although it was not referred to as such [245, 240]. It is also called the pressure Poisson approach. Because of the properties of the velocity-pressure coupling, the method has extremely desirable properties from that point-of-view. While the staggering of the velocities makes the pressure solution well-behaved it greatly complicates the solution of the advection-diffusion portion of the equations. Its most appealing property is that a projection produces a standard seven (for 3-D) or five (for 2-D) point Laplacian for a second-order algorithm. This makes linear algebra extremely simple and the class of methods that can solve the resultant problem efficiently is large. It is for this reason that this discrete Laplacian will be the focus of the next chapter.

One should immediately recognize the intimate connection between this class of methods and projections. Although it was not referred to as a projection, the MAC method was the first projection algorithm. As such the various techniques discussed earlier basically apply to these methods in whole. This includes both exact and approximate projections as discussed in Chaps. 11 and 12.

MAC still enjoys a remarkable amount of use to this day (including in motion picture animations!). For example, see the recent book by Fedkiw and Osher [185] where these algorithms are described and employed in the production of special effects.

For a second-order solution, given a velocity field at a given time level, the algorithm proceeds quite straightforwardly, the divergence of the velocity is

$$\nabla \cdot \tilde{\mathbf{u}}_{i,j,k} = \frac{\tilde{u}_{i+1/2,j,k} - \tilde{u}_{i-1/2,j,k}}{\Delta x} + \frac{\tilde{v}_{i,j+1/2,k} - \tilde{v}_{i,j-1/2,k}}{\Delta y} + \frac{\tilde{w}_{i,j,k+1/2} - \tilde{w}_{i,j,k-1/2}}{\Delta z}, \tag{11.8}$$

coupled to a pressure solution,

$$\Delta t \frac{p_{i+1,j,k} - 2p_{i,j,k} + p_{i-1,j,k}}{\Delta x^2} + \Delta t \, \frac{p_{i,j+1,k} - 2p_{i,j,k} + p_{i,j-1,k}}{\Delta y^2} + \Delta t \, \frac{p_{i,j,k+1} - 2p_{i,j,k} + p_{i,j,k-1}}{\Delta z^2} = \nabla \cdot \tilde{\mathbf{u}}_{i,j,k} \, . \tag{11.9}$$

Finally, the velocity solution is corrected to complete the projection,

$$u_{i+1/2,j,k} = \tilde{u}_{i+1/2,j,k} - \Delta t \, \frac{p_{i+1,j,k} - p_{i,j,k}}{\Delta x}, \tag{11.10a}$$

$$v_{i,j+1/2,k} = \tilde{v}_{i,j+1/2,k} - \Delta t \, \frac{p_{i,j+1,k} - p_{i,j,k}}{\Delta y}, \tag{11.10b}$$

and

$$w_{i,j,k+1/2} = \tilde{w}_{i,j,k+1/2} - \Delta t \frac{p_{i,j,k+1} - p_{i,j,k}}{\Delta z}, \tag{11.10c}$$

with $\nabla \cdot u = 0$ to the accuracy of the solution of the linear system (11.9).

11.3.4 The MAC Projection Used with Godunov-Type Methods

In conjunction with high resolution schemes the MAC projection is a tool to get conservation of quantities as they are advected in a divergence-free flow. In the unsplit Godunov method described in Chap. 14, to complete the computation of time-centered velocities, the pressure at time n must be known. When the Godunov algorithm is initiated, the pressure is only known at time $n - 1/2$. Solving for the time n pressure is desirable for two reasons: if the boundary conditions are a function of time, this step is necessary for accuracy, and the fluxes can be computed conservatively. This is also for stability for a Courant number greater than one-half [46], although this phenomenon does not seem to be present for the approximate projections.

First, the values of $\mathbf{u}$ are computed at cell-edges and advanced to a pseudo-time-centered value with or without the old time pressure. This is directly analogous to the incremental and pressure forms of the projection algorithm. Later, we will test a conjecture related to the pressure form of the predictor step. This is because for the pressure form with time-dependent boundary conditions, the cell-centered upwinding is needed to maintain accuracy. If the old-time pressure is used, it is applied to each cell with a cell-centered derivative (the same as used for the projection).

In the following discussion we will retain the use of two dimensional notation, with the three dimensional implementation being straightforward. The pseudo-time-centered edge velocities are computed from the explicit expressions that will be discussed in more detail in Chap. 14 (Sect. 14.1),

$$\mathbf{u}^{*,n+1/2}_{i+1/2,j} = \mathbf{u}^n_{i,j} - \frac{\Delta t\ \bar{v}^n_{i,j}}{2}\frac{\partial \mathbf{u}^n}{\partial y} + \frac{\Delta t\ \nu}{2} L_{i,j}\mathbf{u}^n + \frac{\Delta t}{2} S_{\mathbf{u}} ,$$

where $S_{\mathbf{u}}$ does not include pressure terms computed at time n, but may include pressure terms from $n-1/2$. Scalar fields can be computed directly

$$\psi^{n+1/2}_{i+1/2,j} = \psi^n_{i,j} - \frac{\Delta t\ \bar{v}^n_{i,j}}{2}\frac{\partial \psi^n}{\partial y} + \frac{\Delta t\ \nu}{2} L_{i,j}\psi^n + \frac{\Delta t}{2} S_{\psi} ,$$

where S_ψ contains terms associated with the derivatives of u as well as source terms (for example, volume forces). Given this edge- and pseudotime-centered velocity field, $\mathbf{u}^{*,n+1/2}$, a divergence is computed

$$D^{\mathrm{MAC}}_{i,j}\mathbf{u}^{*,n+1/2} = \frac{u^{*,n+1/2}_{i+1/2,j} - u^{*,n+1/2}_{i-1/2,j}}{\Delta x} + \frac{v^{*,n+1/2}_{i,j+1/2} - v^{*,n+1/2}_{i,j-1/2}}{\Delta y} ,$$

with the axisymmetric form being

$$D^{\mathrm{MAC}}_{i,j}\mathbf{u}^{*,n+1/2} = \frac{r_{i+1/2}u^{*,n+1/2}_{i+1/2,j} - r_{i-1/2}u^{*,n+1/2}_{i-1/2,j}}{r_i \Delta r} + \frac{v^{*,n+1/2}_{i,j+1/2} - v^{*,n+1/2}_{i,j-1/2}}{\Delta z} .$$

The first discrete Laplacian that we will define is a standard five-point operator in two dimensions

$$\begin{aligned} L_{i,j}\varphi = {} & \frac{\sigma_{i+1/2,j}\left(\varphi_{i+1,j} - \varphi_{i,j}\right) - \sigma_{i-1/2,j}\left(\varphi_{i,j} - \varphi_{i-1,j}\right)}{\Delta x^2} \\ & + \frac{\sigma_{i,j+1/2}\left(\varphi_{i,j+1} - \varphi_{i,j}\right) - \sigma_{i,j-1/2}\left(\varphi_{i,j} - \varphi_{i,j-1}\right)}{\Delta y^2} , \end{aligned} \tag{11.11a}$$

where $\sigma_{i+1/2,j} = \left(\sigma_{i,j} + \sigma_{i+1,j}\right)/2$. In axisymmetric coordinates, the Laplacian is

$$\begin{aligned} L_{i,j}\varphi = {} & \frac{r_{i+1/2}\sigma_{i+1/2,j}\left(\varphi_{i+1,j} - \varphi_{i,j}\right) - r_{i-1/2}\sigma_{i-1/2,j}\left(\varphi_{i,j} - \varphi_{i-1,j}\right)}{r_i \Delta r^2} \\ & + \frac{\sigma_{i,j+1/2}\left(\varphi_{i,j+1} - \varphi_{i,j}\right) - \sigma_{i,j-1/2}\left(\varphi_{i,j} - \varphi_{i,j-1}\right)}{\Delta z^2} . \end{aligned} \tag{11.11b}$$

There are two gradients, one for x-edges, and another for y-edges, in finite difference form. For the x-edges

$$\sigma\mathbf{G}_{i+1/2,j}\varphi = \begin{pmatrix} \dfrac{\varphi_{i+1,j}-\varphi_{i,j}}{\rho^n_{i+1/2,j}\Delta x} \\ \\ \left[\dfrac{\varphi_{i,j+1}-\varphi_{i,j}}{4\rho^n_{i,j+1/2}\Delta y}+\dfrac{\varphi_{i,j}-\varphi_{i,j-1}}{4\rho^n_{i,j-1/2}\Delta y}\right. \\ \left.+\dfrac{\varphi_{i+1,j+1}-\varphi_{i+1,j}}{4\rho^n_{i+1,j+1/2}\Delta y}+\dfrac{\varphi_{i+1,j}-\varphi_{i+1,j-1}}{4\rho^n_{i+1,j-1/2}\Delta y}\right] \end{pmatrix}$$

and for the y-edges

$$\sigma\mathbf{G}_{i,j+1/2}\varphi = \begin{pmatrix} \left[\dfrac{\varphi_{i+1,j}-\varphi_{i,j}}{4\rho^n_{i+1/2,j}\Delta y}+\dfrac{\varphi_{i,j}-\varphi_{i-1,j}}{4\rho^n_{i-1/2,j}\Delta y}\right. \\ \left.+\dfrac{\varphi_{i+1,j+1}-\varphi_{i,j+1}}{4\rho^n_{i+1/2,j+1}\Delta y}+\dfrac{\varphi_{i,j+1}-\varphi_{i-1,j+1}}{4\rho^n_{i-1/2,j+1}\Delta y}\right] \\ \\ \dfrac{\varphi_{i,j+1}-\varphi_{i,j}}{\rho^n_{i,j+1/2}\Delta y} \end{pmatrix}.$$

The edge-centered values of σ are computed from the average values on either side of the edge, i.e., $\rho^n_{i+1/2,j} = (\rho^n_{i,j}+\rho^n_{i+1,j})/2$.

If the divergence operator defined above is applied to a vector field composed of the normal gradients at the edges, the resulting five-point operator is the same (in general form) to the five-point approximate projection Laplacian. This can be used to form an equation for φ,

$$D^{\mathrm{MAC}}\sigma^n\mathbf{G}\varphi = D^{\mathrm{MAC}}\mathbf{u}^*.$$

These equations are given as (11.11a) and (11.11b). This equation is solved with homogeneous Neumann boundary conditions, and the edge velocities are corrected appropriately with the gradients defined above. It is important to construct the ghost cell values using homogeneous Neumann formulas (or more simply, the boundary values of $\mathbf{u}$ can be reset to their proper values). The correction has the form

$$\mathbf{u}^{n+1/2}_{i+1/2,j} = \mathbf{u}^{*,n+1/2}_{i+1/2,j} - \sigma^n_{i+1/2,j}\mathbf{G}_{i+1/2,j}\varphi\,.$$

In either case, the fluxes computed using nonconservative differencing (given in Equations (14.5)–(14.6) of Chap. 14) can be used, but if the MAC projection is used the fluxes can be computed conservatively and because the velocity field is divergence-free quantities will be conserved. This is because the time-centered velocity field on edges is discretely solenoidal. The conservative fluxes have the form

$$\frac{\partial\left(u\psi\right)}{\partial x} = \frac{u_{i+1/2,j}\psi_{i+1/2,j}-u_{i-1/2,j}\psi_{i-1/2,j}}{\Delta x}\,,$$

or in axisymmetric coordinates

$$\frac{1}{r}\frac{\partial}{\partial r}(ru\psi) = \frac{r_{i+1/2}u_{i+1/2,j}\psi_{i+1/2,j} - r_{i-1/2}u_{i-1/2,j}\psi_{i-1/2,j}}{r_i \Delta r},$$

and

$$\frac{\partial(v\psi)}{\partial y} = \frac{v_{i,j+1/2}\psi_{i,j+1/2} - v_{i,j-1/2}\psi_{i,j-1/2}}{\Delta y}.$$

11.3.5 Other Exact Projections

Perhaps the last method to introduce here is the method introduced by Strikwerda [516]. We introduce this method for the purpose of showing how the basic projection methodology can be extended to higher order. This method is distinguished by using third-order accurate finite differences for the gradient and divergence, but otherwise like earlier methods in defining these through a discrete adjoint principle. This method has been tested by and documented with modern high resolution schemes by Almgren [7]. The problem with this method is the larger (nonstandard) stencil which complicates the numerical linear algebra.

For this scheme the gradient is defined by

$$\mathbf{G}_{i,j,k} = \begin{pmatrix} \varphi_{i+1,j,k} - \varphi_{i-1,j,k} \\ \varphi_{i,j+1,k} - \varphi_{i,j-1,k} \\ \varphi_{i,j,k+1} - \varphi_{i,j,k-1} \end{pmatrix}, \tag{11.12}$$

and following the adjoint principle, the divergence is

$$D_{i,j,k} = \varphi_{i+1,j,k} - \varphi_{i-1,j,k} + \varphi_{i,j+1,k} - \varphi_{i,j-1,k} + \varphi_{i,j,k+1} - \varphi_{i,j,k-1}. \tag{11.13}$$

Shu and E [176] use a fourth-order projection that is solved via spectral method. This method can generalize to any order.

11.4 Second-Order Projection Algorithms for Incompressible Flow

Given the aforementioned development, we can prepare the outline of a projection algorithm. The algorithm will consist of two steps as alluded to before: a separate step where the solenoidal nature of the velocity field is ignored and a corrective step using the projection to force the velocity to be solenoidal. This is often referred to as a "fractional step method." In a more general sense, this algorithm can be modified to a first-order method and applied in a sequence of steps through the variety of methods introduced in Chap. 7 (and Chap. 12). In the following algorithm this can be achieved by dropping all evaluations from time level $n+1/2$ to n for explicit schemes. In the parlance of this chapter, the following algorithm is in the Lax-Wendroff tradition as spearheaded by the pioneering work of Bell, Colella and Glaz [45].

In the first step, the velocity is computed solving the motion equations as convection-diffusion equations (the incremental form),

$$\mathbf{u}_{i,j}^{*,n+1} = \mathbf{u}_{i,j}^{n} - \Delta t \left[(\mathbf{u}\cdot\nabla\mathbf{u})_{i,j}^{n+1/2} + \sigma_{i,j}^{n+1/2}\mathbf{G}_{i,j}\phi^{n-1/2} \right. \tag{11.14}$$

$$\left. + \frac{\nu\sigma_{i,j}^{n+1/2}}{2} L_{i,j}\left(\mathbf{u}^n + \mathbf{u}^{*,n+1}\right) + \mathbf{F}_{i,j}^{n+1/2} \right],$$

where $\mathbf{F}$ is the volume force (the subscript V has been dropped for simplicity) and ϕ is the incompressible pressure (see Chap. 2). Another form (the pressure form) has $\mathbf{G}\phi^{n-1/2}$ removed,[3]

$$\mathbf{u}_{i,j}^{*,n+1} = \mathbf{u}_{i,j}^{n} - \Delta t \Big[(\mathbf{u}\cdot\nabla\mathbf{u})_{i,j}^{n+1/2}$$

$$+ \frac{\nu\sigma_{i,j}^{n+1/2}}{2} L_{i,j}\left(\mathbf{u}^n + \mathbf{u}^{*,n+1}\right) + \mathbf{F}_{i,j}^{n+1/2} \Big]. \tag{11.15}$$

where $\sigma_{i,j}^{n+1/2} = \left(\sigma_{i,j}^{n} + \sigma_{i,j}^{n+1}\right)/2$ and $\mathbf{F}_{i,j}^{n+1/2} = \left(\mathbf{F}_{i,j}^{n} + \mathbf{F}_{i,j}^{n+1}\right)/2$, or $\mathbf{F}_{i,j}^{n+1/2} = \mathbf{F}\left(\mathbf{x}_{i,j}, t^{n+1/2}\right)$.

We then take $\mathbf{V} = \left(\mathbf{u}^{*,n+1} - \mathbf{u}^n\right)/\Delta t$ and $\mathbf{V}^d = \left(\mathbf{u}^{n+1} - \mathbf{u}^n\right)/\Delta t$, or $\mathbf{V} = \mathbf{u}^{*,n+1}/\Delta t$ and $\mathbf{V}^d = \mathbf{u}^{n+1}/\Delta t$, and apply the projection. First, we solve the linear system

$$D\sigma^{n+1/2}\mathbf{G}\varphi = D\mathbf{V},$$

then correct the velocity field with (11.2) as

$$\mathbf{u}_{i,j}^{n+1} = \mathbf{u}_{i,j}^{*,n+1} - \frac{\Delta t}{\rho_{i,j}^{n+1/2}}\mathbf{G}_{i,j}\varphi .$$

For the incremental form

$$\phi_{i,j}^{n+1/2} = \phi_{i,j}^{n-1/2} + \varphi_{i,j},$$

or for the pressure form

$$\phi_{i,j}^{n+1/2} = \varphi_{i,j} .$$

Of course the diffusion can also be treated explicitly with a requisite negative impact on the stability condition. In the next chapter we will return to these forms to analyze the discrete errors resulting from the different forms. For

---

[3] Note that the index should be $n-1/2$ if it is a known pre-projection of the pressure, otherwise it should be $n+1/2$.

the exact projections these differences are minor, but when the divergence is a function of the truncation error, these differences are substantiative.

Given in an algorithmic form for an explicit forward Euler step (a good building block for Runge-Kutta for example):

1. Begin with the initial data at time n, velocity, $\mathbf{u}$, density, ρ, and pressure, p.
2. One may or may not want to produce a divergence-free velocity field defined at the edges of the control volumes to perform the advection (i.e., the "MAC projection").
3. Advance the velocity to a pseudo-time $n+1$ ($\mathbf{u}^*$ value) using the equation of motion ignoring the divergence-free constraint.
4. Advance the density equation and any other scalar equations.
5. Form a pressure equation by forming a right hand side from the divergence of velocity, $D\mathbf{u}$, and the Laplacian, $D\mathbf{G}p$.
6. Solve the equation $D\mathbf{G}p^{n+1} = D\mathbf{u}^*$ to some specified accuracy,
7. and correct the velocity field $\mathbf{u}^{n+1} = \mathbf{u}^* - \Delta t\, \mathbf{G}p^{n+1}$.
8. The algorithm is done and ready for the next time step.

11.5 Boundary Conditions

The quality of many results with a given method depend upon the boundary conditions implementation and this is particularly the case with elliptic equations such as the pressure Poisson equation. Here, we will discuss the most essential aspects of defining boundary conditions that improve the efficiency and accuracy of computations.

11.5.1 Solvability

For many common applications in fluid mechanics, solid wall and symmetry boundary conditions are used. At the boundaries this translates to a Neumann condition on φ, where φ is a generic solution variable. Let us first consider the continuous problem

$$\nabla^2 \varphi = f \in \Omega \,,$$

with $\mathbf{n} \cdot \nabla \varphi = b$ on Γ. The subscripts Ω and Γ refer to the interior and boundary of the domain, respectively. Solvability requires that

$$\int f \,\mathrm{d}\Omega = \int b \,\mathrm{d}\Gamma \,.$$

The discrete equations mimic the continuous. Stated more precisely, it is required that for the RHS, $\mathbf{b}$, of the discrete system $\mathbf{Ax} = \mathbf{b}$, $\mathbf{b}^T \mathbf{l} = 0$ where $\mathbf{l}$ is the left eigenvector of $\mathbf{A}$. For the case of a positive semi-definite operator $\mathbf{l}$ is composed of some constant. The above is equivalent to the sum of the right-hand side vector, **b**, summed over all equations being zero. This is a direct consequence of having a conservative divergence operator.

When this condition is violated the system can be "fixed" by either explicitly removing the nonzero amount of divergence [101, 253], or making use of a single Dirichlet boundary condition [415]. In the case of overlapping grids, the loss of conservation is inevitable and this step is essential, but more generally this technique is inadvisable because it covers up an essential discretization error. The Dirichlet boundary-condition-fix is more egregious and while making the system solvable, it renders the conditioning of the system quite poor. Again, this only covers up a more fundamental error.

We will show why the best choice for these boundary conditions are homogeneous Neumann conditions. This topic has been covered in Peyret and Taylor [423] for the case of a projection method on a MAC grid. We will write the equation down in an alternate form with terms defined on the basis of whether or not the edge they are differenced across is on the interior of the domain Ω or on the boundary Γ, i.e.,

$$L\varphi = D\sigma\mathbf{G}\varphi = D_{\Omega}\sigma\mathbf{G}\varphi + D_{\Gamma}\sigma\mathbf{G}\varphi = D_{\Omega}\mathbf{V} + D_{\Gamma}\mathbf{V}\ , \tag{11.16}$$

with our goal to show that $\sum_{i,j} D\mathbf{V} = 0$.

The boundary conditions can be included for $D_{\Gamma}\sigma\mathbf{G}\varphi$ through the use of $\sigma\mathbf{G}\varphi = \mathbf{V} - \mathbf{V}^d$. Substituted into (11.16) and rearranging yields

$$\begin{aligned} D_{\Omega}\sigma\mathbf{G}\varphi + D_{\Gamma}\sigma\left(\mathbf{V} - \mathbf{V}^d\right) = D_{\Omega}\mathbf{V} + D_{\Gamma}\mathbf{V} \rightarrow \\ D_{\Omega}\sigma\mathbf{G}\varphi = D_{\Omega}\mathbf{V} + D_{\Gamma}\mathbf{V}^d\ . \end{aligned} \tag{11.17}$$

Careful examination of $D\mathbf{V}$ reveals that it is conservative, thus sums over the grid cell telescope leaving only boundary terms. If the values of $\mathbf{V}^d$ are chosen so that they are true to the continuous problem then

$$\int (\mathbf{n}\cdot\mathbf{u})\,\mathrm{d}\Gamma \rightarrow \sum_{i,j,k} D_{\Gamma}\mathbf{V}^d = 0\ .$$

Thus, the linear algebra problem is solvable. This is a semi-definite linear algebra problem that has a null space of constants. Simply stated, the solution can have any constant added to it without violating the linear equations. This also means that φ is not unique, but $\mathbf{G}\varphi$ is.

The reason that homogeneous Neumann boundary conditions are chosen is because the value of $D_{\Gamma}\mathbf{V}$ cancels with the discrete boundary condition, leaving only $D_{\Gamma}\mathbf{V}^d$. This can be contributed from the internal discrete divergences, thus the boundary condition on φ can be homogeneous Neumann without affecting the interior solution. The topic of constructing boundary or ghost cell values of φ will be addressed in the following section. The property of conservative differencing for the divergence also allows solvability to be satisfied for periodic problems.

11.5.2 Solid Wall Boundary Conditions

The proper boundary conditions at a solid wall are terribly controversial. There are several valid ways of formulating boundary conditions that are valid for the Navier-Stokes or Euler equations that satisfy the basic governing equations as well as the solvability conditions. Here, we explore these alternative conditions and their impact on the algorithmic performance. These boundary conditions can be differentiated by their relation to the governing equations and numerical order.

The boundary conditions can be stated as

$$\mathbf{u}_\Gamma = \mathbf{w} \ , \ \text{or } \mathbf{n} \cdot \mathbf{u}_\Gamma = \mathbf{w}_n \ , \ \text{when } \nu = 0 \ ,$$

$$\text{with } \int_\Gamma \mathbf{n} \cdot \mathbf{w} \, \mathrm{d}\Gamma = 0 \ ,$$

together with initial conditions

$$\mathbf{u}(\mathbf{x}, 0) = \mathbf{u}_0(\mathbf{x}) \ , \nabla \cdot \mathbf{u}_0 = 0 \ ,$$

and with the equations and geometry comprise a description of the problem to be solved.

Let us assume that the method employed is the predictor-corrector projection with high-resolution Godunov differencing. The standard boundary treatment is a homogeneous Neumann condition. The inhomogeneous boundary conditions can be constructed via either a simple extrapolation (an average on two points) or linear extrapolation using three points.

Let us consider that $\mathbf{u}^*$ is not a physical quantity, then

$$u_t^* + \text{convection} + \text{diffusion} = \text{source} \ .$$

Further, assume that $\mathbf{u}^*$ is not solenoidal

$$\int_\Omega \nabla \cdot \mathbf{u} \, \mathrm{d}\Omega \rightarrow \int_\Gamma \mathbf{n} \cdot \mathbf{u} \, \mathrm{d}\Gamma \ .$$

How should it relate to the physical $\mathbf{u}$? Pressure should enforce $\nabla \cdot \mathbf{u} = 0$ and $\int_\Gamma \mathbf{n} \cdot \mathbf{u} \, \mathrm{d}\Gamma = 0$, i.e., boundary conditions must be numerically efficient.

The boundary conditions are tested using several "standard" problems. First, the basic order and accuracy of the methods without the impact of boundary conditions is established using the doubly periodic shear layer for the Euler equations. The assumed physical/mathematical scenario is quite elementary: the evolution of a 2-D vortex street in a homogeneous incompressible fluid on a doubly periodic unit-square domain, described by the incompressible inviscid (Euler) equations. A detailed description of the problem is given in Sect. 15.1. Further, we have considered the Navier-Stokes solution at $Re = 100$ as well as the Stokes solution, i.e., without the convective terms.

Our goal here is to define the error associated with various algorithmic choices both in terms of the asymptotic convergence rate for smooth problems and a quantitative error estimate. These results in terms of the L_2 error norm on 64^2–256^2 grids for $t = 1.6$ are given in Tables 11.1 and 11.2. Here, the L_2 error is computed via self-convergence where the fine grid solution is used as the standard and is compared with the coarse grid solutions. The convergence rates associated with these errors are shown in Tables 11.3 and 11.4. For all conditions tested (Stokes through Euler) the results indicate that the method produces second-order results. With vertex-staggered pressures the results are more uniform, particularly in relation to the Euler solutions.

Table 11.1. Periodic shear layer results using a cell-centered projection, five different grids and various Reynolds numbers in terms of the L_2 error norms for velocity.

Case	L_2 Norms			
	$16^2 - 32^2$	$32^2 - 64^2$	$64^2 - 128^2$	$128^2 - 256^2$
Euler	8.11×10^{-2}	2.46×10^{-2}	3.57×10^{-3}	1.66×10^{-3}
Re=100	1.08×10^{-2}	2.46×10^{-3}	6.27×10^{-4}	1.58×10^{-4}
Stokes	3.02×10^{-3}	6.78×10^{-4}	1.68×10^{-4}	4.21×10^{-5}

Table 11.2. Periodic shear layer results using a vertex-centered projection, five different grids and various Reynolds numbers in terms of the L_2 error norm for velocity.

Case	L_2 Norms			
	$16^2 - 32^2$	$32^2 - 64^2$	$64^2 - 128^2$	$128^2 - 256^2$
Euler	9.05×10^{-2}	2.30×10^{-2}	5.47×10^{-3}	1.40×10^{-3}
Re=100	1.09×10^{-2}	2.51×10^{-3}	6.37×10^{-4}	1.61×10^{-4}
Stokes	3.02×10^{-3}	6.78×10^{-4}	1.68×10^{-4}	4.21×10^{-5}

The boundary conditions for the pressure Poisson equation were considered by Gresho [229]. The pressure Poisson equation can be projected normal to the boundary producing the following boundary condition,

$$\frac{\partial p}{\partial n} = \mathbf{n} \cdot \left(\nu \nabla^2 \mathbf{u} - u_t - (\mathbf{u} \cdot \nabla)\, \mathbf{u} + \mathbf{F} \right) .$$

This condition is in stark contrast to the standard homogeneous Neumann conditions used either for viscous or inviscid flow. One can consider the ana-

Table 11.3. Periodic shear layer results using a cell-centered projection, five different grids and various Reynolds numbers in terms of the L_2 error norm convergence rates for velocity.

Case	L_2 Convergence Rates		
	$16^2 - 64^2$	$32^2 - 128^2$	$64^2 - 256^2$
Euler	1.72	2.07	1.83
Re=100	2.14	1.97	1.99
Stokes	2.15	2.00	2.00

Table 11.4. Periodic shear layer results using a vertex-centered projection, five different grids and various Reynolds numbers in terms of the L_2 convergence rate for velocity.

Case	L_2 Convergence Rates		
	$16^2 - 64^2$	$32^2 - 128^2$	$64^2 - 256^2$
Euler	1.98	2.07	1.96
Re=100	2.12	1.97	1.99
Stokes	2.15	2.01	2.00

lytic production of vorticity at a solid boundary defined by projection methods

$$-\nu\frac{\partial \omega}{\partial t} = \frac{\partial p}{\partial n} \rightarrow \frac{\partial p}{\partial n} = 0 \ , \ \text{as } Re \rightarrow \infty \ .$$

To formulate this sort of boundary condition for projections one can turn to rather simple approximations. The normal motion equation is accessible through the difference between $\mathbf{u}^*$ and $\mathbf{u}^n$ projected normal to the boundary. The strategy taken is to interpolate this difference onto the boundary in order to produce the normal equation of motion.

The simplest form of extrapolation is to set the boundary velocity (defined in the position denoted by the index 0) with the evolution of the velocity in the first cell (index 1),

$$u_0^* = u_0^n + \delta u_1 \rightarrow u_B^* = u_B^n + \delta u_1 \ ,$$

where

$$\delta u_1 = u_1^* - u_1^n \ .$$

In this and the following discussion we will describe the implementation for a simple constant mesh spacing grid (at least in any given dimension). This provides a poor approximation to the evolution of the velocity at the boundary

$$\frac{\partial \varphi}{\partial n} = -\frac{\partial u_B}{\partial t} - \left[(\mathbf{u} \cdot \nabla)\, \mathbf{u}^{n+1/2} + \mathrm{G}\,(p)^{n-1/2} - \nu \mathrm{L}\,(\mathbf{u})^{n+1/2} - \mathbf{F}^{n+1/2} \right] .$$

Since this also implies that the normal velocity gradient does not change, the solution does not violate the solvability of the resulting system of equations, i.e.,

$$u_1^* - u_0^* = u_1^n - u_0^n \rightarrow \left.\frac{\partial u}{\partial n}\right|^n = \left.\frac{\partial u}{\partial n}\right|^* .$$

We will refer to this as the “simple extrapolation”.

For a better approximation to this condition consider a linear extrapolation for the boundary velocity

$$u_0^* = u_0^n + 2\delta u_1 - \delta u_2 \text{ , remembering that } \delta u_i = u_i^* - u_i^n \text{ ,}$$

where $i = 1$ or 2 denoting cells (or grid points) inside the computational domain. Once the “ghost cell” is averaged with the first cell (a second-order approximation), the following boundary velocity is obtained

$$u_B^* = u_B^n + \frac{3}{2}\delta u_1 - \frac{1}{2}\delta u_2 .$$

Again, this is an approximation to the normal equation of motion at the boundary. Here, it is the second derivative of the normal velocity that does not change over the computational cycle,

$$u_0^* - 2u_1^* + u_2^* = u_0^n - 2u_1^n + u_2^n \rightarrow \left.\frac{\partial^2 u}{\partial n^2}\right|^n = \left.\frac{\partial^2 u}{\partial n^2}\right|^* .$$

We refer to this approach as the “linear extrapolation”. Again, solvability of the resulting linear system is not threatened by this approach.

The second test consists of a vortex placed in a unit square and the boundary conditions were tested on 16^2–256^2 grids for Euler, Navier-Stokes (Re=100), and Stokes flows, for time $t = 0.5$ using $\Delta t = 1/2\Delta x$. The vortex is defined by a solenoidal velocity field given by a streamfunction,

$$\Psi = \frac{1}{\pi} \sin 2\,(\pi x) \sin 2\,(\pi y) .$$

This can be analytically differentiated to produce the desired velocity field. As a start, the results using the standard homogeneous Neumann conditions are given to again set the stage for consideration of alternative boundary conditions. Using the standard boundary conditions the results are more uniformly second-order accurate than the periodic problem. This is shown in the results summarized in Tables 11.5 through 11.8 for both cell- and vertex-centered projections. In this case the convergence rates for the two different projections are very nearly identical. Projections on both cell-centered and vertex-centered grids are used to make sure that conclusions are not specific to the grid topology.

Table 11.5. Standard boundary conditions applied in concert with a cell-centered projection method in terms of L_2 norm errors.

Case	L_2 Norms			
	$16^2 - 32^2$	$32^2 - 64^2$	$64^2 - 128^2$	$128^2 - 256^2$
Euler	1.05×10^{-2}	2.47×10^{-3}	4.88×10^{-4}	1.10×10^{-4}
Re=100	6.16×10^{-3}	1.29×10^{-3}	3.17×10^{-4}	7.97×10^{-5}
Stokes	6.68×10^{-4}	1.52×10^{-4}	3.55×10^{-5}	8.51×10^{-6}

Table 11.6. Standard boundary conditions applied in concert with a cell-centered projection method in terms of L_2 norm convergence rates.

Case	L_2 Norms		
	$16^2 - 64^2$	$32^2 - 128^2$	$64^2 - 256^2$
Euler	2.09	2.33	2.14
Re=100	2.25	2.03	1.99
Stokes	2.13	2.10	2.06

Table 11.7. Standard boundary conditions applied in concert with a vertex-centered projection method in terms of L_2 norm errors.

Case	L_2 Norms			
	$16^2 - 32^2$	$32^2 - 64^2$	$64^2 - 128^2$	$128^2 - 256^2$
Euler	9.83×10^{-3}	2.21×10^{-3}	4.50×10^{-4}	1.00×10^{-4}
Re=100	5.91×10^{-3}	1.35×10^{-3}	3.46×10^{-4}	8.87×10^{-5}
Stokes	1.20×10^{-3}	2.82×10^{-4}	6.85×10^{-4}	1.69×10^{-4}

Table 11.8. Standard boundary conditions applied in concert with a vertex-centered projection method in terms of L_2 norm convergence rates.

Case	L_2 Norms		
	$16^2 - 64^2$	$32^2 - 128^2$	$64^2 - 256^2$
Euler	2.15	2.30	2.16
Re=100	2.13	1.97	1.96
Stokes	2.09	2.04	2.02

Working with the same problem the non-standard boundary conditions can be examined critically. These results are shown in Tables 11.9 through 11.16. Generally speaking the linear extrapolation produces smaller errors than the standard boundary treatment for the cell-centered projections. Likewise the linear extrapolation produces smaller errors for the vertex-centered projections. Both extrapolations are superior to the standard treatment although the improvements are fleeting as the problems considered become less viscous.

Table 11.9. Simple Extrapolation boundary conditions applied in concert with a cell-centered projection method in terms of L_2 norm errors.

Case	L_2 Norms			
	$16^2 - 32^2$	$32^2 - 64^2$	$64^2 - 128^2$	$128^2 - 256^2$
Euler	9.55×10^{-3}	2.33×10^{-3}	4.87×10^{-4}	1.16×10^{-4}
Re=100	5.46×10^{-4}	1.25×10^{-4}	3.14×10^{-5}	7.95×10^{-5}
Stokes	6.38×10^{-4}	1.34×10^{-4}	3.27×10^{-5}	8.13×10^{-6}

Table 11.10. Simple Extrapolation boundary conditions applied in concert with a cell-centered projection method in terms of L_2 norm convergence rates.

Case	L_2 Norms		
	$16^2 - 64^2$	$32^2 - 128^2$	$64^2 - 256^2$
Euler	2.03	2.25	2.07
Re=100	2.12	2.00	1.98
Stokes	2.25	2.25	2.01

Table 11.11. Linear Extrapolation boundary conditions applied in concert with a cell-centered projection method in terms of L_2 norm errors.

Case	L_2 Norms			
	$16^2 - 32^2$	$32^2 - 64^2$	$64^2 - 128^2$	$128^2 - 256^2$
Euler	1.15×10^{-2}	2.48×10^{-3}	5.08×10^{-4}	1.16×10^{-4}
Re=100	5.84×10^{-4}	1.31×10^{-3}	3.17×10^{-4}	7.97×10^{-5}
Stokes	1.42×10^{-3}	1.75×10^{-4}	3.40×10^{-5}	8.18×10^{-6}

Table 11.12. Linear Extrapolation boundary conditions applied in concert with a cell-centered projection method in terms of L_2 norm convergence rates.

Case	L_2 Norms		
	$16^2 - 64^2$	$32^2 - 128^2$	$64^2 - 256^2$
Euler	2.21	2.29	2.13
Re=100	2.15	2.05	1.99
Stokes	3.02	2.36	2.05

Table 11.13. Simple extrapolation boundary conditions applied in concert with a vertex-centered projection method in terms of L_2 norm errors.

Case	L_2 Norms			
	$16^2 - 32^2$	$32^2 - 64^2$	$64^2 - 128^2$	$128^2 - 256^2$
Euler	9.74×10^{-3}	2.19×10^{-3}	4.47×10^{-4}	1.00×10^{-4}
Re=100	5.96×10^{-3}	1.37×10^{-3}	3.48×10^{-4}	8.89×10^{-5}
Stokes	5.18×10^{-3}	1.35×10^{-3}	3.42×10^{-4}	8.59×10^{-5}

Table 11.14. Simple extrapolation boundary conditions applied in concert with a vertex-centered projection method in terms of L_2 norm convergence rates.

Case	L_2 Norms		
	$16^2 - 64^2$	$32^2 - 128^2$	$64^2 - 256^2$
Euler	2.15	2.29	2.16
Re=100	2.12	1.98	1.97
Stokes	1.94	1.98	1.99

Table 11.15. Linear extrapolation boundary conditions applied in concert with a vertex-centered projection method in terms of L_2 norm errors.

Case	L_2 Norms			
	$16^2 - 32^2$	$32^2 - 64^2$	$64^2 - 128^2$	$128^2 - 256^2$
Euler	9.74×10^{-3}	2.19×10^{-3}	4.47×10^{-4}	1.00×10^{-4}
Re=100	5.45×10^{-3}	1.21×10^{-3}	3.30×10^{-4}	8.93×10^{-5}
Stokes	9.17×10^{-4}	2.01×10^{-4}	5.08×10^{-5}	1.28×10^{-5}

Table 11.16. Linear extrapolation boundary conditions applied in concert with a vertex-centered projection method in terms of L_2 norm convergence rates.

Case	L_2 Norms		
	$16^2 - 64^2$	$32^2 - 128^2$	$64^2 - 256^2$
Euler	2.15	2.29	2.16
Re=100	2.18	1.87	1.89
Stokes	2.18	1.99	1.98

12. Approximate Projection Methods

Here, we will introduce the basic aspects and fundamental techniques used in constructing approximate projection methods. This will build upon the theory of exact projections given in the previous chapter. The linear algebra associated with these elliptic operators is still the dominant cost in solving the system of partial differential equations. Many of the details will remain the same, but the removal of the condition that the discrete divergence is exactly zero (to within a small value associated with the solution of a linear system of equations) will extract a price. The benefit is a substantial increase in efficiency of solving standard discrete elliptic operators. The price is that many other auxiliary details in the methods will matter greatly in determining results. These details will occupy much of our attention in the following pages.

The "approximate projections" introduced in [10] do not have the properties of a projection at a discrete level, but rather are discretizations of the continuous projection operator. With approximate projections, the discrete divergence is not exactly zero, instead it is a function of the truncation error. A central focus of this chapter will be the cell-centered approximate projection introduced first by Lai and Colella [312, 313].

The exact discrete projections described in the previous chapter provide a good foundation in the numerical implementation of projection methods, but have some practical difficulties. These problems are commented in [10, 312]. The decoupling of the pressure fields interacts poorly with (chemical, locally varying) source terms leading to instabilities. Additionally, the local decoupling makes efficient linear algebra techniques cumbersome [266] as well as complicates the implementation of adaptive grid techniques [7, 266]. Adaptive grid techniques for incompressible and low-Mach number flows has reached a mature state. The status can be seen in a number of recent publications describing the codes and their capabilities [8, 48, 49, 419, 520].

12.1 Numerical Issues with Approximate Projection Methods

The first issue to dispense with is the reasons for potentially choosing an approximate projection over an exact projection. Philosophically the reason can be simply stated through the realization that the exactness of a projection only holds for the precise divergence stencil used in the projection. If one defines a divergence of velocity on a different stencil it will generally be nonzero, and in fact will be a function of the truncation error of the operators used to project out the non-divergence-free velocity. The reasons for choosing an approximate projection are several fold:

1. The non-standard form of the discrete Laplacian in exact projections allows grid decoupling, which creates a more difficult linear system of equations to solve.
2. This especially complicates the use of highly efficient multi-grid methods.
3. It also complicates the use of more general grids and coordinate systems,
4. as well as deriving effective adaptive mesh refinement algorithms.
5. Finally, it creates difficulties with some physical systems including those with chemical reactions.

In producing an approximate projection algorithm, the principal positive property of the exact projection that should be retained is the stability of the operator.

The approximate projection methods will employ the same discrete divergence, $\mathcal{D}$, and gradient, $\mathcal{G}$, operators as the exact projections. What will change is the composition of these functions into a Laplacian that must be inverted to accomplish the projection. Rather that invert a non-standard Laplacian arising from $\mathcal{D}\sigma\mathcal{G}$, the approximate projection inverts a more standard direct discretization of a Laplacian. It is this aspect of the method that is the key to more efficient numerical linear algebra.

As with exact projections, the basic goal with approximate projection methods is to advance a velocity field, $\mathbf{V} = (V^x, V^y, V^z)^T$ by some convenient means disregarding the solenoidal nature of $\mathbf{V}$, then recover the approximately solenoidal velocity field, $\mathbf{V}^d$ $(\nabla \cdot \mathbf{V}^d \approx 0)$. We use the more general notation of $\mathbf{V}$ because the projection can be applied to a velocity field or its time derivative. The means to this end is an approximate projection, $\tilde{\mathcal{P}}$, which has the effect

$$\tilde{\mathbf{V}}^d = \tilde{\mathcal{P}}(\mathbf{V}) \ .$$

The projection accomplishes this through the approximate decomposition of the velocity field into parts that are approximately divergence and curl-free. This is an approximation to a Hodge or Helmholtz decomposition [109]. The approximately curl-free portion will be denoted by the gradient of a potential, $\nabla\varphi$. This approximate decomposition can be written

$$\mathbf{V} = \tilde{\mathbf{V}}^d + \nabla\varphi \ . \tag{12.1}$$

The above equation holds the key for computing solutions to flow equations.

Taking the divergence of (12.1) gives

$$\nabla \cdot \mathbf{V} = \nabla \cdot \tilde{\mathbf{V}}^d + \nabla^2\varphi \to \nabla \cdot \mathbf{V} = \nabla^2\varphi \ .$$

In the approximate projection the operators $\nabla \cdot \nabla$ has been replaced by a Laplacian ∇^2 not formed through the use of the same operators used for $\nabla\cdot$ and ∇. The new Laplacian should place a premium on compactness and ease of solution. Once φ has been computed, then the solution can be found through

$$\tilde{\mathbf{V}}^d = \mathbf{V} - \nabla\varphi \,. \tag{12.2}$$

For variable density flows, we can proceed as in Chap. 11, but with an approximation of a variable coefficient Laplacian with $\sigma = 1/\rho$.

The projection operators can be written

$$\tilde{\mathcal{P}}_\sigma = \mathbf{I} - \sigma\nabla \left(L_\sigma\right)^{-1} \nabla\cdot \,,$$

where L_σ is the variable coefficient, σ, Laplacian and

$$\tilde{\mathcal{Q}}_\sigma = \mathbf{I} - \tilde{\mathcal{P}}_\sigma \,.$$

with $\sigma\nabla\varphi = \tilde{\mathcal{Q}}_\sigma\left(\mathbf{V}\right)$.

Now the property of exact projections of idempotency has been lost, $\tilde{\mathcal{P}}_\sigma^2 \neq \tilde{\mathcal{P}}_\sigma$. It is still important that the norm of the operator be less than or equal to one for basic stability $\left\|\tilde{\mathcal{P}}_\sigma\left(\mathbf{V}\right)\right\|_2 \leq \|\mathbf{V}\|_2$.

It is important to note that the divergence and gradient operators are exactly the same as with the exact projections. The only aspect of the operators used in the algorithm that changes is the discrete form of the Laplacian, which will take a more usual form to enhance the ease of solving the resultant linear system of equations. Let D be the discrete divergence, and $\mathbf{G}$, the discrete gradient. These discrete operators on a collocated grid are

$$D_{i,j,k}\mathbf{V} = \frac{V^x_{i+1,j,k} - V^x_{i-1,j,k}}{2\Delta x} + \frac{V^y_{i,j+1,k} - V^y_{i,j-1,k}}{2\Delta y} + \frac{V^z_{i,j,k+1} - V^z_{i,j,k-1}}{2\Delta z} \,,$$

and

$$\sigma\mathbf{G}_{i,j}\varphi = \begin{pmatrix} \dfrac{\varphi_{i+1,j,k} - \varphi_{i-1,j,k}}{2\rho_{i,j,k}\Delta x} \\[2ex] \dfrac{\varphi_{i,j+1,k} - \varphi_{i,j-1,k}}{2\rho_{i,j,k}\Delta y} \\[2ex] \dfrac{\varphi_{i,j,k+1} - \varphi_{i,j,k-1}}{2\rho_{i,j,k}\Delta z} \end{pmatrix} .$$

We can analyze a number of properties and stability of the method through Fourier analysis. The Fourier transform and its inverse are

$$\hat{\psi}\left(\alpha_x, \alpha_y, \alpha_z\right) = \frac{1}{2\pi}\int_{-\infty}^{\infty} \psi\left(x, y\right) e^{-\imath\left(\alpha_x x + \alpha_y y + \alpha_z z\right)} \,\mathrm{d}x\,\mathrm{d}y\,\mathrm{d}z \,,$$

and

$$\psi(x,y,z) = \int_{-\infty}^{\infty} \hat{\psi}(\alpha_x, \alpha_y, \alpha_z) \, e^{\imath(\alpha_x x + \alpha_y y + \alpha_z z)} \, d\alpha_x \, d\alpha_y \, d\alpha_z \, ,$$

where $\imath = \sqrt{-1}$ and α_x, α_y, and α_z are the wave numbers in the x-, y- and z-directions, respectively. These will be introduced into the finite-difference equations to produce their symbol.[1] In particular, the stability of the approximate projection is of direct interest. We substitute the discrete operators into the expression for the projection and then replace the grid operations with their Fourier transforms. These expressions then yield the symbol of the operator, which can be interrogated for its amplitude. We confirm that with a standard seven-point Laplacian 3-D (five-points in 2-D),

$$\frac{\varphi_{i-1,j,k} - 2\varphi_{i,j,k} + \varphi_{i+1,j,k}}{\Delta x^2} + \frac{\varphi_{i,j-1,k} - 2\varphi_{i,j,k} + \varphi_{i,j+1,k}}{\Delta y^2} + \frac{\varphi_{i,j,k-1} - 2\varphi_{i,j,k} + \varphi_{i,j,k+1}}{\Delta z^2} \, ,$$

and the above stated collocated divergence and gradient the approximate projection has an amplitude of less than one. The three dimensional case follows with the same result. All of these studies are greatly enabled by the availability of symbolic algebra software such as Mathematica [603].

We can find the truncation error through expanding the Fourier transform of the discrete equations in a Taylor series and compare them with the exact transform. Thus, we can demonstrate that the projection is second-order accurate in space.

The discrete projection operators can be defined as an abstract operator

$$\mathbf{P}_\sigma = \mathbf{I} - \sigma \mathbf{G} \left(L_\sigma\right)^{-1} D \, ,$$

and

$$\mathbf{Q}_\sigma = \sigma \mathbf{G} \left(L_\sigma\right)^{-1} D \, .$$

These are the discrete analogs of $\mathcal{P}_\sigma$ and $\mathcal{Q}_\sigma$. They are also designed to have the desirable properties of the continuous projections carry over as well (thus $\mathbf{P}_\sigma \approx \mathbf{P}_\sigma^2$ and $\|\mathbf{P}_\sigma(\mathbf{V})\|_2 \leq \|\mathbf{V}\|_2$).

We will be defining discrete methods based on the continuous projections rather than demanding that the discrete system algebraically match the conditions for being a projection. Thus, the most straightforward means to discretize each operator ($\nabla\cdot$, ∇ and $\nabla\cdot\sigma\nabla$) will be chosen (not quite true, but nearly). We can visualize the difference between the two approaches by plotting the symbol of an exact projection Laplacian and a more standard

[1] The symbol of the operators are computed by expanding the discrete operators by their Fourier transform, $\exp[\imath(\alpha_x + \alpha_y + \alpha_z)]$, and plotting their absolute value. In the case of these filters this is the divergence of the gradients divided by the magnitude of the diagonal of $\nabla \cdot \nabla$.

five-point Laplacian. This is done in Fig. 12.1. We will stay in two dimensions for clarity and ease of presentation. The standard Laplacian's symbol, its description in terms of Fourier analysis, is larger (in absolute value) than the Laplacian for the exact projection leading to more contractive (dissipative and stable) approximate projection. Only a single constant is in the nullspace (i.e., zero) of the standard operator, rather than the four of the exact projection thus showing that the pressure decoupling has been removed at the price of making the divergence a function of the truncation error. This is seen in the value of zero for the operator in the limit of small wavenumbers (going to zero). The nullspace of an operator are the values where the grid coupling is lost because the operator has no effect as displayed by its zero value. For the exact projection operator this is shown through the values of zero where the wavenumber goes to π.

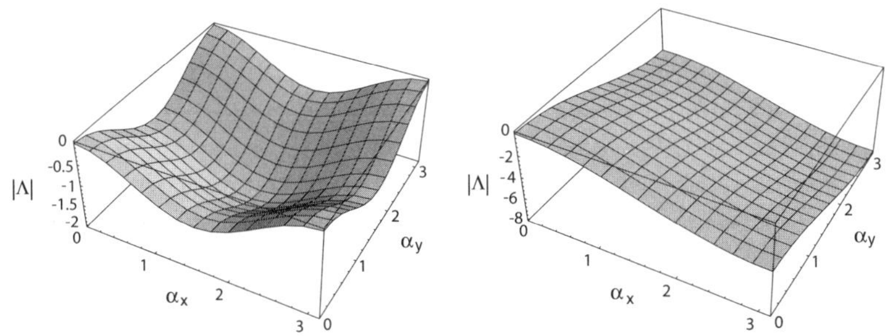

(a) Exact Projection Laplacian

(b) Standard Laplacian

Fig. 12.1. A comparison of the symbol for an exact projection Laplacian and a standard cell-centered Laplacian. The number of zeros shows the dimension of the nullspace of each operator.

A more rigorous derivation for approximate projections has been given by Almgren, Bell and Szymczak [10]. In that work a vertex-staggered pressure grid was used along with a decomposition of the velocity field that allowed an approximate projection with special properties to be defined. The velocity field had two components: the average velocity $\overline{\mathbf{V}}$ and an orthogonal field, $\mathbf{V}^{\perp}$ such that $\left(\overline{\mathbf{V}}, \mathbf{V}^{\perp}\right)_v = 0$. Similarly, divergence and gradient operators are derived for each field. Then an exact projection is defined for this system, with a Laplacian

$$L_\sigma = D\sigma\mathbf{G}\varphi = \overline{D}\sigma\overline{\mathbf{G}}\varphi + D^{\perp}\sigma\mathbf{G}^{\perp}\varphi \, .$$

The approximate projection arises from the algorithm neglecting $\mathbf{V}^\perp$ and solving

$$D\sigma\mathbf{G}\varphi = \overline{D\mathbf{V}} .$$

The convenient and efficient aspect about this projection is that the error in the projection is well defined by $D^\perp\mathbf{V}^\perp$ and the Laplacian $D\sigma\mathbf{G}$ has a usual form.

A similar derivation for the cell-centered projection can be defined. First, some observations about the vertex-staggered approximate projection of [10] are helpful. The operator $D^\perp\sigma\mathbf{G}^\perp$ is identical to a mixed fourth-order spatial derivative.[2] that is equal to the difference between L_σ and the exact projection Laplacian,

$$D^\perp\sigma\mathbf{G}^\perp = L_\sigma - \overline{D}\sigma\overline{\mathbf{G}} .$$

Similarly, this difference can be used with the cell-centered projection to define an approximate projection that has the same properties as the vertex-staggered projections. In two dimensions, the error looks like φ_{xxxx} and φ_{yyyy}. In the cell-centered case, this error is equal to $D^\perp\sigma\mathbf{G}^\perp\varphi$. These difference operators correspond to those in the vertex-staggered case and are orthogonal to the cell-average operators. When scaled by the σ in a weighted inner product, this can produce the proper five-point Laplacian with the appropriately averaged σ at cell edges. This analysis confirms that $\sigma_{i+1/2,j} = (\sigma_{i,j} + \sigma_{i+1,j})/2$ rather than some other average (such as a harmonic mean that the analysis of the continuous operator $\nabla \cdot \sigma\nabla$ would suggest). A simple rule for finding the σ to use in the Laplacian is to follow the interpolation that is implied for the velocity if the divergence is computed through a control volume-based method.

Some distinctions need to be made about the nature of approximate projections as compared with exact projections. To denote these differences, we will refer to approximate projections with the notation, $\widetilde{\mathbf{P}}$. The operator is no longer approximately idempotent (and L_2 stable) (i.e., $\widetilde{\mathbf{P}} \neq \widetilde{\mathbf{P}}^2$), but it is equally important to see that $\widetilde{\mathbf{P}} \approx \widetilde{\mathbf{P}}^2$. This has the effect of making the incremental and pressure versions of the approximate projection substantially different. Stability proofs that relied on the nature of a projection must be revamped. We expect that a successful discretization will converge to the same solutions as the exact projection, and the discrete divergence will converge to zero as the grid is refined.

While these methods cure some problems, they introduce new difficulties that we will address shortly. First, we will describe the basic construction of projection algorithms for incompressible flow.

[2] By mixed spatial derivative we mean that the function is differentiated with respect to more than one direction, e.g., φ_{xy}.

12.2 Projection Algorithms for Incompressible Flow

Given the development above, we can prepare the outline of a projection algorithm. The algorithm will consist of two steps as alluded to before: a separate step where the solenoidal nature of the velocity field is ignored and a corrective step using the projection to force the velocity to be solenoidal. This is often referred to as a "fractional step method." In the first step, the velocity is computed solving the motion equations as convection-diffusion equations

$$\mathbf{u}_{i,j}^{*,n+1} = \mathbf{u}_{i,j}^{n} - \Delta t \left[(\mathbf{u} \cdot \nabla \mathbf{u})_{i,j}^{n+1/2} + \sigma_{i,j}^{n+1/2} \mathbf{G}_{i,j} \phi^{n-1/2} \right.$$
$$\left. - \frac{\nu \sigma_{i,j}^{n+1/2}}{2} L_{i,j} \left(\mathbf{u}^{n} - \mathbf{u}^{*,n+1} \right) + \mathbf{F}_{i,j}^{n+1/2} \right] ,$$

where $\sigma_{i,j}^{n+1/2} = \left(\sigma_{i,j}^{n} + \sigma_{i,j}^{n+1} \right)/2$, $\mathbf{F}_{i,j}^{n+1/2} = \left(\mathbf{F}_{i,j}^{n} + \mathbf{F}_{i,j}^{n+1} \right)/2$, or $\mathbf{F}_{i,j}^{n+1/2} = \mathbf{F}\left(\mathbf{x}_{i,j}, t^{n+1/2} \right)$, ϕ is the incompressible pressure and $L_{i,j}$ is the discrete Laplacian. As mentioned before the advection is discretized with an unsplit high-order Godunov method. The above projection will be denoted as the incremental form for reasons explained below. The other form, the pressure form, has $\sigma_{i,j}^{n+1/2} \mathbf{G}_{i,j} \phi^{n-1/2}$ removed, i.e.,

$$\mathbf{u}_{i,j}^{*,n+1} = \mathbf{u}_{i,j}^{n} - \Delta t \left[(\mathbf{u} \cdot \nabla \mathbf{u})_{i,j}^{n+1/2} \right.$$
$$\left. - \frac{\nu \sigma_{i,j}^{n+1/2}}{2} L_{i,j} \left(\mathbf{u}^{n} + \mathbf{u}^{*,n+1} \right) - \mathbf{F}_{i,j}^{n+1/2} \right] .$$

While this form is obvious, we modify it to control the error in the implicit viscous solution for $\mathbf{u}^{*,n+1}$. To avoid a loss of accuracy this step is modified to a two-step procedure; first,

$$\widetilde{\mathbf{u}}_{i,j}^{*,n+1} = \mathbf{u}_{i,j}^{n} - \Delta t \left[(\mathbf{u} \cdot \nabla \mathbf{u})_{i,j}^{n+1/2} + \sigma_{i,j}^{n+1/2} \mathbf{G}_{i,j} \phi^{n-1/2} \right.$$
$$\left. - \frac{\nu \sigma_{i,j}^{n+1/2}}{2} L_{i,j} \left(\mathbf{u}^{n} + \mathbf{u}^{*,n+1} \right) + \mathbf{F}_{i,j}^{n+1/2} \right] ,$$

then

$$\mathbf{u}_{i,j}^{*,n+1} = \widetilde{\mathbf{u}}_{i,j}^{*,n+1} + \Delta t \, \sigma_{i,j}^{n+1/2} \mathbf{G}_{i,j} \phi^{n-1/2} .$$

This avoids making a large error in the implicit viscous solution. Thus, the pressure error in the velocity will be

$$\left(\mathbf{I} - \frac{\nu \Delta t}{2} L_{i,j} \right)^{-1} \sigma_{i,j}^{n+1/2} \mathbf{G}_{i,j} \left(\phi^{n+1/2} - \phi^{n-1/2} \right) ,$$

rather than the much larger

$$\left(\mathbf{I} - \frac{\nu \Delta t}{2} L_{i,j}\right)^{-1} \sigma_{i,j}^{n+1/2} \mathbf{G}_{i,j} \phi^{n+1/2} ,$$

that the initial formulation yields. In the case where $\nu = 0$ the first (obvious) form is employed. Should other portions of the algorithm be implicit (such as advection), a similar modification is advisable.

12.3 Analysis of Projection Algorithms

12.3.1 Basic Definitions for Analysis

In the following presentation it will be useful to introduce some notation that will allow a more compact representation of the algorithms. We take $\mathbf{V} = \left(\mathbf{u}^{*,n+1} - \mathbf{u}^n\right)/\Delta t$ and $\mathbf{V}^d = \left(\mathbf{u}^{n+1} - \mathbf{u}^n\right)/\Delta t$, or $\mathbf{V} = \mathbf{u}^{*,n+1}/\Delta t$ and $\mathbf{V}^d = \mathbf{u}^{n+1}/\Delta t$, and apply the projection. First, we solve the linear system

$$L_\sigma^{n+1/2} \varphi = D\mathbf{V} ,$$

then correct the velocity field with (11.2) as

$$\mathbf{u}_{i,j}^{n+1} = \mathbf{u}_{i,j}^{*,n+1} - \Delta t\, \sigma_{i,j}^{n+1/2}\, \mathbf{G}_{i,j} \varphi .$$

For the incremental form

$$\phi_{i,j}^{n+1/2} = \phi_{i,j}^{n-1/2} + \varphi_{i,j} ,$$

or for the pressure form

$$\phi_{i,j}^{n+1/2} = \varphi_{i,j} .$$

Symbolically these algorithms can be represented quite compactly. We will use shorthand to describe advection/diffusion, thus we define $\mathbf{N}(\mathbf{u}) = \mathbf{u} \cdot \nabla \mathbf{u}$ or $\mathbf{N}(\mathbf{u}) = \mathbf{u} \cdot \nabla \mathbf{u} - \nu \nabla^2 \mathbf{u} - \mathbf{F}$. The incremental form can be given as

$$\frac{\mathbf{u}^{n+1} - \mathbf{u}^n}{\Delta t} = \mathbf{P}_\sigma \left(\frac{\mathbf{u}^{*,n+1} - \mathbf{u}^n}{\Delta t}\right) ,$$

or

$$\frac{\mathbf{u}^{n+1} - \mathbf{u}^n}{\Delta t} = \mathbf{P}_\sigma \left(-\mathbf{N}^{n+1/2}(\mathbf{u}) - \sigma^{n+1/2} \mathbf{G} \phi^{n-1/2}\right) ,$$

and

$$\sigma^{n+1/2} \mathbf{G} \left(\phi^{n+1/2} - \phi^{n-1/2}\right) = \mathbf{Q}_\sigma \left(\frac{\mathbf{u}^{*,n+1} - \mathbf{u}^n}{\Delta t}\right) ,$$

for the incremental form, or

$$\frac{\mathbf{u}^{n+1} - \mathbf{u}^n}{\Delta t} = \mathbf{P}_\sigma \left(-\mathbf{N}^{n+1/2}(\mathbf{u})\right) ,$$

for the pressure form with

$$\sigma^{n+1/2}\mathbf{G}\phi^{n+1/2} = \mathbf{Q}_\sigma\left(\frac{\mathbf{u}^{*,n+1} - \mathbf{u}^n}{\Delta t}\right) .$$

Another way to write these equations is

$$\mathbf{u}^{n+1} = \mathbf{P}_\sigma\left(\mathbf{u}^{*,n+1}\right) .$$

In both cases, these can be rewritten as before to

$$\mathbf{u}^{n+1} = \mathbf{P}_\sigma\left(-\Delta t\mathbf{N}^{n+1/2}(\mathbf{u}) - \Delta t\sigma^{n+1/2}\mathbf{G}\phi^{n-1/2}\right) ,$$

and

$$\sigma^{n+1/2}\mathbf{G}\left(\phi^{n+1/2} - \phi^{n-1/2}\right) = \mathbf{Q}_\sigma\left(\frac{\mathbf{u}^{*,n+1}}{\Delta t}\right) ,$$

for the incremental form, or for the pressure form

$$\mathbf{u}^{n+1} = \mathbf{P}_\sigma\left(\mathbf{u}^n - \Delta t\mathbf{N}^{n+1/2}(\mathbf{u})\right) ,$$

and

$$\sigma^{n+1/2}\mathbf{G}\left(\phi\right)^{n+1/2} = \mathbf{Q}_\sigma\left(\frac{\mathbf{u}^{*,n+1}}{\Delta t}\right) .$$

For an exact projection there is no difference in writing the operators in these forms, but these differences become substantive in the case where $D\mathbf{u}^n \neq 0$ by either error or design. For approximate projections, the symbols are the same except $\mathbf{P}_\sigma \to \widetilde{\mathbf{P}}_\sigma$ and $\mathbf{Q}_\sigma \to \widetilde{\mathbf{Q}}_\sigma$.

12.3.2 Analysis of Approximate Projection Algorithms

There are also more subtle differences between the exact and approximate projection with regard to the right-hand side of the pressure equation. For exact projections, it makes little difference if we project for pressures or increments in pressure, or if the right-hand side is the divergence of predicted velocity or the divergence of the difference of the predicted and old-time velocities. For approximate projections, these differences are all substantive.

For convenience we will write the equations as

$$\nabla \cdot \mathbf{u} = 0, \text{ and } \frac{\partial \mathbf{u}}{\partial t} + \mathbf{N}(\mathbf{u}) + \sigma\nabla\phi = 0 ,$$

again $\mathbf{N}(\mathbf{u}) = \mathbf{u}\cdot\nabla\mathbf{u}$ or $\mathbf{u}\nabla\mathbf{u} - \nabla^2\mathbf{u} - \mathbf{F}$. Our goal in this section is to understand the nature of the errors present in the approximate projection algorithm (in a generic sense). Other results that are along these lines can be found in Almgren et al. [9] In that work, the additional aspect of characterizing the spatial error and its interaction with the abstract forms are described. The techniques used there are similar as are the results although we will discuss the additional consequences of their analysis.

The following symbols are defined:

- $\delta^n = \phi^{n+1/2} - \phi^{n-1/2}$,
- $\mathbf{V} = \mathbf{u}/\Delta t$, for the pressure form,
- $\mathbf{V}^{*,n+1} = \mathbf{V}^n - \mathbf{N}^{n+1/2}(\mathbf{u})$, and
- for the incremental form $\mathbf{V}^{*,n+1} = \mathbf{V}^n - \mathbf{N}^{n+1/2}(\mathbf{u}) - \sigma^{n+1/2}\mathbf{G}\phi^{n-1/2}$.

We will also analyze the difference between $\mathbf{P}_\sigma\left(\mathbf{V}^{*,n+1}\right)$ or $\widetilde{\mathbf{P}}_\sigma\left(\mathbf{V}^{*,n+1}\right)$, and $\mathbf{P}_\sigma\left(\mathbf{V}^{*,n+1} - \mathbf{V}^n\right)$ or $\widetilde{\mathbf{P}}_\sigma\left(\mathbf{V}^{*,n+1} - \mathbf{V}^n\right)$.

We will begin with the analysis of exact projections as they represent a best case in a sense. As we will see there are cases where this faith is placed a little too cavalierly. For the exact projections, the pressure and incremental forms are identical, but what is being projected matters. We will start with $\mathbf{P}_\sigma\left(\mathbf{V}^{*,n+1} - \mathbf{V}^n\right)$ that leads to a Poisson equation,

$$\begin{aligned} L_\sigma^{n+1/2}\phi^{n+1/2} &= D\left(\mathbf{V}^{*,n+1} - \mathbf{V}^n\right) + \epsilon^{n+1/2} \\ &= -D\mathbf{N}^{n+1/2}(\mathbf{u}) + \epsilon^{n+1/2} . \end{aligned}$$

The quantity ϵ is the tolerance from the linear solver used to invert the Laplacian. The discrete divergence at time $n+1$ is

$$D\mathbf{V}^{n+1} = D\mathbf{V}^n - D\mathbf{N}^{n+1/2}(\mathbf{u}) - D\sigma^{n+1/2}\mathbf{G}\phi^{n+1/2} ,$$

which (using $L_\sigma^{n+1/2} = D\sigma\mathbf{G}$) simplifies to

$$D\mathbf{V}^{n+1} = D\mathbf{V}^n - \epsilon^{n+1/2} .$$

The unfortunate problem is that the old-time divergence contains the sum of previous errors, thus

$$D\mathbf{V}^{n+1} = -\sum_{i=1}^{n} \epsilon^{i+1/2}.$$

This method accumulates error in the discrete divergence and as such is not preferred. In cases where the convergence tolerance is loose this error accumulation can be a problem.

During the early years of solving incompressible flow where numerical linear algebra was not as developed as it is today, convergence tolerances were often much larger than we might choose today ($\mathcal{O}\left(10^{-3}\right)$ instead of $\mathcal{O}\left(10^{-6}\right)$). These decisions were closely associated with the techniques used to solve the linear systems during the 1960's and early 1970's where direct and relaxation methods prevailed. Nowadays Krylov and multigrid methods are available making the linear algebra much more efficient (see Chap. 8).

Now we will look at $\mathbf{P}_\sigma\left(\mathbf{V}^{*,n+1}\right)$. The pressure Poisson equation is

$$\begin{aligned} L_\sigma^{n+1/2}\phi^{n+1/2} &= D\mathbf{V}^{*,n+1} + \epsilon^{n+1/2} \\ &= D\mathbf{V}^n - D\mathbf{N}^{n+1/2}(\mathbf{u}) + \epsilon^{n+1/2} . \end{aligned}$$

This equation can then be simplified by recognizing that $\mathbf{V}^n = \mathbf{V}^{*,n} - \sigma^{n+1/2}\mathbf{G}\phi^{n-1/2}$ and $L_\sigma^{n-1/2}\phi^{n-1/2} = D\mathbf{V}^{*,n} + \epsilon^{n-1/2}$. The pressure equation reduces to

$$L_\sigma^{n+1/2}\phi^{n+1/2} = -D\mathbf{N}^{n+1/2}(\mathbf{u}) + \epsilon^{n+1/2} - \epsilon^{n-1/2}\,.$$

Thus, the pressure solution has the previous times error subtracted from the impact of the current error (normally controlling any growth quite well). The discrete divergence at time $n+1$ is

$$D\mathbf{V}^{n+1} = D\mathbf{V}^{n} - D\mathbf{N}^{n+1/2}(\mathbf{u}) - D\sigma^{n+1/2}\mathbf{G}\phi^{n+1/2}\,,$$

which (using $L_\sigma = D\sigma\mathbf{G}$) simplifies to

$$D\mathbf{V}^{n+1} = -\epsilon^{n+1/2}\,.$$

This method does not accumulate error in the discrete divergence and is preferred.

Moving to approximate projections is more intricate, but the basic ideas still apply. Starting with $\widetilde{\mathbf{P}}_\sigma\left(\mathbf{V}^{*,n+1} - \mathbf{V}^n\right)$ forms, we can see that the same error accumulation will apply, but will now be augmented by a sum of approximation errors.

12.3.3 Incremental Velocity Difference Projection

Writing the incremental form as a "pressure" equation

$$\begin{aligned} L_\sigma^{n+1/2}\delta^n &= D\left(\mathbf{V}^{*,n+1} - \mathbf{V}^n\right) \rightarrow \\ L_\sigma^{n+1/2}\phi^{n+1/2} &= -D\mathbf{N}^{n+1/2}(\mathbf{u}) + \left(L_\sigma^{n+1/2} - D\sigma^{n+1/2}\mathbf{G}\right)\phi^{n-1/2} \\ &\quad + \epsilon^{n+1/2}\,. \end{aligned}$$

The term $\left(L_\sigma^{n+1/2} - D\sigma^{n+1/2}\mathbf{G}\right)\phi^{n-1/2}$ represents the error in the projection from the previous time step which is absorbed in the new-time pressure (really a discrete divergence). Analysis and comparison of the truncation errors for L and $D\mathbf{G}$ indicates that this should be a diffusive (and stabilizing) force in the method. The discrete divergence is computed from

$$D\mathbf{V}^{n+1} = D\mathbf{V}^{n} - D\mathbf{N}^{n+1/2}(\mathbf{u}) - D\sigma^{n+1/2}\mathbf{G}\phi^{n+1/2}\,.$$

As with the exact projection, this method simplifies to

$$D\mathbf{V}^{n+1} = D\mathbf{V}^{n} + \left(L_\sigma^{n+1/2} - D\sigma^{n+1/2}\mathbf{G}\right)\delta^n - \epsilon^{n+1/2}\,,$$

and making observations similar to what we did with the exact projection about $D\mathbf{V}^n$, we get

$$D\mathbf{V}^{n+1} = \sum_{i=1}^{n}\left[\left(L_\sigma^{i+1/2} - D\sigma^{i+1/2}\mathbf{G}\right)\delta^i - \epsilon^{i+1/2}\right].$$

This represents a more serious accumulation error than occurred with the exact projections because of generally larger size of the truncation error than the convergence tolerance.

12.3.4 Pressure Velocity Difference Projection

The pressure form is similar with a pressure equation of

$$L_\sigma^{n+1/2}\phi^{n+1/2} = D\left(\mathbf{V}^{*,n+1} - \mathbf{V}^n\right) = -D\mathbf{N}^{n+1/2}(\mathbf{u}) + \epsilon^{n+1/2}\,.$$

This shows that the previous errors are not absorbed into the subsequent projection. The discrete divergence has the same form as the incremental projection with a critical difference (ϕ instead of the typically smaller δ)

$$D\mathbf{V}^{n+1} = \sum_{i=1}^{n}\left[\left(L_\sigma^{i+1/2} - D\sigma^{i+1/2}\mathbf{G}\right)\phi^{i+1/2} - \epsilon^{i+1/2}\right].$$

This hinders its success.

12.3.5 Incremental Velocity Projection

We now move to $\widetilde{\mathbf{P}}_\sigma\left(\mathbf{V}^{*,n+1}\right)$. The incremental form yields a pressure equation

$$\begin{aligned} L_\sigma^{n+1/2}\delta^n = D\mathbf{V}^{*,n+1} + \epsilon^{n+1/2} &\rightarrow D\mathbf{V}^n - D\mathbf{N}^{n+1/2}(\mathbf{u}) \\ &- D\sigma^{n+1/2}\mathbf{G}\phi^{n-1/2}\,, \end{aligned}$$

which becomes

$$\begin{aligned} L_\sigma^{n+1/2}\phi^{n+1/2} &= D\mathbf{V}^n - D\mathbf{N}^{n+1/2}(\mathbf{u}) \\ &+ \left(L_\sigma^{n+1/2} - D\sigma^{n+1/2}\mathbf{G}\right)\phi^{n-1/2} + \epsilon^{n+1/2}\,. \end{aligned}$$

Using the identities $D\mathbf{V}^n = D\mathbf{V}^{*,n} - D\sigma^{n-1/2}\mathbf{G}\delta^{n-1}$ and $L_\sigma^{n-1/2}\delta^{n-1} = D\mathbf{V}^{*,n} + \epsilon^{n-1/2}$, the above pressure equation becomes

$$\begin{aligned} L_\sigma^{n+1/2}\phi^{n+1/2} &= -D\mathbf{N}^{n+1/2}(\mathbf{u}) + \left(L_\sigma^{n+1/2} - D\sigma^{n+1/2}\mathbf{G}\right)\phi^{n-1/2} \\ &+ \left(L_\sigma^{n-1/2} - D\sigma^{n-1/2}\mathbf{G}\right)\left(\phi^{n-1/2} - \phi^{n-\frac{3}{2}}\right) \\ &+ \epsilon^{n+1/2} - \epsilon^{n-1/2}\,. \end{aligned} \tag{12.3}$$

The discrete divergence follows directly as before to

$$D\mathbf{V}^{n+1} = \left(L_\sigma^{n+1/2} - D\sigma^{n+1/2}\mathbf{G}\right)\delta^n - \epsilon^{n+1/2}\,.$$

The error form for the pressure is actually an estimate of the current error, with $2\phi^{n-1/2} - \phi^{n-3/2}$ being a estimate of the current pressure (in the case of constant density). This turns out to be a poor approximation (especially for variable density flows) and causes the method to be less reliable than the other algorithms as shown later (in Sect. 6.4). A more thorough analysis of this matter is given by Almgren et al. [9] who shed light on this issue in terms of the spatial discretization.

12.3.6 Pressure Velocity Projection

Our last form is the pressure form of $\widetilde{\mathbf{P}}_\sigma\left(\mathbf{V}^{*,n+1}\right)$. The pressure equation is

$$\begin{aligned} L_\sigma^{n+1/2}\phi^{n+1/2} &= D\mathbf{V}^{*,n+1} + \epsilon^{n+1/2} \\ &= D\mathbf{V}^n - D\mathbf{N}^{n+1/2}(\mathbf{u}) + \epsilon^{n+1/2} \,. \end{aligned}$$

Similarly to earlier cases, we use $D\mathbf{V}^n = \left(L_\sigma^{n+1/2} - D\sigma^{n+1/2}\mathbf{G}\right)\phi^{n-1/2} - \epsilon^{n-1/2}$ to get

$$\begin{aligned} L_\sigma^{n+1/2}\phi^{n+1/2} &= -D\mathbf{N}^{n+1/2}(\mathbf{u}) + \left(L_\sigma^{n+1/2} - D\sigma^{n+1/2}\mathbf{G}\right)\phi^{n-1/2} \\ &\quad + \epsilon^{n+1/2} - \epsilon^{n-1/2} \,. \end{aligned}$$

The discrete divergence also follows from our earlier experience as

$$D\mathbf{V}^{n+1} = \left(L_\sigma^{n+1/2} - D\sigma^{n+1/2}\mathbf{G}\right)\phi^{n+1/2} - \epsilon^{n+1/2} \,.$$

This projection has an error form that is quite simple and should be well-behaved numerically. There is no accumulation of error and the method only depends on the data at two time levels (rather than three like the previous method).

The modification made on the pressure projection for viscous flow plays no important role in this error analysis. It does lessen the operator splitting error made in the viscous solution thus improving the velocity field solution, but plays no role in the growth of divergence errors; generally viscosity will help to control divergence errors through better coupling of the velocity field as well as diffuse any errors.

The operator $\left(L_\sigma^{n+1/2} - D\sigma^{n+1/2}\mathbf{G}\right)$ plays a critical role in the error forms for the approximate projections. Because of the truncation errors of the exact and approximate projection Laplacians, this term will act as a stabilizing force in the solution of the pressure equation. It provides a mechanism through which errors from previous time steps are absorbed into the current pressure field. In all methods, the current pressure error is part of the discrete divergence.

12.3.7 Discussion of Analysis Results

As example of these general issues in practice, the basic difference scheme can be replaced. This has shown itself in computations when the second-order differences for the divergence and gradient were replaced with fourth-order central differences. This caused $D\sigma\mathbf{G}$ to be fourth-order, it lacked the second-order terms to cancel with L_σ. Thus, the second-order terms changed sign and destabilized the algorithm.

Another manner of looking at these errors is built upon the analysis in [10]. In that work similar error estimates were made for incremental approximate projections. In [10], the velocity field, $\mathbf{V}^{\perp}$ is updated in the following sequence,

$$\mathbf{V}^{\perp,*,n+1} = \mathbf{V}^{\perp,n} ,$$

then

$$\mathbf{V}^{\perp,n+1} = \mathbf{V}^{\perp,*,n+1} - \sigma^{n+1/2}\mathbf{G}^{\perp}\delta^{n} .$$

Making the observation that $\overline{D\mathbf{V}} = -D^{\perp}\mathbf{V}^{\perp}$, we have an equation for the evolution of the discrete divergence

$$\overline{D\mathbf{V}}^{n+1} = \sum_{i}^{n} D^{\perp}\sigma^{i+1/2}\mathbf{G}^{\perp}\delta^{i} .$$

Recognizing that $D^{\perp}\sigma\mathbf{G}^{\perp} = D\sigma\mathbf{G} - L_{\sigma}$, we see that this error equation is the same as we had for the incremental projection of $D\left(\mathbf{V}^{*,n+1} - \mathbf{V}\right)$.

We can make a similar set of operators for a pressure projection of $D\mathbf{V}^{*,n+1}$. The evolution equation for $\mathbf{V}^{\perp}$ is different

$$\mathbf{V}^{\perp,*,n+1} = \mathbf{V}^{\perp,n} + \sigma^{n-1/2}\mathbf{G}\phi^{n-1/2} ,$$

then

$$\mathbf{V}^{\perp,n+1} = \mathbf{V}^{\perp,*,n+1} - \sigma^{n+1/2}\mathbf{G}^{\perp}\phi^{n+1/2} . \tag{12.4}$$

Taking the divergence of this equation we get

$$\overline{D\mathbf{V}}^{n+1} = D^{\perp}\sigma^{n+1/2}\mathbf{G}^{\perp}\phi^{n+1/2} .$$

Giving us the same relation for error as we had for our earlier analysis of this scheme. In (12.4) we reconciled the difference between this scheme and the incremental scheme thus giving the proper error estimate (look at the difference in the forms for $\mathbf{V}^{*,n+1}$). Just as before, this shows that the divergence error for the pressure projection is completely local in time rather than accumulating.

When the interaction of the spatial discretization is accounted for, the conclusions are somewhat modified. The sort of Fourier analysis employed by Almgren et al. [9] provides the necessary conclusions. Generally speaking, the finite element discretizations favored by Almgren et al. used with vertex-staggered projections provide better numerical properties than the cell-centered finite difference methods. This is countered by the ease of implementation associated with methods that are collocated.

12.4 Pressure Poisson Equation Methods

The formulation given above is functionally close to the pressure Poisson equation (PPE) approach [244]. To make the results from earlier in this chapter have a more general applicability it is important to relate them directly to the PPE methodology. The steps taken with a PPE can be identical to those used for projections, but the pressure equation is explicitly derived. In this approach, a pressure equation is defined by taking the divergence of the motion equation and invoking the solenoidal condition,

$$\nabla \cdot \sigma \nabla \phi = \nabla \cdot \left(\frac{\nu}{\rho} \nabla^2 \mathbf{u} - \mathbf{u} \cdot \nabla \mathbf{u} + \mathbf{F} \right) .$$

This form is common and corresponds to the right hand side of the projection of $D\left(\mathbf{u}^* - \mathbf{u}^n\right)$. This equation can also be given in a incremental pressure form

$$\nabla \cdot \sigma \nabla \varphi = \nabla \cdot \left(\frac{\nu}{\rho} \nabla^2 \mathbf{u} - \mathbf{u} \cdot \nabla \mathbf{u} - \sigma \nabla \phi + \mathbf{F} \right) .$$

Hirt and Harlow [261] suggest a modification to control the growth of errors (given as a pressure equation)

$$\nabla \cdot \sigma \nabla \phi = \nabla \cdot \left(\frac{\nu}{\rho} \nabla^2 \mathbf{u} - \mathbf{u} \cdot \nabla \mathbf{u} + \mathbf{F} + \frac{\mathbf{u}^n}{\Delta t} \right) .$$

Careful examination of this equation confirms that this is equivalent to the projection of $D\mathbf{u}^*/\Delta\mathbf{t}$. The reason is spelled out by Hirt and Harlow [261]. Because in practice $D\mathbf{u}^n \neq 0$, the error from the previous time step(s) should be added back into the equation to suppress its growth. Error is expected even for exact projections because the pressure equation is typically solved iteratively to some error tolerance, ϵ. In the early incompressible solvers like Harlow's MAC method the pressure equation was usually solved to a rather loose tolerance of 10^{-3} [244, 243, 241, 242, 245]. In retrospect the size of the discrete divergence in this case is of the order of that typically seen in approximate projection solutions.

The first form can be derived with the assumption that the pressure Poisson equation is developed by taking the divergence of the motion equation and setting $\nabla \cdot \mathbf{u}^{n+1} = 0$ with $\nabla \cdot \mathbf{u}^n \neq 0$. A similar line of development is given in an Appendix of a paper by Veldman [585]. This form has also been recommended by Henshaw [253] to control the growth of divergence errors on overlapping grids where the pressure solution is much like an approximate projection in the regions of where the grids overlap. This point-of-view is elaborated on extensively by Gresho (and Gresho et al.) in several papers [227, 226, 228] and a book [231].

12.4.1 SIMPLE-Type Methods

Quite often the practical requirements of the application require that only a steady-state solution or long time scales be resolved. In these cases, the methods are different and the advective terms in the equations as well as the pressure-velocity coupling should be handled implicitly. Nonetheless, these methods are quite often taken from the tradition of projection methods introduced in this and previous chapters. The implicit solution of advective (hyperbolic) terms is generally much more difficult than either the elliptic (projection) or parabolic (diffusion) terms. These methods have to deal with the added complexity of implicitly solving for the nonlinear terms in the governing equations. Additional nonlinearity is introduced via the high-resolution methods used to successfully discretize the nonlinear and advective terms. Each of these considerations places a far greater burden on the methods employed for numerical linear algebra.

Because of the focus of applications for these methods, they are quite popular in the engineering community. A typical focus of the methodology are time-independent solution of the flow in a device such as a pipe, channel or heat exchanger. Because the additional difficulties involved in this class of methods much of the research emphasis is on the techniques of numerical linear algebra or acceleration of the nonlinear convergence.

Typical engineering computations have typically relied on two types of methods: implicit or steady-state methods of the sort given in [415], as exemplified by the SIMPLER algorithm an extension of the SIMPLE (Semi-Implicit method for Pressure Linked Equation) method [415], or the Marker-and-Cell (MAC) method [244]. SIMPLE can be thought of as a variant of a projection method although these algorithms are referred to as pressure-correction methods.

Perhaps the best starting point for someone interested in the SIMPLE method is the book by Patankar [415] who along with Spalding invented SIMPLE in 1972. This book is also the reference for SIMPLER or SIMPLE revised, which modifies the method to improve its convergence properties and robustness. The basic idea of the method is an iteration where each equation is solved implicitly in a sequence with the final step being a pressure equation and correction.

Most frequently this class of methods is implemented on a MAC staggered grid, although collocated schemes are increasing in popularity. The general advantages of the MAC grid have briefly been discussed in the previous chapter. Chief among these advantages are the simple form of the elliptic equation associated with pressure, and the lack of decoupling. In the collocated algorithms, the development of a velocity stabilization technique by Rhie and Chow [444] has been important. This step can be viewed as having much in common with some of the filtering techniques discussed in the previous chapter. Indeed, the MAC grid leads to exactly discretely divergence-free flow, and the collocated grid gives approximately divergence-free flows.

The basic SIMPLE algorithm proceeds as follows:

1. Solve the momentum equations holding other quantities constant for the velocity field,
$$N\left(\mathbf{u}^*\right) = \mathbf{F} - \nabla p^n \, . \tag{12.5}$$
2. Solve any other equations separately holding other quantities constant (such as density, species, and temperature).

3. Form the pressure equation and solve, holding everything constant,

$$\nabla \cdot \mathbf{u}^* = \nabla \cdot \mathcal{D}^{-1} \nabla \delta p \,, \tag{12.6}$$

where $\mathcal{D}$ is the diagonal entries for $N(\mathbf{u}^*)$. The pressure equation is formed just as with a projection method using the divergence of existing velocities as the right hand side.
4. Use the results of the pressure equation to correct the velocities and pressure,

$$\mathbf{u}^{n+1} = \mathbf{u}^* - \mathcal{D}^{-1} \nabla \delta p \,, \tag{12.7}$$

and

$$p^{n+1} = p^n + \alpha \delta p \,. \tag{12.8}$$

The constant α is used to under-relax pressure. Without such under-relaxation the iteration sequence often diverges.
5. Check for convergence, and return to step 1 if the not converged, otherwise exit.

In the SIMPLEC variant of this iteration the entries of $\mathcal{D}$ are replaced by absolute value row sums of $N(\mathbf{u}^*)$.

Compare this with SIMPLER, which follows SIMPLE, but with the following modifications: the iteration begins with a pressure solution,

$$\nabla \cdot \mathcal{D}^{-1} \nabla p^{n+1} = \nabla \cdot \mathbf{u}^* \,, \tag{12.9}$$

followed by the solution of the momentum equations using this pressure field,

$$N\mathbf{u}^{**} = \mathbf{F} - \nabla p^{n+1} \,, \tag{12.10}$$

and make the velocity divergence free through

$$\nabla \cdot \mathcal{D}^{-1} \nabla \phi = \nabla \cdot \mathbf{u}^{**} \,, \tag{12.11}$$

and the velocity correction,

$$\mathbf{u}^{n+1} = \mathbf{u}^{**} - \mathcal{D}^{-1} \nabla \phi \,. \tag{12.12}$$

The result of this is an improved rate of convergence as well as robustness. Typically the corrections to the velocity must be under-relaxed to assure convergence. Numerical studies which compare the artificial-compressibility (see Chap. 10) and pressure-Poisson methods can be found in [160, 526].

Methods such as SIMPLE and SIMPLER are limited by several things: spatial accuracy, and nonlinearity in the discrete system of equations destroying the convergence of an iteration. Many of the methodological developments with this type of method involve improving the acceleration and robustness of convergence. Quite often the solution's residual will stagnate at an unacceptably large value for many practical problems. These problems are offset by their simple structure using standard iterative techniques and judicious use of computer memory. The concern with such behavior is that subtle features in the solution will be hidden by the poor convergence. More recent approaches such as Newton-(Krylov) methods that use methods like SIMPLE as a nonlinear preconditioner can overcome these limitations in an efficient manner [421].

Classical MAC methods are often Reynolds number limited and the staggering used for the variables does not allow more modern convective methods to be used effectively without some substantial modifications. Unverdi and Tryggvason [559] use a MAC-like method for variable density flow with front-tracking. Recently, the MAC algorithm has been reformulated in a similar fashion to what is presented here [285]. Dukowicz and Dvinsky [172] also report a similar method for incompressible flow. There are other iterations that are commonly used such as PISO [272], and SIMPLEC [564]. All of these methods are distinguished by differences in the manner in which the velocity and pressure are updated following the solution of the pressure equation.

12.4.2 Implicit High-Resolution Advection

Because of the nature of advective transport, the linear algebra problems encountered in Pressure Poisson Equation Methods are more difficult with either intrinsic asymmetry associated with upwind biased stencils or loss of diagonal dominance for centered methods. As one might expect, a number of clever techniques have been used to overcome these potential efficiency sinks. We will not cover the vast numbers of methods for solving problems in regimes that are not particularly appropriate for high-resolution schemes. Our intention here is not to provide a comprehensive discussion of these methods, but rather introduce them with appropriate connections to other methods discussed in this book.

These methods were originally solved in conjunction with simple advective differencing usually based on upwind differencing. The method is almost identical to the first-order "Godunov" method, but implemented on a staggered grid. Usually, this simply means that the velocity normal to the cell face is used to determine the upwind direction to the flux across that face. Later, the Leonard's QUICK [326] scheme became popular. QUICK is simply a linear high-order upwind method where the upwind stencil is determined by the cell-edge normal velocity. Through the use of higher order methods the quality of solutions could be greatly enhanced [328, 329, 358]. As with the shock-capturing methods, high-order solutions provided an immense improvement in solution quality and opened new vistas of applications.

Inevitably, nonoscillatory versions of QUICK were developed. These methods have acronyms such as SMART [204, 205], SHARP [326, 330, 331] and ULTIMATE [327]; a number of variants of the above have also been proposed [420, 135, 619, 620, 277, 413, 344, 582, 581, 52], to name but a few. These are distinguished from most nonoscillatory methods with their foundation on the MAC staggered grids usually used with these methods. The mechanism of nonlinearity is similar to TVD methods, but specialized to the staggered grid upwinding. Often the method is expressed in the normalized value diagram (NVD), which has rough equivalence to simple TVD methods (see Chaps. 13 and 16) as described by Sweby [523]. As with other high-resolution methods, the nonlinear differencing has superseded the classical methods because of their adaptive accuracy combined with upwinding's inherent robustness. The unfortunate side-effect of these methods is their increased complexity when associated with implicit solutions.

First-order upwind methods can also be used as a means of preconditioning other solution techniques. These methods benefit from the simplicity and advantageous form of the resulting linear systems of equations. First-order upwind differencing results in linear systems that are diagonally dominant (causing them to be well-conditioned). One should note that the nonoscillatory property associated with upwind differencing and its high resolution relatives is closely related to diagonal dominance of the effective linear system of equations. Rather than block-structured systems, the linear equations are scalar in nature and benefit from a vast amount of research on the solution of these systems. The detrimental aspect of these schemes is the often poor nonlinear convergence of the entire system of equations. This aspect is the reason for many variants of these methods often focused on the under-relaxation of the updates in the iterations.

12.4.3 Implicit Direct Methods

The availability of fast efficient numerical linear algebra connected to robust, efficient nonlinear solvers makes the direct solution of the highly nonlinear system possible, in either a time-dependent or time-independent manner. The key problem with an implicit direct solution is the asymmetric nature of the advection operator (as opposed to the positive semi-definite pressure Poisson operator). This concern has been alleviated by the availability of efficient and robust Krylov subspace methods that can be applied to non-symmetric systems. Among these methods the GMRES method [476] is the most robust and theoretically sound.

More recently, efforts have been directed towards efficient methods for solving the incompressible flow equations in a coupled manner. The methods presented earlier often make effective "preconditioners". Central to the capacity to solve the system in an implicit fully coupled manner is efficient numerical linear algebra. Efforts have been focused on Krylov subspace methods (i.e., GMRES, BiCG, etc.) and multigrid or multigrid preconditioned Krylov methods.

A good starting point for this approach is the multigrid algorithm as applied by Vanka [579]. More recently, the lineage of these methods has been improved by using Krylov subspace methods for a Jacobian-free Newton's method [297]. The multigrid method provides the nearly resolution independent cost per mesh cell (ideally at least) while the Krylov method produces a robust nonlinear iteration.

12.5 Filters

The filters given here are a solution to the new problems posed by the approximate satisfaction of the divergence-free condition. Actually, the filters are also related to a problem left over from the exact projection, namely that the divergence operator does not recognize a number of obviously non-divergence-free modes. These modes are commonly referred to as "checkerboard" modes because of the pattern these form on a grid. For more challenging problems these modes can manifest themselves in a variety of ways including anomalous instabilities and solution non-convergence. The checkboard modes if left unchecked can produce effects that render a solution inaccurate or even unstable. The filter is designed to control these errors and allow the algorithm to proceed in a robust manner.

This can also be restated in several ways: these modes are in the null-space of the operator $D\mathbf{V}$, or perhaps more usefully they are not recognized by the divergence operator as being divergence-free (solenoidal). One remedy for this is to use other divergence operators to pick these modes out of solutions and essentially diffuse/remove them. We could certainly define an entire projection to remove these from the solution, but this would be overkill and obviate many of the advantages of the approximate projection such as relative efficiency. The projection apparatus will be used here to define a procedure that will stably diffuse the non-divergence-free modes in the solution as detected by different discrete divergence operators. Another remedy that we will describe is to remove the errors via a more direct velocity filter that is defined by the null space of the original operator.

12.5.1 Classification of Error Modes

First, we will describe the four major types of error modes seen in calculations. While we focus on two dimensional representations, three-dimensional calculations exhibit the same modes. The simplest is primarily one-dimensional in nature and manifests itself as shown in Fig. 12.2a. This is called a line mode. The second is shown in Fig. 12.2b and naturally evokes a vertex-based divergence operator. This is the classic checkerboard mode that prompted the name. The third mode we consider is a diagonal mode as shown in Fig. 12.2c. It is notable that this mode cannot be seen with either the standard cell-centered or vertex-centered divergences discussed in this or the previous chapter. The final mode is shown in Fig. 12.2d. Modes of this type can be seen by the vertex divergence and one of the three edge divergences described below.

The root-cause modes can all be described as linear combinations of the basic one-dimensional modes. These are shown in Fig. 12.3. The modes given in Figs. 12.2a through 12.2d are simply the same pattern for a mode repeated in some region of cells. Different modes than these can exist throughout the flow field. The fundamental filter must recognize the one-dimensional basic modes in order to be completely effective. This argument means that the

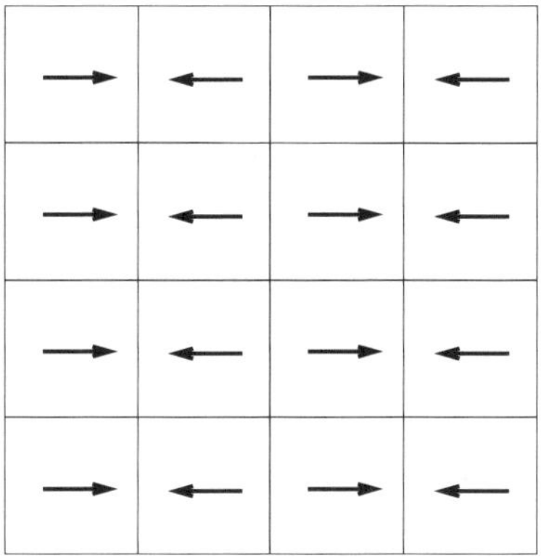

(a) 1-D mode

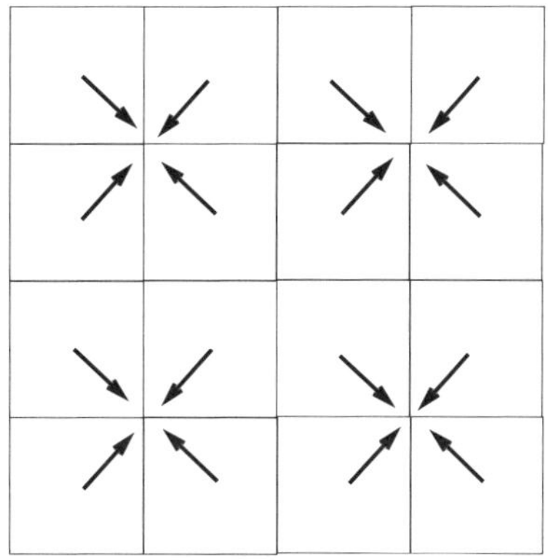

(b) Vertex mode

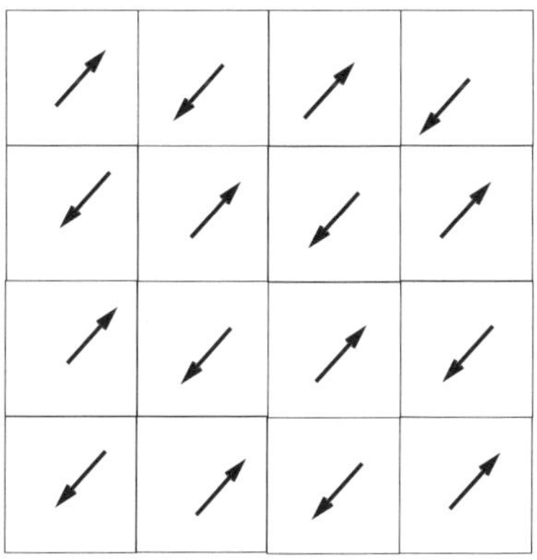

(c) Diagonal mode

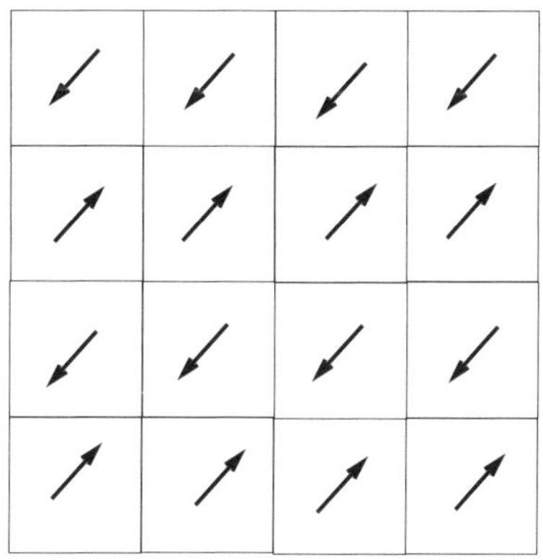

(d) Mode invisible to x-edge divergence

Fig. 12.2. Basic velocity modes that need to be filtered in order to produce a robust algorithm. Each of these modes will evaluate to a zero divergence using the standard divergence stencils discussed in the text.

edge-centered divergences will pick up the error modes at their fundamental level and as such should prove the most successful in improving solution quality.

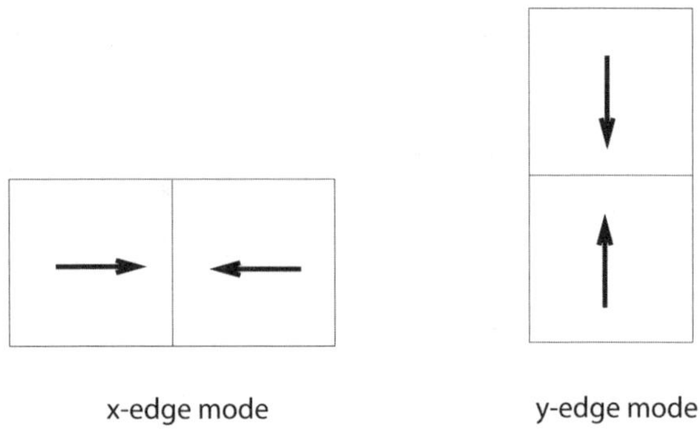

Fig. 12.3. This shows the modes that can be used to define all non-divergence-free modes occurring with cell-centered projection as defined in this chapter.

These modes can be seen in a slightly different light through the application of Fourier analysis. If we plot the symbol of the cell-centered divergence operator and note the number of zeros in the phase plane, we can characterize the null space of the operator. This is done in Fig. 12.4. Its shows that the null space has a dimension of four with three being located in the highest frequency of the grid, thus adjacent grid points are decoupled as we might suspect. To control error modes these decoupled points must be coupled by the action of the filters. This can be done by explicitly damping the decoupling or evaluating the divergence with an alternate stencil and projecting the velocity field with the appropriate potential field.

12.5.2 Projection Filters

It is possible to define a single projection to eliminate all the decoupling modes, but it would be both inaccurate and non-symmetric. This would be highly ill-advised and its impact would cause more damage than good. Maintaining both accuracy and symmetry are important, therefore the filters we define are not as effective as a full projection, but they are also much less expensive. We can apply potential field which only describes the highest frequency modes that are not divergence-free. To accomplish this task the projection will only be solved very approximately keeping the efficiency of the filter at a high priority.

Before moving to the specific discretizations for the filters, a general introduction is in order. The filters are designed to diffuse divergent modes

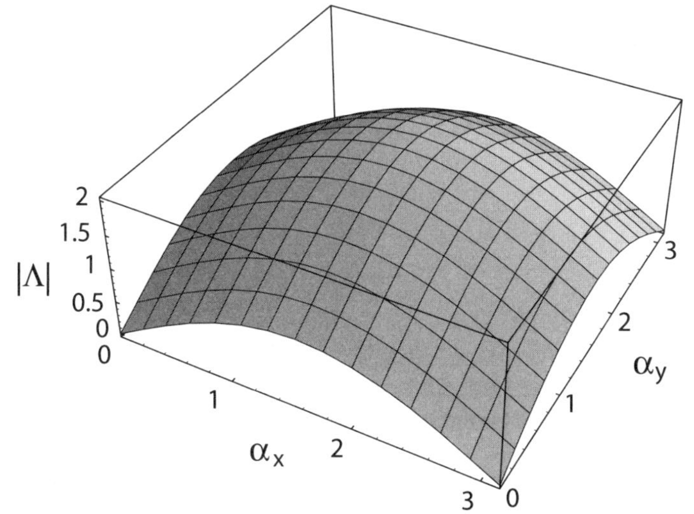

Fig. 12.4. The symbol of the cell-centered divergence operator is plotted for $\alpha_x, \alpha_y \in [0, \pi]$. The nullspace shows itself where the symbol is zero at each of the nearest grid points plus at the origin because the value of the operator is not changed if a constant is added to the vector field. In this and the following plots there is no difference in the wavenumbers between the x- and y-axes. These wavenumbers are interchangable.

described in Figs. 12.2a through 12.2d with the use of a non-converged projection operator. Heuristically, this operation should not threaten the accuracy of the overall scheme because the diffusion operator is second-order thus not impacting the overall order. Taylor series expansions of the operators can be used to confirm this. Thus the truncation error of the projection is consistent with the original algorithm and produces no reduction in the order of accuracy of the overall method.

Symbolically, the filter will act on the advanced time velocity field only,

$$\mathbf{u}^{n+1} = \mathcal{F}\left(\tilde{\mathbf{u}}^{n+1}\right) ,$$

where $\tilde{\mathbf{u}}^{n+1}$ is the product of the preceding projection. The projection is then defined by a divergence–gradient operator pair, D and $\mathbf{G}$. Recalling that the projection can be written as $\mathbf{P} = \mathbf{I} - \sigma\mathbf{G}\left(D\sigma\mathbf{G}\right)^{-1} D$, we will write this as a diffusion operator by replacing $\left(D\sigma\mathbf{G}\right)^{-1}$ by a diagonal operator. This reduces the projection to an explicit diffusion step with a pseudotime-step set by the diagonal of the discretized elliptic operator $D\sigma\mathbf{G}$. We will call this diagonal term $D^d\sigma\mathbf{G}^d$.

Vertex-Divergence Filter. The vertex-divergence based filter was originally developed to solve combustion problems where the divergent mode interacted poorly with source terms [312]. This filter is based on a divergence-gradient operators where the velocities and gradients are cell-centered, but the divergence is vertex-centered. This requires that the pressures be vertex-centered. The discrete divergence is the same as the standard vertex-centered operator, presented in Chap. 11. The diagonal term is defined by the discrete Laplacian and is

$$D^d \sigma \mathbf{G}^d_{i+1/2,j+1/2,k+1/2} = -\frac{\sigma_{i,j+1/2,k+1/2} + \sigma_{i+1,j+1/2,k+1}}{\Delta x^2} - \frac{\sigma_{i+1/2,j,k+1/2} + \sigma_{i+1/2,j+1,k+1/2}}{\Delta y^2} - \frac{\sigma_{i+1/2,j+1/2,k} + \sigma_{i+1/2,j+1/2,k+1}}{\Delta z^2}, \tag{12.13}$$

or in axisymmetric coordinates

$$D^d \sigma \mathbf{G}^d_{i+1/2,j+1/2} = -\frac{r_i \sigma_{i,j+1/2} + r_{i+1} \sigma_{i+1,j+1/2}}{r_{i+1/2} \Delta r^2} - \frac{\sigma_{i+1/2,j} + \sigma_{i+1/2,j+1}}{\Delta z^2}.$$

It is not difficult to establish that this produces a stable algorithm essentially because of the diagonal dominance of the Laplacian. The discrete divergence and gradient are second-order accurate. With this operator specified, the algorithm is straightforward.

Edge-Divergence Filter. The edge-centered filters can be implemented for both the x, y or z-edges of the grid. The values of the velocities on the normal edges of the divergence cell are simple to compute, but the tangential velocities will require averaging. The computed filter pressures will be on edges, thus the normal velocity corrections will be simple compact stencils.

For the x-edges the operators are

$$D_{i+1/2,j,k} \mathbf{V} = \frac{V^x_{i+1,j,k} - V^x_{i,j,k}}{\Delta x} + \frac{V^y_{i,j+1,k} + V^y_{i+1,j+1,k} - V^y_{i,j-1,k} - V^y_{i+1,j-1,k}}{4\Delta y} + \frac{V^z_{i,j,k+1} + V^z_{i+1,j,k+1} - V^z_{i,j,k-1} - V^z_{i+1,j,k-1}}{4\Delta z}, \tag{12.14}$$

or for axisymmetric grids

$$D_{i+1/2,j} \mathbf{V} = \frac{r_{i+1} V^r_{i+1,j} - r_i V^r_{i,j}}{r_{i+1/2} \Delta r} + \frac{V^z_{i,j+1} + V^z_{i+1,j+1} - V^z_{i,j-1} - V^z_{i+1,j-1}}{4\Delta z},$$

and the gradient is

$$\sigma \mathbf{G}_{i,j,k}\varphi = \begin{pmatrix} \dfrac{\varphi_{i+1/2,j,k} - \varphi_{i-1/2,j,k}}{\rho_{i,j,k}\Delta x} \\ \\ \dfrac{\varphi_{i+1/2,j+1,k} + \varphi_{i-1/2,j+1,k} - \varphi_{i-1/2,j-1,k} - \varphi_{i-1/2,j-1,k}}{4\rho_{i,j,k}\Delta y} \\ \\ \dfrac{\varphi_{i+1/2,j,k+1} + \varphi_{i-1/2,j,k+1} - \varphi_{i-1/2,j,k-1} - \varphi_{i-1/2,j,k-1}}{4\rho_{i,j,k}\Delta z} \end{pmatrix} .$$

Above ρ is the density. The diagonal operator is

$$D^d \sigma \mathbf{G}^d_{i+1/2,j,k} = -\frac{\sigma_{i,j,k} + \sigma_{i+1,j,k}}{\Delta x^2} - \frac{\sigma_{i+1/2,j-1/2,k} + \sigma_{i+1/2,j+1/2,k}}{\Delta y^2} - \frac{\sigma_{i+1/2,j,k-1/2} + \sigma_{i+1/2,j,k+1/2}}{\Delta z^2} ,$$

or in axisymmetric coordinates

$$D^d \sigma \mathbf{G}^d_{i+1/2,j} = -\frac{r_i\sigma_{i,j} + r_{i+1}\sigma_{i+1,j}}{r_{i+1/2}\Delta r^2} - \frac{\sigma_{i+1/2,j-1/2} + \sigma_{i+1/2,j+1/2}}{\Delta z^2} .$$

For the y and z-edges the operators are defined similarly via appropriate rotation of the indices.

As with the vertex-based filter, stability is simple to establish via Fourier analysis for both edge-based filters. This analysis replaces the spatial coordinates, x, y, z to wavenumbers $\alpha_x, \alpha_y, \alpha_z$. The wavenumbers are defined as multiples of $\Delta x/\pi$. For the x-edge filter the truncation error (using the Taylor series expansion in the limit as α goes to zero) is

$$\begin{pmatrix} -\frac{1}{24}\alpha_x^2 + \mathcal{O}\left(\alpha_x^4\right) \\ -\frac{1}{6}\alpha_y^2 - \frac{1}{8}\alpha_x^2 - \frac{1}{8}\alpha_z^2 + \mathcal{O}\left(\alpha_z^4, \alpha_y^4, \alpha_z^4, \alpha_x^2\alpha_y^2, \alpha_x^2\alpha_z^2, \alpha_y^2\alpha_z^2\right) \\ -\frac{1}{6}\alpha_z^2 - \frac{1}{8}\alpha_x^2 - \frac{1}{8}\alpha_y^2 + \mathcal{O}\left(\alpha_z^4, \alpha_y^4, \alpha_z^4, \alpha_x^2\alpha_y^2, \alpha_x^2\alpha_z^2, \alpha_y^2\alpha_z^2\right) \end{pmatrix} ,$$

for the divergence and

$$\begin{pmatrix} -\frac{1}{24}\alpha_x^2 + \mathcal{O}\left(\alpha_x^4\right) \\ -\frac{1}{6}\alpha_y^2 - \frac{1}{8}\alpha_x^2 - \frac{1}{8}\alpha_z^2 + \mathcal{O}\left(\alpha_y^4, \alpha_x^2\alpha_y^2, \alpha_y^2\alpha_z^2, \alpha_x^2\alpha_z^2\right) \\ -\frac{1}{6}\alpha_z^2 - \frac{1}{8}\alpha_x^2 - \frac{1}{8}\alpha_y^2 + \mathcal{O}\left(\alpha_z^4, \alpha_x^2\alpha_y^2, \alpha_y^2\alpha_z^2, \alpha_x^2\alpha_z^2\right) \end{pmatrix} ,$$

for the gradient. With a similar expression for the y-edge filter (with appropriate terms switched). This is also a second-order filter. When each of these filters are applied to a velocity field they do not degrade the second-order accuracy of the solution.

In applying the edge-based filter, the correction should only be applied to the velocity in the normal direction of the edge divergence. This is because the symbol of $\sigma \mathbf{G} D$ in the transverse direction is quite similar to the cell-centered divergence, it would not be useful to apply it. We would expect this operation to do little good for the solution (a similar decomposition of the corrections cannot be made for the vertex projection).

By plotting the symbols of the filter divergences we can predict their impact on the solution. This is done in Figs. 12.5 through 12.6. The combination of these filters will couple all grid points, but the application of one of them will leave part of the decoupling intact.

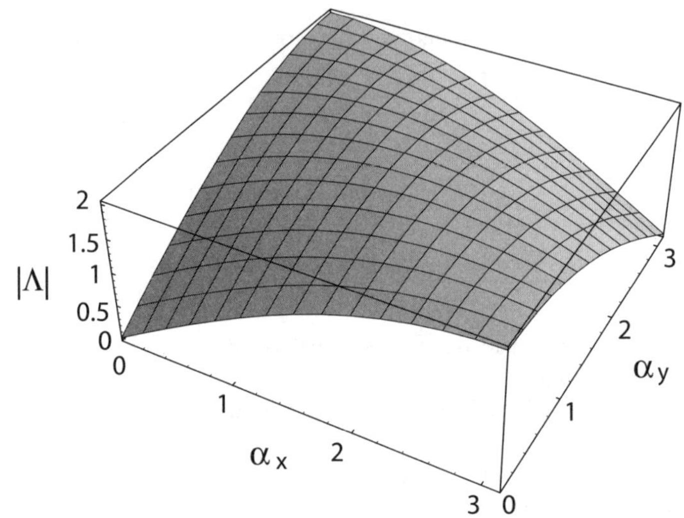

Fig. 12.5. The symbol of the vertex-centered divergence operator is plotted for $\alpha_x, \alpha_y \in [0, \pi]$. This divergence couples the adjacent cells in the x- and y-directions, but leaves the diagonal cell decoupled.

One can also use the exact projection as a filter. Examining the symbol of the approximate and exact projection operators it is obvious that they differ greatly only in the high-frequency portion of the spectrum. It is therefore possible to project out most of the difference between the two operators in the high-frequency regime cheaply using a relaxation like weighted Jacobi. This is close to getting an exact projection for little expense. The issue of grid decoupling will not rear its head because the approximate projection will have chosen the same constant for each of the nullspace components in the exact projection operator.

We will now show the expected impact on the solution through analyzing the constant density projection for one, or four composite weighted Jacobi iterations. The weighted Jacobi method allows us to selectively damp certain high-frequency errors in the solution. Our composite Jacobi iteration consists of one Jacobi sweep with a weight of one followed by a second with a weight of one-half. The weighted Jacobi can be written as

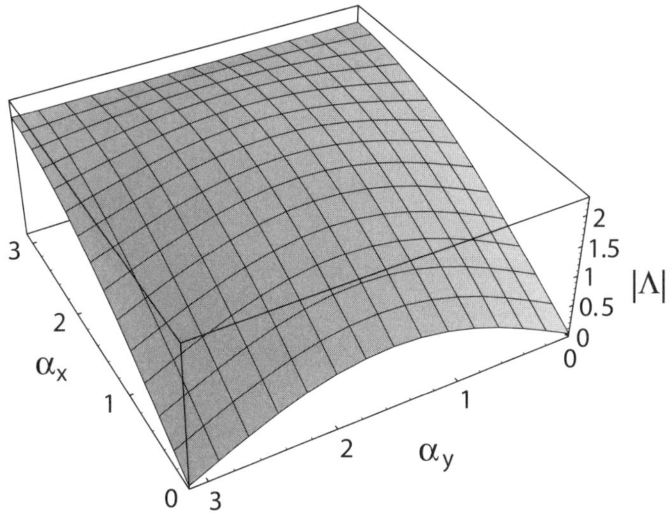

Fig. 12.6. The symbol of the x-edge-centered divergence operator is plotted for $\alpha_x, \alpha_y \in [0, \pi]$. The diagonal and x-direction cells are coupled, but the y-direction cells are decoupled. This also shows the symbol for both x- and y-edge divergences by switching the α's.

$$\varphi_{i,j}^{m+1} = (1 - \omega)\, \varphi_{i,j}^{m} + \omega \sum_{nb} a_{nb,i,j} \varphi_{nb}^{m},$$

where m is the iteration index, ω is the weight and a_{nb} are coefficients for the neighboring cells, nb, in the linear equation for $\varphi_{i,j}$ as given by (11.6a) or for axisymmetric coordinates by (11.6b).

We expect that the iteration will not impact the highest frequency in the solution because of the local decoupling in the exact projection operator, but the $\omega = 1/2$ sweep should be effective in eliminating the next highest frequency error between the approximate and exact projection. Fig. 12.7 shows the relative error in the symbols for the exact and projection operators. When smoothed with one composite Jacobi iteration the errors are significantly reduced as shown by Fig. 12.8. With four composite Jacobi passes, almost all of the high frequency error has been removed at least on $2\Delta x$ scale, errors still persist on Δx scale, but we are not concerned with this (unlike previous filters). The same decoupling is maintained, thus this filter can only be a partial solution.

12.5.3 Velocity Filters

A completely different formulation for filters can be defined on criteria that combine physical intuition and linear vector spaces. These are motivated from

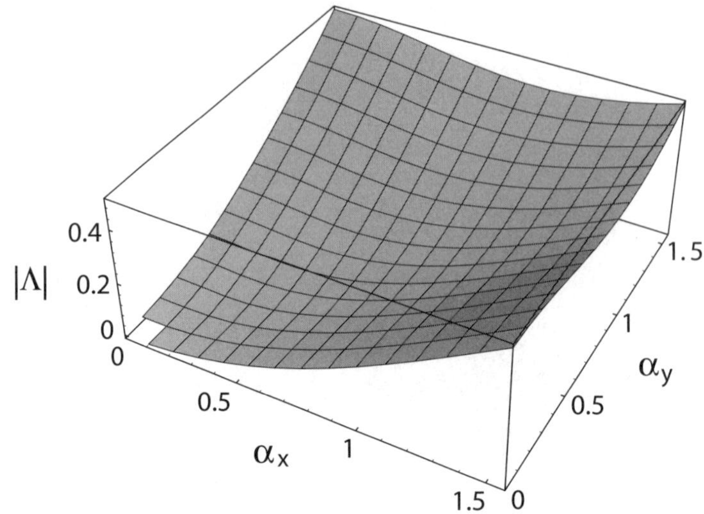

Fig. 12.7. The relative error between the exact and (5-point) approximate projection Laplacians excluding error on the highest frequency scale ($|\alpha_x|$, $|\alpha_y| \leq \pi/2$).

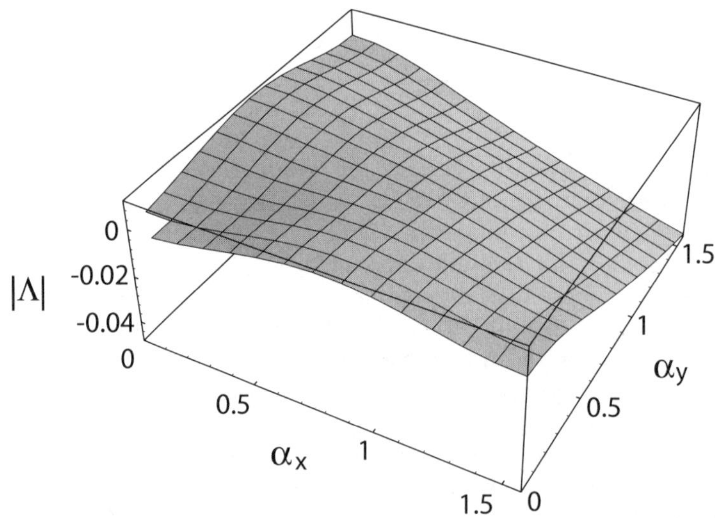

Fig. 12.8. The relative error between the exact and (5-point) approximate projection Laplacians after one composite Jacobi iteration.

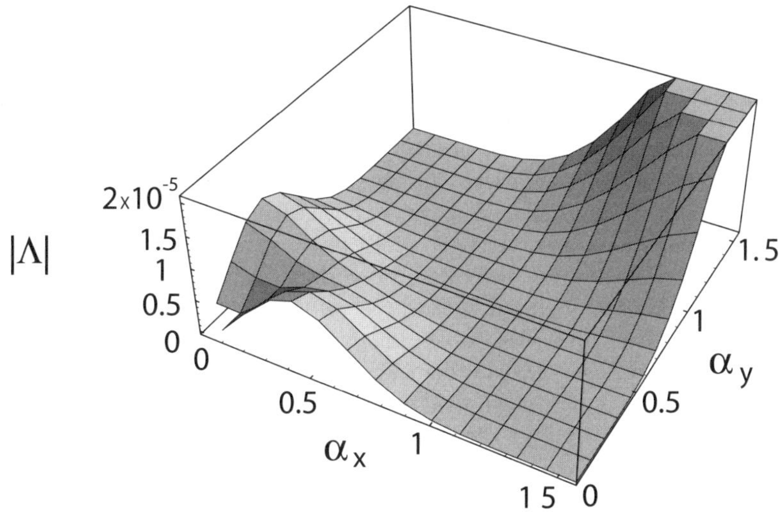

Fig. 12.9. The relative error between the exact and (5-point) approximate projection Laplacians after four composite Jacobi iterations.

work with compressible Lagrangian hydrocodes [95] where elements can be deformed by "hourglass" or checkerboard modes that are not physical.[3] These modes are associated with degrees of freedom in an operator that are not associated with physical forces. This idea is founded on the work of Margolin and Pyun [368] (for a more available lucid discussion of these matters see Benson's review article [53]).

The basic idea is to define a complete linear space from the discrete data used for a stencil that contains both physical and non-physical modes. The non-physical modes are associated with the nullspace of the incomplete space defined by the physical modes. The non-physical modes can then be subtracted from the velocity field to remove these modes from the solution. Ideally speaking, this operation should leave the physical modes undisturbed. The physical modes for two-dimensional flows are two translational modes (average velocities), two shear modes and two stretching modes (the spatial derivatives of velocity). These can be stated as operators, $\bar{u}$, $\bar{v}$, $\partial u/\partial x$, $\partial v/\partial y$, $\partial u/\partial y$, and $\partial v/\partial x$. The finite difference operators defining these operators discretely will define a portion of $\mathbf{S}$. For three-dimensional flows there will

[3] Lagrangian hydrocodes are derived in the tradition of artificial viscosity as defined by Von Neumann and Richtmyer [590]. These codes use a grid that is vertex staggered and the mesh deforms with the material motion. Non-physical modes in the material motion can cause the grid to prematurely cease to be physically realizable and cause the calculation to terminate.

be three translational modes, three shear modes and six stretching modes.[4] Stated differently these are the mean flow, the divergence and rotational or vortical modes.

Vertex or Hourglass Velocity Filter. The vertex filter (projection) has a two-dimensional nullspace that can be interpreted as hourglass modes in the solution. In three dimensions these modes are much more complex with twelve hourglass modes. These modes can be removed with a filter like that defined by [368]. A vertex-based divergence (cell) has four velocity vectors and thus eight degrees of freedom in two dimensions. In three dimensions there are 24 degrees of freedom associated with three velocities at eight nodes of a cell. With only six physical modes in two dimensions, two modes are left to define the hourglass modes. In three dimensions the 24 degrees of freedom and 15 physical modes leave 9 non-physical modes to be removed! These modes can then be removed to improve the solution.

In the cases below, the vector spaces can be found through discovering a linearly independent set of vectors to span the space. These vectors will have physical interpretations such as an average velocity or gradient.

Staying in two dimensions and following [368], we have velocities

$$\mathbf{u}^T = \left(u_{i,j}, u_{i+1,j}, u_{i,j+1}, u_{i+1,j+1}, v_{i,j}, v_{i+1,j}, v_{i,j+1}, v_{i+1,j+1}\right) ,$$

to work with. The average velocity modes are

$$\mathbf{S}_1^T = \frac{1}{4}\left(1,1,1,1,0,0,0,0\right) ,$$

and

$$\mathbf{S}_2^T = \frac{1}{4}\left(0,0,0,0,1,1,1,1\right) .$$

The shearing modes are

$$\mathbf{S}_3^T = \frac{1}{2\Delta x}\left(-1,1,-1,1,0,0,0,0\right) ,$$

and

$$\mathbf{S}_4^T = \frac{1}{2\Delta y}\left(0,0,0,0,-1,-1,1,1\right) .$$

The stretching modes are

$$\mathbf{S}_5^T = \frac{1}{2\Delta x}\left(-1,-1,1,1,0,0,0,0\right) ,$$

and

$$\mathbf{S}_6^T = \frac{1}{2\Delta y}\left(0,0,0,0,-1,1,-1,1\right) .$$

[4] In one-dimension there are two modes: one translation (mean velocity) and the shear (divergence mode). One dimensional incompressible flow is quite dull and we will not consider it further.

The nullspace of this vector space has two modes

$$\mathbf{S}_7^T = \frac{1}{2}(1,-1,-1,1,0,0,0,0) \ ,$$

and

$$\mathbf{S}_8^T = \frac{1}{2}(0,0,0,0,1,-1,-1,1) \ .$$

Examination of the action of these modes on a cell allows one to notice why these are called "hourglass" modes in that they would cause the cell to deform into an hourglass shape.

This gives a complete basis for $\mathbf{u}$. With appropriate normalization the velocity can be expressed as

$$\mathbf{u} = \sum_{i=1}^{8} \mathbf{S}_i^T \mathbf{u} \mathbf{S}_i \ .$$

The filter then can be constructed to remove the hourglass modes

$$\tilde{\mathbf{u}} = \mathbf{u} - C \sum_{i=7}^{8} \mathbf{S}_i^T \mathbf{u} \mathbf{S}_i \ .$$

In [368], C has a maximum value of 0.25 as determined by the explicit stability of the operator associated with the above algorithm. This filter has a differential analog as the fourth-order terms, u_{xyxy} and v_{xyxy}.

Cell-Centered Velocity Filter. Unlike Lagrangian hydrodynamics, a cell-centered grid is most often used for Eulerian computations and do not have a readily defined element on which to define a set of velocities (i.e., degrees of freedom). We will take this set of velocities to be those needed to define a divergence operator. This gives ten degrees of freedom with six physical modes thus leaving four nonphysical modes to be defined. On the vertex grid to move to three dimensions from two, leads to an increase of 10 non-physical modes; the cell-centered case is simpler having the increase from four modes to nine. This arises from the three dimensional stencil of 7 cells, and 21 degrees of freedom coupled with the 15 physical modes thus leaving leaves 9 non-physical modes.

In three dimensions the velocities are

$$\begin{aligned}\mathbf{u}^T = (\ & u_{i-1,j,k} \ , u_{i,j,k}, u_{i+1,j,k}, u_{i,j-1,k}, u_{i,j+1,k}, u_{i,j,k-1}, u_{i,j,k-1}, \\ & v_{i,j,k} \quad , v_{i-1,j,k}, v_{i+1,j,k}, v_{i,j-1,k}, v_{i,j+1,k}, v_{i,j,k-1}, v_{i,j,k+1}, \\ & w_{i,j,k} \quad , w_{i-1,j,k}, w_{i+1,j,k}, w_{i,j-1,k}, w_{i,j+1,k}, w_{i,j,k-1}, w_{i,j,k+1}) \ .\end{aligned}$$

As before, various operators will then be defined by $\mathbf{S}_i^T \mathbf{u}$. The average x-velocity can be given by $\mathbf{S}_1^T \mathbf{u}$ with

$$\mathbf{S}_1^T = \frac{1}{7}(1,1,1,1,1,1,1,0,0,0,0,0,0,0,0,0,0,0,0,0,0) \ .$$

The portion of $\mathbf{S}$ corresponding to the $\bar{v}$ is

$$\mathbf{S}_2^T = \frac{1}{7}(0,0,0,0,0,0,0,1,1,1,1,1,1,1,0,0,0,0,0,0,0) \ .$$

The portion of $\mathbf{S}$ corresponding to the $\bar{w}$ is

$$\mathbf{S}_3^T = \frac{1}{7}(0,0,0,0,0,0,0,0,0,0,0,0,0,0,1,1,1,1,1,1,1) \ .$$

The shearing modes portion of $\mathbf{S}$ are

$$\mathbf{S}_4^T = \frac{1}{2\Delta x}(0,-1,1,0,0,0,0,0,0,0,0,0,0,0,0,0,0,0,0,0,0) \ ,$$

$$\mathbf{S}_5^T = \frac{1}{2\Delta y}(0,0,0,0,0,0,0,0,0,0,-1,1,0,0,0,0,0,0,0,0,0) \ ,$$

and

$$\mathbf{S}_6^T = \frac{1}{2\Delta z}(0,0,0,0,0,0,0,0,0,0,0,0,0,0,0,0,0,0,0,-1,1) \ .$$

The stretching modes portion of $\mathbf{S}$ are

$$\mathbf{S}_7^T = \frac{1}{2\Delta z}(0,0,0,0,0,-1,1,0,0,0,0,0,0,0,0,0,0,0,0,0,0) \ ,$$

$$\mathbf{S}_8^T = \frac{1}{2\Delta y}(0,0,0,-1,1,0,0,0,0,0,0,0,0,0,0,0,0,0,0,0,0) \ ,$$

$$\mathbf{S}_9^T = \frac{1}{2\Delta x}(0,0,0,0,0,0,0,0,-1,1,0,0,0,0,0,0,0,0,0,0,0) \ ,$$

$$\mathbf{S}_{10}^T = \frac{1}{2\Delta z}(0,0,0,0,0,0,0,0,0,0,0,-1,1,0,0,0,0,0,0,0) \ ,$$

$$\mathbf{S}_{11}^T = \frac{1}{2\Delta y}(0,0,0,0,0,0,0,0,0,0,0,0,0,0,0,0,-1,1,0,0) \ ,$$

$$\mathbf{S}_{12}^T = \frac{1}{2\Delta x}(0,0,0,0,0,0,0,0,0,0,0,0,0,0,-1,1,0,0,0,0) \ .$$

Once we find the nullspace of $\mathbf{S}$ we can expand the original velocity field in terms of $\mathbf{S}_i$. Our original set of velocities can be found from

$$\mathbf{u} = \sum_{i=1}^{21} \mathbf{S}_i^T \mathbf{u} \mathbf{S}_i \ ,$$

with $\mathbf{S}$ being normalized so that $\mathbf{S}^T\mathbf{S} = I$ (i.e., $\mathbf{S}^T = \mathbf{S}^{-1}$).

Because the components of velocity corresponding to the nullspace are not physical (motions are not related to physical forces, thus are spurious) we should remove or lessen their participation in the solution. This can be done through subtracting them from the velocity field

$$\tilde{\mathbf{u}} = \mathbf{u} - C \sum_{i=13}^{21} \mathbf{S}_i^T \mathbf{u} \mathbf{S}_i \ . \tag{12.15}$$

For the cell-centered velocity field defined above, the nullspace is

$$\mathbf{S}_{13}^T = (-2,1,,1,0,0,0,0,0,0,0,0,0,0,0,0,0,0,0,0,0,0) \ ,$$

$$\mathbf{S}_{14}^T = (-2,0,0,1,1,0,0,0,0,0,0,0,0,0,0,0,0,0,0,0,0) \ ,$$

$$\mathbf{S}_{15}^T = (-2,0,0,0,0,1,1,0,0,0,0,0,0,0,0,0,0,0,0,0,0) \ ,$$

$$\mathbf{S}_{16}^T = (0,0,0,0,0,0,0,-2,1,1,0,0,0,0,0,0,0,0,0,0,0) \ ,$$

$$\mathbf{S}_{17}^T = (0,0,0,0,0,0,0,0,-2,0,1,1,0,0,0,0,0,0,0,0,0) \ ,$$

$$\mathbf{S}_{18}^T = (0,0,0,0,0,0,0,0,-2,0,0,0,1,1,0,0,0,0,0,0,0) \ ,$$

$$\mathbf{S}_{19}^T = (0,0,0,0,0,0,0,0,0,0,0,0,0,0,-2,1,1,0,0,0,0) \ ,$$

$$\mathbf{S}_{20}^T = (0,0,0,0,0,0,0,0,0,0,0,0,0,0,0,-2,0,1,1,0,0) \ ,$$

and

$$\mathbf{S}_{21}^T = (0,0,0,0,0,0,0,0,0,0,0,0,0,0,0,-2,0,0,0,1,1) \ .$$

If these expressions are substituted into (12.15), we find that it looks like the finite difference expressions for u_{xxxx}, u_{yyyy}, u_{zzzz}, v_{xxxx}, v_{yyyy}, v_{zzzz}, w_{xxxx}, w_{yyyy}, and w_{zzzz}. Thus this operation is equivalent to adding a fourth-order diffusion to the solution. Because of this finite difference form, we find that this modification should not impact the accuracy of the overall scheme. Heuristic analysis of checkerboard errors suggests that a coefficient should multiply the diffusion terms of the form

$$C = \frac{h^4}{32} \ ,$$

where $h = \Delta x$, Δy, and Δz. Fourier analysis confirms that this coefficient is necessary for linear stability of this operation. The symbol of the operator (scaled by C) is shown in Fig. 12.10.

During the development of this filter it was found that applying it to the entire flow field, while effective in controlling the discrete divergence, it also caused the solution quality to be degraded due to its dissipative nature. As a consequence we apply this filter only in the regions of the flow where the dimensionless divergence is large. To accomplish this we then define a second coefficient, C_D, that gives the largest tolerable ratio of divergence to velocity in the flow field and use the edge-based divergences introduced earlier to define the local divergence. The local characteristic velocity is chosen to be the largest velocity in a cell. For the x-, y- and z-direction diffusion, x-, y- and $z-$ divergences are used, respectively. The value of the coefficient is then limited for stability.

The leading coefficient on the velocity filter is then

$$C_{i,j,k}^x = \frac{\Delta x^4}{32} \min\left[1, C_D \frac{\left(\left|D_{i-1/2,j,k}\mathbf{u}\right| + \left|D_{i+1/2,j,k}\mathbf{u}\right|\right)}{\max\left(\left|u_{i,j,k}\right|, \left|v_{i,j,k}\right|, \left|w_{i,j,k}\right|\right)}\right] \ ,$$

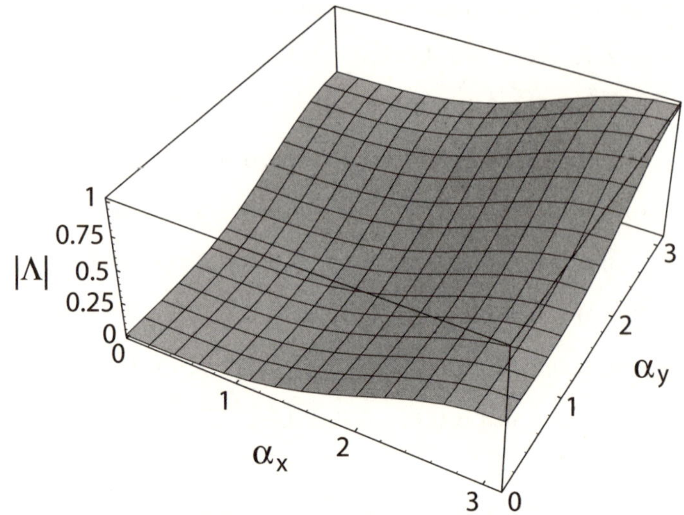

Fig. 12.10. The symbol of the 2-D velocity filter operator is plotted for $\alpha_x, \alpha_y \in [0, \pi]$. The filter effectively couples the adjacent grid points (i.e., it does not have a destructive nullspace associated with them).

$$C^y_{i,j,k} = \frac{\Delta y^4}{32} \min\left[1, C_D \frac{\left(\left|D_{i,j-1/2,k}\mathbf{u}\right| + \left|D_{i,j+1/2,k}\mathbf{u}\right|\right)}{\max\left(\left|u_{i,j,k}\right|, \left|v_{i,j,k}\right|, \left|w_{i,j,k}\right|\right)}\right],$$

and

$$C^z_{i,j,k} = \frac{\Delta z^4}{32} \min\left[1, C_D \frac{\left(\left|D_{i,j,k-1/2}\mathbf{u}\right| + \left|D_{i,j,k+1/2}\mathbf{u}\right|\right)}{\max\left(\left|u_{i,j,k}\right|, \left|v_{i,j,k}\right|, \left|w_{i,j,k}\right|\right)}\right].$$

C_D is taken to be between 1 and 10 typically. Just as the projection based filters, this step is applied at the end of a computational cycle to the advanced time velocities.

We also need to consider the impact of physical viscosity on this filter. Physical viscosity should operate in a similar fashion to this filter through the coupling of adjacent grid cells effectively. When viscous forces dominate, we would like to limit the impact of this velocity filter accordingly. The velocity filter aims to couple cells that are decoupled by the divergence operator. We can compute the degree of damping applied by the viscous forces to the highest frequency errors (i.e., those operating locally) by looking at the symbol of the viscous discretization. For the high frequency (α_x, α_y, $\alpha_z = \pi$), this is

$$\Lambda^{\nu}_{\pi} = \frac{1 - 2\mu_x - 2\mu_y - 2\mu_z}{1 + 2\mu_x + 2\mu_y - 2\mu_z},$$

where $\mu_x = \nu \Delta t / dx^2 \rho$, $\mu_y = \nu \Delta t / \Delta y^2 \rho$, and $\mu_z = \nu \Delta t / \Delta z^2$. We then modify the maximum damping applied by the filter by

$$C^{x,\mathrm{mod}}_{i,j,k} = C^{x}_{i,j,k} \max\left(0, \Lambda^{\nu}_{\pi}\right) \ .$$

With the basic projection algorithm in place terms such as $\mathbf{u} \cdot \nabla \psi$ can be computed. The algorithm is not limited by the cell Reynolds number,[5] and is also applicable to inviscid flows.

12.6 Method Demonstration and Verification

In this section we will give example calculations that will illuminate different strengths and weaknesses of the algorithms described in the preceding sections.

The basic algorithm used to compute the results given here has a number of basic characteristics discussed earlier. It is an approximate projection method using collocated placement of variables, unsplit high-order Godunov advection (see Chap. 14), and Crank-Nicholson (see Chap. 7) diffusion. Unless otherwise stated the method is the cell-centered approximate projection. The linear algebra is solved using a multigrid method (see Chap. 8) that can handle discontinuous large jumps in density.

We use four test problems: a vortex-in-a-box, an inflow problem, an inviscid drop with a 1000:1 density ratio, and a doubly periodic shear-layer. Because we are primarily interested in flows that are at high Reynolds numbers or inviscid, most test problems will not have exact solutions. We can measure convergence rates via Richardson extrapolation [461], (this is also described in Chap. 6). This is done through the use of three grid solutions to compute a single convergence rate. Errors between grids will be reported by considering the finer grid to be exact. The local average of the fine grid is compared with the coarse grid values, with two error estimates yielding a convergence estimate. This approach has been used in a number of other studies (see [45, 50, 10, 514], for example).

12.6.1 Vortex-in-a-Box

Our workhorse test problem will be the vortex-in-a-box problem used in [45]. It uses a stream function

$$\psi = \frac{1}{\pi} \sin^2(\pi x) \sin^2(\pi y) \ ,$$

in a unit square with homogeneous velocity boundary conditions to define the initial conditions. Initial velocities are $u^o = -\psi_y$ and $v^o = \psi_x$. Similarly to [45], we set $\Delta x = \Delta y = 1/2^n$ for $n = 5 - 8$. Our time step is set to $\Delta t = \Delta x$ for an effective CFL number of approximately one.

[5] The cell Reynolds number is defined as $Re^{x}_{\mathrm{cell}} = |u| \, \Delta x / \nu$.

We seek to confirm the earlier analysis of these forms (presented in Sect. 12.3). The convergence results for a vortex-in-a-box are shown in Table 12.1. As expected, the incremental form with the right-hand side $D\left(\mathbf{u}^{*,n+1} - \mathbf{u}^n\right)/\Delta t$, (Sect. 12.3.3),[6] and the pressure form with $D\mathbf{u}^{*,n+1}/\Delta t$, (Sect. 12.3.6), perform well and nearly identically. Both of these are better than the other two options, i.e., the incremental form with the $D\mathbf{u}^{*,n+1}/\Delta t$ (Sect. 12.3.5), with the right-hand side also being second-order, and the pressure projection with $D\left(\mathbf{u}^{*,n+1} - \mathbf{u}^n\right)/\Delta t$ (Sect. 12.3.4), with the right-hand side being only marginally second-order.

Table 12.1. Convergence rates for the vortex-in-a-box problem for different approximate projection formulations. Inc. and Pres. stand for the incremental and pressure projection forms, respectively.

Case	L_2 Norms				
	32^2-64^2	Rate	64^2-128^2	Rate	128^2-256^2
Inc. $D\left(\mathbf{u}^{*,n+1} - \mathbf{u}^n\right)/\Delta t$	2.8×10^{-3}	1.98	7.1×10^{-4}	2.15	1.6×10^{-4}
Inc. $D\mathbf{u}^{*,n+1}/\Delta t$	3.0×10^{-3}	1.99	7.5×10^{-4}	2.10	1.7×10^{-4}
Pres. $D\left(\mathbf{u}^{*,n+1} - \mathbf{u}^n\right)/\Delta t$	1.1×10^{-2}	1.60	3.6×10^{-3}	1.79	1.0×10^{-3}
Pres. $D\mathbf{u}^{*,n+1}/\Delta t$	2.9×10^{-3}	2.00	7.1×10^{-4}	2.13	1.6×10^{-4}

12.6.2 Inflow with Shear

We are also concerned about the convergence behavior of the projections with more general boundary conditions. We have constructed a test problem to examine these concerns. The flow has inflow in shear with a tangentially applied perturbation in a unit square with inflow at one end and outflow at the other. The top and bottom of the domain are set to symmetry conditions. The x-velocity is set by

$$u = 1 + \lambda \tanh\left[\varrho\,(y - 0.5)\right] + 0.05 \sin\left(4\pi t\right) ,$$

where $\lambda = 0.5$, and $\varrho = 30$. The inlet y-velocity is set to

$$v_{\text{inflow}} = 0.05 \sin\left(4\pi t\right) .$$

The results for this test are given in Table 12.2. We report the results for the y-direction velocity as it should be the most sensitive to the boundary

[6] See also Sect. 12.3.1 for basic definitions.

condition. The results reveal that both of the pressure projections are superior to the incremental projections with the projection of $D\mathbf{u}^{*,n+1}/\Delta t$ giving consistently second-order results. The other methods appear to degrade in accuracy as the grid is refined. For the projections of $D\left(\mathbf{u}^{*,n+1} - \mathbf{u}^{n}\right)/\Delta t$, the accumulation of error is the probable explanation. For the worst performer, the incremental projection of $D\mathbf{u}^{*,n+1}/\Delta t$, the likely culprit is the unusual form for the incorporation of previous errors in the projection into the advance time pressure equation.

Table 12.2. Convergence rates for the velocities u and v in the simple inflow problem for different approximate projection formulations.

Case		L_2 Norms				
		32^2-64^2	Rate	64^2-128^2	Rate	128^2-256^2
Inc. $D\left(\mathbf{u}^{*,n+1} - \mathbf{u}^{n}\right)/\Delta t$	u	7.4×10^{-3}	2.17	1.6×10^{-3}	1.71	5.0×10^{-4}
Sect. 12.3.3	v	6.5×10^{-3}	1.96	1.7×10^{-3}	1.75	5.0×10^{-4}
Inc. $D\mathbf{u}^{*,n+1}/\Delta t$	u	8.7×10^{-3}	2.06	2.1×10^{-3}	0.39	1.6×10^{-3}
Sect. 12.3.5	v	7.3×10^{-3}	1.95	1.9×10^{-3}	1.09	8.8×10^{-4}
Pres. $D\left(\mathbf{u}^{*,n+1} - \mathbf{u}^{n}\right)/\Delta t$	u	7.3×10^{-3}	2.10	1.7×10^{-3}	1.77	4.0×10^{-4}
Sect. 12.3.4	v	6.9×10^{-3}	1.97	1.7×10^{-3}	1.88	4.8×10^{-4}
Pres. $D\mathbf{u}^{*,n+1}/\Delta t$	u	7.4×10^{-3}	2.20	1.6×10^{-3}	1.97	4.1×10^{-4}
Sect. 12.3.6	v	6.7×10^{-3}	1.98	1.7×10^{-3}	1.97	4.3×10^{-4}

12.6.3 Doubling Periodic Shear Layer

To further test the types of projections, we turn to the doubly periodic shear flow problem [45] in a unit square box (see also Fig. 12.11). The shear layer problem is set up by defining a velocity field

$$u = \begin{cases} \tanh\left[\varrho\left(y - 0.25\right)\right] & \text{if } y \leq 0.5 \\ \tanh\left[\varrho\left(0.75 - y\right)\right] & \text{if } y > 0.5 \end{cases},$$

and

$$v = 0.05\sin\left(2\pi x\right).$$

The parameter ϱ controls the thickness of the shear-layer and we have chosen $\varrho = 30$.

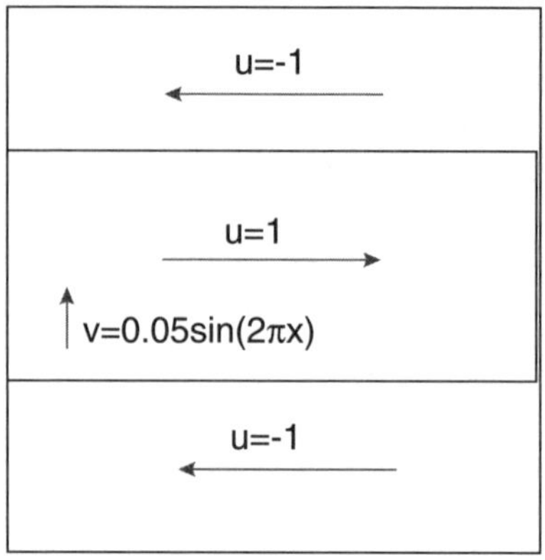

Fig. 12.11. The setup for the doubly periodic shear layer in a unit square box; all boundaries are periodic.

Results are presented for the convergence and a measure of the kinetic energy. The kinetic energy in the solution should be constant for the Euler equations. The decay of kinetic energy is then a measure of the error in the computation.

For the standard convergence test, the error estimated on the finest grid is nearly uniform for all methods except the pressure projection of $D\left(\mathbf{u}^{*,n+1} - \mathbf{u}^n\right)/\Delta t$. On the coarse grids, the error of the projections of $D\mathbf{u}^{*,n+1}/\Delta t$ have less error than the other options, but show lower convergence rates across the range of grids. These results are given in Table 12.3.

The decay of kinetic energy results are given for two methods (Fig. 12.12 and Table 12.4): the incremental projection of $D\left(\mathbf{u}^{*,n+1} - \mathbf{u}^n\right)/\Delta t$ and the pressure projection of $D\mathbf{u}^{*,n+1}/\Delta t$. This test seems to show exactly the opposite of the earlier test. The error is smaller on the coarse grids for the incremental projection, but larger on the fine grid. Consequently, the convergence rate results are also opposite from the other test.

12.6.4 Long Time Integration

We have also found it quite useful to integrate the equations for long periods of time. This will measure how errors build up with various methods and hopefully uncover any long-term numerical instabilities. In order to do this, we integrate the inviscid vortex-in-a-box problem to time 60 with a CFL number of 0.95. This requires approximately 7500–8000 time steps on a 128^2 grid. For the simple convergence tests where this problem is integrated to time one, significant differences do not appear (except with the method exhibiting lower-order convergence).

We have used several of the filters described in [449]. These are used to remove oscillations from the solution that arise from the inability for the

Table 12.3. Convergence rates for the doubly periodic shear flow problem for different approximate projection formulations.

Case		L_2 Norms				
		32^2-64^2	Rate	64^2-128^2	Rate	128^2-256^2
Inc. $D\left(\mathbf{u}^{*,n+1} - \mathbf{u}^n\right)/\Delta t$	u	8.0×10^{-2}	1.91	2.1×10^{-2}	2.16	4.7×10^{-3}
Sect. 12.3.3	v	6.4×10^{-2}	2.09	1.5×10^{-2}	2.02	3.7×10^{-3}
Inc. $D\mathbf{u}^{*,n+1}/\Delta t$	u	6.5×10^{-2}	1.89	1.7×10^{-2}	1.85	4.8×10^{-3}
Sect. 12.3.5	v	6.0×10^{-2}	2.19	1.3×10^{-2}	1.84	3.7×10^{-3}
Pres. $D\left(\mathbf{u}^{*,n+1} - \mathbf{u}^n\right)/\Delta t$	u	9.1×10^{-2}	1.78	2.6×10^{-2}	2.13	6.0×10^{-3}
Sect. 12.3.4	v	7.1×10^{-2}	2.11	1.7×10^{-2}	1.99	4.2×10^{-3}
Pres. $D\mathbf{u}^{*,n+1}/\Delta t$	u	6.1×10^{-2}	1.80	1.8×10^{-2}	1.86	4.8×10^{-3}
Sect. 12.3.6	v	5.6×10^{-2}	2.08	1.3×10^{-2}	1.82	3.7×10^{-3}

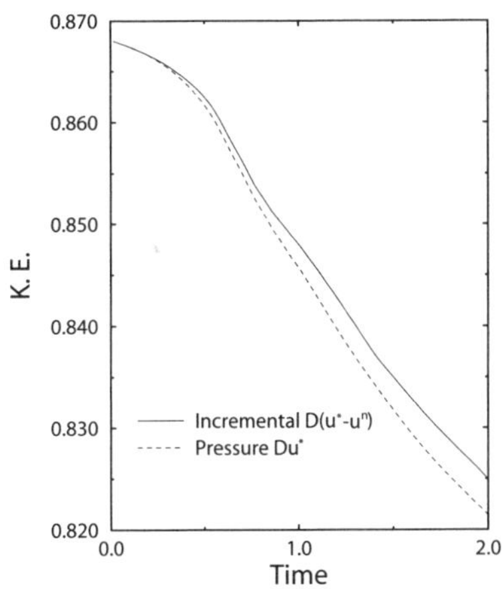

(a) $\|\mathbf{u}\|_2^2$, 32^2 grid

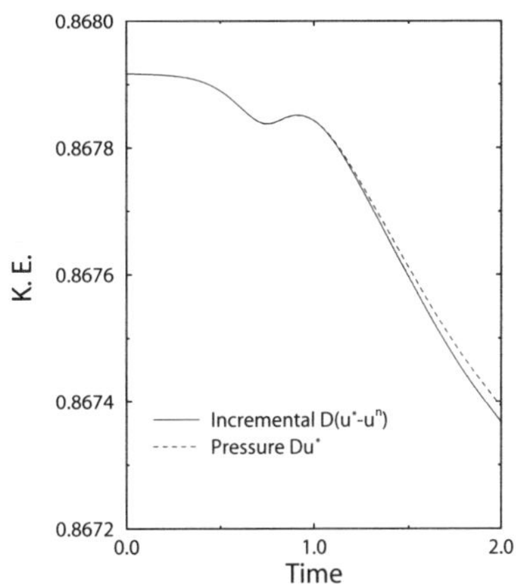

(b) $\|\mathbf{u}\|_2^2$, 256^2 grid

Fig. 12.12. Comparison of projections for the dissipation of kinetic energy for the doubly periodic shear layer problem using the Euler equations.

Table 12.4. Convergence rates for the doubly periodic shear flow problem for the dissipation of kinetic energy, $u^2 + v^2$.

Grid	Inc. $D\left(\mathbf{u}^{*,n+1} - \mathbf{u}^n\right)/\Delta t$		Pres. $D\mathbf{u}^{*,n+1}/\Delta t$	
	$\Vert\mathbf{u}\Vert_2^2$	Rate	$\Vert\mathbf{u}\Vert_2^2$	Rate
32^2	0.824441		0.821216	
		2.12		2.17
64^2	0.857891		0.857533	
		2.15		2.23
128^2	0.865652		0.865699	
		2.04		2.08
256^2	0.867366		0.867392	

discrete divergence to sense non-solenoidal error modes. Based on the results given in [449], we use the cell-centered velocity filter, and the vertex projection filter in tandem. These filters greatly improve the quality of results especially in the cases of long-time integrations and variable density flows with large density jumps or discontinuous density profiles.

The results from computations using the filters and the incremental projection of $D\left(\mathbf{u}^{*,n+1} - \mathbf{u}^n\right)/\Delta t$ the results are shown in Fig. 12.13. Divergence errors are seemingly under control, and the solution is clean. Kinetic energy is decaying in a physical manner and the vorticity norms are behaving physically. Using the incremental projection with the $D\mathbf{u}^{*,n+1}/\Delta t$ right hand side produces results that become unstable at about $t = 15$. The divergence grows greatly and the solutions become noisy and nonphysical.

Fig. 12.14 shows the results for the pressure projection of

$$D\left(\mathbf{u}^{*,n+1} - \mathbf{u}^n\right)/\Delta t\,. \tag{12.16}$$

The results for the vorticity look physical, but the discrete divergence is quite large and has more coherent structure than other projections. This is result of the manner in which the error accumulates in this algorithm and explains the less than second-order convergence experienced with the method.

In Fig. 12.15, the results of the pressure projection of $D\mathbf{u}^{*,n+1}/\Delta t$. This solution behaves quite well and has errors that are lower than any of the other methods (as we might expect from the earlier analysis). The kinetic energy and maximum divergence are compared for the three stable methods in Fig. 12.16. The pressure projection of $D\mathbf{u}^{*,n+1}/\Delta t$ is superior in both kinetic energy dissipation and divergence norms.

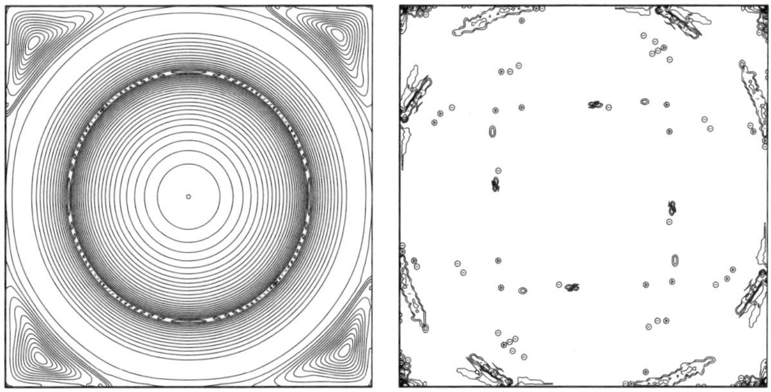

(a) $\log|\psi+1|$ (b) $D\mathbf{u}^{n+1}$

Fig. 12.13. The results of a long-time integration of vortex-in-a-box problem with the incremental $D\left(\mathbf{u}^{*,n+1}-\mathbf{u}^{n}\right)/\Delta t$ projection and vertex (Sect. 12.5.2) and velocity (Sect. 12.5.3), filters for a 128^2 grid.

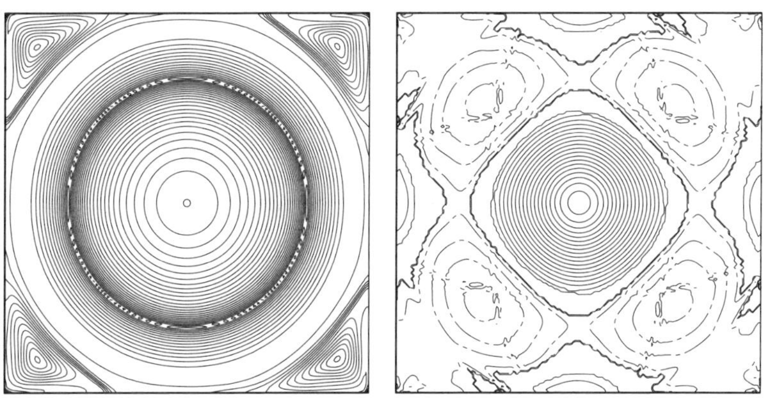

(a) $\log|\psi+1|$ (b) $D\mathbf{u}^{n+1}$

Fig. 12.14. Long-time integration of vortex-in-a-box problem with vertex (Sect. 12.5.2) and velocity (Sect. 12.5.3) filters using the pressure projection of $D\left(\mathbf{u}^{*,n+1}-\mathbf{u}^{n}\right)/\Delta t$ for a 128^2 grid.

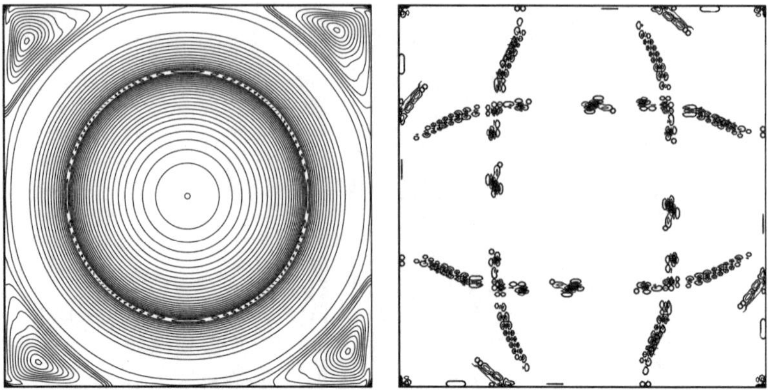

(a) $\log|\psi + 1|$ (b) $D\mathbf{u}^{n+1}$

Fig. 12.15. Long-time integration of vortex-in-a-box problem with vertex and velocity filters using the pressure projection of $D\mathbf{u}^{*,n+1}/\Delta t$ (see Sect. 12.3) for a 128^2 grid.

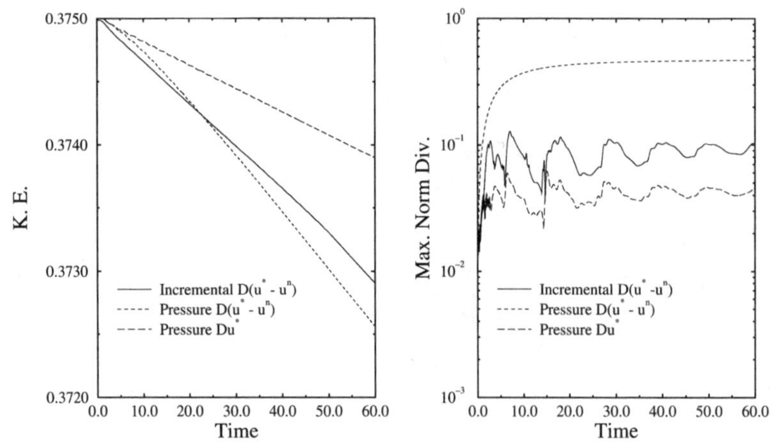

Fig. 12.16. Comparison of global characteristics, kinetic energy (left plot, $\|\mathbf{u}\|_2^2$) and maximum divergence norm (right plot, $D\mathbf{u}^{n+1}/\Delta t$), for different projections on a 128^2 grid.

Another test of the projection form involves introducing more error into the solution at each step of the computation. In our multigrid solution we usually set the error tolerance to 1×10^{-8}, but in this instance we lower it to 1×10^{-4}. For short-time integrations of the vortex-in-a-box problem, this has no impact on the error estimates or the convergence rates. Upon running this problem to $t = 60$ we found that the error tolerance makes a substantial difference. This results in a serious loss of solution quality as time progresses. This is shown in Fig. 12.17a–b. The culprit is a run away divergence error as seen in Fig. 12.18a. Alternately, the projection $L\phi^{n+1/2} = D\mathbf{u}^{*,n+1}/\Delta t$ gives virtually identical results to the higher-error tolerance results. In this case, we made the switch to a composite Jacobi smoother for the multigrid in order to preserve the symmetry of the solution. At the lower tolerance the Gauss-Seidel methods destroyed the symmetry (along with the solution).

12.6.5 Circular Drop Problem

Consider the following test problem, which will illuminate many subtle issues. A circular drop with radius 0.15 is placed at (0.5, 0.75) in a unit square computational domain with solid wall boundary conditions on all sides that is partitioned with a 64×64 grid. Gravity is unity (downward) and all boundaries are frictionless (free-slip). The drop fluid is 1000 times more dense than the background fluid (having unit density). The flow is integrated forward in time to $t = 1$ using the Euler equations. A high-order Godunov method (discussed in Chap. 14) and an unsplit piecewise linear volume tracking algorithm (discussed in Chap. 18) is used to advect the flow. The CFL number is 1/2 unless otherwise stated. The unsteady flow is computed with variations of both the exact and approximate projection methods. Each method demonstrates second-order convergence (in space and time) on sufficiently smooth problems.

Solutions obtained with the standard exact and approximate projection methods (i.e., without filters) are shown in Fig. 12.19.[7] Both solutions exhibit spurious features in the velocity field in the flow above the drop. The exact projection solution (Fig. 12.19) displays some velocity field decoupling and slight asymmetries. Despite the use of a smaller time step in integrating the flow (CFL=1/4), the approximate projection solution in Fig. 12.19 is unacceptable. As discussed later, this solution is compromised in part because projection of $\nabla \cdot \left(\mathbf{u}^{*,n+1} - \mathbf{u}^n\right)$ and solving for a pressure increment (rather than a total pressure). This is consistent with the analysis given in Sect. 12.3.

Without filters, the basic approximate incremental projection of

$$D\left(\mathbf{u}^{*,n+1} - \mathbf{u}^n\right)/\Delta t\,, \tag{12.17}$$

gives exceedingly poor results. This is shown in Fig. 12.20. The resulting velocity field is shown in Figs. 12.20a and 12.20b. Both components of velocity

[7] The standard formulation solves for an increment in pressure and projects $\nabla \cdot \left(\mathbf{u}^{*,n+1} - \mathbf{u}\right)$.

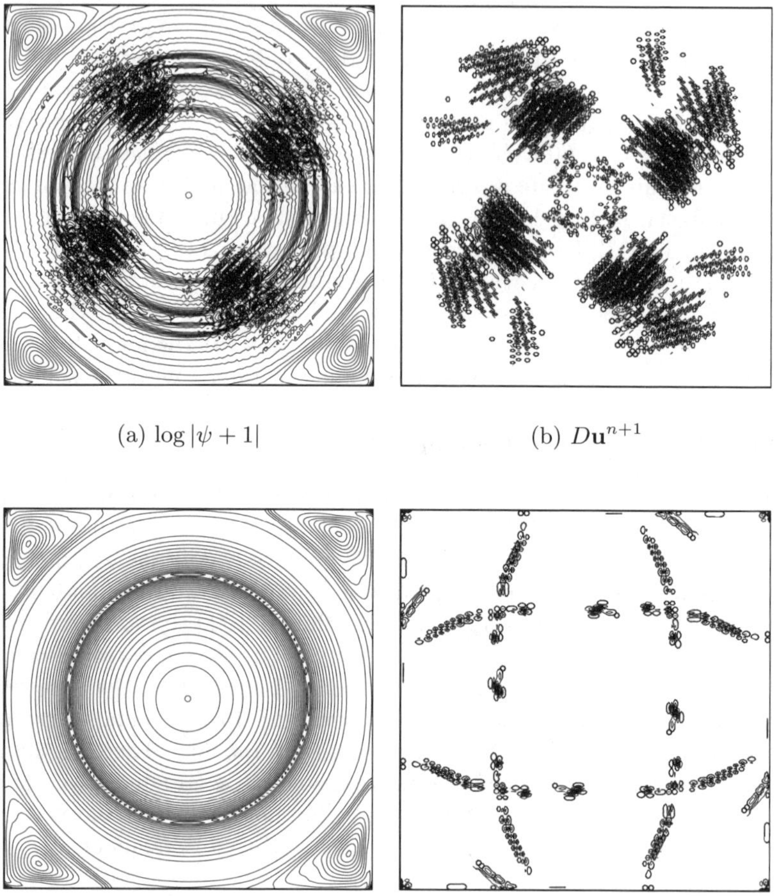

(a) $\log|\psi+1|$ (b) $D\mathbf{u}^{n+1}$

(c) $\log|\psi+1|$ (d) $D\mathbf{u}^{n+1}$

Fig. 12.17. Long-time integration of vortex-in-a-box problem for projections using vertex and velocity filters, $\varepsilon = 1 \times 10^{-4}$ (error tolerance in the multigrid solution) and a composite Jacobi smoother for a 128^2 grid: (a) and (b) plots refer to the projection $L\delta^n = D\left(\mathbf{u}^{*,n+1} - \mathbf{u}^n\right)/\Delta t$; (c) and (d) plots refer to the projection $L\phi^{n+1/2} = D\mathbf{u}^{*,n+1}/\Delta t$.

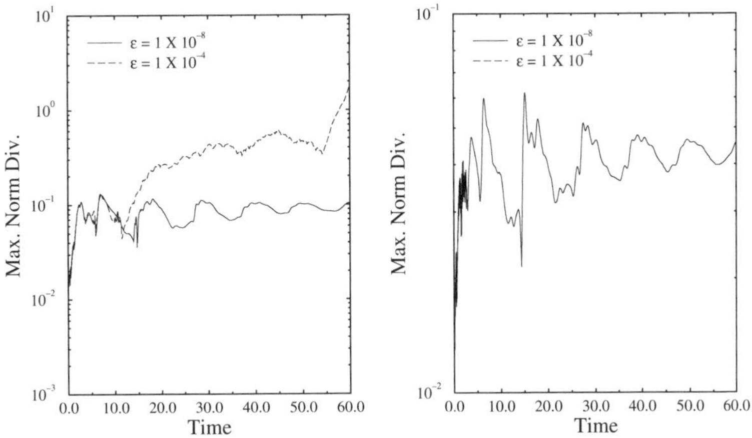

Fig. 12.18. Results for the maximum divergence in the long-time integration of vortex-in-a-box problem with vertex (Sect. 12.5.2) and velocity (Sect. 12.5.3) filters and different projections, $(D\mathbf{u}^{*,n+1} - \mathbf{u}^n)/\Delta t$ (left plot) and $D\mathbf{u}^{*,n+1}/\Delta t$ (right plot) using different convergence criteria (error tolerance ε) for the multigrid algorithm for a 128^2 grid.

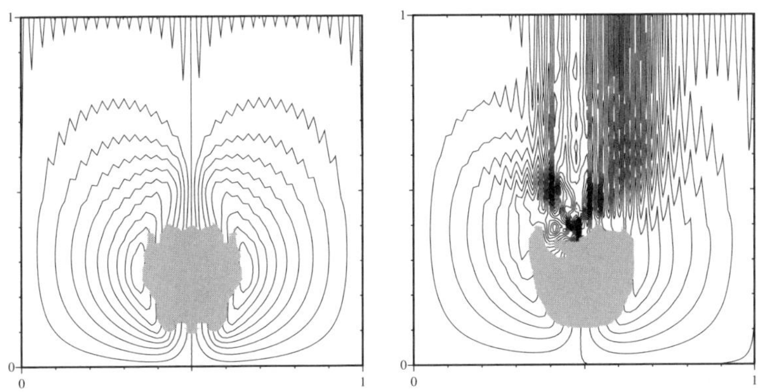

Fig. 12.19. Drop solutions for our standard exact (left plot) and approximate (right plot) pressure increment projections. Both solutions use a grid where all variables are cell-centered. The droplet outline and the velocity streamlines are shown. In both cases the projection equation LHS is given by $\nabla \cdot \left(\mathbf{u}^{*,n+1} - \mathbf{u}^n\right)$ (see Sect. 12.3).

have the appearence of a "halo" effect left behind where the drop was initially (before the density interface had a chance to smear). Both are noisy throughout. The pressure field (shown in Fig. 12.20c) is extremely noisy inside the dense drop. The discrete divergence (i.e., error) is large and concentrated around the periphery of the original bubble position. The discrete divergence is not larger near the velocity field (near the drop's current position). Divergence errors are a combination of checkerboard and line-error modes. These results are all completely unacceptable.

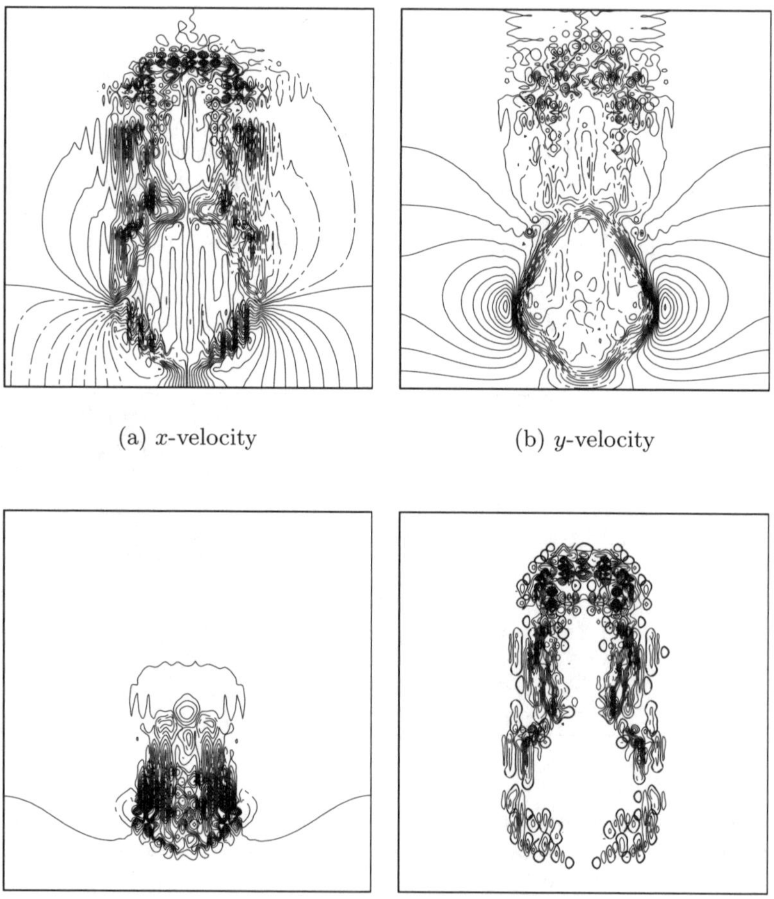

(a) x-velocity

(b) y-velocity

(c) pressure

(d) $D\mathbf{u}^{n+1}$

Fig. 12.20. Inviscid discontinuous drop problem results (isolines) without filters. The incremental, $D\left(\mathbf{u}^{*,n+1} - \mathbf{u}^n\right)/\Delta t$ projection is used (see Sect. 12.3).

The pressure projection of $D\mathbf{u}^{*,n+1}/\Delta t$ itself improves the results as shown in Fig. 12.21. With the exception of the pressure field, these results are as good, or better than anything seen with the incremental projection of $D\left(\mathbf{u}^{*,n+1} - \mathbf{u}^{n}\right)/\Delta t$. We expect the addition of filters can only improve this.

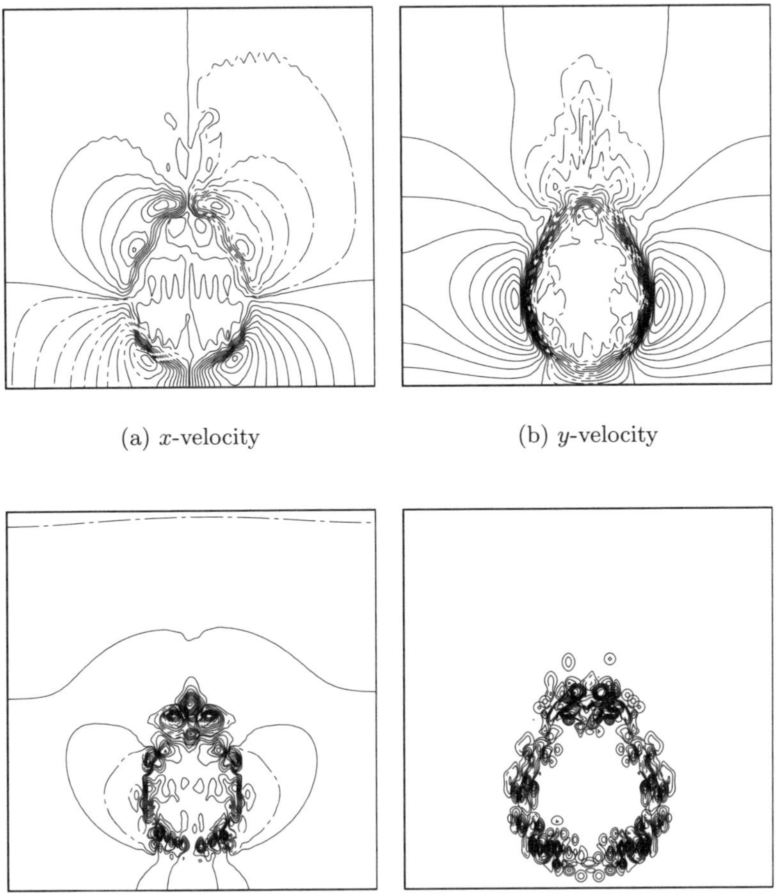

(a) x-velocity

(b) y-velocity

(c) pressure

(d) $D\mathbf{u}^{n+1}$

Fig. 12.21. Inviscid discontinuous drop results without filters using a pressure projection of $D\mathbf{u}^{*,n+1}/\Delta t$ (see Sect. 12.3).

Fig. 12.22 gives the results with filters. Using our earlier experience, we have used the combination of vertex and velocity filters to improve the results. As expected, the results are outstanding and clearly outshine all of the earlier results. The pressure field is still a sore spot with the interior of the drop showing some noise.

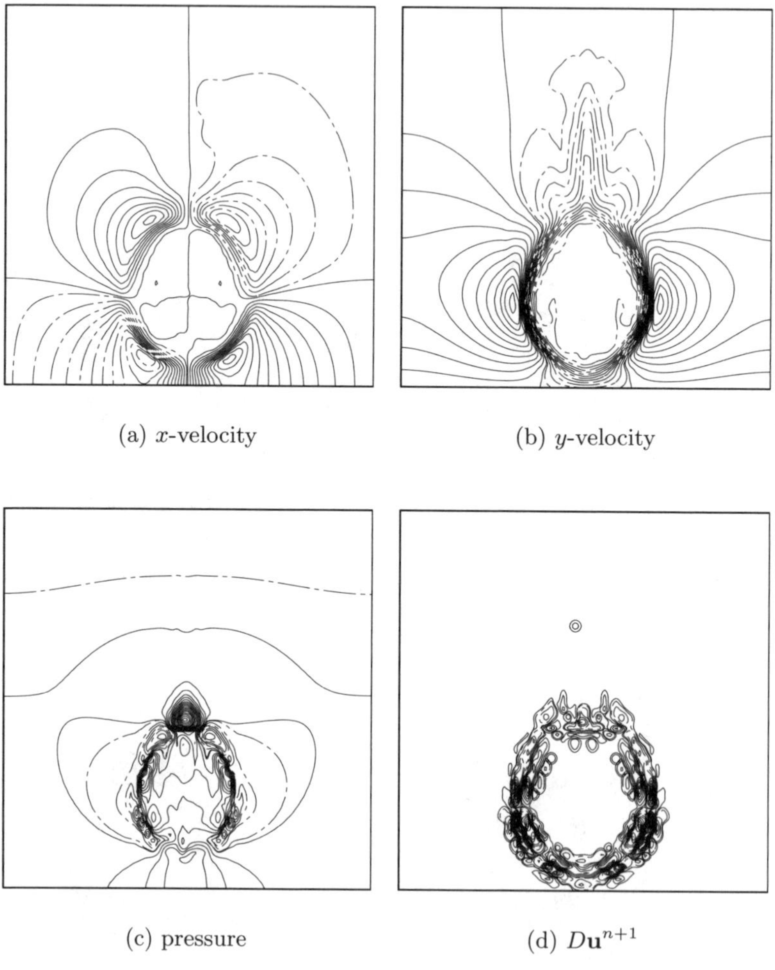

(a) x-velocity (b) y-velocity

(c) pressure (d) $D\mathbf{u}^{n+1}$

Fig. 12.22. Inviscid discontinuous drop results with vertex and velocity based filters using a pressure projection of $D\mathbf{u}^{*,n+1}/\Delta t$ (see Sect. 12.3).

We have shown that when the velocity field is not constrained to be solenoidal, the pressure equation must be carefully constructed for stable, robust behavior of the algorithms. While the test problems presented in this chapter were performed using high-order Godunov methods (Chap. 16) coupled to approximate projection methods on collocated grids, the conclusions should apply to other methods sharing several basic characteristics: a nonsolenoidal velocity field, a divergence operator with a multidimensional nullspace, and second or higher order accuracy. One root cause is that the higher order accuracy fails to provide damping that would otherwise damp the nonsolenoidal modes in the solution.

The form of the pressure equation we prefer based both on analysis and computational evidence is quite clear. The pressure projection of $D\mathbf{u}^{*,n+1}/\Delta t$ is the best method. Coupled with filters this method is quite stable and robust under all circumstances we have tested.

The third major point to emphasize is the utility of using a projection framework to analyze and implement methods. Both from the standpoint of understanding the behavior of the method as well as putting it into action, we have found the framework offered by projections to be superior.

12.6.6 Results Using Various Filters

In this section we will give results that will illuminate different properties of the filters described in the preceding section.

We have uses an approximate projection method using collocated placement of variables, unsplit high-order Godunov advection, and the Crank-Nicholson method for diffusive processes. The linear algebra is solved using a multigrid method.

We will describe a set of calculations that will show the performance and necessity of using the filters defined in Sect. 12.5. Next, we will show their necessity for variable density flows with large density variations and sharp density fronts and long time integrations using the equations of constant density flow. Finally, we will also show that the filters have no negative impact on solution quality or accuracy (for resolved flows, the impact is positive).

Fig. 12.23 shows the impact of the vertex-projection filter. Again, the solution is improved, but the "halo" effect is still prevalent, and the solution quality is poor. The divergence error is again concentrated away from the drop near the initial drop position. The line error modes have been largely suppressed, but other checkerboard modes remain. These modes are related to those not detected by the vertex divergence.

The first problem that we will use is the high density ratio drop problem that was previously used. The cell-centered-velocity filter (Sect. 12.5.3), produces results of improved quality as shown in Fig. 12.24. The "halo" effect is largely alleviated. The flow field away from the "halo" region is not as high quality (noise-free) as the vertex-projection filter solutions. Nevertheless, the value of the discrete divergence is much lower than before, but the quality of the solution is still unacceptable. This is most clearly seen through looking at the pressure field (see Fig. 12.24c) where there is a great deal of noise inside the drop.

Fig. 12.25 shows a comparison of all the methods given for the discontinuous drop problem. The maximum of the discrete divergence is used as

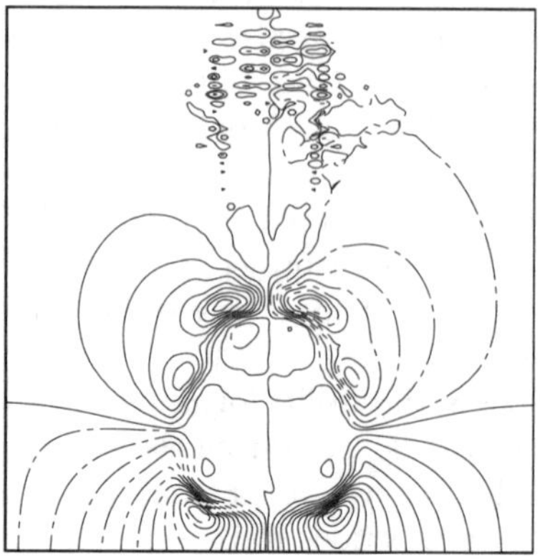

(a) x-velocity

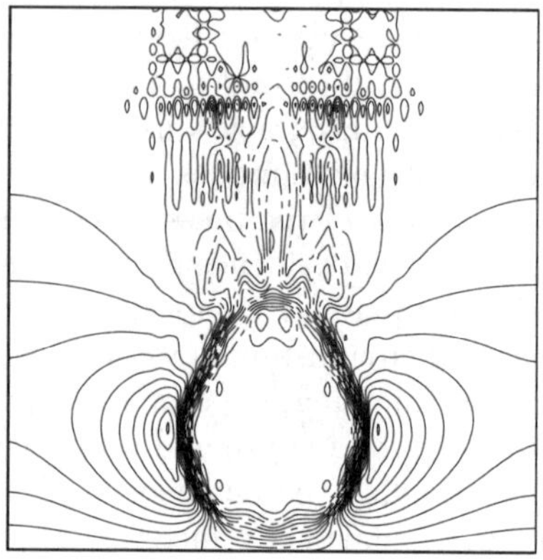

(b) y-velocity

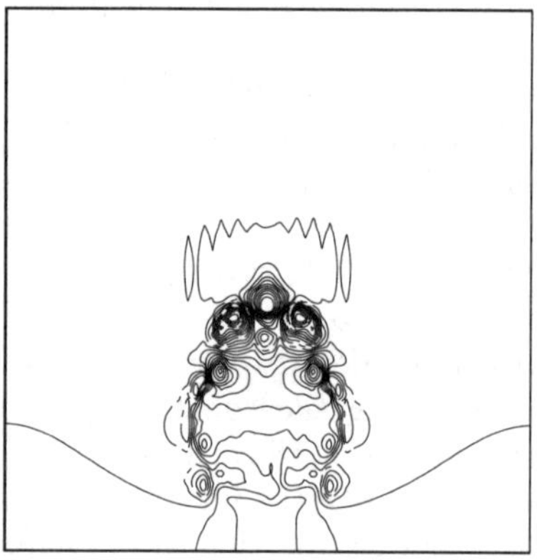

(c) pressure

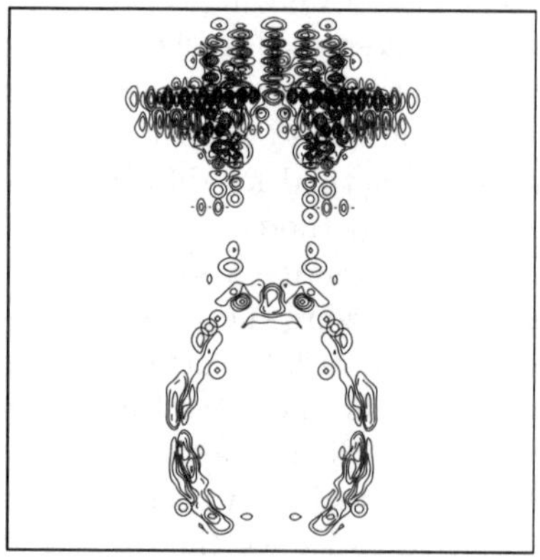

(d) $D\mathbf{u}^{n+1}$

Fig. 12.23. Inviscid discontinuous drop results with a vertex-projection filter. It should be noted that the discrete divergence is not on the same scale as the other figures. Thus the quality and position of this error should be compared with the other figures. The divergences for each of the methods is shown in Fig. 12.25

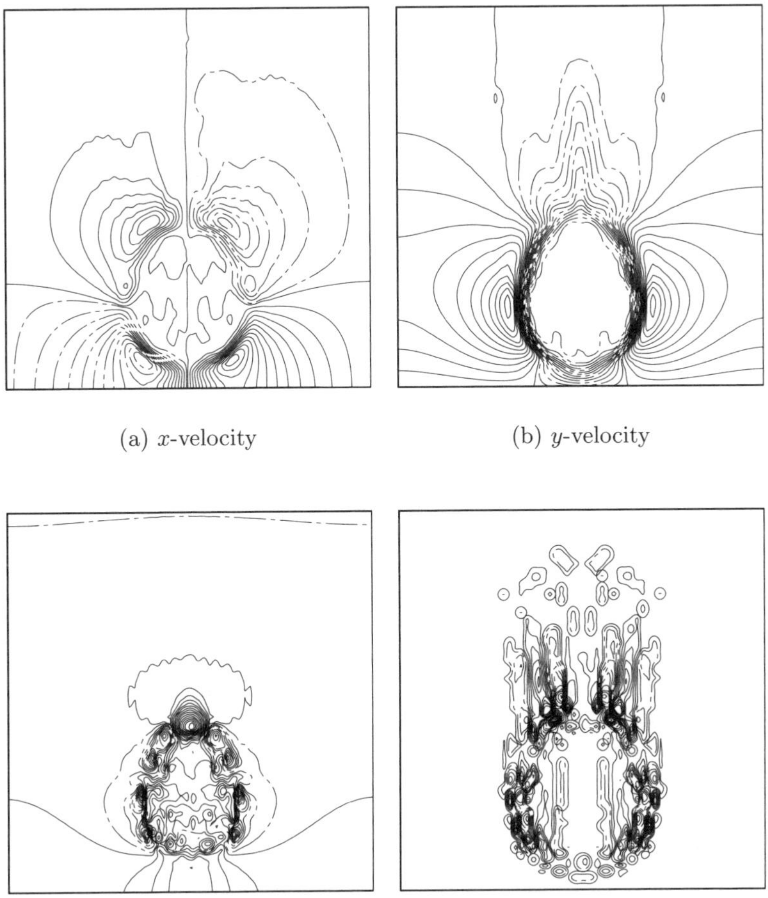

(a) x-velocity (b) y-velocity

(c) pressure (d) $D\mathbf{u}^{n+1}$

Fig. 12.24. Inviscid discontinuous drop results with a cell-centered-velocity filter (Sect. 12.5.3).

a yardstick to measure the quality of each solution. The visual evidence is largely confirmed by these figures. What these figures make abundantly clear is the quantitative magnitude of the improvement achieved through the use of filtering.

We will also apply the seemingly ubiquitous vortex-in-a-box problem to the analysis of the filter's performance. First, the convergence rates and error produced by the projections is tested in the same manner as before. This is followed by a more stringent use of this problem. Still using the Euler

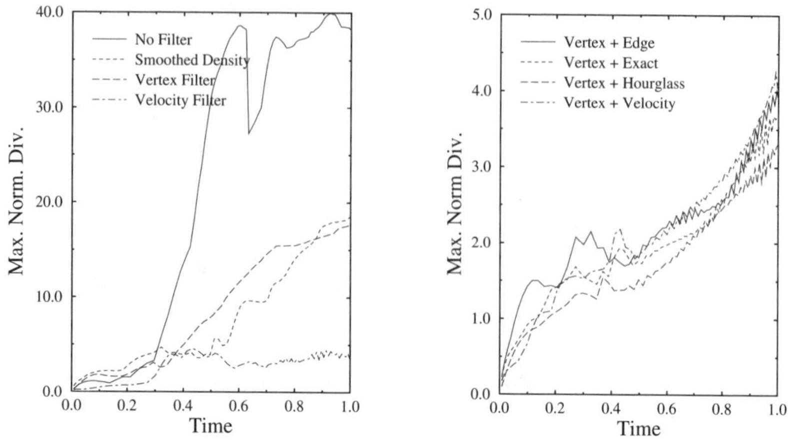

(a) New Filters

(b) Vertex and Velocity Filters

Fig. 12.25. A comparison of method performance for the discontinuous drop problem using the discrete divergence and different filters combinations. The left plot compares various filters and the right plot compares various combinations of the vertex filter (Sect. 12.5.3). We have also included results that are achieved by replacing the discontinuous density field with a smoothed out version. This smoothing is accomplished through the use of three Jacobi iterations (described in Chap. 8).

equations, we will run the vortex-in-a-box out to 60 time units on a 128^2 grid with a CFL number of 0.95. This will require nearly 8000 times steps and should provide a test of the long-term behavior of both the projections and the filter's impact.

For the velocity field the convergence rates and error resulting from using or not using the filters[8] is shown in Table 12.5. Without exception, the filters produce solutions with lower measured error than the unfiltered solution. The convergence rates vary slightly from method to method, but all are second-order accurate. To get more critical results we turn to a longer time integration of the vortex-in-a-box problem.

First, we will look at the results without filters. Fig. 12.26 clearly shows that solution is not acceptable with the error destroying the symmetry of the solution. The break in the symmetry is caused by the red-black Gauss-Seidel relaxation used in the multigrid solver. Fig. 12.26b shows that the problem is with the now familiar line and diagonal modes of discrete divergence. There are other signs of problems such as erratic kinetic energy decay (growth),

[8] The vertex-projection, edge-projection, vertex-velocity and cell-centered velocity filters are given in Sects. 12.5.2, 12.5.2, 12.5.3 and 12.5.3, respectively.

Table 12.5. Convergence rates of the velocity field for the vortex-in-a-box problem for different combinations of filters.

Case	L_2 Norms				
	32^2-64^2	Rate	64^2-128^2	Rate	128^2-256^2
No Filter	4.0×10^{-3}	2.18	8.8×10^{-4}	2.22	1.9×10^{-4}
Vertex-Projection Filter	2.8×10^{-3}	1.97	7.1×10^{-4}	2.14	1.6×10^{-4}
Edge-Projection Filter	3.0×10^{-3}	1.95	7.9×10^{-4}	2.30	1.6×10^{-4}
Vertex- and Edge-Projection Filters	3.3×10^{-3}	2.01	8.3×10^{-4}	2.28	1.7×10^{-4}
Cell-Centered-Velocity Filter	3.7×10^{-3}	2.08	8.7×10^{-4}	2.22	1.9×10^{-4}
Vertex-Projection and Hourglass-Velocity Filters	3.1×10^{-3}	1.96	7.9×10^{-4}	2.30	1.6×10^{-4}
Vertex- and Exact-Projection Filters	3.8×10^{-3}	2.23	8.0×10^{-4}	2.19	1.8×10^{-4}
Vertex-Projection and Cell-Centered-Velocity Filters	2.8×10^{-3}	1.98	7.1×10^{-4}	2.15	1.6×10^{-4}

and jumps in maximum vorticity as well as a large growth in the discrete divergence.

Now that we have seen what we do not want in a solution, we will attempt to repair the algorithm for long-term integrations. Applying the vertex-projection filter we get a distinct improvement in the algorithms performance. This is made clear by Fig. 12.27. The kinetic energy and vorticity behave in physical fashions with the discrete divergence under seeming control. Fig. 12.27 also makes it clear that our work is not done. Small errors near the wall dominate the divergence error and visibly distort the vorticity profile. The discrete divergence is also rising constantly thus signaling the eventual demise of the solution.

Alone the cell-centered-velocity filter (Sect. 12.5.3) should suffice to clean up the solution. This proposition is clearly shown to be correct with Fig. 12.28. The solutions are visually pleasing and the solutions are physically relevant. The solutions are not as high in quality as the combination of vertex- and edge-projection or hourglass-velocity, (Sect. 12.5.3) filters. As the reader may already guessed, the combination of vertex-projection and cell-centered-velocity filters gives outstanding results.

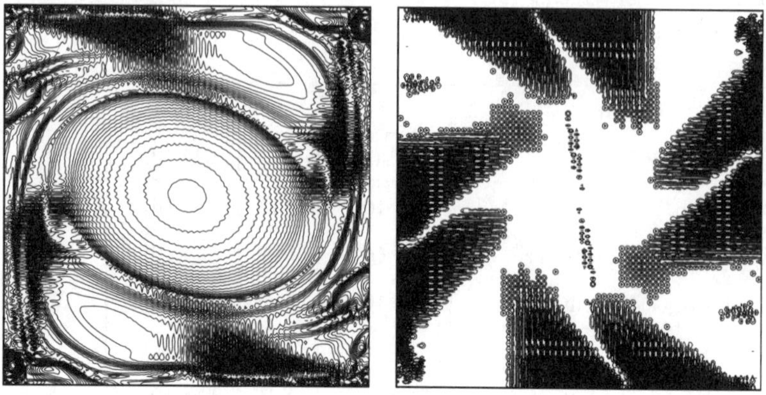

(a) $\log|\psi + 1|$ (b) $D\mathbf{u}^{n+1}$

Fig. 12.26. Long time integration of vortex-in-a-box problem without filters for a 128^2 grid.

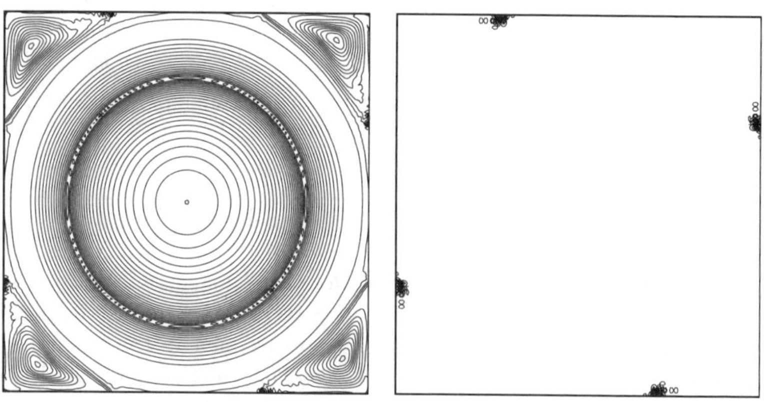

(a) $\log|\psi + 1|$ (b) $D\mathbf{u}^{n+1}$

Fig. 12.27. Long time integration of vortex-in-a-box problem with the vertex-projection filter (Sect. 12.5.3) for a 128^2 grid.

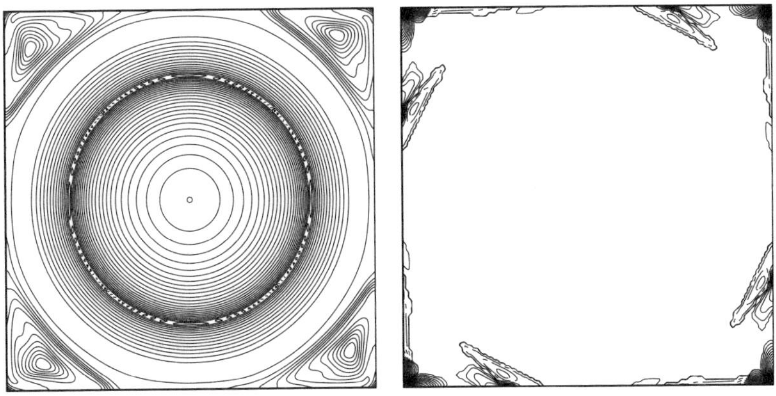

(a) $\log|\psi + 1|$ (b) $D\mathbf{u}^{n+1}$

Fig. 12.28. Long time integration of vortex-in-a-box problem with a cell-centered-velocity filter (Sect. 12.5.3), for a 128^2 grid.

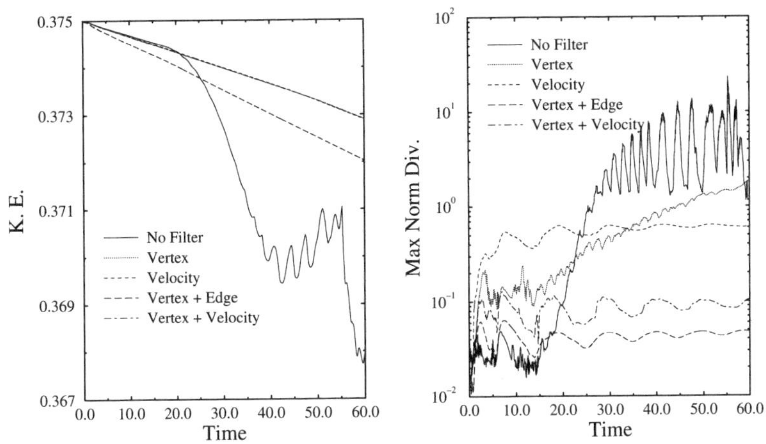

Fig. 12.29. Long time integration of vortex-in-a-box problem comparison of results for the kinetic energy (left plot) and maximum divergence norms (right plot) for a 128^2 grid.

The reason why these two filters were chosen for use in tandem can be gleaned from Fig. 12.29. The vertex- and edge-projection filters, Sect. 12.5.2, in tandem produce the lowest divergence errors, but at the cost of increased loss of kinetic energy. The vertex-projection plus cell-centered-velocity filter duo produces nearly as low divergence error, but at no cost in dissipation of kinetic energy.

The presence of filters is shown to be useful in the case of variable density flows or long time integrations, where viscous forces are small. The combination of vertex based projection filters with a velocity filter produced the best solutions in most applications.

If we base our algorithm on a vertex-staggered grid, we would likely find that a combination of edge-centered and/or the hourglassing filter is effective in removing unwanted oscillations. It should be noted that results from vertex-staggered approximate projection methods are less susceptible to oscillations because of lower dimension of the null space of the discrete divergence operator.

Although these methods were presented in the framework of an approximate projection method, we feel that they are applicable to PPE based methods. The important work involves identifying the null space of the divergence operator and then applying the proper filter to reduce the decoupling. This should provide PPE-based approaches (where $D\mathbf{u}^{n+1} \neq 0$) a means to improving their solution.

Part III

Modern High-Resolution Methods

13. Introduction to Modern High-Resolution Methods

In this chapter we will introduce the history and most basic concepts associated with high-resolution methods (see also introduction in Chap. 9). These methods got their start through the parallel efforts of Boris in developing the flux-corrected transport (FCT) method and van Leer's direct extension of Godunov's work to higher order.[1] These efforts were then formalized principally through the numerical analysis of Harten whose total variation diminishing (TVD) methods codified the techniques we refer to as high-resolution methods.

13.1 General Remarks about High-Resolution Methods

Before discussing Godunov-type methods and their history it is useful to clarify what exactly a high-resolution method is (and is not). Simply put, high-resolution schemes employ some sort of nonlinear "recipe" to control oscillations in the solution. This is opposed to methods that are linear using the same differencing stencil everywhere regardless of the solution. Thus, high-resolution combines two elements: nonlinear differencing where the stencil is dependent on the local solution and the use of this nonlinearity to control oscillations. Some high-resolution schemes attempt to totally eliminate oscillations while others simply minimize them.

Historically speaking, one might consider high-resolution methods to be the second generation of numerical methods for hyperbolic PDEs. The first generation of methods were linear. They were typified by either making the choice of an oscillatory solution such as that produced by the Lax-Wendroff scheme, or a diffusive solution such as that produced by upwind or Lax-Friedrichs schemes. The second generation methods achieve high-resolution by adaptively using the first generation's methods where they are most appropriate. Perhaps we are now witnessing the beginnings of the next generation of methods where the high-resolution methods themselves are adaptively chosen.

[1] We will also discuss a largely unknown extension of Godunov's work that took place in the Soviet Union at roughly the same time as Boris and van Leer's work.

Just what constitutes a high-resolution method and distinguishes this from a non-high-resolution approach?

This is both a simple and complex question to answer. We can begin with a simple yet somewhat misleading answer: high-resolution methods are intrinsically nonlinear. In fact they are nonlinear even if the equations being approximated are linear. Stated differently, the effective finite difference stencil employed by a high-resolution method is a function of both space and time and dependent upon the nature of the local solution.

The catch with this simple answer is that it is not entirely correct. A method can be nonlinear and adaptive while not being high-resolution. For example a nonlinear combination of two or three first-order methods is not high-resolution. The nonlinear principle used must remove any significant spurious oscillations and allow higher than first-order of accuracy when the solution is smooth (at least in the case of a linear discontinuous problem).

The second answer is that high-resolution methods are nonoscillatory. When a normal linear finite-difference method encounters an under-resolved solution it reacts by allowing oscillatory solutions. By under-resolved, we mean that the dissipative mechanisms in the physics being solved or the numerics being used are insufficient to control the stability and behavior of the resulting solution. This is usually a consequence of not having enough computational grid points to simulate the dissipative effects (directly) and the choice of numerical method.

Thus a well-behaved solution will become ill-behaved and if the equation being solved is nonlinear, the results are unpredictable. High-resolution methods use the intrinsic nonlinearity to choose local differencing that will not produce the uncontrolled oscillations.

This leads us to consider a better definition of high-resolution: these methods select the "best" technique for approximating the solution given the evidence provided by the local solution. Thus, high-resolutions method adapt themselves to their circumstances so that the solutions produced have predictable properties such as accuracy and some reasonable guarantee of physically meaningful results.

Among the most important high-resolution methods are the Godunov-type methods based on the work of S. K. Godunov [215] and the extensions due to van Leer [571]. The distinguishing characteristic of Godunov-type methods is their basis in interpolation of the dependent variables in a control volume and the resolution of the resulting multivalued edge values through a (approximate) Riemann solver. The basic procedure has a physical appeal. The method is usually divided in two steps, with the first being interpolation, that is often referred to as "reconstruction". For higher than first-order methods this produces a subgrid distribution for the dependent variables. The second step is the Riemann solution which produces a physically relevant flux from the dependent variable profile used in the first step. This separation of functionality has added much of the methods appeal, along with its impressive results.

Recently, van Leer has provided an overview of the early development of high-resolution methods [575]. In that paper the efforts of V. P. Kolgan are discussed [298]. Kolgan's work has been largely ignored perhaps due to his untimely death or the whims of fate, but simply looking at the title of his paper, "Application of the principle of minimum values of the derivative to the construction of finite-difference schemes for calculating discontinuous solutions of Gas Dynamics" hints at how ahead of his time he was. Closer examination shows that the "limiter" Kolgan proposed was in fact a second-order method equivalent to the ENO scheme (also known as mineno). Nonetheless, his efforts were not widely known even when van Leer presented his work in the Soviet Union in the late 1970's. This demonstrates the fickle nature of discovery, success and fame in science that has played out over-and-over historically.

As alluded to above, the order of a Godunov-type method is determined by the order of the interpolation. A first-order method arises from the piecewise constant sub-cell distribution. The second-order method associated with a linear distribution and is the most commonly associated with the name Godunov these days. Third-order methods are defined through parabolic profiles are also successful (see PPM below). Arbitrary order methods are available via ENO methods (these are covered in more detail in Chap. 17).

The order of Godunov-type methods can be classified in several ways for general nonlinear problems. One manner is to use a finite difference definition where the dependent variables are point values. This can be used to derive high-order flux methods. If the dependent variables are viewed as averages over a cell then a mean preserving high-order interpolation can produce high-order methods. Note that these definitions of the order are equivalent for linear problems. The difference between these approaches, however, is potentially fleeting, in many cases, as *general* nonlinear problems develop discontinuities in all but pathological cases, and once the discontinuity forms, the solution is nominally first-order accurate in any case [364].

The other aspect of Godunov-type methods are Riemann solvers. Initially, Riemann solvers were of an exact nature as the solution to the Riemann initial value problem for ideal gases. Due to the cost of this procedure under the best of circumstances and the relatively small practical value of an exact Riemann solution, approximate Riemann solvers have supplanted exact Riemann solutions in the vast majority of cases thus are most commonly applied. Moreover, approximate Riemann solvers are more reasonable for general circumstances (complicated physics, equations of state) encountered in most applications. Most of these issues are significantly simpler in the case of incompressible flow, but it is the enhanced performance of Godunov-type methods for a variety of more challenging circumstances that spurred the development of such methods for incompressible flow. The use of a Riemann solution imbues this method with a certain semi-analytic character and combined with the sub-cell reconstruction makes the method philosophically and mathematically appealing. Coupled with its performance these factors have led to its adoption as one of the pre-eminent methods of the past several decades.

The basic Godunov scheme was introduced in 1959 by Godunov [215] as an alternative to von Neumann-Richtmyer schemes [590]. The method models flow problems as a series of piecewise constant slabs that then interact through solving a Riemann problem [457]. The Riemann problem is the shock tube solution that arises from the instantaneous motion between two discontinuous fluid states. Originally, Godunov's method was set in Lagrangian coordinates, but in 1961 an Eulerian method was published [217]. The basic algorithm is explained geometrically in Fig. 13.1. Godunov has recently recounted that the method would not have arisen had he had access to the Lax-Friedrichs [317] method [216]! Godunov's 1959 paper also had a significant result in the field of numerical analysis, a theorem that states that no linear method can be both second-order and monotone. Overcoming the limitations imposed by this theorem is the foundation of modern high-resolution methods.

Godunov's method and result remained relatively obscure for approximately a decade. Interest was re-invigorated from two sources: Jay Boris (with David Book) who focused on overcoming the theorem by introducing nonlinear methods, and Bram van Leer who brought his attention on both the theorem and ultimately an upgrade of Godunov's method that will be discussed shortly. Boris and Book introduced an algorithm known as flux-corrected transport (FCT) [65, 68, 66, 67, 62] that overcame Godunov's theorem by ignoring it. Another parallel research path taken by Bram van Leer produced methods with the same goal as FCT, but using a different perspective based on increasing the order of interpolation in a Godunov-style algorithm (from piecewise constant to piecewise linear or quadratic) [568, 570, 569]. Unlike Boris, van Leer was acutely aware of the limitations imposed by Godunov's theorem [566, 567]. The basic idea is displayed in Fig. 13.2 for a piecewise linear interpolant. We note that this heuristic is only based on spatial considerations, and through consideration of the time-dependent nature of a flow these bounds can be relaxed. Because of the piecewise interpolation, that is cell-by-cell, the problem retains the need for a Riemann solver to rectify the jump discontinuity at cell interfaces. Furthermore, the nonlinear differencing introduced by van Leer will cause the interpolation to degenerate to first-order (piecewise constant) near large jumps or extrema in the data. This idea is the key to overcoming Godunov's theorem, namely the nonlinearity of the basic differencing method (its data dependence).

In 1979, the true successor to Godunov's method was unveiled [571]. This method used piecewise linear interpolation, and a two step algorithm with a Lagrangian step followed by a re-map (or advection step) back to the orig-

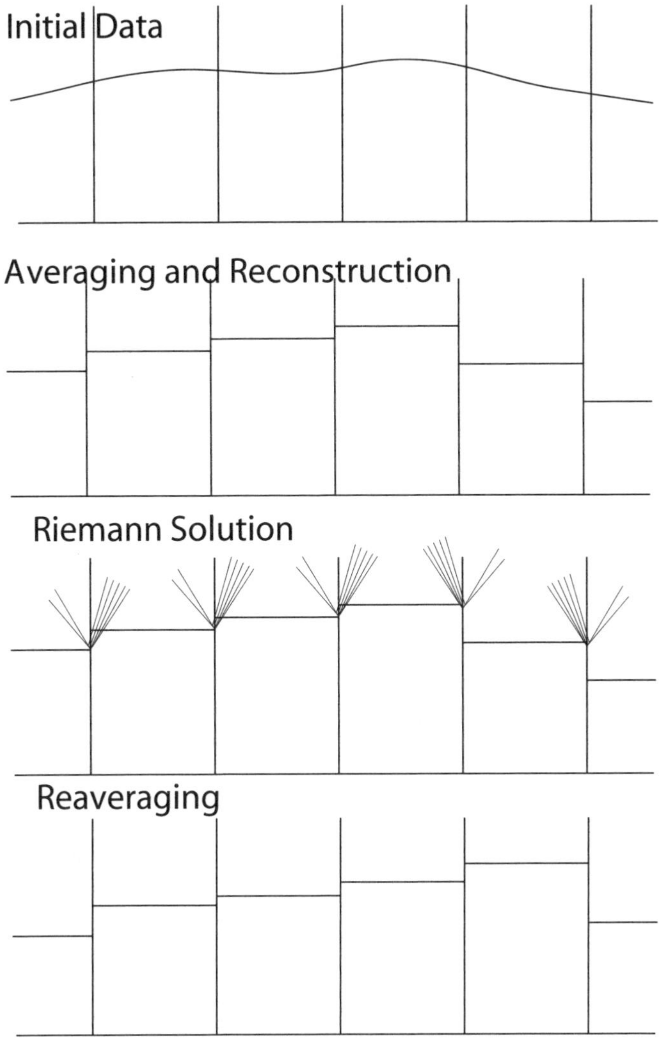

Fig. 13.1. The basic geometric picture of Godunov's method showing the steps of the algorithm. The piecewise constant reconstruction, the evolution via the Riemann solution and the averaging associated with the finite volume update.

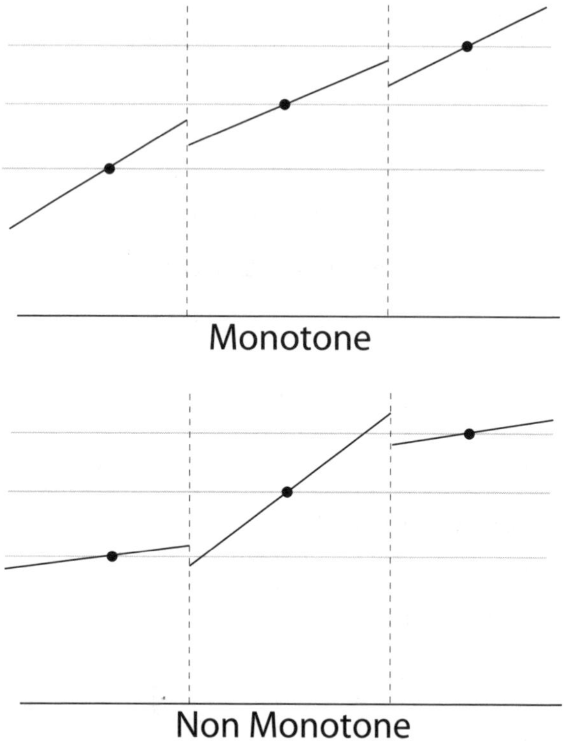

Fig. 13.2. The geometric idea of monotone limiters as introduced by van Leer. The piecewise profile is constrained to lie between the cell average values of the adjacent cells. If this constraint is violated, the interpolant is modified so that it is not violated.

inal grid. The differences between this method and the original Godunov method can be seen in Fig. 13.3 in the linear profiles rather than the piecewise constant. This work was done with the assistance of Paul Woodward who then created the next algorithm in the progression with Phil Colella. This method is called the piecewise parabolic method (PPM), which like the name says is based on a piecewise parabolic profile (first introduced by van Leer as method IV [569]). This method is still actively used especially in the astrophysics community [200].

Then in the early 1980's a floodgate of creative energy was opened by these methods and their stunning results. This produced a veritable phalanx of methods: total variation diminishing (TVD), essentially nonoscillatory (ENO), total variation bounding (TVB), weighted ENO (WENO) and others, in addition a wide variety of Riemann solvers, limiters and auxiliary techniques; many of the above are discussed in the following chapters. Colella went on to refine the piecewise linear, high-resolution Godunov method in a series of papers in the mid-to-late 1980s [116, 117]. This creative explosion left the scientific community with a quantum leap in computational capability to simulate highly nonlinear phenomena dominated by hyperbolic terms. These methods also bred the inception of ENO methods (covered in more depth in Chap. 17), although more recent developments have produced methods that are somewhat distinct from the tradition started by Godunov. We only mean that the ENO methods are not currently based explicitly on the interpolation or reconstruction of a function in a computational zone followed by a Riemann solution.

Eventually, high-resolution, Godunov-type, methods were applied to incompressible flows in conjunction with the projection [45] and artificial compressibility approaches [156, 252, 310, 411, 469], respectively. High-resolution methods were motivated by the desire to compute flows with high accuracy and robust results. There are a wide range of incompressible flows that can benefit from these capabilities.

13.2 The Concept of Nonoscillatory Methods and Total Variation

The fundamental aspect of all the methods discussed in the following five chapters is the goal of producing a nonoscillatory solution. What does this mean?

The method does not produce (significant) unphysical oscillations in the numerical solution.

Following Harten's definition [247], we classify as high-resolution methods those with the following properties:

- provide at least second order of accuracy in smooth areas of the flow,
- produce numerical solutions (relatively) free from spurious oscillations, and
- in the case of discontinuities, the number of grid points in the transition zone containing the shock wave is smaller in comparison with that of first-order monotone methods.

As was aforementioned, these methods are nonlinear even when the equation being solved is linear. This manifests itself by having the finite difference stencil used by the method to adapt to the character of the local solution. The methods are designed with the use of the following two concepts: total variation and monotonicity.

The total variation of a function $U(x)$ is defined as

$$\mathrm{TV}(U) = \lim_{\epsilon \to 0} \sup \frac{1}{\epsilon} \int_{-\infty}^{+\infty} |U(x+\epsilon) - U(x)| \, \mathrm{d}x \, . \tag{13.1}$$

If $U(x)$ is smooth then (13.1) can be written

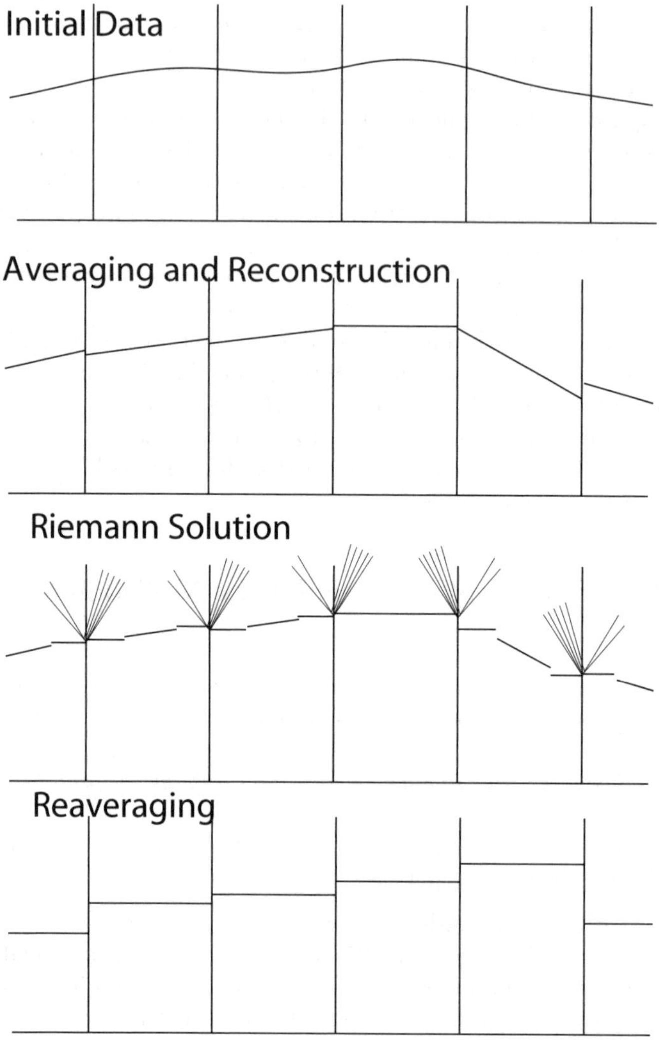

Fig. 13.3. The basic geometric picture of van Leer's high-resolution Godunov's method showing the steps of the algorithm. The piecewise constant reconstruction, the evolution via the Riemann solution and the averaging associated with the finite volume update.

$$\mathrm{TV}(U) = \int_{-\infty}^{+\infty} |U'(x)| \, \mathrm{d}x \,. \tag{13.2}$$

If U is a function of space and time, $U(x,t)$, then we define the total variation of U at fixed time, t. In a discretized domain, U is a function of the mesh and its total variation at a time instant indicated by the index n, is defined as

$$\mathrm{TV}(U^n) \equiv \mathrm{TV}(U(t)) = \sum_{j=-\infty}^{+\infty} |U_{j+1}^n - U_j^n| \,. \tag{13.3}$$

The function U is assumed to be either 0 or constant as the index j approaches the infinity, in order to obtain finite total variation.

13.3 Monotonicity

Let us consider the scalar conservation law

$$\begin{aligned} &\frac{\partial U}{\partial t} + \frac{\partial E(U)}{\partial x} = 0 \,, \\ &\qquad \text{or} \\ &\frac{\partial U}{\partial t} + \alpha(U)\frac{\partial U}{\partial x} = 0 \,, \end{aligned} \tag{13.4}$$

where

$$\alpha(U) = \frac{\mathrm{d}E}{\mathrm{d}U} \,, U(x,0) = \phi(x) \,, -\infty < x < \infty \,, \tag{13.5}$$

and $\phi(x)$ is assumed to be of bounded total variation.

An important property of the weak solution of the scalar initial value problem is the *monotonicity property* according to which:

- No new local extrema in x may be created.
- The value of a local minimum increases, i.e., it is a non-decreasing function [247], and the value of a local maximum decreases, i.e., it is a non-increasing function [247].

Thus the total variation, $\mathrm{TV}(U(t))$, is a decreasing function of time

$$\mathrm{TV}(U(t_2)) \leq \mathrm{TV}(U(t_1)) \quad \forall \quad t_2 \geq t_1 \,. \tag{13.6}$$

Let us now consider the explicit discretization of (13.4) which can be written in the shorter form

$$U_j^{n+1} = H(U_{j-k}^n, U_{j-k+1}^n, ..., U_{j+k}^n) = L \cdot U_j^n \,, \tag{13.7}$$

where L is an operator. We say that the scheme (13.7) is *total variation non-increasing* (TVNI) or *total variation diminishing* (TVD) if for all w

$$\mathrm{TV}(L \cdot U) \leq \mathrm{TV}(U) . \tag{13.8}$$

The scheme (13.7) is monotonicity preserving if the finite difference operator L is monotonicity preserving, that is, if w is a monotone mesh function, so is $L \cdot U$. Moreover, the scheme (13.7) is a monotone scheme if H is a monotone increasing function function of each of its $2k+1$ arguments. The hierarchy of these properties can be stated as follows: the set of monotone schemes is contained in the set of TVD schemes and this is in turn contained in the set of monotonicity preserving schemes.

For a constant coefficient $\alpha(U) = \alpha$, (13.4) yields the linear advection equation. Well known schemes such as the Godunov first-order upwind scheme and the Lax-Wendroff method, among others, can cast in the general form

$$U_j^{n+1} = \sum_{l=-k_L}^{l=k_R} b_l U_{j+l}^n , \tag{13.9}$$

where k_L and k_R are two non-negative integers and b_l are constant coefficients. Harten [247] has shown that the linear finite difference approximation (13.9) is monotonicity preserving if the coefficients b_l are non-negative, i.e.,

$$b_l \geq 0 , \quad -k_L \leq l \leq k_R . \tag{13.10}$$

Thus any linear monotonicity preserving scheme is a monotone, first-order, accurate scheme.

Second-order accurate three- and five-point (nonlinear) schemes of the form (13.7) can be rewritten in the form [247]

$$U_{j+1}^n = U_j^n - C_{j-1/2} \Delta U_{j-1/2} + D_{j+1/2} \Delta U_{j+1/2} , \tag{13.11}$$

where $\Delta U_{j+1/2} = U_{j+1}^n - U_j^n$, $\Delta U_{j-1/2} = U_j^n - U_{j-1}^n$ and

$$\left.\begin{aligned} C_{j-1/2} &\equiv C(U_{j-2}, U_{j-1}, U_j, U_{j+1}) \\ D_{j+1/2} &\equiv D(U_{j-1}, U_j, U_{j+1}, U_{j+2}) \end{aligned}\right\} . \tag{13.12}$$

Harten [247] has proved that any scheme (13.12) satisfying the inequalities

$$\left.\begin{aligned} &C_{j+1/2} \geq 0 \quad , \quad D_{j+1/2} \geq 0 \\ &0 \leq C_{j+1/2} + D_{j+1/2} \leq 1 \end{aligned}\right\} , \tag{13.13}$$

is a TVNI scheme. Hereafter, we shall refer to the above as *Harten's theorem.*

An alternative criterion to Harten's theorem is the data compatibility condition as proposed by Roe [464]. Roe's idea was to circumvent Godunov's theorem by constructing adaptive algorithms that would adjust themselves to the local nature of the solution. This leads to the design of schemes with variable coefficients (which are functions of the data), in other words, in nonlinear schemes even for linear PDEs such as the linear advection equation. A scheme is compatible if the solution U_{j+1}^n at each point j is bounded by the pair (U_{j-s}^n, U_j^n), where $s \equiv \text{sign}(\alpha)$ [543]. The data compatibility condition is satisfied by the inequality

$$0 \leq \frac{U_j^{n+1} - U_j^n}{U_{j-s}^{n+1} - U_j^n} \leq 1 \, . \tag{13.14}$$

By applying the data compatibility condition for specific sets of data, we can construct combination of schemes which satisfy the whole set of data. This would eventually result in adaptive, nonlinear schemes that are monotone and second-order accurate. Different approaches for building monotone schemes have been discussed in [532] (see also Chap. 16).

A hierarchy of numerical methods is schematically shown in Fig. 13.4 [609]. Monotone schemes can be either upwind or centered schemes so can TVD and ENO[2] schemes. The set of monotone schemes is the smallest set of schemes and is a subset of the set of TVD schemes.

13.4 General Remarks on Riemann Solvers

Above we discussed the concept of a Riemann solver. The Riemann solution is the solution of the initial value problem resulting from the jump discontinuity between two semi-infinite slabs of fluid. Because Godunov-type methods rely upon Riemann solvers it is useful to understand a few basic concepts regarding these techniques (see also introductory discussion of Riemann solvers in Chap. 9).

In its simplest form a Riemann solver is upwind differencing, or using the physical direction of transport to bias a finite-difference stencil. Consider, the scalar wave equation $U_t + aU_x$, with $a > 0$. Using our usual indexing conventions, the flow is from low indices to higher ones, if we want to update the cell indexed as j, the simplest approximation would be $U_j^{n+1} = U_j^n - a\Delta t \left(U_j^n - U_{j-1}^n\right)/\Delta x$. The first order derivative $\left(U_j^n - U_{j-1}^n\right)/\Delta x$ has been biased by the direction of a or upwinded. This is the simplest form of Riemann solver. This choice of differencing scheme is stable and produces physically realizable results (i.e., ones that dissipate energy appropriately).

The challenge is to use the same basic concept for nonlinear equations and systems of equations. For nonlinear equations the speed of transport becomes a function of the solution. Approaches to dealing with this include making the problem locally linear for any point of interest. More complex Riemann solvers encode analytical knowledge of the solution into the numerical approach. If the exact numerical solution to the Riemann problem is

[2] Essentially Nonoscillatory Schemes are discussed in Chap. 17.

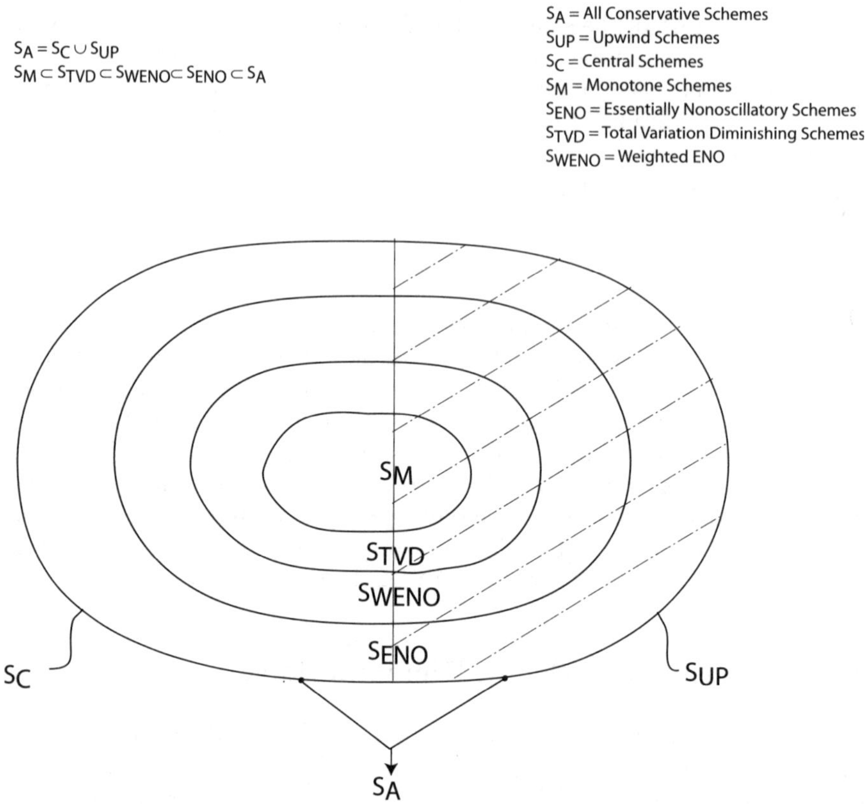

Fig. 13.4. Schematic representation of the hierarchy of conservative schemes for hyperbolic conservation laws.

known (and used for the Riemann solver) then the method is called an *exact Riemann solver*. If the solution is approximated using some sort of linearization, the Riemann solver is called an *approximate Riemann solver*. Generally speaking, the more analytical knowledge that is used in constructing the Riemann solver, the greater resolution can be achieved when that Riemann solver is implemented. Because the analytical knowledge of the Riemann solution is scarce for nonlinear problems, approximate Riemann solvers are more common than exact Riemann solvers.

Similarly, systems of equations are often dealt with by transforming the problem into a set of linear scalar equations. The same naming convention described above is used to describe the Riemann solvers.

All of these basic concepts introduced in this chapter are expanded upon in the coming chapters.

14. High-Resolution Godunov-Type Methods for Projection Methods

High-resolution Godunov-type methods are most commonly associated with compressible flow solutions. Their introduction to incompressible flow solutions was most directly impacted by their use with a projection method[1] by Bell, Colella and Glaz (BCG) [45] in 1987. BCG incorporated an exact projection with second-order Godunov methods that had then begun to reach a state of maturity with respect to compressible shock dynamic calculations. This parallels the current time where these methods are commonly used in incompressible flow calculations as well.

This chapter will describe the salient aspects of high-resolution Godunov-type methods used in conjunction with projection methods. We will begin by describing the fundamental approximations in the form of the first-order accurate time integration scheme. This will include the Riemann solutions specialized for the velocity field associated with the incompressible flow. This is followed by extensions necessary for the first-order method to produce formally second-order approximations using monotonicity limited derivatives (slopes) in one dimension. This approach is expanded upon with the discussion of genuinely multidimensional derivative approximations that are similarly limited using extensions of the one dimensional monotonicity criteria to two and three dimensions. Finally, we close with a discussion and analysis of the numerical stability of these methods for the advection-diffusion equation.

14.1 First-Order Algorithm

The first issue to emphasize is that incompressible flow is intrinsically multidimensional and this character should be directly reflected in the integration methods applied. A multidimensional advection scheme thus allows the impact of $\nabla \cdot \mathbf{u} = 0$ to be felt directly in the advection. This is difficult to accomplish through a split algorithm. As such, "unsplit" time integration methods are highly recommended. Operator splitting is quite common and involves the solution of multidimensional problems with a sequence of one-dimensional solutions. This can be arranged to not effect linear (second-order)

[1] Approximate and exact projection methods are discussed in Chaps. 11 and 12, respectively.

accuracy through the use of Strang splitting [515]. Operator splitting is often used effectively with compressible flow where unsplit methods show little added benefit in most cases.

To simplify the exposition, we first describe a first-order multidimensional algorithm as given by Colella [117]. This method provides corner coupling, and alleviates the necessity for operator split implementations of high-order methods. A geometric depiction of the method can be seen in Fig. 14.1. We will also cover the Hancock algorithm [573, 517] as an alternative corner upwind method because it is somewhat simpler and cheaper per time step although it decreases the allowable time step size. The importance of multidimensional algorithms is explored in [47]. Incompressible flows are intrinsically multidimensional because of the solenoidal constraint.

With the first-order Godunov algorithm, the profile of the dependent variables is taken to be constant in a computational cell. For example, given $U^n_{i,j,k}$ we can set the edge values,

$$U^n_{i+1/2,j,k} = U^n_{i-1/2,j,k} = U^n_{i,j+1/2,k} = U^n_{i,j-1/2,k} = U^n_{i,j,k} \ .$$

After applying this rule to each cell, the value of U at the edge is double valued. This necessitates the use of some mechanism to determine which value is valid for transport through the cell-edge. This mechanism is upwinding, or more generally a Riemann solver to be discussed in the pages that follow. Given these values, they must be time centered. This is accomplished with a Taylor series expansion

$$U^{n+1/2}_{i+1/2,j,k} = U^n_{i,j,k} + \frac{\Delta x}{2}\frac{\partial U^n_{i,j,k}}{\partial x} + \frac{\Delta t}{2}\frac{\partial U^n_{i,j,k}}{\partial t} \ ,$$

and

$$U^{n+1/2}_{i+1/2,j,k} = U^n_{i+1,j,k} - \frac{\Delta x}{2}\frac{\partial U^n_{i+1,j,k}}{\partial x} + \frac{\Delta t}{2}\frac{\partial U^n_{i+1,j,k}}{\partial t} \ ,$$

with the other edges defined analogously. By assuming the profile of U is constant (first-order, piecewise constant), the spatial derivative drops out of the series, but we must still deal with the temporal derivative, which is done in a Lax-Wendroff fashion (discussed in Chaps. 7 and 9). This is given by the governing equation for U at time n.

Let us assume that the evolution of a scalar U is governed by the following equation

$$\frac{\partial U}{\partial t} + \mathbf{u}\cdot\nabla U = \nu\nabla^2 U + S \ , \tag{14.1}$$

where S is a source term such as gravitational acceleration. Because we have assumed that the profile of U is constant, the convective term normal to the face of the cell drops out, but the transverse derivative remains. This is a consequence of stability considerations and it turns out that the stability is the same if one includes the normal part of the time derivative or no time derivative at all. This stability is less than the corner transport method

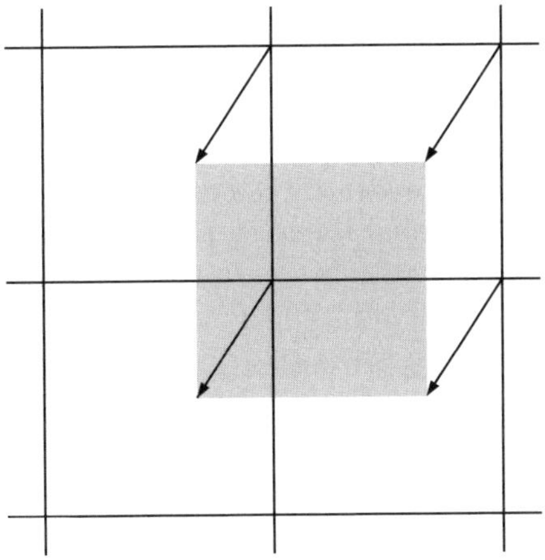

Fig. 14.1. A geometric representation of the first-order version of the multidimensional advection algorithm. The shaded region will become the new cell average value in the cell outlined by the arrows at its corners.

(CTU) which is defined by the maximum of the one-dimensional CFL numbers, $\max(\Delta t\,|u_k|\,/h_k)$, while the alternative is the sum of the CFL numbers, $\mathcal{C} \leq \sum_{k=1}^{d} \Delta t\,|u_k|\,/h_k$, where h is the spatial step, u is the velocity (locally) on the grid cell and k denotes a spatial direction. Thus, we are left with

$$\frac{\partial U}{\partial t} = -\left(\boldsymbol{\tau}\cdot\mathbf{u}\cdot\nabla U^n\right) + \nu\nabla^2 U^n + S^n\ ,$$

where $\boldsymbol{\tau}$ is the (tangent) vector of transverse derivatives. For first-order the Hancock method is even simpler, there is no time-centering [573]. The initial values are used to produce the values to advance the solution in time. The difference between the two methods is the stability constraint. Define the dimensional Courant number, $\mathcal{C} = |\mathbf{u}|\,\Delta t/\Delta x$, then the CTU scheme is stable for $\mathcal{C} \leq 1$ for the largest of each direction, while Hancock is stable for $\mathcal{C} \leq 1$ for the sum of each direction.

As noted, the variables are double valued at cell-edges. In order to solve this dilemma we use a Riemann solver to find the single value transported through edges. This is equivalent to upwinding the velocity, but generalizes well to higher-order differencing. The Riemann solver often used is based on one of those employed for the Burgers' equation in one dimension (see [573] for a more complete discussion of this).

We will return to our dependent variables and describe the algorithm for an x-edge $i+1/2, j, k$, with other edges following in an analogous fashion. We will denote the values supplied by cell i, j, k by the subscript L (for left), and the values from $i+1, j, k$ by the subscript R (for right). The final values will be denoted by no additional subscript. For CTU, this Riemann solver will have to be applied twice, once to supply transverse convective terms, and a second time to find the cell-edge, time-centered values of the dependent variables.

Let us look at the most straightforward approach to getting an upwind value in keeping with a Burgers' equation solution. This is the method introduced by BCG [45]. First, we identify the normal velocity at the edge, which is u in this case. Then a normal Riemann problem is solved for u

$$u_{i+1/2,j,k} = \begin{cases} u_{i+1/2,j,k,L} \quad \text{if} \\ u_{i+1/2,j,k,L} > 0 \;\text{ and }\; u_{i+1/2,j,k,L} + u_{i+1/2,j,k,R} > 0 \\ 0 \text{ if } u_{i+1/2,j,k,L} < 0 \;\text{ and }\; u_{i+1/2,j,k,R} > 0 \\ u_{i+1/2,j,R} \quad \text{otherwise} \end{cases} . \tag{14.2}$$

With this solution in hand, we can then solve for the other variables (treating them as scalar quantities)

$$\phi_{i+1/2,j,k} = \begin{cases} \phi_{i+1/2,j,k,L} \text{ if } u_{i+1/2,j,k} > 0 \\ 1/2\left(\phi_{i+1/2,j,k,L} + v_{i+1/2,j,k,R}\right) \text{ if } u_{i+1/2,j,k} = 0 \\ \phi_{i+1/2,j,k,R} \quad \text{otherwise} \end{cases} \tag{14.3}$$

where ϕ stands for the velocity, v, density, ρ, or temperature θ. One issue that can become particularly serious in some applications is the preservation of symmetry [146]. Generally, symmetry is desirable for any algorithm, but if the flow is unstable the results are sensitive to small changes such as deviations from symmetry. Although symmetry is only broken only around a normal velocity of zero, this can have a profound effect near stagnation points and instabilities [146]. The manner to treat this is to make decisions symmetrically near zero normal velocities in the solution, then (14.2) becomes

$$u_{i+1/2,j,k} = \begin{cases} u_{i+1/2,j,k,L} \text{ if } \; u_{i+1/2,j,k,L} > \varepsilon \;\text{ and} \\ u_{i+1/2,j,k,L} + u_{i+1/2,j,k,R} > \varepsilon \\ \left(u_{i+1/2,j,k,L} + u_{i+1/2,j,k,R}\right)/2 \\ \text{if } \; u_{i+1/2,j,k,L} < -\varepsilon \;\text{ and }\; u_{i+1/2,j,k,R} > \varepsilon \\ u_{i+1/2,j,k,R} \quad \text{otherwise,} \end{cases} \tag{14.4}$$

where ε is a small number near the square root of machine varepsilon. For double precision (8 bit reals), $\varepsilon = 10^{-6}$ to 10^{-8} seems to work well. One might argue to scale this constant with a variable like velocity. This can be achieved by replacing $\varepsilon := \Delta x/\Delta t\ \varepsilon$. Similar modifications are recommended for most Riemann solvers so that symmetry is not disturbed under these conditions.

Other options are possible; after all Riemann solvers have been called a "cottage industry" [435]. Here, we can also substitute another algorithm based on Roe's Riemann solver [463]. For the normal Riemann problem we solve

$$u_{i+1/2,j,k} = \frac{1}{2}\left[u_{i+1/2,j,k,L} + u_{i+1/2,j,k,R} - \mathrm{Sign}\left(\bar{u}\right)\left(u_{i+1/2,j,k,R} - u_{i+1/2,j,k,L}\right)\right] ,$$

where Sign(x) returns the sign of x, and $\bar{u} = \left(u_{i+1/2,j,k,L} + u_{i+1/2,j,k,R}\right)/2$. The symmetry "fix" can be implemented by making Sign$(x) = 0$ for $|x| < \varepsilon$. The scalar values can then be found easily

$$v_{i+1/2,j,k} = \frac{1}{2}\left[v_{i+1/2,j,k,L} + v_{i+1/2,j,k,R} - \mathrm{Sign}\left(u_{i+1/2,j,k}\right)\left(v_{i+1/2,j,k,R} - v_{i+1/2,j,k,L}\right)\right] ,$$

$$\rho_{i+1/2,j,k} = \frac{1}{2}\left[\rho_{i+1/2,j,k,L} + \rho_{i+1/2,j,k,R} - \mathrm{Sign}\left(u_{i+1/2,j,k}\right)\left(\rho_{i+1/2,j,k,R} - \rho_{i+1/2,j,k,L}\right)\right] ,$$

and

$$\theta_{i+1/2,j,k} = \frac{1}{2}\left[\theta_{i+1/2,j,k,L} + \theta_{i+1/2,j,k,R} - \mathrm{Sign}\left(u_{i+1/2,j,k}\right)\left(\theta_{i+1/2,j,k,R} - \theta_{i+1/2,j,k,L}\right)\right] .$$

Riemann solvers can be put into a different form that defines a flux rather than a primitive value. These even simpler Riemann solvers can be used with Lax-Friedrichs (LF), local Lax-Friedrichs (LLF) and Harten-Lax-van Leer (HLL) versions. In LF, the wave speed is replaced by the mesh ratio, $\mathcal{C} = \Delta x/\Delta t$,

$$(uu)_{i+1/2,j,k} = \frac{1}{2}\left[(uu)_{i+1/2,j,k,L} + (uu)_{i+1/2,j,k,R} - \mathcal{C}\left(u_{i+1/2,j,k,R} - u_{i+1/2,j,k,L}\right)\right] .$$

In LLF, the wave speed is chosen from the largest of the local values, $\mathcal{C} = \max(\left|u_{i+1/2,j,k,L}\right|, \left|u_{i+1/2,j,k,R}\right|)$. The HLL solvers replaces the full Riemann fan with a simple three wave model (totally needless[2] in this case, but it does work). A HLL solver can be defined as follows: define $\mathcal{C}_L = \min\left(u_{i+1/2,j,k,L}, u_{i+1/2,j,k,R}, 0\right)$, and $\mathcal{C}_R = \max\left(u_{i+1/2,j,k,L}, u_{i+1/2,j,k,R}, 0\right)$ then the flux can be simply expressed

$$(uu)_{i+1/2,j,k} = \frac{\mathcal{C}_R\,(uu)_{i+1/2,j,k,L} - \mathcal{C}_L\,(uu)_{i+1/2,j,k,R}}{\mathcal{C}_R - \mathcal{C}_L} + \frac{\mathcal{C}_L\mathcal{C}_R\left(u_{i+1/2,j,k,R} - u_{i+1/2,j,k,L}\right)}{\mathcal{C}_R - \mathcal{C}_L} .$$

[2] The HLL solver is desirable because it simplifies the Riemann problem to only two waves, for incompressible flow there is only one wave.

Each of these Riemann solvers is more dissipative and thus safer to use than the first set of methods introduced. This is the consequence of the additional dissipation that results from the conservative estimate of the local wave speeds.

It may be desirable (this makes a difference for higher order methods) to use the old time cell-centered data to solve the Riemann problem rather than the extrapolated data at edges. The cell-centered old-time data has the effects of pressure properly applied to it through the action of the previous cycle's pressure solution.[3] The Riemann solver has nearly the same form as the previous method for scalar data, but the normal solver changes to

$$u_{i+1/2,j,k} = \begin{cases} u_{i+1/2,j,k,L} & \text{if } u^n_{i,j,k} > 0 \text{ and } u^n_{i,j,k} + u^n_{i+1,j,k} > 0 \\ 0 & \text{if } u^n_{i,j,k} < 0 \text{ and } u^n_{i+1,j,k} > 0 \\ u_{i+1/2,j,k,R} & \text{otherwise.} \end{cases}$$

Roe's solver for velocities u is

$$u_{i+1/2,j,k} = \frac{1}{2}\left[\, u_{i+1/2,j,k,L} + u_{i+1/2,j,k,R} - \operatorname{Sign}(\bar{u})\left(u_{i+1/2,j,k,R} - u_{i+1/2,j,k,L}\right)\right],$$

where $\bar{u} = \left(u^n_{i,j,k} + u^n_{i+1,j,k}\right)/2$. Again, the scalar expressions are unchanged.

After the Riemann solver has been run on all the edges in the domain, we can construct the transverse fluxes to enable us to achieve time-centered values on the edges. The fluxes are computed with

$$u\frac{\partial U}{\partial x} = \bar{u}_{i,j,k}\frac{U_{i+1/2,j,k} - U_{i-1/2,j,k}}{\Delta x}, \tag{14.5}$$

or in axisymmetric coordinates

$$\frac{u}{r}\frac{\partial}{\partial r}(rU) = \bar{u}_{i,j,k}\frac{r_{i+1/2,j,k}U_{i+1/2,j,k} - r_{i-1/2,j,k}U_{i-1/2,j,k}}{r_i \Delta r},$$

and

$$u\frac{\partial U}{\partial y} = \bar{v}_{i,j,k}\frac{U_{i,j+1/2,k} - U_{i-1/2,j,k}}{\Delta y}, \tag{14.6}$$

where

$$\bar{u}_{i,j,k} = \frac{1}{2}\left(u_{i+1/2,j,k} + u_{i-1/2,j,k}\right),$$

and

$$\bar{v}_{i,j,k} = \frac{1}{2}\left(v_{i,j+1/2,k} + v_{i,j-1/2,k}\right).$$

[3] The MAC projection has not been applied to the data at this point, thus the cell-centered data is better in some sense from a pressure coupling point-of-view.

When the time-centered values are achieved, then the Riemann solver is used again to resolve the double-valued time-centered edge values. This basic step is identical whether the method is a CTU or Hancock solver (the difference is the time-centering of the transverse terms in CTU). Note that for the first-order Hancock scheme $U^{n+1/2} = U^n$. The non-conservative update is

$$\begin{aligned} U^{n+1}_{i,j,k} = U^n_{i,j,k} &- \Delta t \bar{u}_{i,j,k} \frac{U^{n+1/2}_{i+1/2,j,k} - \Delta t U^{n+1/2}_{i-1/2,j,k}}{\Delta x} \\ &- \Delta t \bar{v}_{i,j,k} \frac{U^{n+1/2}_{i,j+1/2,k} - \Delta t U^{n+1/2}_{i,j-1/2,k}}{\Delta y} \\ &- \Delta t \bar{w}_{i,j,k} \frac{U^{n+1/2}_{i,j,k+1/2} - U^{n+1/2}_{i,j,k-1/2}}{\Delta z} + \nu \Delta t \, LU \ , \end{aligned} \tag{14.7}$$

where $\bar{u}_{i,j,k}$, $\bar{v}_{i,j,k}$ and $\bar{w}_{i,j,k}$ are the average velocities, e.g.,

$$\bar{u}_{i,j,k} = 1/2 \left(u^{n+1/2}_{i-1/2,j,k} + u^{n+1/2}_{i+1/2,j,k} \right) .$$

In conservation or divergence form the equation is

$$\begin{aligned} U^{n+1}_{i,j,k} = U^n_{i,j,k} &- \Delta t \frac{(uU)^{n+1/2}_{i+1/2,j,k} - \Delta t \, (uU)^{n+1/2}_{i-1/2,j,k}}{\Delta x} \\ &- \Delta t \frac{(vU)^{n+1/2}_{i,j+1/2,k} - \Delta t \, (vU)^{n+1/2}_{i,j-1/2,k}}{\Delta y} \\ &- \Delta t \frac{(wU)^{n+1/2}_{i,j,k+1/2} - \Delta t \, (wU)^{n+1/2}_{i,j,k-1/2}}{\Delta z} + \nu \Delta t \, LU \ . \end{aligned} \tag{14.8}$$

where the discretization of the Laplacian LU is discussed below. The point of the time-centering for CTU is the additional stability gained (see Sect. 14.7 for a more complete discussion of stability).

At this point in the algorithm, either the fluxes are computed for the update of the cell-centered variables, or the process is put off until a pressure solution is completed. This intermediate pressure solution (the MAC projection) is discussed later in Chap. 11. In a nutshell, this method produces a divergence-free version of the edge-centered velocity field. This is particularly useful in applications where the conservation of quantities is particularly important and deviations in conservation can be tied directly to the departure of the velocity field from being divergence-free. In addition, providing a pressure solution at the middle time level produces a more stable algorithm [266].

Diffusive processes are also important in many flows. These terms can all be stably approximated at the old time n. The diffusive term can be approximated by a standard five (or seven)-point Laplacian stencil

$$LU = \frac{U_{i+1,j,k} - 2U_{i,j,k} + U_{i-1,j,k}}{\Delta x^2} + \frac{U_{i,j+1,k} - 2U_{i,j,k} + U_{i,j-1,k}}{\Delta y^2}$$

$$+ \frac{U_{i,j,k+1} - 2U_{i,j,k} + U_{i,j,k-1}}{\Delta z^2} ,$$

or in axisymmetric coordinates

$$LU = \frac{r_{i+1/2}\left(U_{i+1,j,k} - U_{i,j,k}\right) - r_{i-1/2}\left(U_{i,j,k} - U_{i-1,j,k}\right)}{r_i \Delta r^2} + \frac{U_{i,j+1,k} - 2U_{i,j,k} + U_{i,j-1,k}}{\Delta z^2} ,$$

and the source S (14.1) can be dealt with appropriately as was discussed in Sect. 11.4.

14.2 High-Resolution Algorithms

Now that the first-order algorithm has been established, we will show how it can be improved in spatial accuracy (when the solution remains smooth). Time accuracy can also be improved through a Lax-Wendroff technique. These methods are covered in Chaps. 7 and 15. High-resolution is typically accomplished through allowing the variables to be interpolated by linear or quadratic functions in a cell as opposed to the piecewise constant approach used for first-order algorithms. This will also allow the normal convective derivatives to come into play for the CTU method, but these still will not involve Riemann solutions. Other techniques such as Runge-Kutta or Adams-Bashforth can be used to provide a multidimensional update of a different form (see Chap. 7 for a discussion of various time integration techniques).

14.2.1 Piecewise Linear Methods (PLM)

In general, we will interpolate in each grid direction some polynomial that describes the profile of a variable. As an example, we will use Fromm's scheme. The polynomial is linear,

$$U^n(x) = U^n_{i,j,k} + \frac{\partial U^n}{\partial x}\left(x - x_{i,j,k}\right) ,$$

which we will rewrite for convenience as

$$U(\xi) = U^n_{i,j,k} + \delta_{i,j,k} U^n \xi ,$$

where $\xi = \left(x - x_{i,j,k}\right)/\Delta x \in [-1/2, 1/2]$.

For Fromm's scheme, $\delta_{i,j,k} U^n = 1/2\left(U^n_{i+1,j,k} - U^n_{i-1,j,k}\right)$. To include the normal convective term, we can work with this one-dimensional form, recognizing that it is

$$u \frac{\partial U}{\partial x} ,$$

we can interpolate to an one-dimensional edge- and time-centered value of U in an one-dimensional sense,

$$\tilde{U}_{i+1/2,j,k} = U^n_{i,j,k} + \frac{1}{2}\left(1 - \mathcal{C}^x_{i,j,k}\right)\delta_{i,j,k}U^n \ ,$$

or

$$\tilde{U}_{i+1/2,j,k} = U^n_{i+1,j,k} - \frac{1}{2}\left(1 + \mathcal{C}^x_{i+1,j,k}\right)\delta_{i+1,j,k}U^n \ ,$$

where $\mathcal{C} = u^n_{i,j,k}\Delta t/\Delta x$ is the local Courant number. When working with parabolic (piecewise parabolic method - PPM) or higher order functions, it is important to recognize that this is really

$$\tilde{U}_{i+1/2,j,k} = -\frac{1}{\mathcal{C}^x_{i,j,k}}\int_{1/2}^{1/2-\mathcal{C}^x_{i,j,k}} U_{i,j,k}\left(\xi\right)\mathrm{d}\xi \ ,$$

or

$$\tilde{U}_{i+1/2,j,k} = -\frac{1}{\mathcal{C}^x_{i,j,k}}\int_{-1/2}^{-1/2-\mathcal{C}^x_{i,j,k}} U_{i+1,j}\left(\xi\right)\mathrm{d}\xi \ .$$

This scheme gives quite better results when compared with other classical second-order methods. It has nice amplitude and dispersion properties, but it also produces nonphysical oscillations in the solution near discontinuous data (or data that looks discontinuous on a grid). To alleviate this problem we must apply a slope limiter to rid of the oscillations [570, 523, 465, 448]. Stated heuristically, a slope limiter assures that the interpolant for a variable in a grid cell is bounded above and below by the neighboring grid cells values, or the profile is set to zero if the cell is a minima or maxima. Properly applied, this removes oscillations from the solution (true in one-dimension, but for multidimensional cases Saltzman investigated the necessary conditions [479, 47, 517]).

For Fromm's scheme, monotonicity can be obtained through the use of the following expression for the slope

$$\delta^M_{i,j,k}U^n = \mathrm{S}\max\left[0, \min\left(\left|\delta^C_{i,j,k}U^n\right|, 2\mathrm{S}\delta^L_{i,j,k}U^n, 2\mathrm{S}\delta^R_{i,j,k}U^n\right)\right] , \qquad (14.9)$$

where $\delta^C_{i,j,k}U^n$ is the slope from the original Fromm's scheme, $S = \mathrm{Sign}(\delta^C_{i,j,k}U)$, $\delta^L_{i,j,k}U^n = U^n_{i,j,k} - U^n_{i-1,j,k}$, and $\delta^R_{i,j,k}U^n = U^n_{i+1,j,k} - U^n_{i,j,k}$. Thus, the limited Fromm's scheme becomes a TVD method. More generally, slope limiters can be expressed as one parameter functions of $r = \delta^L U/\delta^R U$. Thus, the above limiter can be expressed as

$$\psi\left(r\right) = \max\Big(0, \min\left(2, 2r, 1/2\left(1+r\right)\right)\Big) \ . \qquad (14.10)$$

This form of describing a limiter was introduced by Sweby [523]. The slope is then given by $\delta^M U = \psi\left(r\right)\delta^R U$. As we will show later, this form is useful for describing other second-order monotone methods similar to slope-limited Fromm's scheme.

It is useful to recognize that the TVD bounds can be extended through the consideration of the local Courant number of the flow. Because the flux of material that is ultimately used to evolve the quantities on the grid are found through time averaging, it is these time integrated values that determine the monotonicity preserving character of a method. Through viewing the limiter as a modification on a linear function and integrating this linear function over a time step, the bounds on the linear function can be relaxed. For example with Fromm's scheme, (14.10), the bounding values 2 and $2r$ can be changed to $2\max\left(1/\left(1-\mathcal{C}\right),-1/\mathcal{C}\right)$ and $2\max\left(1/\left(1+\mathcal{C}\right),1/\mathcal{C}\right)r$ for the local Courant number $\mathcal{C}$. This expression takes both positive and negative values of $\mathcal{C}$ into account.

Colella [116] defined an improved version of this scheme based on a fourth-order prescription for the central slope (PLMI method). This method is defined using the previous expression and substituting

$$\delta^C_{i,j,k}U^n = \frac{2}{3}\left(U^n_{i+1,j,k} - U^n_{i-1,j,k}\right) - \frac{1}{6}\left(\delta^M_{i-1,j,k}U^n + \delta^M_{i+1,j,k}U^n\right) .$$

If the limiting has not modified the Fromm's slopes this gradient is fourth-order accurate,

$$\delta^C_{i,j,k}U^n = \frac{U_{i-2,j,k} - 8U_{i-1,j,k} + 8U_{i+1,j,k} - U_{i+2,j,k}}{12} ,$$

which has better accuracy and resolution in smooth regions of the flow. In Chap. 17 we will extend this slope to one based on the primitive function (one dimensional conservative polynomial being interpolated).

Another modification that can improve results is to selectively apply the normal convective term so that it only describes propagation of information into a cell. For cell edge $i+1/2,j,k$ in the x-direction, this modification leads to

$$\tilde{U}_{i+1/2,j,k} = U^n_{i,j,k} + \frac{1}{2}\Big(1 - \max\left(0, \mathcal{C}^x_{i,j,k}\right)\Big)\delta_{i,j,k}U^n,$$

and

$$\tilde{U}_{i+1/2,j,k} = U^n_{i+1,j,k} - \frac{1}{2}\Big(1 + \min\left(0, \mathcal{C}^x_{i,j,k}\right)\Big)\delta_{i,j,k}U^n.$$

The remainder of the algorithm follows as before with one modification. Because terms only need to be computed to first-order in the Taylor expansion in order to realize a second-order algorithm in time for the entire time step, the transverse convective terms can be computed with first-order upwind differences. These terms can also follow from the high-order interpolation given above (1-D time-centered with respect to material motion) for slightly better, but slightly more expensive results.

Other second-order methods based on a slope can be described by replacing the slope limiter in Fromm's method with some other expression.[4] A diffusive second-order method is the "minmod" method [247]

$$\psi_{\text{MM}}(r) = \max[0, \min(1, r)] \ .$$

An alternative to this is a second-order ENO scheme (mineno) given by [248]

$$\psi_{\text{ENO}}(r) = \begin{cases} r & \text{if } |r| < 1 \ , \\ 1 \ . & \text{otherwise} \end{cases}$$

This ENO scheme can be taken to higher order via a uniformly high-order (UNO) method [251]. This entails computing a limited value of the second derivative at each edge of a cell using limiters. This second derivative is used to improve the estimate of the slope at the cell center through extrapolation. Thus, two extrapolated cell-centered slopes are used instead of cell-edged slopes. The second derivatives are computed as "left" and "right" estimates at each cell edge. As an algorithm, this would work in the following fashion: Algorithm 1: UNO Limiter

1. Compute second derivatives in each cell $D_{i,j,k}U$.
2. Limit the above using $D^L_{i-1/2,j,k}U = \psi(r)\, D_{i,j,k}U$.
3. Compute high-order, first-order derivative estimates: $\delta^{R,UNO}U = \delta^R U - 1/2\ D_{i+1/2,j,k}U$ and $\delta^{L,UNO}U = \delta^L U + 1/2\ D_{i-1/2,j,k}U$.
4. Limit the above using $\delta^{UNO}_{i-1/2,j,k}U = \psi(r)\, \delta^{R,UNO}_{i+1/2,j,k}U$.

We will stop here with ENO schemes as further developments beyond UNO are part of a distinct method that is worth its own complete discussion as is given in Chap. 17, where a different algorithmic implementation will be presented.

The basic limited Fromm's approach is a good robust choice, but to round this discussion out other "TVD" limiters can be used effectively. The steepest second-order method is given by the "superbee" limiter [465]

$$\psi_{\text{SB}}(r) = \max[0, \min(2, r), \min(1, 2r)] \ .$$

A modified harmonic mean limiter is an alternative (van Leer limiter) [570]

$$\psi_{\text{vL}}(r) = \frac{r + |r|}{1 + |r|} \ .$$

A method that is similar to the harmonic mean limiter (van Albada limiter) [561]

$$\psi_{\text{vA}}(r) = \frac{r + r^2}{1 + r^2} \ .$$

[4] The Lax-Wendroff method is obtained by $\psi(r) = r$. A standard second-order upwind method is obtained by $\psi(r) = 1$.

Each of these limiters is strongly related to a data-dependent weighted least squares approach where the weights are chosen to accentuate smooth data. Plots of $\psi(r)$ for each limiter is given in Fig. 14.2.

14.2.2 Piecewise Parabolic Methods (PPM)

As noted earlier, we can go beyond linear reconstruction to higher order polynomials. Parabolas are the obvious next step. We will now explore two methods based on parabolic polynomial interpolation. The first of these methods is the piecewise parabolic method [120]. It is defined by a polynomial built from the cell-average value of a variable and cell-edge values. The polynomial has the form

$$\begin{aligned} U(\xi) &= U^n_{i,j,k} + \left(\Delta_{i,j,k}U_L + \Delta_{i,j,k}U_R\right)\xi \\ &\quad + 3\left(\Delta_{i,j,k}U_R - \Delta_{i,j,k}U_L\right)\left(\xi^2 - \frac{1}{12}\right), \end{aligned}$$

where

$$\Delta_{i,j,k}U_L = \frac{1}{2}\left(U^n_{i,j,k} - U^n_{i-1,j,k}\right) + \frac{1}{6}\left(\delta^M_{i,j,k}U^n - \delta^M_{i-1,j,k}U^n\right),$$

and

$$\Delta_{i,j,k}U_R = \frac{1}{2}\left(U^n_{i+1,j,k} - U^n_{i,j,k}\right) - \frac{1}{6}\left(\delta^M_{i+1,j,k}U^n - \delta^M_{i,j,k}U^n\right).$$

This polynomial can be made monotone through the use of the following slope limiters

$$\Delta^M_{i,j,k}U_L = \mathrm{S}\max\Big(0, \min\left(\left|\Delta_{i,j,k}U_L\right|, 2\mathrm{S}\Delta_{i,j,k}U_R\right)\Big),$$

and

$$\Delta^M_{i,j,k}U_R = \mathrm{S}\max\Big(0, \min\left(2\left|\Delta_{i,j,k}U_L\right|, \mathrm{S}\Delta_{i,j,k}U_R\right)\Big),$$

where $\mathrm{S} = \mathrm{Sign}\left(\Delta_{i,j,k}U_L\right)$. With the monotone polynomial, we can then integrate at an edge over the time step to obtain

$$\begin{aligned} \tilde{U}_{i+1/2,j,k} &= U^n_{i,j,k} + \frac{1}{2}\left(1 - \max\left(0, C^x_{i,j,k}\right)\right)\left(\Delta^M_{i,j,k}U_L + \Delta^M_{i,j,k}U_R\right) \\ &\qquad + 3\left(\Delta^M_{i,j,k}U_R - \Delta^M_{i,j,k}U_L\right) \\ &\qquad \left(\frac{1}{3}[\max\left(0, C^x_{i,j,k}\right)]^2 - \frac{1}{2}\max\left(0, C^x_{i,j,k}\right) + \frac{1}{6}\right). \end{aligned} \tag{14.11}$$

This method is third-order without the limiters (and fourth-order in the limit of vanishing time step size). With limiters it produces convergence that is better than second-order for a linear advection equation ($U_t + aU_x = 0$).

Another nominally third-order method can be found by finding a polynomial similar to the one above, but derived on the basis of cell average data

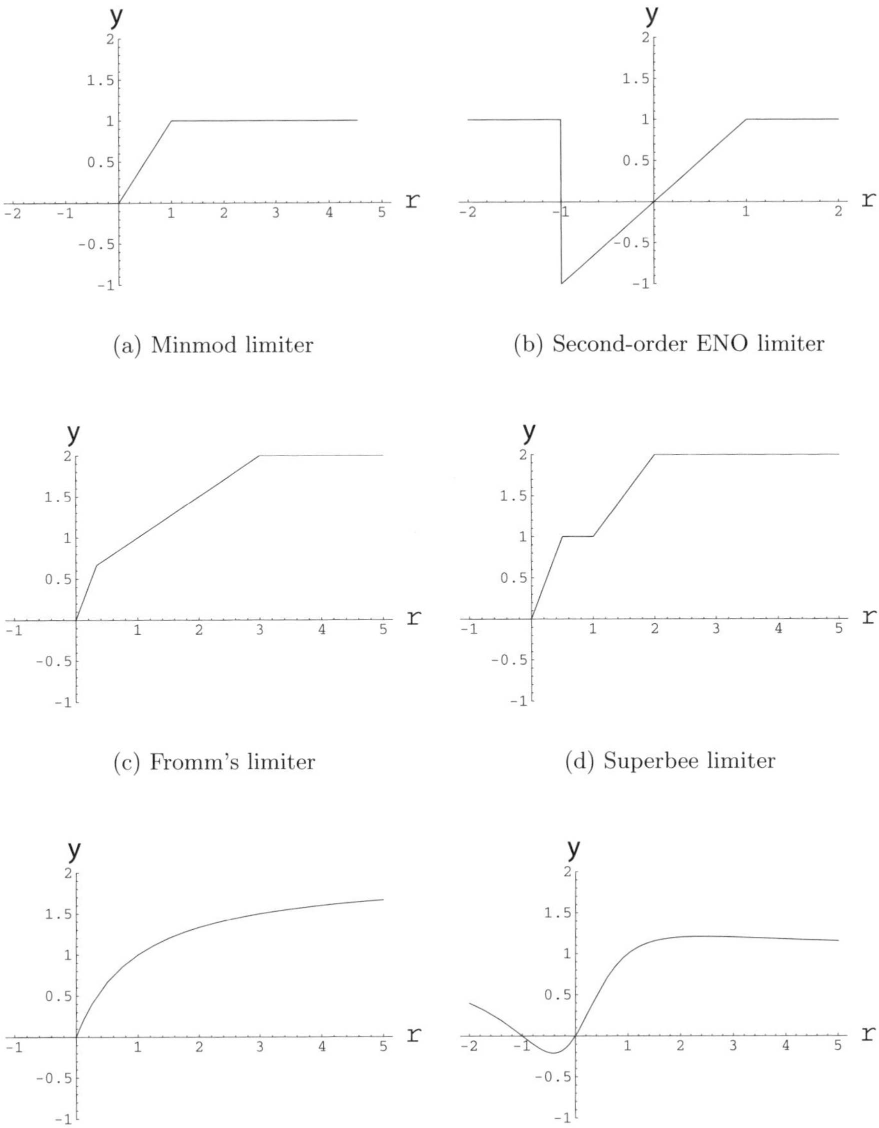

(a) Minmod limiter

(b) Second-order ENO limiter

(c) Fromm's limiter

(d) Superbee limiter

(e) Harmonic mean or van Leer limiter

(f) van Albada limiter

Fig. 14.2. Plots of $\psi(r)$ for the limiters introduced in the text.

and the cell-edge derivatives. The resulting method (see van Leer [574] for more details) is third-order, but on a smaller stencil than the above method. The stencil width is the same as Fromm's scheme, but the accuracy is higher. Defining $S_L = U^n_{i,j,k} - U^n_{i-1,j,k}$ and $S_R = U^n_{i+1,j,k} - U^n_{i,j,k}$, the polynomial can be written as

$$U(\xi) = U^n_{i,j,k} + \frac{1}{2}(S_L + S_R)\xi + \frac{1}{2}(S_R - S_L)\left(\xi^2 - \frac{1}{12}\right) .$$

The monotonicity algorithm is applied as follows:

1. Compute the time n edge values

$$\Delta_L = \frac{1}{3}S_L + \frac{1}{6}S_R ,$$

and

$$\Delta_R = \frac{1}{6}S_L + \frac{1}{3}S_R .$$

2. Compute sign factors (monotonicity problem flags),

$$\chi_L = \mathrm{Sign}\left(|\Delta_L| - |S_L|\right) ,$$

and

$$\chi_R = \mathrm{Sign}\left(|\Delta_R| - |S_R|\right) ,$$

3. Look for local maxima/minima

$$\mathrm{s}_L = \mathrm{Sign}\left(S_L\right) ,$$

and

$$\mathrm{s}_R = \mathrm{Sign}\left(S_R\right) .$$

4. Correct S_L and S_R for presence of local minima or maxima

$$S_L = \mathrm{s}_L \max\left(0, \mathrm{s}_R S_L\right) ,$$

and

$$S_R = \mathrm{s}_R \max\left(0, \mathrm{s}_L S_R\right) .$$

5. Compute monotone slopes

$$S_L^M = \chi_L S_L + 4\chi_R S_R + (1 - \chi_L)(1 - \chi_R) S_L ,$$

and

$$S_R^M = \chi_R S_R + 4\chi_L S_L + (1 - \chi_L)(1 - \chi_R) S_R .$$

The time-centered values at cell-edges can now be computed as with the PPM,

$$\tilde{U}_{i+1/2,j,k} = U^n_{i,j,k} + \frac{1}{4}\left(1 - \max\left(0, \mathcal{C}^x_{i,j,k}\right)\right)\left(S^M_L + S^M_R\right) + \frac{1}{2}\left(S^M_R - S^M_L\right)\left[\frac{1}{3}\left(\max\left(0, \mathcal{C}^x_{i,j,k}\right)\right)^2 - \frac{1}{2}\max\left(0, \mathcal{C}^x_{i,j,k}\right) + \frac{1}{6}\right]. \quad (14.12)$$

14.2.3 Algorithm Verification Tests

Next, we turn our attentions to testing the performance of advection schemes in association with incompressible flow. For the purposes of this demonstration, the details of the projection method matter little. Nonetheless, the method is a second-order approximate projection that was discussed in detail in Chap. 12. A workhorse test problem will be the vortex-in-a-box problem introduced in [45]. It uses a streamfunction

$$\Psi = \frac{1}{\pi}\sin^2\left(\pi x\right)\sin^2\left(\pi y\right) ,$$

in a unit square with homogeneous velocity boundary conditions to define the initial conditions. Initial velocities are $u^o = -\partial\Psi/\partial y$ and $v^o = \partial\Psi/\partial x$. Similarly to [45], we set $\Delta x = \Delta y = 1/2^n$ for $n = 5 - 8$. Our time step is set to $\Delta t = \Delta x$ for an effective CFL number of approximately one. This is the maximum CFL number in the problem and locally the value will be smaller. This was done because we felt it was important to demonstrate the algorithms ability to integrate the flow equations with a CFL number of unity.

The numerical results are given in Table 14.1. As expected, first-order upwinding reduces the algorithm to first-order overall with the remainder of the methods producing at least nominally second-order results. The only other linear method, Fromm's scheme produces some of the best results, but the inherently non-monotonic results are a major drawback in terms of robustness. Another notable scheme uses the superbee limiter, the least dissipative of the TVD methods. This produces poor results from the overcompression of the nonlinear advection terms (its performance with Burger's equation foreshadows this behavior). More advanced methods such as the UNO, PPM, and PQM schemes do not yield any noteworthy improvements over the second-order methods (in particular Fromm's (PLM) or fourth-order slopes (PLMI)). One should note that UNO does produce nearly the lowest error solution on all grids, and by a large degree on the coarser grids.

To fully assess the behavior of the various advection methods, we examined them through another test problem, the Green-Gauss vortices, which has the benefit of an exact solution. The equations are solved in a periodic domain of a unit square with initial conditions

Table 14.1. Convergence rates for vortex-in-a-box problem with various advection schemes.

Case	L_2 Norms				
	32-64	Rate	64-128	Rate	128-256
First-Order	2.5e-2	0.93	1.3e-2	0.92	6.9e-3
Fromm's	2.8e-3	2.06	6.8e-4	2.15	1.5e-4
PLM	3.0e-3	1.96	7.7e-4	2.22	1.7e-4
PLMI	2.8e-3	1.98	7.1e-4	2.15	1.6e-4
Minmod	6.6e-3	1.78	1.9e-3	1.82	5.4e-4
Mineno	6.7e-3	1.81	1.9e-3	1.82	5.4e-4
van Leer	3.6e-3	2.02	8.8e-3	2.21	1.9e-4
van Albada	4.6e-3	2.07	1.1e-3	2.24	2.3e-4
Superbee	5.1e-3	1.51	1.8e-3	1.73	5.3e-4
UNO	2.7e-3	2.07	6.5e-4	2.05	1.6e-4
PPM	3.0e-3	2.04	7.3e-4	2.16	1.6e-4
PQM	3.1e-3	1.97	7.9e-4	2.23	1.7e-4

$$u(x,y,0) = -\cos(2m\pi x)\sin(2m\pi y) \ ,$$

and

$$v(x,y,0) = \sin(2m\pi x)\cos(2m\pi y) \ .$$

The pressure is also given by

$$p(x,y,0) = -\frac{1}{4}\Big[\cos(4m\pi x) + \cos(4m\pi y)\Big].$$

This is a steady-state solution thus should remain constant for all time. If $\nu \neq 0$, the solution will decay so that

$$u(x,y,t) = u(x,y,0)\exp\left[-2\nu(2m\pi)^2 t\right] \ ,$$

$$v(x,y,t) = v(x,y,0)\exp\left[-2\nu(2m\pi)^2 t\right] \ ,$$

and

$$p(x,y,t) = p(x,y,0)\exp\left[-4\nu(2m\pi)^2 t\right] \ .$$

We will carry six methods from the prior test problem for further consideration: PLMI, van Leer, van Albada, UNO, PQM, and PPM. We will study the test problem for values of $m = 2$ or 4. As m grows, the solution has higher frequency (and more) vortices (16 for $m = 2$, and 64 for $m = 4$). Each problem encompasses a finer structure and will challenge the advective schemes more.

Using these relations, we can see that for $\nu = 0$, the kinetic energy, $\|\mathbf{u}\|_2^2 = 1/2$, should remain constant. We can use this to estimate the effective Reynolds number for each method. This is a measure of the dissipation in the scheme; higher is better. The effective Reynolds number uses the exact maximum velocity (of one) and the mesh spacing as reference velocity and length, respectively. The effective value of ν is found by examining the numerical solution for the value that best reproduces the data from the "exact" solution. The results for the convergence rates and effective Reynolds number are given in Tables (14.2 and 14.3).

Table 14.2 gives the results for $m = 2$. The first thing to notice is that all methods converge at least second-order with the UNO scheme converging at third-order. The PPM scheme gives the lowest intrinsic dissipation (followed by UNO and van Albada). The PLMI scheme also does quite well. The differences in dissipation shrinks with smaller grids, with the van Albada scheme giving relatively poor results for the 64×64 and 32×32 grids. It is worth noting that the PPM, UNO and PLMI schemes have the widest stencils of the schemes examined.

The results for $m = 4$ are summarized in Table 14.3. These are generally the same as for $m = 2$, but the rates of dissipation have increased because of the smaller, and less well resolved vortices. This highlights the frequency dependent dissipation from the high-resolution Godunov methods. Features of the flow that are less resolved will experience more dissipation than well-resolved features. This time, the PPM scheme outperforms the other methods with UNO a close second. The van Leer and van Albada limiters also fail to produce convergence rates that are greater than second-order for the finest grids.

14.3 Staggered Grid Spatial Differencing

The MAC staggered grid often favored for pressure-velocity coupling provides some relatively significant challenges to the Godunov-type algorithms (see Fig. 14.3, for an example). Most of the previous discussion applies to MAC staggered grid, but certain portions of the algorithm require some delicate choices [528]. In particular, the chief issue is the choice of velocities at directions transverse to the direction that the edge-velocity is normal to. This velocity is not uniquely defined as the center of the transverse edge (face in 3-D) lies equidistant from four (eight in 3-D) velocities normal to that edge (face in 3-D). The simplest choice is to simply average the four (eight in 3-D) values. Less stable choices are to make an upwind choice of velocity that is quite susceptible to symmetry breaking.

Table 14.2. Convergence rates and effective dissipation for the periodic vortices problem (the Green-Gauss vortices), for $m = 2$ (see text for details).

Method	Measure	Grid			
		32	64	128	256
PLMI	Convergence		2.53	2.52	2.29
	Effective *Re*	1.29e4	1.82e5	1.66e6	1.37e7
van Leer	Convergence		2.63	2.09	2.22
	Effective *Re*	9.61e3	1.52e5	1.62e6	1.48e7
van Albada	Convergence		2.28	2.04	2.57
	Effective *Re*	9.15e3	1.66e5	1.93e6	1.73e7
UNO	Convergence		2.87	2.90	2.98
	Effective *Re*	1.36e4	1.97e5	1.83e6	1.54e7
PPM	Convergence		2.32	2.41	2.12
	Effective *Re*	1.31e4	2.22e5	2.19e6	1.83e7
PQM	Convergence		2.64	2.33	2.14
	Effective *Re*	1.01e4	1.65e5	1.62e6	1.34e7

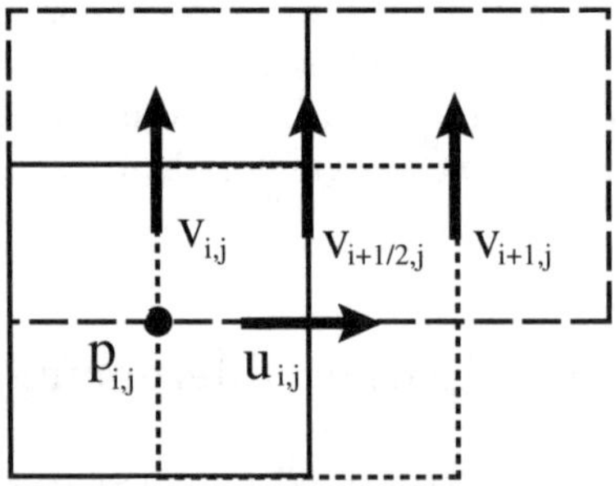

Fig. 14.3. A two-dimensional MAC grid showing the indices in our example. The prototypical problem is to find $v_{i,j+1/2}$ from the surrounding data.

Table 14.3. Convergence rates and effective dissipation for the periodic vortices problem (the Green-Gauss vortices), for $m = 4$.

Method	Measure	Grid			
		32	64	128	256
PLMI	Convergence		2.65	2.82	2.48
	Effective Re	1.42e3	2.82e4	3.95e5	3.46e6
van Leer	Convergence		2.39	3.06	1.81
	Effective Re	1.39e3	2.09e4	3.14e5	3.25e6
van Albada	Convergence		2.49	2.50	1.79
	Effective Re	1.39e3	1.87e4	3.36e5	3.76e6
UNO	Convergence		2.77	2.93	2.84
	Effective Re	1.56e3	3.05e4	4.34e5	3.81e6
PPM	Convergence		2.78	2.49	2.37
	Effective Re	1.39e3	2.86e4	4.96e5	4.58e6
PQM	Convergence		2.37	3.04	2.27
	Effective Re	1.38e3	2.25e4	3.51e5	3.320e6

Once these decisions are made the algorithm is an extremely straightforward extension of the cell-centered version. One will construct a piecewise polynomial interpolant in each cell, and then choose a time-advancement technique. Aside from the fashion of choosing transverse-normal velocities, the Riemann solvers can be identical (the same as the cell-centered case once this decision is made). The need to provide divergence-free velocity fields is the same as in the cell-centered case although, generally speaking, the pressure-velocity coupling is less problematic on the MAC grid.

Because of these various issues differencing on MAC staggered grids has a greater difficulty with symmetry. This is countered by the extremely good pressure-velocity coupling and the ability to achieve this with a standard five (seven in 3-D) point Poisson operator. This makes the numerical linear algebra quite straightforward and as efficient as possible.

14.4 Unsplit Spatial Differencing

It is often useful to move the interpolation/differencing associated with the spatial differencing to a less dimensionally split approach. This is necessary on more unstructured grids, but is also useful on the sort of structured grids used thus far in the discussion. The sorts of methods described here were originally developed by Barth for use in aerospace applications on unstructured triangular or tetrahederal meshes [35, 36, 38]. There are several forms to be considered, but each is distinguished by producing the complete polynomial basis in a single (but more expensive) step. The objective of this approach is to justify the expense with better results from both an operational and philosophical point-of-view.

With the one-dimensional high-resolution monotone schemes well established the extension to multidimensional constructions seems an obvious next step. Most multidimensional applications of monotone schemes are derived from a serial (e.g., operator-split or Runge-Kutta) application of the basic one-dimensional method. Although the design and application of truly multidimensional monotone methods has attracted much attention recently, their understanding and widespread use and reliability remains elusive. This is especially true for multidimensional schemes that must reconstruct and integrate data on complex grid topologies.

Our starting point is the class of multidimensional "k-exact" methods devised by Barth [34, 35, 36, 37, 38]. These schemes are based on the minimum number of degrees of freedom required to reconstruct a polynomial from discrete data. Barth's extended approach derives a reconstruction based upon a least-squares methodology [61]. This loosens the requirements from the "k-exact" methods by having the reconstruction based on a broader set of discrete values. This invokes a minimization approach that produces and "optimal" difference for the choice of stencil. Further, it offers a high degree of flexibility to the user in constructing the overall method. One can also apply a line integral approach to computing a gradient, and while this method is faster than the least squares method, it is also more restrictive in its implementation. For a specific set of conditions within the least squares approach the two are equivalent.

Barth applies monotonicity after the minimization process, following principles similar to those set forth by Dukowicz [166, 168] and Zalesak in his multidimensional FCT algorithm [615]. The algorithm is an obvious extension of the one-dimensional monotonicity which is defined by the minimum and maximum local values, and the same for the reconstruction. One then modifies the reconstruction by a scalar value that makes the reconstruction monotonicity satisfying. The issue is that this algorithm is based on heuristic arguments that are extended from van Leer's original idea.

The problem is that the definition and application of monotonicity is not part of minimization process, and therefore remains quasi-one-dimensional. One can show that if monotonicity considerations are recast as constraints in the minimization process, the resulting reconstruction is truly multidimensional, i.e., the difference between a constrained (monotonic) and unconstrained (non-monotonic) reconstruction can be interpreted as a geometric "limiter" that is in general a vector or tensor. This takes the form of a constraint that is activated when the constraint is violated by the reconstruction.

Rather than necessarily providing a practical algorithm this point-of-view provides a window into the standard algorithm.

14.4.1 Least Squares Reconstruction

Given discrete values of a scalar quantity u_i at grid points i, a polynomial reconstruction u_i^{R} at an arbitrary point (x, y) (for two dimensions) near point i is defined by

$$U_i^{\mathrm{R}}(x,y) = U_i + \nabla_{x,i} U\,(x - x_{1,i}) + \nabla_{y,i} U\,(y - y_{1,i}) + \nabla_{x,i}^2 U\left(x^2 - x_{2,i}\right) + \cdots\,, \tag{14.13}$$

where the constants $x_{q,i}$ are defined so that

$$\int_\Omega U_i(x,y)\,\mathrm{d}\Omega = U_i\,,$$

i.e., mean-preserving. We wish to minimize the weighted L_2 error in the reconstructed value of U at points j near i (U_j^{R}) relative to the discrete values U_j, i.e.,

$$\min \sum_j \left\| w_j \left(U_j^{\mathrm{R}} - U_j\right)\right\|_2\,, \tag{14.14}$$

where w_j are weights for the reconstruction at point j that arise from geometric or data considerations. As example of a geometric weight, points j at greater distances from the reference point i will be weighted less than those closer (e.g., an inverse-distance weight). Data weights can arise from monotonic considerations where many one-dimensional limiters can be recast as weighted least squares (van Leer, van Albada, and second-order ENO in the limit as the weights go to infinity). For the polynomial reconstruction given earlier, (14.14) results in a least squares problem given by (for two dimensions)

$$\min \left\|\mathbf{w}^T\left(\mathbf{A}\mathbf{x} - \mathbf{b}\right)\right\|_2 \rightarrow \left(\mathbf{w}\mathbf{A}\right)^T\left(\mathbf{w}\mathbf{A}\right)\mathbf{x} = \left(\mathbf{w}\mathbf{A}\right)^T \mathbf{w}\mathbf{b}\,, \tag{14.15}$$

where the matrix $\mathbf{A}$ is given by

$$\mathbf{A} = \begin{pmatrix} (x_{1,k} - x_{1,i}) & (y_{1,k} - y_{1,i}) & (x_{2,k} - x_{2,i}) & \cdots \\ \vdots & \vdots & \vdots & \ddots \\ (x_{1,n} - x_{1,i}) & (y_{1,n} - y_{1,i}) & (x_{2,n} - x_{2,i}) & \cdots \end{pmatrix},$$

which can be rewritten as

$$\mathbf{A} = \begin{pmatrix} \Delta x_{1,k} & \Delta y_{1,k} & \Delta x_{2,k} & \cdots \\ \vdots & \vdots & \vdots & \ddots \\ \Delta x_{1,n} & \Delta y_{1,n} & \Delta x_{2,n} & \cdots \end{pmatrix} .$$

The vector $\mathbf{b}$ is given by

$$\mathbf{b} = \begin{pmatrix} U_k - U_i \\ \vdots \\ U_n - U_i \end{pmatrix} = \begin{pmatrix} \Delta U_k \\ \vdots \\ \Delta U_n \end{pmatrix} .$$

The weights are specified as those which act as a multiplier for each row in the matrix or vector,

$$\mathbf{w} = \begin{pmatrix} w_k \\ \vdots \\ w_n \end{pmatrix} ,$$

and the solution vector $\mathbf{x}$ is given by

$$\mathbf{x} = \begin{pmatrix} \nabla_x U_i \\ \nabla_y U_i \\ \nabla_x^2 U_i \\ \vdots \end{pmatrix} .$$

The number of columns (n) in $\mathbf{A}$ and the length of vectors $\mathbf{w}$, $\mathbf{b}$, and $\mathbf{x}$ depends upon the number of terms taken in the reconstruction, and the number of rows (m) in $\mathbf{A}$ depends upon the number of neighbors j considered for each point i. The system is generally over-determined, i.e., $m > n$.

Caution must be taken in constructing the local linear system of equations. The system can become quite ill-conditioned with certain choices of weights. The construction and solution of this system of equations requires careful attention to any errors that might be introduced in the solution process. For example, a solution via the normal equations with a standard method is usually sufficient on regular grids, but in the more general case this approach can be prone to failure. A QR-factorization circumvents this problem in cases where the system is ill-conditioned.[5] In addition, if the minimization problem becomes rank-deficient, it can be regularized or solved

[5] The QR factorization is a numerically stable manner of solving least square problems using an orthogonal decomposition [61, 136]. See Chap. 8 for more explanation.

outright with a SVD algorithm.[6] A simple and effective choice is to regularize the solution with a Tikhonov-type method in which a small parameter, ε ($\approx 1.0 \times 10^{-6}$), is added to the entry (and retains the entry's sign). This provides a solution that retains the fundamental structure of the well-posed problem.

Consider the example of two-dimensional discrete data for $\mathbf{U}$ given in Fig. 14.4. We wish to find a linear reconstruction of this data (i.e., retain only the first derivative terms in the polynomial expansion), using the least squares system given by (14.15). Two least squares solutions for the gradient of U are shown in Fig. 14.5, a "centered" gradient obtained without weighting (i.e., $\mathbf{w} = 1$) and "distance squared" gradient obtained with inverse distance squared weights. One inequality constraint is active, $\nabla_x U + \nabla_y U = 8$ (see Sect. 14.4.3 for further discussion on inequality constraints). These two gradients, which are similar, lie inside of a gradient space formed by points resulting from "raw" (finite difference) gradients computed from a set of eight nearest-neighbor triangles.

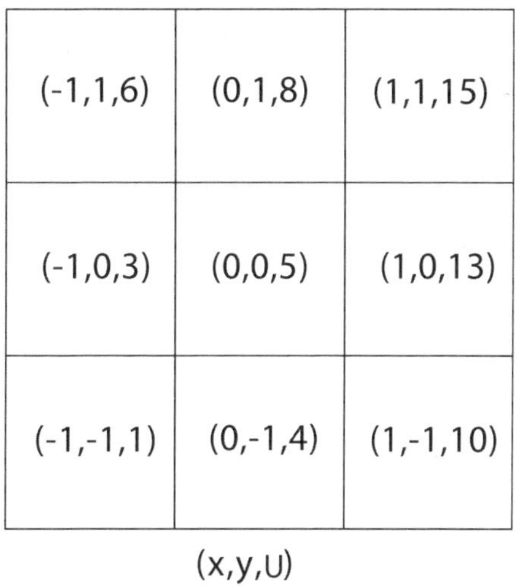

Fig. 14.4. The raw data for the example problem for least squares.

On a fixed two-dimensional grid ($\Delta x = \Delta y$), we can write down the solution to the least squares gradient explicitly. With no weighting, the gradients are

$$\frac{\partial U}{\partial x} \approx \frac{U_{i+1,j+1} + U_{i+1,j} + U_{i+1,j-1} - U_{i-1,j+1} - U_{i-1,j} - U_{i-1,j-1}}{6\Delta x},$$

[6] The singular value decomposition (SVD) solves ill-posed problems producing the best solution in the same sense as a least square problem [61, 136].

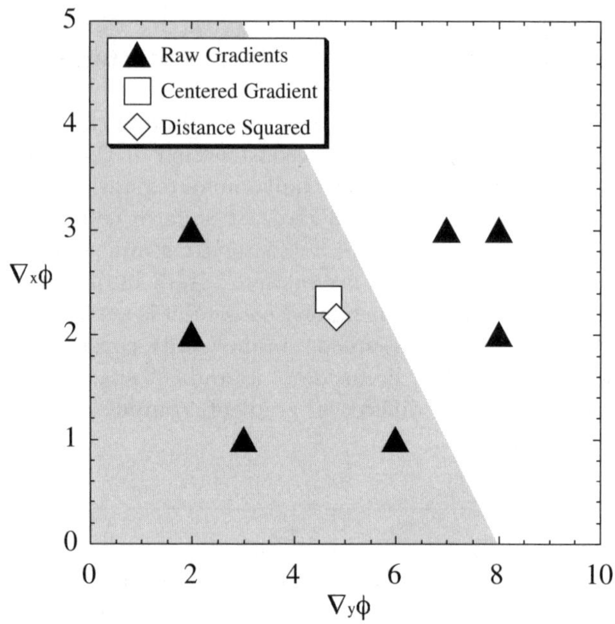

Fig. 14.5. Gradients computed from various pairs of raw data as well as from the regular and distance squared weighted for the least squares problem. The shaded region is defined by the inequality constraint $\nabla_x\phi + \nabla_y\phi = 8$, where ϕ is a generic variable, e.g., $\phi = U$.

and

$$\frac{\partial U}{\partial y} \approx \frac{U_{i+1,j+1} + U_{i,j+1} + U_{i-1,j+1} - U_{i-1,j-1} - U_{i,j-1} - U_{i-1,j-1}}{6\Delta y} .$$

With a distance weighting applied to the problem, the gradients become

$$\frac{\partial U}{\partial x} \approx \frac{U_{i+1,j+1} + 4U_{i+1,j} + U_{i+1,j-1} - U_{i-1,j+1} - 4U_{i-1,j} - U_{i-1,j-1}}{12\Delta x} ,$$

and

$$\frac{\partial U}{\partial y} \approx \frac{U_{i+1,j+1} + 4U_{i,j+1} + U_{i-1,j+1} - U_{i-1,j-1} - 4U_{i,j-1} - U_{i-1,j-1}}{12\Delta y} .$$

If the weighting is the distance-squared the gradient is

$$\frac{\partial U}{\partial x} \approx \frac{U_{i+1,j+1} + 8U_{i+1,j} + U_{i+1,j-1} - U_{i-1,j+1} - 8U_{i-1,j} - U_{i-1,j-1}}{20\Delta x} ,$$

and

$$\frac{\partial U}{\partial y} \approx \frac{U_{i+1,j+1} + 8U_{i,j+1} + U_{i-1,j+1} - U_{i-1,j-1} - 8U_{i,j-1} - U_{i-1,j-1}}{20\Delta y} .$$

Note that the changes in the weighting induce changes in the relative weighting of the nearest grid points in the direction of the gradient.

14.4.2 Monotone Limiters and Extensions

In this section we will introduce the basic heuristic limiter and summarize some simple extensions of this limiter to provide different properties for pure monotone advection [456].

First, consider the steps necessary to modify a reconstruction so that it gives a monotone representation of the local data. This is defined in the following way: *the reconstruction in a cell is bounded by the minimum and maximum of the surrounding cells.* A typical implementation of this algorithm is as follows:

1. Find the nearby minimum and maximum of the dependent data $U^{\min}$ and $U^{\max}$.
2. Find the reconstruction minimum and maximum in the computational zone, Ω, $U_\Omega^{\min}$ and $U_\Omega^{\max}$.
3. If $U_\Omega^{\min} < U^{\min}$ compute α such that $U_\Omega^{\min} = U^{\min}$, otherwise $\alpha = 1$,
$$\alpha_{\min} = \min\left(1, \frac{U - U^{\min}}{U - U_\Omega^{\min}}\right) .$$
4. If $U_\Omega^{\max} > U^{\max}$ compute α such that $U_\Omega^{\max} = U^{\max}$, otherwise $\alpha = 1$.
$$\alpha_{\max} = \min\left(1, \frac{U^{\max} - U}{U_\Omega^{\max} - U}\right) .$$
5. Choose α to be the smallest of the available choices from steps 3 and 4,
$$\alpha = \min\left(\alpha_{\min}, \alpha_{\max}\right) .$$

This algorithm can be modified to produce an algorithm that is only bounded by specific (nonlocal) values, or to preserve the sign of the data.

If one desires for various values to be maintained as absolute bounds for a method, then the local minimum or maximum found in step 1 above is replaced by global values. Another approach might be to include the global values with the local values.

An important special case is when one desires sign-preserving methods, or more generally positive definite solutions. This is desirable for a variety of physical quantities such a mass fraction, concentration, or density. In this case, the maximum part of α is disregarded, and only the minimum is found. Another manner of looking at this is that a global minimum is set to zero, and the maximum is not disregarded. The same set of steps can be inverted to set a global maximum or define methods as negative definite.

Note further, that these algorithms can be applied to any one-dimensional discretization. In order to apply this, one has to choose the high-order stencil desired then apply these limiters only in the direction of the reconstruction (dimension-by-dimension). The basic principles remain the same.

14.4.3 Monotonic Constrained Minimization

It is useful from a pedagogical standpoint to consider monotonicity as a constraint in a minimization process of the reconstruction for the over-determined discrete data in L_2. As noted earlier, this process is not computationally competitive for the improvement in the results, but provides a unique perspective on the algorithm.

When designing monotonic reconstruction methods, the monotonic constraint can be considered as a *discrete data concept*, in contrast to a high-order reconstruction method, which is *accuracy-driven*. In short, reconstruction follows from consideration of nearby discrete data and their physical location, whereas monotonicity constraints follow from consideration of nearby discrete data (not their location), as well as reference cell geometry.

First consider a scalar monotonicity constraint from the previous section, in which a constant α is found to multiply the higher order terms in the interpolant

$$\begin{aligned} U_i\left(x,y\right) = U_i + \alpha[&\nabla_x U\left(x - x_{1,i}\right) + \nabla_y U\left(y - y_{1,i}\right) \\ &+ \nabla_x^2 U\left(x^2 - x_{2,i}\right)] + \cdots\ , \end{aligned} \tag{14.16}$$

such that the above expansion (reconstruction) is monotone with respect to the local data. Next, consider the algorithm from the previous section which produces a scalar that modifies the reconstruction to meet the constraints.

By relying upon the extensive work done in solving least squares problems [61] we can extend the utility of this approach. The formulation of the monotonicity satisfying interpolation with an inequality constraint is quite similar. Basically, the least squares problem is modified by any of the constraining inequalities that is violated by the base solution. This recasts the minimization as

$$\min\left\|\mathbf{w}^T\left(\mathbf{A}\mathbf{x} - \mathbf{b}\right)\right\|_2 \text{ subject to } \mathbf{C}\mathbf{x} = \mathbf{d}\ . \tag{14.17}$$

If none of the inequalities are violated then $\mathbf{C}$ is null. At most, the rank of $\mathbf{C}$ is equal to $\mathbf{x}$. For example, if there are two unknowns then up to two constraints can be active and their solution will determine the system. When one constraint is active then a minimization would take place.

We find the active constraint in a process similar to scalar algorithm. The difference is that more than one constraint can be used to determine the overall limiter. The process proceeds as follows:

1. Find the local minimum and maximum of the dependent data $U^{\min}$ and $U^{\max}$.

2. Compute the smallest positive difference $\Delta U^{\min} = \min(U - U^{\min}, U^{\max} - U)$.
3. Find the reconstruction minimum and maximum in Ω, $U_\Omega^{\min}$ and $U_\Omega^{\max}$.
4. Check to see if the reconstruction maxima and minima violate the monotonicity constraint. If $U - U_\Omega^{\min} > \Delta U^{\min}$ this is an active constraint, and if $U_\Omega^{\max} - U > \Delta U^{\min}$ this is also an active constraint.

For simple rectangular mesh cells several simplifications can be made. The constraints that can be active are

$$h_x \eta_x \nabla_x U + h_y \eta_y \nabla_y U \leq 2\Delta U^{\min} ,$$

$$h_x \eta_x \nabla_x U - h_y \eta_y \nabla_y U \leq 2\Delta U^{\min} ,$$

and

$$-h_x \eta_x \nabla_x U + h_y \eta_y \nabla_y U \leq 2\Delta U^{\min} ,$$

where $\eta_x = \text{sign}\,(\nabla_x U)$ and $\eta_y = \text{sign}\,(\nabla_y U)$, with h_x and h_y being the grid spacing in $x-$ and $y-$direction, respectively.

The equation can be applied to the minimization problem in the following way: suppose we are computing the following interpolant,

$$U_i + \nabla_x U\,(x - x_{1,i}) + \nabla_y U\,(y - y_{1,i}) \ ,$$

and we have a constraint

$$h_x \nabla_x U + h_y \nabla_y U = \Delta U^{\text{lim}} .$$

We have two options in applying this constraint to the interpolation problem: the method of weighting and algebraic elimination. The method of weighting would take some constant C and add the constraint equation to the matrix $\mathbf{A}$ as a new row and an entry in $\mathbf{b}$ as

$$C\left(h_x \nabla_x U + h_y \nabla_y U - \Delta U^{\text{lim}}\right) \ ,$$

where ΔU^{lim} is the difference in U to produce a monotone result. The algebraic elimination would solve for one of the unknowns in terms of the other(s) and then a reduced system would be minimized. In this example we could take

$$\nabla_x U = \frac{\Delta U^{\text{lim}} - h_y \nabla_y U}{h_x} \ .$$

The result of this process is a vector scaling of the interpolant such that the constraint is satisfied and the error is minimized.

Within the constrained least-squared minimization framework, several variations can be constructed. One can take the view that we are now filtering the stencil in order to accentuate certain properties in the data. The practical effect will be to either remove or add dissipation from the solution. In this way, much of the flexibility inherent in one-dimensional high-resolution methods can be applied to multidimensional methods on arbitrary grids.

For incompressible flow there is an obvious and powerful constraint that pervades the physical character of the flow, the divergence-free condition. This requires that the values of normal derivatives of the velocity have a specific relation. This can be applied as a constraint on the derivatives.

14.4.4 Divergence-Free Reconstructions

In a two dimensional Cartesian geometry the divergence-free condition is

$$\frac{\partial u}{\partial x} + \frac{\partial v}{\partial y} = 0 \rightarrow \frac{\partial u}{\partial x} = -\frac{\partial v}{\partial y},$$

and in three dimensions it can be written as

$$\frac{\partial u}{\partial x} + \frac{\partial v}{\partial y} + \frac{\partial w}{\partial z} = 0 \rightarrow \frac{\partial u}{\partial x} = -\frac{\partial v}{\partial y} - \frac{\partial w}{\partial z},$$

or in other permutations of this equation. This reduces the reconstruction by one degree of freedom, but includes some physical conditioning to the result. Care must be taken so that the utilization of the above does not preclude nonphysical results. Generally speaking, the application of this technique will increase the amount of limiting that needs to be applied to the raw reconstruction results.

Let us examine the modifications that might simply be applied in order to not violate the nonoscillatory nature of a reconstruction while appealing to the divergence-free constraint. This will occur most commonly when the computed slopes do not have an appropriate relation with respect to their sign. In two dimensions $\text{sign}(\partial u/\partial x) = -\text{sign}(\partial v/\partial y)$, if this is not true, the safest thing to do is set both derivatives to zero (assuming that a divergence-free reconstruction is desired).

One can apply to idea of monotonicity in more than one dimension. The general idea and principles are much the same, but a number of technical caveats apply.

14.4.5 Extending Classical TVD Limiters

We begin with the statement that the procedures given above produce multi-dimensional extensions of a monotone Fromm's (or centered) method limiter. By making the weights data dependent, that is dependent on the data being interpolated, these limiters can easily be implemented. Each of these multi-dimensional extensions reduces to the corresponding one-dimensional limiter if the flow becomes one-dimensional (mesh aligned on an orthogonal grid).

The minmod limiter in one dimension chooses the minimum of the available slopes, and the superbee chooses the maximum with the condition that the choice be monotonicity preserving. One way to generalize this is to make the choice based on the average slope. Others [170, 41] have done the same, but with the interpretation of superbee as the largest of the available slopes.

These choices are implemented by setting the weights to zero if the data does not meet the selection criteria. Thus the minmod limiter generalizes by choosing the data that is smaller than the average and the superbee through choosing data that is larger. For a two-dimensional linear reconstruction the algorithm would be organized as follows:

1. Solve (14.15) with centered data.
2. For each cell in the stencil compute
$$\delta_k U = (x_{1,i} - x_{1,k}) \nabla_x U + (y_{1,i} - y_{1,k}) \nabla_y U \ .$$
3. For the minmod limiter if $|\Delta_k U| > |\delta_k U|$ then $w_k = 0$, otherwise it is unchanged.
4. For the superbee limiter if $|\Delta_k U| < |\delta_k U|$ then $w_k = 0$, otherwise it is unchanged.
5. Solve the re-weighted least square problem again.
6. Impose the monotonicity constraints on the resulting solution.

We show the solution for the minmod and superbee limiters as applied in multidimensional least squares fashion in an example below (see also Fig. 14.6). The minmod gradient uses the following four data points, (-1,-1), (-1,0), (-1,1), (0,-1). The superbee gradient uses the following four data points, (0,1), (1,-1), (1,0), (1,1). In this case the minmod limiter results in a gradient that is smaller in magnitude than any of the raw gradients. The superbee limiter provides a gradient that is near the size of the largest raw gradients.

In Fig. 14.7 we show the result of applying the monotonicity constraints to the superbee limiter. One inequality constraint is active, $\nabla_x U + \nabla_y U = 8$, although a second constraint is shown, the superbee gradient does not violate it. The constraint is defined by the values in the interpolant that produce cell maximum values equal to the maximum of the surrounding zones. The other constraint is defined by cell minimum values. As can be observed the scalar limiter simply uniformly scales the gradient onto the line formed by the equality form of the constraint and differs significantly from the minimized constrained solution.

Another important concept is slope steepening (related strongly to artificial compression [246, 607]). One-dimensional piecewise linear schemes have been expressed in a rather complete way by Huynh [271] who introduced an interesting and extensible slope steepener. We first discuss this in a simplified setting from that given by Huynh, but then move the concept to a purely multidimensional implementation. Huynh defines the scheme using a constant κ to determine the compression. The left and right slopes are differenced and multiplied by κ then made monotone. In the standard form

$$\psi(r) = \mathrm{minmod}\left[\max\left(1/2\left(1+r\right), \kappa\left|1-r\right|\right), 2, 2r\right] \ .$$

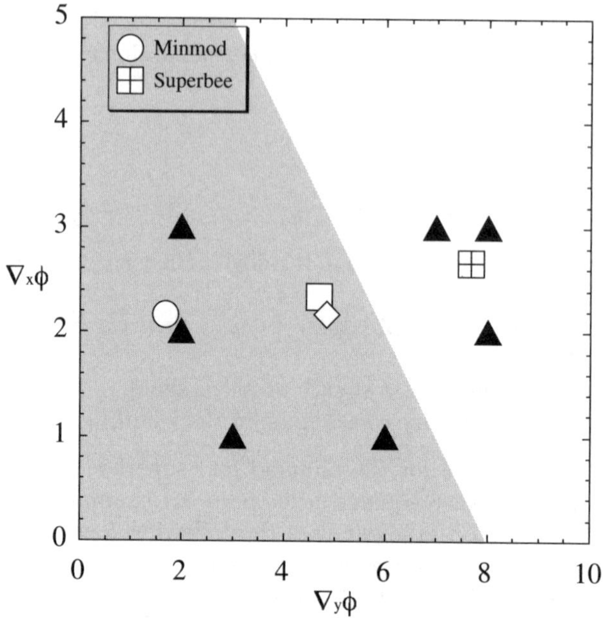

Fig. 14.6. Computed minmod and superbee least squares gradients for the given data. The closed symbols show the data while the open symbols show the interpolated values.

In multidimensions we replace the term, $\kappa\left|1-r\right|$, by $\kappa\left(U^{\max}-2U+U^{\min}\right)$. The other terms are also substituted as the implementation of the van Leer scheme in multiple dimensions suggests.

Two other important limiters are the harmonic mean limiter [569][7] and van Albada's limiter [561].[8] By choosing the weights to be inversely proportional to the absolute value of size of the data, the harmonic mean limiter is generalized. This simply requires the weight vector be scaled by $w_k := w_k / \left|\Delta_k U\right|$. If this choice is the inverse of the size of the data squared, then the van Albada limiter is generated. Again, this requires the weight vector be scaled, in this case by $w_k := w_k / \left(\Delta_k U\right)^2$. These limiters may be useful in fine-tuning the resolution and characteristics of the interpolant to a given application or situation.

[7] $\psi\left(r\right) = \left(r + \left|r\right|\right) / \left(1 + \left|r\right|\right)$.
[8] $\psi\left(r\right) = \left(r + r^2\right) / \left(1 + r^2\right)$.

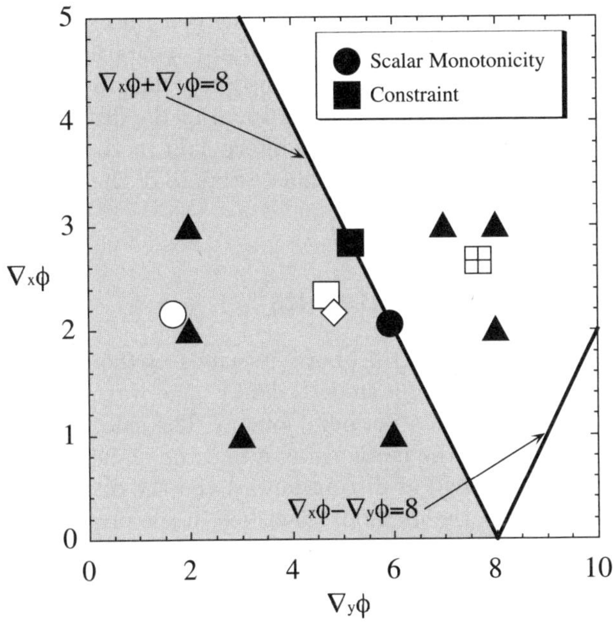

Fig. 14.7. The standard scalar and constrained limiter applied to the non-monotone superbee gradients. Two of the four lines formed by the constraints are shown. The other symbols are the same as Fig. 14.6.

Yet, another small modification of the weighting used to extend the harmonic mean and van Albada limiters can be used to implement a L_1 minimization. This is accomplished through using an iteratively re-weighted least squares calculation where the weights are inversely proportional to the residual. Thus, the van Leer limited scheme can be used to compute the first guess, then the inverse of the residual can be used to weight the data and the solution is found again.

Let now discuss weighting in a broader sense. The use of least squares methods in computing the functional dependence of data is well known. It seems sensible to apply more or less weight to data points depending on their reliability. The above generalization of limiters could be viewed in this context. For the purposes that we have employed the least squares machinery at a given point in space, the appropriate weighting is inversely proportional to the distance. In [206] the form is discussed with relation to the expected error. There a linear interpolant was determined for some data using the model,

$$y = a + bx + \varepsilon \Delta y \ ,$$

where the last term, $\varepsilon \Delta y$, is the expected error. Since we are essentially approximating a Taylor series expansion, the general form of the expected error is known (where the function is smooth). For a linear expansion the error terms scale with the square of the distance, thus the geometric portion of the weight should be inversely proportional to the distance squared. This relation will change as the order of the interpolant increases. As a result the norm of the residual will be a reasonable estimate of the truncation error of the method.

14.5 Multidimensional Results

We will present the results of the above described methods on two test problems: the scalar advection of a smooth double sine wave and a circular discontinuous region, both on a periodic domain. The sine wave test will show the order of accuracy of the methods used and the circular region will show the behavior of the method at discontinuities and the distortion of a smooth body. We will examine the methods described in the previous section.

Table 14.4 shows the error for the sine wave test on a series of grids 16×16 through 64×64. The fine grid error and the order of convergence are shown. These results reveal that the multidimensional reconstructions are superior and constraint-based limiters improve the solution.

Table 14.4. Order of accuracy for various reconstruction methods for a double sine wave advection. Errors are given on the fine grid.

Method	Error L_1	Error L_∞	Order L_1	Order L_∞
1-D Fromm	1.03×10^{-3}	1.19×10^{-2}	1.97	1.42
2-D Fromm	4.28×10^{-4}	1.13×10^{-2}	2.68	1.53
2-D Constrained	4.02×10^{-4}	1.13×10^{-2}	2.72	1.51
2-D Minmod	6.30×10^{-3}	3.81×10^{-2}	1.68	1.14
2-D Harmonic	7.91×10^{-3}	3.64×10^{-2}	1.60	1.25
2-D L_1	7.24×10^{-4}	1.72×10^{-2}	2.36	1.38

We show the results using the methods described above in Figs. 14.8 through 14.10. The solutions are computed with a monotone unsplit differencing method similar to the one in [47]. Each is computed at a CFL number of one-half on a 50×50 grid. We believe the close-ups of the details of the solution provide critical details to judge the fidelity of the calculations.

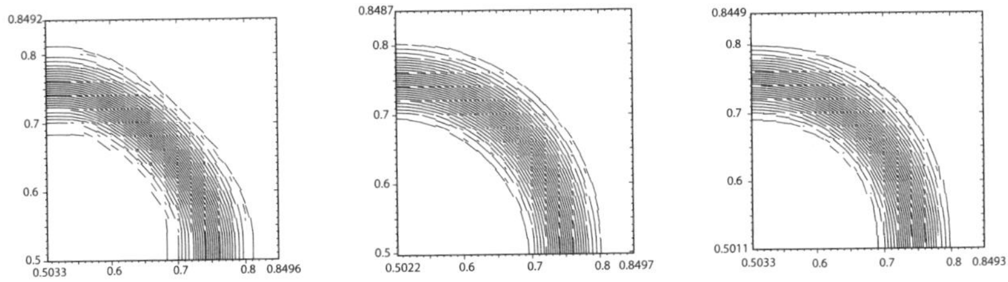

(a) 1-D Fromm's Reconstructions

(b) 2-D Fromm's Reconstructions

(c) 2-D Constrained Fromm's Reconstructions

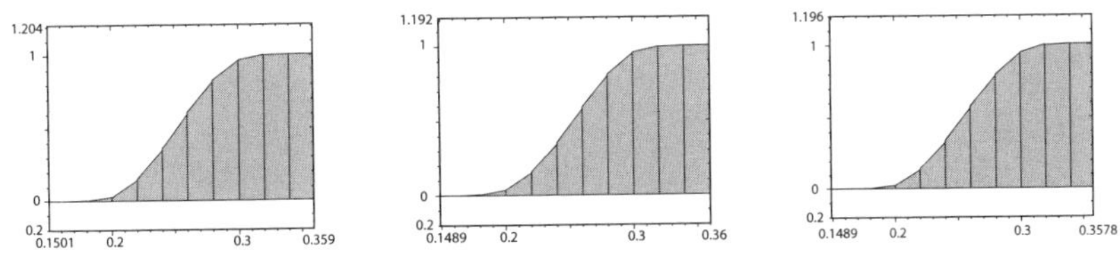

(d) 1-D Fromm's Front

(e) 2-D Fromm's Front

(f) 2-D Constrained Fromm's Front

Fig. 14.8. Advection of a circle by an unsplit advection scheme using various Fromm-van Leer methods.

In general the purely multidimensional methods provide a significantly less grid-dependent solution. The general features of the one-dimensional limiters have been replicated in a consistent two-dimensional manner. The superbee limiter shows small 3-4 cell transitions and the Huynh compressive limiter is even sharper. While the constraint-based limiting is somewhat superior to scalar limiting, its superbee implementation shows some evidence of increased grid dependence. This may be due to a decrease in dissipation, and the interface thickness, while consistent in thickness in both methods, is sharper with the constraint-based limiting.

Here, we have presented several extensions of existing methods for reconstructing functions for the purpose of developing a Godunov-type algorithm. We have demonstrated that these methods are genuinely multidimensional and naturally extend to arbitrary grids. Furthermore, the procedures improve the accuracy and quality of solutions. These methods are also more flexible than existing multidimensional methods.

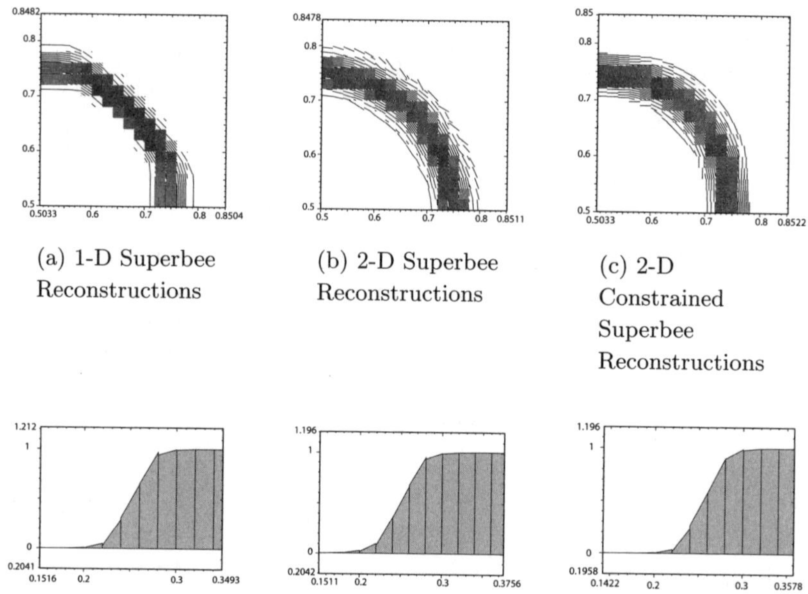

(a) 1-D Superbee Reconstructions

(b) 2-D Superbee Reconstructions

(c) 2-D Constrained Superbee Reconstructions

(d) 1-D Superbee Front

(e) 2-D Superbee Front

(f) 2-D Constrained Superbee Front

Fig. 14.9. Advection of a circle by an unsplit advection scheme using various Superbee methods.

14.6 Viscous Terms

Here, we briefly discuss the discretization of the viscous terms. The discretization of these terms in conjunction with curvilinear coordinates has been presented in Chap. 4. The discrete form will use a second-order accurate form, and the same spatial operators will be used for implicit and explicit time integrations. For incompressible flows the form is especially simple using a standard Laplacian operator (in two dimensions),

$$\nabla^2 p^{n-1/2} = \frac{\mathbf{u}_{i+1,j} - 2\mathbf{u}_{i,j} + \mathbf{u}_{i-1,j}}{\Delta x^2} + \frac{\mathbf{u}_{i,j+1} - 2\mathbf{u}_{i,j} + \mathbf{u}_{i,j-1}}{\Delta y^2} . \quad (14.18)$$

If the viscosity, ν is a function of position (e.g., in non-Newtonian flows described in Chap. 3) then the differencing is slightly more complex. For example, the second-order derivative of the velocity component u is given by

$$\frac{\partial^2 u}{\partial x^2} = \frac{\nu_{i+1/2,j}\left(u_{i+1,j} - u_{i,j}\right) - \nu_{i-1/2,j}\left(u_{i,j} - u_{i-1,j}\right)}{\Delta x^2} . \quad (14.19)$$

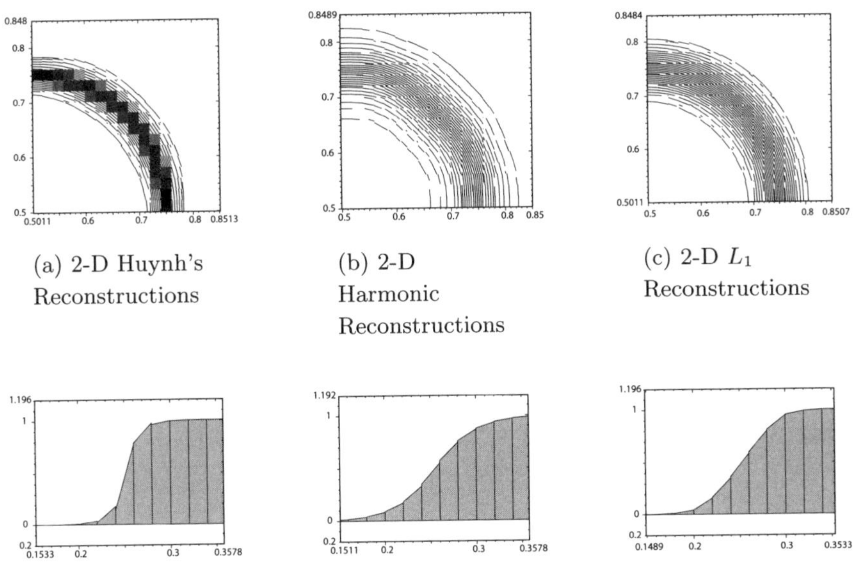

(a) 2-D Huynh's Reconstructions

(b) 2-D Harmonic Reconstructions

(c) 2-D L_1 Reconstructions

(d) 2-D Huynh's Front

(e) 2-D Harmonic Front

(f) 2-D L_1 Fromm's Front

Fig. 14.10. Advection of a circle by an unsplit advection scheme using various multidimensional slope reconstruction methods.

For variable density flows cross derivative terms are introduced, for example,

$$\frac{\partial^2 u}{\partial x \partial y} = \frac{u_{i+1,j+1} - u_{i-1,j+1} - u_{i+1,j-1} + u_{i-1,j-1}}{4\Delta x \Delta y} \, . \tag{14.20}$$

The formula can simply be evaluated for an explicit time integration. An implicit time integration will require that a linear system of equations be solved.

14.7 Stability

We will now show that the CTU algorithm presented earlier in this chapter has its stability limited by the material CFL condition. A similar result can be proven for Hancock's method with a suitable change in the definition of the CFL number. In one-dimension our model equation is

$$\frac{\partial U}{\partial t} + a\frac{\partial U}{\partial x} = D\frac{\partial^2 U}{\partial x^2} \, .$$

with $D \geq 0$. We conduct Fourier analysis on this problem using Fromm's and the predictor-corrector methods assuming that $a > 0$ and the method is

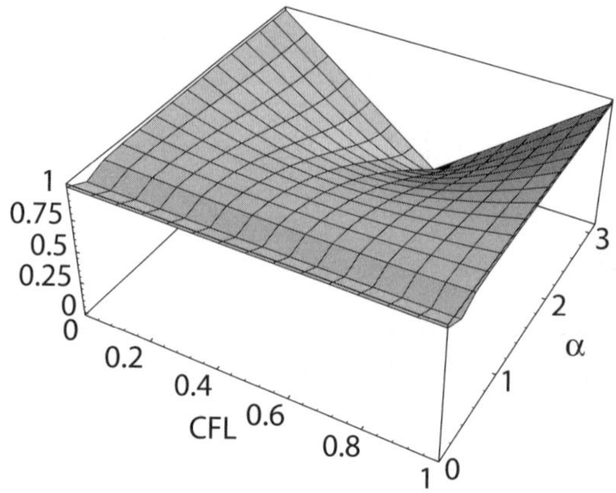

Fig. 14.11. The amplification factor is shown for a constant Fourier number of 100 with varying CFL number and angle.

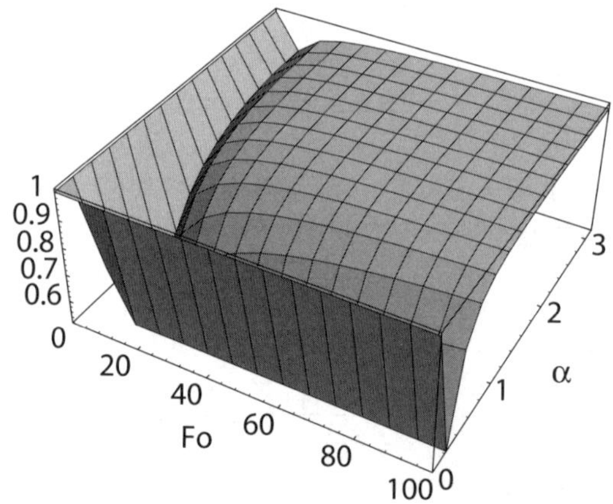

Fig. 14.12. The amplification factor is shown for a constant CFL number of 1 and varying Fourier number and angle.

implemented in the fashion described in this chapter. For a linear problem where Fourier analysis applies, the details of the upwind treatment simplify to the same algebraic form. The predictor will use an explicit expression for the viscous term and Fromm's scheme for advection. These are combined to advance the solution along with a Crank-Nicholson diffusion solution. Two dimensionless quantities easily appear in the analysis, i.e., the CFL number

$$\mathcal{C} = \frac{a\Delta t}{\Delta x} ,$$

and the Fourier number

$$Fo = \frac{D\Delta t}{\Delta x^2} \, .$$

The amplification factor is plotted for two cases: in Fig. 14.11 where the Fourier number is 100 and in Fig. 14.12 where the CFL number is 1. In both cases, the amplification factor is less than one indicating that the algorithm is at least linearly stable.

For the full multidimensional algorithm, stability is limited by the maximum velocity [117, 478],

$$\Delta t_{stab} = \min_{i,j,k} \left(\frac{\Delta x}{|u|}, \frac{\Delta y}{|v|} \right) \, .$$

For cases where source terms are present this is modified [50] to

$$\Delta t_{stab} = \min_{i,j,k} \left[\left(\frac{\min(\Delta x, \Delta y)}{|\nabla p - \mathbf{F}|} \right)^{1/2}, \frac{\Delta x}{|u|}, \frac{\Delta y}{|v|} \right] \, .$$

For the pressure form of the projection this can be further simplified to

$$\Delta t_{stab} = \min_{i,j,k} \left[\left(\frac{\min(\Delta x, \Delta y)}{|\mathbf{F}|} \right)^{1/2}, \frac{\Delta x}{|u|}, \frac{\Delta y}{|v|} \right] \, .$$

Each of these time steps is further modified by some safety factor, c to $\Delta t = c\Delta t_{stab}$ where $0 < c \leq 1$.

15. Centered High-Resolution Methods

Numerical schemes which do not involve the sign of the characteristic speeds in the discretization of the spatial derivatives can be classified as centered schemes. In contrast to the first-order upwind discretization, centered schemes make use of points from the left and the right of the center of the stencil (Fig. 15.1). Examples of centered schemes include the Lax-Friedrichs [322], Lax-Wendroff [321], Toro's first order centered scheme (FORCE) [544], and variants of second and third-order nonoscillatory schemes for hyperbolic conservation laws by Tadmor and collaborators [281, 279, 343, 357, 396].

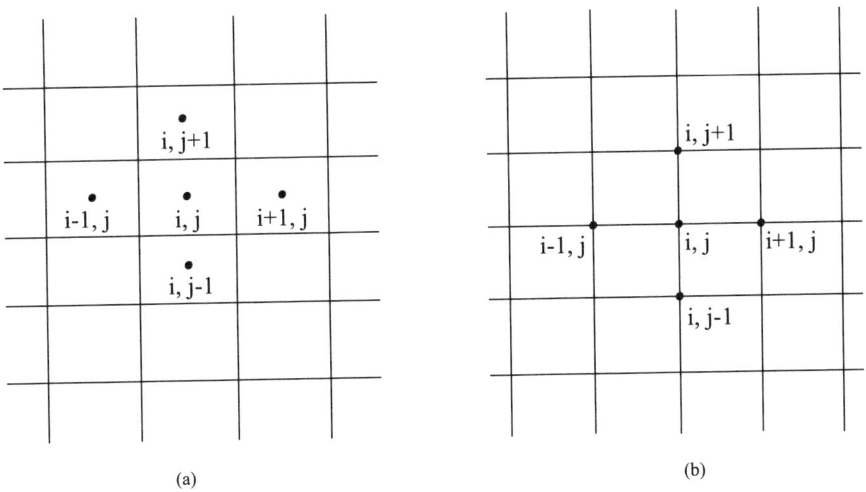

Fig. 15.1. Stencil for centered discretization based on (a) the cell centers and (b) grid vertices.

15.1 Lax-Friedrichs Scheme

We consider the one-dimensional, linear advection equation $U_t + \alpha U_x = 0$. Using centered discretization in space and first-order explicit discretization in time, we obtain

$$U_i^{n+1} = U_i^n - \frac{\alpha \Delta t}{2\Delta x} \left(U_{i+1}^n - U_{i-1}^n\right) . \tag{15.1}$$

Perfofming von Neumann stability analysis using the solution $U_i^n = E^n e^{\imath k i \Delta x}$, where E^n is the amplitude, k is the wave number[1], and $\imath = \sqrt{-1}$ is the unit complex number, we obtain

$$E^n = 1 - \imath \frac{\alpha \Delta t}{\Delta x} \sin(k\Delta x) . \tag{15.2}$$

The stability requirement is $||E^n|| \leq 1$. However,

$$||E^n|| = 1 + (\alpha \Delta t/\Delta x)^2 \sin^2(k\Delta x) \geq 1 ,$$

thus the discretization (15.1) is unconditionally unstable.

The Lax-Friedrichs scheme [322] aims to rectify the above problem by replacing U_i^n by $(U_{i-1}^n + U_{i+1}^n)/2$. The linear advection equation is written as

$$U_i^{n+1} = \frac{U_{i-1}^n + U_{i+1}^n}{2} - \frac{\alpha \Delta t}{2\Delta x} \left(U_{i+1}^n - U_{i-1}^n\right) . \tag{15.3}$$

Equation (15.3) can also be written

$$U_i^{n+1} = U_i^n - \frac{\alpha \Delta t}{2\Delta x} \left(U_{i+1}^n - U_{i-1}^n\right) + \frac{1}{2}\left(U_{i+1}^n - 2U_i^n + U_{i-1}^n\right) . \tag{15.4}$$

The last term in (15.4) can be considered as the discretization of the dissipative term $\nu_{num} \cdot u_{xx}$, where $\nu_{num} = \Delta x^2/2\Delta t$ is the numerical viscosity [256].

Let us consider now the nonlinear system of conservation laws

$$\frac{\partial \mathbf{U}}{\partial t} + \frac{\partial \mathbf{E}}{\partial x} = 0 , \tag{15.5}$$

where $\mathbf{E} \equiv \mathbf{E}(\mathbf{U})$. A conservative discretization of (15.5) can be obtained as

$$\mathbf{U}_i^{n+1} = \mathbf{U}_i^n - \frac{\Delta t}{\Delta x} \left(\mathbf{E}_{i+1/2}^* - \mathbf{E}_{i-1/2}^*\right) , \tag{15.6}$$

where $\mathbf{E}_{i-1/2}^*$ and $\mathbf{E}_{i+1/2}^*$ are numerical approximations of the fluxes $\mathbf{E}_{i-1/2}$ and $\mathbf{E}_{i+1/2}$, respectively. The Lax-Friedrichs numerical approximations of the fluxes can be defined by

$$\mathbf{E}_{i+1/2}^* = \frac{1}{2}(\mathbf{E}_i^n + \mathbf{E}_{i+1}^n) + \frac{1}{2}\frac{\Delta x}{\Delta t}(\mathbf{U}_i^n - \mathbf{U}_{i+1}^n) , \tag{15.7}$$

[1] The wave length is $\lambda = 2\pi/k$.

$$\mathbf{E}^*_{i-1/2} = \frac{1}{2}(\mathbf{E}^n_{i-1} + \mathbf{E}^n_i) + \frac{1}{2}\frac{\Delta x}{\Delta t}(\mathbf{U}^n_{i-1} - \mathbf{U}^n_i) \,. \tag{15.8}$$

The fluxes given by (15.7) and (15.8) are labeled as $\mathbf{E}^{\mathrm{LF}}_{i+1/2}$ and $\mathbf{E}^{\mathrm{LF}}_{i-1/2}$, respectively. Substitution of (15.7) and (15.8) into (15.6) yields

$$\mathbf{U}^{n+1}_i = \frac{1}{2}(\mathbf{U}^n_{i-1} + \mathbf{U}^n_{i+1}) + \frac{1}{2}\frac{\Delta t}{\Delta x}(\mathbf{E}^n_{i-1} - \mathbf{E}^n_{i+1}) \,. \tag{15.9}$$

Note that (15.9) derived for non-linear conservation laws is similar to (15.3) derived for the linear advection equation.

Toro [544] has shown that the Lax-Friedrichs scheme can also be seen as an integral average within a cell. For example, the variable U^{n+1}_i in the linear advection equation can be written

$$U^{n+1}_i = \frac{1}{\Delta x}\int_{x_{i-1/2}}^{x_{i+1/2}} \tilde{U}(x, \frac{1}{2}\Delta t)\, \mathrm{d}x \,, \tag{15.10}$$

where $\tilde{U}(x,t)$ is the solution of the Riemann problem

$$\tilde{U}(x/t) = \begin{cases} U^n_{i-1} \text{ if } x/t < \alpha \,, \\ U^n_{i+1} \text{ if } x/t > \alpha \,. \end{cases} \tag{15.11}$$

Similarly, for the system of conservation laws (15.5) the variable $\mathbf{U}^{n+1}_i$ can be written as

$$\mathbf{U}^{n+1}_i = \frac{1}{\Delta x}\int_{x_{i-1/2}}^{x_{i+1/2}} \tilde{\mathbf{U}}(x, \frac{1}{2}\Delta t)\mathrm{d}x \,, \tag{15.12}$$

where $\tilde{\mathbf{U}}(x,t)$ is the solution obtained by a Riemann solver.

For the linear advection equation the Lax-Friedrichs scheme (15.3) can also be written as [544]

$$U^{n+1}_i = \frac{(1+c)}{2}\, U^n_{i-1} + \frac{(1-c)}{2}\, U^n_{i+1} \,, \tag{15.13}$$

where $c = \alpha\Delta t/\Delta x$. Because the term U^n_{i-1} has the larger weight, the Lax-Friedrichs scheme could also be viewed as an upwind biased scheme.

The generalization of the Lax-Friedrichs scheme in two dimensions is straightforward. Consider the two-dimensional system of conservation laws

$$\frac{\partial \mathbf{U}}{\partial t} + \frac{\partial \mathbf{E}}{\partial x} + \frac{\partial \mathbf{F}}{\partial y} = 0 \,, \tag{15.14}$$

where $\mathbf{E} \equiv \mathbf{E}(\mathbf{U})$ and $\mathbf{F} \equiv \mathbf{F}(\mathbf{U})$. A conservative (explicit) discretization of (15.14) can be obtained as

$$\mathbf{U}_{i,j}^{n+1} = \mathbf{U}_{i,j}^{n} - \frac{\Delta t}{\Delta x}\Big(\mathbf{E}_{i+1/2,j}^{*} - \mathbf{E}_{i-1/2,j}^{*}\Big) - \frac{\Delta t}{\Delta y}\Big(\mathbf{F}_{i,j+1/2}^{*} - \mathbf{F}_{i,j-1/2}^{*}\Big)\,, \tag{15.15}$$

where $\mathbf{E}_{i-1/2,j}^{*}$, $\mathbf{E}_{i+1/2,j}^{*}$, $\mathbf{F}_{i,j-1/2}^{*}$, and $\mathbf{F}_{i,j+1/2}^{*}$ are numerical approximations of the intercell fluxes $\mathbf{E}_{i-1/2,j}$, $\mathbf{E}_{i+1/2,j}$, $\mathbf{F}_{i,j-1/2}$, and $\mathbf{F}_{i,j+1/2}$, respectively. The Lax-Friedrichs numerical approximations of the fluxes can be defined by

$$\mathbf{E}_{i+1/2,j}^{*} = \frac{1}{2}(\mathbf{E}_{i,j}^{n} + \mathbf{E}_{i+1,j}^{n}) + \frac{1}{2}\frac{\Delta x}{\Delta t}(\mathbf{U}_{i,j}^{n} - \mathbf{U}_{i+1,j}^{n})\,, \tag{15.16}$$

$$\mathbf{E}_{i-1/2,j}^{*} = \frac{1}{2}(\mathbf{E}_{i-1,j}^{n} + \mathbf{E}_{i,j}^{n}) + \frac{1}{2}\frac{\Delta x}{\Delta t}(\mathbf{U}_{i-1,j}^{n} - \mathbf{U}_{i,j}^{n})\,. \tag{15.17}$$

$$\mathbf{F}_{i,j+1/2}^{*} = \frac{1}{2}(\mathbf{F}_{i,j}^{n} + \mathbf{F}_{i,j+1}^{n}) + \frac{1}{2}\frac{\Delta y}{\Delta t}(\mathbf{U}_{i,j}^{n} - \mathbf{U}_{i,j+1}^{n})\,, \tag{15.18}$$

$$\mathbf{F}_{i,j-1/2}^{*} = \frac{1}{2}(\mathbf{F}_{i,j-1}^{n} + \mathbf{F}_{i,j}^{n}) + \frac{1}{2}\frac{\Delta y}{\Delta t}(\mathbf{U}_{i,j-1}^{n} - \mathbf{U}_{i,j}^{n})\,. \tag{15.19}$$

The computational stencil for the two-dimensional version of the Lax-Friedrich scheme is shown in Fig. 15.2.

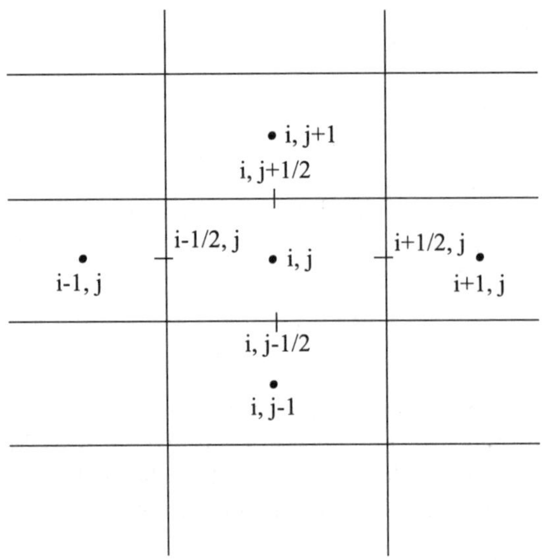

Fig. 15.2. Stencil for the two-dimensional version of the Lax-Friedrichs scheme.

The Lax-Friedrichs scheme given by (15.16)-(15.19) is very dissipative. An improved version of the scheme can be obtained by high order reconstruction of the variable $\mathbf{U}$ [163]. In this case, the Lax-Friedrichs intercell fluxes are defined by

$$\mathbf{E}^*_{i+1/2,j} = \frac{1}{2}\Big[\mathbf{E}^n(\mathbf{U}_{L,i+1/2}) + \mathbf{E}^n(\mathbf{U}_{R,i+1/2})\Big] + \frac{1}{2}\frac{\Delta x}{\Delta t}\left(\mathbf{U}^n_{L,i+1/2} - \mathbf{U}^n_{R,i+1/2}\right), \tag{15.20}$$

$$\mathbf{E}^*_{i-1/2,j} = \frac{1}{2}(\mathbf{E}^n(\mathbf{U}_{L,i-1/2}) + \mathbf{E}^n(\mathbf{U}_{R,i-1/2})) + \frac{1}{2}\frac{\Delta x}{\Delta t}(\mathbf{U}^n_{L,i-1/2} - \mathbf{U}^n_{R,i-1/2}), \tag{15.21}$$

$$\mathbf{F}^*_{i,j+1/2} = \frac{1}{2}\Big[\mathbf{F}^n(\mathbf{U}_{L,j+1/2}) + \mathbf{F}^n(\mathbf{U}_{R,j+1/2})\Big] + \frac{1}{2}\frac{\Delta y}{\Delta t}(\mathbf{U}^n_{L,j+1/2} - \mathbf{U}^n_{R,j+1/2}), \tag{15.22}$$

$$\mathbf{F}^*_{i,j-1/2} = \frac{1}{2}\Big[\mathbf{F}^n(\mathbf{U}_{L,j-1/2}) + \mathbf{F}^n(\mathbf{U}_{R,j-1/2})\Big] + \frac{1}{2}\frac{\Delta y}{\Delta t}(\mathbf{U}^n_{L,j-1/2} - \mathbf{U}^n_{R,j-1/2}). \tag{15.23}$$

The intercell variables $\mathbf{U}_L$ and $\mathbf{U}_R$ can be calculated by second or third order interpolation schemes (see, for example, Sect. 16.4.5). In Fig. 15.3 we compare the solutions obtained by (15.16)-(15.19) (first-order-based interpolation) and (15.20)-(15.23) (in conjunction with the MUSCL interpolation from Sect. 16.4.5) for the double mixing layer problem. The assumed physical/mathematical scenario is quite elementary: the evolution of a 2-D vortex street in a homogeneous incompressible fluid on a doubly periodic unit-square domain, described by the standard incompressible Navier-Stokes equations. The initial condition consists of a double shear layer

$$u = \begin{cases} \tanh((y-0.25)\delta) \text{ if } y \leq 0.5\,, \\ \tanh((0.75-y)\delta) \text{ if } y > 0.5\,, \end{cases} \tag{15.24}$$

where δ determines the shear layer thickness that is weakly perturbed in the spanwise direction through a sinusoidal perturbation of the spanwise velocity

$$v = v'\sin(2\pi x)\,, \tag{15.25}$$

where v' is the perturbation amplitude. As defined, the problem has the converged solution that takes the familiar form of a regular vortex street (see Fig. 15.3 second row of plots)—an apparent manifestation of the unstable wave-number-one mode. The results in Fig. 15.3 are plotted at dimensionless times $t = 0.6$ and $t = 1$, for computations performed using a 64×64 grid.

First order of accuracy for defining the intercell variables leads to very diffusive solutions as manifested by the thick shear layers on the upper and lower sides of the domain (Fig. 15.3 first row of plots). The excessive numerical diffusion does not allow the leg of the shear layer to form as it happens in the case when MUSCL interpolation is used for the left and right intercell variables.

Finally, we mention that McDonald [363] has presented a modified version of the Lax-Friedrichs scheme known as corrected viscosity scheme [257, 565].

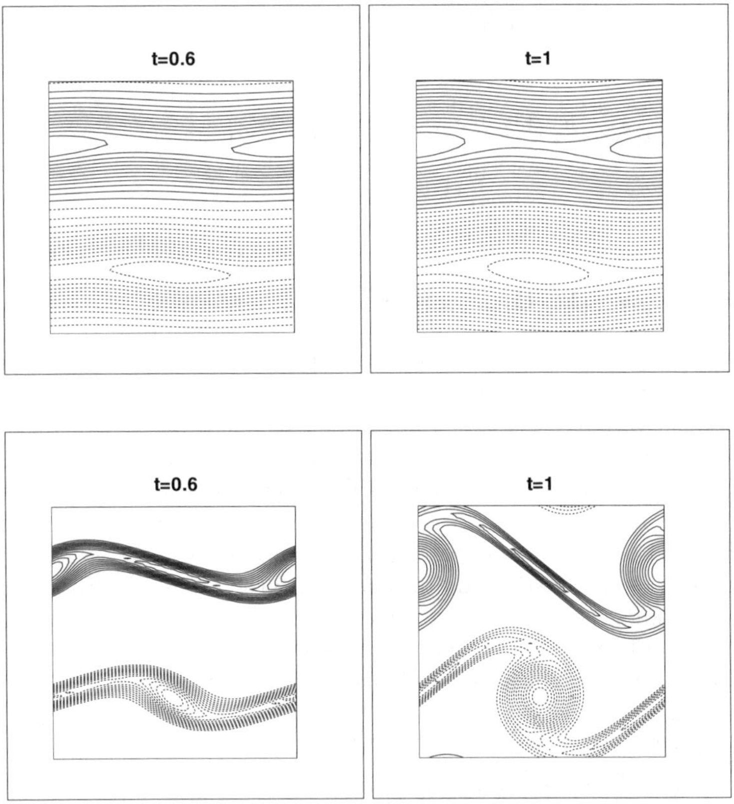

Fig. 15.3. Solutions for the double mixing layer problem as obtained using the Lax-Friedrichs scheme in conjunction with first-order (top) and second-order MUSCL (bottom) interpolation.

15.2 Lax-Wendroff Scheme

One of the most popular space-centered schemes is the second-order Lax-Wendroff scheme [321]. The original two-step Lax-Wendroff was presented in a report by Richtmyer in 1963 [446]. The scheme was introduced in Chap. 7 (and Chap. 9) and is extended here to 2/3-D. The scheme was proposed for solving hyperbolic conservation laws and has gained popularity due to its second-order of accuracy and simplicity. Variants of the schemes include the two-step procedure introduced by Richtmyer and Morton [447] also known as the Richtmyer scheme, and the MacCormack's predictor-corrector scheme [362]. A generalization of the MacCormack's scheme on the basis of a quasi-explicit extension has been proposed by Casier et al. [97], while a systematic investigation of variants of space-centered predictor-corrector two-step structure was presented by Lerat and Peyret [333]. A comprehensive review of the Lax-Wendroff family of schemes can be found in [257]. The basic formulation of the Lax-Wendroff scheme is presented below.

We consider the non-linear system of conservation laws (15.5) and develop $\mathbf{U}^{n+1}$ in a Taylor (time) series

$$\mathbf{U}^{n+1} = \mathbf{U}^n + \Delta t\, U_t + \frac{\Delta t^2}{2}\,\mathbf{U}_{tt} + \frac{\Delta t^3}{6}\,\mathbf{U}_{ttt}\,. \tag{15.26}$$

Replacing $\mathbf{U}_{tt}$ by the space derivative term

$$\frac{\partial^2 \mathbf{U}}{\partial t^2} = -\frac{\partial^2 \mathbf{E}}{\partial x \partial t}\,, \tag{15.27}$$

and introducing the Jacobian $\mathbf{A} = \partial \mathbf{E}/\partial \mathbf{U}$ we can write

$$\frac{\partial^2 \mathbf{U}}{\partial t^2} = -\frac{\partial}{\partial x}\left(\mathbf{A}\frac{\partial \mathbf{U}}{\partial t}\right) = \frac{\partial}{\partial x}\left(\mathbf{A}\frac{\partial \mathbf{E}}{\partial x}\right)\,. \tag{15.28}$$

Using (15.28) and discretizing the spatial derivatives with central differences around the grid point i, (15.26) can be written

$$\begin{aligned}\mathbf{U}_i^{n+1} &= \mathbf{U}_i^n - \frac{1}{2}\frac{\Delta t}{\Delta x}\left(\mathbf{E}_{i+1}^n - \mathbf{E}_{i-1}^n\right) \\ &+\frac{1}{2}\left(\frac{\Delta t}{\Delta x}\right)^2\left[\mathbf{A}_{i+1/2}^n(\mathbf{E}_{i+1}^n - \mathbf{E}_i^n) - \mathbf{A}_{i-1/2}^n(\mathbf{E}_i^n - \mathbf{E}_{i-1}^n)\right]\,,\end{aligned} \tag{15.29}$$

where the intercell Jacobian can be calculated as $\mathbf{A}_{i+1/2} = \mathbf{A}(\mathbf{U}_{i+1/2})$.

The conservative form of (15.29) is given by

$$\mathbf{U}_i^{n+1} = \mathbf{U}_i^n - \frac{\Delta t}{\Delta x}\left(\mathbf{E}_{i+1/2}^* - \mathbf{E}_{i-1/2}^*\right)\,, \tag{15.30}$$

where

$$\mathbf{E}^*_{i+1/2} = \mathbf{E}_{i+1/2} - \frac{1}{2}\frac{\Delta t}{\Delta x}\,\mathbf{A}_{i+1/2}\,(\mathbf{E}_{i+1} - \mathbf{E}_i)\,,$$
$$\mathbf{E}_{i+1/2} = \frac{\mathbf{E}_i + \mathbf{E}_{i+1}}{2}\,. \tag{15.31}$$

The formulation given by (15.30) and (15.31) requires the calculation of the Jacobian **A**. The latter can be calculated analytically or numerically. The numerical calculation of the Jacobian has been proposed by Roe [463] and Harten [247] in connection with the implementation of the Lax-Wendroff scheme for the compressible Euler equations. For example,

$$\mathbf{A}_{i+1/2} = \begin{cases} \dfrac{\mathbf{E}_{i+1} - \mathbf{E}i}{\mathbf{U}_{i+1} - \mathbf{U}_i} & \text{if } \mathbf{U}_{i+1} - \mathbf{U}_i \neq 0\,, \\ \\ \mathbf{A}(\mathbf{U}_i) & \text{if } \mathbf{U}_{i+1} = \mathbf{U}_i\,. \end{cases} \tag{15.32}$$

In multi-dimensional computations the evaluation of the Jacobian matrices can take significant part of the computing time. To avoid it, Richtmyer and Morton [447] and McCormack [362] proposed alternative formulations. Richtmyer and Morton introduced an intermediate state at $t = (n+1/2)\Delta t$

$$\mathbf{U}^{n+1/2}_{i+1/2} = \frac{1}{2}(\mathbf{U}^n_i + \mathbf{U}^n_{i+1}) - \frac{1}{2}\frac{\Delta t}{\Delta x}(\mathbf{E}^n_{i+1} - \mathbf{E}^n_i)\,. \tag{15.33}$$

The flux $\mathbf{E}^{\mathrm{RI}}_{i+1/2} = \mathbf{E}(\mathbf{U}^{n+1/2}_{i+1/2})$, where $\mathbf{U}^{n+1/2}_{i+1/2}$ is defined by (15.33) is designated as the Richtmyer flux. The flux $\mathbf{E}^{n+1/2}_{i+1/2}$ is used at the final state to compute $\mathbf{U}^{n+1}_i$

$$\mathbf{U}^{n+1}_i = \mathbf{U}^n_i - \frac{\Delta t}{\Delta x}\,(\mathbf{E}^{n+1/2}_{i+1/2} - \mathbf{E}^{n+1/2}_{i-1/2})\,. \tag{15.34}$$

Note that the intermediate state (15.33) is equivalent to the Lax-Friedrichs scheme at the intercell point $(i+1/2)$ between the times n and $n+1/2$, while the final state (15.34) is a leapfrog scheme applied at $n+1/2$ [256].

Another formulation has been proposed by MacCormack [362]. This is a two-step predictor-corrector scheme. The predictor step is defined by

$$\bar{\mathbf{U}}_i = \mathbf{U}^n_i - \frac{\Delta t}{\Delta x}\,(\mathbf{E}^n_{i+1} - \mathbf{E}^n_i)\,, \tag{15.35}$$

followed by a corrector step

$$\mathbf{U}^{n+1}_i = \frac{1}{2}(\mathbf{U}^n_i + \bar{\mathbf{U}}_i) - \frac{1}{2}\frac{\Delta t}{\Delta x}\,(\bar{\mathbf{E}}_i - \bar{\mathbf{E}}_{i-1})\,. \tag{15.36}$$

The fluxes $\bar{\mathbf{E}}_i$ and $\bar{\mathbf{E}}_{i-1}$ are evaluated using $\bar{\mathbf{U}}_i$ and $\bar{\mathbf{U}}_{i-1}$, respectively. Note that the predictor and corrector steps lead to unstable schemes, if considered separately, for positive and negative characteristic speeds, respectively. The predictor-corrector combined scheme, however, leads to a stable scheme due to the cancellations of the truncation errors.

Other formulations of the scheme are given by [257]

$$\bar{\mathbf{U}}_i = \mathbf{U}_i^n - \frac{\Delta t}{\Delta x}\left(\mathbf{E}_{i+1}^n - \mathbf{E}_i^n\right), \tag{15.37}$$

$$\bar{\bar{\mathbf{U}}}_i = \mathbf{U}_i^n - \frac{\Delta t}{\Delta x}\left(\bar{\mathbf{E}}_i - \bar{\mathbf{E}}_{i-1}\right), \tag{15.38}$$

$$\mathbf{U}_i^{n+1} = \frac{1}{2}(\bar{\mathbf{U}}_i + \bar{\bar{\mathbf{U}}}_i). \tag{15.39}$$

or, alternatively,

$$\bar{\mathbf{U}}_i = \mathbf{U}_i^n - \frac{\Delta t}{\Delta x}\left(\mathbf{E}_i^n - \mathbf{E}_{i-1}^n\right), \tag{15.40}$$

$$\bar{\bar{\mathbf{U}}}_i = \mathbf{U}_i^n - \frac{\Delta t}{\Delta x}\left(\bar{\mathbf{E}}_{i+1} - \bar{\mathbf{E}}_i\right), \tag{15.41}$$

$$\mathbf{U}_i^{n+1} = \frac{1}{2}(\bar{\mathbf{U}}_i + \bar{\bar{\mathbf{U}}}_i). \tag{15.42}$$

The formulations given by (15.35)-(15.36), (15.37)-(15.39) and (15.40)- (15.42) are identical for linear problems, but they lead to different results for non-linear problems.

The Richtmyer and MacCormack schemes can also be generalized to two and three dimensions. In two dimensions the Richtmyer scheme can be written as

$$\begin{aligned}\mathbf{U}_{i,j}^{n+1/2} &= \frac{1}{4}(\mathbf{U}_{i+1,j}^n + \mathbf{U}_{i-1,j}^n + \mathbf{U}_{i,j+1}^n + \mathbf{U}_{i,j-1}^n)\\ &\quad -\frac{1}{2}\frac{\Delta t}{\Delta x}\left(\mathbf{E}_{i+1,j}^n - \mathbf{E}_{i-1,j}^n\right)\\ &\quad -\frac{1}{2}\frac{\Delta t}{\Delta y}\left(\mathbf{F}_{i,j+1}^n - \mathbf{F}_{i,j-1}^n\right),\end{aligned} \tag{15.43}$$

$$\begin{aligned}\mathbf{U}_{i,j}^{n+1} &= \mathbf{U}_{i,j}^n - \frac{\Delta t}{\Delta x}\left(\mathbf{E}_{i+1,j}^{n+1/2} - \mathbf{E}_{i-1,j}^{n+1/2}\right)\\ &\quad -\frac{\Delta t}{\Delta y}\left(\mathbf{F}_{i,j+1}^{n+1/2} - \mathbf{F}_{i,j-1}^{n+1/2}\right).\end{aligned} \tag{15.44}$$

The computational stencil for the above scheme is shown in Fig. 15.4.

Another formulation has been proposed by Zwas [623]

$$\begin{aligned}\mathbf{U}_{i+1/2,j+1/2}^{n+1/2} &= \frac{1}{4}\left(\mathbf{U}_{i+1,j+1}^n + \mathbf{U}_{i+1,j}^n + \mathbf{U}_{i,j+1}^n + \mathbf{U}_{i,j}^n\right)\\ &\quad -\frac{1}{2}\frac{\Delta t}{\Delta x}\left(\mathbf{E}_{i+1,j+1/2}^n - \mathbf{E}_{i,j+1/2}^n\right)\end{aligned}$$

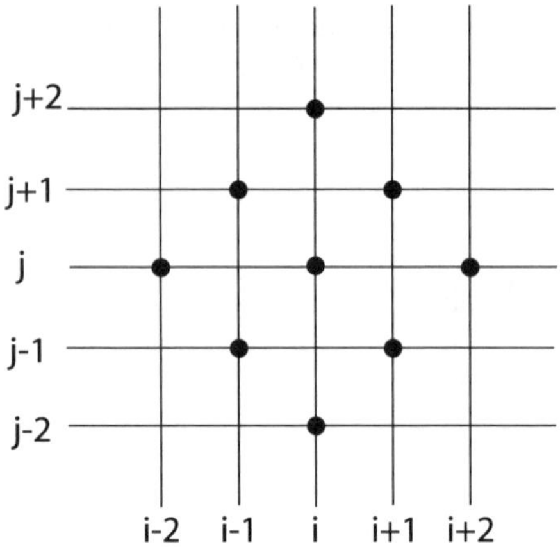

Fig. 15.4. Computational stencil for the Richtmyer's scheme.

$$-\frac{1}{2}\frac{\Delta t}{\Delta y}\left(\mathbf{F}^n_{i+1/2,j+1}-\mathbf{F}^n_{i+1/2,j}\right), \tag{15.45}$$

$$\begin{aligned}\mathbf{U}^{n+1}_{i,j} &= \mathbf{U}^n_{i,j}-\frac{\Delta t}{\Delta x}\left(\mathbf{E}^{n+1/2}_{i+1/2,j}-\mathbf{E}^{n+1/2}_{i-1/2,j}\right)\\ &-\frac{\Delta t}{\Delta y}\left(\mathbf{F}^{n+1/2}_{i,j+1/2}-\mathbf{F}^{n+1/2}_{i,j-1/2}\right),\end{aligned} \tag{15.46}$$

where

$$\mathbf{E}^n_{i+1,j+1/2}=\mathbf{E}\Bigl(\frac{\mathbf{U}^n_{i+1,j+1}+\mathbf{U}^n_{i+1,j}}{2}\Bigr), \tag{15.47}$$

$$\mathbf{E}^{n+1/2}_{i+1/2,j}=\mathbf{E}\Bigl(\frac{\mathbf{U}^{n+1/2}_{i+1/2,j+1/2}+\mathbf{U}^{n+1/2}_{i+1/2,j-1/2}}{2}\Bigr), \tag{15.48}$$

$$\mathbf{F}^n_{i+1/2,j+1}=\mathbf{F}\Bigl(\frac{\mathbf{U}^n_{i+1,j+1}+\mathbf{U}^n_{i,j+1}}{2}\Bigr), \tag{15.49}$$

$$\mathbf{F}^{n+1/2}_{i+1/2,j}=\mathbf{F}\Bigl(\frac{\mathbf{U}^{n+1/2}_{i,j}+\mathbf{U}^{n+1/2}_{i+1,j}}{2}\Bigr), \tag{15.50}$$

and the computational stencil is shown in Fig. 15.5.

For the two step MacCormack's scheme in two dimensions there are four different variants depending how the forward and backward differences are combined in the predictor and corrector steps, for example, one can choose forward or backward differences in both the predictor and corrector steps or forward differences at one step and backward at another. The forward-backward version of the scheme in two dimensions is given by

$$\bar{\mathbf{U}}_{i,j}=\mathbf{U}^n_{i,j}-\frac{\Delta t}{\Delta x}\left(\mathbf{E}^n_{i+1,j}-\mathbf{E}^n_{i,j}\right)-\frac{\Delta t}{\Delta y}\left(\mathbf{F}^n_{i,j+1}-\mathbf{F}^n_{i,j}\right), \tag{15.51}$$

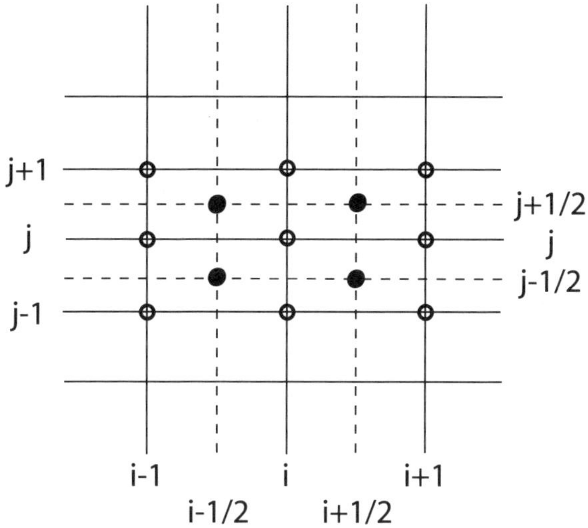

Fig. 15.5. Computational stencil for the Zwas's scheme.

$$\bar{\bar{\mathbf{U}}}_{i,j} = \mathbf{U}^n_{i,j} - \frac{\Delta t}{\Delta x}\left(\bar{\mathbf{E}}_{i,j} - \bar{\mathbf{E}}_{i-1,j}\right) - \frac{\Delta t}{\Delta y}\left(\bar{\mathbf{F}}_{i,j} - \bar{\mathbf{F}}_{i,j-1}\right) , \tag{15.52}$$

$$\mathbf{U}^{n+1}_{i,j} = \frac{1}{2}(\bar{\mathbf{U}}_{i,j} + \bar{\bar{\mathbf{U}}}_{i,j}) . \tag{15.53}$$

In the forward-forward version (15.52) is replaced by

$$\bar{\bar{\mathbf{U}}}_{i,j} = \mathbf{U}^n_{i,j} - \frac{\Delta t}{\Delta x}(\bar{\mathbf{E}}_{i+1,j} - \bar{\mathbf{E}}_{i,j}) - \frac{\Delta t}{\Delta y}(\bar{\mathbf{F}}_{i,j+1} - \bar{\mathbf{F}}_{i,j}) . \tag{15.54}$$

Other variants of the Lax-Wendroff family of schemes are discussed below:

1. Lerat and Peyret [332, 333] have defined a family of schemes on the basis of the predictor-corrector, space-centered discretization. The numerical flux can be written [257]

$$\mathbf{E}^*_{i+1/2} = \frac{1}{2}(\mathbf{E}_{i+1} + \mathbf{E}_i) - \frac{1}{2\alpha}\Big[\beta\mathbf{E}_{i+1} + (1-\beta)\mathbf{E}_i\Big] + \frac{1}{2\alpha}\bar{\mathbf{E}}_i\ , \tag{15.55}$$

where $\bar{\mathbf{E}}$ is defined as

$$\bar{\mathbf{E}} = \mathbf{E}(\mathbf{U}^{n+\alpha}_{i+\beta}) .$$

Equation (15.55) defines a family of schemes, called S^β_α, for different values of the coefficients α and β. Specifically, we obtain

- For $\alpha = 1$ and $\beta = 0$ the MacCormack's scheme (15.37)-(15.39);
- for $\alpha = 1$ and $\beta = 1$ the version (15.40)-(15.42) of the MacCormack's scheme;

- for $\alpha = \beta = 1/2$ the Richtmyer version of the Lax-Wendroff scheme;
- for $\beta = 1/2$ the family of α-schemes as considered by McGuire and Morris [380];
- for $\alpha = 1$ and $\beta = 1/2$ the version proposed by Rubin and Burstein [471];
- for $\beta = 0$ or $\beta = 1$ another family of α-schemes as proposed by Warming et al. [593].

2. All versions of the Lax-Wendroff schemes can be consolidated in a general form as follows [321]

$$\mathbf{E}^*_{i+1/2} = \frac{1}{2}(\mathbf{E}_{i+1} + \mathbf{E}_i) + \frac{1}{2}\frac{\Delta t}{\Delta x}\,\mathbf{A}_{i+1/2} \\ (\mathbf{E}_{i+1} - \mathbf{E}_i) - \mathcal{D}\,(\mathbf{U}_{i+1} - \mathbf{U}_i)\,, \tag{15.56}$$

where $\mathcal{D}$ is a positive function that plays the role of artificial dissipation[2] in order to control spurious oscillations. The numerical flux (15.56) is essentially the original Lax-Wendroff flux (15.31) modified by the addition of numerical dissipation. The artificial dissipation term can also be written as

$$\Delta x\,\mathcal{D}\,\frac{\partial \mathbf{U}}{\partial x}\,.$$

Note that the last term is similar to the viscous terms appeared in the Navier-Stokes equations.

15.3 First-Order Centered Scheme

The first-order centered scheme (FORCE) was developed by Toro [541, 542] for systems of hyperbolic conservation laws. The scheme is based on a reinterpretation of the random choice method (RCM) [214] on a staggered grid. Toro developed and implemented the scheme for the compressible Euler equations. More recently Drikakis and Smolarkiewicz [163] presented the implementation of the scheme in incompressible flows, in conjunction with the artificial compressibility approach, and conducted several numerical experiments for the double mixing layer problem using variants of the schemes. In the following paragraphs we provide an introduction to the RCM, followed by the description of the FORCE scheme as developed by Toro [541, 542] and implemented in incompressible flows by Drikakis and Smolarkiewicz [163].

[2] Lax and Wendroff [321] call it artificial viscosity.

15.3.1 Random Choice Method

The RCM was introduced by Glimm [214] as a proof of existence of solutions to a class of nonlinear systems of hyperbolic conservation laws and was further implemented by Chorin [107] to solve the compressible Euler equations. An excellent review of the method is given by Toro [544]. The implementation of RCM requires the solutions of local Riemann problems and random sampling of these solutions within a cell in order to assign a state to the next time level.

The starting point is the nonlinear system of hyperbolic conservation laws (15.5). Assuming a piecewise constant distribution of the data, for example, by defining cell averages, $\bar{\mathbf{U}}$, the solution consists of two steps:

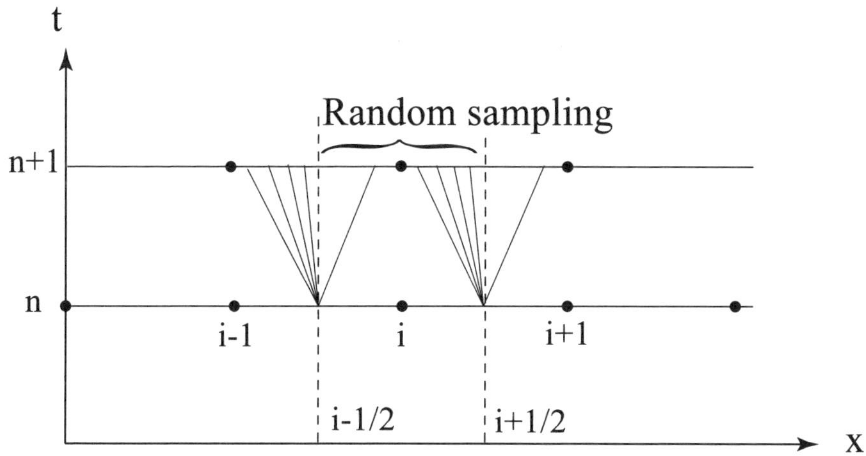

Fig. 15.6. The Random Choice Method on a non-staggered grid.

- The first step is the solution of Riemann problems for pairs of neighboring states around the cell i (Fig. 15.6); according to the stencil of this figure we define the non-staggered version of the RCM. Using the pairs $\bar{\mathbf{U}}_{i-1}$, $\bar{\mathbf{U}}_i$, and $\bar{\mathbf{U}}_i$, $\bar{\mathbf{U}}_{i+1}$ we solve Riemann problems to find the solutions for $\mathbf{U}_{i-1/2}(x/t)$ and $\mathbf{U}_{i+1/2}(x/t)$, respectively.
- The second step is to pick up a state and asssign it to a cell. This step depends on a random number δ^n in the interval $[0,1]$. The solution at $n+1$ for each cell is given by

$$\mathbf{U}_i^{n+1} = \begin{cases} \mathbf{U}_{i-1/2}(\delta^n \Delta x/\Delta t)\ , & \text{if} \quad 0 \le \delta^n \le 1/2\ , \\ \\ \mathbf{U}_{i+1/2}\Big[(\delta^n - 1)\Delta x/\Delta t\Big]\ , & \text{if} \quad 1/2 \le \delta^n \le 1\ . \end{cases} \tag{15.57}$$

The choice of the random numbers is crucial for the accuracy of the solution. Chorin [107] proposed that one random number is sufficient for each time level. Colella [115] proposed pseudo-random numbers based on a van der Corput sequence [239] as follows

$$\delta^n = \sum_{i=0}^{m} A_i k_1^{-(i+1)} , \tag{15.58}$$

$$A_i = \mathrm{mod}(k_2 a_i, k_1) , \tag{15.59}$$

$$n = \sum_{i=0}^{m} a_i k_1^i , \tag{15.60}$$

where a_i are coefficients, for example, taking values 0 or 1.

On a staggered grid (Fig. 15.7), the RCM also consists of two steps

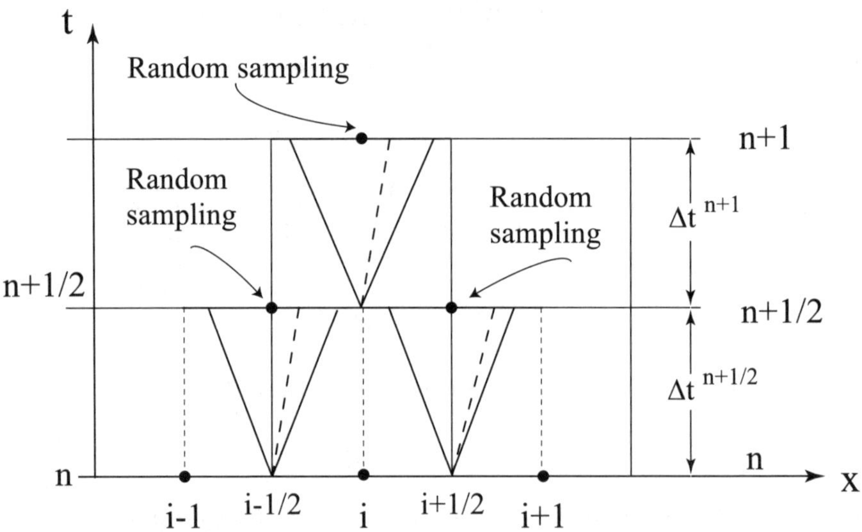

Fig. 15.7. The Random Choice Method on a staggered grid.

- Using the pairs $\bar{\mathbf{U}}_{i-1}$, $\bar{\mathbf{U}}_i$, and $\bar{\mathbf{U}}_i$, $\bar{\mathbf{U}}_{i+1}$ we solve Riemann problems to find the solutions for $\tilde{\mathbf{U}}^{n+1/2}_{i-1/2}(x,t)$ and $\tilde{\mathbf{U}}^{n+1/2}_{i+1/2}(x,t)$, respectively. Then, we random sample the above solutions at $\Delta t^{n+1/2}$

$$\mathbf{U}^{n+1/2}_{i-1/2}(x,t) = \tilde{\mathbf{U}}^{n+1/2}_{i-1/2}(\delta^n \Delta x, \Delta t^{n+1/2}) , \tag{15.61}$$

and

$$\mathbf{U}^{n+1/2}_{i+1/2}(x,t) = \tilde{\mathbf{U}}^{n+1/2}_{i+1/2}(\delta^n \Delta x, \Delta t^{n+1/2}) . \tag{15.62}$$

- In the second step we solve a Riemann problem for $\mathbf{U}_{i-1/2}^{n+1/2}$ and $\mathbf{U}_{i+1/2}^{n+1/2}$ to find a solution $\tilde{\mathbf{U}}_i^{n+1}(x,t)$ and subsequently random sample it at a time Δt^{n+1} to obtain

$$\mathbf{U}_i^{n+1}(x,t) = \tilde{\mathbf{U}}_i^{n+1}(\delta^{n+1}\Delta x, \Delta t^{n+1}) \,. \tag{15.63}$$

15.3.2 FORCE

According to the FORCE scheme [541, 542, 544], (15.61) and (15.62) are replaced by deterministic integrals. Assuming

$$\Delta t^{n+1/2} = \Delta t^{n+1} = \frac{1}{2}\Delta t$$

the deterministic integrals are written

$$\mathbf{U}_{i-1/2}^{n+1/2}(x,t) = \frac{1}{\Delta x}\int_{-\frac{1}{2}\Delta x}^{\frac{1}{2}\Delta x} \tilde{\mathbf{U}}_{i-1/2}^{n+1/2}(x, \frac{\Delta t}{2})\,\mathrm{d}x \,, \tag{15.64}$$

$$\mathbf{U}_{i+1/2}^{n+1/2}(x,t) = \frac{1}{\Delta x}\int_{-\frac{1}{2}\Delta x}^{\frac{1}{2}\Delta x} \tilde{\mathbf{U}}_{i+1/2}^{n+1/2}(x, \frac{\Delta t}{2})\,\mathrm{d}x \,. \tag{15.65}$$

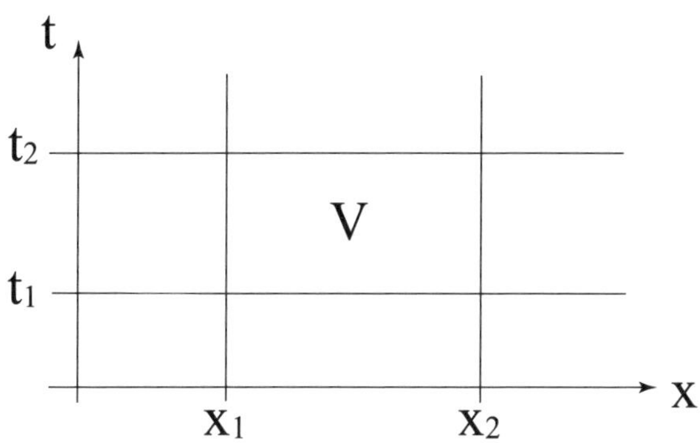

Fig. 15.8. Control volume definition in space-time coordinates.

The integrals (15.64) and (15.65) can be calculated by applying the integral form of the conservation laws [544]. For a control volume defined by the coordinates $[x_1, x_2] \times [t_1, t_2]$ in a $x - t$ plane (Fig. 15.8), the integral form of conservation laws can be written as

$$\frac{\mathrm{d}}{\mathrm{d}t}\int_{x_1}^{x_2} \mathbf{U}(x,t)\,\mathrm{d}x = \mathbf{E}(\mathbf{U}(x_1,t)) - \mathbf{E}(\mathbf{U}(x_2,t)) \,. \tag{15.66}$$

An alternative integral form can be obtained by integrating (15.66) in time from t_1 to t_2. This yields

$$\int_{x_1}^{x_2} \mathbf{U}(x,t_2)\,\mathrm{d}x = \int_{x_1}^{x_2} \mathbf{U}(x,t_1)\,\mathrm{d}x + \int_{t_1}^{t_2} \mathbf{E}(\mathbf{U}(x_1,t))\,\mathrm{d}t - \int_{t_1}^{t_2} \mathbf{E}(\mathbf{U}(x_2,t))\,\mathrm{d}t\,. \quad (15.67)$$

Applying (15.67) to (15.64) and (15.65), we obtain [544]

$$\mathbf{U}_{i-1/2}^{n+1/2} = \frac{1}{2}(\mathbf{U}_{i-1}^n + \mathbf{U}_i^n) + \frac{\Delta t}{2\Delta x}\left(\mathbf{E}_{i-1}^n - \mathbf{E}_i^n\right), \quad (15.68)$$

$$\mathbf{U}_{i+1/2}^{n+1/2} = \frac{1}{2}(\mathbf{U}_{i}^n + \mathbf{U}_{i+1}^n) + \frac{\Delta t}{2\Delta x}\left(\mathbf{E}_{i}^n - \mathbf{E}_{i+1}^n\right). \quad (15.69)$$

The values $\mathbf{U}_{i-1/2}^{n+1/2}$ and $\mathbf{U}_{i+1/2}^{n+1/2}$ are used to solve a Riemann problem and find a solution $\tilde{\mathbf{U}}_i^{n+1}$ which is integrated within the cell to calculate $\mathbf{U}_i^{n+1}$ as follows

$$\mathbf{U}_i^{n+1} = \frac{1}{\Delta x}\int_{-\frac{1}{2}\Delta x}^{\frac{1}{2}\Delta x} \tilde{\mathbf{U}}_i(x, \frac{1}{2}\Delta t)\,\mathrm{d}x\,. \quad (15.70)$$

Applying (15.67) to (15.70) yields

$$\mathbf{U}_i^{n+1} = \frac{1}{2}(\mathbf{U}_{i-1/2}^{n+1/2} + \mathbf{U}_{i+1/2}^{n+1/2}) - \frac{\Delta t}{2\Delta x}\left(\mathbf{E}_{i+1/2}^{n+1/2} - \mathbf{E}_{i-1/2}^{n+1/2}\right). \quad (15.71)$$

where $\mathbf{E}_{i+1/2}^{n+1/2}$ is calculated according to the Richtmyer flux $\mathbf{E}_{i+1/2}^{\mathrm{RI}} = \mathbf{E}(\mathbf{U}_{i+1/2}^{n+1/2})$, where $\mathbf{U}_{i+1/2}^{n+1/2}$ is defined by (15.33).

The above are equivalent with the conservation form (15.6) if the numerical flux is defined by

$$\mathbf{E}_{i+1/2}^{*} = \frac{1}{2}\left[\mathbf{E}_{i+1/2}^{n+1/2} + \frac{1}{2}(\mathbf{E}_i^n + \mathbf{E}_{i+1}^n)\right] + \frac{1}{4}\frac{\Delta x}{\Delta t}\left(\mathbf{U}_i^n - \mathbf{U}_{i+1}^n\right). \quad (15.72)$$

As shown by Toro [544] the FORCE flux (15.72) is the arithmetic mean of the Richtmyer (15.33) and Lax-Friedrichs (15.7) schemes, that is

$$\mathbf{E}_{i+1/2}^{\mathrm{force}} = \frac{1}{2}\left(\mathbf{E}_{i+1/2}^{\mathrm{LF}} + \mathbf{E}_{i+1/2}^{\mathrm{RI}}\right) \quad (15.73)$$

Toro [544] and Billet [59] have shown that for the linear advection equation $U_t + \alpha U_x = 0$ the scheme is monotone and stable with stability condition

$$0 \le |\mathcal{C}| \le 1\,, \quad (15.74)$$

where

$$\mathcal{C} = \frac{\Delta t \alpha}{\Delta x}\,. \quad (15.75)$$

15.3.3 Variants of the FORCE Scheme

The FORCE scheme has been implemented by Drikakis and Smolarkiewicz [163] to solve the incompressible Navier-Stokes equations. In [163] the artificial-compressibility formulation was employed to couple the continuity and momentum equations in a pseudotime, τ. The dual-time stepping in conjunction with multigrid acceleration was used to iterate the solution in real time, t. In the context of the pseudo-compressible system, $n + 1/2$ in (15.68), (15.69), (15.71) and (15.72) denote an intermediate pseudotime level between τ and $\tau + \Delta\tau$.

Eq. (15.73) offers the possibility to experiment with different variants of the Lax-Friedrichs and Richtmyer fluxes thus obtaining different versions of the FORCE scheme. For the Lax-Friedrichs flux, $\mathbf{E}^{\mathrm{LF}}_{i+1/2}$ (15.7), we aforementioned that the variables $\mathbf{U}_i$ and $\mathbf{U}_{i+1}$ can be replaced by $\mathbf{U}_L$ and $\mathbf{U}_R$, respectively, where $\mathbf{U}_L$ and $\mathbf{U}_R$ can be calculated by a second or higher order extrapolation schemes. Similar considerations can be made for the Richtmyer flux (15.33).

In [163] experiments were conducted using two different interpolators: the "third-order" Lagrangian interpolator[3]

$$\mathbf{U}_L = \frac{1}{6}(5\mathbf{U}_i - \mathbf{U}_{i-1} + 2\mathbf{U}_{i+1}) \; , \tag{15.76}$$

$$\mathbf{U}_R = \frac{1}{6}(5\mathbf{U}_{i+1} - \mathbf{U}_{i+2} + 2\mathbf{U}_i) \; , \tag{15.77}$$

and the MUSCL scheme [571] (see Sect. 16.4.5).

The different variants of the FORCE scheme are obtained if the MUSCL or "third-order" interpolators are implemented in both the $\mathbf{E}^{\mathrm{RI}}$ and $\mathbf{E}^{\mathrm{LF}}$ fluxes, or in only one of them. Numerical experiments have revealed that the first term (at least), on the right hand side of (15.7), of the flux $\mathbf{E}^{\mathrm{LF}}$ in (15.73) must be calculated by higher-order interpolation, otherwise the solutions become overly diffusive (Figs. 15.9 and 15.10). In [163] numerical experiments were conducted with and without high-order interpolation in the flux $\mathbf{E}^{\mathrm{RI}}$. All FORCE schemes that use higher-order interpolation in the flux $\mathbf{E}^{\mathrm{RI}}$ in (15.73) are (slightly) less diffusive (Fig. 15.11) than the equivalent schemes with the first-order interpolation employed for $\mathbf{E}^{\mathrm{RI}}$ (Fig. 15.12).

[3] Note that the interpolation in (15.76) and (15.77) is not third-order-accurate per se, but it assures third-order accuracy of the wave-speed dependent term $(\mathbf{U}_R - \mathbf{U}_L)$, [178, 177].

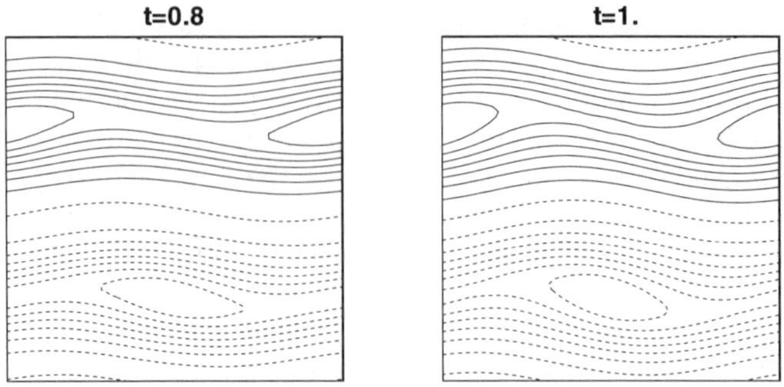

Fig. 15.9. Variant of the FORCE scheme using first-order interpolation for $\mathbf{U}$ in the fluxes $\mathbf{E}^{\mathrm{LF}}$ and $\mathbf{E}^{\mathrm{RI}}$.

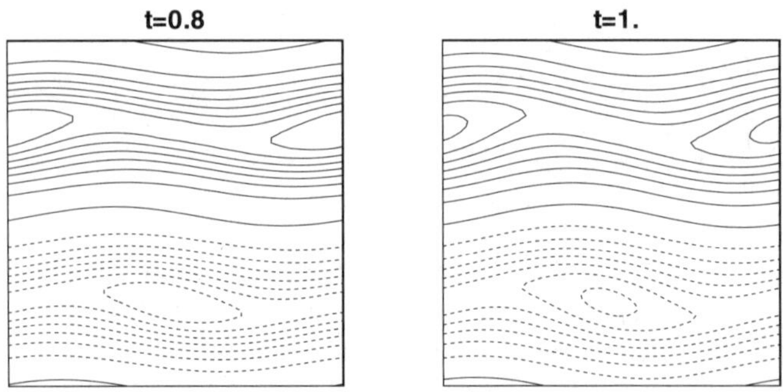

Fig. 15.10. Variant of the FORCE scheme using first-order interpolation for $\mathbf{U}$ in the flux $\mathbf{E}^{\mathrm{LF}}$ and the interpolator (15.76) and (15.77) in the flux $\mathbf{E}^{\mathrm{RI}}$.

15.4 Second- and Third-Order Centered Schemes

15.4.1 Nessyahu-Tadmor Second-Order Scheme

Nessyahu and Tadmor [396] developed second-order centered schemes for conservation laws as extensions of the Lax-Friedrichs scheme. The Lax-Friedrichs scheme is simpler than the Godunov scheme because it does not require the solution of a Riemann problem since it integrates over the entire Riemann fan. Essentially, it can be seen as projection of successive non-interacting

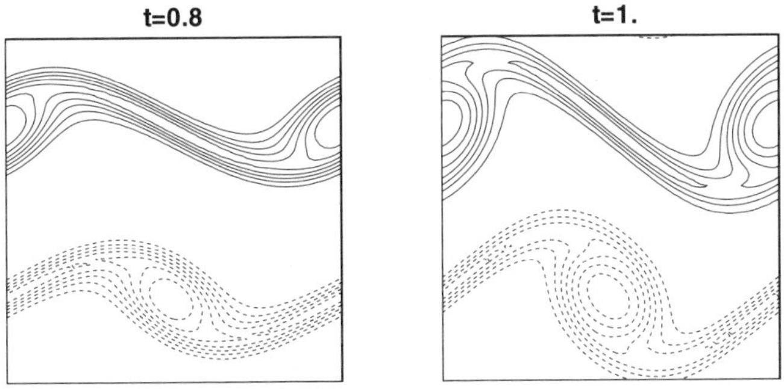

Fig. 15.11. Variant of the FORCE scheme using the interpolator (15.76) and (15.77) in the fluxes $\mathbf{E}^{\mathrm{LF}}$ and $\mathbf{E}^{\mathrm{RI}}$.

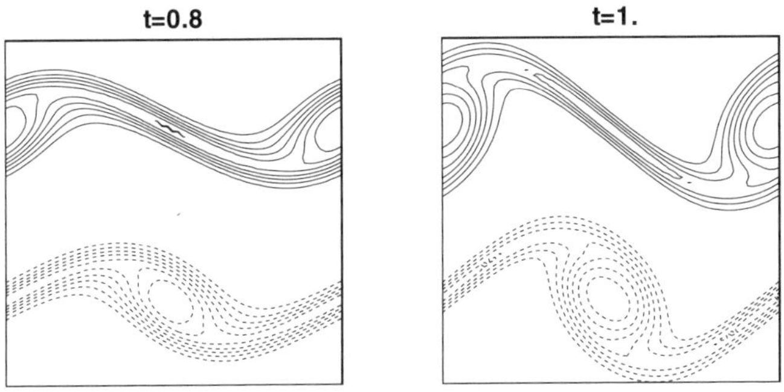

Fig. 15.12. Variant of the FORCE scheme using the interpolator (15.76) and (15.77) in the flux $\mathbf{E}^{\mathrm{LF}}$ and first-order interpolation in $\mathbf{E}^{\mathrm{RI}}$.

Riemann problems integrated over a staggered grid (Fig. 15.13) [396]. The Lax-Friedrichs scheme, however, encompasses significant numerical viscosity as we already demonstrated in Sect. 15.1 for the double mixing layer flow problem.

In [396], it was proposed to overcome the problem of the excessive numerical dissipation of the Lax-Friedrichs scheme by using high-resolution MUSCL interpolation instead of the first-order piecewise constant ones. For the one-dimensional system of conservation laws (15.5)

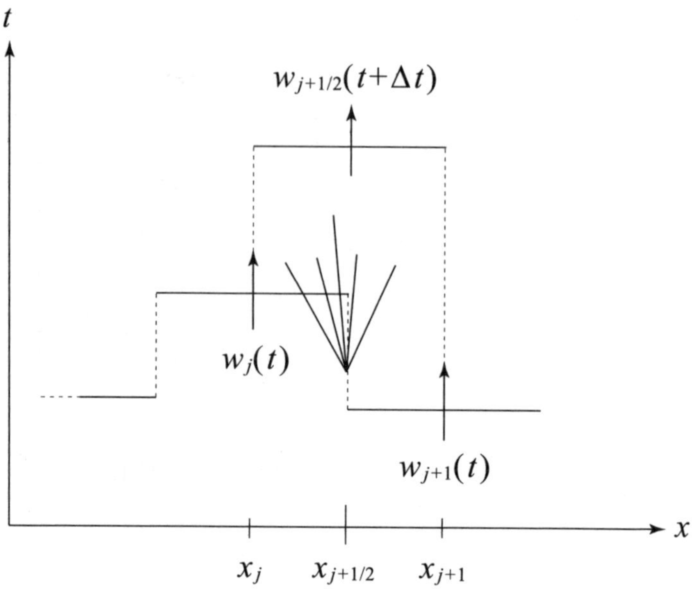

Fig. 15.13. Interpretation of the Lax-Friedrichs scheme as a piecewise constant projection of successive non-interacting Riemann problems integrated over a staggered grid, where w is a generic variable.

$$\frac{\partial \mathbf{U}}{\partial t} + \frac{\partial \mathbf{E}}{\partial x} = 0 , \tag{15.78}$$

an approximate solution $\mathbf{U}^{n+1}$ can be found by

$$\mathbf{U}^n_{i+1/2} = \frac{1}{2}\Big(\mathbf{U}^n_i + \mathbf{U}^n_{i+1}\Big) - \frac{\Delta t}{\Delta x}\Big(\mathbf{E}_{i+1} - \mathbf{E}_i\Big) . \tag{15.79}$$

The modified numerical flux $\mathbf{E}_i$ is given by [396]

$$\mathbf{E}_i = \mathbf{E}\Big(\mathbf{U}^{n+1/2}_i\Big) + \frac{\Delta x}{8\Delta t}\,\mathbf{U}^x_i , \tag{15.80}$$

where

$$\mathbf{U}^{n+1/2}_i = \mathbf{U}^n_i - \frac{1}{2}\frac{\Delta t}{\Delta x}\,\mathbf{E}^x_i . \tag{15.81}$$

The calculation of the numerical derivatives $\mathbf{U}^x_i$ and $\mathbf{E}^x_i$ takes place as follows [396]. One can use the minmod (,) limiter which takes the form

$$\text{minmod}\,(a,b) = \frac{1}{2}[\text{sign}(a) + \text{sign}(b)]\,\min(|a|,|b|) . \tag{15.82}$$

Using (15.82) the numerical derivatives can be calculated by

$$\mathbf{U}_i^x = \text{minmod}\left(\Delta\mathbf{U}_{i+1/2,k}, \Delta\mathbf{U}_{i-1/2,k}\right) , \tag{15.83}$$

or

$$\mathbf{U}_i^x = \text{minmod}\Big(\alpha\Delta\mathbf{U}_{i+1/2,k}, \frac{\mathbf{U}_{i+1,k} - \mathbf{U}_{i-1,k}}{2}, \alpha\Delta\mathbf{U}_{i-1/2,k}\Big) , \tag{15.84}$$

where α is a constant and $k = 1, ..., N$ is the dimension of the vector $\mathbf{U}$. Alternatively,

$$\begin{aligned}\mathbf{U}_i^x = \text{minmod}\Big(&\Delta\mathbf{U}_{i-1/2,k} + \frac{1}{2}\text{minmod}(\Delta^2\mathbf{U}_{i-1,k}, \Delta^2\mathbf{U}_{i,k}),\\ &\Delta\mathbf{U}_{i+1/2,k} - \frac{1}{2}\text{minmod}(\Delta^2\mathbf{U}_{i,k}, \Delta^2\mathbf{U}_{i+1,k})\Big) .\end{aligned} \tag{15.85}$$

The differences $\Delta\mathbf{U}_{i+1/2}$ are defined by $\Delta\mathbf{U}_{i+1/2,k} = \mathbf{U}_{i+1,k} - \mathbf{U}_{i,k}$.

For the numerical derivative of the flux, $\mathbf{E}_i^x$, one can use the Jacobian matrix $\mathbf{A} = \partial\mathbf{E}/\partial\mathbf{U}$ and define

$$\mathbf{E}_i^x = \mathbf{A}\mathbf{U}_i^x . \tag{15.86}$$

The estimation of the Jacobian matrix is time consuming, especially in multi-dimensional computations. Alternatively, one can define $\mathbf{E}_i^x$ by

$$\mathbf{E}_i^x = \text{minmod}(\Delta\mathbf{E}_{i+1/2,k}, \Delta\mathbf{E}_{i-1/2,k}) , \tag{15.87}$$

or

$$\mathbf{E}_i^x = \text{minmod}\Big(\alpha\Delta\mathbf{E}_{i+1/2,k}, \frac{\mathbf{E}_{i+1,k} - \mathbf{E}_{i-1,k}}{2}, \alpha\Delta\mathbf{E}_{i-1/2,k}\Big) . \tag{15.88}$$

Below we utilise the one-dimensional formulation to present the extension of the method to two-dimensional problems.

15.4.2 Two-Dimensional Formulation

We consider the two-dimensional system of conservation laws

$$\frac{\partial\mathbf{U}}{\partial t} + \frac{\partial\mathbf{E}}{\partial x} + \frac{\partial\mathbf{F}}{\partial y} = 0 , \tag{15.89}$$

and introduce a piecewise polynomial approximate solution U_k^n for each element k of $\mathbf{U}$ at the time level t^n as follows [279]

$$U_k^n(x, y) = \sum_{i,j} p_{i,j}^{(k)}(x, y)\chi_{i,j} , \tag{15.90}$$

where $\chi_{i,j}$ is the characteristic function of the cell $I_{i,j}$ (Fig. 15.14) and $p_{i,j}(x, y)$ are polynomials supported at the cells $I_{i,j} = \{(\xi, \eta) | |\xi - x_i| \leq \Delta x/2, |\eta - y_j| \leq \Delta y/2\}$. The second-order centered scheme can be obtained by

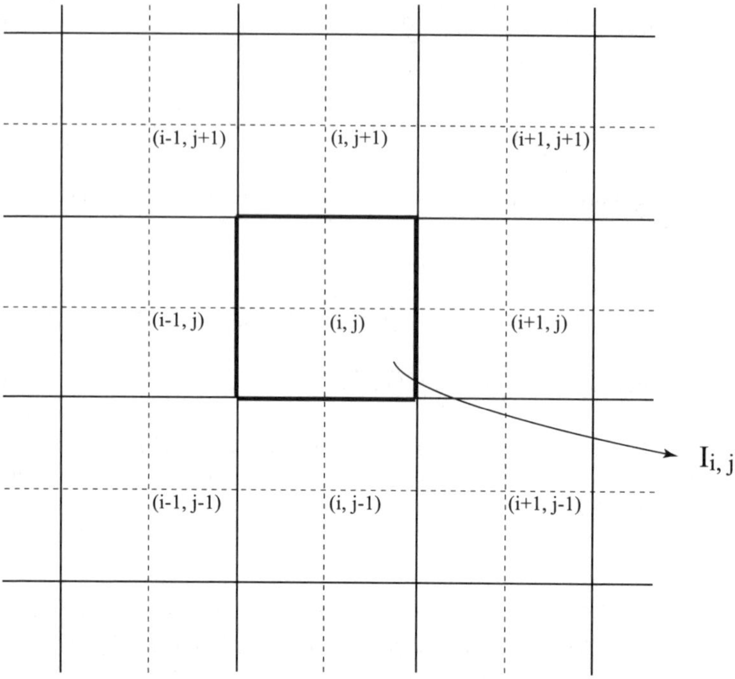

Fig. 15.14. Two-dimensional stencil for the for the Nessyahu's and Tadmor's scheme.

a reconstructed piecewise-linear MUSCL approximation according to which $p_{i,j}(x,y)$ are given by (for the sake of simplicity we drop the index k),

$$p_{i,j}(x,y) = \bar{U}_{i,j} + U^x_{i,j}\left(\frac{x-x_i}{\Delta x}\right) + U^y_{i,j}\left(\frac{y-y_j}{\Delta y}\right) , \tag{15.91}$$

where $U^x_{i,j}$ and $U^y_{i,j}$ are second-order accurate discrete slopes in the x- and y-direction, respectively, and $\bar{U}_{i,j} = U_{i,j}$ (i.e., the given available solution). The exact staggered averages of the reconstructed solution at $t = t^n$ are given by

$$\begin{aligned}\bar{\mathbf{U}}^n_{i+1/2,j+1/2} = & \frac{1}{4}\left(\bar{\mathbf{U}}^n_{i,j} + \bar{\mathbf{U}}^n_{i+1,j} + \bar{\mathbf{U}}^n_{i,j+1} + \bar{\mathbf{U}}^n_{i+1,j+1}\right) \\ & + \frac{1}{16}\left(\mathbf{U}^x_{i,j} - \mathbf{U}^x_{i+1,j} + \mathbf{U}^x_{i,j+1} - \mathbf{U}^x_{i+1,j+1}\right) \\ & + \frac{1}{16}\left(\mathbf{U}^y_{i,j} - \mathbf{U}^y_{i,j+1} + \mathbf{U}^y_{i+1,j} - \mathbf{U}^x_{i+1,j+1}\right) .\end{aligned} \tag{15.92}$$

The reconstruction (15.90) is evolved in time and projected on staggered cell averages to yield cell average values for $\bar{\mathbf{U}}^{n+1}_{i+1/2,j+1/2}$ as follows

$$\begin{aligned}\bar{\mathbf{U}}^{n+1}_{i+1/2,j+1/2} = &< \frac{1}{4}(\bar{\mathbf{U}}^n_{i,.} + \bar{\mathbf{U}}^n_{i+1,.}) + \frac{1}{8}(\mathbf{U}^x_{i,.} - \mathbf{U}^x_{i+1,.}) \\ &- \frac{\Delta t}{6\Delta x}\left(\mathbf{E}^n_{i+1,.} - \mathbf{E}^n_{i,.}\right) >_{j+1/2} \\ &+ < \frac{1}{4}(\bar{\mathbf{U}}^n_{.,j} + \bar{\mathbf{U}}^n_{.,j+1}) + \frac{1}{8}(\mathbf{U}^y_{.,j} - \mathbf{U}^y_{.,j+1}) \\ &- \frac{\Delta t}{6\Delta y}\left(\mathbf{F}^n_{.,j+1} - \mathbf{F}^n_{.,j+1}\right) >_{i+1/2} ,\end{aligned} \tag{15.93}$$

where the brackets stand for staggered-averaging

$$< \mathbf{U}_{i,.} >_{j+1/2} = \frac{1}{2}(\mathbf{U}_{i,j} + \mathbf{U}_{i,j+1}) , \tag{15.94}$$

$$< \mathbf{U}_{.,j} >_{i+1/2} = \frac{1}{2}(\mathbf{U}_{i,j} + \mathbf{U}_{i+1,j}) . \tag{15.95}$$

The mid-values $\mathbf{U}^{n+1/2}$ are evaluated by

$$\mathbf{U}^{n+1/2}_{i,j} = \bar{\mathbf{U}}_{i,j} - \frac{\Delta t}{2\Delta x}\mathbf{E}^x_{i,j} - \frac{\Delta t}{2\Delta y}\mathbf{F}^y_{i,j} , \tag{15.96}$$

where $\mathbf{E}^x_{i,j}$ and $\mathbf{F}^y_{i,j}$ are one-dimensional, second-order accurate, discrete slopes in the x- and y-directions, respectively.

In summary, the method is a predictor-corrector scheme that uses the cell averages (15.92) to calculate the mid-values $\mathbf{U}^{n+1/2}_{i,j}$ from (15.96) and, subsequently, is followed by the second-order corrector (15.93) for the calculation of the new cell averages, $\bar{\mathbf{U}}^{n+1}_{i+1/2,j+1/2}$. The discrete slopes (numerical derivatives) $\mathbf{E}^x_{i,j}$ and $\mathbf{F}^y_{i,j}$ can be calculated as described in Sect. 15.4.1.

15.4.3 Third-Order Centered Scheme

Liu and Tadmor [357] developed a third-order accurate Godunov-type scheme for the approximate solution of hyperbolic systems of conservation laws. The scheme consists of a nonoscillatory piecewise-quadratic reconstruction of pointvalues from their cell averages and central differencing based on staggered evolution of the reconstructed cell averages. Similarly with the second-order scheme, the third-order extension does not require the use of Riemann solvers. Liu and Tadmor [357] showed that the scheme is nonoscillatory by proving that it satisfies the Number of Extrema Diminishing Property. This property states that the number of extrema $N(U(\cdot,t^n))$ of $U(x,t^n)$ does not exceed that of its piecewise-constant projection, $N(\Sigma \bar{U}^n_i \chi_i(\cdot))$,

$$N(U(\cdot,t^n)) \leq N(\Sigma \bar{U}^n_i \chi_i(\cdot)) . \tag{15.97}$$

In [357] they presented the formulation of the scheme for one dimensional problems including numerical experiments for the linear and nonlinear, one-dimensional, advection equations as well as for the one-dimensional Euler equations of gas dynamics.

The method was extended by Levy and Tadmor [343] to two dimensions. The starting point is a piecewise parabolic reconstruction, $U^n(x,y) = \sum_{i,j} p_{i,j}(x,y)\chi_{i,j}$, where $p_{i,j}(x,y)$ consists of quadratic terms

$$\begin{aligned} p_{i,j}(x,y) = \bar{U}_{i,j} + U^x_{i,j}\left(\frac{x-x_i}{\Delta x}\right) + \frac{1}{2}U^{xx}_{i,j}\left(\frac{x-x_i}{\Delta x}\right)^2 \\ +U^y_{i,j}\left(\frac{y-y_j}{\Delta y}\right) + \frac{1}{2}U^{yy}_{i,j}\left(\frac{y-y_j}{\Delta y}\right)^2 , \end{aligned} \tag{15.98}$$

with the mixed terms being ignored. The terms $U^{xx}_{i,j}$ and $U^{yy}_{i,j}$ denote second-order discrete slopes in the $x-$ and $y-$direction, respectively. Levy and Tadmor [343] placed the following constraints:

- The cell average of $p_{i,j}(x,y)$ coincides with the given average $\bar{U}^n_{i,j}$, i.e., $\bar{p}_{i,j}(x,y) = \bar{U}^n_{i,j}$.
- The cell averages of $p_{i,j}(x,y)$ over the four neighboring cells coincide with the underlying given averages $\bar{U}^n_{i\pm1,j\pm1}$.

Taking into account the above, the third-order scheme first calculates the point values $U^n_{i,j}$

$$U^n_{i,j} = \bar{U}^n_{i,j} - \frac{1}{24}U^{xx}_{i,j} - \frac{1}{24}U^{yy}_{i,j} . \tag{15.99}$$

The first-order discrete slopes $U^x_{i,j}$ and $U^y_{i,j}$ are reconstructed as follows [343]

$$U^x_{i,j} = \theta_{i,j}\Delta^x_0\bar{U}^n_{i,j} , \tag{15.100}$$

$$U^y_{i,j} = \theta_{i,j}\Delta^y_0\bar{U}^n_{i,j} . \tag{15.101}$$

The second-order discrete slopes are given by

$$U^{xx}_{i,j} = \theta_{i,j}\Delta^x_+\Delta^x_-\bar{U}^n_{i,j} , \tag{15.102}$$

$$U^{yy}_{i,j} = \theta_{i,j}\Delta^y_+\Delta^y_-\bar{U}^n_{i,j} . \tag{15.103}$$

The parameters $\theta_{i,j}$ $(0 < \theta \leq 1)$ are limiters designed to avoid spurious oscillations. One-dimensional limiters have been proposed in [357]. The operators $\Delta^x_\pm$ and $\Delta^y_\pm$ are defined by

$$\Delta^x_\pm = \pm\Big[w(x \pm \Delta x) - w(x)\Big] , \tag{15.104}$$

$$\Delta^y_\pm = \pm\Big[w(y \pm \Delta y) - w(y)\Big] , \tag{15.105}$$

and

$$\Delta_0^x = \frac{1}{2}(\Delta_+^x - \Delta_-^x)\,, \tag{15.106}$$

$$\Delta_0^y = \frac{1}{2}(\Delta_+^y - \Delta_-^y)\,. \tag{15.107}$$

The piecewise-parabolic reconstruction can be evolved in time using Taylor series expansion in conjunction with the Simpson's rule or the second-order Runge-Kutta. Using Taylor series expansion the values $U_{i,j}^{n+1/2}$ and $U_{i,j}^{n+1}$ are calculated by

$$U_{i,j}^{n+1/2} = U_{i,j}^n + \frac{\Delta t}{2}\,\dot{U}_{i,j}^n + \frac{(\Delta t)^2}{8}\ddot{U}_{i,j}^n \tag{15.108}$$

$$U_{i,j}^{n+1} = U_{i,j}^n + \Delta t\,\dot{U}_{i,j}^n + \frac{(\Delta t)^2}{2}\ddot{U}_{i,j}^n\,, \tag{15.109}$$

where $\dot{U}_{i,j}^n$ and $\ddot{U}_{i,j}^n$ denote the first and second time derivatives; these are replaced by discrete derivatives. Using (15.108) and (15.109) in conjunction with the Simpson rule, we can proceed with the corrector step to calculate $\bar{\mathbf{U}}_{i+1/2,j+1/2}^{n+1}$ for the system of equations,

$$\begin{aligned}
\bar{\mathbf{U}}_{i+1/2,j+1/2}^{n+1} = &< \frac{1}{4}(\bar{\mathbf{U}}_{i,.}^n + \bar{\mathbf{U}}_{i+1,.}^n) + \frac{1}{8}(\mathbf{U}_{i,.}^x - \mathbf{U}_{i+1,.}^x) >_{j+1/2} \\
&+ < \frac{1}{4}(\bar{\mathbf{U}}_{.,j}^n + \bar{\mathbf{U}}_{.,j+1}^n) + \frac{1}{8}(\mathbf{U}_{.,j}^y - \mathbf{U}_{.,j+1}^y) >_{i+1/2} \\
&- \frac{\Delta t}{6\Delta x}\Big(< \mathbf{E}_{i+1,.}^n - \mathbf{E}_{i,.}^n >_{j+1/2} \\
&+4 < \mathbf{E}_{i+1,.}^{n+1/2} - \mathbf{E}_{i,.}^{n+1/2} >_{j+1/2} \\
&+ < \mathbf{E}_{i+1,.}^{n+1} - \mathbf{E}_{i,.}^{n+1} >_{j+1/2}\Big) \\
&- \frac{\Delta t}{6\Delta y}\Big(< \mathbf{F}_{.,j+1}^n - \mathbf{F}_{.,j+1}^n >_{i+1/2} \\
&+4 < \mathbf{F}_{.,j+1}^{n+1/2} - \mathbf{F}_{.,j}^{n+1/2} >_{i+1/2} \\
&+ < \mathbf{F}_{.,j+1}^{n+1} - \mathbf{F}_{.,j}^{n+1} >_{i+1/2}\Big)\,.
\end{aligned} \tag{15.110}$$

The second- and third-order centered schemes have been implemented by Levy and Tadmor [343] for the solution of the incompressible Euler equations in their vorticity formulation. The vorticity formulation poses difficulties regarding the implementation of boundary conditions and the extension of the second-order Nessyahu-Tadmor [396] to three dimensions. The scheme has been implemented by Kupferman and Tadmor [309] for the solution of incompressible flows using the velocity formulation and the projection method.

Further, the second-order Nessyahu-Tadmor [396] scheme has been applied by Kupferman [308] to study the axisymmetric Couette-Taylor flow.

16. Riemann Solvers and TVD Methods in Strict Conservation Form

In this chapter, we will present the construction of Riemann solvers and total variation diminishing methods via flux limiters primarily in the context of the artificial compressibility formulation of the incompressible flow equations, which allows the equations to be written in a strict conservation form. Further, the methods discussed here are also applicable to a variety of problems where artificial compressibility is not used such as general advection-diffusion problems.

16.1 The Flux Limiter Approach

The construction of high-order TVD schemes can be achieved using the flux limiter approach [65, 66, 68, 525, 523, 524, 543]. Consider the linear advection equation

$$\frac{\partial U}{\partial t} + \frac{\partial E(U)}{\partial x} = 0 \, , \quad E(U) = aU \, , \tag{16.1}$$

where a is a constant. Explicit discretization of this equation yields

$$U_j^{n+1} = U_j^n - \frac{\Delta t}{\Delta x}(E_{j+1/2} - E_{j-1/2}) \, . \tag{16.2}$$

The flux limiter approach discretizes the intercell fluxes $E_{j\pm1/2}$ by combining first-order (low-order), $E^{\mathrm{LO}}_{j\pm1/2}$, and second- or higher-order, $E^{\mathrm{HI}}_{j\pm1/2}$, fluxes. The TVD flux is defined as

$$E^{\mathrm{TVD}}_{j\pm1/2} = E^{\mathrm{LO}}_{j\pm1/2} + \psi_{j+1/2}(E^{\mathrm{HI}}_{j\pm1/2} - E^{\mathrm{LO}}_{j\pm1/2}) \, , \tag{16.3}$$

where $\psi_{j+1/2}$ is a flux limiter function that needs to be defined.

Let us consider the low and high order fluxes of the form

$$\left.\begin{aligned} E^{\mathrm{LO}}_{j+1/2} &= \alpha_0 a U_j^n + \alpha_1 a U_{j+1}^n \\ E^{\mathrm{HI}}_{j+1/2} &= \beta_0 a U_j^n + \beta_1 a U_{j+1}^n \end{aligned}\right\} \, . \tag{16.4}$$

Using (16.4) the TVD flux (16.3) is written as

$$E_{j+1/2}^{\text{TVD}} = \left[\alpha_0 + (\beta_0 - \alpha_0)\psi_{j+1/2}\right](aU_j^n) + \left[\alpha_1 + (\beta_1 - \alpha_1)\psi_{j+1/2}\right](aU_{j+1}^n) \,. \tag{16.5}$$

Substitution of (16.5) into (16.2) gives (see also Chap. 13, Eq. (13.11)

$$U_j^{n+1} = U_j^n - C\Delta U_{j-1/2} + D\Delta U_{j+1/2} \,, \tag{16.6}$$

where the coefficients C and D are given by

$$\left.\begin{aligned} C &= \mathcal{C}\left[\alpha_0 + (\beta_0 - \alpha_0)\psi_{j-1/2}\right] \\ D &= -\mathcal{C}\left[\alpha_1 + (\beta_1 - \alpha_1)\psi_{j+1/2}\right] \end{aligned}\right\} \,, \tag{16.7}$$

where $\mathcal{C} = a\Delta t/\Delta x$, $\Delta U_{j-1/2} = U_j^n - U_{j-1}^n$ and $\Delta U_{j+1/2} = U_{j+1}^n - U_j^n$. Depending on the choice of the coefficients different numerical schemes can be obtained [543]. For the low order flux choices include the following fluxes:

- Godunov's first order upwind flux

$$\alpha_0 = \frac{1}{2}(1+s) \,, \quad \alpha_1 = \frac{1}{2}(1-s) \,, \quad s = \text{sign}(a) \,. \tag{16.8}$$

- The FORCE flux (see Chap. 15)

$$\alpha_0 = \frac{1}{4}(1+\mathcal{C})^2 \,, \quad \alpha_1 = \frac{1}{4}(1-\mathcal{C})^2 \,. \tag{16.9}$$

- The Lax-Friedrichs flux

$$\alpha_0 = \frac{1}{2\mathcal{C}}(1+\mathcal{C}) \,, \quad \alpha_1 = \frac{1}{2\mathcal{C}}(1-\mathcal{C}) \,. \tag{16.10}$$

For the high-order flux the Lax-Wendroff flux can be chosen for which

$$\beta_0 = \frac{1}{2}(1+\mathcal{C}) \,, \quad \beta_1 = \frac{1}{2}(1-\mathcal{C}) \,. \tag{16.11}$$

16.2 Construction of Flux Limiters

Following the flux limiters analysis as presented by Toro [543], in the following sections we discuss two examples from the construction of flux limiters in conjunction with the advection equation.

16.2.1 Flux Limiter for the Godunov/Lax-Wendroff TVD Scheme

Consider the first order Godunov flux with coefficients (16.8) and the Lax-Wendroff flux with coefficients (16.11), in conjunction with (16.1) for $a > 0$. For $a > 0$ the coefficients of the Godunov scheme are $\alpha_0 = 1$ and $\alpha_1 = 0$. The coefficients C and D (16.7) are accordingly written

$$\left.\begin{aligned} C &= \mathcal{C}[1 + (\beta_0 - 1)\psi_{j-1/2}] \\ D &= -\mathcal{C}\beta_1\psi_{j+1/2} \end{aligned}\right\} . \tag{16.12}$$

Eq. (16.6) can also be written as

$$U_j^{n+1} = U_j^n - \hat{C}\Delta U_{j-1/2} , \tag{16.13}$$

where $\hat{C}$ is a function of the local data

$$\hat{C} = C - \frac{D}{r} \quad , \quad r = \frac{\Delta U_{j-1/2}}{\Delta U_{j+1/2}} . \tag{16.14}$$

Application of Harten's theorem (13.13) gives

$$0 \leq \mathcal{C}\Big[1 + (\beta_0 - 1)\psi_{j-1/2} + \beta_1\psi_{j+1/2}\frac{1}{r}\Big] \leq 1 . \tag{16.15}$$

Following [543] we can bound the limiter $\psi_{j-1/2}$ between a lower, ψ_B, and an upper bound, ψ_T

$$\psi_B \leq \psi_{j-1/2} \leq \psi_T . \tag{16.16}$$

Multiplying (16.16) by $(\beta_0 - 1)$, add 1 and multiply by $\mathcal{C}$, one obtains

$$\begin{aligned} \mathcal{C}[1 + (\beta_0 - 1)\psi_T] \leq \mathcal{C}[1 + (\beta_0 - 1)\psi_{j-1/2}] \\ \leq \mathcal{C}[1 + (\beta_0 - 1)\psi_B] . \end{aligned} \tag{16.17}$$

To retrieve Harten's TVD constraint we add to (16.17) the following inequalities

$$-\mathcal{C}[1 + (\beta_0 - 1)\psi_T] \leq \mathcal{C}\beta_1\psi_{j+1/2}\frac{1}{r} \leq 1 - \mathcal{C}[1 + (\beta_0 - 1)\psi_B] . \tag{16.18}$$

The left inequality of (16.18) yields

$$\psi_{j+1/2} = \begin{cases} \geq \psi_L \text{ for } r \geq 0 , \\ \leq \psi_L \text{ for } r < 0 . \end{cases} \tag{16.19}$$

where

$$\psi_L = (\psi_T - \frac{1}{\beta_1})r = \psi_T - \frac{2}{1 - \mathcal{C}} . \tag{16.20}$$

The right inequality of (16.18) yields

$$\psi_{j+1/2} = \begin{cases} \leq \psi_R \text{ for } r \geq 0\,, \\ \geq \psi_L \text{ for } r < 0\,. \end{cases} \tag{16.21}$$

where

$$\psi_R = \left(\psi_B + \frac{1-\mathcal{C}}{\mathcal{C}\beta_1}\right) r = \psi_B + \frac{2}{\mathcal{C}}\,. \tag{16.22}$$

Eqs. (16.19)-(16.22) determine the limiter $\psi_{j+1/2}$ for different choices of the bottom and top bounds. The above equations are also valid for $a < 0$ if $\mathcal{C}$ is replaced by $|\mathcal{C}|$. The general TVD region as defined by the above equations is shown in Fig. 16.1.

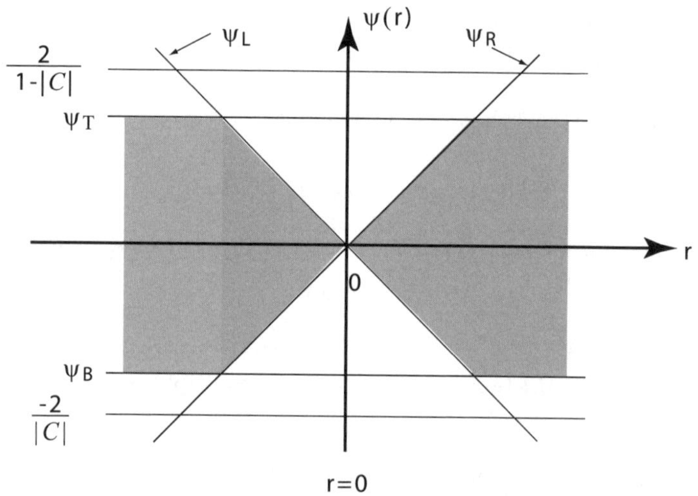

Fig. 16.1. The general TVD region (dark region) as defined by (16.19)-(16.22).

16.2.2 Flux Limiter for the Characteristics-Based/Lax-Friedrichs Scheme

In this section we will show the implementation of the flux limiter approach for a characteristics-based high-order flux [156, 149] and the low-order flux by the Lax-Friedrichs scheme [317]. For a one-dimensional stencil the characteristics-based flux is written[1]

[1] The multidimensional version of the scheme is presented in detail in Sec. 16.4.

$$E_{j+1/2}^{\mathrm{HI}} = \beta_{-1}E_{j-1} + \beta_0 E_j + \beta_1 E_{j+1} + \beta_2 E_{j+2}\,, \tag{16.23}$$

where $\beta_{-1} = -(1+s)/12$, $\beta_0 = (7+3s)/12$, $\beta_1 = (7-3s)/12$, $\beta_2 = -(1-s)/12$, and $s = \mathrm{sign}(a)$.

We examine first the order of accuracy of the characteristics-based scheme in conjunction with the linear advection equation (16.1). In this case the accuracy can be derived by using Roe's theorem [464, 543]. For $U_t + (aU)_x = U_t + E_x = 0$, where a is a constant and $E = aU$, the theorem states that any scheme written in the form

$$U_j^{n+1} = \sum_{k=-k_L}^{k_R} c_k U_{i+k}^n\,, \tag{16.24}$$

is p-th order accurate *in space and time* if

$$\sum_{k=-k_L}^{k_R} k^q c_k = (-\mathcal{C})^q, \quad 0 \le q \le p\,, \tag{16.25}$$

where k_L and k_R are two non-negative integers, c_k are scheme dependent coefficients.

The second-order Runge-Kutta discretization of the linear advection equation yields

$$\left.\begin{aligned} K_1 &= -\Delta t\; E_x(t^n, U^n) \\ K_2 &= -\Delta t\; E_x(t^n + \Delta t, U^n + K_1) \\ U^{n+1} &= U^n + \tfrac{1}{2}\Big(K_1 + K_2\Big) \end{aligned}\right\} . \tag{16.26}$$

Implementation of (16.23) into (16.26) yields

$$K_1 = b_{-2}U_{j-2}^n + b_{-1}U_{j-1}^n + b_0 U_j^n + b_1 U_{j+1}^n + b_2 U_{j+2}^n \tag{16.27}$$

and

$$\left.\begin{aligned} K_2 = {} & b_{-2}^2 U_{j-4}^n + b_{-2}b_{-1}U_{j-3}^n + b_{-2}(1+b_0)U_{j-2}^n + b_{-2}b_1 U_{j-1}^n + \\ & b_{-2}b_2 U_j^n + b_{-1}b_{-2}U_{j-3}^n + b_{-1}^2 U_{j-2}^n + b_{-1}(1+b_0)U_{j-1}^n + \\ & b_{-1}b_1 U_j^n + b_{-1}b_2 U_{j+1}^n + b_0 b_{-2} U_{j-2}^n + b_0 b_{-1} U_{j-1}^n + \\ & b_0(1+b_0)U_j^n + b_0 b_1 U_{j+1}^n + b_0 b_2 U_{j+2}^n + b_1 b_{-2} U_{j-1}^n + \\ & b_1 b_{-1} U_j^n + b_1(1+b_0)U_{j+1}^n + b_1^2 U_{j+2}^n + b_1 b_2 U_{j+3}^n + \\ & b_2 b_{-2} U_j^n + b_2 b_{-1} U_{j+1}^n + b_2(1+b_0)U_{j+2}^n + b_2 b_1 U_{j+3}^n + \\ & b_2^2 U_{j+4}^n \end{aligned}\right\} , \tag{16.28}$$

where

$$\left.\begin{aligned} b_{-2} &= -\frac{\mathcal{C}}{12}(1+s) \\ b_{-1} &= \frac{\mathcal{C}}{12}(8+4s) \\ b_0 &= -\frac{\mathcal{C}}{2}s \\ b_1 &= -\frac{\mathcal{C}}{12}(8-4s) \\ b_2 &= \frac{\mathcal{C}}{12}(1-s) \end{aligned}\right\} . \tag{16.29}$$

Using (16.27) and (16.28), (16.26) yields

$$\begin{aligned} U_j^{n+1} = {} & c_{-4}U_{j-4}^n + c_{-3}U_{j-3}^n + c_{-2}U_{j-2}^n + c_{-1}U_{j-1}^n + \\ & c_0U_j^n + c_1U_{j+1}^n + c_2U_{j+2}^n + c_3U_{j+3}^n + c_4U_{j+4}^n , \end{aligned} \tag{16.30}$$

where

$$\left.\begin{aligned} c_{-4} &= \tfrac{1}{2}b_{-2}^2 \\ c_{-3} &= b_{-2}b_{-1} \\ c_{-2} &= \tfrac{1}{2}(2b_0b_{-2} + 2b_{-2} + b_{-1}^2) \\ c_{-1} &= b_1b_{-2} + b_0b_{-1} + b_{-1} \\ c_0 &= \tfrac{1}{2}(2b_2b_{-2} + 2b_1b_{-1} + b_0^2 + 2b_0) + 1 \\ c_1 &= b_2b_{-1} + b_0b_1 + b_1 \\ c_2 &= \tfrac{1}{2}(2b_0b_2 + 2b_2 + b_1^2) \\ c_3 &= b_2b_1 \\ c_4 &= \tfrac{1}{2}b_2^2 \end{aligned}\right\} . \tag{16.31}$$

The coefficients (16.31) satisfy (16.25) for second-order of accuracy.

The Lax-Friedrichs flux is given by

$$E^{\mathrm{LO}} = \alpha_0E_j + \alpha_1E_{j+1} , \tag{16.32}$$

where $\alpha_0 = (1+\mathcal{C})/2\mathcal{C}$ and $\alpha_1 = -(1-\mathcal{C})/2\mathcal{C}$ [543].
Using (16.28) and (16.32), the TVD version (16.3) of the flux E at the cell faces $j-1/2$ and $j+1/2$ is written

$$\begin{aligned} E_{j+1/2}^{\mathrm{TVD}} = {} & \beta_{-1}\psi_{j+1/2}E_{j-1} + \beta_2\psi_{j+1/2}E_{j+2} + \\ & \Big[\alpha_0 + \psi_{j+1/2}(\beta_0 - \alpha_0)\Big]E_j + \Big[\alpha_1 + \psi_{j+1/2}(\beta_1 - \alpha_1)\Big]E_{j+1} , \end{aligned} \tag{16.33}$$

$$E^{\rm TVD}_{j-1/2} = \beta_{-1}\psi_{j-1/2}E_{j-2} + \beta_2\psi_{j-1/2}E_{j+1} +$$
$$\Big[\alpha_0 + \psi_{j-1/2}(\beta_0 - \alpha_0)\Big]E_{j-1} + \Big[\alpha_1 + \psi_{j-1/2}(\beta_1 - \alpha_1)\Big]E_j \ . \tag{16.34}$$

Using (16.33) and (16.34) the discretized linear advection equation is written

$$U_j^{n+1} = U_j^n - C\Delta U_{j-1/2} + D\Delta U_{j+1/2} - E\Delta U_{j-3/2} + F\Delta U_{j+3/2} \ , \tag{16.35}$$

where $\Delta U_{j-1/2} = U_j - U_{j-1}$, $\Delta U_{j+1/2} = U_{j+1} - U_j$, $\Delta U_{j-3/2} = U_{j-1} - U_{j-2}$, and $\Delta U_{j+3/2} = U_{j+2} - U_{j+1}$ and

$$\left.\begin{aligned} C &= \mathcal{C}\Big[\alpha_0 + \psi_{j-1/2}(\beta_0 - \alpha_0)\Big] \\ D &= -\mathcal{C}\Big[\alpha_1 + \psi_{j+1/2}(\beta_1 - \alpha_1)\Big] \\ E &= \mathcal{C}\beta_{-1}\psi_{j-1/2} \\ F &= -\mathcal{C}\beta_2\psi_{j+1/2} \end{aligned}\right\} \ . \tag{16.36}$$

To derive limiter functions such that the scheme will be TVD, we apply the data compatibility condition [543, 464]. We present the derivation for $a > 0$ (a similar analysis can be applied for $a < 0$), (16.1) can be written as

$$\frac{U_j^{n+1} - U_j^n}{U_{j-1}^n - U_j^n} = C - \frac{D}{r} + \frac{E}{\tilde{r}} - \frac{F}{\hat{r}} \ , \tag{16.37}$$

where $r = \Delta U_{j-1/2}/\Delta U_{j+1/2}$ and $\tilde{r} = \Delta U_{j-1/2}/\Delta U_{j-3/2}$. Depending on the sign of a the ratio of upwind change r should be correctly interpreted, that is

$$r = \frac{\Delta_{upw}}{\Delta_{loc}} = \begin{cases} \dfrac{U_j^n - U_{j-1}^n}{U_{j+1}^n - U_j^n} \ , & a > 0 \ , \\ \dfrac{U_{j+2}^n - U_{j+1}^n}{U_{j+1}^n - U_j^n} \ , & a < 0 \ . \end{cases} \tag{16.38}$$

Eqs. (16.35) and (16.37) yield

$$0 \le \mathcal{C}\Big[\alpha_0 + \psi_{j-1/2}(\beta_0 - \alpha_0)\Big] + \mathcal{C}\Big[\alpha_1 + \psi_{j+1/2}(\beta_1 - \alpha_1)\Big]\frac{1}{r} - \psi_{j-1/2}\frac{\mathcal{C}}{6\tilde{r}} \le 1 \ . \tag{16.39}$$

We impose a global constraint

$$\psi_B \le \psi_{j-1/2} \le \psi_T \ , \tag{16.40}$$

where ψ_T and ψ_B are the top and bottom bounds of the flux limiter, which are considered to be independent of r and $\tilde{r}$. Then, (16.40) gives

$$\mathcal{C}\Big[\alpha_0 + \psi_T(\beta_0 - \alpha_0 - \frac{1}{6\tilde{r}})\Big] \leq \mathcal{C}\Big[\alpha_0 + \psi_{j-1/2}(\beta_0 - \alpha_0 - \frac{1}{6\tilde{r}})\Big] \\ \leq \mathcal{C}\Big[\alpha_0 + \psi_B(\beta_0 - \alpha_0 - \frac{1}{6\tilde{r}})\Big] , \quad (16.41)$$

if $\beta_0 - \alpha_0 - \dfrac{1}{6\tilde{r}} \leq 0$, which is equivalent to

$$\tilde{r} = \frac{U_{j-1} - U_j}{U_{j-1} - U_{j-2}} < \frac{\mathcal{C}}{2\mathcal{C} - 3} . \quad (16.42)$$

For $\beta_0 - \alpha_0 - \dfrac{1}{6\tilde{r}} > 0$ the analysis will be the same if in (16.41) we swap ψ_B with ψ_T and vice versa.

To satisfy (16.39) and (16.41), the following inequality should be satisfied

$$-\mathcal{C}\Big[\alpha_0 + \psi_T(\beta_0 - \alpha_0 - \frac{1}{6\tilde{r}})\Big] \leq \mathcal{C}\Big[\alpha_1 + \psi j + 1/2(\beta_1 - \alpha_1)\Big]\frac{1}{r} \\ \leq 1 - \mathcal{C}\Big[\alpha_0 + \psi_B(\beta_0 - \alpha_0 - \frac{1}{6\tilde{r}})\Big] . \quad (16.43)$$

Analysis of the left inequality gives

$$\psi_{j+1/2} = \begin{cases} \geq \psi_L \text{ for } r \geq 0 , \\ \leq \psi_L \text{ for } r < 0 , \end{cases} \quad (16.44)$$

where

$$\psi_L = \frac{1}{\beta_1 - \alpha_1}\left\{-r\Big[\alpha_0 + \psi_T(\beta_0 - \alpha_0 - \frac{1}{6\tilde{r}})\Big\} - \alpha_1\right) . \quad (16.45)$$

Analysis of the right inequality gives

$$\psi_{j+1/2} = \begin{cases} \leq \psi_R \text{ for } r \geq 0 , \\ \geq \psi_R \text{ for } r < 0 , \end{cases} \quad (16.46)$$

where

$$\psi_R = \frac{1}{\beta_1 - \alpha_1}\left\{\frac{r}{\mathcal{C}} - r\Big[\alpha_0 + \psi_B(\beta_0 - \alpha_0 - \frac{1}{6\tilde{r}})\Big] - \alpha_1\right\} . \quad (16.47)$$

Eqs. (16.45) and (16.47) are functions of the ratios r and $\tilde{r}$. For large flow gradients occurring within the stencil $i-1$, i and $i+1$, $\Delta U_{j-3/2} < \Delta U_{j+1/2}$ thus $r/\tilde{r} < 1$. As a result, $\psi_B r/\tilde{r}$ and $\psi_T r/\tilde{r}$ can be considered small compared to the rest of the terms in (16.45) and (16.47) thus can be dropped from (16.45) and (16.47), respectively. Numerical experiments for the one-dimensional Burgers' equation and for the incompressible Navier-Stokes equations have shown that these terms do not alter the accuracy of computations [149]. Alternatively, one can attempt to approximate $\psi_B r/\tilde{r}$ and $\psi_T r/\tilde{r}$ as functions of the ratio r.

We substitute $\beta_0, \beta_1, \alpha_0$ and α_1 into (16.45) and (16.47) and obtain

$$\left.\begin{aligned} \psi_R &= \frac{3(1-\mathcal{C}) - (2\mathcal{C}-3)\psi_B}{3-\mathcal{C}} r + \frac{3(1-\mathcal{C})}{3-\mathcal{C}} \\ \psi_L &= -\frac{3+3\mathcal{C}+(2\mathcal{C}-3)\psi_T}{3-\mathcal{C}} r + \frac{3(1-\mathcal{C})}{3-\mathcal{C}} \end{aligned}\right\} . \tag{16.48}$$

The construction of the limiter functions ψ_L and ψ_R is completed after defining the top and bottom boundaries of the flux limiter. Concerning this point one has flexibility to construct different TVD schemes depending on the definitions of ψ_B and ψ_T.

For $\psi_B = 0$ and $\psi_T = 3(1+\mathcal{C})/(3-2\mathcal{C})$, we obtain

$$\psi_R = \frac{3(1-\mathcal{C})}{3-\mathcal{C}}(r+1) , \tag{16.49}$$

$$\psi_L = \frac{3(1-\mathcal{C})}{3-\mathcal{C}} . \tag{16.50}$$

Using (16.49) and (16.50), we define the flux limiter (henceforth labeled *CB-limiter*) (Fig. 16.2)

$$\psi = \begin{cases} \dfrac{3(1-\mathcal{C})}{3-\mathcal{C}} & \text{if} \quad r \le 0 , \\[2ex] \dfrac{3(1-\mathcal{C})}{3-\mathcal{C}}(r+1) & \text{if} \quad 0 \le r \le \dfrac{\mathcal{C}(7-3\mathcal{C})}{(3-2\mathcal{C})(1-\mathcal{C})} , \\[2ex] \dfrac{3(1+\mathcal{C})}{3-2\mathcal{C}} & \text{if} \quad r > \dfrac{\mathcal{C}(7-3\mathcal{C})}{(3-2\mathcal{C})(1-\mathcal{C})} . \end{cases} \tag{16.51}$$

To calculate any of the flux limiter functions the following procedure can be used

- The ratios r^L and r^R are calculated at the cell faces

$$\left.\begin{aligned} r^L_{j+\frac{1}{2}} &= \frac{\Delta U_{j-1/2}}{\Delta U_{j+1/2}} \\ r^R_{j+\frac{1}{2}} &= \frac{\Delta U_{j+3/2}}{\Delta U_{j+1/2}} \end{aligned}\right\} , \tag{16.52}$$

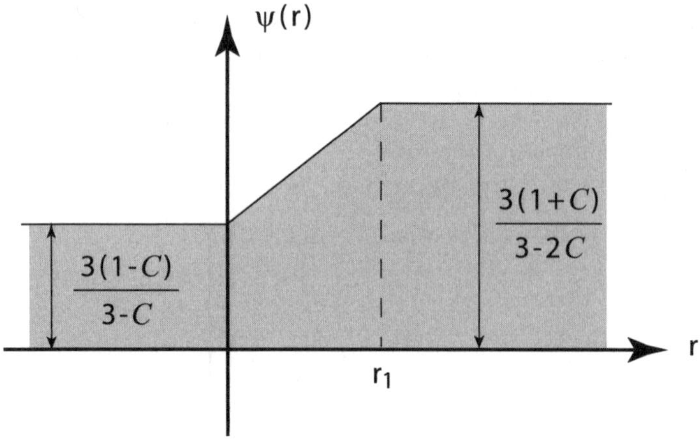

Fig. 16.2. The TVD region for the CB-limiter as defined by (16.51).

where $\Delta(\cdot)$ denotes u-velocity differences at the cell faces. The pressure or velocity can be used to calculate the intercell slopes.

- Then, we compute the flux limiter as

$$\psi_{j+1/2} = \min\Big(\psi(r^L_{j+1/2}), \psi(r^R_{j+1/2})\Big) . \tag{16.53}$$

The flux limiter $\psi_{j+1/2}$ should be applied to all flux components.

16.3 Other Approaches for Constructing Advective Schemes

Other approaches than Harten's theorem and the data compatibility condition to limiting advective fluxes include the positive schemes [270] and the universal limiter [331]. The approach based on positive schemes allows different time stepping schemes to be used by treating the space and time discretization separately [270]. Thuburn [532] has shown that different approaches for constructing limiters can lead to equivalent schemes at least in the context of the one-dimensional linear advection equation. However, the differences between the aforementioned approaches still remain important since each of these approaches can be extended and utilized in different ways. For example, the TVD approach can be extended to conservation laws other than the advection equation and the universal limiter can be implemented to multiple advection on arbitrary meshes [533].

16.3.1 Positive Schemes

A scheme is called positive if, and only if, satisfies

$$U_j = 0 \quad \text{and} \quad U_i \geq 0 \quad \forall \quad i \neq j \quad \text{then} \quad \frac{dU}{dt} \geq 0 \,. \tag{16.54}$$

A discretization of the linear advection equation is defined as positive (see also [270]) if

- the spatial discretization is positive, and
- no positive value of U can become negative during one time step when all U values are initially greater than or equal to zero.

We examine now how the above constraints apply to the linear advection equation. We consider the discretized equation (16.1) with the intercell fluxes defined by

$$\left. \begin{aligned} E_{j+1/2} &= E_j + \frac{1}{2}\psi_{j+1/2}(U_j - U_{j-1}) \\ E_{j-1/2} &= E_{j-1} + \frac{1}{2}\psi_{j-1/2}(U_{j-1} - U_{j-2}) \end{aligned} \right\} . \tag{16.55}$$

Using (16.55), (16.1) gives

$$U_j^{n+1} = U_j^n - \mathcal{C}(U_j^n - U_{j-1}^n)\Big(1 + \frac{1}{2}\psi_{j+1/2} - \frac{1}{2}\psi_{j-1/2}r_{j-1/2}\Big) \,, \tag{16.56}$$

where r is defined by (16.38). For $U_j^n = 0$ and $U_{j-1}^n = 0$, the spatial discretization will be positive if

$$0 \leq 1 + \frac{1}{2}\psi_{j+1/2} - \frac{1}{2}\psi_{j-1/2}r_{j-1/2} \leq \epsilon \quad \forall\, i \tag{16.57}$$

where $\epsilon \geq 1$ defines an upper bound. The above inequalities are satisfied by the following conditions

$$\left. \begin{aligned} 0 \leq \psi_{j+1/2} \qquad & \leq \tilde{\epsilon} \qquad \forall\, i \\ 0 \leq \psi_{j+1/2}r_{j+1/2} & \leq 2 \qquad \forall\, i \end{aligned} \right\} , \tag{16.58}$$

where $\tilde{\epsilon} = 2(\epsilon - 1)$.

For the case $U_j^n > 0$ and $U_{j-1}^n = 0$, the condition for a positive scheme will be satisfied if in addition to (16.57) the following inequality is satisfied

$$\mathcal{C}\epsilon \leq 1 \,. \tag{16.59}$$

The inequalities (16.58) and (16.59) provide a way to define flux limiters depending on the choice of the upper bound ϵ. The inequality (16.58) implies that

$$U_{j-1}^n \leq U_j^{n+1} \leq U_j^n \quad \forall\, i \,. \tag{16.60}$$

The inequality (16.60) is the local bounding property [467], which can be used itself to construct advective schemes [568].

16.3.2 Universal Limiter

The universal limiter approach [327] aims to satisfy the local bounding property formulated as [532]

$$\min(U_{j-1}^n, U_j^n) \le U_j^{n+1} \le \max(U_{j-1}^n, U_j^n) \quad \forall\, j \;, \tag{16.61}$$

by posing constraints on the intercell values $U_{j-1/2}$ according to

$$\min(U_{j-1}^n, U_j^n) \le U_{j-1/2}^n \le \max(U_{j-1}^n, U_j^n) \quad \forall\, j \;. \tag{16.62}$$

The inequalities (16.61), (16.62) in conjunction with (16.1) result in the following condition

$$U_{\min,j} \le U_{j+1/2}^n \le U_{\max,j} \quad \forall\, j \;, \tag{16.63}$$

where

$$\left.\begin{aligned} U_{\min,j} &= \frac{1}{\mathcal{C}}\Big[U_j^n + (\mathcal{C}-1)\max(U_{j-1}^n, U_j^n)\Big] \\ U_{\max,j} &= \frac{1}{\mathcal{C}}\Big[U_j^n + (\mathcal{C}-1)\min(U_{j-1}^n, U_j^n)\Big] \end{aligned}\right\} \;. \tag{16.64}$$

The inequalities (16.62) and (16.63) provide a way to construct advective schemes which satisfy the local bounding property.

Finally, we note that normalized value diagram (NVD) schemes, which has rough equivalence to simple TVD methods, has been pursued in conjunction with pressure-Poisson equation methods, e.g., [204, 344, 413, 596, 619, 620, 526] (see also Sect. 12.4.2).

16.4 The Characteristics-Based Scheme

16.4.1 Introductory Remarks and Basic Formulation

The characteristics-based (CB) scheme [145, 148, 156] is a Riemann solver for incompressible flows. The CB scheme discretizes the advective fluxes by defining the primitive variables as functions of their values on the characteristics. Unlike compressible flows where shock wave solutions may make sense even in the one-dimensional context, incompressible flows need to be studied at least in two dimensions. The CB scheme has been developed for two [156] as well as for three-dimensional flows [153, 145, 157, 148] including its implementation in complex geometries in the context of block-structured [222] and unstructured grids [617]. We first discuss the basic idea of the CB scheme in the context of (nonlinear) hyperbolic conservation laws

$$\frac{\partial U}{\partial t} + \frac{\partial E(U)}{\partial x} = 0 \;, \tag{16.65}$$

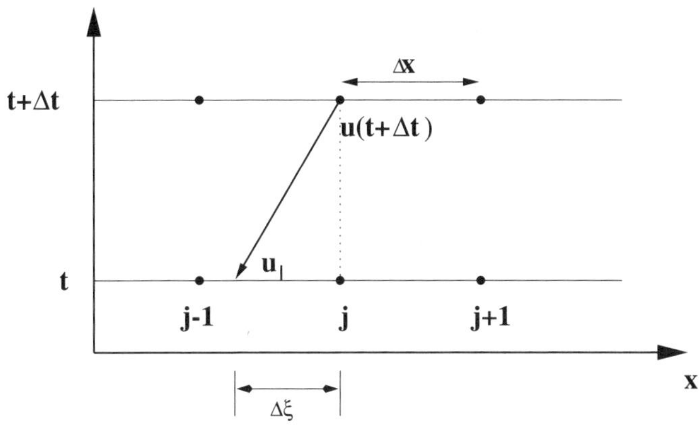

Fig. 16.3. Schematic representation of the definition of variable $U(t + \Delta t) \equiv \widetilde{U}$ as function of the characteristic variable U_l.

where E is a nonlinear function of U. The update of the solution is given by

$$U(t + \Delta t) = U(t) - \int_t^{t+\Delta t} \frac{\partial E(U)}{\partial x} \, \mathrm{d}t \; . \tag{16.66}$$

In the last equation the spatial derivative of the flux is only known at the initial time level t which is exactly at the lower bound of the interval of integration. However, the fundamental theorem of integration requires the integrand to be known inside the limits of integration for stable numerical update. For example, the trapezoidal rule would require the integrand to be known at $t + \Delta t/2$. Therefore, we need to make use of a device which will allow us to propagate the integrand in time and perform a stable numerical update. Consider the propagation of the solution from time t to $t + \Delta t$ as shown in Fig. 16.3. In order to define the solution at the point $(x, t + \Delta t)$, we perform a linear backward Taylor series expansion in the neighborhood of that point

$$U(x, t + \Delta t) = U(x - \Delta\xi, t) + \Delta\xi \frac{\partial U}{\partial x} + \Delta t \frac{\partial U}{\partial t} \; , \tag{16.67}$$

where higher order terms have been neglected. By denoting $U_l \equiv U(x - \Delta\xi, t)$ and $\widetilde{U} \equiv U(x, t + \Delta t)$, and by introducing the wave speed, $\dot{\xi} = \partial\xi/\partial t$, such that $\Delta\xi = \dot{\xi}\Delta t$, (16.67) is written

$$\widetilde{U} = U_l + \left(\frac{\partial U}{\partial t} + \frac{\partial U}{\partial x}\dot{\xi}\right) \Delta t \, . \tag{16.68}$$

All terms in the above equation are unknown since neither the spatial and time derivatives nor the value U_l are known. Yet, the position from which the initial value U_l should be taken is also unknown. From the hyperbolic conservation law (16.65) the term $\partial U/\partial t \equiv \dot{U}$ is given by

$$\dot{U} = -\frac{\partial E(U)}{\partial x} \, . \tag{16.69}$$

Substituting $\dot{U}$ from the last relation into (16.68), one obtains

$$\widetilde{U} = U_l + \frac{\partial U}{\partial x}\left[\dot{\xi} - \frac{\partial E(U)}{\partial U}\right] \Delta t \, , \tag{16.70}$$

where we made use of

$$\frac{\partial E(U)}{\partial x} = \frac{\partial E(U)}{\partial U}\frac{\partial U}{\partial x} \, . \tag{16.71}$$

If, by definition, the wave speed is an eigenvalue, i.e., $\dot{\xi} = \partial E(U)/\partial U$, then the coefficient of the uknown spatial derivative on the right hand side of the above equation can be eliminated, thus, obtaining

$$\widetilde{U} = U_l \, . \tag{16.72}$$

In other words, the solution $\widetilde{U}$ can be calculated using U_l (henceforth called *characteristic variable*) where the latter corresponds to the point with co-ordinates $(x - \dot{\xi}\Delta t, t)$ and belongs to the line with slope $1/\dot{\xi}$ (characteristic line). The variable U_l can subsequently be calculated by high-order interpolation as we will present later on. Once the value $\widetilde{U} = U_l$ is known, the flux $E(\widetilde{U})$ can be calculated. In the case where more than one characteristic lines exist, for example, in multi-dimensional flow problems, the flux $E(\widetilde{U})$ would then be defined as $E(g(U_l))$ since the variable $\widetilde{U}$ will be a function of all characteristic variables lying on the lines with slopes $1/\dot{\xi}_l$ ($l = 0 \div 2$), i.e., $\widetilde{U} = g(U_l)$.

In the next section we present how $E(g(U_l))$ can be constructed for the case of the three-dimensional incompressible equations.

16.4.2 Dimensional Splitting

We recall from Chap. 10 the inviscid counterpart of the incompressible equations in conjunction with the artificial compressibility formulation, written in a matrix form for a three-dimensional curvilinear co-ordinates system:

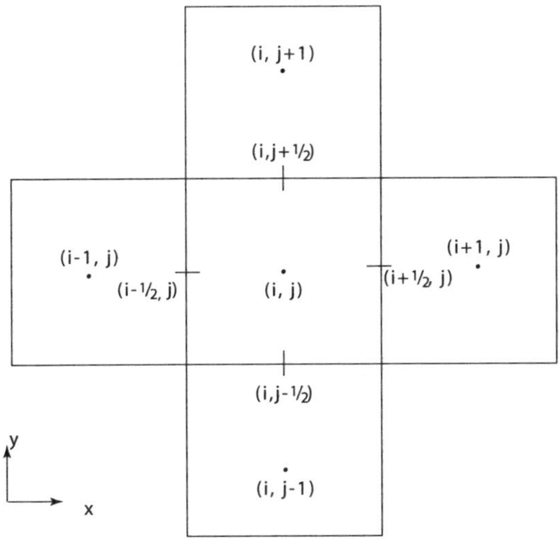

Fig. 16.4. Two-dimensional cell notation for the CB scheme.

$$\frac{\partial \bar{\mathbf{U}}}{\partial t} + \frac{\partial \bar{\mathbf{E}}}{\partial \xi} + \frac{\partial \bar{\mathbf{F}}}{\partial \eta} + \frac{\partial \bar{\mathbf{G}}}{\partial \zeta} = 0 , \tag{16.73}$$

where

$$\bar{\mathbf{U}} = J\mathbf{U} = J(p/\beta, u, v, w)^T , \tag{16.74}$$

$$\bar{\mathbf{E}} = J(\mathbf{E}\xi_x + \mathbf{F}\xi_y + \mathbf{G}\xi_z) , \tag{16.75}$$

$$\bar{\mathbf{F}} = J(\mathbf{E}\eta_x + \mathbf{F}\eta_y + \mathbf{G}\eta_z) , \tag{16.76}$$

$$\bar{\mathbf{G}} = J(\mathbf{E}\zeta_x + \mathbf{F}\zeta_y + \mathbf{G}\zeta_z) , \tag{16.77}$$

and β is the artificial compressibility parameter. The matrices $\bar{\mathbf{E}}, \bar{\mathbf{F}}$ and $\bar{\mathbf{G}}$ are given by

$$\boldsymbol{\mathcal{E}} = J \begin{pmatrix} \beta Q \\ uk_t + pk_x + uQ \\ vk_t + pk_y + vQ \\ wk_t + pk_z + wQ \end{pmatrix} , \tag{16.78}$$

where $\boldsymbol{\mathcal{E}} = \bar{\mathbf{E}}, \bar{\mathbf{F}}, \bar{\mathbf{G}}$ for $k = \xi, \eta, \zeta$, respectively, and

$$Q = uk_x + vk_y + wk_z . \tag{16.79}$$

For the sake of simplicity, we consider the presentation for a non-moving grid, i.e., $k_t = 0$. The cell notation in two and three dimensions is shown in Figs. 16.4 and 16.5, respectively. The advective fluxes in (16.73) are discretized using the intercell values $(i \pm 1/2, j, k), (i, j \pm 1/2, k), (i, j, k \pm 1/2)$

$$\frac{\partial \bar{\mathbf{U}}}{\partial t} + \frac{\bar{\mathbf{E}}_{i+1/2,j,k} - \bar{\mathbf{E}}_{i-1/2,j,k}}{\Delta x} + \frac{\bar{\mathbf{F}}_{i,j+1/2,k} - \bar{\mathbf{F}}_{i,j-1/2,k}}{\Delta \eta} + \frac{\bar{\mathbf{G}}_{i,j,k+1/2} - \bar{\mathbf{G}}_{i,j,k-1/2}}{\Delta \zeta} = 0\,. \tag{16.80}$$

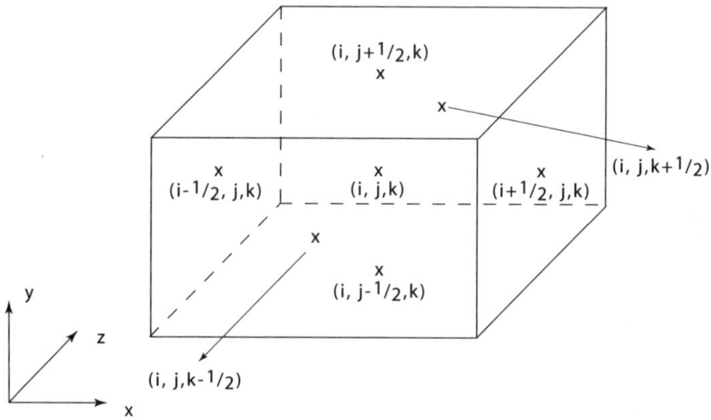

Fig. 16.5. Three-dimensional cell notation for the CB scheme.

To derive solutions to the Riemann problem in each of the directions ξ, η and ζ one can consider the equations[2]

$$\left.\begin{aligned} \frac{\partial \widetilde{\mathbf{U}}}{\partial t} + \frac{\partial \mathbf{E}(\widetilde{\mathbf{U}})}{\partial \xi} &= 0 \\ \frac{\partial \widetilde{\mathbf{U}}}{\partial t} + \frac{\partial \mathbf{F}(\widetilde{\mathbf{U}})}{\partial \eta} &= 0 \\ \frac{\partial \widetilde{\mathbf{U}}}{\partial t} + \frac{\partial \mathbf{G}(\widetilde{\mathbf{U}})}{\partial \zeta} &= 0 \end{aligned}\right\}, \tag{16.81}$$

[2] Dimensional splitting is used *only* for analytically deriving characteristics-based solution for the intercell variables according to which the advective fluxes are calculated. The numerical time integration is obtained for the complete system (Euler or Navier-Stokes) of equations after adding all the discretized fluxes (inviscid and viscous), using a time integration scheme (see Chaps. 7, 9 and 10).

where $\widetilde{\mathbf{U}}$ is the vector of the variables for which characteristics-based solutions will be derived, and $\mathbf{E}(\widetilde{\mathbf{U}})$ is the invisicid flux which will be calculated using $\widetilde{\mathbf{U}}$.

It suffices to present the derivation of the Riemann solution for the ξ direction. Similarly, one can derive the the formulas for the other two directions. The Riemann solutions derived for each direction will subsequently be used in the calculation of the intercell fluxes.

16.4.3 Characteristics-Based Reconstruction in Three Dimensions

For the system of equations

$$\frac{\partial \widetilde{\mathbf{U}}}{\partial t} + \frac{\partial \mathbf{E}(\widetilde{\mathbf{U}})}{\partial \xi} = 0 \,, \tag{16.82}$$

we consider the non-conservative form

$$\left.\begin{array}{l}
\dfrac{\widetilde{p}_\tau}{\beta\sqrt{\xi_x^2+\xi_y^2+\xi_z^2}} + \widetilde{u}_\xi\tilde{x} + \widetilde{v}_\xi\tilde{y} + \widetilde{w}_\xi\tilde{z} = 0 \\[2ex]
\dfrac{\widetilde{u}_\tau}{\sqrt{\xi_x^2+\xi_y^2+\xi_z^2}} + \widetilde{u}_\xi(\widetilde{u}\tilde{x} + \widetilde{v}\tilde{y} + \widetilde{w}\tilde{z}) + \\
\widetilde{u}(\widetilde{u}_\xi\tilde{x} + \widetilde{v}_\xi\tilde{y} + \widetilde{w}_\xi\tilde{z}) + \widetilde{p}_\xi\tilde{x} = 0 \\[2ex]
\dfrac{\widetilde{v}_\tau}{\sqrt{\xi_x^2+\xi_y^2+\xi_z^2}} + \widetilde{v}_\xi(\widetilde{u}\tilde{x} + \widetilde{v}\tilde{y} + \widetilde{w}\tilde{z}) + \\
\widetilde{v}(\widetilde{u}_\xi\tilde{x} + \widetilde{v}_\xi\tilde{y} + \widetilde{w}_\xi\tilde{z}) + \widetilde{p}_\xi\tilde{y} = 0 \\[2ex]
\dfrac{\widetilde{w}_\tau}{\sqrt{\xi_x^2+\xi_y^2+\xi_z^2}} + \widetilde{w}_\xi(\widetilde{u}\tilde{x} + \widetilde{v}\tilde{y} + \widetilde{w}\tilde{z}) + \\
\widetilde{w}(\widetilde{u}_\xi\tilde{x} + \widetilde{v}_\xi\tilde{y} + \widetilde{w}_\xi\tilde{z}) + \widetilde{p}_\xi\tilde{z} = 0
\end{array}\right\} , \tag{16.83}$$

where

$$\tilde{k} = \frac{\xi_k}{\sqrt{\xi_x^2+\xi_y^2+\xi_z^2}}, \quad k = x, y, z \,. \tag{16.84}$$

The indices τ and ξ denote derivatives in pseudo-time τ and spatial direction ξ (in the computational plane). We develop $\widetilde{\mathbf{U}}$ in Taylor series expansion around the time level τ.

$$\widetilde{\mathbf{U}}(\tau + \Delta\tau) = \widetilde{\mathbf{U}}_l(\tau) + \widetilde{\Delta\xi}\widetilde{\mathbf{U}}_\xi + \widetilde{\mathbf{U}}_\tau\Delta\tau \,, \tag{16.85}$$

where $\widetilde{\mathbf{U}}_l$ $(l = 0, 1, 2)$ are the variables along the characteristics, l, and the the interval $\widetilde{\Delta\xi}$ is defined by introducing a wave speed λ such that:

$$\widetilde{\Delta\xi} = \lambda\sqrt{\xi_x^2 + \xi_y^2 + \xi_z^2}\Delta\tau. \tag{16.86}$$

Eq. (16.85) can be solved with respect to $\widetilde{\mathbf{U}}(\tau)$

$$\widetilde{\mathbf{U}}_\tau = \frac{\widetilde{\mathbf{U}} - \widetilde{\mathbf{U}}_l}{\Delta\tau} - \lambda\sqrt{\xi_x^2 + \xi_y^2 + \xi_z^2}\widetilde{\mathbf{U}}_\xi \,. \tag{16.87}$$

Using (16.87), (16.83) yields

$$\left.\begin{array}{l}
\dfrac{\widetilde{p} - p_l}{\beta\Delta\tau\sqrt{\xi_x^2 + \xi_y^2 + \xi_z^2}} + \dfrac{1}{\beta}\widetilde{p}_\xi\lambda + \\
\widetilde{u}_\xi\tilde{x} + \widetilde{v}_\xi\tilde{y} + \widetilde{w}_\xi\tilde{z} = 0 \\
\\
\dfrac{\widetilde{u} - u_l}{\Delta\tau\sqrt{\xi_x^2 + \xi_y^2 + \xi_z^2}} + \widetilde{u}_\xi(\lambda_0 - \lambda) + \\
\widetilde{u}(\widetilde{u}_\xi\tilde{x} + \widetilde{v}_\xi\tilde{y} + \widetilde{w}_\xi\tilde{z}) + \widetilde{p}_\xi\tilde{x} = 0 \\
\\
\dfrac{\widetilde{v} - v_l}{\Delta\tau\sqrt{\xi_x^2 + \xi_y^2 + \xi_z^2}} + \widetilde{v}_\xi(\lambda_0 - \lambda) + \\
\widetilde{v}(u_\xi\tilde{x} + \widetilde{v}_\xi\tilde{y} + \widetilde{w}_\xi\tilde{z}) + \widetilde{p}_\xi\tilde{y} = 0 \\
\\
\dfrac{\widetilde{w} - w_l}{\Delta\tau\sqrt{\xi_x^2 + \xi_y^2 + \xi_z^2}} + \widetilde{w}_\xi(\lambda_0 - \lambda) + \\
\widetilde{w}(\widetilde{u}_\xi\tilde{x} + \widetilde{v}_\xi\tilde{y} + \widetilde{w}_\xi\tilde{z}) + \widetilde{p}_\xi\tilde{z} = 0
\end{array}\right\} , \tag{16.88}$$

where the eigenvalue λ_0 is defined by

$$\lambda_0 = \widetilde{u}\tilde{x} + \widetilde{v}\tilde{y} + \widetilde{w}\tilde{z} \,. \tag{16.89}$$

To eliminate the spatial derivatives from (16.88), we make use of the idea presented in the book of Courant and Hilbert [124] regarding elimination of unknowns in a system of linear equations.[3] According to [124] we first multiply each from (16.88) with arbitrary non-zero coefficients a, b, c and d, respectively, and after summation we obtain

[3] In [124] this is referred to as Riemann method.

$$\frac{1}{\Delta\tau\sqrt{\xi_x^2+\xi_y^2+\xi_z^2}}\Big[\frac{1}{\beta}a(\widetilde{p}-p_l)+b(\widetilde{u}-u_l)+c(\widetilde{v}-v_l)+d(\widetilde{w}-w_l)\Big]$$
$$+\widetilde{p}_\xi(-\frac{a}{\beta}\lambda+b\tilde{x}+c\tilde{y}+d\tilde{z})+$$
$$\widetilde{u}_\xi\Big[a\tilde{x}+b(\lambda_0-\lambda+\widetilde{u}\tilde{x})+c\widetilde{v}\tilde{x}+d\widetilde{w}\tilde{x}\Big]+$$
$$\widetilde{v}_\xi\Big[a\tilde{y}+b\widetilde{u}\tilde{y}+c(\lambda_0-\lambda+\widetilde{v}\tilde{y})+d\widetilde{w}\tilde{y}\Big]+$$
$$\widetilde{w}_\xi\Big[a\tilde{z}+b\widetilde{u}\tilde{z}+c\widetilde{v}\tilde{z}+d(\lambda_0-\lambda+\widetilde{w}\tilde{z})\Big]=0 \tag{16.90}$$

An ordinary set of differential equations can be defined by setting the coefficients of the partial spatial derivatives, i.e., the terms into brackets in (16.90) to be zero

$$\frac{1}{\beta}a(\widetilde{p}-p_l)+b(\widetilde{u}-u_l)+c(\widetilde{v}-v_l)+d(\widetilde{w}-w_l)=0\,, \tag{16.91}$$

$$-\frac{a}{\beta}\lambda+b\tilde{x}+c\tilde{y}+d\tilde{z}=0\,, \tag{16.92}$$

$$a\tilde{x}+b(\lambda_0-\lambda+\widetilde{u}\tilde{x})+c\widetilde{v}\tilde{x}+d\widetilde{w}\tilde{x}=0\,, \tag{16.93}$$

$$a\tilde{y}+b\widetilde{u}\tilde{y}+c(\lambda_0-\lambda+\widetilde{v}\tilde{y})+d\widetilde{w}\tilde{y}=0\,, \tag{16.94}$$

$$a\tilde{z}+b\widetilde{u}\tilde{z}+c\widetilde{v}\tilde{z}+d(\lambda_0-\lambda+\widetilde{w}\tilde{z})=0\,. \tag{16.95}$$

The eigenvalues of the system of the above equations are

$$\left.\begin{aligned}\lambda_0&=\widetilde{u}\tilde{x}+\widetilde{v}\tilde{y}+\widetilde{w}\tilde{z}\\ \lambda_1&=\lambda_0+s\\ \lambda_2&=\lambda_0-s\end{aligned}\right\}\,, \tag{16.96}$$

where $s=\sqrt{{\lambda_0}^2+\beta}$.

A non-trivial solution of (16.91)-(16.95) can be found for each of the eigenvalues. Specifically, for $\lambda_0=u\tilde{x}+v\tilde{y}+w\tilde{z}$, where u,v,w are the velocities calculated at the previous time step, we obtain

$$\tilde{x}(\widetilde{w}-w_0)-\tilde{z}(\widetilde{u}-u_0)=0\,, \tag{16.97}$$

$$\tilde{x}(\widetilde{v}-v_0)-\tilde{y}(\widetilde{u}-u_0)=0\,. \tag{16.98}$$

For $\lambda_1=\lambda_0+s$:

$$\widetilde{p}=p_1-\lambda_1\Big[\tilde{x}(\widetilde{u}-u_1)+\tilde{y}(\widetilde{v}-v_1)+\tilde{z}(\widetilde{w}-w_1)\Big]\,. \tag{16.99}$$

For $\lambda_2=\lambda_0-s$:

$$\widetilde{p}=p_2-\lambda_2\Big[\tilde{x}(\widetilde{u}-u_2)+\tilde{y}(\widetilde{v}-v_2)+\tilde{z}(\widetilde{w}-w_2)\Big]\,. \tag{16.100}$$

Eqs. (16.97)-(16.100) can be solved to obtain the values of $\widetilde{p}, \widetilde{u}, \widetilde{v}$ and $\widetilde{w}$ as functions of the characteristic values p_l, u_l, v_l and w_l $(l = 0, 1, 2)$. After some algebra we obtain

$$\widetilde{\mathbf{U}} = \begin{pmatrix} \widetilde{p} \\ \widetilde{u} \\ \widetilde{v} \\ \widetilde{w} \end{pmatrix} = \begin{pmatrix} \frac{1}{2s}(\lambda_1 k_2 - \lambda_2 k_1) \\ R\tilde{x} + u_0(\tilde{y}^2 + \tilde{z}^2) - v_0\tilde{x}\tilde{y} - w_0\tilde{x}\tilde{z} \\ R\tilde{y} + v_0(\tilde{x}^2 + \tilde{z}^2) - w_0\tilde{z}\tilde{y} - u_0\tilde{x}\tilde{y} \\ R\tilde{z} + w_0(\tilde{y}^2 + \tilde{x}^2) - v_0\tilde{z}\tilde{y} - u_0\tilde{x}\tilde{z} \end{pmatrix}, \tag{16.101}$$

where

$$R = \frac{1}{2s}\Big[p_1 - p_2 + \tilde{x}(\lambda_1 u_1 - \lambda_2 u_2) + \tilde{y}(\lambda_1 v_1 - \lambda_2 v_2) + \tilde{z}(\lambda_1 w_1 - \lambda_2 w_2)\Big], \tag{16.102}$$

$$k_1 = p_1 + \lambda_1(u_1\tilde{x} + v_1\tilde{y} + w_1\tilde{z}), \tag{16.103}$$

$$k_2 = p_2 + \lambda_2(u_2\tilde{x} + v_2\tilde{y} + w_2\tilde{z}). \tag{16.104}$$

The variables (16.101) are the *reconstructed characteristics-based* variables which are used in the calculation of the advective intercell fluxes $\bar{\mathbf{E}}_{i\pm1/2,j,k}(\widetilde{\mathbf{U}})$.

Note that the eigenvalues are functions of u, v and w which have been calculated at the previous pseudotime level τ. Alternatively, the eigenvalues can also be calculated using the reconstructed variables $\widetilde{u}, \widetilde{v}$ and $\widetilde{w}$ by performing the following procedure (as an iterative loop):

1. Initially calculate the eigenvalues using u, v, w from the pseudotime level τ.
2. Proceed with the calculation of the "tilde" variables using the CB scheme.
3. Recalculate the eigenvalues using the "tilde" variables.
4. Recalculate the "tilde" variables using the new eigenvalues.
5. Calculate the advective flux.

The above procedure increases the computing time per time step, but its advantages in terms of accuracy yet remain to be investigated.

16.4.4 Reconstructed Characteristics-Based Variables in Two Dimensions

For a two-dimensional flow (16.101) to (16.104) are written

$$\widetilde{\mathbf{U}} = \begin{pmatrix} \widetilde{p} \\ \widetilde{u} \\ \widetilde{v} \end{pmatrix} = \begin{pmatrix} \frac{1}{2s}(\lambda_1 k_2 - \lambda_2 k_1) \\ R\tilde{x} + \tilde{y}(u_0\tilde{y} - v_0\tilde{x}) \\ R\tilde{y} - \tilde{x}(u_0\tilde{y} - v_0\tilde{x}) \end{pmatrix}, \tag{16.105}$$

where

$$R = \frac{1}{2s}\Big[p_1 - p_2 + \tilde{x}(\lambda_1 u_1 - \lambda_2 u_2) + \tilde{y}(\lambda_1 v_1 - \lambda_2 v_2)\Big] , \tag{16.106}$$

$$k_1 = p_1 + \lambda_1(u_1\tilde{x} + v_1\tilde{y}) , \tag{16.107}$$

$$k_2 = p_2 + \lambda_2(u_2\tilde{x} + v_2\tilde{y}) . \tag{16.108}$$

and $\lambda_0 = u\tilde{x} + v\tilde{y}$.

The reconstructed variables (16.101) and (16.105) can be used to calculate the intercell advective fluxes in three and two dimensions, respectively, if the characteristic variables p_l, u_l, v_l and w_l ($l = 0, 1, 2$) are known. In the next section a high-order interpolation procedure [178, 177] for the characteristic variables is presented.

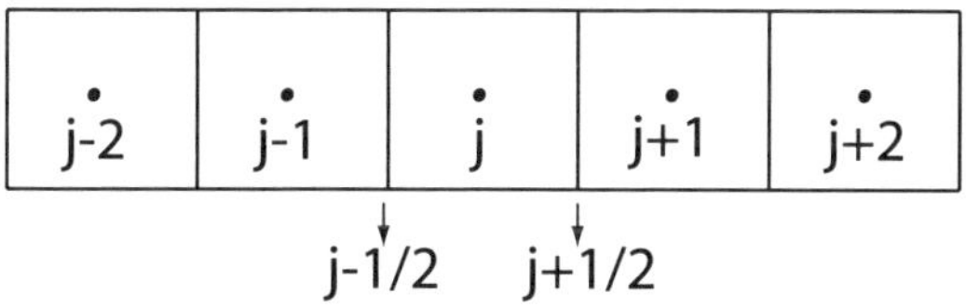

Fig. 16.6. One-dimensional stencil used to define the high-order interpolation.

16.4.5 High-Order Interpolation

To complete the presentation of the characteristics-based scheme, here we discuss different orders of interpolation for the calculation of the primitive variables. Further discussion about high-order schemes follows in Chap. 17.

Consider the one-dimensional stencil (equidistant grid in the computational space) shown in Fig. 16.6 and define two states, left and right, for the intercell characteristic variables, as follows

$$\mathbf{U}_{L,j+1/2} = a\mathbf{U}_j - b\mathbf{U}_{j-1} + c\mathbf{U}_{j+1} + d\mathbf{U}_{j+2} , \tag{16.109}$$

for the left state, and

$$\mathbf{U}_{R,j+1/2} = a\mathbf{U}_{j+1} - b\mathbf{U}_{j+2} + c\mathbf{U}_j + d\mathbf{U}_{j-1} , \tag{16.110}$$

for the right state. The coefficients a, b, c and d need to be determined.

The derivative of the characteristic variable at the cell center for the case of a positive eigenvalue - the result will be analogous if a negative eigenvalue is considered - yields

$$\left(\frac{\partial \mathbf{U}}{\partial \xi}\right)_j = \mathbf{U}_{L,j+1/2} - \mathbf{U}_{L,j-1/2} =$$
$$a\mathbf{U}_j - b\mathbf{U}_{j-1} + c\mathbf{U}_{j+1} + d\mathbf{U}_{j+2} - =$$
$$(a\mathbf{U}_{j-1} - b\mathbf{U}_{j-2} + c\mathbf{U}_j + d\mathbf{U}_{j+1})$$
$$(a-c)\mathbf{U}_j - (a+b)\mathbf{U}_{j-1} + b\mathbf{U}_{j-2}$$
$$+(c-d)\mathbf{U}_{j+1} + d\mathbf{U}_{j+2} \,. \tag{16.111}$$

By developing all variables in a Taylor series expansion around the cell center j, (16.111) yields

$$\left(\frac{\partial \mathbf{U}}{\partial \xi}\right)_j = (a-c)\mathbf{U}_j -$$
$$(a+b)\Big[\mathbf{U}_j - \mathbf{U}^{(1)} + \mathbf{U}^{(2)} - \mathbf{U}^{(3)} + \mathbf{U}^{(4)}\Big]$$
$$+b\Big[\mathbf{U}_j - 2\mathbf{U}^{(1)} + 4\mathbf{U}^{(2)} - 8\mathbf{U}^{(3)} + 16\mathbf{U}^{(4)}\Big]$$
$$+(c-d)\Big[\mathbf{U}_j + \mathbf{U}^{(1)} + \mathbf{U}^{(2)} + \mathbf{U}^{(3)} + \mathbf{U}^{(4)}\Big]$$
$$+d\Big[\mathbf{U}_j + 2\mathbf{U}^{(1)} + 4\mathbf{U}^{(2)} + 8\mathbf{U}^{(3)} + 16\mathbf{U}^{(4)}\Big] \,. \tag{16.112}$$

The superscripts denote order of derivatives. The denominators in the Taylor series expansion have been omitted and can be considered to be part of the unknown coefficients which are yet to be determined. Further, the grid spacing is considered to be equal to one since we are working in the computational space. Equation (16.112) can be written as

$$\left(\frac{\partial \mathbf{U}}{\partial \xi}\right)_j = (a-b+c+d)\mathbf{U}^{(1)} + \Big[c-a+3(b+d)\Big]\mathbf{U}^{(2)}$$
$$\Big[c+a+7(d-b)\Big]\mathbf{U}^{(3)} + \Big[c-a+15(b+d)\Big]\mathbf{U}^{(4)} \,. \tag{16.113}$$

Using (16.113) schemes of different order of accuracy can be derived.

- First-order upwind scheme for

$$a = 1 \quad \text{and} \quad b = c = d = 0 \,. \tag{16.114}$$

The left and right states of the variables at the cell face are accordingly defined by

$$\mathbf{U}_{L,j+1/2} = \mathbf{U}_j \quad , \quad \mathbf{U}_{R,j+1/2} = \mathbf{U}_{j+1} \,. \tag{16.115}$$

- The second-order scheme is obtained for $c = d = 0$,

$$a - b = 1 \,, \tag{16.116}$$

for satisfying the CFL like restriction [177], i.e., having the coefficient of the first-order derivative equal to one, and

$$3b - a = 0\,, \tag{16.117}$$

for eliminating the second-order derivative term from (16.113). From (16.116) and (16.117) the values $a = 3/2$ and $b = 1/2$ are obtained. The left and right states are accordingly defined by

$$\left.\begin{aligned} \mathbf{U}_{L,j+1/2} &= \frac{3}{2}\mathbf{U}_j - \frac{1}{2}\mathbf{U}_{j-1} \\ \mathbf{U}_{R,j+1/2} &= \frac{3}{2}\mathbf{U}_{j+1} - \frac{1}{2}\mathbf{U}_{j+2} \end{aligned}\right\}\,. \tag{16.118}$$

- The third-order scheme is obtained for $d = 0$, the CFL-like restriction

$$a - b + c = 1\,, \tag{16.119}$$

and the following conditions for eliminating the second- and third-order derivative terms from (16.113)

$$\left.\begin{aligned} 3b - a + c &= 0 \\ a - 7b + c &= 0 \end{aligned}\right\}\,. \tag{16.120}$$

Eqs. (16.119) and (16.120) give the values $a = 5/6, b = 1/6$ and $c = 1/3$. The left and right states are accordingly defined

$$\left.\begin{aligned} \mathbf{U}_{L,j+1/2} &= \frac{5}{6}\mathbf{U}_j - \frac{1}{6}\mathbf{U}_{j-1} + \frac{1}{3}\mathbf{U}_{j+1} \\ \mathbf{U}_{R,j+1/2} &= \frac{5}{6}\mathbf{U}_{j+1} - \frac{1}{6}\mathbf{U}_{j+2} + \frac{1}{3}\mathbf{U}_j \end{aligned}\right\}\,. \tag{16.121}$$

- Similarly, one can obtain the fourth-order scheme

$$\mathbf{U}_{R,L,j+1/2} = \frac{1}{12}(7\mathbf{U}_j + 7\mathbf{U}_{j+1} - 7\mathbf{U}_{j-1} - 7\mathbf{U}_{j+2})\,. \tag{16.122}$$

The interpolation formulas (16.115), (16.118), (16.121) and (16.122) can be used for calculating the characteristic variables p_l, u_l, v_l and w_l $(l = 0, 1, 2)$ for each of the three eigenvalues. The decision on the selection of the left or right state can be made according to the sign of the local (intercell) eigenvalue according to the formula

$$\mathbf{U}_{l,j+1/2} = \frac{1}{2}\Big\{\big[1+\mathrm{sign}(\lambda_l)\big]\mathbf{U}_{R,j+1/2} + \big[1-\mathrm{sign}(\lambda_l)\big]\mathbf{U}_{L,j+1/2}\Big\}\,. \tag{16.123}$$

Eq. (16.123) completes the calculation of the characteristic variables and consequently the calculation of the advective fluxes in the context of the CB scheme.

An alternative interpolation formula for the intercell values along the characteristics can be obtained by using the MUSCL scheme [568]. The MUSCL scheme defines the left and right states by

$$\left.\begin{aligned} \mathbf{U}_{L,j+1/2} &= \mathbf{U}_j + \frac{g_j}{4}\Big[(1-kg_j)\nabla + (1+kg_j)\Delta\Big]\mathbf{U}_j \\ \mathbf{U}_{R,j+1/2} &= \mathbf{U}_{j+1} - \frac{g_{j+1}}{4}[(1+kg_{j+1})\nabla \\ &\quad +(1-kg_{j+1})\Delta]\mathbf{U}_{j+1} \end{aligned}\right\}, \tag{16.124}$$

where the parameter k controls different MUSCL realizations: fully upwind for $k = -1$, third-order for $k = 1/3$ (for $k = 1/3$ the scheme is strictly third-order only for one-dimensional problems), and centered for $k = 1$; g_j is the van Albada limiter [561] (see also Sect. 14.2.1)

$$g_j = \frac{r + r^2}{1 + r^2 + \epsilon}, \tag{16.125}$$

where $r = \nabla\mathbf{U}_j/\Delta\mathbf{U}_j$, $\nabla\mathbf{U}_j = \mathbf{U}_j - \mathbf{U}_{j-1}$, $\Delta\mathbf{U}_j = \mathbf{U}_{j+1} - \mathbf{U}_j$, and ϵ is a small positive constant preventing division by zero.

Remark 16.4.1. In the implementation of high-order interpolation, one or two fictitious values (depending on the order of interpolation) need to be specified inside the computational boundaries. The authors' experience is that extrapolation of these values from the interior of the domain works satisfactorily. In the case of wall boundaries, these values can still be extrapolated without violating the no-slip and no-penetration condition (see Chap. 10) if the flux component normal to the wall boundary is set equal to zero (see the advective flux subroutine in Appendix 2).

16.4.6 Advective Flux Calculation

We summarize the steps for the calculation of the advective flux:

- **Step 1**: Calculate the three eigenvalues λ_l for $l = 0, 1, 2$ using the velocities u, v and w from the previous time step.
- **Step 2:** Use one set from the interpolation formulas (16.115), (16.118), (16.121) or (16.124) to calculate the left and right states of the characteristic variables. The selection depends on the order of accuracy defined by the numerical analyst.
- **Step 3:** Use (16.123) to calculate the characteristic variables. For each eigenvalue, i.e., for each characteristic, we calculate one set of primitive variables p_l, u_l, v_l and w_l.
- **Step 4:** Use (16.101) to (16.104) for three-dimensional, or (16.105) to (16.108) for two-dimensional, computations to calculate the reconstructed variables $\widetilde{\mathbf{U}}$.

- **Step 5**: Use the reconstructed variables to calculate the intercell advective flux, $\bar{\mathbf{E}}_{j+1/2} \equiv [\bar{\mathbf{E}}(\widetilde{\mathbf{U}})]_{j+1/2}$.

The above five steps are also performed for the calculation of the advective fluxes in η and ζ directions. Then, the discretized flux derivatives are added (including the viscous fluxes in the case of the Navier-Stokes equations) and the system of equations is integrated in time using a time integration scheme (see Chaps. 7 and 10). An example of a FORTRAN subroutine for the discretization of the advective flux is in the Appendix B.1

16.4.7 Results

We demonstrate the accuracy of the characteristics-based method by presenting results from computations of three-dimensional flows in straight and curved channels, respectively [157]. The flow in a straight channel is considered at Reynolds number of 100, based on the centerline velocity and channel width. The calculations are performed on a single quadrant of the channel due to the symmetry. The grid contains 58 points in the streamwise direction and 39×39 in the quadrant cross section. The grid is slightly clustered in the $x-$direction near the channel entrance, while it is uniform in the other directions. The computed pressure coefficient along the channel centerline is compared with the experimental data of Beavers et al. [43] (Fig. 16.7).

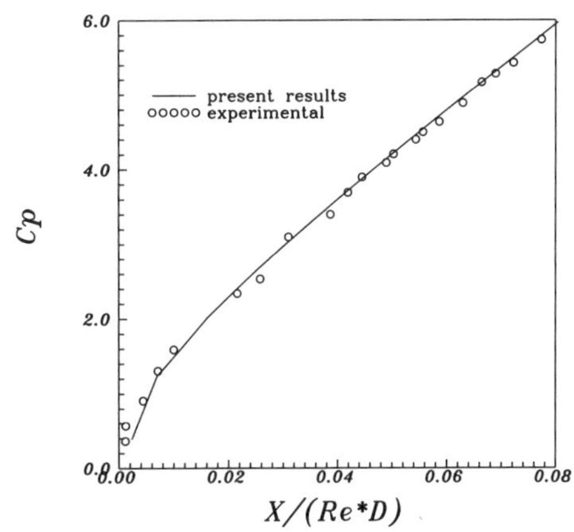

Fig. 16.7. Pressure distribution along the centerline of a straight channel for the flow at $Re = 100$. Comparison of computational results, using the CB scheme, with the experimental data of Beavers et al. [43].

The axial development of the streamwise velocity at the channel centerline as well as the velocity profile at $X/(D * Re) = 0.02$, where D is the channel width, are compared with the corresponding laser Doppler velocimetry measurements of Goldstein and Kreid [218] (Fig. 16.8). These computations have been performed using the third-order interpolation formula (16.121) for the left and right states of the characteristic variables.

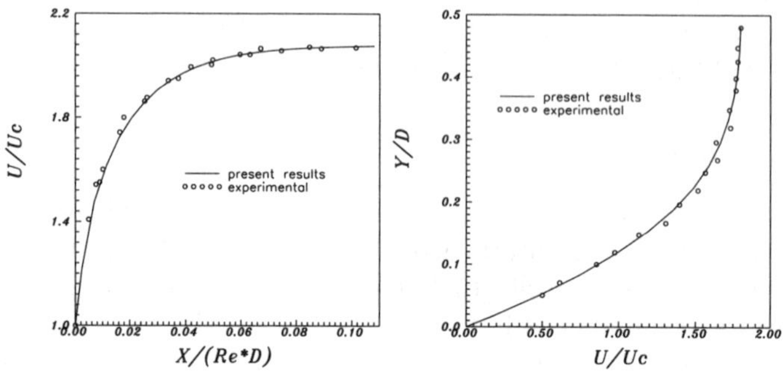

Fig. 16.8. Comparison of computational results, using the CB scheme, with the experimental data of Goldstein and Kreid [218] for the three-dimensional flow in a straight channel.

Further, computations for the flow in a 90° bend (Fig. 16.9) at $Re = 790$ have been performed in [157] using the CB scheme (in conjunction with (16.121)).

This flow has been studied experimentally by Humphrey et al. [269]. The mean radius of the bend is 92 mm attached to the end of rectangular channel of 40×40 mm cross-section. A straight extension section is attached upstream of the bend entrance. The parameters of the experiment are such that the bend has large enough turning angle and small enough mean radius to generate severe distortion and significant secondary flow. The fine grid used in [157] has 80 nodes in the streamwise direction, and 80×40 in the transverse plane, i.e., a total 256,000 grid points. Computations were carried out using the half section of the duct in the $z-$direction because of symmetry. As inflow conditions the corresponding developed flow in a straight duct at $Re = 790$ was imposed at the inlet. In Fig. 16.10 the formation of secondary flow at $\theta = 90^o$ is shown.

In Figs. 16.11 and 16.12 comparisons of the computations with the experimental results of [269] are shown. The comparisons are presented at two different radial locations, $\tilde{R} = 0.3$ and $\tilde{R} = 0.7$, for angles $\theta = 60^o$ and $\theta = 90^o$. The $\tilde{R}$ is defined by: $\tilde{R} = (R - R_o)/(R_j - R_o)$, where R_j and R_o are the inner and outer radius, respectively.

A challenging area for implementation of high-resolution and high-order schemes (these are discussed in more detail in a subsequent chapter) is when a flow exhibits transitional flow features. At its early stages, flow transition

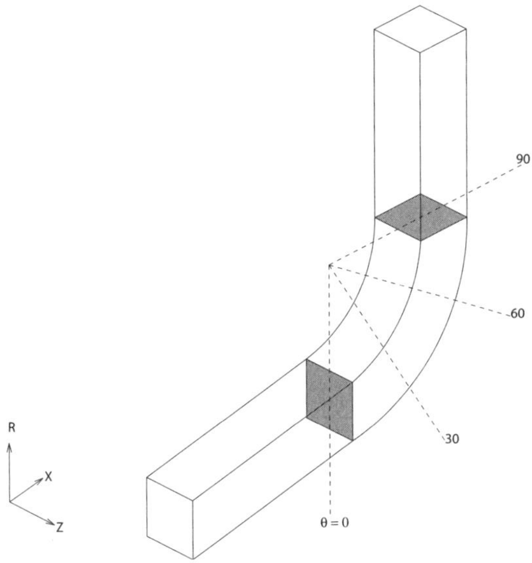

Fig. 16.9. Schematic of the 90° curved channel.

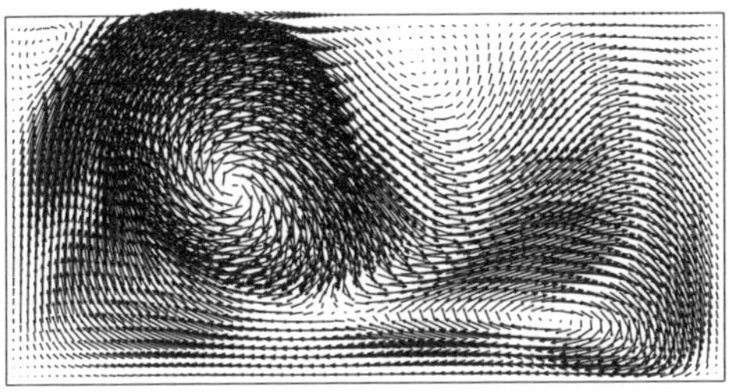

Fig. 16.10. Formation of the secondary flow at $\theta = 90°$ for the curved channel.

is associated with instabilities and flow bifurcations [146]. Under such conditions flows may become unstable leading to turbulence, or may return to stable laminar flow conditions. Computation of flows featuring instabilities is a challenging task for any numerical scheme because numerical dissipation can suppress the instabilities, thus incorrectly appearing the flow to be laminar. Similarly, numerical dispersion can lead to spurious fluctuations, which can subsequently trigger the flow to become falsely unstable. Here, we show an example from implementation of the third-order variant of the characteristics-based scheme in pulsatile flow through a pipe with a stenosis.

In the context of biofluid mechanics, the computational study of flow through a stenosis is motivated by the need to obtain a better understand-

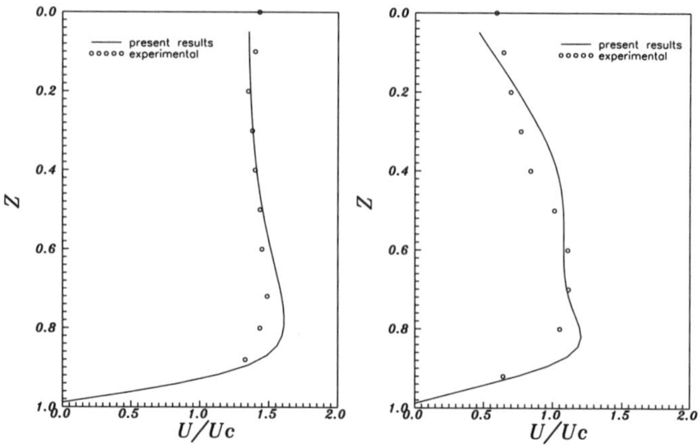

Fig. 16.11. Comparison of computational results using the CB scheme with the experimental data of Humphrey et al. [269] at $\theta = 60°$, $\bar{R} = 0.3$ (left plot) and $\bar{R} = 0.7$ (right plot).

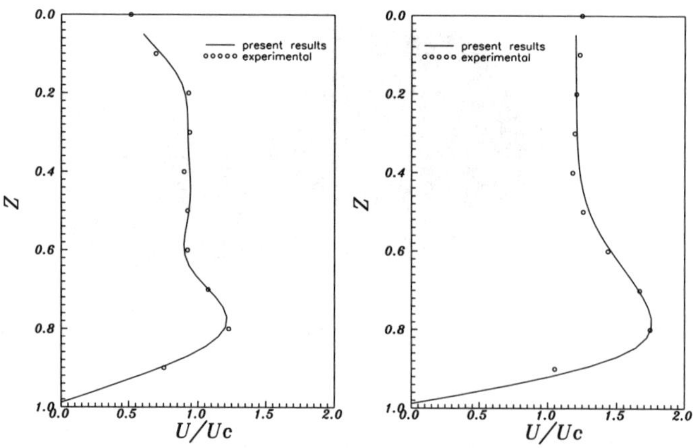

Fig. 16.12. Comparison of computational results using the CB scheme with the experimental data of Humphrey et al. [269] at $\theta = 90°$, $\bar{R} = 0.3$ (left plot) and $\bar{R} = 0.7$ (right plot).

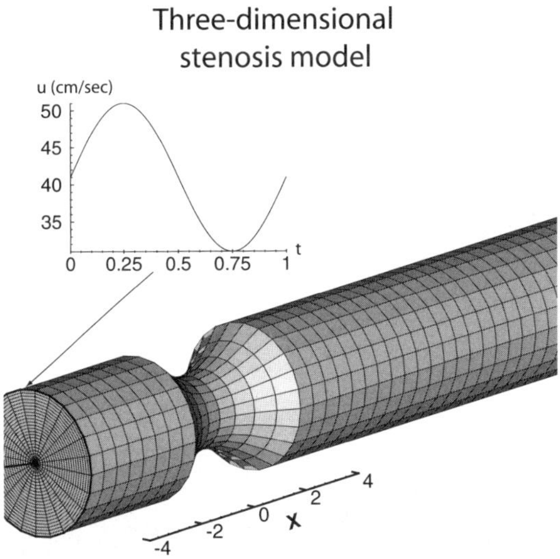

Fig. 16.13. Inlet velocity profile and stenosis geometry for the computation of three-dimensional flow using the characteristics-based scheme.

ing of the impact of flow phenomena on diseases such as atherosclerosis and stroke. The flow phenomena occurring in stenotic arteries include asymmetric flow separation, instabilities and laminar-to-turbulent transition. These phenomena may have significant effects on the wall shear stress (WSS). Experimental studies have shown that in pulsatile flows both high and low WSS values have important hemodynamic effects [306, 96]; the former because of their magnitude and the latter because of their rapid variations in space and time.

In [366, 367] the characteristics-based scheme was employed to simulate the three-dimensional, pulsatile flow through a stenosis. The geometry consists of an axisymmetric stenosis with 75% reduction in the cross-sectional area, i.e., the stenotic area is 25% of the pipe area (Fig. 16.13). Several numerical experiments have been conducted using different pipe lengths upstream and downstream of the stenosis in order to ensure independence of the results from the position of the inflow and outflow boundaries; it was found that $2D$ and $70D$ lengths upstream and downstream, respectively, are sufficient. Time-dependent, three-dimensional flow computations were carried out on a grid containing $250,000$ ($400 \times 25 \times 25$) cells; the grid was non-uniform in the radial direction with a clustering of grid lines in the near wall region. The dimensions of the pipe were $2D$ and $70D$ lengths upstream and downstream, respectively. The instantaneous Reynolds number, based on the centreline temporally-averaged streamwise velocity and the pipe radius, has minimum and maximum values of 760 and 1245, respectively. The flow pa-

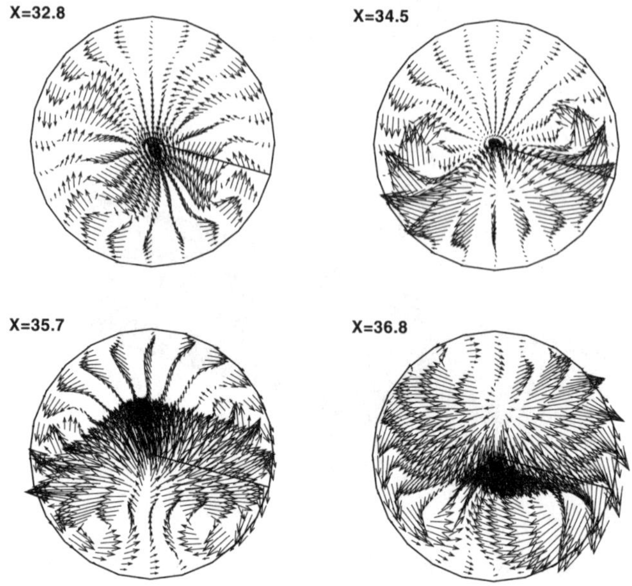

Fig. 16.14. Velocity vectors at different cross sections showing the disturbed secondary flow downstream of the stenosis. The results have been obtained using the characteristics-based scheme.

rameters correspond to a pulsatile frequency number (Womersley number) $\alpha = R(\omega/\nu)^{1/2} = 9.87$, where R is the pipe radius, ω is the frequency of the pulsatile velocity profile and ν is the kinematic viscosity.

In the numerical simulation the onset of instability can be detected as asymmetric flow in the cross-sectional planes. The instability results in substantial asymmetries within the separated flow region downstream of the stenosis. These asymmetries will lead to disturbed secondary flows as can be seen from the velocity vectors plotted at the different positions downstream of the stenosis (Fig. 16.14). Note that in the case of stable flow the solution will remain axisymmetric throughout the domain, even when random perturbations are imposed on the initial velocity profiles. This is the case when a first or a second-order (artificial-viscosity-type) numerical scheme is utilized in the computations.

Here, the separated flow is extended up to 60 radii downstream of the stenosis. With respect to the spatial growth of the instability, the simulations reveal that important flow changes occur within two downstream regions. These can be observed in the isosurfaces of the streamwise velocity at different time instants (Fig. 16.15). The first region is closer to the stenosis and encompasses the fluid jet arising from the stenotic region. In this re-

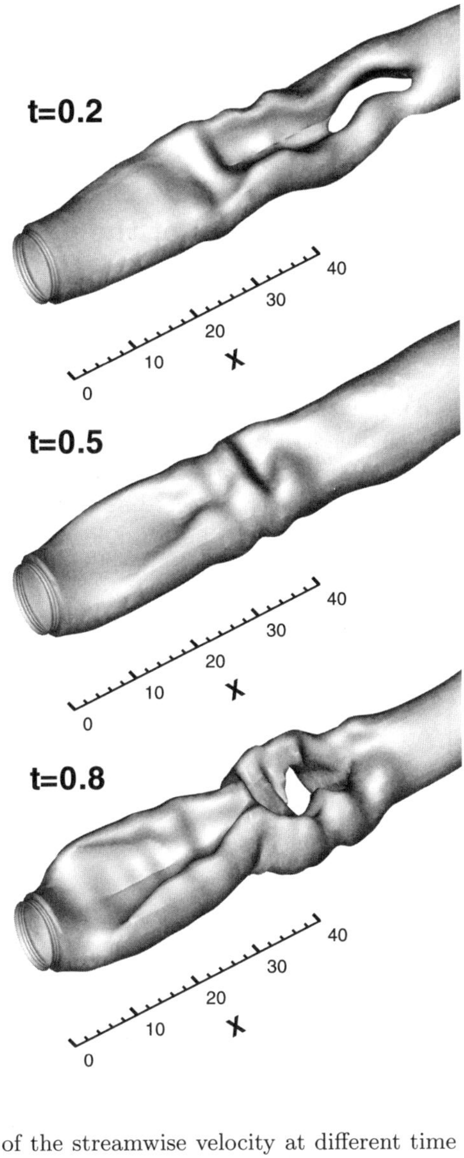

Fig. 16.15. Isosurfaces of the streamwise velocity at different time instants: $u = 0.96$ at $t = 0.2$; $u = 0.66$ at $t = 0.5$; and $u = 0.62$ at $t = 0.8$. The results have been obtained using the characteristics-based scheme.

gion, the instability has been established, but has not broken the coherence of the fluid jet. The second region is further downstream where substantial variations of the flowfield occur. In this region, the coherence of the fluid jet cannot be maintained due to the swirling motion of the fluid. Although the lengths of these regions vary among different time instants, the largest flow variations take place within $x = 30 - 50$ radii downstream of the stenosis.

16.5 Flux Limiting Version of the CB Scheme

The flux limiting approach can be combined with the CB scheme using

$$\mathbf{E}^{\mathrm{TVD}}_{j\pm1/2} = \mathbf{E}^{\mathrm{LF}}_{j\pm1/2} + \psi_{j\pm1/2}(\mathbf{E}^{\mathrm{CB}}_{j\pm1/2} - \mathbf{E}^{\mathrm{LF}}_{j\pm1/2}) , \tag{16.126}$$

where $\mathbf{E}^{\mathrm{LF}}$ is the Lax-Friedrich flux - given by (15.20) and (15.21) in Chap. 15 - and $\mathbf{E}^{\mathrm{CB}}$ is the characteristics-based flux defined using the characteristics-based variables (16.101) and the reconstruction steps described in (16.4.6). The limiter $\psi_{j\pm1/2}$ can be defined according to the CB-limiter (16.51).

The TVD scheme (16.126) has been implemented in [149] using the CB-limiter as well as using the superbee limiter[4] [466]

$$\psi_{\mathrm{SB}} = \begin{cases} 0 & \text{if} \quad r \le 0 , \\ 2r & \text{if } 0 \le r \le \frac{1}{2} , \\ 1 & \text{if } \frac{1}{2} \le r \le 1 , \\ \min\left[2, \psi_g + (1-\psi_g) r\right] & r > 1 , \end{cases} \tag{16.127}$$

where $\psi_g = (1 - \mathcal{C})/(1 + \mathcal{C})$.

16.6 Implementation of the Characteristics-Based Method in Unstructured Grids

In the preceding paragraphs we presented the development of the characteristics-based method in conjunction with structured grids. The method has also been implemented in unstructured grids for incompressible flow simulations [617]. In [617], the incompressible Navier-Stokes equations were discretized on an unstructured tetrahedral grid using a cell-vertex approach according to which the primitive variables are stored at the vertices of the tetrahedral cells. For each vertex, a control volume is constructed as shown in Fig. 16.16.

The convective terms are discretized using an edge-based procedure

[4] In Sect. 14.2.1, the superbee limiter is presented in its original form without taking into account the time-dependent effect. The formula here is equivalent with that of Sect. 14.2.1 except for the inclusion of the CFL number.

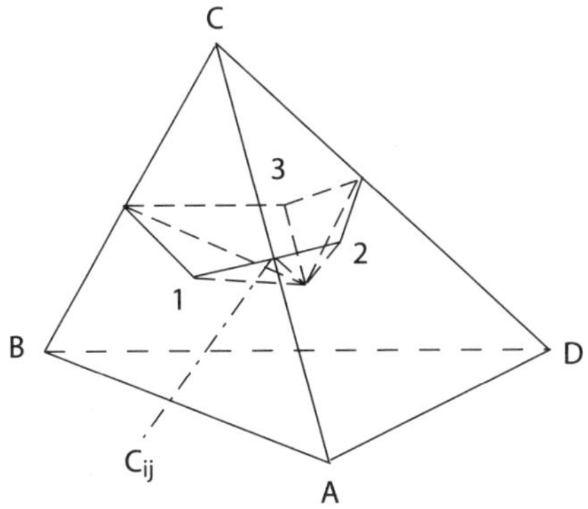

Fig. 16.16. Structure of control volume within a tetrahedron.

$$\int\int_V\int \nabla \cdot \mathbf{E} \, \mathrm{d}V = \int\int_S \mathbf{E} \cdot \mathbf{n} \, \mathrm{d}S = \sum_{n=1}^{nth-edge} (\mathbf{E} \cdot \mathbf{n} \, \Delta S)_n \,, \tag{16.128}$$

where ΔS_n is the surface associated with the egde n, for example, the edge CA in Fig. 16.16.

For example, the viscous terms, $\mathbf{R}$ (in $x-$direction), are calculated according to a cell-based approach.

$$\int\int_V\int \nabla \mathbf{R} \, \mathrm{d}V = \int\int_S \mathbf{R} \cdot \mathrm{d}\mathbf{S} = \sum_{j}^{nth-cell} (\mathbf{R} \cdot \mathbf{n} \, \Delta S)_j \,, \tag{16.129}$$

where ΔS_j is the surface of cell i, for example, the cell C_{ij} in Fig. 16.16. Using the relation

$$\int\int_S \mathrm{d}\mathbf{S} = 0 \,, \tag{16.130}$$

the total vector surface of the control volume in a cell i is calculated by

$$\Delta S_j = \frac{1}{3} \Delta S_{Ci} \,, \tag{16.131}$$

where ΔS_{Ci} is the surface vector of the face opposite node C of the tetrahedron under consideration. Using (16.131), the viscous terms are calculated by

$$\int\int_S \mathbf{R} \cdot \mathrm{d}\mathbf{S} = \sum_{j}^{nth-cell} (\mathbf{R} \cdot \mathbf{n} \, \Delta S)_j = \frac{1}{3} \sum_{j}^{nth-C-cell} (\mathbf{R} \cdot \mathbf{n} \, \Delta S_C)_j \,, \tag{16.132}$$

The gradient of a flow variable at the centre of a tetrahedron is

$$\text{grad } \psi_j = -\frac{\sum_{j=1}^{4} 9\psi_j \mathbf{S}_j}{27V} = -\frac{1}{3}\frac{\sum_{j=1}^{4} \psi_j \mathbf{S}_j}{V} \tag{16.133}$$

where ψ_j is the flow variable at the vertex j of the tetrahedron, $\mathbf{S}_j$ is the surface vector that is opposite to node j, and V is the volume of the tetrahedron.

Using the above definitions the characteristics-based scheme can be implemented in an unstructured grid environment. Results for incompressible flows around a cylinder and a sphere, as well as in a three-dimensional cavity have been presented in [617]. Zhao and Tai [617] have shown that the characteristics-based method is simpler in the implementation than Roe's method and less diffusive than a second-order central scheme with artificial viscosity.

16.7 The Weight Average Flux Method

16.7.1 Basic Formulation

The weighted average flux (WAF) method [543, 537, 538, 540, 58] is a generalization of the Lax-Wendroff and the Godunov first-order upwind method. The origin of the method lies in the random flux scheme [543]. The WAF method is deterministic and lead to second-order explicit schemes. Here, we discuss the WAF method in the context of non-linear systems, for example, the incompressible Euler and Navier-Stokes equations.

We consider the system of conservation laws (16.222) discretized by an explicit scheme (16.223). According to the WAF method the intercell flux is defined by

$$\mathbf{E}_{j+1/2} = \frac{1}{\Delta x}\int_{-\Delta x/2}^{\Delta x/2} \mathbf{E}[\mathbf{U}_{i+/2}(x, \frac{\Delta t}{2})]\, dx \,, \tag{16.134}$$

where $\mathbf{U}_{j+1/2}$ is the solution of the Riemann problem with piecewise constant data $\mathbf{U}_j^n$, $\mathbf{U}_{j+1}^n$ at the cell face $i + 1/2$. For the solution of the Riemann problem using approximate Riemann solvers different approaches have been proposed in the literature and are reviewed in [543, 596]. For the three wave speeds the integral (16.134) can be evaluated by

$$\mathbf{E}_{j+1/2} = \sum_{k=1}^{N} b_k \mathbf{E}_{j+1/2}^{(k)} \,, \tag{16.135}$$

where $\mathbf{E}_{j+1/2}^{(k)} \equiv \mathbf{E}(\mathbf{U}^{(k)})$, N is the number of waves involved in the solution of the Riemann problem and b_k are the (normalized) lengths of the segments $A_{k-1}A_k$ as shown in Fig. 16.17, that is

$$b_k = \frac{|A_{k-1}A_k|}{\Delta x} \,. \tag{16.136}$$

The simplest way to calculate the wave speeds S_k (Fig. 16.17) is directly by the eigenvalues similar to David's suggestion [131] for the estimation of wave speeds for the case of the Euler equations of gas dynamics. The weight coefficients can also be defined in terms of the wave speeds S_k as

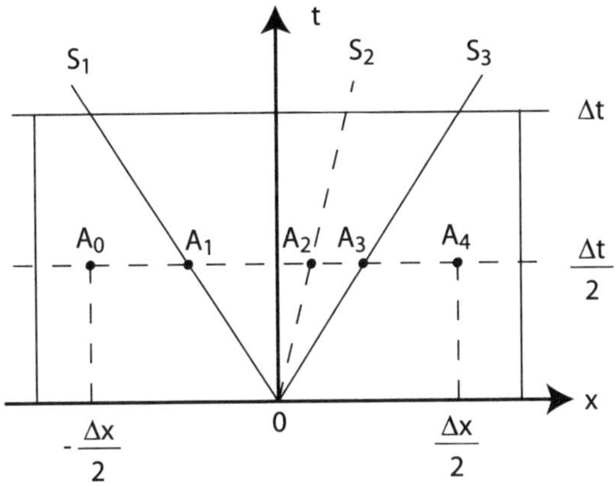

Fig. 16.17. Schematic for the evaluation of the WAF intercell flux.

$$b_k = \frac{1}{2}(\mathcal{C}_k - \mathcal{C}_{k-1}) \,, \tag{16.137}$$

where

$$\mathcal{C}_k = \frac{\Delta t S_k}{\Delta x} \,, \quad \mathcal{C}_0 = -1 \,, \quad \mathcal{C}_{N+1} = 1 \,. \tag{16.138}$$

Substitution of (16.137) into (16.135) gives the WAF flux

$$\mathbf{E}_{j+1/2} = \frac{1}{2}(\mathbf{E}_j + \mathbf{E}_{j+1}) - \frac{1}{2}\sum_{k=1}^{N} \mathcal{C}_k \Delta\mathbf{E}^{(k)}_{j+1/2} \,, \tag{16.139}$$

where

$$\Delta\mathbf{E}^{(k)}_{j+1/2} = \mathbf{E}^{(k+1)}_{j+1/2} - \mathbf{E}^{(k)}_{j+1/2} \,. \tag{16.140}$$

An alternative way to formulate the WAF method is by defining a weighted average state

$$\bar{\mathbf{U}}_{j+1/2} = \sum_{k=1}^{N} b_k \mathbf{U}^{(k)}_{j+1/2} \,, \tag{16.141}$$

with $\mathbf{U}^{(k)}_{j+1/2}$ being the value of $\mathbf{U}_{j+1/2}$ in region k (Fig. 16.17). Using (16.137), (16.141) yields

$$\bar{\mathbf{U}}_{j+1/2} = \frac{1}{2}(\mathbf{U}^n_j + \mathbf{U}^n_{j+1}) - \frac{1}{2}\sum_{k=1}^{N} \mathcal{C}_k \Delta \mathbf{U}^{(k)}_{j+1/2} , \tag{16.142}$$

where

$$\Delta \mathbf{U}^{(k)}_{j+1/2} = \mathbf{U}^{(k+1)}_{j+1/2} - \mathbf{U}^{(k)}_{j+1/2} . \tag{16.143}$$

The intercell flux is subsequently defined by

$$\mathbf{E}_{j+1/2} = \mathbf{E}(\bar{\mathbf{U}}_{j+1/2}) . \tag{16.144}$$

16.7.2 TVD Version of the WAF Schemes

The WAF scheme is a second-order method in space and time. For linear systems there is one wave family, and the TVD condition (property) can be exactly implemented, whereas for non-linear systems such as the Euler or the Navier-Stokes equations there are three wave families. Therefore, the computation of three limiter functions per intercell boundary is required. The TVD version of the WAF flux (16.139) is given by

$$\mathbf{E}_{j+1/2} = \frac{1}{2}(\mathbf{E}_j + \mathbf{E}_{j+1}) - \frac{1}{2}\sum_{k=1}^{N} \operatorname{sign}(\mathcal{C}_k)\phi^{(k)}_{j+1/2}\Delta \mathbf{E}^{(k)}_{j+1/2} , \tag{16.145}$$

where $\phi^{(k)}_{j+1/2}(r^{(k)})$ is the WAF limiter. The parameter $r^{(k)}$ refers to the wave k in the solution $\mathbf{U}_{j+1/2}$ of the Riemann problem and is defined by

$$r = \begin{cases} \dfrac{\Delta q^{(k)}_{j-1/2}}{\Delta q^{(k)}_{j+1/2}} , & \mathcal{C}_k > 0 , \\[2em] \dfrac{\Delta q^{(k)}_{j+3/2}}{\Delta q^{(k)}_{j+1/2}} , & \mathcal{C}_k < 0 , \end{cases} \tag{16.146}$$

where $\Delta q_{j-1/2} = q^{(k)}_j - q^{(k)}_{j-1}$, $\Delta q_{j+1/2} = q^{(k)}_{j+1} - q^{(k)}_j$, and $\Delta q_{j+3/2} = q^{(k)}_{j+2} - q^{(k)}_{j+1}$. The quantity q can be any of the primitive variables p, u, v or w.

The TVD version of the average state (16.142) is defined by

$$\bar{\mathbf{U}}_{j+1/2} = \frac{1}{2}(\mathbf{U}^n_j + \mathbf{U}^n_{j+1}) - \frac{1}{2}\sum_{k=1}^{N} \operatorname{sign}(\mathcal{C}_k)\, \phi_k \, \Delta \mathbf{U}^{(k)}_{j+1/2} , \tag{16.147}$$

and the intercell flux

$$\mathbf{E}_{j+1/2} = \mathbf{E}(\bar{\mathbf{U}}_{j+1/2}) \,. \tag{16.148}$$

Some limiter functions can be any of those given below [543]:

- The van Leer limiter [568]

$$(\phi_{vl})_{j+1/2} = \begin{cases} 1 & \text{if } r \leq 0 \,, \\ 1 - \dfrac{(1-|\mathcal{C}|)2r}{1+r} & \text{if } r \geq 0 \,. \end{cases} \tag{16.149}$$

- The van Albada limiter [561]

$$(\phi_{va})_{j+1/2} = \begin{cases} 1 & \text{if } r \leq 0 \,, \\ 1 - \dfrac{(1-|\mathcal{C}|)r(1+r)}{1+r^2} & \text{if } r \geq 0 \,. \end{cases} \tag{16.150}$$

- The minmod limiter [465]

$$(\phi_{mb})_{j+1/2} = \begin{cases} 1 & \text{if } \quad r \leq 0 \,, \\ 1-(1-|\mathcal{C}|)r & \text{if } \; 0 \leq r \leq 1 \,, \\ |\mathcal{C}| & \text{if } \quad r \geq 1 \,. \end{cases} \tag{16.151}$$

The van Albada's van Leer's limiters were also defined in Sect. 14.2.1 for the case where the time-dependent effect is not taken into account, thus the equivalent formula except for the inclusion of the CFL number.

16.8 Roe's Method

We consider again the one-dimensional hyperbolic conservation law

$$\frac{\partial \mathbf{U}}{\partial t} + \frac{\partial \mathbf{E}}{\partial x} = 0 \,, \tag{16.152}$$

accompanied by an appropriate set of initial and boundary conditions. The Godunov intercell numerical flux is defined by

$$\mathbf{E}_{j+1/2} = \mathbf{E}(\tilde{\mathbf{U}}_{j+1/2}) \,, \tag{16.153}$$

where $\tilde{\mathbf{U}}_{j+1/2}$ is the exact similarity solution $\mathbf{U}_{j+1/2}(x/t)$ of the Riemann problem formulated by (16.152) and the set of initial conditions

$$\mathbf{U}_{j+1/2} = \begin{cases} \mathbf{U}_L & \text{if } x < 0\,, \\ \\ \mathbf{U}_R & \text{if } x > 0\,, \end{cases} \tag{16.154}$$

evaluated at $x/t = 0$. The wave structure of the solution of the Riemann problem for (16.152) is shown in Fig. 16.18.

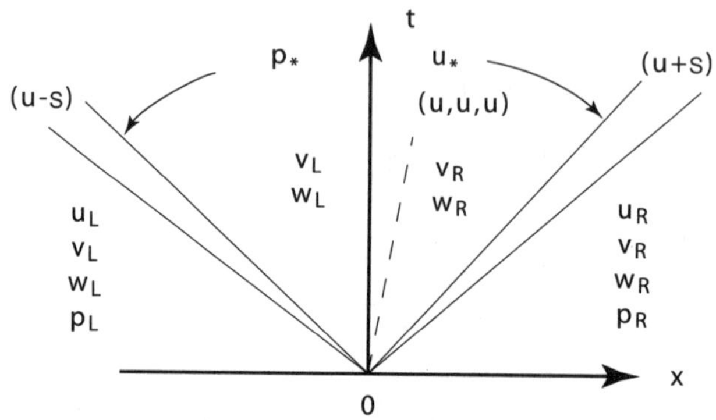

Fig. 16.18. Wave structure of the solution of the Riemann problem for (16.152). For the artificial compressibility method the pseudo-sound speed is given by $s = \sqrt{u^2 + \beta}$.

Roe [463] solved the Riemann problem (16.152) and (16.154) approximately. By introducing the Jacobian matrix, $\mathbf{A}(\mathbf{U})$, of the flux $\mathbf{E}(\mathbf{U})$, (16.152) is written

$$\frac{\partial \mathbf{U}}{\partial t} + \mathbf{A}(\mathbf{U})\frac{\partial \mathbf{U}}{\partial x} = 0\,. \tag{16.155}$$

According to Roe's method, the Jacobian $\mathbf{A}(\mathbf{U})$ is replaced by a Jacobian matrix, which is a function of some intermediate state $\widetilde{\mathbf{U}}$, i.e.,

$$\widetilde{\mathbf{A}} \equiv \widetilde{\mathbf{A}}(\widetilde{\mathbf{U}}) = \widetilde{\mathbf{A}}(\mathbf{U}_L, \mathbf{U}_R)\,, \tag{16.156}$$

thus (16.155) is replaced by

$$\frac{\partial \mathbf{U}}{\partial t} + \widetilde{\mathbf{A}}(\mathbf{U})\frac{\partial \mathbf{U}}{\partial x} = 0\,. \tag{16.157}$$

Subsequently, the Riemann problem is solved for (16.157) with the set of initial conditions (16.154). The Roe's Jacobian matrix is required to satisfy the properties

1. The matrix $\widetilde{\mathbf{A}}$ is required to have real eigenvalues and a complete set of linearly independent right eigenvectors.

2. Consistency with the exact Jacobian, i.e.,

$$\widetilde{\mathbf{A}} = \mathbf{A} \text{ if } \mathbf{U}_L = \mathbf{U}_R = \widetilde{\mathbf{U}} = \mathbf{U} \, . \tag{16.158}$$

3. Conservation across discontinuities

$$\widetilde{\mathbf{A}}(\mathbf{U}_R - \mathbf{U}_L) = \mathbf{E}(\mathbf{U}_R) - \mathbf{E}(\mathbf{U}_L) \, . \tag{16.159}$$

Roe's method [463] has been very popular among the CFD community for computing compressible flows featuring shock waves and other discontinuities. The method has also been implemented in the context of incompressible flows [375, 411]. The numerical intercell flux is written as

$$\begin{aligned}\mathbf{E}_{j+1/2} = \frac{1}{2}\left[\mathbf{E}(\mathbf{U}_{R,j+1/2}) + \mathbf{E}(\mathbf{U}_{L,j+1/2})\right] \\ -\frac{1}{2}\widetilde{\mathbf{A}}(\mathbf{U}_{R,j+1/2} - \mathbf{U}_{L,j+1/2}) \, .\end{aligned} \tag{16.160}$$

For incompressible flows the only average state, $\widetilde{\mathbf{U}}$ that satisfies the aforementioned properties is obtained by a simple average of $\mathbf{U}_R$ and $\mathbf{U}_L$ [252, 411]

$$\widetilde{\mathbf{U}} = \frac{\mathbf{U}_R + \mathbf{U}_L}{2} \, . \tag{16.161}$$

The variables $\mathbf{U}_R$ and $\mathbf{U}_L$ can be calculated by an interpolation scheme such as the MUSCL (16.124) or the characteristics-based (interpolation) scheme (16.123).

The first-order version of the flux (16.160) is given by

$$\mathbf{E}_{j+1/2} = \frac{1}{2}\left[\mathbf{E}(\mathbf{U}_{j+1}) + \mathbf{E}(\mathbf{U}_j)\right] - \frac{1}{2}\widetilde{\mathbf{A}}(\mathbf{U}_{j+1} - \mathbf{U}_j) \, . \tag{16.162}$$

Replacing the Jacobian matrix $\widetilde{\mathbf{A}}$ by the eigenvector and eigenvalue matrices, the Roe's flux is written

$$\begin{aligned}\mathbf{E}_{j+1/2} = \frac{1}{2}\left[\mathbf{E}(\mathbf{U}_{R,j+1/2}) + \mathbf{E}(\mathbf{QU}_{L,j+1/2})\right] \\ -\frac{1}{2}\sum_{k=1}^{n} \mathbf{r}_k |\lambda_k| \mathbf{l}_k (\mathbf{U}_{R,j+1/2} - \mathbf{U}_{L,j+1/2}) \, ,\end{aligned} \tag{16.163}$$

where $\mathbf{r}_k$ and $\mathbf{l}_k$ are columns of the right and left eigenvectors of $\widetilde{\mathbf{A}}$ and λ_k are the eigenvalues. These matrices have been defined in Chap. 10. Their calculation is obtained for the average value (16.161).

Note that in the original Roe's method the Jacobian matrix $\widetilde{\mathbf{A}}$ is not explicitly required for the calculation of the numerical flux. The flux can be written as

$$\mathbf{E}_{j+1/2} = \frac{1}{2}\left[\mathbf{E}(\mathbf{U}_{R,j+1/2}) + \mathbf{E}(\mathbf{U}_{L,j+1/2})\right] - \frac{1}{2}\sum_{k=1}^{n} \alpha_k |\lambda_k| \mathbf{r}_k \, , \tag{16.164}$$

where $\mathbf{r}_k$ are the right eigenvector elements and $\alpha_k \equiv \alpha_k(\mathbf{U}_L, \mathbf{U}_R)$ are the *wave strengths*, which are found by the solution of the system

$$\Delta\mathbf{U} = \mathbf{U}_R - \mathbf{U}_L = \sum_{k=1}^{n} \alpha_k \mathbf{r}_k \,, \tag{16.165}$$

Alternatively, the flux (16.164) can be given by

$$\mathbf{E}_{j+1/2} = \mathbf{E}(\mathbf{U}_{L,j+1/2}) + \sum_{\lambda_k \leq 0, k=1}^{n} \alpha_k \lambda_k \mathbf{r}_k \,, \tag{16.166}$$

$$\mathbf{E}_{j+1/2} = \mathbf{E}(\mathbf{U}_{R,j+1/2}) - \sum_{\lambda_k \geq 0, k=1}^{n} \alpha_k \lambda_k \mathbf{r}_k \,. \tag{16.167}$$

The above equations are applicable for negative and positive eigenvalues, respectively.

16.9 Osher's Method

The approximate Riemann solver of Osher was first presented in [180] (and [410]). The scheme was applied to the compressible Euler's equations by Osher and Chakravarthy [409] and since then a number of papers have been presented regarding the implementation of the scheme in aerodynamic applications [543]. The intercell flux is defined by integration in phase space and depends on the choice of integration paths, which are associated with the set of right eigenvectors. Two different approaches have been proposed for the integration. The P-ordering that is based on relations across waves in physical space and the O-ordering that is based on the original Osher's scheme.

The Osher's approach considers the flux splitting scheme (Chap. 9) according to which the flux is decomposed into positive, $\mathbf{E}^+(\mathbf{U})$, and negative fluxes, $\mathbf{E}^-(\mathbf{U})$, that satisfy

$$\mathbf{E}(\mathbf{U}) = \mathbf{E}^+(\mathbf{U}) + \mathbf{E}^-(\mathbf{U}) \,. \tag{16.168}$$

The positive and negative fluxes are associated with the positive and negative eigenvalues. For a hyperbolic system of conservation laws the following relations are also valid for the Jacobian and the fluxes:

$$\mathbf{A} = \mathbf{A}^+ + \mathbf{A}^- \,, \tag{16.169}$$

$$\mathbf{E} = \mathbf{E}^+ + \mathbf{E}^- \,, \tag{16.170}$$

where

$$\mathbf{A}^+ = \frac{\partial \mathbf{E}^+}{\partial \mathbf{U}} \,, \tag{16.171}$$

$$\mathbf{A}^- = \frac{\partial \mathbf{E}^-}{\partial \mathbf{U}} , \tag{16.172}$$

are the Jacobian matrices of the positive and negative fluxes, respectively. The positive and negative fluxes (for a hyperbolic system $m \times m$) are defined as follows:

- $\mathbf{E}^+ = \mathbf{E}$ and $\mathbf{E}^- = 0$ if $\lambda_j \geq 0$ for $j = 1, ..., m$.
- $\mathbf{E}^+ = 0$ and $\mathbf{E}^- = \mathbf{E}$ if $\lambda_j \leq 0$ for $j = 1, ..., m$.

For a set of initial data $\mathbf{U}_L$ and $\mathbf{U}_R$ the flux (16.168) is written

$$\mathbf{E}(\mathbf{U}) = \mathbf{E}^+(\mathbf{U}_L) + \mathbf{E}^-(\mathbf{U}_R) , \tag{16.173}$$

Using the integral relations

$$\int_{\mathbf{U}_L}^{\mathbf{U}_R} \mathbf{A}^+(\mathbf{U})\, \mathrm{d}\mathbf{U} = \mathbf{E}^+(\mathbf{U}_R) - \mathbf{E}^+(\mathbf{U}_L) , \tag{16.174}$$

and

$$\int_{\mathbf{U}_L}^{\mathbf{U}_R} \mathbf{A}^-(\mathbf{U})\, \mathrm{d}\mathbf{U} = \mathbf{E}^-(\mathbf{U}_R) - \mathbf{E}^-(\mathbf{U}_L) , \tag{16.175}$$

The intercell flux (16.173) can be written in the following forms

$$\mathbf{E}_{j+1/2}(\mathbf{U}) = \mathbf{E}(\mathbf{U}_L) + \int_{\mathbf{U}_L}^{\mathbf{U}_R} \mathbf{A}^-(\mathbf{U})\, \mathrm{d}\mathbf{U} , \tag{16.176}$$

$$\mathbf{E}_{j+1/2}(\mathbf{U}) = \mathbf{E}(\mathbf{U}_R) - \int_{\mathbf{U}_L}^{\mathbf{U}_R} \mathbf{A}^+(\mathbf{U})\, \mathrm{d}\mathbf{U} , \tag{16.177}$$

or

$$\mathbf{E}_{j+1/2}(\mathbf{U}) = \frac{1}{2}\left[\mathbf{E}(\mathbf{U}_L) + \mathbf{E}(\mathbf{U}_R)\right] - \frac{1}{2}\int_{\mathbf{U}_L}^{\mathbf{U}_R} |\mathbf{A}|\, \mathrm{d}\mathbf{U} . \tag{16.178}$$

The integration is obtained along integration paths. Fig. 16.19 shows a choice of integration paths (P-ordering) for a 3×3 hyperbolic system, for example, the two-dimensional incompressible Euler equations in conjunction with the artificial compressibility formulation. The vectors $\mathbf{U}_0$ (=$\mathbf{U}_L$), $\mathbf{U}_{1/3}$, $\mathbf{U}_{2/3}$ and $\mathbf{U}_1$ (=$\mathbf{U}_R$) are constant states arising in the exact solution to the Riemann problem in physical space $x - t$ (see also [543], Chap. 4). For the P-ordering (16.176) can be written

$$\mathbf{E}_{j+1/2}(\mathbf{U}) = \mathbf{E}(\mathbf{U}_0) + \int_{\mathbf{U}_0}^{\mathbf{U}_{1/3}} \mathbf{A}^-(\mathbf{U})\,\mathrm{d}\mathbf{U} + \int_{\mathbf{U}_{1/3}}^{\mathbf{U}_{2/3}} \mathbf{A}^-(\mathbf{U})\,\mathrm{d}\mathbf{U} + \int_{\mathbf{U}_{2/3}}^{\mathbf{U}_1} \mathbf{A}^-(\mathbf{U})\,\mathrm{d}\mathbf{U}\,. \quad (16.179)$$

The integration is then obtained by taking into account the sign of the eigenvalues along each path. Osher originally presented the O-ordering of integration paths that follows exactly the invert part. Fig. 16.20 shows a choice of integration paths according to O-ordering. A detailed description of the method for compressible flows can also be found in [543] for the compressible Euler equations.

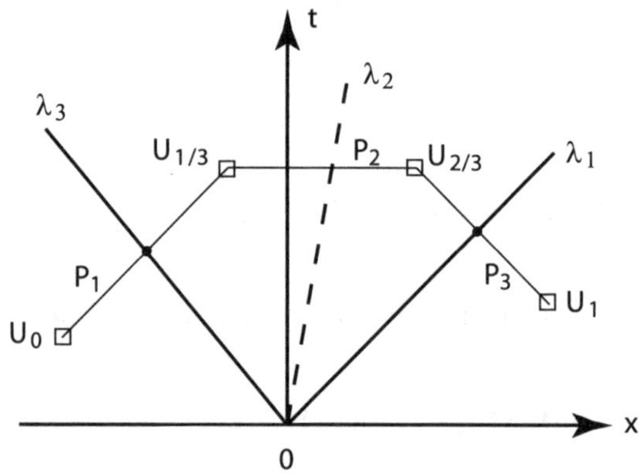

Fig. 16.19. Configuration of integration paths (P-ordering) and intersection points $\mathbf{U}_{1/3}$, $\mathbf{U}_{2/3}$ for a 3×3 system.

16.10 Chakravarthy-Osher TVD Scheme

Chakravarthy and Osher [99] have developed a high-order TVD scheme that consists of a first-order upwind scheme and correction terms that lead to higher-order discretization. The general form of the Chakravarthy and Osher (numerical) flux is given by

$$\widetilde{\mathbf{E}}_{j+1/2}(\mathbf{U}) = \mathbf{E}^{(1)}_{j+1/2}(\mathbf{U}) - \frac{1-k}{4}(\Delta_{j+3/2}\mathbf{E}^-) - \frac{1+k}{4}(\Delta_{j+1/2}\mathbf{E}^-)$$

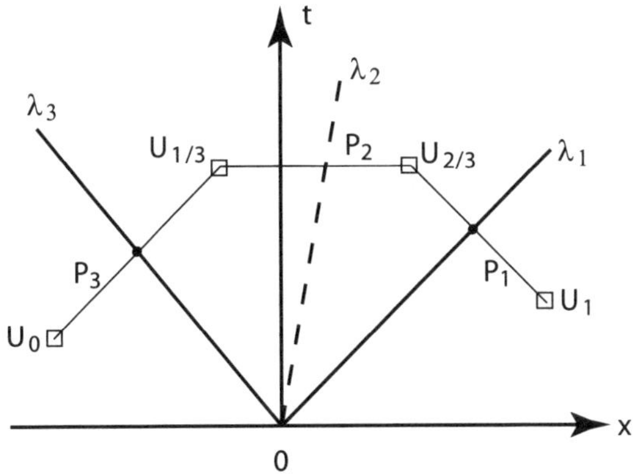

Fig. 16.20. Configuration of integration paths (O-ordering) and intersection points $\mathbf{U}_{1/3}$, $\mathbf{U}_{2/3}$ for a 3×3 system.

$$+\frac{1+k}{4}(\varDelta_{j+1/2}\mathbf{E}^+) + \frac{1-k}{4}(\varDelta_{j-1/2}\mathbf{E}^+)\,, \tag{16.180}$$

where $\varDelta_{j+3/2} = (\cdot)_{j+2} - (\cdot)_{j+1}$, $\varDelta_{j+1/2} = (\cdot)_{j+1} - (\cdot)_j$, and $\varDelta_{j-1/2} = (\cdot)_j - (\cdot)_{j-1}$. The flux, $\mathbf{E}^{(1)}_{j+1/2}(\mathbf{U})$, on the right-hand-side of (16.180) is a first-order upwind flux, for example, the first-order Roe's flux or the characteristics-based flux in conjunction with first-order upwind interpolation (16.115). The flux (16.180) does not guarantee the TVD properties per se. The TVD properties can be recovered if the high order terms are bounded by using flux limiters. Accordingly, the flux (16.180) is written

$$\begin{aligned}\widetilde{\mathbf{E}}_{j+1/2}(\mathbf{U}) = \mathbf{E}^{(1)}_{j+1/2}(\mathbf{U}) &\\ &-\frac{1-k}{4}(\widetilde{\varDelta}_{j+3/2}\mathbf{E}^-) - \frac{1+k}{4}(\widetilde{\widetilde{\varDelta}_{j+1/2}}\mathbf{E}^-)\\ &+\frac{1+k}{4}(\widetilde{\varDelta}_{j+1/2}\mathbf{E}^+) + \frac{1-k}{4}(\widetilde{\widetilde{\varDelta}_{j-1/2}}\mathbf{E}^+)\,,\end{aligned} \tag{16.181}$$

where

$$\widetilde{\varDelta}_{j+3/2}\,\mathbf{E}^- = \text{minmod}[\varDelta_{j+3/2}\,\mathbf{E}^-, \omega\varDelta_{j+1/2}\,\mathbf{E}^-]\,, \tag{16.182}$$

$$\widetilde{\widetilde{\varDelta}_{j+1/2}}\,\mathbf{E}^- = \text{minmod}[\varDelta_{j+1/2}\,\mathbf{E}^-, \omega\varDelta_{j+3/2}\,\mathbf{E}^-]\,, \tag{16.183}$$

$$\widetilde{\varDelta}_{j+1/2}\,\mathbf{E}^+ = \text{minmod}[\varDelta_{j+1/2}\,\mathbf{E}^+, \omega\varDelta_{j-1/2}\,\mathbf{E}^+]\,, \tag{16.184}$$

$$\widetilde{\widetilde{\varDelta}_{j-1/2}}\,\mathbf{E}^+ = \text{minmod}[\varDelta_{j-1/2}\,\mathbf{E}^+, \omega\varDelta_{j+1/2}\,\mathbf{E}^+]\,, \tag{16.185}$$

where the slope of the limiter is controlled by the coefficient ω that satisfies the inequality

$$1 \leq \omega \leq \frac{3-k}{1-k} \, . \tag{16.186}$$

The minmod limiter is defined by

$$\text{minmod}(x, \omega y) = \text{sign}(x) \cdot \max\{0, \min[|x|, \omega y \text{sign}(x)]\} \, . \tag{16.187}$$

The truncation error of the flux (16.180) is

$$\frac{1}{12}(3k-1)\Delta x^2 \frac{\partial^3 \mathbf{E}}{\partial x^3} + \mathcal{O}(\Delta x^3) \, . \tag{16.188}$$

Therefore, the scheme without limiting is third-order accurate for $k = 1/3$ and becomes second-order accurate for all the other values of k. For $k = -1$, $k = 0$, $k = 1/2$ and $k = 1$ the scheme becomes a second-order accurate upwind scheme, Fromm's scheme [199], QUICK scheme [325], and central difference scheme, respectively.

Arakawa et al. [18] have also presented implementation of the same scheme for complex turbulent flows through a Francis water runner. Shin [492] has also presented steady turbulent flow as well as large eddy simulations of turbulent flow in a 90-deg bend using the Chakravarthy-Osher TVD scheme. Their computations showed that the third-order upwind TVD scheme was more stable and exhibited higher computational efficiency than the QUICK scheme.

16.11 Harten, Lax and van Leer (HLL) Scheme

Hartex, Lax and Van Leer [250] have proposed a Riemann solver (henceforth labeled HLL) for the direct approximation of the intercell fluxes. The method was also introduced in Chap. 14 in connection with projection methods. We consider the one-dimensional system of hyperbolic conservation laws

$$\frac{\partial \mathbf{U}}{\partial t} + \frac{\partial \mathbf{E}}{\partial x} = 0 \, , \tag{16.189}$$

for a set of initial data

$$\mathbf{U}(x,0) = \begin{cases} \mathbf{U}_L & \text{if} \quad x < 0 \, , \\ \mathbf{U}_R & \text{if} \quad x > 0 \, . \end{cases} \tag{16.190}$$

For a control volume $[x_l, x_r] \times [0, \tau]$ as shown in Fig. 16.21, the integral form of (16.189) reads

$$\begin{aligned} \int_{x_l}^{x_r} \mathbf{U}(x,\tau) \, \mathrm{d}x = \int_{x_l}^{x_r} \mathbf{U}(x,0) \, \mathrm{d}x &+ \int_0^\tau \mathbf{E}(\mathbf{U}(x_l,t)) \, \mathrm{d}t \\ &- \int_0^\tau \mathbf{E}(\mathbf{U}(x_r,t)) \, \mathrm{d}t \, . \end{aligned} \tag{16.191}$$

We develop the integrals on the right hand side of (16.190) and obtain

$$\int_{x_l}^{x_r} \mathbf{U}(x,\tau)\,\mathrm{d}x = x_r\mathbf{U}_R - x_l\mathbf{U}_L + \tau[\mathbf{E}(\mathbf{U}_L) - \mathbf{E}(\mathbf{U}_R)] \,. \tag{16.192}$$

The integral on the left hand side can be written

$$\begin{aligned}\int_{x_l}^{x_r} \mathbf{U}(x,\tau)\,\mathrm{d}x = \int_{x_l}^{\tau S_L} \mathbf{U}(x,\tau)\,\mathrm{d}x + \int_{\tau S_L}^{\tau S_R} \mathbf{U}(x,\tau)\,\mathrm{d}x \\ + \int_{\tau S_R}^{x_R} \mathbf{U}(x,\tau)\,\mathrm{d}x \,,\end{aligned} \tag{16.193}$$

where S_L and S_R are the fastest signal velocities at time τ. The integral (16.193) can further be written

$$\begin{aligned}\int_{x_l}^{x_r} \mathbf{U}(x,\tau)\,\mathrm{d}x = \int_{\tau S_L}^{\tau S_R} \mathbf{U}(x,\tau)\,\mathrm{d}x + (\tau S_L - x_l)\mathbf{U}_L \\ +(x_R - \tau S_R)\mathbf{U}_R \,.\end{aligned} \tag{16.194}$$

Comparing (16.192) and (16.194) we obtain

$$\int_{\tau S_L}^{\tau S_R} \mathbf{U}(x,\tau)\,\mathrm{d}x = \tau(S_R\mathbf{U}_R - S_L\mathbf{U}_L + \mathbf{E}_L - \mathbf{E}_R) \,, \tag{16.195}$$

which after division with $\tau(S_R - S_L)$ gives the velocity $\mathbf{U}_{hll}$ in the intermediate state between the fastest signal velocities

$$\mathbf{U}^{hll} = \frac{S_R\mathbf{U}_R - S_L\mathbf{U}_L + \mathbf{E}_L - \mathbf{E}_R}{S_R - S_L} \,. \tag{16.196}$$

By applying the integral form of conservation laws to the control volumes $[x_l, 0] \times [0, \tau]$ and $[0, x_r] \times [0, \tau]$ one obtains the fluxes $\mathbf{E}_{LR}$ and $\mathbf{E}_{LR}^{'}$

$$\mathbf{E}_{LR} = \mathbf{E}_L - \frac{1}{\tau}\int_{\tau S_L}^{0} \mathbf{U}(x,\tau)\mathrm{d}x - S_L\mathbf{U}_L \tag{16.197}$$

$$\mathbf{E}_{LR}^{'} = \mathbf{E}_R + \frac{1}{\tau}\int_{0}^{\tau S_R} \mathbf{U}(x,\tau)\mathrm{d}x - S_R\mathbf{U}_R \tag{16.198}$$

Harten et al. [250] defined an approximate Riemann solver by

$$\mathbf{U}_{LR} = \begin{cases} \mathbf{U}_L & \text{if} \quad x/t \le S_L \,, \\ \mathbf{U}^{hll} & \text{if} \quad S_L \le x/t \le S_R \,, \\ \mathbf{U}_R & \text{if} \quad x \ge S_R \,. \end{cases} \tag{16.199}$$

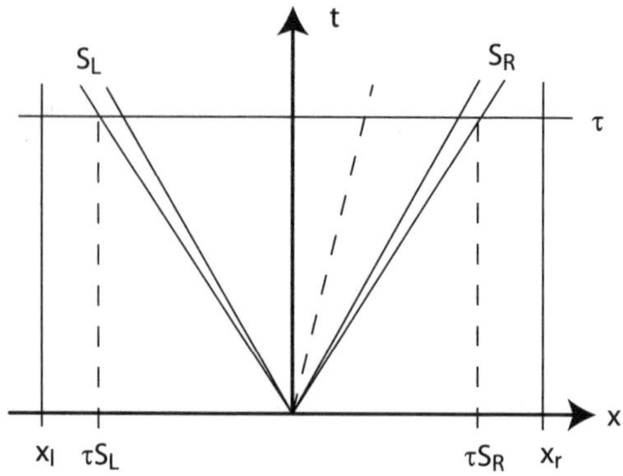

Fig. 16.21. Signal velocities S_L and S_R and control volume $[x_l, x_r] \times [0, \tau]$ on $x-t$ plane.

Substituting (16.196) into (16.197) we obtain the corresponding HLL intercell flux

$$\mathbf{E}_{j+1/2}^{\mathrm{HLL}} = \begin{cases} \mathbf{E}_L & \text{if} \quad 0 \le S_L \, , \\ \widetilde{\mathbf{E}}^{\mathrm{HLL}} & \text{if} \quad S_L \le 0 \le S_R \, , \\ \mathbf{E}_R & \text{if} \quad S_R \le 0 \, . \end{cases} \tag{16.200}$$

where

$$\widetilde{\mathbf{E}}_{j+1/2}^{\mathrm{HLL}} = \frac{S_R}{S_R - S_L}\mathbf{E}_L - \frac{S_L}{S_R - S_L}\mathbf{E}_R + \frac{S_L S_R}{S_R - S_L}(\mathbf{U}_R - \mathbf{U}_L) \, . \tag{16.201}$$

The above can be combined in a single formula [250]

$$\mathbf{E}_{j+1/2}^{\mathrm{HLL}} = \frac{S_R^- - S_L^-}{S_R - S_L}\,\mathbf{E}_R + \frac{S_R^+ - S_L^+}{S_R - S_L}\,\mathbf{E}_L - \frac{1}{2}\frac{S_R|S_L| - S_L|S_R|}{S_R - S_L}(\mathbf{U}_R - \mathbf{U}_L) \, . \tag{16.202}$$

where $S_{L,R}^- = \min(0, S_{L,R})$ and $S_{L,R}^+ = \max(0, S_{L,R})$.

Various procedures to estimate the wave speeds are discussed in the Sect. 16.13.

16.12 HLLC Scheme

Toro et al. [547] have proposed an efficient (and simple) modification of the HLL scheme according to which the missing contact and shear waves are restored. The schemes was named HLLC ("C" stands for the contact wave) This modification is obtained by including an additional (middle) wave of speed S_*.

For the derivation of the flux value, we split in (16.195) the LHS of the integral (after division with $\tau(S_R - S_L)$) into two terms

$$\frac{1}{\tau(S_R - S_L)} \int_{\tau S_L}^{\tau S_R} \mathbf{U}(x,\tau)\,\mathrm{d}x = \frac{1}{\tau(S_R - S_L)} \int_{\tau S_L}^{\tau S_*} \mathbf{U}(x,\tau)\,\mathrm{d}x + \frac{1}{\tau(S_R - S_L)} \int_{\tau S_*}^{\tau S_R} \mathbf{U}(x,\tau)\,\mathrm{d}x\,. \tag{16.203}$$

Using (16.196), we can write

$$\frac{S_* - S_L}{S_R - S_L}\mathbf{U}_{*L} + \frac{S_R - S_*}{S_R - S_L}\mathbf{U}_{*R} = \mathbf{U}^{hll}\,, \tag{16.204}$$

where

$$\mathbf{U}_{*L} = \frac{1}{\tau(S_R - S_L)} \int_{\tau S_L}^{\tau S_*} \mathbf{U}(x,\tau)\,\mathrm{d}x\,, \tag{16.205}$$

$$\mathbf{U}_{*R} = \frac{1}{\tau(S_R - S_L)} \int_{\tau S_*}^{\tau S_R} \mathbf{U}(x,\tau)\,\mathrm{d}x\,. \tag{16.206}$$

The HLLC Riemann solver yields

$$\mathbf{U}^{hllc} = \begin{cases} \mathbf{U}_L & \text{if} \quad x/t \le S_L\,, \\ \mathbf{U}_{*L} & \text{if} \quad S_L \le x/t \le S_*\,, \\ \mathbf{U}_{*R} & \text{if} \quad S_* \le x/t \le S_R\,, \\ \mathbf{U}_R & \text{if} \quad x \ge S_R\,. \end{cases} \tag{16.207}$$

By applying Rankine-Hugoniot conditions across each of the wave speeds one can obtain the HLLC intercell flux

$$\mathbf{E}^{\mathrm{HLLC}}_{j+1/2} = \begin{cases} \mathbf{E}_L & \text{if} \quad 0 \le S_L\,, \\ \mathbf{E}_{*L} & \text{if} \quad S_L \le 0 \le S_*\,, \\ \mathbf{E}_{*R} & \text{if} \quad S_* \le 0 \le S_R\,, \\ \mathbf{E}_R & \text{if} \quad S_R \le 0\,, \end{cases} \tag{16.208}$$

where

$$\mathbf{E}_{*L} = \mathbf{E}_L + S_L(\mathbf{U}_{*L} - \mathbf{U}_L) , \tag{16.209}$$

$$\mathbf{E}_{*R} = \mathbf{E}_R + S_R(\mathbf{U}_{*R} - \mathbf{U}_R) . \tag{16.210}$$

The speeds $\mathbf{U}_{*L}$ and $\mathbf{U}_{*R}$ are defined in a similar fashion as in the compressible flow case [547]

$$\mathbf{U}_{*L,*R} = \frac{S_{L,R} - U_{L,R}}{S_{L,R} - S_*} \begin{pmatrix} 1 \\ S_* \\ v_{L,R} \\ w_{L,R} \end{pmatrix} . \tag{16.211}$$

16.13 Estimation of the Wave Speeds for the HLL and HLLC Riemann Solvers

Various estimates of the wave speeds for the compressible flow case have been reviewed in [543]. Simple estimates for the compressible case have been suggested by Davis [131], which for the case of the artificial compressibility approach can be written as (assuming the flux $\mathbf{E}$ in the $x-$direction)

$$S_L = u_L - \sqrt{u_L^2 + \beta} , \quad S_R = u_R + \sqrt{u_R^2 + \beta} . \tag{16.212}$$

If the wave speeds in the HLL flux are defined as $S_L = -S^+$ and $S_R = S^+$, we obtain the Rusanov flux

$$\mathbf{E}_{j+1/2}^{\text{Rusanov}} = \frac{1}{2}(\mathbf{E}_L + \mathbf{E}_R) - \frac{1}{2}S^+(\mathbf{U}_R - \mathbf{U}_L) . \tag{16.213}$$

Following Davis's [131] definition for S^+ (compressible case), for the case of the artificial compressibility method Drikakis and Smolarkiewicz [163] proposed a similar formula

$$\begin{aligned} S^+ = \max\big(&|u_L - \sqrt{U_L^2 + \beta}|, |u_R - \sqrt{u_R^2 + \beta}|, \\ &|u_L + \sqrt{u_L^2 + \beta}|, |u_R + \sqrt{u_R^2 + \beta}|\big) . \end{aligned} \tag{16.214}$$

The wave speed S_* can be estimated for the incompressible case in a similar way as proposed by Batten et al. [41] for the compressible flow case. For the artificial compressibility formulation one obtains

$$S_* = \frac{p_R/\beta - p_L/\beta + u_L(S_L - u_L) - u_R(S_R - u_R)}{S_L - S_R + u_R - u_L} . \tag{16.215}$$

16.14 HLLE Scheme

Einfeldt [179] has derived an approximate Riemann solver for compressible gas flow. In contrast to other Riemann solvers, where a numerical approximation for velocities and pressure at contact discontinuities is computed, Einfeldt derived a numerical approximation for the largest and smallest signal velocity in the Riemann problem. Using the numerical signal velocities, he used theoretical results of Harten, Lax and Van Leer [250] to obtain the numerical flux.

According to Einfeldt [179] the numerical intercell flux (henceforth labeled HLLE) is given by

$$\mathbf{E}_{j+1/2}^{\mathrm{HLLE}} = \frac{b_{j+1/2}^{+}}{b_{j+1/2}^{+} - b_{j+1/2}^{-}} \mathbf{E}_L - \frac{b_{j+1/2}^{-}}{b_{j+1/2}^{+} - b_{j+1/2}^{-}} \mathbf{E}_R + \frac{b_{j+1/2}^{+} b_{j+1/2}^{-}}{b_{j+1/2}^{+} - b_{j+1/2}^{-}} (\mathbf{U}_R - \mathbf{U}_L) , \tag{16.216}$$

where $b_{j+1/2}^{+} = \max(0, b_{j+1/2}^{r})$ and $b_{j+1/2}^{-} = \min(0, b_{j+1/2}^{l})$. In contrast to the original version of the HLL Riemann solver the numerical wave speeds $b_{j+1/2}^{r}$ and $b_{j+1/2}^{l}$ are not the lower and upper bounds of the physical wave speeds. Approximations for the numerical wave speeds in the context of compressible and incompressible flow equations have been given by Einfeldt [179] and Drikakis et al. [149, 163], respectively. For the case of the artificial compressibility method the numerical waves speeds are defined by [149, 163]

$$b_{j+1/2}^{+} = \max\Big((\lambda_1)_j, (\lambda_1)_{j+1} \Big) , \tag{16.217}$$

$$b_{j+1/2}^{-} = \min\Big((\lambda_2)_j, (\lambda_2)_{j+1} \Big) . \tag{16.218}$$

The eigenvalues λ_1 and λ_2 are defined by

$$\lambda_1 = u + \sqrt{u^2 + \beta}, \quad \lambda_2 = u - \sqrt{u^2 + \beta} . \tag{16.219}$$

16.15 Comparison of CB and HLLE Schemes

We present results from numerical experiments [149], using the HLLE and CB schemes, for the case of a two-dimensional temporal mixing layer defined by a velocity profile $U(y) = \tanh(y)$, where $U(y)$ and y are made dimensionless by the free-stream velocity, U_∞, and half of the initial vorticity thickness, $\delta/2$ (δ is the initial vorticity thickness). The dimensionless time is $t = 2TU_\infty/\delta$, where T is the time with dimensions. The Reynolds number is defined as $Re = 0.5U_\infty\delta/\nu$, where ν is the kinematic viscosity. The

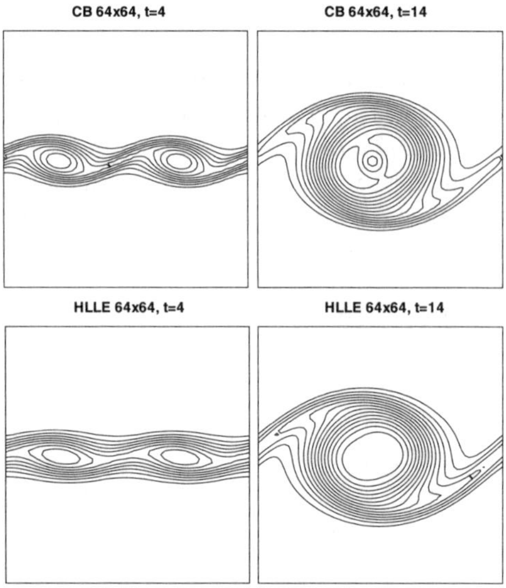

Fig. 16.22. Isovorticity contours on the 64×64 grid for the CB and HLLE schemes.

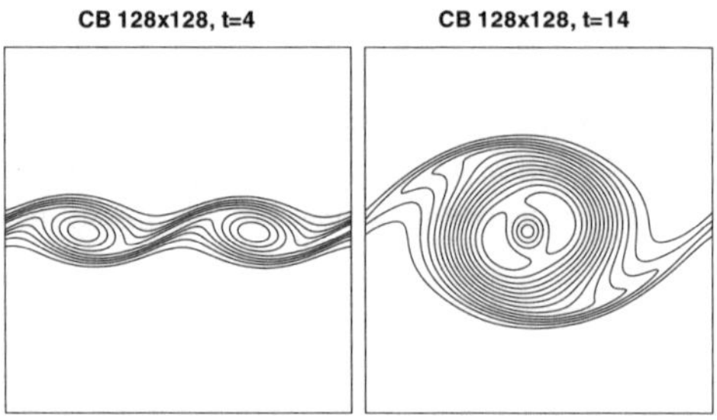

Fig. 16.23. Isovorticity contours on the 128×128 grid for the CB scheme.

numerical experiments have been conducted for $Re = 200$. The equations are solved in a square domain $[0, L] \times [-L/2, L/2]$, imposing periodic boundary conditions in the $x-$direction and free slip walls in the y-direction, i.e., $v = (\partial p/\partial y) = (\partial u/\partial y) = 0$ at the boundaries. A sine wave superimposed by a solenoidal white-noise perturbation, $E(x, y)$, of small amplitude is added to the basic flow [591, 338]. The initial conditions for u and v are given by

$$U_{in} = U(y) + d_1 \; E(x, y) \exp(-y^2) + d_2 \sin(\pi x/\lambda_U) \; , \tag{16.220}$$

$$v_{in} = d_2 \sin(\pi x/\lambda_U) \; , \tag{16.221}$$

where $E(x, y)$ is the perturbation function having values in the interval $[0, 1]$ and λ_U is the most unstable wavelength which, according to the theory [388],

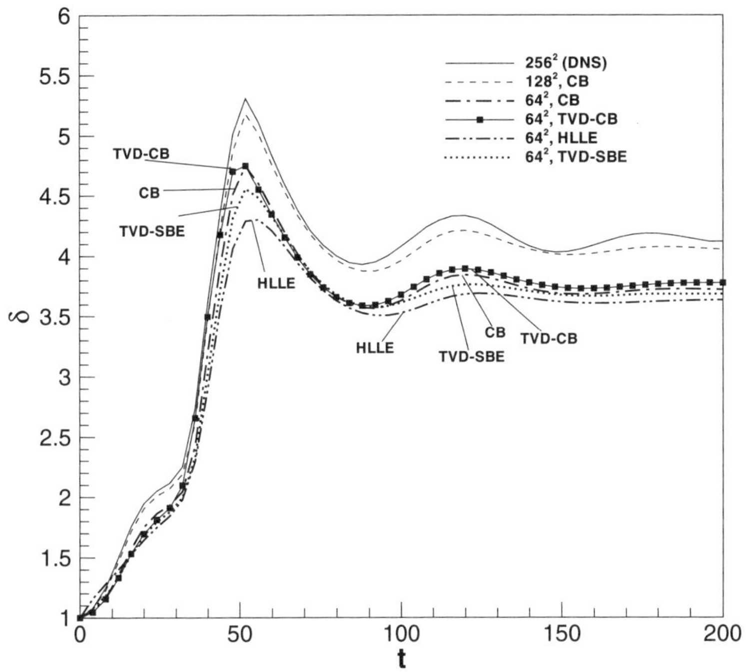

Fig. 16.24. Growth of the vorticity thickness computed by the HLLE and variants of the CB scheme. TVD-CB and TVD-SBE are the TVD (16.126) variants of the CB scheme in conjunction with the CB (16.51) and Superbee (16.127) limiters, respectively. The solution on the 256 × 256 grid is labeled as DNS.

is defined as $\lambda_U = 7\delta$. In the present experiments $d_1 = 0.1$ and $d_2 = 0.05$. The length of the computational domain should be taken equal to $L_n/(0.5\delta) = 14n$ (dimensionless) for obtaining n Kelvin-Helmoltz vortices [338]. For $n = 2$ two vortices are initially formed and later merge to form one large vortex (Fig. 16.22). Computations have been conducted using the CB and HLLE schemes, in conjunction with (16.118) and (16.123), on a sequence of increasingly finer grids containing 64×64, 128×128 and 256×256 grid points. The results on 256×256 and 512×512 grids exhibit differences in the u values about 1%.

Fig. 16.22 shows the results for the CB and HLLE schemes on the 64×64 grid. The same contour values have been plotted for both schemes at two different time instants. The HLLE scheme results in thicker - more diffusive - shear layers at $t = 4$. Additionally, the details of the core of the vortex at $t = 14$ are missing. The CB scheme provides very similar results on the

64×64 and 128×128 grids. The solution for the CB scheme on the 128×128 grid is shown in Fig. 16.23.

Comparisons of the various solutions can be obtained on the basis of the vorticity thickness. This is defined as $\delta = 2U_\infty/(\mathrm{d}\bar{u}/\mathrm{d}y)_{max}$, where the "bar" denotes an average in the $x-$direction. In Fig. 16.24 we plot the growth of vorticity thickness for the HLLE and CB schemes as well as for the variants TVD-CB and TVD-SBE of the CB scheme obtained by (16.126) in conjunction with (16.51) and (16.127) limiters, respectively, and third-order interpolation (16.118) for the left and right states of the variables. The CB scheme provides better results than the HLLE scheme on the 64×64 grid, while the TVD-CB scheme is slightly less diffusive than the CB scheme.

16.16 "Viscous" TVD Limiters

We close this chapter by presenting how in the derivation of flux limiters the physical viscosity can also be taken into account as proposed in [539]. This approach can be demonstrated using the advection-diffusion equation as a model of conservation law[5]

$$\frac{\partial U}{\partial t} + \frac{\partial E(U)}{\partial x} = \nu \frac{\partial^2 U}{\partial x^2} , \tag{16.222}$$

where $E(U) = aU$, a is a constant coefficient, and ν is a viscosity coefficient.

Discretization of (16.222) using an explicit scheme and second-order discretization for the derivative on the right hand side of (16.222), yields

$$U_j^{n+1} = U_j^n - \frac{\Delta t}{\Delta x}(E_{j+1/2} - E_{j-1/2}) + d(U_{j+1}^{n+1} - 2U_j^n + U_{j-1}^n) , \tag{16.223}$$

where $d = \nu \Delta t/\Delta x^2$. Similar to the case of the linear advection equation (16.1), one can define a high-order TVD flux as

$$E_{j\pm1/2}^{\mathrm{TVD}} = E_{j\pm1/2}^{\mathrm{LO}} + \psi_{j\pm1/2}(E_{j\pm1/2}^{\mathrm{HI}} - E_{j\pm1/2}^{\mathrm{LO}}) , \tag{16.224}$$

where $\psi_{j\pm1/2}$ is a flux limiter function yet to be determined. To preserve some generality we assume that $E_{j\pm1/2}^{\mathrm{HI}}$ and $E_{j\pm1/2}^{\mathrm{LO}}$ are respectively of the form (16.4). Substituting (16.4) into (16.224) we obtain

$$\begin{aligned} E_{j+1/2}^{\mathrm{TVD}} = & \left[\alpha_0 + (\beta_0 - \alpha_0)\psi_{j\pm1/2}\right](aU_j^n) + \\ & \left[\alpha_1 + (\beta_1 - \alpha_1)\psi_{j\pm1/2}\right](aU_{j+1}^n) , \end{aligned} \tag{16.225}$$

which is the same with the TVD flux (16.5) obtained for the linear advection equation. The coefficients α_0, α_1, β_1 and β_2 depend on the scheme chosen to

[5] Work regarding the implementation of viscous limiters for the incompressible and compressible Navier-Stokes equations is in progress [151].

define the low and high-order fluxes. In the presentation below we will assume that these coefficients correspond to Godunov's first-order upwind flux

$$\alpha_0 = \frac{1}{2}(1+s) \quad , \quad \alpha_1 = \frac{1}{2}(1-s) \, , \quad s = \mathrm{sign}(a) \, , \tag{16.226}$$

and Lax-Wendroff flux

$$\beta_0 = \frac{1}{2}(1+\mathcal{C}) \quad , \quad \beta_1 = \frac{1}{2}(1-\mathcal{C}) \, , \tag{16.227}$$

where $\mathcal{C} = a\Delta t/\Delta x$.

Up to this point, the presence of the viscous terms on the right hand side of (16.222) are not taken into account in the definition of the TVD flux. Substitution of (16.225) into (16.222) gives

$$\begin{aligned} U_j^{n+1} &= U_j^n - C\Delta U_{j-1/2} + D\Delta U_{j+1/2} \\ &\quad + d(U_{j+1}^n - 2U_j^n + U_{j-1}^n) \, , \end{aligned} \tag{16.228}$$

where the coefficients C and D are defined as in (16.7).

We denote by ψ flux limiters associated with upwind directions. To derive limiter functions such that the scheme will be TVD, we apply (for $a > 0$) the data compatibility condition [543]

$$0 \le \frac{U_j^{n+1} - U_j^n}{U_{j-1}^n - U_j^n} \le 1 \, . \tag{16.229}$$

Using (16.228), the ratio $(U_j^{n+1} - U_j^n)/(U_{j-1}^n - U_j^n)$ is written

$$\frac{U_j^{n+1} - U_j^n}{U_{j-1}^n - U_j^n} = C - D\frac{1}{r} + d(1 - \frac{1}{r}) \, , \tag{16.230}$$

where $r = \Delta U_{j-1/2}/\Delta U_{j+1/2}$.

Substituting the coefficients C and D from (16.7) we obtain

$$\begin{aligned} 0 \le \mathcal{C}\Big[\alpha_0 + \psi_{j-1/2}(\beta_0 - \alpha_0)\Big] + \mathcal{C}\Big[\alpha_1 + \psi_{j+1/2}(\beta_1 - \alpha_1)\Big]\frac{1}{r} \\ + d(1 - \frac{1}{r}) \le 1 \, . \end{aligned} \tag{16.231}$$

We impose a global constraint

$$\psi_B \le \psi_{j-1/2} \le \psi_T \, , \tag{16.232}$$

where ψ_T and ψ_B are the top and bottom boundaries of the flux limiter, which are considered to be independent of the slope r; this constraint yields[6]

[6] Note that for the Godunov and Lax-Wendroff fluxes as low and high-order fluxes, respectively: $\beta_0 - \alpha_0 = (\mathcal{C} - 1)/2 \le 0$ and $\beta_1 - \alpha_1 = (1 - \mathcal{C})/2 > 0$.

$$\mathcal{C}\Big[\alpha_0 + \psi_T(\beta_0 - \alpha_0)\Big] \le \mathcal{C}\Big[\alpha_0 + \psi_{j-1/2}(\beta_0 - \alpha_0)\Big] \le \mathcal{C}\Big[\alpha_0 + \psi_B(\beta_0 - \alpha_0)\Big] . \qquad (16.233)$$

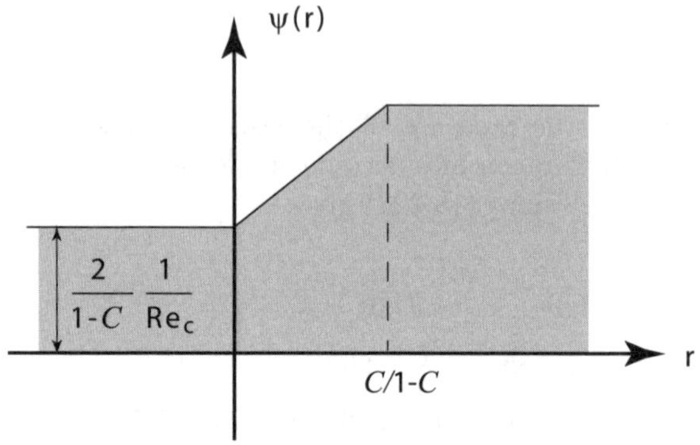

Fig. 16.25. The TVD region for the viscous limiter as defined by (16.243).

To satisfy (16.229) and (16.233), the following inequality should be satisfied

$$-\mathcal{C}\Big[\alpha_0 + \psi_T(\beta_0 - \alpha_0)\Big] \le \mathcal{C}\Big[\alpha_1 + \psi_{j+1/2}(\beta_1 - \alpha_1)\Big]\frac{1}{r} + d(1 - \frac{1}{r}) \le 1 - \mathcal{C}\Big[\alpha_0 + \psi_B(\beta_0 - \alpha_0)\Big] \qquad (16.234)$$

Analysis of the left inequality yields

$$\psi_{j+1/2} = \begin{cases} \ge \psi_L \text{ for } r \ge 0 \, , \\ < \psi_L \text{ for } r < 0 \, , \end{cases} \qquad (16.235)$$

where

$$\psi_L = \frac{1}{\beta_1 - \alpha_1}\left\{ -r\Big[\alpha_0 + \psi_T(\beta_0 - \alpha_0)\Big] - \alpha_1 - \frac{1}{Re_c}(r-1)\right\} , \qquad (16.236)$$

and $Re_c = \mathcal{C}/d$ is the cell Reynolds number. Analysis of the right inequality gives

$$\psi_{j+1/2} = \begin{cases} \le \psi_R \text{ for } r > 0 \, , \\ \ge \psi_R \text{ for } r < 0 \, , \end{cases} \qquad (16.237)$$

where

$$\psi_R = \frac{1}{\beta_1 - \alpha_1}\left\{\frac{r}{\mathcal{C}} - r\left[\alpha_0 + \psi_B(\beta_0 - \alpha_0) - \frac{1}{Re_c}(r-1)\right] - \alpha_1\right\} . \quad (16.238)$$

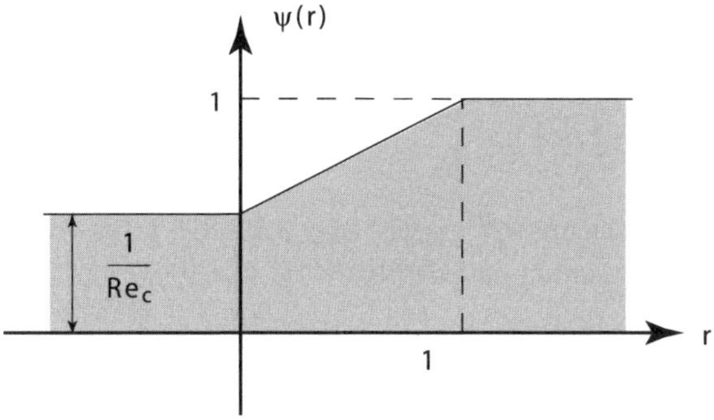

Fig. 16.26. The TVD region for the viscous limiter as defined by (16.244).

The equations (16.235)-(16.238), are valid for both $a > 0$ and $a < 0$ if r is defined by

$$r = \frac{\Delta_{upw}}{\Delta_{loc}} = \begin{cases} \dfrac{U_j^n - U_{j-1}^n}{U_{j+1}^n - U_j^n}, & a > 0\,, \\ \\ \dfrac{U_{j+2}^n - U_{j+1}^n}{U_{j+1}^n - U_j^n}, & a < 0\,. \end{cases} \quad (16.239)$$

Substituting the coefficients α_0, α_1 and β_0, β_1, for the Godunov and Lax-Wendroff schemes, respectively, into (16.236) and (16.238) we obtain

$$\psi_L = r\left[\psi_T - \frac{2}{1-|\mathcal{C}|}(1 + \frac{1}{Re_c})\right] + \frac{2}{(1-|\mathcal{C}|)Re_c}\,, \quad (16.240)$$

$$\psi_R = r\left(\frac{2}{|\mathcal{C}|} + \psi_B - \frac{2}{1-|\mathcal{C}|}\frac{1}{Re_c}\right) + \frac{2}{(1-|\mathcal{C}|)Re_c}\,. \quad (16.241)$$

The choice of the top, ψ_T, and bottom, ψ_B, boundaries produces different TVD schemes. For example, if we define

$$\begin{aligned} \psi_B &= \frac{2}{1-|\mathcal{C}|}\frac{1}{Re_c}\,, \\ \psi_T &= \frac{2}{1-|\mathcal{C}|}\left(1 + \frac{1}{Re_c}\right)\,, \end{aligned} \quad (16.242)$$

the limiter is obtained as (Fig. 16.25)

$$\psi_{j+1/2} = \begin{cases} \dfrac{2}{1-|\mathcal{C}|}\dfrac{1}{Re_c} & \text{if} \quad r \le 0\,, \\[2ex] \dfrac{2}{|\mathcal{C}|}r + \dfrac{2}{1-|\mathcal{C}|}\dfrac{1}{Re_c} & \text{if} \quad 0 \le r \le \dfrac{|\mathcal{C}|}{1-|\mathcal{C}|}\,, \\[2ex] \dfrac{2}{1-|\mathcal{C}|}\left(1+\dfrac{1}{Re_c}\right) & \text{if} \quad r > \dfrac{|\mathcal{C}|}{1-|\mathcal{C}|}\,. \end{cases} \tag{16.243}$$

For $\psi_B = 0$ and replacing $2/(1-|\mathcal{C}|)$ and $2/|\mathcal{C}|$ by 1 (in order to produce a limiter similar to minmod) one obtains (Fig. 16.26)

$$\psi_{j+1/2} = \begin{cases} \dfrac{1}{Re_c} & \text{if} \quad r \le 0\,, \\[2ex] r\left(1-\dfrac{1}{Re_c}\right)+\dfrac{1}{Re_c} & \text{if} \quad 0 \le r \le 1\,, \\[2ex] 1 & \text{if} \quad r > 1\,. \end{cases} \tag{16.244}$$

The physical viscosity is not sufficient to remove any spurious solutions from the numerical solutions. In fact, Toro [539] found that for certain TVD discretization schemes one does not require artificial viscosity only if

$$Re_c \le \frac{2}{1-|\mathcal{C}|}\,. \tag{16.245}$$

The "viscous" limiters approach has not yet been investigated in the context of multidimensional problems.

17. Beyond Second-Order Methods

The successful high-resolution methods discussed in earlier chapters are predominantly second-order in formal accuracy. As one might guess there have been focused efforts to extend these methods to higher order accuracy. In this chapter we will introduce several approaches to achieving this end.

The development of accurate and efficient numerical methods for time-dependent transport phenomena, both for practical applications as well as for fundamental studies, remains a challenging task for numerical analysts and practitioners. This chapter is concerned with the development and application of numerical methods of *very* high-order of accuracy; in the context of this chapter this means at least third and possibly fourth, fifth or higher. Note that several developments discussed in this chapter are applicable to the broader field of computational physics. Obvious advantages of high-order methods are:[1]

- The desired accuracy is attained with reduced memory and storage requirements. This is of paramount importance for any three-dimensional time dependent problems as well as for multiphysics problems involving many equations, such as in multiphase flows.
- In conjunction with current computing resources available, very high-order methods may be the only way of providing reliable solutions for practical engineering problems, which are presently studied using under-resolved simulations, i.e., using coarse computational grids. Low-order methods, first or second-order, can be completely inadequate without resorting to very fine meshes that are not affordable with current computers.
- High-order methods may offer higher accuracy in fundamental studies concerned with the modeling of associated processes not yet fully understood, for example, turbulence.

[1] This depends upon the smoothness of the solution defined by the number of derivatives that can be computed accurately. This is measured using the verification techniques discussed in Chap. 6.

17.1 General Remarks on High-Order Methods

High-order methods which are applicable to complex flows, in terms of both geometry and physics, should satisfy the following requirements:

1. For model problems (e.g., one dimensional advection-diffusion equation) they can be of arbitrary accuracy, in the sense of being free of theoretical accuracy barriers.
2. The design accuracy of the method should be directly confirmed in these cases using mesh refinement.
3. For realistic problems they should be of high-order of accuracy (uniform), with third-order accuracy being commonly used, both in space and time.
4. They should be conservative where this is appropriate given the governing equations. This does not mean that the methods are not applicable to non-conservative equations. More generally, this is in keeping with the principle that the numerical solution should have the same character as the physics itself.
5. They should be (essentially) nonoscillatory. Such schemes are to be nonlinear even when applied to linear problems, so as to resolve the paradoxical requirements embodied in Godunov's theorem.

With respect to the time discretization, the schemes can be either explicit or implicit, but explicit implementation ensures simplicity and the schemes can be combined efficiently with various boundary conditions methods for the case of two and three dimensional flows.

The benefits obtained from developing high-order methods would be important for several applications, especially those involving time-dependent phenomena, including: flows in aerospace applications, for example, flows around helicopter, turbomachinery and wind turbine blades; environmental flows, for example, weather and climate forecasting; biological flows, for example, blood flow, flows in artificial organs such artificial heart and heart valves; flows in internal combustion engines, inside valve devices (spool valves in injection systems); aeroacoustics; multiphase flows; free surface flows; fluid-structure interaction.

There are different computational frameworks within which one can pursue the development of high-order methods. These include:

1. Spectral methods, e.g., [94, 284].
2. Finite element methods [231], particularly the more recently emphasized discontinuous Galerkin approach [112, 441, 562].
3. Compact difference schemes [121, 323, 423, 536].
4. Schemes based on flux reconstruction such as the essentially nonoscillatory schemes (ENO) [248, 495, 496] and uniformly high-order scheme (UHO) [148].
5. The arbitrary-order (in space and time) nonoscillatory advection schemes, ADER [535, 546].

6. Methods based on preserving accuracy and monotonicity [271, 451, 518].

The following preliminary remarks need to be made:

- Although both spectral and finite element approaches have many attractive features, the finite volume approach remains the most broadly employed in the development of both academic and industrial codes.[2]
- At present, a large number of successful methods within the finite volume framework have their origin in the Godunov approach.
- The compact difference methods, spectral methods and Galerkin finite element methods are linear and thus oscillatory.
- Compact, spectral and finite element methods match the requirement of spatial accuracy, but satisfying time accuracy, i.e., be high-order both in *space and time*, remains an open issue.
- Compact and spectral methods have both been the subject of extensions involving nonoscillatory techniques, but these will not be covered here.
- The weighted variant of the ENO schemes (called WENO) [280] uses up to eleventh-order spatial discretization when it is implemented in conjunction with Runge-Kutta methods for time integration. The Runge-Kutta methods must, however, be TVD in order to avoid spurious oscillations. This leads to accuracy limitation in terms of both spatial and time accuracy since the order of accuracy of TVD Runge-Kutta methods cannot be higher than fifth. Therefore, even though the spatial accuracy can go beyond fifth order the overall accuracy will be decreased because of the accuracy of the time integration method. Fifth-order TVD Runge-Kutta schemes are complicated in the implementation and have reduced stability range. As a result, most researchers currently adopt the third-order implementation.
- Methods that preserve or relax monotonicity incorporate elements of ENO/WENO methods upon sensing violations of monotonicity.

When considering the construction of high-order methods (higher than second-order) a certain subtlety comes into play for nonlinear problems. The construction of a numerical scheme can proceed along two different lines of thinking. It is important to consider the various collateral effects of the choice made in which line of thinking to follow (we also discussed this basic issue in Chap. 9):

1. The reconstruction of the dependent variables as the average of the variable in the computational cells. This implies that the fluxes used to evolve the cells will be evaluated at the cell walls, and the differences between neighboring cells will be reconciled through the application of a Riemann solver. In this case the average of the cell value is evolved in time[3]

[2] Finite element methods can be viewed a generalized finite volume approach.

[3] Note that the indices (i) and (j) are equivalently used throughout the chapter with respect to one-dimensional stencil discretization.

$$\frac{\partial \mathcal{A}(\mathbf{U})}{\partial t} + \left[\mathbf{E}\left(\mathbf{U}_{j+1/2}\right) - \mathbf{E}\left(\mathbf{U}_{j-1/2}\right)\right] = 0\,,$$

where

$$\mathcal{A}\left[\mathbf{U}(x)\right] = \int_{x_j-\Delta x/2}^{x_j+\Delta x/2} \mathbf{U}(x)\,\mathrm{d}x\,.$$

These methods can use either a method-of-lines or generalized Lax-Wendroff procedure for time integration.

2. The other approach is the reconstruction of the flux of dependent variables. In this case the fluxes are usually split into their directional components, the information propagating in negative and positive directions in place of the Riemann solution prior to interpolation. Here, the cell values of the dependent variables are viewed as point values. The relation of the point value evolution to the cell averages has been explained by Merriman [385] as

$$\frac{\partial \mathbf{U}}{\partial t} + \left[\mathcal{A}^{-1}\mathbf{E}\left(\mathbf{U}_{j+1/2}\right) - \mathcal{A}^{-1}\mathbf{E}\left(\mathbf{U}_{j-1/2}\right)\right] = 0\,,$$

where $\mathcal{A}^{-1}$ is the (abstract) inverse of the integral averaging operator and $\mathbf{U}$ denotes approximate point values at x_j. Generally, these methods have used method-of-lines time integration, but more recently Qia et al [434] have defined Lax-Wendroff time integrators for these methods.

The truly wonderful aspect of these approaches is that a given interpolation scheme once chosen will work to produce high-order results for both choices. The additional complication is that any interpretation of the results then follows the precepts of the choice. For example, if the dependent variables are reconstructed any accuracy consideration must be examined with the cell integral values for the fiducial solution. Likewise, if the flux reconstruction is chosen, point values of the dependent variables must be examined. This comes about because point values and cell averages begin to deviate from one another starting with second order,

$$\bar{W} = U_j + \frac{1}{24}U_{xx}\Delta x^2 + \frac{7}{5760}U_{xxxx}\Delta x^4 + \text{H.O.T.}, \tag{17.1}$$

where H.O.T. means higher order terms which are an infinite sequence of even derivatives.

The key concept is that the reconstruction choices all preserve the value of mean when integrated over a cell. For example, third-order methods that satisfy this requirement have the generic form of,

$$U_j(x) = \bar{U} + U_x(x - x_j) + U_{xx}\left[(x - x_j)^2 - \frac{\Delta x^2}{24}\right]. \tag{17.2}$$

Note, that the quadratic term has a correction due to the disparity between the point and cell average values. As a practical matter the choice of reconstructing dependent variables or their flux is largely a matter of style. Where this is not stylistic is in the formal achievement of accuracy in multiple space dimensions where the flux-based interpolation is a superior approach. In this case reconstruction of variables requires complex quadratures for accuracy.

This point becomes pragmatic in the presentation below. Originally, ENO schemes were developed for the reconstruction of dependent variables. More recently with the popular WENO methods flux reconstruction has been used. The bottom line is that the basic techniques associated with either method can be used interchangeably once the type of reconstruction is chosen. The reader should keep this in mind while reading the remainder of the chapter. We will present each method as it was originally introduced, but the underlying algorithms can be used with either basic design philosophy.

The development and implementation of high-order methods is currently a topic of active research.

17.2 Essentially Nonoscillatory Schemes (ENO)

The development of ENO schemes was initiated by Harten et al. [248] aiming at constructing high-order schemes in smooth regions and through the use of adaptive stencils achieving high accuracy up to discontinuities (provided multiple discontinuities are adequately separated). At the same time, the schemes aim at providing high-order solutions that are free of spurious oscillations. Since the development of these schemes there were a number of research studies concerned with the implementation of the methods for flows encompassing discontinuities [495, 496]. Further, ENO schemes as well as their variant weighted ENO (WENO) [280] have been implemented in incompressible flows. At present, ENO schemes and their variants consist a well established advanced numerical framework for constructing high-order methods for multidimensional problems, which provide solutions "theoretically" (almost) free from spurious oscillations.

The schemes have been developed to provide solutions to systems of hyperbolic conservation laws. Let us consider the hyperbolic conservation law

$$\frac{\partial \mathbf{U}}{\partial t} + \frac{\partial \mathbf{E}(\mathbf{U})}{\partial x} = 0 \,, \tag{17.3}$$

in conjunction with an initial condition $\mathbf{U}(x,0) \equiv \mathbf{U}_0$; the vectors $\mathbf{U}$ and $\mathbf{E}(\mathbf{U})$ have m components. The system is hyperbolic considering that the $m \times m$ Jacobian matrix $\mathbf{A}(\mathbf{U}) = \partial \mathbf{E}/\partial \mathbf{U}$ has m real eigenvalues

$$\lambda_1(\mathbf{U}) \leq \lambda_2(\mathbf{U}) \leq ... \leq \lambda_m(\mathbf{U}) \,, \tag{17.4}$$

and a complete set of m linearly independent right eigenvectors $\mathbf{T}_k$, where $k = 1 \ldots m$. If $\mathbf{T}_k^{-1}$ are the left eigenvectors then $\mathbf{T}_k^{-1} \cdot \mathbf{T}_k = \mathbf{I}$ ($\mathbf{I}$ is the unit matrix).

The discretization of (17.3) can be written as

$$\bar{\mathbf{U}}^{n+1} = \bar{\mathbf{U}}^n - \frac{\Delta t}{\Delta x}(\bar{\mathbf{E}}_{i+1/2} - \bar{\mathbf{E}}_{i-1/2}) = [\Re(\tau) \cdot \bar{\mathbf{U}}^n]_i \ , \tag{17.5}$$

where $\bar{\mathbf{E}}$ is the numerical flux, $\bar{\mathbf{U}}$ is the numerical approximation of $\mathbf{U}$ and $\Re(\tau)$ denotes the numerical solution operator. The numerical flux is a function of $2k$ variables, i.e.,

$$\bar{\mathbf{E}}_{i+1/2} = \bar{\mathbf{E}}(\bar{\mathbf{U}}^n_{i-k+1}, ..., \bar{\mathbf{U}}^n_{i+k}) \ , \tag{17.6}$$

and is consistent with the flux $\mathbf{E}$ according to

$$\bar{\mathbf{E}}(\mathbf{U}, ..., \mathbf{U}) = \mathbf{E} \ . \tag{17.7}$$

Harten and Osher [251] presented a second-order accurate scheme[4] which is strictly nonoscillatory (known as uniformly nonoscillatory scheme or UNO) in the scalar case ($m = 1$), that is

$$N_0(U^{n+1}) \leq N_0(U^n) \ , \tag{17.8}$$

where $N_0(U)$ denotes the number of local extrema in U. The scheme presented by Harten and Osher [251] is a modification of the MUSCL scheme, which is a TVD scheme for the scalar case. In the second-order extension of the Godunov's scheme, the piecewise linear reconstruction of the data $R(x; \bar{U})$ is obtained by

$$R(x; \bar{U}) = \bar{U}_i + S_i(x - x_i) \quad \text{for} \quad x_{i-1/2} < x < x_{i+1/2} \ , \tag{17.9}$$

with the slope S_i being defined as

$$S_i = \frac{\partial U(x_i)}{\partial x} + \mathcal{O}(\Delta x) \ . \tag{17.10}$$

In the MUSCL scheme the slope is function of a limiter. The limiter is used to "guarantee" the TVD condition by switching the scheme to be first-order accurate at local extrema. However, the limiter results in a discontinuous Taylor series expansion (at local extrema), thus leading to a loss of accuracy. Harten and Osher [251] modified the second-order accurate piecewise linear reconstruction of the data $R(x; \bar{u})$ by using a second-order slope S_i that satisfies

$$S_i = \frac{\partial U(x_i)}{\partial x} + \mathcal{O}(\Delta x^2) \ . \tag{17.11}$$

Further, they introduced the condition of nonoscillatory schemes according to which in the scalar case

$$\mathrm{TV}(\Re(\tau) \cdot \bar{U}) \leq \mathrm{TV}(U) + \mathcal{O}(\Delta x^2) \ , \tag{17.12}$$

where $\bar{U}$ is a cell average of a piecewise smooth function $U(x)$.

---

[4] This was also introduced in Chap. 14.

The condition (17.12) limits the accuracy to second order. To construct schemes of accuracy higher than second order, they introduced the condition

$$\mathrm{TV}(\Re(\tau)\cdot\bar{U}) \le \mathrm{TV}(R(x;\bar{U})) \le \mathrm{TV}(U) + \mathcal{O}(\Delta x^r)\,, \tag{17.13}$$

where the reconstruction $R(x;\bar{U})$ is $(r+1)$th-order accurate, i.e.,

$$R(x;\bar{U}) = U(x) + e(x)(\Delta x)^r + \mathcal{O}(\Delta x^{r+1})\,, \tag{17.14}$$

where $e(x)$ is a coefficient. The above condition leads to ENO schemes that allow the production of spurious oscillations only on the level of truncation error.

Unlike TVD schemes, ENO schemes do not use monotonicity limiters. In the case of ENO schemes, uniform order of accuracy ($r-$th order accurate schemes) is obtained by controlling any increase of the total variation of the numerical solution through an adaptive stencil in such a way that each (grid) point attempts to use the smoothest information available. The number of points in the stencil should be $r+1$.

According to [248], the information about smoothness of $U(x)$ can be obtained by a table of divided differences of $U(x)$. This is achieved by a recursive operation as follows:

$$U[x_i] = W(x_i)\,, \tag{17.15}$$

and

$$W[x_i,...,x_{i+k}] = \frac{W[x_{i+1},...,x_{i+k}] - W[x_i,...,x_{i+k-1}]}{x_{i+k}-x_i}\,. \tag{17.16}$$

If $W(x)$ is a continuous function in (x_i, x_{i+k}) then

$$W[x_i,...,x_{i+k}] = \frac{1}{k!}\frac{\mathrm{d}^k}{\mathrm{d}x^k}U(\xi_{i,k})\,, \quad x_i \le \xi_{i,k} \le x_{i+k}\,. \tag{17.17}$$

If W has a jump discontinuity in the pth derivative $(0 \le p \le k)$ then

$$W[x_i,...,x_{i+k}] = \mathcal{O}\Big(\Delta x^{-k+p}[U^p]\Big)\,, \tag{17.18}$$

where $[W^p]$ denotes the jump in the pth derivative [248]. The above equations provide an asymptotic measure of the smoothness of U, which is equivalent to finding an interval in which U has the smallest divided differences.

In [248], implementation of the ENO schemes was obtained via a Lax-Wendroff-type time discretization. This is, however, complicated to program, especially for multidimensional problems and problems described by partial differential equations with source terms. For the efficient implementation of ENO schemes, Shu and Osher [495] have proposed the use of TVD Runge-Kutta type methods to discretize in time. These methods are presented in Chap. 7 and discussed further in Chap. 9. Further, the original version of the

ENO schemes involved cell averages as well as point values. As a result a reconstruction procedure is required to recover point values from cell averages to the correct order of accuracy. This complicates the algorithm especially in multidimensional problems. In [495], it was proposed to use the adaptive stencil directly on fluxes to get ENO schemes without using cell averages. This approach is discussed below.

17.3 ENO Schemes Using Fluxes

The objective of this approach is to derive ENO schemes using only fluxes. For the sake of simplicity, we consider the hyperbolic conservation law for the scalar case

$$\frac{\partial U}{\partial t} + \frac{\partial E(U)}{\partial x} = 0 \,, \tag{17.19}$$

which can be discretized as

$$\bar{U}^{n+1} = \bar{U}^n - \frac{\Delta t}{\Delta x}(\bar{E}_{i+1/2} - \bar{E}_{i-1/2}) \,. \tag{17.20}$$

The numerical flux $\bar{E}$ in (17.20) can be split into positive and negative fluxes[5] as

$$\bar{E}_{i+1/2} = \bar{E}_i^+ + \bar{E}_{i+1}^- \,, \tag{17.21}$$

where the fluxes $\bar{E}_i^+$ and $\bar{E}_{i+1}^-$ are the numerical fluxes which are associated with the positive and negative eigenvalues, respectively, of the flux E,

$$\frac{\partial E^+}{\partial U} \geq 0 \,, \quad \frac{\partial E^-}{\partial U} \leq 0 \,, \tag{17.22}$$

$$E(U) = E^+(U) + E^-(U) \,. \tag{17.23}$$

The positive and negative fluxes can be calculated by a first-order monotone scheme, for example, the Lax-Friedrichs scheme (15.7). The numerical flux (17.21) can be developed in a Taylor series expansion as follows

$$\bar{E}_{i+1/2} = E_{i+1/2} + \sum_{k=1}^{m-1} a_{2k} \Delta x^{2k} \left(\frac{\partial^{2k}}{\partial x^{2k}} E \right)_{i+1/2} + \mathcal{O}(\Delta x^{2m+1}) \,, \tag{17.24}$$

where $a_2, a_4, ..., a_{2m-2}, ...,$ are constant coefficients. Following (17.21) we can also require

$$\bar{E}_{i+1/2} = \bar{E}_{i+1/2}^+ + \bar{E}_{i+1/2}^- \,. \tag{17.25}$$

The positive $\bar{E}^+$ and negative $\bar{E}^-$ fluxes can be defined to satisfy (17.23), i.e.,

[5] Flux splitting schemes were discussed in Chaps. 9 and 13.

$$\bar{E}^{\pm}_{i+1/2} = E^{\pm}_{i+1/2} + \sum_{k=1}^{m-1} a_{2k} \Delta x^{2k} \Big(\frac{\partial^{2k}}{\partial x^{2k}} E^{\pm} \Big)_{i+1/2} + \mathcal{O}(\Delta x^{2m+1}) \, . \quad (17.26)$$

Shu and Osher [495] proposed to use polynomial interpolants $p^{\pm}_{i+1/2}$ of $E^{\pm}$ such that

$$p^{\pm}_{i+1/2}(x) = E^{\pm}(U(x)) + \mathcal{O}\big(\Delta x^{2m+1}\big) \, , \quad (17.27)$$

and defined the positive and negative fluxes as

$$\bar{E}^{\pm}_{i+1/2} = p^{\pm}_{i+1/2} + \sum_{k=1}^{m-1} a_{2k} \Delta x^{2k} \Big(\frac{\partial^{2k}}{\partial x^{2k}} p^{\pm}_{i+1/2} \Big)_{i+1/2} \, . \quad (17.28)$$

The interpolating polynomials are constructed by using the adaptive stencil ENO procedure according to which we choose the $2m+1$ points automatically from the smoothest possible region of the stencil, starting with the one obtained by (17.21). We present below the (procedure for the polynomial $p^{+}_{i+1/2}$):

1. We initiate the procedure by defining the indices

$$j^{(0)}_{\min} = j^{(0)}_{\max} = i \, , \quad \text{and} \quad C^{(0)}_{+} = \bar{E}^{+}(U_i) \, . \quad (17.29)$$

2. Consider that $j^{(n-1)}_{\min}$, $j^{(n-1)}_{\max}$ and $C^{(n-1)}_{+}$ are known (from the $(n-1)$th divided difference), then we calculate the nth divided difference of $\bar{E}^{+}(U(x))$ using (17.16)

$$a^{(n)} = \bar{E}^{+} \Big[U\big(x_{k^{(n-1)}_{\min}}\big), ..., U\big(x_{k^{(n-1)}_{\max}+1}\big) \Big] \, , \quad (17.30)$$

$$b^{(n)} = \bar{E}^{+} \Big[U\big(x_{k^{(n-1)}_{\min}-1}\big), ..., U\big(x_{k^{(n-1)}_{\max}}\big) \Big] \, . \quad (17.31)$$

Then, we add a a point to the stencil according to the smallest nth divided difference as follows

- If $|a^{(n)}| \geq |b^{(n)}|$, then

$$c^{(n)} = b^{(n)} \, , \quad (17.32)$$

$$j^{(n)}_{\min} = j^{(n-1)}_{\min} - 1 \, , \quad j^{(n)}_{\max} = j^{(n-1)}_{\max} \, . \quad (17.33)$$

- If $|a^{(n)}| < |b^{(n)}|$, then

$$c^{(n)} = a^{(n)} \, , \quad (17.34)$$

$$j^{(n)}_{\min} = j^{(n-1)}_{\min} \, , \quad j^{(n)}_{\max} = j^{(n-1)}_{\max} + 1 \, . \quad (17.35)$$

Using the above we write

$$C_+^{(n)}(x) = C_+^{(n-1)}(x) + c^{(n)} \prod_{k=k_{\min}^{(n-1)}}^{k=k_{\max}^{(n-1)}} (x - x_k) . \tag{17.36}$$

3. Finally, the positive interpolant $p^+_{i+1/2}$ is defined by

$$p^+_{i+1/2} = C_+^{(2m)}(x) \tag{17.37}$$

The construction of the negative interpolant $p^-_{i+1/2}$ follows a similar procedure:

1. We initiate the procedure by defining the indices $j_{\min}^{(0)} = j_{\max}^{(0)} = i+1$ and $C_-^{(0)} = \bar{E}^-(U_{i+1})$.
2. Then, we repeat the step 2 as above with $\bar{E}^+$ replaced by $\bar{E}^-$ and C_+ replaced by C_-.
3. Finally,

$$p^-_{i+1/2} = C_-^{(2m)}(x) . \tag{17.38}$$

For smooth solutions, the divided differences (17.30) and (17.31) should be bounded by the maximum norm of the nth derivative of $E^\pm$ multiplied by a constant, for example [495],

$$d^{(n)} = \min(|c^{(n)}|, M^{(n)})\text{sign}(c^{(n)}) , \tag{17.39}$$

or

$$d^{(n)} = \min(|c^{(n)}|, M^{(n)} \Delta x^{n-2})\text{sign}(c^{(n)}) . \tag{17.40}$$

Accordingly, in (17.36) we replace $c^{(n)}$ by $d^{(n)}$, where $M^{(n)}$ are constants preferably related to the maximum norm of the nth derivative of $E^\pm$ in (initially) smooth regions.

Remark 17.3.1. In multiple dimensions the procedure described by equations (17.21)-(17.38) is applied to each of the flux derivatives appearing in the conservation laws. The solution of the equations in time is then obtained by applying the TVD Runge-Kutta type discretization.

For nonlinear systems of equations, the polynomial interpolants are obtained using the left, $(\mathbf{T}_k)_{i+1/2}$, and right, $(\mathbf{T}_k^{-1})_{i+1/2}$, eigenvectors: (i) Firstly, we interpolate $(\mathbf{T}_k)_{i+1/2} \cdot \mathbf{E}^\pm$ to obtain $(\mathbf{T}_k)_{i+1/2} \cdot p^\pm_{i+1/2}$ following the procedure (17.29)-(17.38). (ii) Secondly, we define the polynomial interpolants as

$$p^\pm_{i+1/2}(x) = \sum_{k=1}^{m} \left[(\mathbf{T}_k)_{i+1/2} \cdot p^\pm_{i+1/2}\right](\mathbf{T}_k^{-1})_{i+1/2} . \tag{17.41}$$

The fluxes $\bar{\mathbf{E}}^\pm_{i+1/2}$ are then calculated by (17.28).

17.4 Weighted ENO Schemes

The ENO schemes select an interpolating stencil in which the solution is supposed to be the smoothest one. If a cell is near a discontinuity (or large gradient), then the smoothest possible solution is assigned to this cell and so spurious oscillations (Gibbs phenomenon) is avoided. Liu et al. [356] developed a new version of the ENO schemes called weighted ENO schemes or WENO. The WENO schemes use a convex combination of all the corresponding interpolating polynomials on the stencil in order to compute an approximate polynomial for each cell. The interpolating polynomials are combined by assigning weights to the convex combination.

The WENO schemes satisfy the essentially nonoscillatory property by combining only the interpolating polynomials on the smoothest stencils. The interpolating polynomials on the discontinuous stencil do not contribute to the convex combination. The advantage of WENO schemes compared to the original ENO schemes are:

- Spurious oscillations near discontinuities are avoided because the cells near discontinuities are assigned stencils from the smooth part of the solution.
- The convex combination of interpolating polynomials results in the cancellation of truncation errors thus improving the order of accuracy by one.
- Reduction of ENO's oscillatory behavior near convergence.

The starting point for the reconstruction procedure are the cell average values $\bar{U}_i$ of the solution at each cell i from which the point values at the interface $i+1/2$ are calculated, for example, by simple averaging. Then, we select a stencil

$$S_i = (x_{i-r+1/2}, x_{i-r+3/2}, ..., x_{i+1/2}) , \tag{17.42}$$

and interpolate $\bar{U}_{i+1/2}$ to obtain a polynomial p_i. In (17.42) r denotes the order of accuracy.

For each stencil S_i, a *smoothness indicator* $(IS)_i$ is evaluated by first computing a table of differences

$$\begin{array}{ccc} \Delta[\bar{U}_{i-r+1}], & \Delta[\bar{U}_{i-r+2}], \quad ..., & \Delta[\bar{U}_{i-1}] \\ \Delta^2[\bar{U}_{i-r+1}], & \Delta^2[\bar{U}_{i-r+2}], \quad ..., & \Delta^2[\bar{U}_{i-1}] \\ & \vdots & \\ \Delta^{r-1}[\bar{U}_{i-r+1}], & \Delta^{r-1}[\bar{U}_{i-r+2}], ..., & \Delta^{r-1}[\bar{U}_{i-1}] , \end{array} \tag{17.43}$$

where

$$\left. \begin{array}{ll} \Delta[\bar{U}_l] & = \bar{U}_{l+1} - \bar{U}_l \\ \Delta^k[\bar{U}_l] & = \Delta^{k-1}[\bar{u}_{l+1}] - \Delta^{k-1}[\bar{U}_l] . \end{array} \right\} , \tag{17.44}$$

The smoothness indicator is defined by the summation of all averages of square values of the same order differences [356]

$$IS_i = \frac{\sum_{l=1}^{r-1}\left(\sum_{k=1}^{l}(\Delta^{r-l}[\bar{U}_{i-r+k}])^2\right)}{l} . \tag{17.45}$$

The convex combination is obtained as follows: For each cell we define r stencils

$$(S_{i+k})_{k=0}^{r-1} = \left\{x_{i+k-r+1/2}, x_{i+k-r+3/2}, ..., x_{i+k+1/2}\right\}_{k=0}^{r-1} , \tag{17.46}$$

and r corresponding interpolating polynomials $\left(p_{i+k}\right)_{k=0}^{r-1}$. The WENO schemes use a convex combination of all the interpolating polynomials $\left(p_{i+k}\right)_{k=0}^{r-1}$ to obtain a new polynomial $P_i(x)$

$$P_i(x) = \sum_{k=0}^{r-1} \frac{a_k^i}{\sum_{l=0}^{r-1} a_l^i} p_{i+k}(x) . \tag{17.47}$$

The polynomial $P_i(x)$ is the reconstructed solution for $U(x)$, i.e.,

$$P_i(x) = U(x,t) + \mathcal{O}(\Delta x^r) . \tag{17.48}$$

In (17.47) a_k^i are positive coefficients $(k = 0, 1, 2, ..., r-1)$ defined as

$$a_k^i = \frac{C_k^i}{(\epsilon + IS_{i+k})^r} , \quad k = 0, 1, ..., r-1 , \tag{17.49}$$

where ϵ is a small positive number (e.g., $\epsilon = 10^{-5}$) to prevent division by zero, and $C_k^i = \mathcal{O}(1)$ ($C_k^i > 0$). The coefficients C_k^i are defined by

$$C_k^i = \begin{cases} 1 & \text{if} \quad b_k^i(x_i) = 0 , \\[2ex] \dfrac{\Delta x^r}{\eta_p |b_k^i(x_i)|} & \text{if} \quad b_k^i(x_i) > 0 , \\[2ex] \dfrac{\Delta x^r}{\eta_n |b_k^i(x_i)|} & \text{if} \quad b_k^i(x_i) < 0 . \end{cases} \tag{17.50}$$

where

$$b_k^i(x) = \sum_{s=0}^{r} \left\{ \prod_{l=0, l\neq s}^{r} (x - x_{i+k-l+1/2}) \right\} , \tag{17.51}$$

and η_p is the number of positive terms in $b_k^i(x)$ and η_n is the number of negative terms in $b_k^i(x)$.

The advective flux derivative in (17.19) is calculated using the reconstructed solution (17.47), i.e.,

$$\frac{\partial E(U)}{\partial x} = \frac{1}{\Delta x}\Big\{\bar{E}(P_i(x_{i+1/2}), P_{i+1}(x_{i+1/2})) - \bar{E}(P_{i-1}(x_{i-1/2}), P_i(x_{i-1/2}))\Big\}, \tag{17.52}$$

where $\bar{E}$ denotes a numerical flux, which can be approximated by any of the numerical fluxes we have discussed in previous chapters, for example, Lax-Friedrichs (Chap. 15) or characteristic-based flux (i.e., Chap. 16).

17.4.1 Third-Order WENO Reconstruction

The third-order reconstruction is obtained for $r = 2$ [356]. Note that using this reconstruction Liu et al. [356] achieved fourth-order of accuracy in their numerical experiments. The key points of the reconstruction procedure are listed below:

1. For the cell defined by $[x_{i-1/2}, x_{i+1/2}]$ we define the stencils

$$\left.\begin{array}{ll} S_i & = (x_{i-3/2}, x_{i-1/2}, x_{i+1/2}) \\ S_{i+1} & = (x_{i-3/2}, x_{i-1/2}, x_{i+1/2}) \end{array}\right\} . \tag{17.53}$$

2. For each stencil, the linear interpolation polynomial are given by

$$\left.\begin{array}{ll} p_i(x) & = \bar{U}_i + \dfrac{\bar{U}_i - \bar{U}_{i-1}}{\Delta x}(x - x_i) \\ p_{i+1}(x) & = \bar{U}_i + \dfrac{\bar{U}_{i+1} - \bar{U}_i}{\Delta x}(x - x_i) \end{array}\right\} . \tag{17.54}$$

   where $\bar{U}$ are numerical values of $U(x)$.
3. The convex combination is defined by

$$P_i = \frac{a_0^i}{a_0^i + a_1^i} p_i(x) + \frac{a_1^i}{a_0^i + a_1^i} p_{i+1}(x) , \tag{17.55}$$

   where

$$a_0^i = \frac{C_0^i}{(\epsilon + (IS)_i)^2} , \quad a_1^i = \frac{C_1^i}{(\epsilon + (IS)_{i+1})^2} . \tag{17.56}$$

   The smoothness indicators are calculated by $(IS)_i = (\bar{U}_i - \bar{U}_{i-1})^2$ and $(IS)_i = (\bar{U}_{i+1} - \bar{U}_i)^2$. The coefficients C_0^i and C_1^i are calculated as follows:

- For $\partial E(U)/\partial U > 0$, we specify $b_0^i(x_{i+1/2}) = 2\Delta x^2$ and $b_1^i(x_{i+1/2}) = -\Delta x^2$, and obtain $\eta_p = 1$ and $\eta_n = 1$, that gives $C_0^i = 1/2$ and $C_1^i = 1$. The coefficients a_0^i and a_1^i are calculated by

$$\left.\begin{aligned} a_0^i &= \frac{1}{2(\epsilon + (IS)_i)^2} \\ a_1^i &= \frac{1}{(\epsilon + (IS)_{i+1})^2} \end{aligned}\right\} . \tag{17.57}$$

- For $\partial E(U)/\partial U < 0$, we specify $b_0^i(x_{i-1/2}) = -\Delta x^2$ and $b_1^i(x_{i-1/2}) = 2\Delta x^2$, and obtain $\eta_p = 1$ and $\eta_n = 1$, that gives $C_0^i = 1$ and $C_1^i = 1/2$. The coefficients a_0^i and a_1^i are calculated by

$$\left.\begin{aligned} a_0^i &= \frac{1}{(\epsilon + (IS)_i)^2} \\ a_1^i &= \frac{1}{2(\epsilon + (IS)_{i+1})^2} \end{aligned}\right\} . \tag{17.58}$$

17.4.2 Fourth-Order WENO Reconstruction

The fourth-order reconstruction is obtained for $r = 3$ [356]. The key points of the reconstruction procedure are listed below:

1. For the cell defined by $[x_{i-1/2}, x_{i+1/2}]$ we define the stencils

$$\left.\begin{aligned} S_i &= (x_{i-5/2}, x_{i-3/2}, x_{i-1/2}, x_{i+1/2}) \\ S_{i+1} &= (x_{i-3/2}, x_{i-1/2}, x_{i+1/2}, x_{i+3/2}) \\ S_{i+2} &= (x_{i-1/2}, x_{i+1/2}, x_{i+3/2}, x_{i+5/2}) \end{aligned}\right\} . \tag{17.59}$$

2. For each stencil, the linear interpolation polynomials are given by

$$\begin{aligned} p_i(x) = {} & \frac{\bar{U}_i - 2\bar{U}_{i-1} + \bar{U}_{i-2}}{\Delta x^2}(x - x_{i-1})^2 + \frac{\bar{U}_i - \bar{U}_{i-2}}{2\Delta x}(x - x_{i-1}) \\ & + \bar{U}_{i-1} - \frac{\bar{U}_i - 2\bar{U}_{i-1} + \bar{U}_{i-2}}{24} , \end{aligned} \tag{17.60}$$

$$\begin{aligned} p_{i+1}(x) = {} & \frac{\bar{U}_{i+1} - 2\bar{U}_i + \bar{U}_{i-1}}{\Delta x^2}(x - x_i)^2 + \frac{\bar{U}_{i+1} - \bar{U}_{i-1}}{2\Delta x}(x - x_i) \\ & + \bar{U}_i - \frac{\bar{U}_{i+1} - 2\bar{U}_i + \bar{U}_{i-1}}{24} , \end{aligned} \tag{17.61}$$

$$\begin{aligned} p_{i+2}(x) = {} & \frac{\bar{U}_{i+2} - 2\bar{U}_{i+1} + \bar{U}_i}{\Delta x^2}(x - x_{i+1})^2 + \frac{\bar{U}_{i+2} - \bar{U}_i}{2\Delta x}(x - x_{i+1}) \\ & + \bar{U}_{i+1} - \frac{\bar{U}_{i+2} - 2\bar{U}_{i+1} + \bar{U}_i}{24} . \end{aligned} \tag{17.62}$$

3. The convex combination is defined by

$$P_i = \frac{a_0^i}{a_0^i + a_1^i + a_2^i} p_i(x) + \frac{a_1^i}{a_0^i + a_1^i + a_2^i} p_{i+1}(x) + \frac{a_2^i}{a_0^i + a_1^i + a_2^i} p_{i+2}(x) \;, \tag{17.63}$$

where

$$\left. \begin{aligned} a_0^i &= \frac{C_0^i}{(\epsilon + IS_i)^3} \\ a_1^i &= \frac{C_1^i}{(\epsilon + IS_{i+1})^3} \\ a_2^i &= \frac{C_2^i}{(\epsilon + IS_{i+2})^3} \end{aligned} \right\} \;. \tag{17.64}$$

The smoothness indicators are calculated by

$$(IS)_i = \frac{1}{2}\left[(\bar{U}_{i-1} - \bar{U}_{i-2})^2 + (\bar{U}_i - \bar{U}_{i-1})^2\right] + (\bar{U}_i - 2\bar{U}_{i-1} + \bar{U}_{i-2})^2 \;, \tag{17.65}$$

$$(IS)_{i+1} = \frac{1}{2}\left[(\bar{U}_i - \bar{U}_{i-1})^2 + (\bar{U}_{i+1} - \bar{U}_i)^2\right] + (\bar{U}_{i+1} - 2\bar{U}_i + \bar{U}_{i-1})^2 \;, \tag{17.66}$$

$$(IS)_{i+2} = \frac{1}{2}\left[(\bar{U}_{i+1} - \bar{U}_i)^2 + (\bar{U}_{i+2} - \bar{U}_{i+1})^2\right] + (\bar{U}_{i+2} - 2\bar{U}_{i+1} + \bar{U}_i)^2 \;. \tag{17.67}$$

The coefficients C_0^i, C_1^i and C_2^i in (17.64) are calculated as follows:

- For $\partial E(U)/\partial U > 0$, we specify $b_0^i(x_{i+1/2}) = 6\Delta x^3$, $b_1^i(x_{i+1/2}) = -2\Delta x^3$ and $b_2^i(x_{i+1/2}) = 2\Delta x^3$. For these values, we obtain $\eta_p = 2$ and $\eta_n = 1$, that gives $C_0^i = 1/12$, $C_1^i = 1/2$ and $C_2^i = 1/4$. The coefficients a_0^i, a_1^i and a_2^i are calculated by

$$\left. \begin{aligned} a_0^i &= \frac{1}{12(\epsilon + (IS)_i)^3} \\ a_1^i &= \frac{1}{2(\epsilon + (IS)_{i+1})^3} \\ a_2^i &= \frac{1}{4(\epsilon + (IS)_{i+2})^3} \end{aligned} \right\} \;. \tag{17.68}$$

- For $\partial E(U)/\partial U < 0$, we specify $b_0^i(x_{i-1/2}) = -2\Delta x^3$, $b_1^i(x_{i+1/2}) = 2\Delta x^3$ and $b_2^i(x_{i+1/2}) = -6\Delta x^3$. For these values, we obtain $\eta_p = 1$ and $\eta_n = 2$, that gives $C_0^i = 1/4$, $C_1^i = 1/2$ and $C_2^i = 1/12$. The coefficients a_0^i, a_1^i and a_2^i are calculated by

$$\left.\begin{aligned} a_0^i &= \frac{1}{4(\epsilon + (IS)_i)^3} \\ a_1^i &= \frac{1}{2(\epsilon + (IS)_{i+1})^3} \\ a_2^i &= \frac{1}{12(\epsilon + (IS)_{i+2})^3} \end{aligned}\right\} . \tag{17.69}$$

Remark 17.4.1. For the same order of accuracy, WENO schemes are slightly more expensive than ENO schemes for serial computations. In parallel computations, WENO schemes will be (substantially) more expensive than their ENO counterparts because the former involve more data transfer between computational cells. WENO schemes lead to a smoother flux than that of ENO schemes.

17.5 A Flux-Based Version of the WENO Scheme

In WENO schemes the weight on a stencil has to vary according to the relative smoothness of this stencil to the other candidate stencils. The definition of the weight depends strongly on the way of evaluating the smoothness of a stencil. Jiang and Shu [280] proposed a way of measuring the smoothness of the numerical solution that is based on minimizing the L_2 error norm of the derivatives of the reconstruction polynomials. Jiang's and Shu's WENO scheme is approximately twice faster than the original WENO scheme. First we present the modified WENO scheme for the scalar case.

The numerical flux $\bar{E}$ is defined by

$$\bar{E}_{i+1/2} = \bar{E}^+_{i+1/2} + \bar{E}^-_{i+1/2} , \tag{17.70}$$

where the positive and negative fluxes can be defined from the Lax-Friedrichs, characteristic-based, or Roe flux. The positive and negative fluxes can by evaluated by an rth order ENO approximation.[6]

$$\bar{E}^+_{i+1/2} = \sum_{k=0}^{r-1} \omega_k q_k^{(r)}(E_{i+k-r+1}, ..., E_{i+k}) , \tag{17.71}$$

where

$$\omega_k = \frac{a_k}{a_0 + a_1 + ... + a_{r-1}} , \tag{17.72}$$

[6] We consider the positive flux only. The negative flux can be symmetrically written with respect to $x_{i+1/2}$.

$$a_k = \frac{C_k^{(r)}}{(\epsilon + (IS)_k)^p} \,, \quad k = 0, 1, ..., r-1 \,, \tag{17.73}$$

and

$$q_k^{(r)}(E^+_{i+k-r+1}, ..., E^+_{i+k}) = \sum_{l=0}^{r-1} \hat{a}_{k,l}^{(r)} g_l(E^+_{i+k-r+1}, ..., E^+_{i+k}) \,. \tag{17.74}$$

The coefficients $\hat{a}_{k,l}^{(r)}$ and optimal weights $C_k^{(r)}$ are defined in Tables 17.1 and 17.2, respectively. In [280], the value $p = 2$ is suggested to obtain essentially nonoscillatory approximations at least for $r = 2$ and $r = 3$.

Table 17.1. Values of coefficients $\hat{a}_{k,l}^{(r)}$ used in (17.74).

r	k	l=0	l=1	l=2
2	0	-1/2	3/2	-
	1	1/2	1/2	-
3	0	1/3	-7/6	11/6
	1	-1/6	5/6	1/3
	2	1/3	5/6	-1/6

Table 17.2. Values of optimal weights $C_k^{(r)}$ used in (17.73).

C_k^r	$k=0$	$k=1$	$k=2$
r=2	1/3	2/3	-
r=3	1/10	6/10	3/10

Using the above, the positive flux in (17.71) is written

$$\begin{aligned} \bar{E}^+_{i+1/2} = & \,\omega_0^+ \left(\frac{2}{6} E^+_{i-2} - \frac{7}{6} E^+_{i-1} + \frac{11}{6} E^+_i \right) \\ & + \omega_1^+ \left(-\frac{1}{6} E^+_{i-1} + \frac{5}{6} E^+_i + \frac{2}{6} E^+_{i+1} \right) \\ & + \omega_2^+ \left(\frac{2}{6} E^+_i + \frac{5}{6} E^+_{i+1} - \frac{1}{6} E^+_{i+2} \right) . \end{aligned} \tag{17.75}$$

The original WENO scheme defines the smoothness by the smoothness indicators (17.45). In the modified WENO scheme, new weights are defined, which replace (17.45). In the derivation of new weights, Jiang and Shu [280] explored the idea of the total variation as a good measure for smoothness, thus aiming at minimizing the total variation for the approximation.

Let us denote the r candidate stencils (r is the order of accuracy) by S_k, where $k = 0, 1..., r-1$, and the interpolation polynomial on stencil S_k by p_k. According to [280] the smoothness indicators are given by

$$(IS)_k = \sum_{l=1}^{r-1} \int_{x_{i-1/2}}^{x_{i+1/2}} \Delta x^{2l-1} \, (p_k^{(l)})^2 \, \mathrm{d}x \,, \tag{17.76}$$

where $p_k^{(l)}$ is the lth derivative of $p_k(x)$. The right-hand-side of (17.76) is the sum of the L_2 norms of all the derivatives of the interpolation polynomial $p_k(x)$ over the interval $[x_{i-1/2}, x_{i+1/2}]$. The term $(\Delta x)^{2l-1}$ is included to remove the dependence on Δx in the derivatives of the polynomials. For $r = 2$, (17.76) results in the same smoothness indicators as in the original WENO scheme. For $r = 3$ the result is different, i.e.,

$$\left.\begin{aligned} (IS)_0 &= \tfrac{13}{12}(E_{i-2} - 2E_{i-1} + E_i)^2 + \\ &\quad \tfrac{1}{4}(E_{i-2} - 4E_{i-1} + 3E_i)^2 \\ (IS)_1 &= \tfrac{13}{12}(E_{i-1} - 2E_i + E_{i+1})^2 + \\ &\quad \tfrac{1}{4}(E_{i-1} - E_{i+1})^2 \\ (IS)_2 &= \tfrac{13}{12}(E_i - 2E_{i+1} + E_{i+2})^2 + \\ &\quad \tfrac{1}{4}(3E_i - 4E_{i+1} + E_{i+2})^2 \end{aligned}\right\} \,. \tag{17.77}$$

The extension of the scheme to multiple dimensions and nonlinear systems $(m \times m)$ can be obtained similarly to the implementation of the ENO schemes. Let $\mathbf{l}_n$ and $\mathbf{r}_n$ be the nth row and column vectors of the left $\mathbf{T}^{-1}$ and right $\mathbf{T}$ eigenvector matrices, respectively, of the Jacobian matrix $\mathbf{A}_{i+1/2}$, where n denotes the characteristic field. For each characteristic field one first obtains the flux

$$\tilde{\mathbf{E}}_{i+1/2,n} = \sum_{k=0}^{r-1} \omega_{k,n} q_k^{(r)}(\mathbf{l}_n \cdot \bar{\mathbf{E}}_{i+k-r+1}, ..., \mathbf{l}_n \cdot \bar{\mathbf{E}}_{i+k}) \,, \tag{17.78}$$

where

$$\omega_{k,n} = \omega_k(\mathbf{l}_n \cdot \bar{\mathbf{E}}_{i-r+1}, ..., \mathbf{l}_n \cdot \bar{\mathbf{E}}_{i+r-1}) \,, \tag{17.79}$$

for $k = 0, 1, ..., r-1$, are the weights in the nth characteristic field, and ω_k is given by (17.72). The numerical fluxes in each characteristic field are projected back to the physical space by multiplying the flux $\tilde{\mathbf{E}}$ with the nth column vector of the right eigenvector matrix

$$\bar{\mathbf{E}}_{i+1/2} = \sum_{n=1}^{m} \tilde{\mathbf{E}}_{i+1/2,n} \cdot \mathbf{r}_n \ . \tag{17.80}$$

The higher order WENO methods of this type were discussed by Balsara and Shu [28]. These methods include a seventh-, ninth- and eleventh-order extensions of the fifth-order method described above. They were developed in conjunction with the monotonicity-preserving (MP) methods [518] and are known as MPWENO (monotonicity preserving WENO) schemes. MP methods are described and discussed in Sect. 17.8 in detail.

Yang et al. [608] have implemented the third-order modified WENO scheme in incompressible flows in conjunction with the artificial compressibility approach. Their computations for flow through a ninety degree bend (square duct) as well as for three-dimensional lid-driven cavity flow have shown that the third-order version of the WENO scheme provides both more accurate and efficient results than the second-order ENO scheme.

17.6 Artificial Compression Method for ENO and WENO

In some cases the ENO and WENO methods can be quite diffusive especially for linear discontinuities. In the case of gas dynamics this often accompanies contact discontinuities which do not have the self-steepening mechanism accompanying shock waves. The same sort of dynamics can be found operating in incompressible shear layers. In order to combat these problems a technique known as the artificial compression method (ACM) was developed by Yang [607]. Yang's method built upon an earlier method derived by Harten [246].

In ACM, a correction is added to the flux defined by the ENO or WENO algorithm which reduces the amount of diffusion (hence is anti-diffusive in nature),

$$E_{j+1/2} := E_{j+1/2} + C_{j+1/2} \ . \tag{17.81}$$

The issue is to apply this correction in a stable manner. This is accomplished by making the correction consistent with monotonicity through a trigger that finds discontinuous profiles and only applies the correction at discontinuities. Experience has shown that applying a correction of this sort in smooth regions often introduces unpleasant side-effects such as turning a smooth transition into a series of steps.

The correction can be compactly defined by

$$
\begin{aligned}
C_{j+1/2} = \text{minmod}\Big[& \\
& \frac{\alpha_j}{2}\text{minmod}\left(\hat{E}_{j+1/2} - E_{j+1/2}, \hat{E}_{j-1/2} - E_{j-1/2}\right) \\
& \text{minmod}\left(E_{j+1} - E_{j+1/2}, \hat{E}_{j-1/2} - E_{j-1}\right)\Big] ,
\end{aligned}
\tag{17.82}
$$

where $\hat{E}$ is the flux defined with the opposite direction as the physical flux. The minmod limiter has been defined in Sect. 14.2.1. The last two terms are defined by monotonicity and are essential for the stability of the resulting method. The last ingredient for the method is the discontinuity detector,

$$
\alpha_j = \alpha\left(\frac{|E_{j+1} - 2E_j + E_{j-1}|}{|E_{j+1} - E_j| + |E_j - E_{j-1}|}\right)^2 , \tag{17.83}
$$

with α being some constant. Yang found that $\alpha = 33$ worked well.

Finally, we note that the correction that Yang defined is of the same order as the truncation error and, therefore, the accuracy of the method is not affected. Nevertheless, the method should be used with caution due to its potentially harmful side-effects in smooth flows.

17.7 The ADER Approach

The ADER approach [546] utilizes a modified GRP (Generalized Riemann Problem) scheme to construct schemes of very high order of accuracy both in space and time. The modified GRP scheme of Toro is based on the GRP scheme of Ben-Artzi and Falcovitz [51], which is a second-order Godunov-type method. The ADER approach consists of a series of steps aiming at producing high-order discretizations for the equations. These steps are summarized below:

1. Proceed with high-order reconstruction of the initial data and pose a generalized Riemann problem with the initial data being piecewise smooth.[7]
2. Development of the GRP solution to Taylor series expansion in time to any order of accuracy.
3. Obtain an average ADER state by integrating the Taylor series expansion.
4. Substitution of time derivatives by spatial derivatives.
5. Derivation of evolution equations for all spatial derivatives.
6. Define and solve Riemann problems for the spatial derivatives.
7. Calculation of the advective flux.

The above steps are described in following sections. We first present the development of this approach for a linear scalar equation.

[7] Since the ADER scheme has been developed to capture discontinuities, in [546] a discontinuity at the intercell edge is considered.

17.7.1 Linear Scalar Case

The high-order reconstruction of the initial data is obtained by using the ENO interpolation procedure [248]. ENO interpolation is adaptive and leads to non-linear schemes even when the schemes are applied to linear problems.

The computation of the ADER flux of m-th order of accuracy at an intercell position is obtained by solving the generalized Riemann problem,

$$\frac{\partial U}{\partial t} + \frac{\partial E}{\partial x} = 0 \, , \tag{17.84}$$

$$U(x,0) = \begin{cases} U_i & \text{for } x < 0 \, , \\ U_{i+1} & \text{for } x > 0 \, , \end{cases} \tag{17.85}$$

where U_i and U_{i+1} are $(m-1)$-th order polynomial functions. The calculation of the numerical flux can be obtained by a series expansion [535, 546] as follows:
We write a Taylor expansion of the intercell state in time

$$U^n_{i+1/2} = U^{(0)}_{i+1/2} + \sum_{k=1}^{m-1} \frac{\tau^k}{k!} \frac{\partial^k}{\partial t^k} [U^{(0)}_{i+1/2}] \, , \tag{17.86}$$

where $U^{(0)}_{i+1/2}$ denotes the value of U at $t = 0$.

The next step is to replace all time derivatives by spatial derivatives by means of the Lax-Wendroff procedure [321]. For the scalar linear advection equation

$$\frac{\partial U}{\partial t} + \lambda \frac{\partial U}{\partial x} = 0 \, , \tag{17.87}$$

where λ is the *constant* characteristic speed, the expansion is written

$$U^{\text{ader}}_{i+1/2} = U^{(0)}_{i+1/2} + \sum_{k=1}^{m-1} \frac{(-\lambda \Delta t)^k}{(k+1)!} \frac{\partial^k}{\partial x^k} (U_{i+1/2}) \, , \tag{17.88}$$

where $U^{\text{ader}}_{i+1/2}$ is the ADER average state. The average state (17.88) has the leading term $U^{(0)}_{i+1/2}$, which corresponds to the Godunov first-order upwind method. In (17.88) the spatial derivatives can be evaluated by solving a a linearized Riemann problem for the spatial derivatives $U^{(k)} \equiv \partial^k U / \partial x^k$ according to the following theorem [546]:

Theorem: Let $q \equiv U^{(k)}$ be the k-th order spatial derivative of $U(x,t)$. Then q obeys the linearized equation

$$\frac{\partial q}{\partial t} + \hat{\lambda} \frac{\partial q}{\partial x} = 0 \, , \tag{17.89}$$

where $\hat{\lambda} = \lambda(U^{(0)}_{i+1/2})$. The proof of the theorem is easily obtained by differentiation of (17.87) with respect to x. Note that the above result is also valid for all time derivatives.

The GRPs for the derivatives are defined as

$$\frac{\partial q}{\partial t} + \lambda \frac{\partial q}{\partial x} = 0 \,, \tag{17.90}$$

$$q(x,0) = \begin{cases} \dfrac{\partial^k}{\partial x^k}[U_L(x_{i+1/2})] \text{ for } x < x_{i+1/2} \,, \\ \\ \dfrac{\partial^k}{\partial x^k}[U_R(x_{i+1/2})] \text{ for } x > x_{i+1/2} \,. \end{cases} \tag{17.91}$$

The initial condition for solving the GRP described by (17.90) and (17.91) is obtained by differentiating the ENO (or WENO) reconstruction for U with respect to x. Because optimal weights for derivatives do not always exist we can use the same weights and smoothness indicators for the function and for all its derivatives.
With the spatial derivatives evaluated by the solution of the GRP (17.90) and (17.91), the intercell variable (17.88) can be calculated. The ADER flux is then defined by

$$E^{\text{ader}}_{i+1/2} = E(U^{\text{ader}}_{i+1/2}) \,. \tag{17.92}$$

Another possibility to obtain the ADER flux is by using the following theorem:

Theorem: For the scalar conservation law (17.84), the flux function $E(U)$ obeys the evolution equation

$$\frac{\partial E(U)}{\partial t} + \lambda(U)\frac{\partial E(U)}{\partial x} = 0 \,, \tag{17.93}$$

where $\lambda(U)$ is the characteristic speed. The proof is obtained if we multiply (17.84) by $\lambda(U)$ and use the chain rule to obtain $\partial E(U)/\partial t = \lambda(U)\partial U/\partial t$. The theorem is also applicable to nonlinear systems. In this case the characteristic speed should be replaced by the Jacobian matrix. Using the above property the ADER flux can be written as follows

$$E^{\text{ader}}_{i+1/2} = E^{(0)}_{i+1/2} + \sum_{k=1}^{m-1} \frac{(-\lambda \Delta t)^k}{(k+1)!} \frac{\partial^k}{\partial x^k}(E_{i+1/2}) \,. \tag{17.94}$$

In (17.94) one can use Riemann solvers to calculate directly the flux.

17.7.2 Multiple Dimensions: Scalar Case

Toro et al. [546] has extended the scheme for the scalar two-dimensional linear advection equation

$$\frac{\partial U}{\partial t} + \frac{\partial E}{\partial x} + \frac{\partial F}{\partial y} = 0 \,. \tag{17.95}$$

The numerical fluxes $\bar{E}$ and $\bar{F}$ are defined by $\bar{E} = \lambda_1 \bar{U}$ and $\bar{F} = \lambda_2 \bar{U}$, respectively, where λ_1 and λ_2 are constant velocity components.

The unsplit discretization of (17.95) using an explicit (one-step) finite volume scheme is written

$$\begin{aligned} \bar{U}^{n+1}_{i,j} = \bar{U}^{n}_{i,j} &- \frac{\Delta t}{\Delta x}(\bar{E}_{i+1/2,j} - \bar{E}_{i-1/2,j}) \\ &- \frac{\Delta t}{\Delta y}(\bar{F}_{i+1/2,j} - \bar{F}_{i-1/2,j}) \,. \end{aligned} \tag{17.96}$$

Similarly with the one-dimensional scalar case, we consider the truncated Taylor series expansion around $t = 0$

$$\bar{U}(x, y, \tau) = \sum_{k=0}^{m-1} \frac{\tau^k}{k!} \frac{\partial^k}{\partial t^k}[U(x, y, 0)] \,. \tag{17.97}$$

The time average of (17.97) in the interval $0 - \Delta t$ is obtained as

$$\bar{U} = \frac{1}{\Delta t} \int_0^{\Delta t} \left\{ \sum_{k=0}^{m-1} \frac{(\Delta t)^k}{k!} \frac{\partial^k}{\partial t^k}[U(x, y, 0)] \right\} \mathrm{d}t \,, \tag{17.98}$$

which gives

$$\bar{U} = \sum_{k=0}^{m-1} \frac{(\Delta t)^k}{(k+1)!} \frac{\partial^{(k)}}{\partial t^k}[U(x, y, 0)] \,. \tag{17.99}$$

According to the ADER approach, the time derivatives need to be replaced by spatial derivatives. The time derivatives are given by

$$\begin{aligned} \frac{\partial^k}{\partial t^k} &= \left[-\lambda_1 \frac{\partial}{\partial x} - \lambda_2 \frac{\partial}{\partial y}\right]^k \\ &= \sum_{A_k} \binom{k}{\xi} (-\lambda_1)^{\xi} (-\lambda_2)^{\eta} \frac{\partial^{(\xi)}}{\partial x^{(\xi)}} \frac{\partial^{(\eta)}}{\partial y^{(\eta)}} \,, \end{aligned} \tag{17.100}$$

where

$$A_k = \left\{(\xi, \eta) : \xi + \eta = k, \xi, \eta \geq 0\right\} \,. \tag{17.101}$$

Equations (17.99) and (17.100) yield

$$\bar{U} = \sum_{k=0}^{m-1} \frac{(\Delta t)^k}{(k+1)!} \sum_{A_k} \binom{k}{\xi} (-\lambda_1)^{\xi} (-\lambda_2)^{\eta}\, U^{\xi\eta}(x, y, 0) \;, \qquad (17.102)$$

where

$$U^{\xi\eta} = \frac{\partial^{(\xi)}}{\partial x^{(\xi)}} \frac{\partial^{(\eta)}}{\partial y^{(\eta)}} U \;. \qquad (17.103)$$

Using the above, the average value over a cell boundary, for example at $x = x_{i+1/2}$, is calculated as

$$\bar{U}^{\mathrm{ader}}_{i+1/2,j} = \sum_{k=0}^{m-1} \frac{(\Delta t)^k}{(k+1)!} \sum_{A_k} \frac{(-\lambda_1)^{\xi}(-\lambda_2)^{\eta}}{\xi!\eta!} \Delta y \int_j U^{(\xi\eta)}(x_{i+1/2}, y, 0)\, \mathrm{d}y \;. \qquad (17.104)$$

The numerical fluxes are evaluated as

$$\bar{E}_{i+1/2,j} = E(\bar{U}^{\mathrm{ader}}_{i+1/2,j}) \quad , \quad \bar{F}_{i+1/2,j} = F(\bar{U}^{\mathrm{ader}}_{i+1/2,j}) \;, \qquad (17.105)$$

where $\bar{U}_{i+1/2,j}$ is given by (17.104).

In a similar fashion the method can be extended to three dimensions. In this case the intercell variables are calculated by

$$\bar{U}^{\mathrm{ader}}_{i+1/2,j,k} = \sum_{k=0}^{m-1} \frac{(\Delta t)^k}{(k+1)!} \sum_{A_k} \frac{(-\lambda_1)^{\xi}(-\lambda_2)^{\eta}(-\lambda_3)^{\zeta}}{\xi!\;\eta!\;\zeta!\;\Delta y\;\Delta z} \int_j \int_k U^{(\xi\eta\zeta)}(x_{i+1/2}, y, z, 0)\, \mathrm{d}y\, \mathrm{d}z \;, \qquad (17.106)$$

where

$$U^{\xi\eta\zeta} = \frac{\partial^{(\xi)}}{\partial x^{(\xi)}} \frac{\partial^{(\eta)}}{\partial y^{(\eta)}} \frac{\partial^{(\zeta)}}{\partial z^{(\zeta)}} U \;, \qquad (17.107)$$

and

$$A_k = \Big\{ (\xi, \eta, \zeta) : \xi + \eta + \zeta = k, \xi, \eta, \zeta \geq 0 \Big\} \;. \qquad (17.108)$$

The intercell fluxes are then calculated using the values from (17.106).

17.7.3 Extension to Nonlinear Hyperbolic Systems

The extension of the ADER scheme to nonlinear hyperbolic systems has been presented by Titarev and Toro [535] including application to the Burgers' equation and one-dimensional Euler equation.

The starting point is the conservation law

$$\frac{\partial \mathbf{U}}{\partial t} + \frac{\partial \mathbf{E}}{\partial x} = 0 , \tag{17.109}$$

which after integration in a control volume in $x - t$ space, defined by the intervals $[x_{i-1/2}, x_{i+1/2}]$ and $[t^n, t^{n+1}]$, yields

$$\begin{aligned} \bar{\mathbf{U}}_i^{n+1} = \bar{\mathbf{U}}_i^n - \frac{\Delta t}{\Delta x}\Big[\int_{t^n}^{t^{n+1}} & \mathbf{E}(\mathbf{U}_{i+1/2})\, \mathrm{d}t \\ & - \int_{t^n}^{t^{n+1}} \mathbf{E}(\mathbf{U}_{i-1/2})\, \mathrm{d}t\Big] , \end{aligned} \tag{17.110}$$

where $\bar{\mathbf{U}}_i^n$ is the average of the solution in the cell at time t^n. Equation (17.110) can be approximated as

$$\mathbf{U}_i^{n+1} = \mathbf{U}_i^n - \frac{\Delta t}{\Delta x}(\bar{\mathbf{E}}_{i+1/2} - \bar{\mathbf{E}}_{i-1/2}) , \tag{17.111}$$

where $\mathbf{U}_i^n$ is a high-order approximation of the cell average $\bar{\mathbf{U}}_i^n$ [535] and $\bar{\mathbf{E}}_{i+1/2}$, $\bar{\mathbf{E}}_{i-1/2}$ are numerical fluxes.

The first step of the ADER approach is to obtain a high-order reconstruction of the initial data, at t^n, using the ENO interpolation procedure [248]. ENO interpolation is adaptive and leads to non-linear schemes even when the schemes are applied to linear problems. Using ENO reconstruction, we replace the conservative variables by polynomials $\mathbf{p}_i(x)$. Then, at each cell face we have to solve the following generalized Riemann problem described by (17.109) and the initial conditions

$$\mathbf{U}(x,0) = \begin{cases} \mathbf{U}_L = \mathbf{p}_i(x) & \text{for } x < x_{i+1/2} , \\ \\ \mathbf{U}_R = \mathbf{p}_{i+1}(x) & \text{for } \quad x > 0 . \end{cases} \tag{17.112}$$

The approximate solution $\mathbf{U}(x_{i+1/2}, \tau)$ at a local time $\tau = t - t^n$ is evaluated by

$$\mathbf{U}(x_{i+1/2}, \tau) = \mathbf{U}(x_{i+1/2}, 0) + \sum_{k=1}^{m-1} \frac{\tau_n^k}{k!} \frac{\partial^{(k)}}{\partial t^{(k)}}[\mathbf{U}(x_{i+1/2}, 0)] , \tag{17.113}$$

where $\mathbf{U}(x_{i+1/2}, 0)$ is obtained by an exact or approximate Riemann solver.

The next step is to replace all time derivatives in (17.113) by spatial derivatives as we did for the scalar case (17.88). Then, the spatial derivatives are evaluated by solving a generalized Riemann problem for the spatial derivatives $\mathbf{U}^{(k)} \equiv \partial^{(k)}\mathbf{U}/\partial x^{(k)}$, that is

$$\frac{\partial \mathbf{U}^{(k)}}{\partial t^{(k)}} + \mathbf{A}(\mathbf{U}(x_{i+1/2}, 0))\frac{\partial \mathbf{U}^{(k)}}{\partial x^{(k)}} = 0 \; , \tag{17.114}$$

$$\mathbf{U}^{(k)}(x, 0) = \begin{cases} \dfrac{\partial^{(k)}}{\partial x^{(k)}}[\mathbf{U}_L(x_{i+1/2})] \text{ for } x < x_{i+1/2} \; , \\ \\ \dfrac{\partial^{(k)}}{\partial x^{(k)}}[\mathbf{U}_R(x_{i+1/2})] \text{ for } x > x_{i+1/2} \; . \end{cases} \tag{17.115}$$

The initial condition for solving the GRP described by (17.114) and (17.115) is obtained by differentiating the ENO (or WENO) reconstruction for $\mathbf{U}$ with respect to x [535]. Because optimal weights for derivatives do not always exist we can use the same weights and smoothness indicators for the function and for all its derivatives. The coefficient matrix $\mathbf{A}$ is the same for all derivatives and, therefore, has to be evaluated only once.

Having calculated all the spatial derivatives through the solution of the Riemann problem (17.114) and (17.115), we can calculate the average ADER state using (17.88). This yields an m-th order accurate average state

$$\mathbf{U}^{\text{ader}}(x_{i+1/2}, \tau) = A_0 + A_1\tau + A_2\tau^2 + \ldots + A_{m-1}\tau^{m-1} \; , \qquad 0 \le \tau \le \Delta t \; , \tag{17.116}$$

where the coefficients A_l $(l = 0, \ldots, (m-1))$ encompass the spatial derivatives terms (see (17.88)). The numerical (intercell) ADER flux is calculated by a Gaussian rule [535]

$$\bar{\mathbf{E}}_{i+1/2} = \sum_{p=0}^{p_n} \mathbf{E}(\mathbf{U}^{\text{ader}}(x_{i+1/2}, \delta_p \Delta t)) \, \omega_p \; , \tag{17.117}$$

where δ_p and ω_p are scaled nodes and weights of the Gaussian rule, and p_n is the number of nodes.

Remark 17.7.1. The main difference between the ADER and ENO schemes is how high-order time accuracy is obtained. The key ingredient in the ADER approach is the development of the state expansion in time using the concept of generalized Riemann problem followed by integration of the flux using this expansion. The ADER scheme can be extended directly to PDEs with source terms. In [534] the extension of the scheme to handle source terms for the scalar case is discussed. Toro et al. [546] have shown that for linear equations the ADER scheme is two to three times faster than the corresponding

ENO scheme for the same CFL number. Numerical experiments using the Burgers' equation [535] have shown that the gap in accuracy between fifth-order WENO and ADER schemes increases as the grid is further refined. The ADER scheme has not yet been extended to nonlinear systems in two and three dimensions, but the fact that the scheme is of unbounded accuracy both in space and time motivates further studies on this approach.

17.8 Extending and Relaxing Monotonicity in Godunov-Type Methods

Monotonicity is not the end-all principle for limiters, other bounds can produce useful nonlinear schemes (for example, the discussion in Chap. 14 for the UNO limiter demonstrates this). The essential aspect of limiters is that they provide a bounding principle that produces nonlinear stability for a method. This stability is founded on linear stability, but takes this further by introducing monotonicity, smoothness or positivity as well, and produces schemes that can converge to both strong and weak solutions of the underlying differential equations.

For example, a sign-preserving method can be produced, or a method that defines only lower or upper bounds. A method can be limited to defined lower and upper bounds. Another approach is to enforce monotonicity where it is absolutely necessary, at discontinuities, while carefully allowing more accurate differencing to be used away from the discontinuities. In all cases, the key solution characteristics is to be careful with the discontinuities (shocks in general, but also shear layers), and extrema (the local minima and maxima). Note that at discontinuities the usual notions of numerical accuracy are meaningless and limit the utility of higher order methods.

Below, we will examine several alternatives that relax monotonicity in differing manners. Accuracy and monotonicity preserving schemes as a class of methods that preserve monotonicity at discontinuities and accuracy away from them is discussed.[8]

17.8.1 Accuracy and Monotonicity Preserving Limiters

Recent work has produced a set of techniques that can produce monotone results at discontinuities and sharp gradients while preserving accuracy in smooth flows. This is in response to the negative characteristics of monotone limiters at extrema. If these extrema are part of a smooth solution rather than spurious oscillations, this limiting is unappealing. The necessary elements for the less restrictive limiters are the characterization and detection of smooth flow structures. Different algorithms are discussed below.

---

[8] This section will only use one-dimensional notation. With the proper rotation of indices, multidimensional methods can be defined using dimensional splitting.

This section will focus on two incarnations of this basic idea: application as a slope limiter [271] and a general algorithm for bounding edge values reconstructed via polynomial interpolation [518]. Each of these techniques uses some nonstandard techniques in their development. A key function to define is the minmod function which returns a value with the minimum size in absolute terms unless the arguments differ in sign then zero is returned. It is commonly written as

$$\mathrm{minmod}\left(a,b\right) = \mathrm{sign}\left(a\right) \max\left[0, \min\left|a\right|, \mathrm{sign}\left(a\right) b\right] \; .$$

With a bit of engineering it can be rewritten algebraically as

$$\mathrm{minmod}\left(a,b\right) = \frac{1}{2}\left[\mathrm{sign}\left(a\right) + \mathrm{sign}\left(b\right)\right]\left[\left|a+b\right| - \left|a-b\right|\right] \; .$$

Another function that can be written algebraically is the “minbar” or “mineno” function which returns the smallest absolute value regardless of whether or not the values have the same sign,

$$\mathrm{mineno}\left(a,b\right) = \frac{1}{2}\left[\mathrm{sign}\left(a+b\right)\right]\left(\left|a+b\right| - \left|a-b\right|\right) .$$

Another key function is the median which will return the value of three arguments that is bounded below and above by the other two. It can be constructed from the two argument minmod function as

$$\mathrm{median}\left(a,b,c\right) = a + \mathrm{minmod}\left(b-a, c-a\right) \; .$$

A key aspect of the median limiter is that if two of the three arguments are accurate to a certain order, the result of the median will be the same order (even if the third argument is chosen). Huynh [271] defines a smoothest slope through a limiter, called “extended minmod”, defined by the median as,

$$\mathrm{xm}\left(a,b\right) = \mathrm{median}\left(a,b,-a-b\right) .$$

Slope Limiting Algorithm. The limiter begins by defining several variables, left and right undivided differences,

$$S_- = U_j - U_{j-1} \; , \quad S_+ = U_{j+1} - U_j \; .$$

First, we will give the standard monotone method as a reference point. The bound for a monotone slope in terms of these variables is

$$Q^* = \mathrm{median}\left(0, 2S_-, 2S_+\right) \; ,$$

and the monotone slope (with the initial unlimited slope, S_j) is

$$S_j := \mathrm{median}\left(0, S_j, Q^*\right) \; .$$

This approach can be modified to account for the Courant number in Lax-Wendroff type differencing. Take a signed value of the CFL number, $\mathcal{C}$, and compute the following,

$$C_- = 2\max\left[1/\mathcal{C}, 1/\left(1+\mathcal{C}\right)\right] \; , \quad C_+ = 2\max\left[1/\left(1-\mathcal{C}\right), -1/\mathcal{C}\right] \; ,$$

and the bounding slope

$$Q^* = \mathrm{median}\left(0, C_- S_-, C_+ S_+\right).$$

Finish as before with $S_j := \mathrm{median}\left(0, S_j, Q^*\right)$. Next, this method will be modified to loosen the restrictive monotonicity bounds on the slope.

Huynh's algorithm for limiting a slope can now be defined. The whole process is centered around estimating the smoothest slope at each edge of a zone (this is equivalent to the UNO method given in Chap. 14).

Define a set of directionally biased second-order slopes

$$P_- = \frac{1}{2}\left(3U_j - 4U_{j-1} + U_{j-2}\right), \quad P_+ = \frac{1}{2}\left(-U_{j+2} + 4U_{j+1} - 3U_j\right),$$

and a centered one,

$$P_0 = \frac{1}{2}\left(U_{j+1} - U_{j-1}\right).$$

These values are used to define bounding second-order slopes using the median function

$$Q_- = \mathrm{median}\left(P_-, S_-, P_0\right) \text{ and } Q_+ = \mathrm{median}\left(P_0, S_+, P_+\right),$$

finally an accurate second-order smoothest slope can be found with

$$Q_* = \mathrm{xm}\left(Q_-, Q_+\right).$$

Other forms can be used in this last step in place of the extended minmod such as the minmod or mineno limiters, as well as van Albada's limiter in abstract form. Furthermore, this process could be extended to third-order differences to further improve results.

In terms of these variables the monotone bounding flux is easily defined,

$$Q^* = \mathrm{median}\left(Q_*, 2S_-, 2S_+\right).$$

Huynh also defines a means to provide a steeper slope representation. This fulfills the same role as an ACM algorithm (we introduced the concept for multidimensional slopes was introduced in Chap. 14). If the field should be steepened the high-order slope is replaced near discontinuities through a simple means. Huynh defines a constant $\kappa \in [0, 20]$ to provide a measure of steepening (higher κ means steeper). The method is simple and involves replacing the high-order slope with

$$S_j := \mathrm{sign}\left(S_j\right) \max\left(\left|S_j\right|, \kappa \left|Q_+ - Q_-\right|\right). \tag{17.118}$$

The basic idea is that if the change in second-order differences is large enough it denotes a discontinuity (or large gradient) and the high-order slope can be replaced. This is done prior to testing the slope for monotonicity preservation.

Everything is in place except for the high-order slope that will be used as much as possible. Huynh uses a fourth-order slope,

$$S_j = \frac{1}{12}\left(-U_{j+2} + 8U_{j+1} - 8U_{j-1} + U_{j-2}\right),$$

and other slopes third-order and higher could be substituted. For the solution of equations based on a conservation principle a slope based on a conservative interpolation may be advisable; for fourth-order this is

$$S_j = \frac{1}{48}\left(-5U_{j+2} + 34U_{j+1} - 34U_{j-1} + 5U_{j-2}\right) ,$$

where this slope will produce methods with a very mild anti-diffusive character. The final limited value that will be used in the solution is

$$S_j := \text{median}\left(Q_*, S_j, Q^*\right) . \tag{17.119}$$

Accuracy and Monotonicity Preserving Edge Limiting Algorithm. Another approach for producing a limited scheme can be defined by modifying the edge values produced by the reconstruction. Such a method is described by Suresh and Huynh [518]. It begins by defining a monotonicity-preserving scheme. Working with the assumption that velocities are positive, one defines an upwind extrapolation to the edge; we are concerned with $j+1/2$,

$$U^{UL}_{j+1/2} = \text{median}\Big(U_j, (U_j + \alpha(U_j - U_{j-1})), U_{j+1}\Big) , \tag{17.120}$$

where α is a constant usually set to 2 or 4. Note, that the upstream interpolation associated with van Leer limiting would give $\alpha = 1$. A bit of further examination shows that these constants can be determined more generally by making their values CFL number dependent. Since these limiters are being defined under the assumption that characteristic velocities are positive, the limit is for $\alpha = 1/\mathcal{C}$ (the inverse of the CFL number).

The monotone cell edge value is determined by

$$U^{M}_{j+1/2} = \text{median}\left(U_j, U_{j+1/2}, U^{UL}\right) . \tag{17.121}$$

Suresh and Huynh were specifically interested in Runge-Kutta time integrators where the CFL number is usually below one-half (rather than one).

This algorithm can be modified to account for the CFL number where Lax-Wendroff methods are used. The setting of the value α can be redefined as $\alpha = 1/\mathcal{C}$. Likewise, the value U_{j+1} is really defined as $U_j + \beta\left(U_{j+1} - U_j\right)$ with $\beta = 1$. For a CFL number based algorithm, $\beta = 1/\left(1-\mathcal{C}\right)$. Thus, the algorithm with CFL number dependence is

$$\begin{aligned} U^{UL}_{j+1/2} = \text{median}\Big(&U_j, U_j + (U_j - U_{j-1})/\mathcal{C}, \\ &U_j + (U_{j+1} - U_j)/\left(1-\mathcal{C}\right)\Big) , \end{aligned} \tag{17.122}$$

and the monotone cell edge value is determined by (17.121).

To take this approach toward one that can preserve accuracy, valid (smooth enough) extrema must be differentiated from discontinuities (or large gradients) and oscillations. First, an accurate value for the edge value needs to be estimated. For example, a fifth-order upwind value for the edge (with a positive velocity) is

$$U_{j+1/2} = \frac{2}{60}U_{j-2} - \frac{13}{60}U_{j-1} + \frac{47}{60}U_j + \frac{27}{60}U_{j+1} - \frac{3}{60}U_{j+2} .$$

The goal is to use this value as much as possible so that monotonicity is not violated and spurious extrema are not created. Extrema can be identified from two reference points, one centered about the edge being approximates, and a second extrapolated to that edge from upwind-oriented data. This approach has also been used in conjunction with very high-order WENO methods (seventh through eleventh order) where they are known as MPWENO methods [28] as mentioned in Sect. 17.5.

To estimate reasonable extremas from the edge in question $(j + 1/2)$, values are extrapolated to second-order from cell j and $j + 1$,

$$U^{EX,1}_{j+1/2} = \frac{1}{2}\left(U_j + U_{j+1}\right) ,$$

$$U^{EX,2}_{j+1/2} = \frac{1}{2}\left(3U_j - U_{j-1}\right) ,$$

and

$$U^{EX,3}_{j+1/2} = \frac{1}{2}\left(3U_{j+1} - U_{j+2}\right) .$$

Using an extrapolation from the upwind direction is less strict and is conducted in a bounding sense. The choice used is the following,

$$U^{LC}_{j+1/2} = \frac{1}{2}\left(3U_j - U_{j-1}\right) + \frac{4}{3}(d4)_j ,$$

where LC stands for largest curvature and the curvature $(d4)$ is chosen using a minmod limiter,

$$(d4)_j = \text{minmod}\left(d_j, d_{j+1}\right) ,$$

or

$$(d4)_j = \text{minmod}\Big(\text{minmod}(4d_j - d_{j+1}, 4d_{j+1} - d_j)\\ \text{minmod}(d_j, d_{j+1})\Big) , \tag{17.123}$$

with $d_j = U_{j+1} - 2U_j + U_{j-1}$. The logic behind the choices is described in [518].

The final choice of limiter is made to bound the solution taking allowable extrema and monotonicity into account. The minimum allowable value is

$$U^{\min} = \max\left(\min\left(U_j, U^{LC}, U^{UL}\right), \min\left(U_j, U_{j+1}, U^{EX}\right)\right) .$$

The maximum allowable value is

$$U^{\max} = \min\left(\max\left(U_j, U^{LC}, U^{UL}\right), \max\left(U_j, U_{j+1}, U^{EX}\right)\right) ,$$

where $U^{EX} = \text{median}\left(U^{EX,1}, U^{EX,2}, U^{EX,3}\right)$. The final value is the median of these values and the original accurate interface value,

$$U^{AMP}_{j+1/2} = \text{median}\left(U^{\min}, U_{j+1/2}, U^{\max}\right) .$$

17.8.2 Extrema and Monotonicity Preserving Methods

In this section the above algorithm is extended so that an accurate stencil can be used as often as possible. This is accomplished by using a nonoscillatory stencil in place of the bounding method used in the previous section. By the properties of the median function used to make the final stencil selection, the result will inherit the order of the nonoscillatory method used.

One can easily define a high-order stencil based purely on linear design concepts. The key is to use this high-order stencil whenever possible as long as it is bounded by two values: the monotonicity lower (upper) limits that keep the solution stable, and the "smoothest" local stencil that provides good representation of most extrema. Huynh [271] and Suresh and Huynh [518] used this approach to good effect as discussed in the previous section. We will show how to get the extrema-preserving behavior of (W)ENO into Godunov-type methods while not inducing too much dissipation on deterministic shock (or shock-like) flow structures. When one is at neither an extrema nor a discontinuity, the high-order treatment should be used. The basic idea is to combine the smooth estimate of the differencing where it does not threaten monotonicity in such a manner that the high-order treatment is exploited as much as possible.

Here, the earlier approaches are taken a step further by both clarifying the various limiters, extending the order and unifying the edge value limiting in keeping with Huynh's approach (less like that of Suresh and Huynh where the bounding estimate are not necessarily high-order themselves). All of these details figure in providing exceptional, robust and efficient results. The details of the implementation matter a great deal in contributing to the accuracy of the method.

Extrema and Accuracy Preserving Slopes. The form of the high-order differencing is important and, therefore, basing the fundamental differencing on the conservation of quantities is essential. Interpolating what is known as the primitive function one can derive these approximations,

$$\xi_k = \sum_{j=0}^{k} \Delta x_j U_j \ , \tag{17.124}$$

and then differentiate it appropriately (where k depends on the order of accuracy or size of the stencil). In computing the first derivative of a function to fourth-order on a constant mesh the differences are small, but evident upon testing. The standard fourth-order first derivative is

$$S_j = \frac{-U_{j+2} + 8U_{j+1} - 8U_{j-1} + U_{j-2}}{12\Delta x} \ , \tag{17.125}$$

and the primitive function variant $(\partial^2 \xi / \partial x^2)$ is

$$S_j = \frac{-5U_{j+2} + 34U_{j+1} - 34U_{j-1} + 5U_{j-2}}{48\Delta x} \ . \tag{17.126}$$

The spatial differencing is accomplished often through a second-order linear interpolation,

$$U^n_{j+1/2} = U^n_j + \frac{1}{2}\Delta x S_j \; . \qquad (17.127)$$

Second-order time differencing can be directly combined with this

$$U^{n+1/2}_{j+1/2} = U^n_{j+1/2} - \frac{1}{2}\Delta x \mathcal{C} S_j \; . \qquad (17.128)$$

Slope and Edge Limiters. When one wants accurate computations several approaches are taken each with advantages and disadvantages based on the character of the solution. This character-dependent behavior leads to nonlinear methods that dominate the current choices. Before we discuss the choices high-order centered differencing is worth a mention. These methods, while highly accurate and low in dissipation, are not stable without some sort of additional dissipation and/or the full resolution of those dissipative effects, be natural, modeled or numerical. Among these are high-order monotone methods that preserve monotone solutions, but are limited to second-order accuracy for smooth solutions. These methods can be put into a generic form using slope "limiters". For example, take the high-order slopes, (17.125), applying monotonicity will produce

$$S_j := \text{minmod}\left(S_j, Q^*_j\right) \; , \qquad (17.129)$$

this can be put in a median (bounding) form as

$$S_j := \text{median}\left(0, S_j, Q^*_j\right) \; ,$$

with

$$Q^*_j = 2 \;\text{minmod}\left(\left(U_j - U_{j-1}\right), \left(U_{j+1} - U_j\right)\right) \; ,$$

or

$$Q^*_j = 2 \;\text{median}\left(0, \left(U_j - U_{j-1}\right), \left(U_{j+1} - U_j\right)\right) \; .$$

The PLM extrema and monotoncity preserving algorithm can be best laid out in discrete steps:

1. Monotonicity is defined by

   $$Q^* = \text{median}\left(0, 2S_-, 2S_+\right) \; ,$$

   and

   $$S_j := \text{median}\left(0, S_j, Q^*\right) \; .$$

2. If the slope is unchanged, return, otherwise return to the initial value of S_j and produce a UNO slope with the following steps:

3. Define the base differences, first-order

$$S_- = U_j - U_{j-1} \; ,$$

$$S_+ = U_{j+1} - U_j \; ,$$

and second-order

$$P_- = (3U_j - 4U_{j-1} + U_{j-2})/2 \; ,$$

$$P_0 = (U_{j+1} - U_{j-1})/2 \; ,$$

$$P_+ = (-U_{j+2} + 4U_{j+1} - 3U_j)/2 \; .$$

4. Conduct a stencil selection that is equivalent to the UNO limiter for up to second-order,

$$Q_- = \text{median}(S_-, P_0, P_-) \; ,$$

$$Q_+ = \text{median}(S_+, P_0, P_+) \; .$$

5. Chose the "smoothest" accurate slope $Q_* = \text{xm}(Q_-, Q_+)$. A number of limiters can be used, the extended minmod is shown, but mineno, minmod or van Albada, which in particular work as well.
6. For the PLM scheme the slope is $S_j := Q_*$. For the xPLM scheme the final slope is $S_j := \text{median}(Q_*, S_j, Q^*)$. The xPLM scheme is identical to Huynh's method [271] in many details, but also has a number of enhancements.

Thus, the accurate value is used except where monotonicity is violated. There the UNO slope is used; consequently, *the first-order value is never used.*

Extrema and Accuracy Preserving Edge Values. The same principle can be applied with piecewise parabolic (PPM) schemes [120] where the parabolic interpolant is defined by the triple (the average and two edge values), $\left(\langle U_j \rangle, U_{j\pm1/2}\right)$. The parabola

$$P_0 + P_1\theta + P_2\theta^2 \; , \quad \theta = (x - x_j)/\Delta x \; ,$$

has coefficients,

$$P_0 = \frac{3}{2}\langle U_j \rangle - \left(U_{j-1/2} + U_{j+1/2}\right) \; ,$$

or $\langle U_j \rangle - P_2/6$;

$$P_1 = U_{j+1/2} - U_{j-1/2};$$

and

$$P_2 = 3\left(U_{j+1/2} - 2\langle U_j \rangle + U_{j-1/2}\right) \; .$$

Monotonicity is assured if

$$U_{j\pm1/2} \in [U_j, U_{j\pm1}] \; ,$$

and

$$U_{j\pm1/2} \in \left[U_j, 3U_j - 2U_{j\mp1/2}\right] \ .$$

This can be accomplished through the use of two median functions,

$$U_{j\pm1/2} := \text{median}\left(U_j, U_{j\pm1/2}, U_{j\pm1}\right) \ ,$$

followed by

$$U_{j\pm1/2} := \text{median}\left(U_j, U_{j\pm1/2}, 3U_j - 2U_{j\mp1}\right) \ .$$

The same ideas can be used to construct a third-order (spatially) piecewise parabolic method. The parabola is determined by three values of U_j, the integral average of U in cell j and the edge values, $U_{j\pm1/2}$, which is approximated by a linear combination of neighboring zone data. Making these approximations nonlinear will endow the approximations with accuracy and/or monotonicity preserving character.

In answer to these issues weighted ENO methods were designed to alleviate some of these problems. Here, weights are chosen to recover a high-order upwind-biased stencil in smooth regions of the flow from a group of upwind-biased lower order methods. For example, the fifth-order WENO method is constructed from three third-order stencils.

The basic algorithm works as follows: if monotonicity is not threatened, then move on with the high-order stencil. If it is threatened, then construct an ENO/WENO approximation and set the return value to be the median of the high-order, ENO/WENO and monotone-preserving values. This general approach gives the L_1-norm-convergence of the high-order stencil, and infinity norm convergence of the ENO/WENO stencil. The relaxation of monotonicity is not complete: monotonicity is preserved at discontinuities, but not at smooth extrema and features. This can be accomplished by alternating stencils as in the ENO methods.

Other aspects of the method are also important for compressible (shocked) flows. This includes the basic high-order scheme, the details of the Riemann solution and the characteristic decomposition. The combination of these factors will lead to solutions that are two to four times more accurate for problems with non-trivial flow structures at the same mesh spacing. In terms of efficiency, the effort required to get the same accuracy, the gains are even larger.

The WENO methods can be used for general reconstruction of variables, and not only for constructing fluxes, and also in conjunction with providing states to input into a Riemann solution. As noted earlier, the principal approximation in weighted ENO methods is the use of a weighted combination of high-order stencils based on local smoothness to produce a higher order stencil. For a fifth-order WENO scheme a combination of three third-order stencils are used,

$$U_{j+1/2} = \omega_1 U_{j+1/2,1} + \omega_2 U_{j+1/2,2} + \omega_3 U_{j+1/2,3} \ , \tag{17.130}$$

where

$$U_{j+1/2,1} = \frac{(11U_j - 7U_{j-1} + 2U_{j-2})}{6} ,$$

$$U_{j+1/2,2} = \frac{(2U_{j+1} + 5U_j - U_{j-1})}{6} ,$$

and

$$U_{j+1/2,3} = \frac{(-U_{j+2} + 5U_{j+1} + 2U_j)}{6} ,$$

all for a positive wave-speed. These are combined with the following weights

$$\omega_k = \frac{\alpha_k}{\sum_{k=1}^{k=3} \alpha_k} ,$$

with

$$\alpha_1 = \frac{1}{(IS_1 + \epsilon)^2} ,$$

$$\alpha_2 = \frac{6}{(IS_2 + \epsilon)^2} ,$$

and

$$\alpha_3 = \frac{3}{(IS_3 + \epsilon)^2} ,$$

where ϵ is a small positive number (e.g., $\epsilon = 10^{-5}$). The smoothness measures, IS are defined in Sect. 17.5 and in [280]. This procedure can be used in place of ENO interpolations.

The same basic idea as for the PLM method can be applied to the piecewise parabolic method (PPM). The basic steps are like those for slope limiters, but applied to edge values. Again, the key is to get the ENO type values into use when monotonicity is violated. The various steps are listed below:

1. Check the monotonicity with the high-order edge values, $U_{j\pm1/2}$, in two steps

   $$U_{j\pm1/2} := \text{median}\left(U_j, U_{j\pm1/2}, U_{j\pm1}\right) ,$$

   and

   $$U_{j\pm1/2} := \text{median}\left(U_j, U_{j\pm1/2}, 3U_j - 2U_{j\mp1/2}\right) .$$

2. If the high-order edge values are unmodified the algorithm is finished and return (move to testing the next zone or field). Should either edge be modified, return to the initial edge values $U_{j\pm1/2}$ and use the following algorithm:

3. Define edge values second, and third-order accurate; for example, the second-order edges are $U_{j\pm1/2,2,+} = (U_j + U_{j\pm1})/2$ and $U_{j\pm1/2,2,-} = (3U_j - U_{j\mp1})/2$. The index 2 (or 3 used later) refers to the order of the edge approximation. The third-order edges are given by the fifth-order WENO candidate stencils.
4. Begin selecting the smooth slope as before using a median function, but one can use the full ENO method where the high-order stencils are selected on the basis of the next lower order. The function to use in the selection is the extended median,

$$\text{xmedian}(a, b, c) = a + \text{mineno}(b - a, c - a) .$$

This will return either b or c based on which one is closer to a. Thus, the overall procedure is

$$U_{j\pm1/2,-} = \text{xmedian}\left(U_{j\pm1/2,2,-}, U_{j\pm1/2,3,-}, U_{j\pm1/2,0}\right) ,$$

and

$$U_{j\pm1/2,+} = \text{xmedian}\left(U_{j\pm1/2,2,+}, U_{j\pm1/2,0}, U_{j\pm1/2,3,+}\right) ,$$

using up to third-order edge values ($U_{j\pm1/2,0}$ is first-order edge value, e.g., U_j).
5. Choose the smoothest edge value using one of the limiters (using a difference), e.g.,

$$U_{j\pm1/2,*} := U_j + \text{vanAlbada}\left(U_{j\pm1/2,-} - U_j, U_{j\pm1/2,+} - U_j\right) ,$$

and set $U_{j\pm1/2} := U_{j\pm1/2,*}$. The van Albada limiter was introduced earlier in Chap. 14 as a slope limiter in a Godunov-type method. Complete the xPPM scheme with one final step producing an accurate value from the original edge value, $U_{j\pm1/2}$, the monotone limit for the edge, $U^*_{j\pm1/2}$, and the (W)ENO edge value, $U_{j\pm1/2,*}$,

$$U_{j\pm1/2} := \text{median}\left(U_{j\pm1/2,*}, U_{j\pm1/2}, U^*_{j\pm1/2}\right) ,$$

where $U^*_{j\pm1/2}$ is calculated by (17.121) using (17.120) for the upwind extrapolation to the edge (or (17.122) for a CFL based algorithm).

The ENO part of the algorithm can be replaced with other procedures, most notably a WENO based selection of the smoothest stencils.

17.8.3 Steepened Transport Methods

Just as the ENO and WENO methods can be modified to reduce the numerical diffusion, regular Godunov-type methods with monotone and enhanced monotonicity limiters can be as well. We will highlight two approaches to accomplishing this one based upon the contact steepeners introduced earlier and a second one based on the PPM algorithm [120].

The use of ACM method prior to applying monotonicity makes things simpler than using ENO and WENO methods where the monotonicity is included in the flux correction (as discussed previously in Sect. 17.6). We will present the steepened transport method in the context of a slope reconstruction method. Like the flux correction a value will be added to the original slope value in the neighborhood of a slope. Thus, we define a slope correction

$$S_j := S_j + \delta S_j \ ,$$

by

$$\delta S_j = \alpha_j \text{minmod}\left(U_{j+1/2,R} - U_{j+1/2,L}, U_{j-1/2,R} - U_{j-1/2,L}\right) \ .$$

The edge values are defined by a diffusive interpolation

$$U_{j-1/2,L} = U_{j-1} + \frac{1}{2}\text{minmod}\left(U_{j-1} - U_{j-2}, U_j - U_{j-1}\right) \ ,$$

$$U_{j-1/2,R} = U_j - \frac{1}{2}\text{minmod}\left(U_j - U_{j-1}, U_{j+1} - U_j\right) \ ,$$

$$U_{j+1/2,L} = U_j + \frac{1}{2}\text{minmod}\left(U_j - U_{j-1}, U_{j+1} - U_j\right) \ ,$$

and

$$U_{j+1/2,R} = U_{j+1} - \frac{1}{2}\text{minmod}\left(U_{j+1} - U_j, U_{j+2} - U_{j+1}\right) \ .$$

We note that ENO interpolations also do well for higher order schemes because this algorithm preserves order-of-accuracy. Like before α_j is a discontinuity detector similar to the flux correction detector

$$\alpha_j = \alpha\left(\frac{|U_{j+1} - 2U_j + U_{j-1}|}{|U_{j+1} - U_j| + |U_j - U_{j-1}|}\right)^2 \ ,$$

where the value of $\alpha = 33$ works well.

For defining steep edge quantities used in a PPM (or (W)ENO) algorithm a similar approach will suffice. The advantage is that the the approach will only have to work one edge at a time as opposed to the two edges used in the slope modification. For example, take the edge, $j + 1/2$,

$$U_{j+1/2,S} = U_{j+1} - \frac{1}{2}\text{minmod}\Big(\frac{1}{2}\left(U_{j+1} - U_{j-1}\right) \\ 2\text{minmod}\left(U_{j+1} - U_j, U_{j+2} - U_{j+1}\right)\Big) \ . \quad (17.131)$$

In both cases the slopes can be replaced by a less diffusive interpolant,

$$U_{j+1/2,S} = U_{j+1} - \frac{1}{2}\text{minmod}\left(U_{j+1} - U_j, U_{j+2} - U_{j+1}\right) \ .$$

The last issue is the detection of discontinuities. The idea is to use a smooth monotone interpolation from a cell outside the discontinuity.

In the case of edges for PPM, Colella and Woodward [120] defined a test that examines the local solution for discontinuities while excluding small jumps,

$$\xi = \max\left[0, \min\left(1, 20\left(\zeta - 0.01\right)\right)\right] ,$$

which is combined with the steep edges using

$$U_{j+1/2} = \xi U_{j+1/2,S} + \left(1 - \xi\right) U_{j+1/2} .$$

If the sign of the second derivative changes sign then steepening is turned off, $d_{j-1} d_{j+1} < 0$, with $d_j = U_{j+1} - 2U_j + U_{j-1}$. The detector ζ is defined as the ratio of U_{xxx}/U_x

$$\zeta = -\frac{d_{j+1} - d_{j-1}}{6\left(U_{j+1} - U_{j-1}\right)} .$$

17.9 Discontinuous Galerkin Methods

Combining Godunov methods and finite element methods is an attractive approach for solving hyperbolic conservation laws. Such a combination is found with discontinuous Galerkin (DG) methods where a discontinuous basis is used in the Galerkin approximation. This method is rather natural for approximating weak solutions. As a necessity for resolving the discontinuities at element boundaries Riemann solvers are employed to determine a unique inter-element flux. In addition, DG methods are more compact than typical high resolution Godunov methods not requiring extensive memory accesses to surrounding cells or elements. For the purpose of nonlinear limiting and flux evaluations only nearest neighbors are generally required.

Discontinuous Galerkin methods were first introduced by Reed and Hill [441] for neutron transport. Subsequently, the method has found far greater use in the fluid dynamics community, although Morel and coworkers have revitalized its use in radiation transport [390]. Key developments were made by Cockburn and Shu (see [112] for example). Recent work has culminated in a robust, high-resolution method for conservation laws [113]. Liu and Shu [354] applied the discontinuous Galerkin methods to solving the stream-function-vorticity form of the incompressible flow equations.

In the following, we will use a semi-discrete form for the method in keeping with Cockburn's and Shu's recent work. Fully discrete forms for discontinuous Galerkin methods were introduced earlier first by van Leer ([569], as Scheme III) and later by White and Woodward [604]. Van Leer's method uses linear expansions while the later method uses parabolic expansions.

Generally, we are interested in solving a conservation law, $U_t + E_x = 0$. In its simplest form the finite difference equation for updating the conservation law is

$$\frac{\partial U_j}{\partial t} = -\frac{1}{\Delta x}\left[E\left(U_{j+1/2}\right) - E\left(U_{j-1/2}\right)\right] , \tag{17.132}$$

where we have explicitly employed a semi-discrete form. One can write the linear basis for a DG(1) scheme as

$$U_j + S_j\left(x - x_j\right) \ , \quad x \in \left[x_j - \Delta x/2, x_j + \Delta x/2\right]$$

For DG(1) the slope, S_j, is updated using the following,

$$\begin{aligned}\frac{\partial S_j}{\partial t} &= -\frac{6}{\Delta x^2}\left[E\left(U_{j-1/2}\right) + E\left(U_{j+1/2}\right)\right] \\ &\quad + \frac{12}{\Delta x^2}\int_{-\Delta x/2}^{\Delta x/2} E(U)\,\mathrm{d}x \ .\end{aligned} \tag{17.133}$$

The DG(2) scheme uses a quadratic basis

$$\begin{aligned}&U_j + S_j\left(x - x_j\right) + Q_j\left[\left(x - x_j\right)^2 - \Delta x^2/12\right] \ , \\ &\qquad\qquad x \in \left[x_j - \Delta x/2, x_j + \Delta x/2\right] \ .\end{aligned}$$

Here, we have made use of a Legendre polynomial basis resulting in a diagonal mass matrix. The form for updating S_j is retained and the quadratic term, Q_j, is updated using the following equation,

$$\begin{aligned}\frac{\partial Q_j}{\partial t} &= -\frac{30}{\Delta x^3}\left[E\left(U_{j+1/2}\right) - E\left(U_{j-1/2}\right)\right] \\ &\quad + \frac{360}{\Delta x^3}\int_{-\Delta x/2}^{\Delta x/2} E\left(U\right)\left(x - x_o\right)\mathrm{d}x \ .\end{aligned} \tag{17.134}$$

The TVD Runge-Kutta methods can be used as integrators (Chap. 7), also referred to as strong stability preserving scheme [493, 220]. Typically, we will use a second-order integrator with DG(1), thus this method is referred to as RK2-DG(1). It provides second-order accuracy for sufficiently smooth flows. The DG(2) method is used with a third-order integrator, (RK3-DG(2) method) and provides third-order of accuracy.

Riemann solvers can also be used in the context of DG methods. Because the quest for correct physical solutions depends crucially on satisfying an entropy condition, sufficiently dissipative Riemann solutions are important. Exact Riemann solvers (upwind in the scalar case) are only marginally entropy satisfying [407]. More dissipation can be entertained to the limit of a Lax-Friedrichs (LF) method (Sect. 15.1).

The importance of the dissipative Riemann solvers is in the design of robust numerical methods. In difficult circumstance one can use more dissipation via the LF Riemann solvers to achieve robustness. This, combined with nonlinear spatial differencing, provides a reliable numerical method for a variety of problems including some of the most challenging in existence. The Godunov flux is the least dissipative flux that satisfies an entropy condition (i.e., an E-flux, $\left(E\left(U_l, U_r\right) - E\left(U\right)\right)\left(U_r - U_l\right) \le 0, U \in \left[U_l, U_r\right]$), and for typical schemes LF is the most dissipative flux that leads to a stable scheme. LF method provides the ability to securely produce adequate entropy through the numerical flux. However, Rider and Lowrie [455] showed that the LF flux results in instabilities in the DG methods and must be used with care.

17.10 Uniformly High-Order Scheme for Godunov-Type Fluxes

An alternative framework to develop high-order schemes for incompressible flows has been presented in [148]; the family of schemes was designated [148] as uniformly high order (UHO) schemes (for non-periodic boundaries the accuracy at the boundaries is reduced down to second order). The first step is to obtain approximated intercell fluxes through a numerical reconstruction procedure and then in a second step interpolate these fluxes to obtain high-order approximations. In [148], the characteristics-based presented in Chap. 16 was used to reconstruct the intercell fluxes.

We consider the one dimensional hyperbolic conservation law

$$\frac{\partial \mathbf{U}}{\partial t} + \frac{\partial \mathbf{E}}{\partial x} = 0 , \tag{17.135}$$

and define the spatial operator

$$\mathbf{L} = -\frac{1}{\Delta x}\left(\widetilde{\mathbf{E}}_{i+1/2} - \widetilde{\mathbf{E}}_{i-1/2}\right) , \tag{17.136}$$

where $\widetilde{\mathbf{E}}$ is the characteristic-based approximation of the flux $\mathbf{E}$ (or any other approximation of the physical flux using a Riemann solver or a reconstruction procedure) at the cell faces. The flux $\widetilde{\mathbf{E}}$ can be split into positive and negative parts

$$\widetilde{\mathbf{E}} = \widetilde{\mathbf{E}}^{+} + \widetilde{\mathbf{E}}^{-} , \tag{17.137}$$

where $\mathrm{d}\widetilde{\mathbf{E}}^{+}/\mathrm{d}\widetilde{\mathbf{U}} \geq 0$ and $\mathrm{d}\widetilde{\mathbf{E}}^{-}/\mathrm{d}\widetilde{\mathbf{U}} \leq 0$; $\widetilde{\mathbf{U}}$ is an approximation of the unknown variables that can be obtained through the characteristics-based scheme (Chap. 16). The positive and negative parts can be calculated using any flux splitting scheme, e.g., the Lax-Friedrichs or Roe flux formulas. The Lax-Friedrichs positive and negative fluxes are given by

$$\widetilde{\mathbf{E}}^{\pm} = \frac{1}{2}(\widetilde{\mathbf{E}}(\widetilde{\mathbf{U}}) \pm \alpha\widetilde{\mathbf{U}}) , \tag{17.138}$$

where $\alpha = \max|\partial\widetilde{\mathbf{E}}(\widetilde{\mathbf{U}})/\partial\widetilde{\mathbf{U}}|$.

The UHO scheme aims to find high-order approximations of the fluxes $\widetilde{\mathbf{E}}^{\pm}$. Below we present the analysis only for the positive flux, similarly one can obtain the analysis for the negative flux. For the sake of simplicity, in the presentation, we drop the "+" sign from the superscript. We define the fluxes $\widetilde{\mathbf{E}}_{i+1/2}$ and $\widetilde{\mathbf{E}}_{i-1/2}$ through r-th order polynomial approximations as follows

$$\widetilde{\mathbf{E}}_{i+1/2} = \sum_{k=-r+3-n}^{r-2} \alpha_k^r \widetilde{\mathbf{E}}_{i+k} , \tag{17.139}$$

$$\widetilde{\mathbf{E}}_{i-1/2} = \sum_{k=-r+2-n}^{r-3} \alpha_{k+1}^r \widetilde{\mathbf{E}}_{i+k} , \tag{17.140}$$

where

$$n = 0 \quad \forall \quad r > 3 \quad \text{and} \quad n = 1 \quad \text{if} \quad r = 3 .$$

α_k^r are weight coefficients which need to be defined. For the case $r > 3 \quad (n = 0)$, the derivative of the flux at i is calculated by (to simplify the analysis, the grid is considered uniform and the spacing between the cell centers equal to one)

$$\begin{aligned}(\widetilde{\mathbf{E}}_x)_i &= \widetilde{\mathbf{E}}_{i+1/2} - \widetilde{\mathbf{E}}_{i-1/2} \\ &= \sum_{k=-r+3}^{r-2} \alpha_k^r \widetilde{\mathbf{E}}_{i+k} - \sum_{k=-r+2}^{r-3} \alpha_{k+1}^r \widetilde{\mathbf{E}}_{i+k} ,\end{aligned} \tag{17.141}$$

which can be expanded to give

$$\begin{aligned}(\widetilde{\mathbf{E}}_x)_i = {} & \alpha_{-r+3}^r \widetilde{E}_{i-r+3} + \alpha_{-r+4}^r \widetilde{\mathbf{E}}_{i-r+4} + \cdots + \alpha_0^r \widetilde{\mathbf{E}}_i \\ & + \cdots + \alpha_{r-3}^r \widetilde{\mathbf{E}}_{i+r-3} + \alpha_{r-2}^r \widetilde{E}_{i+r-2} \\ & -\alpha_{-r+3}^r \widetilde{\mathbf{E}}_{i-r+2} - \alpha_{-r+4}^r \widetilde{\mathbf{E}}_{i-r+3} - \cdots \\ & -\alpha_1^r \widetilde{\mathbf{E}}_i - \cdots - \alpha_{r-3}^r \widetilde{\mathbf{E}}_{i+r-4} - \alpha_{r-2}^r \widetilde{\mathbf{E}}_{i+r-3} = \\ & -\alpha_{-r+3}^r \widetilde{\mathbf{E}}_{i-r+2} + (\alpha_{-r+3}^r - \alpha_{-r+4}^r) \widetilde{E}_{i-r+3} + \cdots + \\ & (\alpha_0^r - \alpha_1^r) \widetilde{\mathbf{E}}_i + \cdots + (\alpha_{r-3}^r - \alpha_{r-2}^r) \widetilde{\mathbf{E}}_{i+r-3} + \alpha_{r-2}^r \widetilde{\mathbf{E}}_{i+r-2} .\end{aligned}$$

By developing the fluxes $\widetilde{\mathbf{E}}_{i+k}$ $(k = -r+2, \cdots, r-2)$ in Taylor series expansion around the point i, up to $r-$th order of accuracy, we obtain

$$
\begin{aligned}
(\widetilde{\mathbf{E}}_x)_i = & -\alpha^r_{-r+3}\Big(\widetilde{\mathbf{E}}_i - \mid -r+2 \mid \cdot\widetilde{\mathbf{E}}^{(1)} + \frac{\mid -r+2 \mid^2}{2!}\cdot\widetilde{\mathbf{E}}^{(2)} \\
& - \frac{\mid -r+3 \mid^3}{3!}\cdot\widetilde{\mathbf{E}}^{(3)} + \cdots + (-1)^r\frac{\mid -r+2 \mid^r}{r!}\cdot\widetilde{\mathbf{E}}^{(r)}\Big) \\
& + (\alpha^r_{-r+3} - \alpha^r_{-r+4})\Big(\widetilde{\mathbf{E}}_i - \mid -r+3 \mid \cdot\widetilde{\mathbf{E}}^{(1)} + \frac{\mid -r+3 \mid^2}{2!}\cdot\widetilde{\mathbf{E}}^{(2)} - \\
& \frac{\mid -r+3 \mid^3}{3!}\cdot\widetilde{\mathbf{E}}^{(3)} \cdots + (-1)^r\frac{\mid -r+3 \mid^r}{r!}\cdot\widetilde{\mathbf{E}}^{(r)}\Big) + \cdots + (\alpha^r_0 - \alpha^r_1)\widetilde{\mathbf{E}}_i \\
& + \cdots + (\alpha^r_{r-3} - \alpha^r_{r-2})\Big(\widetilde{\mathbf{E}}_i + (r-3)\cdot\widetilde{\mathbf{E}}^{(1)} + \frac{(r-3)^2}{2!}\cdot\widetilde{\mathbf{E}}^{(2)} \\
& + \cdots + \frac{(r-3)^r}{r!}\cdot\widetilde{\mathbf{E}}^{(r)}\Big) + \\
& \alpha^r_{r-2}\Big(\widetilde{\mathbf{E}}_i + (r-2)\cdot\widetilde{\mathbf{E}}^{(1)} + \frac{(r-2)^2}{2!}\cdot\widetilde{\mathbf{E}}^{(2)} \\
& + \cdots + \frac{(r-2)^r}{r!}\cdot\widetilde{\mathbf{E}}^{(r)}\Big),
\end{aligned}
$$

where the terms $\widetilde{\mathbf{E}}^{(\cdot)}$ represent high-order flux derivatives. After some rearrangement of the coefficients, we can write the term $(\widetilde{\mathbf{E}}_x)_i$

$$
\begin{aligned}
(\widetilde{\mathbf{E}}_x)_i = & \widetilde{\mathbf{E}}^{(1)}\Big(\alpha^r_{-r+3}|-r+2| - (\alpha^r_{-r+3} + \alpha^r_{-r+4})|-r+3| + \\
& \cdots (\alpha^r_{r-3} - \alpha^r_{r-2})(r-3) + \alpha^r_{r-2}(r-2)\Big) + \\
& \widetilde{\mathbf{E}}^{(2)}\Big(-\alpha^r_{-r+3}\frac{\mid -r+2 \mid^2}{2!} - (\alpha^r_{-r+3} - \alpha^r_{-r+4})\frac{\mid -r+3 \mid^2}{2!} + \cdots \\
& + (\alpha^r_{r-3} - \alpha^r_{r-2})\frac{(r-3)^2}{2!} + \alpha^r_{r-2}\frac{(r-2)^2}{2!}\Big) + \cdots + \\
& \widetilde{\mathbf{E}}^{(r)}\Big((-1)^{r+1}\alpha^r_{-r+3}\frac{\mid -r+2 \mid^r}{r!} + (-1)^r(\alpha^r_{-r+3} - \alpha^r_{-r+4})\frac{\mid -r+3 \mid^r}{r!} \\
& + \cdots + \alpha^r_{r-2}\frac{(r-2)^r}{r!}\Big).
\end{aligned}
$$

The above relation can be used to determine the unknown coefficients $\alpha^r_{(\cdot)}$ by posing certain numerical conditions. For example, we know that the even and odd derivatives are responsible for the dissipation and dispersion of the numerical scheme, respectively. Therefore, to construct schemes with minimum dissipation and dispersion errors one can set the coefficients of the derivatives $\widetilde{\mathbf{E}}^{(2)}, \widetilde{\mathbf{E}}^{(3)}, ...\widetilde{\mathbf{E}}^{(r)}$ equal to zero. Further, the CFL condition requires the coefficient of the derivative $\widetilde{\mathbf{E}}^{(1)}$ to be equal to 1. Thus, the following system of algebraic equations can be obtained

$$
\begin{aligned}
& \alpha^r_{-r+3}|-r+2| - (\alpha^r_{-r+3} + \alpha^r_{-r+4})|-r+3| + \cdots \\
& + (\alpha^r_{r-3} - \alpha^r_{r-2})(r-3) + \alpha^r_{r-2}(r-2) = 1\,,
\end{aligned} \tag{17.142}
$$

$$
\begin{aligned}
& -\alpha^r_{-r+3}\frac{\mid -r+2 \mid^2}{2!} - (\alpha^r_{-r+3} - \alpha^r_{-r+4})\frac{\mid -r+3 \mid^2}{2!} + \cdots \\
& + (\alpha^r_{r-3} - \alpha^r_{r-2})\frac{(r-3)^2}{2!} + \alpha^r_{r-2}\frac{(r-2)^2}{2!} = 0\,,
\end{aligned} \tag{17.143}
$$

. . . .

$$(-1)^{r+1}\alpha^r_{-r+3}\frac{|-r+2|^r}{r!}+(-1)^r(\alpha^r_{-r+3}-\alpha^r_{-r+4})\frac{|-r+3|^r}{r!}+\cdots$$
$$+(\alpha^r_{r-3}-\alpha^r_{r-2})\frac{(r-3)^r}{r!}+\alpha^r_{r-2}\frac{(r-2)^r}{r!}=0\,. \qquad (17.144)$$

Solution of the above system provides the values of the coefficients $\alpha^r_{(\cdot)}$ of (17.139) and (17.140).

In the case of the fourth-order scheme we obtain

$$\left.\begin{aligned} \alpha_{-1}+\alpha_0+\alpha_1+\alpha_2 &= 1\\ \alpha_1-\alpha_0+3(\alpha_2-\alpha_{-1}) &= 0\\ \alpha_1+\alpha_0+7(\alpha_2+\alpha_{-1}) &= 0\\ \alpha_1-\alpha_0+15(\alpha_2-\alpha_{-1}) &= 0 \end{aligned}\right\}, \qquad (17.145)$$

and the solution of the above gives gives $\alpha_0=\alpha_1=7/12$, $\alpha_{-1}=1/12$, and $\alpha_2=-1/12$. For the case of third-order reconstruction: $\alpha_0=5/6, \alpha_{-1}=-1/6, \alpha_1=1/3$ and $\alpha_2=0$. The third-order version of the scheme was employed in [163] to carry out calculations for a double mixing layer in a periodic box.

17.11 Flux-Corrected Transport

The flux-corrected transport scheme was the first algorithm developed that overcame the limitation of Godunov's theorem. It is notable that Godunov's work is not referenced with the earliest papers on FCT [63]. Perhaps not knowing what you cannot do, allows one to try something different. Some of the flux limiters (notably the minmod limiter) seem to have their genesis with the FCT method.

The original FCT was defined in a series of papers which gave analysis and results of using the scheme. The best recent reference is the book by Oran and Boris [403]. This method blends a high-order flux with a low-order monotone flux in such a way as to prevent the creation of new extrema. The FCT has been used extensively in turbulent [64], MHD [140] and reactive flow problems [404].

Zalesak [615] redefined the FCT in such a way as to make it more general. A standard low-order solution, similar to that obtained by a first-order upwind monotonic solution (also called donor-cell differencing), is used to define a monotonic solution. This solution is then used to limit an antidiffusive flux, which is defined as the difference between a high-order and low-order flux. As with the earlier versions of the FCT, the limiter is designed to give a no antidiffusive flux when an extrema or a discontinuity is reached. This prescription of the FCT can allow the user to specify a wide range of low-order fluxes as well as a large variety of high-order fluxes. These have included central differencing of second- or higher-order, Lax-Wendroff, and spectral fluxes [379].

The steps for FCT algorithms are listed as follows:

1. Find low-order monotonic cell-edge fluxes, $\hat{E}^L_{j+1/2}$;
2. find the diffused solution, $\tilde{U}_j$;
3. find a high order flux, $\hat{E}^H_{j+1/2}$;
4. define an antidiffusive flux, $\hat{E}^A_{j+1/2} = \hat{E}^H_{j+1/2} - \hat{E}^L_{j+1/2}$;
5. limit the antidiffusive flux to $\hat{E}^C_{j+1/2}$, and
6. apply the corrected antidiffusive flux to the diffused solution to find U^{n+1}_j.

The Boris's and Book's algorithm differs to Zalesak's algorithm only in few points. The high-order flux can be defined rather generally. A good example is provided in Zalesak's paper [615] while in one case Boris and Book provide a spectral scheme as a high-order flux. In fact, Boris and Book measure the error on a square wave and the spectral high-order flux gives the smallest error of any method they tested.

For the linear advection equation $U_t + [E(U)]_x = 0$ ($E(U) = aU$, where a is the characteristic speed), the Boris and Book algorithm uses a monotonic flux for the predictor stage defined by

$$\hat{E}^L_{j+1/2} = \frac{1}{2}\left(E_{j+1} - E_{j-1}\right) - \left(\frac{\nu}{\lambda} + \frac{1}{2}\mathcal{C}^2\right)\left(U_{j+1} - 2U_j + U_{j-1}\right) ,\quad (17.146)$$

where $\mathcal{C}$ is the CFL number and $\lambda = \Delta t/\Delta x$. The value of $\nu = 0.125$ can be optimally used to minimize both diffusion and dispersion errors. For $\nu = 0$ the Lax-Wendroff scheme is obtained. In Zalesak's algorithm, a simple donor-cell flux may be used (or any other monotone method) as the low-order flux.

In the Boris and Book algorithm, the antidiffusion corrector stage is defined by the monotone flux

$$\hat{E}^A_{j+1/2} = \mu\left(\tilde{U}_{j+1} - \tilde{U}_j\right) , \quad (17.147)$$

where the tilde variables are the ones calculated by the predictor stage, i.e., $\tilde{U}^{n+1}_j = U^n_j - \lambda(E_{j+1/2} - E_{j-1/2})$. The optimal values of the coefficient μ (and ν) will depend on the underlying scheme. One choice is $\mu = \nu$. Alternative choices for optimal minimization of dispersion have been proposed [66]

$$\nu = \frac{1}{6} + \frac{\mathcal{C}^2}{3} , \quad \mu = \frac{1}{6} - \frac{\mathcal{C}^2}{6} .$$

Because the antidiffusion stage should not create new maxima or minima in the solution, the antidiffusion fluxes should be corrected by limiting the fluxes as follows

$$E^C_{j+1/2} = \text{sign}(\Delta_{j+1/2})\max\{0, \min(\Delta_{j-1/2}\text{sign}(\Delta_{j+1/2}),\\ \mu|\Delta_{j+1/2}|, \Delta_{j+3/2}\text{sign}(\Delta_{j+1/2}))\}\ , \qquad (17.148)$$

where $\Delta_{j+1/2} = \tilde{U}_{j+1} - \tilde{U}_j$ and $\text{sign}(\cdot) = (\cdot)/|(\cdot)|$.

In Zalesak's algorithm the antidiffusive flux (17.147) could be a Lax-Wendroff flux or another higher order flux minus the monotone flux used in the predictor stage. Zalesak's FCT has been classified as a hybrid method that is applied in two steps. By being hybrid, the algorithm is based on the blending of high- and low-order difference schemes together. Step one is accomplished with a donor-cell differencing plus some additional diffusion (the entropy fix discussed in the previous section adds such dissipation). This could be accomplished with other first-order algorithms such as Godunov's [215] or Engquist and Osher's [180]. These fluxes are used to produce a transported diffused solution $\tilde{U}$ as follows:

$$\tilde{U}_j = U^n_j - \lambda\left(\hat{E}^{DC}_{j+1/2} - \hat{E}^{DC}_{j-1/2}\right)\ , \qquad (17.149)$$

where $\hat{E}^{DC}$ is the first-order upwind (donor-cell differencing) flux. A high-order flux, E^H, is defined in some way and then the low-order flux is subtracted from the high-order flux to define the antidiffusive flux as

$$\hat{E}^A_{j+1/2} = \hat{E}^H_{j+1/2} - \hat{E}^L_{j+1/2}\ .$$

The antidiffusive flux is then limited with respect to the local gradients of the conserved variable computed with the transported and diffused solution. Zalesak defined his limiter as a prelude to a truly multidimensional limiter, but also defined an equivalent limiter as

$$\hat{E}^C_{j+1/2} = \text{median}\left[\lambda\Delta_{j-1/2}\tilde{U}, \hat{E}^A_{j+1/2}, \lambda\Delta_{j+\frac{3}{2}}\tilde{U}\right]\ , \qquad (17.150)$$

This limiter is identical to the limiter defined by Boris and Book [65], but with a different definition of $\hat{E}^A$. The FCT generally carries a stability limit on its time step of $\mathcal{C} \leq 1$.

Steinle and Morrow [513] have also introduced an implicit FCT algorithm; however, this algorithm is limited to small multiples of the CFL number. This is because the low-order solution is produced by multiple sub-cycles with an explicit donor-cell (or other monotonic) solution and an implicit high-order solution. The high-order solution is only stable for small multiples of the CFL number, thus limiting the applicability of this algorithm. The FCT has also been extended for use with a finite-element solution method with great success [360]. The use of adaptive unstructured grids is another key part of the success of this work.

The FCT method does not extend to systems in the same manner as other schemes. Some schemes have used an equation-by-equation synchronization of flux limiters [360].

17.12 MPDATA

The MPDATA (Multidimensional Positive Definite Advection Transport Algorithm) scheme [501] is another manner of defining a scheme using nonlinear properties to achieve a nonoscillatory result. By nonoscillatory we mean that the scheme has the property of sign or monotonicity preservation or more generally is nonlinearly stable. Although most schemes are based on the idea of flux limiting, MPDATA is formulated more directly on the particular properties of upwind (upstream) differencing. In its most basic form, MPDATA is sign-preserving (but not monotonicity preserving), and second-order accurate. MPDATA is a two-time level or forward-in-time algorithm. It is a multidimensional scheme, and its implementation does not involve spatial (Strang) splitting. The scheme is constructed using two or more passes of upwind differencing, but with well-chosen "velocities" based on the truncation error to correct the numerical errors in the initial physical upwind pass.

A basic tool in deriving MPDATA is Taylor series expansions, leading naturally to the concept of the modified equation. Here, we describe the derivation of the basic MPDATA algorithm to simulate the simple case of one-dimensional advection of a scalar $U(x,t)$ by a velocity field u

$$\frac{\partial U}{\partial t} = -u\,\frac{\partial U}{\partial x}\,. \tag{17.151}$$

The first step is an upwind scheme; the scheme depends on the sign of the velocity

$$U_j^{n+1} = U_j^n - \frac{\Delta t}{\Delta x}(E_{j+1/2} - E_{j-1/2}) \tag{17.152}$$

where the flux is:

$$E_{j+1/2} = \frac{\mathcal{C}}{2}\left(U_j^n + U_{j+1}^n\right) - \frac{|\mathcal{C}|}{2}\left(U_{j+1}^n - U_j^n\right)\,. \tag{17.153}$$

In common notation, the subscript j identifies the computational cell, the superscript n the time, and $\mathcal{C} = u\Delta t/\Delta x$ is the Courant number. Note that the flux (i.e., the spatial derivative) has been estimated one-half cell upstream, where the upstream direction is determined by the sign of u.

The scheme described by (17.152) and (17.153) is stable and sign preserving when the Courant number is bounded: $\mathcal{C} \in [-1, 1]$. However, this is only first-order accurate. That is, expanding the discrete field U_j^n in a Taylor series, one finds that (17.152) and (17.153) more accurately approximate the advection-diffusion equation

$$\frac{\partial U}{\partial t} = -u\,\frac{\partial U}{\partial x} + \frac{\partial}{\partial x}\left(K\frac{\partial U}{\partial x}\right)\,, \tag{17.154}$$

where the diffusion coefficient $K = \Delta x^2/(2\Delta t)(|\mathcal{C}| - \mathcal{C}^2)$. Under the assumed bounds on the Courant number, the diffusion coefficient K is positive thus insuring stability. We say the scheme is first-order accurate, meaning that the error is of order $\mathcal{O}(\Delta x^2)$ relative to U itself. We refer to (17.154) as the modified equation of the scheme (17.152) and (17.153).

To derive a more accurate algorithm, one can compensate the second-order (i.e., diffusional) error, by estimating the error and subtracting it in the algorithm. The important character of MPDATA is how to estimate that error while preserving the nonoscillatory properties of the solution. An upstream estimate of the error will have this property. The error term can be written in advective form

$$\frac{\partial}{\partial x}\left(K\frac{\partial U}{\partial x}\right) \simeq \frac{\partial}{\partial x}\left(U^{(1)}U\right) , \tag{17.155}$$

where

$$U^{(1)} \equiv \frac{\Delta x^2}{2\Delta t}(|\mathcal{C}| - \mathcal{C}^2)\frac{1}{U}\frac{\partial U}{\partial x} , \tag{17.156}$$

is called a pseudovelocity. To complete the basic MPDATA algorithm, a second step is taken, repeating (17.152) using the pseudovelocity, $U^{(1)}$ in (17.153). Note that if U_j is defined at the centers of computational cells, then the pseudovelocity is defined at the cell edges halfway between the cell centers, and varies in space and time even when U is constant. It is easy to show that the bounds on the physical CFL number imply the same bounds on the pseudo-velocity. Each step of the algorithm is stable and sign-preserving and therefore the overall scheme also has these properties. The error terms in the modified equation of basic MPDATA (not shown) now appear at the third order, implying that MPDATA is a second-order algorithm.

With the application of another correction step the method can be raised to third-order of accuracy. The algorithm is formulated just as before, but now the pseudovelocities are constructed from the third-order truncation errors. Through Taylor series analysis the next step can be derived through replacing time-derivatives by space-derivatives in a Lax-Wendroff procedure. The derivation is given in [370] resulting in a “pseudo-velocity” in the third-order case of

$$U^{(2)} \equiv \frac{\Delta x^3}{6\Delta t^2}\left(\mathcal{C} - 3\mathcal{C}\,|\mathcal{C}| + 2\mathcal{C}^3\right)\frac{1}{U}\frac{\partial^2 U}{\partial x^2} . \tag{17.157}$$

This is applied just as before with another donor cell pass. The time step is restricted just as the first corrective step, but with an important caveat: The sum of $U^{(1)} + U^{(2)}$ should satisfy $U^{(1)} + U^{(2)} \leq 1$ to insure stability of the correction.

Part IV

Applications

18. Variable Density Flows and Volume Tracking Methods

The flow of incompressible fluids with large (discontinuous) density variations (interfaces) occurs in widespread applications. Water/air free surface flow is a classical example, e.g., a water drop falling into a pool of water. Other important examples are the filling of a cast metal mold with a molten metal alloy; the production and transport of micron-sized ink drops during inkjet printer operation; environmental and combustion problems, as well as many applications in mechanical, aerospace and chemical engineering industries. Reliable simulation of these types of flows demands a numerical model with accuracy, fidelity, and robustness (see Chap. 6 for an introduction).

In this chapter we will focus on a particularly useful application of high-resolution methods to incompressible flows, that is flows with large density variations. As the numerical methods have become more robust and economical, their use in analyzing industrial processes has increased. With this increased emphasis it is worthwhile to briefly review the details of methods that are useful in simulating such flows.

18.1 Multimaterial Mixing Flows

There are several prototypical examples of variable density mixing flows. These flows have different character starting with a shear layer including a density difference across the shear. Many examples are driven by gravity, for example, either a rising bubble (a hot less dense fluid) below a cooler one, and a Rayleigh-Taylor instability where a higher density fluid lies over a lower density fluid in gravity (or some other acceleration).

The solutions in Sects. 18.1.1 to 18.1.3 are computed using a cell-centered approximate projection method as described in Chap. 12. This includes the use of vertex- and edge-projection filters. The velocity field is projected in order to compute the pressure. The nonlinear terms are discretized using an unsplit CTU method described in Chap. 14. The spatial differences use the improved PLM method with monotone limiting. Where the interface is tracked, it is computed with the piecewise linear interface calculation method described later in this chapter.

18.1.1 Shear Flows

We will now show several flows that are driven by a shear at inflow. These shear flows are patterned after the famous Brown-Roshko shear layer experiment [85]. The first example has a constant density and the second has a variable density with the difference in velocity being caused by density variations (equal mass flow rate across the inflow). Both computations are shown in a domain that measures four units in the x-direction and one unit in the y-direction with outflow boundary conditions and no gravity. For the constant density flow, the flow is initialized by

$$u = 1 + \lambda \tanh\left[\varrho\left(y - 0.5\right)\right] , \tag{18.1}$$

where $\lambda = 0.5$ ($\lambda \in [0, 1]$), and $\varrho = 60$. This is also the inlet x-velocity for all time. The passive scalar is set by

$$\psi = 0.5 + 0.5 \tanh\left[\varrho\left(y - 0.5\right)\right].$$

The inlet y-velocity is used to perturb the flow and is set to

$$v_{\text{inflow}} = \sum_{k=0}^{2} \frac{0.05}{2^k} \sin\left[4\left(2^k\right)\pi t\right].$$

Fig. 18.1 shows two snapshots of the constant density shear duct flow. By this time a regular pattern has established itself that has the same period as the lowest frequency perturbation of the inlet flow. At short distance into the flow vortices begin to appear. Pairs of vortices near the inlet merge then are advected toward the end of the domain without interacting strongly with other vortices.

The variable density flow is similarly set up. We use $\tilde{\beta} = \rho_{\text{high}}/\rho_{\text{low}} = 7$, thus for the inlet mass flow rates to be equal, where λ in (18.1) is 0.75. This problem highlights the differences between constant and variable density flows. The only other major difference with the constant density flow is the density profile which is set by

$$\rho = \frac{1}{2}\left(\tilde{\beta} + 1\right) + \frac{1}{2}\left(\tilde{\beta} - 1\right) \tanh\left[\varrho\left(0.5 - y\right)\right].$$

The results shown in Fig. 18.2 are similar to the constant density case, but the initial vortex formation is more vigorous. This is caused by the baroclinic generation of vorticity.[1] Vortex merging does not occur until the vortices have nearly passed out of the domain. This process is occurring in the vortex that is second from the right, and as before, the solution is periodic as the next two vortices are about to go through the same process.

[1] Baroclinic vorticity generation occurs whenever the pressure and density gradients are misaligned. Therefore, $\nabla(1/\rho) \times \nabla p$ will be non-zero.

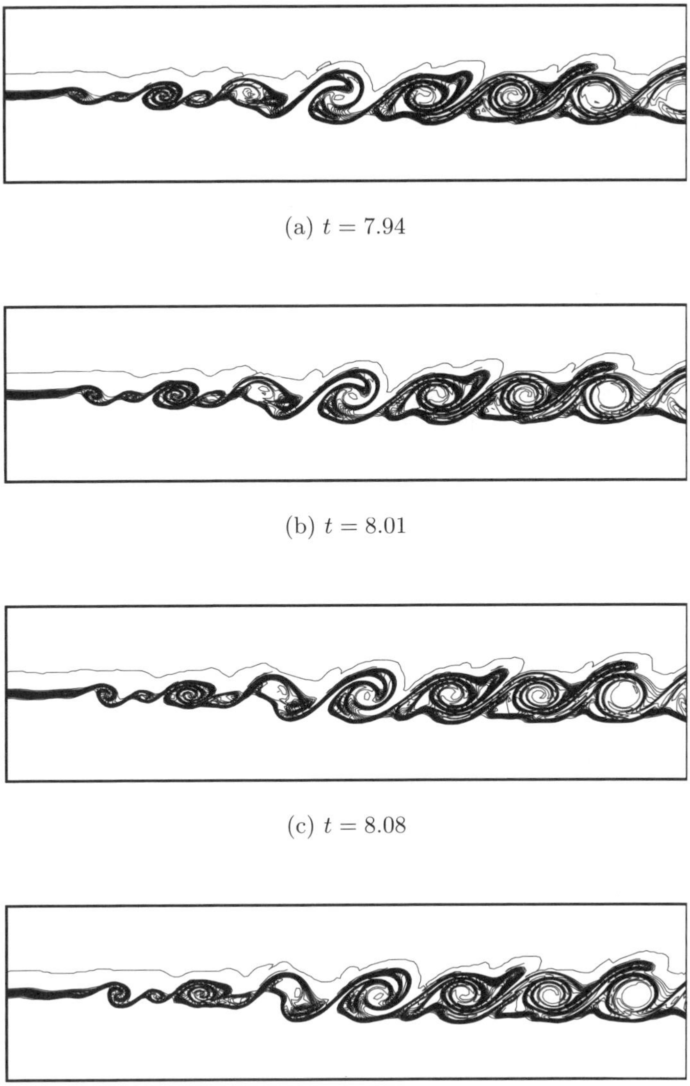

(a) $t = 7.94$

(b) $t = 8.01$

(c) $t = 8.08$

(d) $t = 8.15$

Fig. 18.1. Incompressible shear-layer with inflow from the left and outflow at the right using a 512×128 grid. The flow is visualized via a passively advected scalar field. This is a marker that moves with the fluid velocity, but does not effect its evolution; this is governed by $Z_t + \nabla \cdot (\mathbf{u}Z) = 0$.

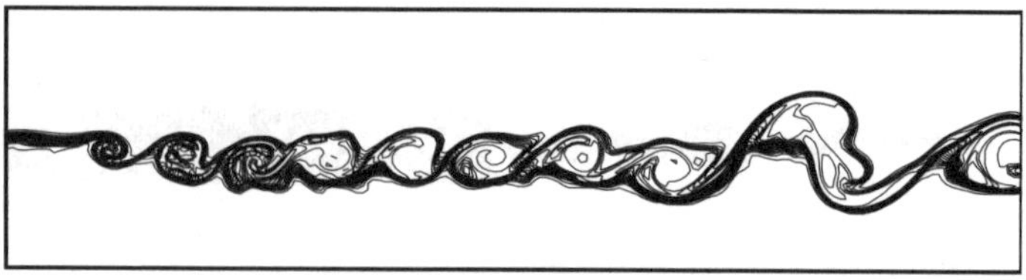

(a) $t = 7.15$

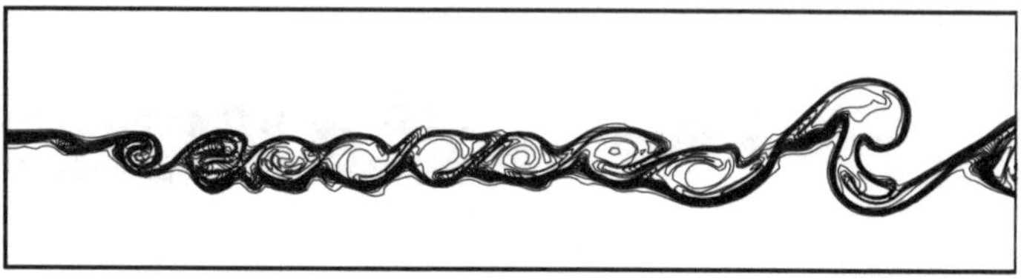

(b) $t = 7.36$

Fig. 18.2. Variable density ($\tilde{\beta} = 7$) shear-layer with inflow from the left and outflow at the right using a 512×128 grid. The isodensity lines are shown; the time is dimensionless.

18.1.2 Rising Bubbles

The next two examples concern bubbles rising through a denser medium. Both examples employ axisymmetric coordinates. The first one is of a Boussinesq flow, where the ambient flow field is cold, $\theta = 0$, and embedded in this is a hot bubble, $\theta = 1$ at the axis of symmetry. The bubble has a radius of 0.25 and is centered at $(r_o, z_o) = (0, 0.5)$. A distance function $d = \sqrt{(r - r_o)^2 + (z - z_o)^2}$ is used allowing us to define the temperature field as

$$\theta = \frac{1}{2} + \frac{1}{2} \tanh \left[\varrho \left(0.25 - d \right) \right],$$

where $\varrho = 60$ with the flow field initially at rest and the (dimensionless) viscosity is $\nu = 0.001$. We have carried out calculation for $Gr = 15.6$ (Eq. 3.88) in a computational domain with width of one unit radius and height of two units radius, using a 128×256 equally spaced grid.

For $Gr = 7.6 \times 10^9$ the computational domain has width of one unit radius and height of three units radius (Fig. 18.3). The computations were carried out on a 128×384 equally spaced grid. In this case, the variable density formulation is used. The density ratio is $\tilde{\beta} = 10$ and the density field is set by

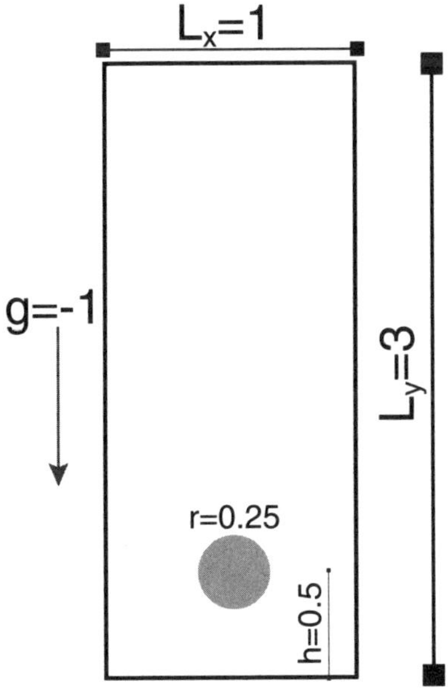

Fig. 18.3. The setup for the rising bubble problem where the grey patch is the bubble's initial position. All boundaries are solid walls meaning no slip boundary conditions.

$$\rho = \frac{1}{2}\left(\frac{1}{\tilde{\beta}} + 1\right) + \frac{1}{2}\left(1 - \frac{1}{\tilde{\beta}}\right) \tanh\left[\varrho\left(d - 0.25\right)\right] ,$$

again with $\varrho = 60$ and $\nu = 0.001$. The grid is the same as with the first Boussinesq bubble problem.

Fig. 18.4 shows the Boussinesq results. The effects of heat conduction are quite evident throughout when contrasted with the qualitative features of the variable density, high Grashof number case, (Fig. 18.5). Both bubbles form the typical mushroom-cap-shaped structure. Computations with no explicit diffusion were also carried out for $Gr = 15.6$ and as can be seen in Fig. 18.6, the generation of baroclinic vorticity is much more intense than for the Boussinesq case.

18.1.3 Rayleigh-Taylor Instability

The final demonstration concerns Rayleigh-Taylor instabilities. We begin with a single mode perturbation. The single mode problem is patterned after problems described by Bell and Marcus [50] (see also [432]). We set up the problem with density ratio $\tilde{\beta} = 5$ in a rectangular domain with a width of one and a height of four dimensionless units (Fig. 18.7). The boundaries are solid walls above and below with periodic conditions in the x-direction.

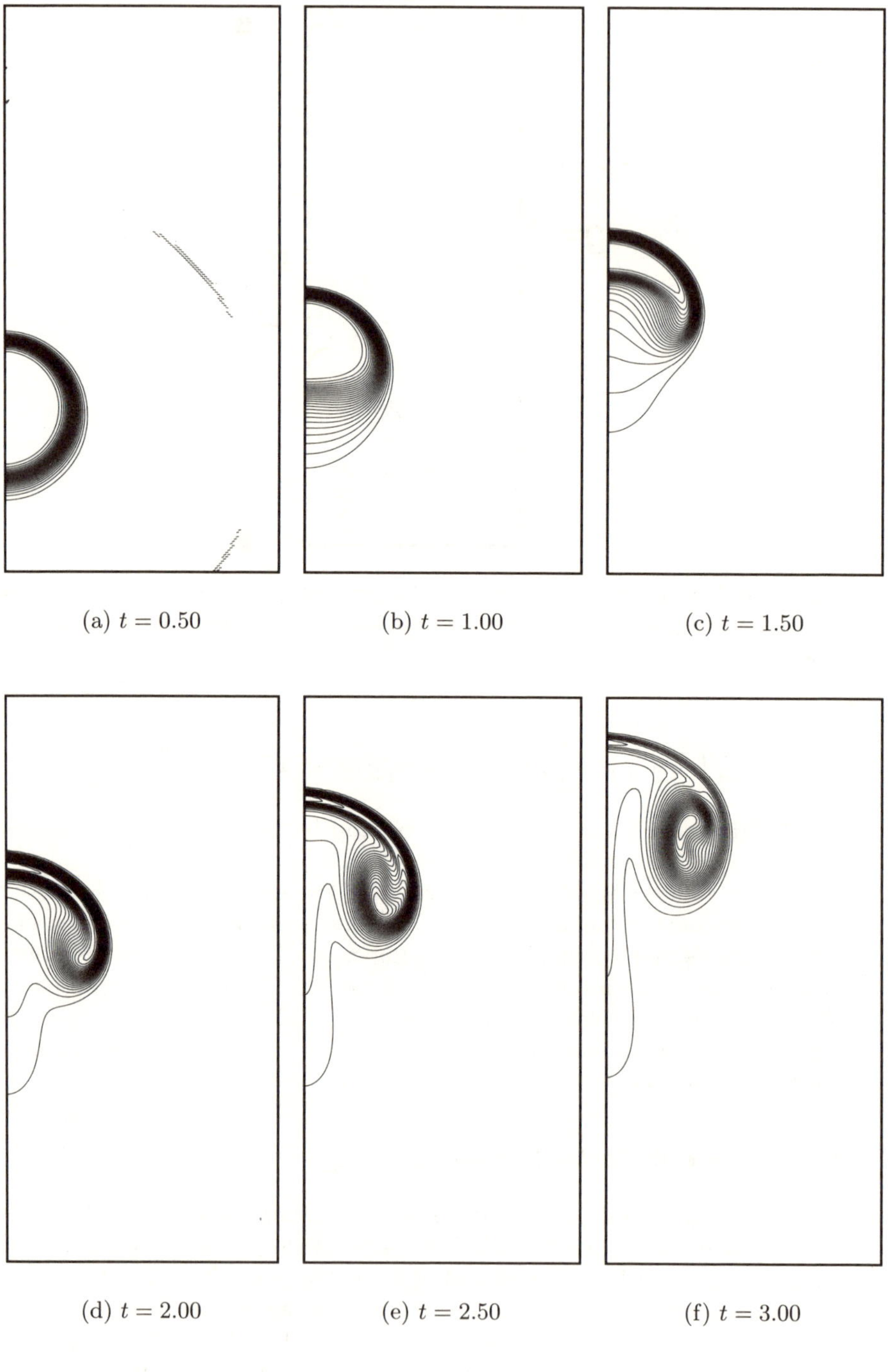

Fig. 18.4. Boussinesq bubble rise results ($Gr = 15.6$ and $\nu = 0.001$). The temperature field is plotted.

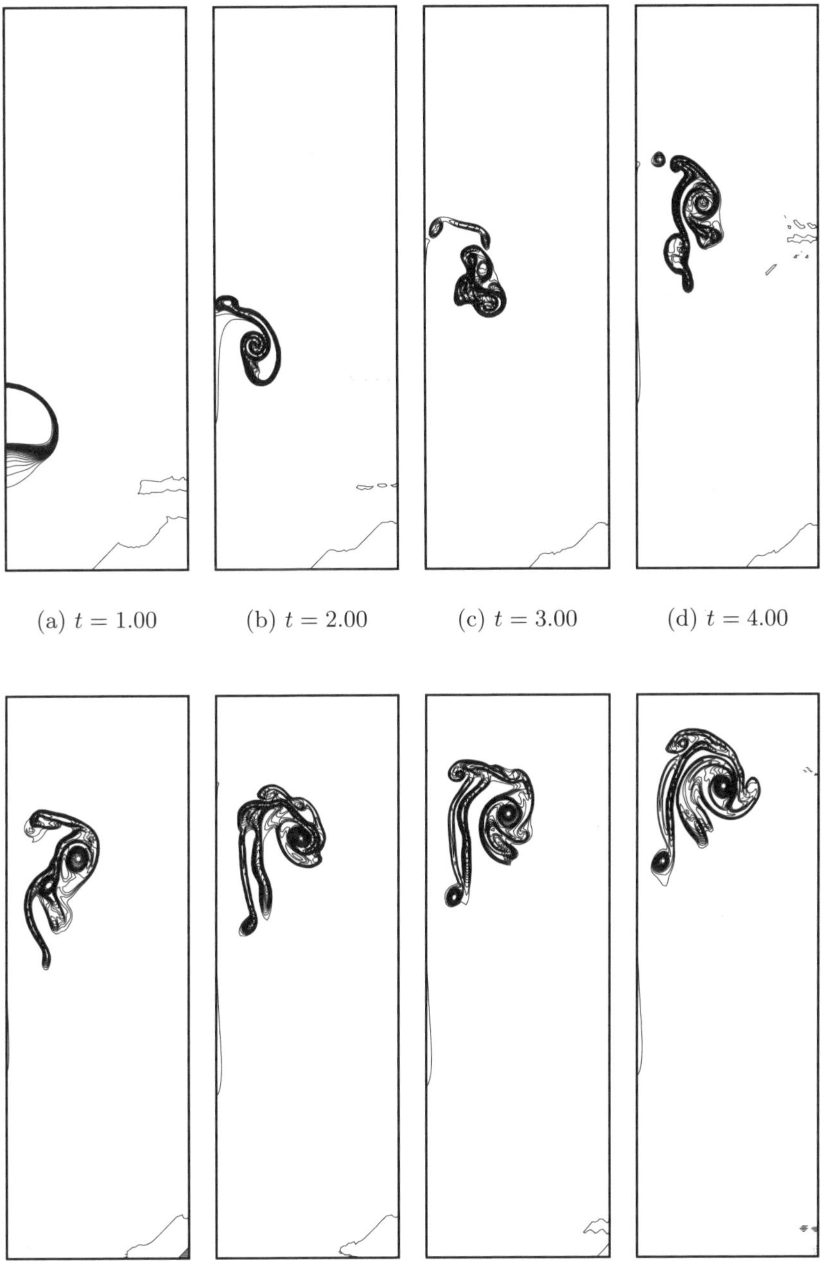

(a) $t = 1.00$ (b) $t = 2.00$ (c) $t = 3.00$ (d) $t = 4.00$

(e) $t = 4.50$ (f) $t = 5.00$ (g) $t = 5.50$ (h) $t = 6.00$

Fig. 18.5. Variable density bubble-rise results ($Gr = 7.6 \times 10^9$ and $\nu = 0.001$). The temperature field is plotted.

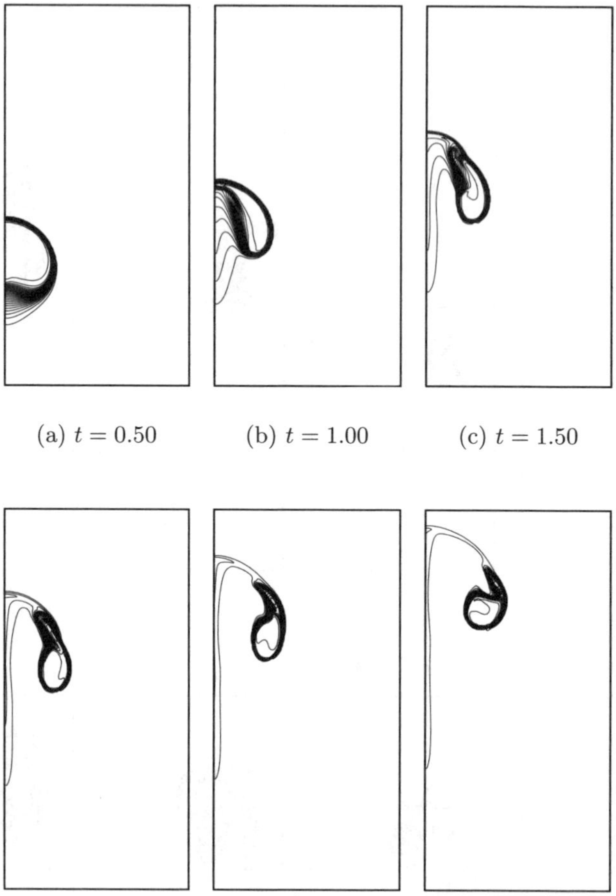

(a) $t = 0.50$ (b) $t = 1.00$ (c) $t = 1.50$

(d) $t = 2.00$ (e) $t = 2.50$ (f) $t = 3.00$

Fig. 18.6. Variable density bubble-rise results ($Gr = 15.6$ and $\nu = 0$). The flow is visualized with the density field.

The computational grid is 128×512. The x-velocity is initially zero with a perturbation applied to the y-velocity of

$$v = 0.005\left[\cos(2\pi x) + 1\right] .$$

The density field is initially set by

$$\rho = \frac{1}{2}\left(\tilde{\beta} + 1\right) + \frac{1}{2}\left(\tilde{\beta} - 1\right) \tanh\left[\varrho\left(y - 2\right)\right] ,$$

with $\varrho = 30$ and no explicit diffusion ($\nu = 0$).

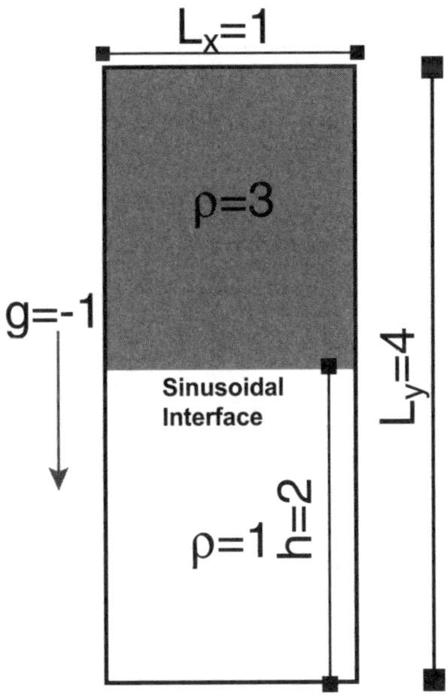

Fig. 18.7. The setup for the Rayleigh-Taylor instability. No-slip boundary conditions are used on the upper and lower boundaries. Periodic conditions are used on the sides.

The late-time evolution of the instability is shown in Fig. 18.8. At $t = 4.00$ (Fig. 18.8b), the flow is still relatively simple with the initial mode beginning to roll up into two counter rotating vortices. By $t = 5.00$ (Fig. 18.8b), the flow has developed a significant nonlinear structure with secondary instabilities breaking away from the initial spike. This continues to evolve and tertiary instabilities (instabilities developing from the secondary instability, Kelvin-Helmholtz) develop as shown in Fig. 18.8d.

One problem that this test makes evident can be seen in the last frame. Small asymmetries in the solution have grown large enough to be noticeable. These are driven by the non-symmetric nature of the Gauss-Seidel relaxation used in the multigrid algorithm. Replacing this with a symmetric relaxation cures this problem.

The next example is based on Rayleigh-Taylor experiments [141] (see also [128] for further experimental investigations). In this experiment often referred to as the linear electric motor (LEM) experiments, a box is accelerated using electromagnetic rails. A number of cameras are stationed along the path of the box containing the mixing fluids to image the evolution of the flow. The LEM allows the time dependent acceleration of the interface to be tailored in a variety of ways. We will consider comparison with the "constant" acceleration history where water and hexane are the mixing fluids. These flu-

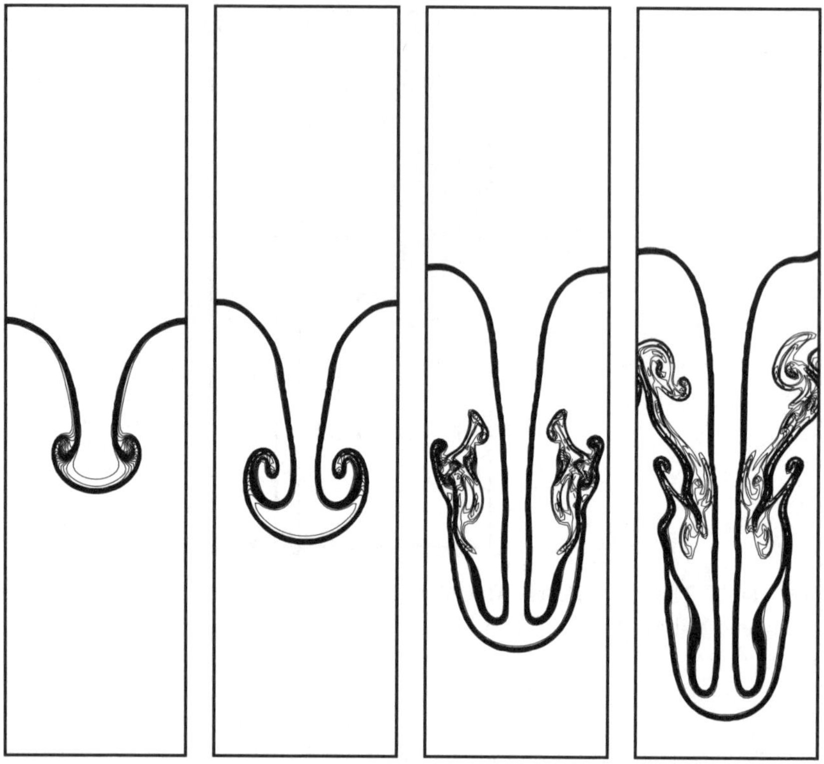

(a) $t = 3.50$ (b) $t = 4.00$ (c) $t = 5.00$ (d) $t = 5.50$

Fig. 18.8. Evolution of the Rayleigh-Taylor instability. The density profile is shown at each time instant.

ids give an Atwood number, $A = (\rho_{\max} - \rho_{\min}) / (\rho_{\max} + \rho_{\min}) = 0.5$ [141]. Time and depth of mixing are measured by the expected self-similar profile expected for the growth of the bubble dome (the edge of the light fluid mixing into the heavy) from its initial interface position, $h(t) = \alpha g t^2$, where α is a proportionality constant that is one principal object of the experimental investigation (in LEM $\alpha = 0.054$). The Reynolds number of the experiment is strongly time-dependent (with a t^3 dependence) with a value of $\approx 10^5$ at the end of the acceleration. This dependence is a function of the quadratic growth of the integral scale and the linear growth rate of the large scale velocity field. The setup of the problem is shown in Fig. 18.9.

The flow is essentially incompressible and best approximated using methods for incompressible flow with immiscible interfaces. The flow solver is

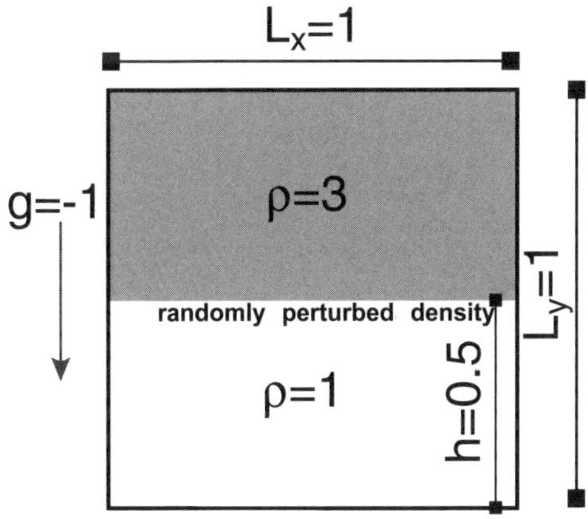

Fig. 18.9. The setup for the random mode Rayleigh-Taylor instability. The upper boundaries are solid walls meaning no slip conditions and the sides are periodic.

a high resolution Godunov solver using an approximate projection for the pressure-velocity coupling combined with interface tracking [431]. The time integration method is a genuinely multidimensional "Hancock" method [543, 573] described in Chap. 14. The volume-of-fluid (VOF) interface tracking method covered at the end of this chapter is most appropriate for fluids that behave immiscibly. We use an ensemble small amplitude ($\ll \Delta x$) multimode perturbations with random phase in the material interface to initialize the instability. It then evolves via constant acceleration through approximately five generations of bubble merger. Lattice Boltzmann results were provided by Tim Clark (LANL)[111].

First, we give the comparison with the integral scale (the integral scale is the overall size of the mixing layer) growth rate for the bubble height that is measured experimentally. This is shown in Fig. 18.10. As most existing calculations (the "alpha" group[2] [141]), the computed α is less than the experimentally measured value (see caption of Fig. 18.10). The present results here are no different in this respect. With LEM we have detailed two-dimensional slices of experimental structure. This is shown in Fig. 18.11) along side the numerical simulations. Using the fractal dimension as a statistical measurement tool we directly compare the calculations (Fig. 18.12). The fractal dimension measures the complexity of a shape, or surface. An example of this would be the extent to which a two-dimensional surface (such as a mountain range) fills the third dimension. Jagged mountains would have a higher fractal dimension (more space filling) than smooth rolling hills (less space filling). In this case the qualitative and quantitative results are best

[2] The "alpha group" is a group of scientists who have simulated the same experiment using a wide variety of methods.

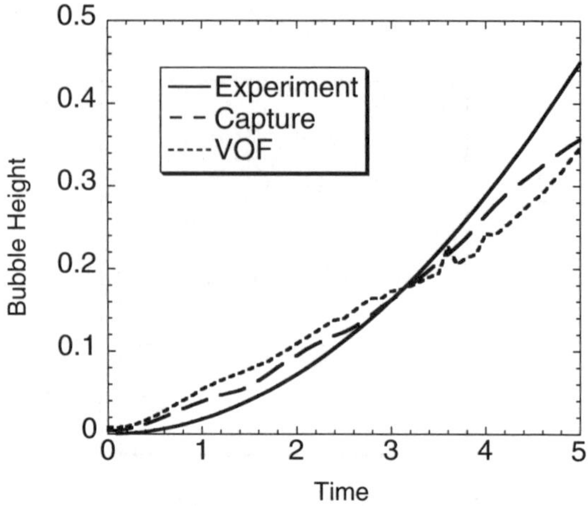

Fig. 18.10. The comparison of the integral scale growth from the LEM experiment and simulations for an idealized version of this experiment. At early times the computed α for the bubble height is higher than the experiment, but it becomes less than the experimentally measured value at late times. The 99% integral volume fraction was used to compute the bubble height in each case. The experimental value is $\alpha = 0.054$, the captured interface (labeled as "captured" that is simply using the shock capturing as described in Chap. 14) yields $\alpha = 0.043$, and the volume of fluid (labeled "VOF" or volume tracking) gives $\alpha = 0.042$. These methods are discussed later in this chapter.

computed with the high resolution Godunov method with interface tracking (i.e., numerically immiscible). Indeed, the results seem to show that the data can be considered to be fractal at large scales as indicated by the fractal dimension at large scales (greater than 0.1 of the integral scale).

18.2 Volume Tracking

In solving a broad spectrum of variable density flows, the dynamics of the interface between materials is the essence of the flow. When this interface is sharp, the key to the successful numerical solution is to maintain this nature in the approximation. A number of techniques exist to accomplish this end, but we will focus on the volume tracking or volume-of-fluid class of method. These methods have the important feature of intrinsically conserving volume (mass). Many other methods in common use are not based on conservation principles.

Other techniques such as the projection and numerical linear algebra will be strained by the presence of a numerically sharp interface. In the case of projection methods the use of filters provides for robust solutions. Techniques

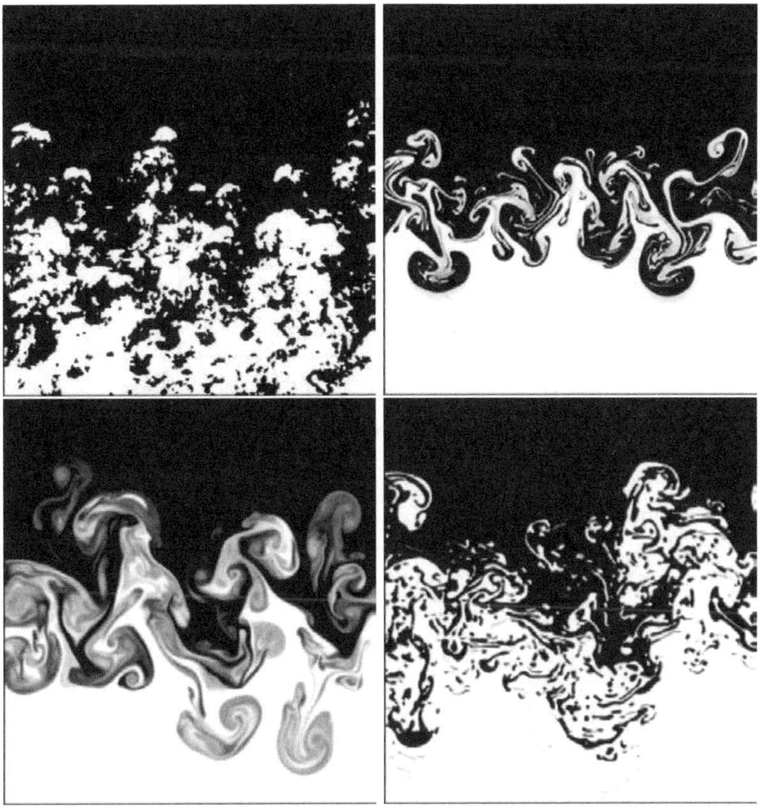

Fig. 18.11. The comparison of the mixing calculations of Rayleigh-Taylor instability starting top-to-bottom, left-to-right, experimental image, lattice Boltzmann, interface capturing, using a high-resolution, Godunov-type, method (Chap. 14), and volume tracking (volume of fluid, VOF). The volume tracking method will be discussed in much greater detail later in this chapter.

such as Krylov (conjugate gradient) methods preconditioned by multigrid can make these projections efficient.

Volume tracking methods and its variants have enjoyed widespread use and success since the mid-1970's, yet they possess solution algorithms that are too often perceived as being heuristic and without mathematical formalism. Partly, this perception stems from the difficulty in applying standard hyperbolic PDE analysis, which assumes algebraic formulations, to a method that is largely geometric in nature (hence the more appropriate term *volume tracking*). To some extent the lack of formalism in volume tracking methods, manifested as an obscure underlying methodology, has impeded progress in evolutionary algorithmic improvements.

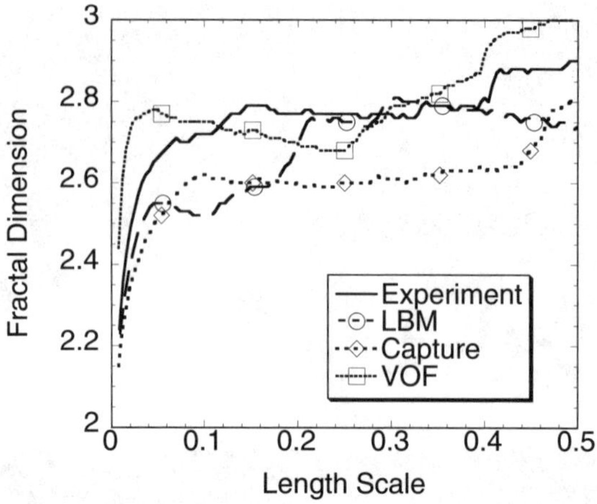

Fig. 18.12. The comparison of the local fractal dimension of the experimental data and computations. LBM, "Capture" and VOF stand for the lattice Boltzmann method, captured interface (using a high-resolution method) and volume of fluid (volume tracking) method, respectively.

Here, the methodology underlying modern volume tracking methods is discussed systematically. The two-dimensional algorithm constructed from purely geometric constructs. The algorithms are second-order in space through the use of a linearity-preserving, piecewise-linear interface geometry approximation. Second-order temporal accuracy is realized with a multidimensional unsplit time integration scheme. The method described uses volume fluxes computed with a set of straightforward geometric tasks.

18.2.1 Fluid Volume Evolution Equations

We first derive material volume evolution equations in the presence of an incompressible flow field. This is a straightforward manipulation of standard consistency relations. We begin by defining V^k, the volume of material k,

$$V^k = \int \alpha^k (V)\, \mathrm{d}V \,, \tag{18.2a}$$

where $\alpha (V)$ is an indicator function, given by,

$$\alpha (V) = \begin{cases} 1 \text{ if fluid } k \text{ is present;} \\ 0 \text{ otherwise.} \end{cases} \tag{18.2b}$$

Given V^k, the volume fraction f^k is defined as

$$f^k = \frac{V^k}{V} \,, \tag{18.3}$$

where the total volume V is $\int \mathrm{d}V$. We require that the material volumes fill all space:

$$V = \sum_k V^k ,$$

or, equivalently, $\sum_k f^k = 1$.

It is readily apparent in the following that volume tracking methods are naturally control volume methods and the volume fractions f^k are integrally-averaged quantities. Given a flow field $\mathbf{u}$, a standard advection equation governs the evolution of f^k,

$$\frac{\mathrm{d}f^k}{\mathrm{d}t} = 0 \rightarrow \frac{\partial f^k}{\partial t} + \mathbf{u} \cdot \nabla f^k = 0. \tag{18.4a}$$

If the flow field is incompressible, i.e., $\nabla \cdot \mathbf{u} = 0$, the f^k advection equation can be easily recast in conservative form:

$$\frac{\partial f^k}{\partial t} + \mathbf{u} \cdot \nabla f^k = 0 \rightarrow \frac{\partial f^k}{\partial t} + \nabla \cdot \left(\mathbf{u} f^k\right) = 0. \tag{18.4b}$$

An equivalent statement for (18.4b) is that material volumes remain constant on streamlines. Incompressibility allows this statement to be further expressed as a conservation law. This confirms our intuition that incompressible flow conserves volume allowing the evolution of the volume fraction to be written as the divergence of fluxes. As such the method abides by the Lax-Wendroff theorem although the imposition of an entropy condition necessary for weak solutions in an open problem. This issue is elaborated further by Sethian [489] with specific application to level sets methods. Interestingly, level sets methods provide an entropy satisfying solution through the motion of their implicit interface using high-resolution methods, but do not conserve volume discretely.

18.2.2 Basic Features of Volume Tracking Methods

First, we review the basic features of most volume tracking methods. To begin, fluid volumes are initialized in each computational cell from a specified interface geometry. This task requires computing fluid interface volumes in each cell containing the interface (hereafter referred to as *mixed* cells). Exact interface information is then discarded in favor of the discrete volume data. Conservation of volume is the essence of the method. The volume data is traditionally retained as volume fractions (denoted as f hereafter), whereby mixed cells will have a volume fraction f between zero and one, and cells without interfaces (*pure* cells) will have a volume fraction f equal to zero or unity. Since a unique interface configuration does not exist once the exact interface location is replaced with discrete volume data, detailed interface information cannot be extracted until an interface is *reconstructed.* The principal reconstruction constraint is local volume conservation, i.e., the reconstructed interface must truncate cells with a volume equal to the discrete fluid volumes.

Interfaces are "tracked" in volume tracking methods by evolving fluid volumes forward in time with solutions of an advection equation, (18.4b), using the inferred interface positions At any time in the solution, exact interface locations are not known, i.e., a given distribution of volume data does not guarantee a unique interface. Interface geometry must be inferred, based on local volume data and the assumptions of the particular algorithm, before interfaces can be reconstructed. The reconstructed interface is then used to compute the volume fluxes necessary to integrate the volume evolution equations. Typical implementations of these algorithms are one-dimensional, with multidimensionality traditionally acquired through operator splitting [515].

A flow chart of the principal algorithmic steps discussed in is shown in Fig. 18.13. It is highly advisable to "interrogate" a method in problems with gross interface topology changes, whereby an initially simple interface configuration is subjected to flows with appreciable vorticity.

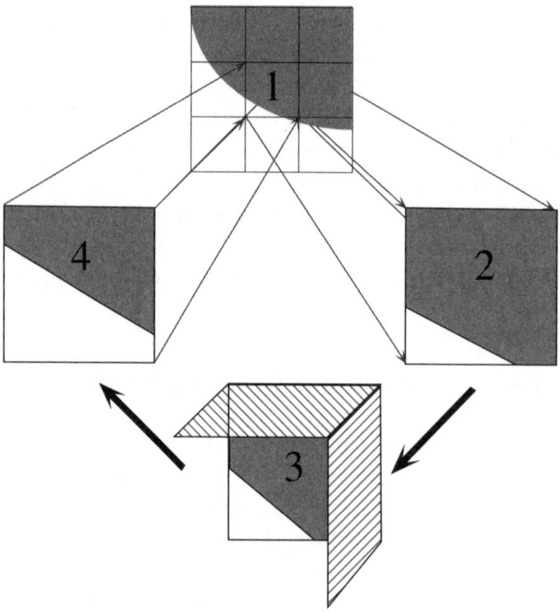

Fig. 18.13. Flow chart of the four basic steps comprising the volume tracking method. First, (1) discrete material volume data is provided on the computational domain (the shaded region), then (2) a piecewise linear interface is reconstructed, next (3) material volume fluxes are computed as truncation volumes (the hatched regions); and, finally, (4) the volumes are integrated to a new time level.

18.3 The History of Volume Tracking

Table 18.1 summarizes the salient features of notable volume tracking methods published since 1974. Listed for each method are important aspects of the interface reconstruction and volume advection algorithms. Identifiable reconstruction features include the assumed or implied interface geometry, which tends to be either piecewise constant, piecewise constant/"stair-stepped", or piecewise linear; and the method used for computing the interface normal, which is either one-dimensional (operator split) or multidimensional. Similarly, time integration of the volume advection equation can be constructed in an operator split or multi-dimensional fashion. Below, we summarize briefly the chronology and impact of these developments.

Beginning in the late 1990's these methods came into wide use especially in commercial CFD codes used for industrial processing. Thus, the methods are now an important part of the mainstream CFD methodology. Furthermore, they have spawned a plethora of papers including an important review [480] and continued development in ever more complex computational geometries. Recently, a number of authors have both refined and improvised the basic volume-of-fluid methods [22, 193, 192, 481, 482, 490]. In particular, the adaptive approach used in [98] utilizes the VOF method as a shock-capturing method in the regions where the interface is under-resolved. Yet, another important improvised method is Sussman and Puckett's combined VOF and level sets methods [521, 519].

Author(s)	**Reconstructed Geometry**	**Integrator**
DeBar [133]	Linear, Split	Split
Noh and Woodward [398]	Constant, Split	Split
Hirt and Nichols [263]	Constant, Multi-Dimensional	Split
Chorin [108]	Constant, Multi-Dimensional	Split
Barr and Ashurst [32]	Constant, Multi-Dimensional	Split
Ashgriz and Poo [20]	Linear, Split	Split
Youngs [610]	Linear, Multi-Dimensional	Split
Pilliod and Puckett [425, 426]	Linear, Multi-Dimensional	Unsplit
Rider and Kothe [454]	Linear, Multidimensional	Unsplit
Scardovelli and Zaleski [480]	Linear, Multidimensional	Unsplit

Table 18.1. Reconstructed interface geometry and time integration method used in a variety of published volume tracking algorithms.

Within a short period of time in the early 1970s, the first three volume tracking methods were introduced: DeBar's method [133], Hirt and Nichols' VOF method [262, 397], and Noh and Woodward's SLIC (for Simple Line Interface Calculation) method [398]. Each of these methods chose a different reconstructed interface geometry: the DeBar's algorithm used a piecewise linear approximation, the VOF method used a piecewise constant/"stair-stepped" approximation, and the SLIC algorithm invoked a piecewise constant approximation. As shown in Table 18.1, most volume tracking algorithms published to date fall into one of these three interface reconstruction categories: piecewise constant, piecewise constant (producing stair-stepped interface in appearance), or piecewise linear. DeBar's piecewise linear choice for the reconstructed interface geometry is still generally preferred in modern volume tracking algorithms.

The SLIC method approximates interfaces as piecewise constant, where interfaces within each cell are assumed to be lines (or planes in three dimensions) aligned with one of the logical mesh coordinates. This choice of a simpler interface geometry (relative to Debar's piecewise linear choice) appears to have been made to facilitate treatment of multiple (> 2) materials within a given mixed cell. In piecewise constant/stair-stepped methods such as VOF, interfaces are also forced to align with mesh coordinates, but are additionally allowed to "stair-step" (align with more than one mesh coordinate) within each cell, depending upon the local distribution of discrete volume data.

Interface normals are acquired in DeBar's method with one-dimensional volume fraction differences, i.e., by considering only those cells sharing a face across which volume fluxes are to be estimated. In this sense the reconstruction can be considered "operator split" since the interface normal follows from one-dimensional differences based upon the current advection sweep direction. The SLIC method, as in DeBar's method, estimates interface normals with operator split differences. Modern SLIC implementations have improved slightly via use of multidimensional operators (3×3 stencil in two dimensions) for the normal and center-of-mass coordinates to aid in placing the interface within the cell [453]. The VOF algorithm also uses a multi-dimensional operator in determining interface orientation. This information helps positioning the reconstructed stair-stepped interface within each cell.

Another pioneering development is attributed to James LeBlanc. His method, as described in [72, 71], has an algebraic (rather than geometric) basis. However, LeBlanc's scheme for the transport of volumes to and from mixed cells (the so-called "mixed-to-mixed" cell transport) also has a geometric interpretation. Close scrutiny of this mode of transport reveals that LeBlanc's "area" factors are nearly identical to those used in a subsequent piecewise linear scheme devised by Youngs (discussed later) [610]. This demonstrates the ability to derive equivalent methods from fairly different bases (algebraic for LeBlanc and geometric for Youngs). A method similar to

DeBar's method is also described in the work of Norman and Winkler [400], where it is attributed to earlier work of LeBlanc (see [72, 71] for further discussion).

Most early volume tracking methods were devised and formulated algebraically (i.e., using combinations of upwind and downwind fluxes) rather than geometrically. In fact, some low-order methods (e.g., SLIC or VOF) can be derived in a number of ways: algebraically, geometrically, or heuristically. The chosen derivation, for these cases, is largely a matter of taste. For piecewise lincar methods, however, a geometric framework is preferable because concise algebraic descriptions can be difficult, especially in three dimensions. To facilitate comparisons with the piecewise linear method, we will interpret volume tracking methods geometrically where possible, while keeping in mind this interpretation might be contrary to the original authors' philosophy. For a reference on algebraic approaches to volume tracking methods, see [472]. Moreover, the geometric approach has all but completely supplanted the algebraic approach during the last half of the 1990's.

Many piecewise constant volume tracking algorithms have been published subsequent to the VOF and SLIC algorithms, e.g., see Chorin [108], and Barr and Ashurst [32]. As seen in Table 18.1, newer versions of these methods offered improvements such as a multidimensional reconstruction algorithm, but operator-split time integration techniques have still been relied upon. In general, these methods have been evolutionary, but still retained the simplistic piecewise constant geometry assumption.

A notable feature of the VOF method is that its volume fluxes can be formulated algebraically, i.e., without needing an exact reconstructed interface position. The volume fluxes can be expressed as a weighted sum of upwind and downwind contributions, depending upon the orientation of the interface relative to the local flow direction. If the reconstructed interface is parallel to the flow, an upwind flux is used; otherwise a portion of downwind flux is added to counteract numerical diffusion brought about by the piecewise constant upwinding. This approach, which falls into the general family of flux-corrected transport (FCT) methods [615, 65] (simplifying its analysis), was the underlying theme behind the design of the VOF method. This flux-limiting methodology has also been used recently to define modern variants, e.g., see [311, 472].

As discussed in Chap. 17, various slope steepeners or artificial compression methods can also be used to retain compact interfaces computationally. Another approach is to use a "limited" anti-upwind method to retain compact interfaces algebraically [139, 311]. These approaches are almost identical to the FCT-based tracking approach.

A feature characteristic of piecewise constant volume tracking methods (with or without stair-stepping) is the unphysical creation of what Noh and Woodward termed [398] *flotsam* ("floating wreckage") and *jetsam* ("jettisoned goods"). These terms are appropriate for isolated, sub-mesh-size material bodies that separate from the main material body because of errors induced by the volume tracking algorithm. These material remnants tend to be ejected from interfaces in piecewise constant volume tracking methods when the flow has significant vorticity and/or shear near the interface. An example of this behavior is demonstrated clearly with the SLIC results presented later. The presence of *flotsam* near interfaces can severely compromise the overall interfacial flow solution, especially when interface dynamics (e.g., surface tension and phase change) are also being modeled.

In the early 1980s, volume tracking methods were advanced significantly by the new piecewise linear schemes of Youngs and coworkers [610, 611]. Youngs' methods positioned each reconstructed interface line, defined by a slope and intercept, within the volume fraction control volume (cell). This is in contrast to DeBar's method, where the reconstructed interface was positioned across cell faces. The slope of the line is given by the interface normal (gradient of the volume fractions), and the intercept follows from invoking volume conservation. The interface normal is determined with a multidimensional algorithm (9-point stencil in two dimensions, 27-point stencil in three) that does not depend upon the sweep direction. The methods of Youngs, formulated for both two [610] and three [611] dimensions on orthogonal meshes, were subsequently adopted in many high-speed hydrocodes involving material interfaces [2, 264, 301, 422].

The two-dimensional (2-D) and three-dimensional (3-D) piecewise linear methods developed by Youngs differed by more than just dimensionality. Although the time integration scheme for both methods was identical (operator splitting), the normal used in interface reconstruction was more accurate in the 2-D algorithm than the 3-D algorithm. The interface normal computed in Youngs' 2-D algorithm will reproduce a line regardless of its orientation on an orthogonal mesh, and is therefore second-order accurate (according to Pilliod and Puckett's criteria [425, 431]). The 3-D normal will reproduce a plane for certain simple orientations, hence the algorithm is not formally second-order accurate. We will refer to Youngs' 2-D and 3-D methods as "Youngs' first method" and "Youngs' second method", respectively, because of this important accuracy difference. It is obviously desirable to retain second-order accuracy in a piecewise linear volume tracking method, otherwise a line (or plane in 3-D) will not be preserved after simple translation.

Many extensions and enhancements to the significant work of Youngs have occurred since its introduction. Johnson extended Youngs' 2-D method to nonorthogonal meshes [2]. The first use of adaptive mesh refinement (AMR) in a volume tracking method can be found in [118]. Puckett and Saltzman [431, 433] coupled an AMR algorithm [44, 55] to the 3-D method. Pilliod and Puckett have recently refined Youngs' algorithm in two dimensions with an unsplit, "corner-coupled" time integration scheme extension that has second-order accuracy through the use of an improved interface normal [425, 426]. More recently, a simpler version of this method has been discussed by Colella et al. [225]. It uses a wider stencil (5x3 or 3x5 in 2-D based on interface orientation) to simplify the method. A similar method is used in [453] for comparison with other interface tracking methodologies. Mosso [392] has recently introduced new methods for second-order interface normal approximations on irregular meshes, and devised a new second-order time integration scheme based on the concept of remapping a displaced mesh. This approach couples nicely with arbitrary Lagrangian-Eulerian (ALE) schemes. Also, Kothe and coworkers have extended Youngs' 3-D method to unstructured meshes [302] by introducing a second order Runge-Kutta method for time integration and a robust method for plane truncation of arbitrary polyhedra.

We will refer to the family of piecewise methods introduced by Youngs (and its extensions) as PLIC (for Piecewise Linear Interface Calculation) methods. Details and capabilities of many PLIC volume tracking methods unfortunately remain obscure because of insufficient widespread publication. Despite this fact, PLIC methods have been used successively for high speed hydrodynamic calculations during the 1980's and early 1990's by a host of researchers. Most recent progress has been made in applying PLIC to incompressible multiphase flows often in an industrial setting [480].

18.4 A Geometrically Based Method of Solution

We now describe an algorithm for the geometric solution of the volume evolution equations given in Sect 18.2.1. This algorithm appeals to geometry because material volume fluxes, defined as material volumes passing through a given cell face over one time step, are the n-sided polygons formed by interface line segments passing through total volume flux polygons. These fluxes are computed in a straightforward and systematic manner using algorithms for lines intersecting n-sided polygons. The algorithm is constructed from a "geometric toolbox", as described in Sect. 18.4.1. By using this toolbox, heuristic, "case-by-case" logic is not required to find solutions. This is a common manner of implementation that leads to rather complex coding and difficulty in extension to 3-D, mapped grids or more complex integration schemes. We also discuss an unsplit method for the time integration of the volume evolution equations. Time integration must be accurate to at least second-order to maintain interface integrity, otherwise an interface propagating at an arbitrary orientation to the mesh might distort (and potentially breakup) unphysically sooner. Results show that all interfaces eventually breakup, but higher order methods reach this point later in calculations.

Next, we describe a geometric toolbox in Sect. 18.4.1, the piecewise linear interface (PLIC) reconstruction in Sect. 18.4.2, the calculation of material volume fluxes in Sect. 18.4.3, and time advancement in Sect. 18.4.4. Examples of an interface tracking method's basic performance on translation and rotation tests are then presented in Sect. 18.4.5.

18.4.1 A Geometric Toolbox

This 2-D PLIC method requires the following geometric functions to be defined:

1. Line–Line Intersection.
2. Point Location.
3. Polygon Collection.
4. Polygon Area.

These functions are simple, well-defined, and widely used, for example, in the field of computational geometry [405]. Concise algorithm design and implementation is made possible with these functions, as is evident later in this section. The capabilities of this toolbox can also be extended to accommodate a 3-D PLIC method if additional functions such as plane/surface intersection are incorporated [302].

In these geometric functions a line is defined by the equation

$$\mathbf{n} \cdot \mathbf{x} + \varrho = 0 \,, \tag{18.5}$$

where $\mathbf{n}$ is the normal to the line, $\mathbf{x}$ is a point on the line, and ϱ is the line constant. Computational cells are defined (in 2-D) as n-sided polygons given by a set of n vertices $\mathbf{X}_v = (x_v, y_v)$. Any cell having volume fractions f between zero and one will possess an interface defined by (18.5). The interface line equation will in general be different in each interface cell, i.e., values of $\mathbf{n}$ and ϱ will vary (from cell to cell) since the overall interface geometry is approximated as piecewise linear. As discussed later in this section, values of $\mathbf{n}$ and ϱ in (18.5) result from a volume fraction gradient and enforcement of volume conservation, respectively. For each interface cell, the interface line divides space into regions inside the fluid and outside the fluid, depending upon the convention chosen for $\mathbf{n}$. We choose the coordinate convention that $\mathbf{n}$ points lie into the fluid, hence (18.5) will be positive for any point $\mathbf{x}$ lying within the fluid, zero for any point $\mathbf{x}$ lying on the line, and negative for any point $\mathbf{x}$ lying outside of the fluid.

Below we summarize each of the four necessary geometric functions.

Function 1: Line/Line Intersection. The most basic geometric function is one that locates the point of intersection between two lines. If the lines are actually line segments, as in the PLIC method, steps must be taken to determine if the intersection point is valid (i.e., lies within segments). The intersection point is found from a simultaneous solution of the two line equations (with checks for parallel lines). By invoking this function only in cases where there must be a valid intersection point, costly checks for validity are avoided. A valid intersection point will result when one end of one line segment lies "across" the intersection point. This function is needed to find the points of intersection between the interface line and any cell edge.

Function 2: Point Location. Given a line defined by (18.5), a point location function returns true if a point (x_v, y_v) lies inside the fluid, which is true if (18.5) is positive for (x_v, y_v). Given a point (x_i, y_i) on the line, the point (x_v, y_v) lies within the fluid if

$$n_x\,(x_v - x_i) + n_y\,(y_v - y_i) > 0\ .$$

This function is needed to determine which of the cell vertices $\mathbf{X}_v$ lie inside the fluid.

Function 3: Polygon Collection. A polygon collection function collects the vertices (x_v, y_v) of a n-sided polygon in counterclockwise order. The polygon vertices collected for each interface cell are those cell vertices lying inside the fluid and interface line/cell edge intersection points. A four-sided polygon example is shown in Fig. 18.14. By definition, two adjacent cell vertices are on opposite sides of the interface line if one cell vertex lies inside the fluid and the other lies outside. The resultant n-sided polygon surrounds the fluid in that cell, as seen in Fig. 18.14. This function is needed before the area inside the n-sided polygon can be computed.

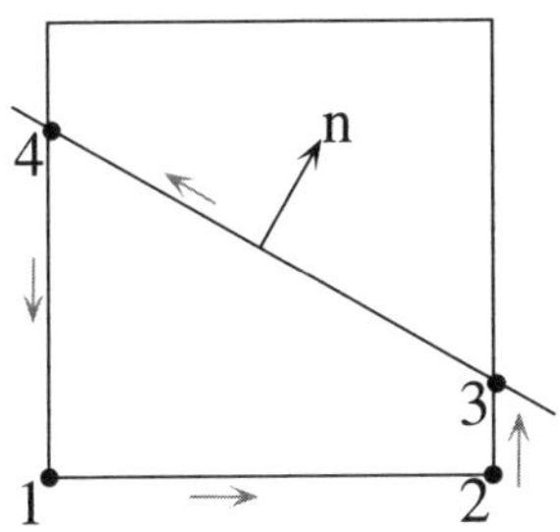

Fig. 18.14. Typical example of a 4-sided polygon formed when an interface line truncates a computational cell. Functions 1 to 4 in the text are performed sequentially to compute the area surrounded by this polygon. Arrows indicate the line integration path.

The polygon collection function uses the output of Function 2, which initializes a boolean variable that describes the "state" of cell vertices with respect to the interface (i.e., inside or outside the fluid). A vertex state variable minimizes source code logic and enables robust procedures. The benefits of this implementation are easily realized for the (fairly typical) case of an interface line passing nearby a cell vertex (see Fig. 18.15). In this case a decision regarding the identity of this cell vertex is forced (based on some prescribed tolerance), thereby avoiding numerical difficulties. This problem and its solution are illustrated by the two relevant cases in Fig. 18.15. The cost of cell vertex ambiguity situations is an extra vertex that becomes associated with the truncated polygon.

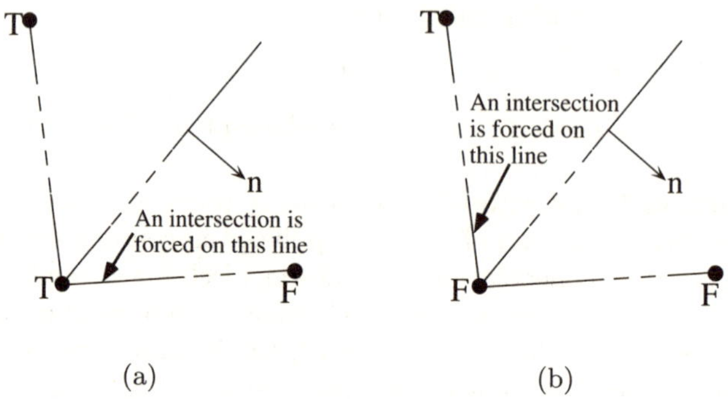

Fig. 18.15. Two examples of an interface line passing nearby a cell vertex. This (critical) vertex is "ambiguous" in that it can be considered inside (a) or outside (b) the fluid, depending upon the prescribed tolerance of the boolean variable that identifies the vertex as being inside or outside the fluid.

Function 4: Polygon Area. A polygon area function takes the vertices (x_v, y_v) of an n-sided polygon, collected in counterclockwise order (the result of Function 3), and computes the exact area enclosed by the polygon. In Cartesian geometry, the area enclosed by the polygon is given by

$$A = \frac{1}{2}\sum_{v=1}^{n}(x_v y_{v+1} - x_{v+1} y_v), \tag{18.6}$$

where the vertex $v = n + 1$ is assumed to coincide with the vertex $v = 1$. In cylindrical geometry having azimuthal (θ) symmetry, the area (an azimuthally-symmetric volume) is given by

$$A = \frac{\pi}{6}\sum_{v=1}^{n}(r_v + r_{v+1})(r_v z_{v+1} - r_{v+1} z_v). \tag{18.7}$$

The single algorithmic change required for the method to perform correctly in 2-D cylindrical rather than Cartesian geometry is the use of (18.7) instead of (18.6) for the polygon area computation.

18.4.2 Reconstructing the Interface

Given the functions provided by the geometric toolbox, linear interface segments must be *reconstructed* in each mixed cell. This reconstruction step requires the line (18.5) to be defined for each interface segment, therefore an interface normal **n** and a line constant ϱ must be determined. The line constant ϱ follows from enforcement of volume conservation and the interface normal **n** follows from volume fraction gradients. Interface reconstruction examples of simple volume fraction distributions (circles and squares) are also presented in this section. The examples illustrate the importance of an accurate, linearity-preserving estimate for **n**.

Finding the Interface Constant. Determining the line constant ϱ is the most difficult reconstruction task because the value of ϱ is constrained by volume conservation. In other words, the value of ϱ is constrained such that the resulting line passes through the cell with a truncation volume equal to the cell material volume V. This determination requires inverting a $V(\varrho)$ relation, in which V can vary linearly, quadratically, or cubically with ϱ, depending upon the coordinate system and the shape of the n-sided polygon formed by the interface segment truncating the cell. It is tempting to construct an algorithm for determining ϱ based on a direct solution of the $V(\varrho)$ relation. But, since this relation is often nonlinear and varies in each mixed cell, due to its dependence upon local data, a "case-by-case" implementation results not to be efficient (vectorizable or parallel), general, concise, or easily maintained and understood. The algorithm inverts the $V(\varrho)$ relation iteratively in each mixed cell. The resulting algorithm is independent of the mixed-cell properties and data. The algorithm is simple, robust, general, efficient, and easily understood.

The line constant ϱ is found when the generally nonlinear function

$$f(\varrho) = V(\varrho) - V\,, \tag{18.8}$$

becomes zero. Here $V(\varrho)$ is the material volume in the cell bounded by the interface segment (with line constant of ϱ) and the portion of the cell edges within the material. When these two volumes are equal (to within some tolerance), the interface segment is declared "reconstructed" in that cell. A host of root-finding algorithms are available to find the zero of this function, but we have found Brent's method [430, Chap. 9] to give the best results in practice. Bisection will converge, but is slow, and Newton's method may diverge, but Brent's method invokes a combination of bisection and inverse quadratic interpolation to find a near-optimal next guess for ϱ.

An intelligent initial guess for ϱ is essential for efficient convergence (< 4-6 iterations). One can initialize ϱ prior to the iteration in the following manner. Lines possessing the interface normal **n** are passed through each vertex of the polygonal cell, and the resulting truncation volumes are computed. Those two lines forming truncation volumes that bound the actual material volume in that cell provide upper ($\varrho_{\max}$) and lower ($\varrho_{\min}$) bounds for ϱ. Our initial guess for ϱ is then a linear average of $\varrho_{\min}$ and $\varrho_{\max}$.

Fig. 18.16 illustrates the iterative progression of Brent's algorithm in finding ϱ for an example reconstruction. Convergence of the algorithm occurs when the line is properly placed, or when (18.8) becomes zero.

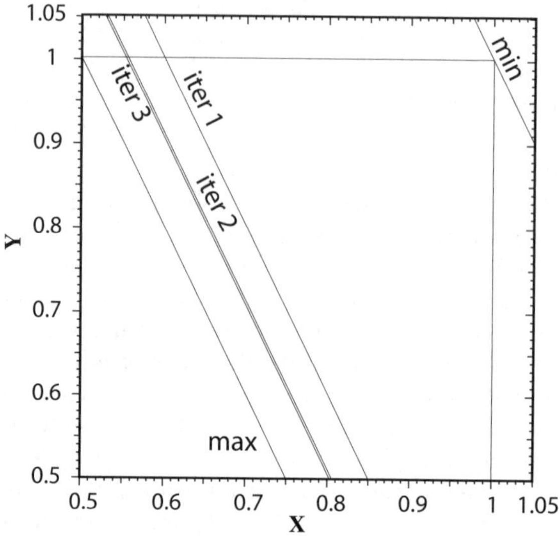

Fig. 18.16. An example of the iterative placement of an interface segment within a mixed cell using Brent's method, which prescribes a new value for interface line constant ϱ at each iteration. The cell size is 1×1. The line is properly placed after five iterations, although only the first three are shown here (after which progress is indistinguishable). Initial bounding guesses ($\varrho_{\min}$ and $\varrho_{\max}$) for the line are also shown.

The final algorithm for finding the interface constant ϱ is summarized below:

Interface Reconstruction Algorithm.

1. Given an interface line segment, described by (18.5), truncating a mixed cell:
2. Find and assemble the n vertices of the polygon formed by the those cell vertices inside the fluid and the interface line/cell edge intersection points.
3. Compute the volume bounded by this polygon.
4. Determine if the polygon volume differs from the known fluid volume by some prescribed tolerance.
5. If the volumes differ, use Brent's method to find a new estimate for ϱ in (18.5) and go back to step one.
6. If the volumes do not differ, the interface line is declared reconstructed.

If this algorithm is applied only to mixed cells, an iterative search for ϱ is computationally efficient.

This approach is also attractive because mixed cells usually comprise <10% of the total cells in the computational domain, even for the most topologically complex cases. Also, because volume tracking methods are local in nature, the increase in mixed cells number over a given time step is bounded because only those cells that are mixed or directly adjacent to mixed cells experience a volume fraction change. An example of the behavior of this algorithm is shown in Fig. 18.17, where the line constant ϱ and iteration-count as a function of fluid volume are shown for a given interface normal $\mathbf{n}$. Brent's method is well suited for the $V(\varrho)$ function since V often varies quadratically with ϱ. Newton's method typically converges in half the iterations of Brent's method, but Brent's method requires less computational effort because an evaluation of the derivative of V with respect to ϱ is not needed.

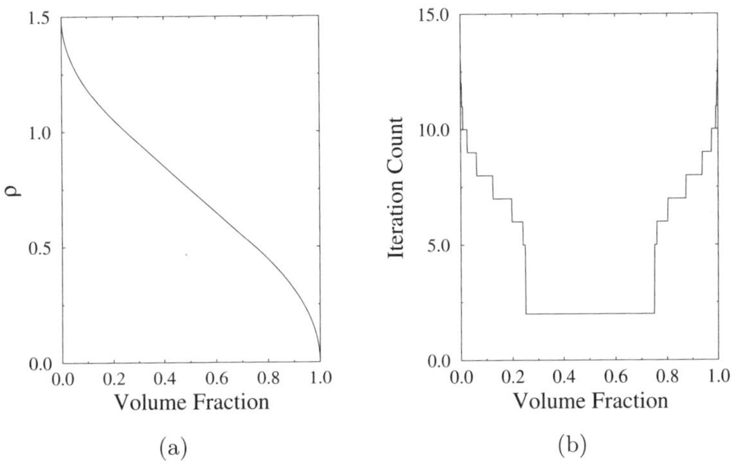

Fig. 18.17. An example of the variation of line constant ϱ (a) and Brent's method iteration-count (b) as a function of fluid volume for a given interface normal $\mathbf{n}$. Note the linear and quadratic dependence of ϱ as a function of fluid volume.

Finding the Interface Normal. Unlike the line constant ϱ, the method used in determining the interface normal $\mathbf{n}$ is arbitrary, i.e., it is not constrained by volume conservation considerations. The algorithm used in determining $\mathbf{n}$ is nonetheless crucial, being a key distinguishing aspect of volume tracking algorithms. Simple estimations for $\mathbf{n}$ can cause a volume tracking algorithm to exhibit overall first-order tendencies, i.e., an inability to reproduce linear interfaces. For example, a method for $\mathbf{n}$ that forces $\mathbf{n}$ to align with mesh logical coordinates yields a low-order volume tracking algorithm reminiscent of the SLIC method. Other than the temporal integration method (discussed in Sects. 18.4.3 and 18.4.4), the method used for estimation of **n** plays a principal role in the overall accuracy of a volume tracking algorithm. In the following, we will first discuss three methods for computing **n**, then illustrate the impact of using **n** resulting from these methods on the reconstruction of simple circular and square volume distributions.

Our first method is an extension of Youngs' second method [611] to general grids [302], the second is an extension of Pilliod and Puckett's minimization principles [425, 426] (there is also an algorithm inspired by the recent work of Swartz and Mosso [392] described in [454]). The first method extends Youngs' approach of using finite difference stencils, yielding an explicit expression for **n**. This method is shown to be first-order accurate, as measured with the least squares criteria [425, 426]. In the second method **n** does not result from an explicit expression, but rather from a least squared error minimization procedure that is iterative (but local). This method is shown to be second-order accurate. Despite our ability to demonstrate second-order accuracy with our current algorithms for **n**, important outstanding issues remain, such as whether tractable and efficient (non-iterative) 3-D extensions are possible.

Youngs' second method for estimation of **n** was originally developed for 3-D tensor-product meshes in which cells are orthogonal bricks having generally different widths. The basic algorithm invokes straightforward (yet wide stencil) finite-difference approximations for the volume fraction gradient, ∇f, to define the interface normal **n** [611]. It is robust and simple, but unfortunately first-order accurate, capable reproducing a linear interface only in certain isolated cases.

One can extend Youngs' finite difference approximations for ∇f to 2-D or 3-D arbitrary-connectivity (fully unstructured) meshes. The method is based on the work of Barth [35], who has devised innovative least squares algorithms for the linear and quadratic reconstruction of discrete data on unstructured meshes. Second (and higher) order accuracy has been demonstrated on highly irregular (e.g., random triangular) meshes. In this approach, volume fraction Taylor series expansions f_i^{TS} are formed from each reference cell volume fraction f_i at point $\mathbf{x}_i$ to each cell neighbor f_k at point $\mathbf{x}_k$, where k is an arbitrary index. The sum $(f_i^{\mathrm{TS}} - f_k)^2$ over all n immediate neighbors is then minimized in the least squares sense using the normal equations. This L_2 norm minimization will yield the volume fraction gradient ∇f_i as solutions to the following linear system (in 2-D):

$$\mathbf{A}^T\mathbf{A}\mathbf{x} = \mathbf{A}^T\mathbf{b}\,, \tag{18.9}$$

where

$$\mathbf{A} = \begin{pmatrix} \omega_k\,(x_k - x_i) & \omega_k\,(y_k - y_i) \\ \vdots & \vdots \\ \omega_n\,(x_n - x_i) & \omega_k\,(y_n - y_i) \end{pmatrix},$$

$$\mathbf{b} = \begin{pmatrix} \omega_k (f_k - f_i) \\ \vdots \\ \omega_n (f_n - f_i) \end{pmatrix},$$

where $\omega_k = 1/|\mathbf{x}_k - \mathbf{x}_i|^t$ (we take $t = 2$), n is the number of equations used and

$$\mathbf{x} = \begin{pmatrix} \nabla_x f_i \\ \nabla_y f_i \end{pmatrix}.$$

The normal is then computed as

$$\mathbf{n} = \begin{pmatrix} \nabla_x f_i / |\nabla f_i| \\ \nabla_y f_i / |\nabla f_i| \end{pmatrix}, \tag{18.10}$$

where

$$|\nabla f_i| = \sqrt{(\nabla_x f_i)^2 + (\nabla_y f_i)^2}.$$

This is just like the multidimensional Godunov method described at the beginning of Chap. 14. Least squares reconstruction methods, such as that implied by the linear system above, are quite powerful and attractive. They are not married to any particular mesh topology or dimensionality, hence are easily amenable to any (unstructured) 2-D or 3-D mesh. All that is required is a set of discrete data points described by their data values and their physical location.

Despite the attractive characteristics of least squares methods, we have yet to construct a linear system whose solution yields interface normals $\mathbf{n}$ that are generally second-order accurate. This methodology offers perhaps the best possibility (over other methods) for a generally second-order accurate $\mathbf{n}$ estimation in 3-D. Careful construction of data-dependent weights might provide a linear system that yields such a second-order discretization.

The second method for estimation of $\mathbf{n}$ is based upon the recent volume tracking work of Pilliod [425], in which particular scrutiny is paid to the interface reconstruction step. They conclude that methods for estimating $\mathbf{n}$ based on a minimization of error (by some measure) are optimal. Developed as a result of this analysis was an algorithm for $\mathbf{n}$ that, like our first method, is also guided by a least squares prescription for error. In the least squares procedure, reconstructed interface line segments are extrapolated into nearest-neighbor cells (all cells bordering the reference mixed cell), and the nearest-neighbor truncation volumes are computed (in addition to the reference mixed cell). Differences between nearest-neighbor truncation volumes and their actual material volumes are minimized in the least squares sense by iteratively changing the reference mixed cell value of $\mathbf{n}$ until convergence is achieved. This method is second-order in the sense that it is able to reconstruct linear interfaces exactly.

Unfortunately, the least squares method can be prohibitively expensive, especially in 3-D, because interfaces must be reconstructed in all mixed cells bordering the reference mixed cell at each iteration required to find **n**. For 2-D tensor-product Cartesian meshes, however, this expensive iterative procedure can be circumvented with an innovative "fast least squares" (FLS) procedure [425], making the method viable.

The accuracy of these interface normal algorithms is most easily assessed by analyzing estimates of **n**, operating on discrete volume data that replicates known (exact) interface geometries, such as a circle or square. If the various **n** estimates are used to perform piecewise linear interface reconstructions on the data, the differences and errors can also be visualized.

Reconstruction Examples. We now assess the accuracy of the interface normal algorithms detailed in the previous section by applying them to known circular and square distributions of volume data. To avoid artificial regularity of the results, we offset the circle and square centers (relative to mesh logical coordinates). For all tests, we partition a unit square domain from 10^2 to 160^2 orthogonal, uniform cells. Discrete volume data is initialized on the domain to replicate either a circle (radius 0.368) centered at $(0.525, 0.464)$ (or, alternatively, a square with side length of 0.512) centered at $(0.468, 0.541)$. The discrete volume initialization is exact in every cell, so we expect computational values of the interface normals **n** to converge to the analytical values.

In this case the precise location of the interface is known for the object being reconstructed thus the detailed interface error in each zone can be computed using this knowledge. Quantitative error assessments are possible for these known interface geometries if one defines the reconstruction error as the volume integral of the difference between the actual interface and the linear representation. With errors quantified, convergence rates can be computed, as shown in Tables 18.2–18.3. For the circular distribution, which has no curvature singularities, we expect each interface normal method to converge. For the square, on the other hand, curvature singularities (the corners) will preclude convergence. Along with inhibiting convergence, curvature singularities will also be the site of the most flagrant reconstruction errors.

The errors measured on these simple tests give rise to some interesting observations. Youngs' second method (which is first order) surprisingly exhibited the lowest errors on the coarsest grids, but its lower convergence rate eventually led to errors that exceeded the other methods as the grid is refined. In general, for large curvatures (relative to mesh spacing), Youngs' second method is quite robust. Results for Youngs' second method are given in Table 18.2.

Finally, consider the reconstruction errors resulting from the least squares error method. This method is also second-order, as seen in Table 18.3. The

Table 18.2. L_1 (sum of absolute errors) circle reconstruction error norms using Young's second method for the interface normal $\mathbf{n}$.

Grid	Error	Order-of-Accuracy
10^2	2.04×10^{-3}	
		1.59
20^2	6.79×10^{-4}	
		1.38
40^2	2.60×10^{-4}	
		1.36
80^2	1.01×10^{-4}	
		0.93
160^2	5.29×10^{-5}	

reconstruction is more continuous for the circle and the singularities in the corners are more localized.

Table 18.3. L_1 (sum of absolute errors) circle reconstruction error norms using the least squares method for the interface normal $\mathbf{n}$.

Grid	Error	Order-of-Accuracy
10^2	3.21×10^{-3}	
		2.18
20^2	7.08×10^{-4}	
		2.29
40^2	1.45×10^{-4}	
		1.94
80^2	3.78×10^{-5}	
		2.01
160^2	9.38×10^{-6}	

18.4.3 Material Volume Fluxes

Integration of the material volume evolution equations discussed in Section 18.2.1 requires the evaluation of material volume fluxes δV^k at each cell face. The fluxes δV^k represent the volume of material k passing through a given face during the current time step. The sum of all material volume fluxes must equal the total volume flux, which for face $(i+1/2, j)$ is given by:

$$\sum_k \delta V^k_{i+1/2,j} = \delta V_{i+1/2,j} = \Delta t\, \mathbf{u}_{i+1/2,j} \cdot \mathbf{H}_{i+1/2,j}\,, \tag{18.11}$$

where $\mathbf{u}$ is the velocity vector and $\mathbf{H}$ is the edge area vector.

To compute δV^k, two steps are required: (1) the polygons bounding the volume swept by the velocity field over the time step must be constructed, and (2) the amount of material k must be found for these polygons. The machinery to accomplish the second step has been discussed earlier (the geometric toolbox given in Sect. 18.4.1), but the first step requires definition.

Here, we discuss both operator split and unsplit time integration schemes, which are depicted schematically in Fig. 18.18. The operator split method constructs the solution in a series of one-dimensional sweeps shown in Fig. 18.18(a–b). Usually Strang's splitting [515] is employed to lessen the symmetry breaking effects of the operator splitting. In the unsplit method (Fig. 18.18d), the flow is considered to be completely multidimensional, but the method can be constructed from the operator split method via systematic corrections shown below.

As an example, consider the formation of edge flux volumes for an x-sweep of an operator split method. A sample result is shown in Fig. 18.19a. The steps to form the edge flux polygon for face $(i+1/2, j)$ are outlined in the following algorithm.

Edge Flux Polygon Algorithm.

1. Identify the vertices of the edge being evaluated, which form the first two vertices in the flux polygon:
$\left(x_{i+1/2,j-1/2},\ y_{i+1/2,j-1/2}\right),\ \left(x_{i+1/2,j+1/2},\ y_{i+1/2,j+1/2}\right)$.
2. Given the velocity normal to this edge ($u_{i+1/2,j}$) and the time step (Δt), trace characteristically to locate the final vertices:
$\left(x_{i+1/2,j-1/2} - u_{i+1/2,j}\Delta t,\ y_{i+1/2,j-1/2}\right)$,
$\left(x_{i+1/2,j+1/2} - u_{i+1/2,j}\Delta t,\ y_{i+1/2,j+1/2}\right)$.

The unsplit algorithm requires that correction be made to the edge (operator split) of the flux volume defined above. The corrections are defined by the velocity field assumed in the algorithm. In our implementation, the corner velocity transverse to the direction of the edge under consideration is "donored" given the sign of the normal edge velocity. For example, if $u_{i+1/2,j}$ is greater than zero, the velocities $v_{i,j\pm 1/2}$ will define the corner corrections

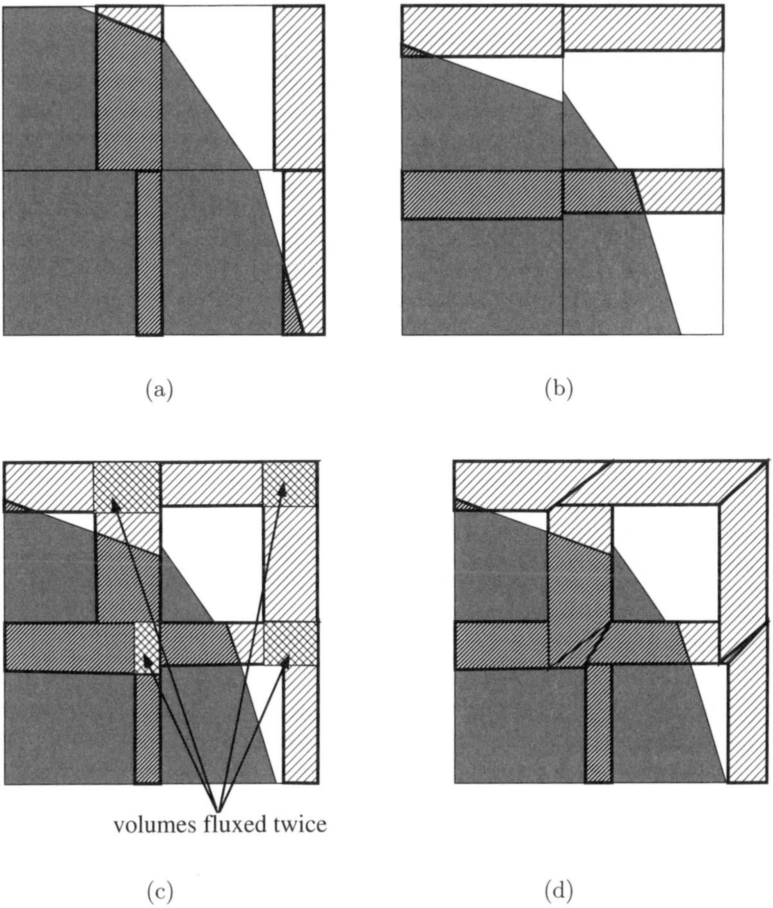

Fig. 18.18. Face volume fluxes δV and their bounding polygons for three different time-integration schemes in two dimensions. For operator-split integration, an x-direction sweep is made in (a), followed by a y-direction sweep in (b). A naive (multidimensional) unsplit method, shown in (c), is not recommended because face-adjacent polygons overlap, hindering conservative, monotonic volume advection. In the unsplit (multidimensional) method of (d), the polygons are constructed by tracing along upwind characteristics, yielding a multi-dimensional, "corner-coupled" fluxing of material volumes.

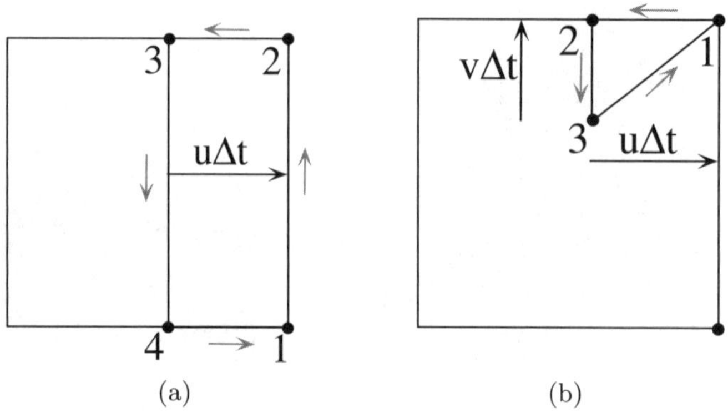

Fig. 18.19. (a) Quadrilateral and (b) corner flux polygons formed at face $(i+1/2, j)$ when tracing back along a positive velocity field. The edge polygon that encloses the volume flux $\delta V_{i+1/2,j}$ is defined by the four points in (a) for the x-sweep of an operator-split time integration. For multidimensional unsplit time integration, the edge polygon in (a) must be modified by corner fluxes such as the triangle in (b).

to the flux volume at the $i + 1/2, j$ edge. The corner correction polygon is depicted in Fig. 18.19b. The correction polygons are computed according to the following corner flux polygon algorithm.

Corner Flux Polygon Algorithm.

1. Identify the vertex of the corner being evaluated, which forms the first vertex in the flux polygon:
   $\left(x_{i+1/2,j+1/2},\ y_{i+1/2,j+1/2}\right)$.
2. Using the velocity normal to this edge, $u_{i+1/2,j}$, and the time step, Δt, trace characteristically to the next vertex:
   $\left(x_{i+1/2,j+1/2} - u_{i+1/2,j}\Delta t,\ y_{i+1/2,j+1/2}\right)$.
3. Using the donored transverse velocity at the corner, trace characteristically to the final vertex:
   $\left(x_{i+1/2,j+1/2} - u_{i+1/2,j}\Delta t,\ y_{i+1/2,j+1/2} - v_{i+1/2,j+1/2}\Delta t\right)$.

Given the edge and corner flux polygon algorithms, unsplit edge flux polygons at face $(i + 1/2, j)$ are found according to the following algorithm.

Unsplit Edge Flux Polygon Algorithm.

1. Apply the *edge flux polygon algorithm*.
2. Apply the *corner flux polygon algorithm* at $i + 1/2, j + 1/2$.

3. If the donored transverse velocity at $(x_{i+1/2,j+1/2}, y_{i+1/2,j+1/2})$ is positive, subtract the corner flux volume, otherwise add it.
4. Apply the *corner flux polygon algorithm* at $i + 1/2, j - 1/2$.
5. If the donored transverse velocity at $(x_{i+1/2,j-1/2}, y_{i+1/2,j-1/2})$ is positive, add the corner flux volume, otherwise subtract it.

18.4.4 Time Integration

We now consider the time integration of (18.4b) for incompressible flows, whereby the material k volume fractions $f^{k,n}$ at discrete time level n are marched forward in time to step $n + 1$. In performing this time integration, care must be taken to ensure that material volumes V^k remain volume-filling in each cell, i.e., material volumes must sum to the total cell volume V $(= \sum V^k)$. For incompressible flow, an additional global volume-filling constraint applies, namely material volume summed over the entire mesh must also be conserved (this applies locally as well). For compressible flows, any local discrete volume-filling errors can be absorbed by the fluid compressibility, but for incompressible flow, this can lead to unacceptable errors. If compressibility (velocity divergence) is used to absorb local volume-filling errors for incompressible flows, the integration accuracy of the right hand side of (18.4b) degrades accordingly.

A key point in our time integration scheme for the incompressible flow volume evolution equations is the recognition that discrete velocity divergences are not necessarily zero, but rather a function of the convergence tolerance used in obtaining the linear system solutions. To employ this "divergence correction" $\nabla \cdot \mathbf{u}$ is added to both sides of (18.4b), giving:

$$\frac{\partial f}{\partial t} + \nabla \cdot (\mathbf{u} f) = f \nabla \cdot \mathbf{u} \, . \tag{18.12}$$

By applying this correction, even for incompressible flows, we find that local and global volume-filling constraints are adhered to much more closely. We integrate (18.12) forward in time to advance volume fractions from $f^{k,n}$ to $f^{k,n+1}$ in the presence of incompressible flows.

Our time integration schemes operate on two basic sets of cells: those that are *mixed* and those that are *active*. Before integration proceeds, those cells defined as mixed and active are flagged appropriately. Mixed cells are labeled as such if:

1. Their volume fractions f satisfy $\epsilon \le f \le 1 - \epsilon$; or
2. Their volume fractions f satisfy $f > 1 - \epsilon$, *and* a face is shared with an empty cell having $f < \epsilon$.

In flagging mixed and active cells, ϵ is a small number relative to zero, for example, taken to be 1×10^{-12}. Active cells are labeled as such if at least one cell in its domain of dependence is mixed. In two dimensions, the domain of dependence for operator-split time integration is three cells along the current sweep direction, while it is a 3×3 array of cells for unsplit time integration. This cell categorization localizes algorithm computational work in the proximity of interface. For 2-D computations of topologically simple interfaces, we have found the work required to time-integrate the volume evolution equations scales like the square root of the total grid size.

For an operator-split time integration scheme, (18.12) must be integrated twice in two dimensions (thrice in three dimensions), one integration per each sweep, and one sweep per dimension. The volume fractions $f^{k,n+1}$ are therefore constructed from multiple, sequential solutions to (18.12). In two dimensions, volume fractions $f^{k,n}$ are advanced in the first sweep to $\tilde{f}^k$ according to

$$\tilde{f}^k_{i,j} = f^{k,n}_{i,j} - \frac{\delta V^k_{i+\delta i,j+\delta j} - \delta V^k_{i-\delta i,j-\delta j}}{V_{i,j}} + \Delta t\, f^{k,n}_{i,j}\, \nabla \cdot \mathbf{u}_{i,j}\,, \tag{18.13}$$

and in the second sweep $\tilde{f}^k$ are advanced to their final values $f^{k,n+1}$:

$$f^{k,n+1}_{i,j} = \tilde{f}^k_{i,j} - \frac{\delta V^k_{i+\delta i,j+\delta j} - \delta V^k_{i-\delta i,j-\delta j}}{V_{i,j}} + \Delta t\, f^{k,n+1}_{i,j}\, \nabla \cdot \mathbf{u}_{i,j}\,. \tag{18.14}$$

Since this method is implemented on structured, orthogonal meshes in Cartesian (or azimuthally-symmetric cylindrical) geometry, each sweep is associated with either an x or y (r or z in axisymmetric co-ordinates) direction. For sweeps in the x (r) direction, $(\delta i, \delta j) = (1/2, 0)$, and for sweeps in the y (z) direction, $(\delta i, \delta j) = (0, 1/2)$. Sweep-direction-order is alternated every time step to minimize asymmetries induced by the sequential sweeping process.

Note the RHS rightmost terms in (18.13) and (18.14), which are the important divergence correction terms. These terms need to enforce volume-filling constraints, containing volume fractions f at differing time levels, explicit (time level n) for the first sweep and implicit (time level $n+1$) for the second sweep. This form of the correction is found to be optimal in practice, giving a net divergence correction that employs a volume fraction having an intermediate time level. We now summarize the operator-split time integration algorithm.

Operator-Split Time Integration Algorithm.

1. Flag all mixed, active, and isolated cells.
2. Compute the discrete velocity divergence $\nabla \cdot \mathbf{u}_{i,j}$ in all flagged cells.
3. Reconstruct interfaces in all mixed cells according to the Interface Reconstruction Algorithm given in Section 18.4.2.
4. Compute edge volume fluxes δV^k in all flagged cells according to the Edge Flux Polygon Algorithm given in Section 18.4.3. If this is an x (r in axisymmetric co-ordinates) direction sweep, the fluxes will be right-face

fluxes; if this is a y (z in axisymmetric co-ordinates) direction sweep, they will be top-face fluxes.
5. Advance volume fractions f^k in time using (18.13) if this is the first sweep or (18.14) if this is the second sweep.
6. Look for and conservatively redistribute any volume fraction undershoots ($f^k < 0$) or overshoots ($f^k > 1$).

As a final note, regarding operator-split time integration, if $\nabla \cdot \mathbf{u} = 0$ *is assumed*, then employing the identity (for 2-D Cartesian geometry)

$$\frac{\partial u}{\partial x} = -\frac{\partial v}{\partial y} ,$$

is often useful, especially for operator-split time integration discretizations of (18.4b). This approach appears to improve the discrete conservation properties of operator-split incompressible flow time integration [426, 432].

As opposed to an operator-split time integration of (18.12), an unsplit time integration scheme advances time level n volume fractions $f^{k,n}$ to $f^{k,n+1}$ with one equation, given for two dimensions as

$$f^{k,n+1}_{i,j} = f^{k,n}_{i,j} - \frac{\delta V^k_{i+1/2,j} - \delta V^k_{i-1/2,j}}{V_{i,j}} - \frac{\delta V^k_{i,j+1/2} - \delta V^k_{i,j-1/2}}{V_{i,j}} + \Delta t \frac{f^{k,n}_{i,j} + f^{k,n+1}_{i,j}}{2} \nabla \cdot \mathbf{u}_{i,j} . \tag{18.15}$$

Note the centered time level $(n + 1/2)$ used for the volume fraction in the rightmost divergence correction term on the RHS above. We now summarize the unsplit time integration algorithm.

Unsplit Time Integration Algorithm.

1. Flag all mixed, active and isolated cells.
2. Compute the discrete velocity divergence $\nabla \cdot \mathbf{u}_{i,j}$ in all flagged cells.
3. Reconstruct the interface in all mixed cells according to the Interface Reconstruction Algorithm given in Sect. 18.4.2.
4. Compute volume fluxes δV^k in all flagged cells according to the Unsplit Edge Flux Polygon Algorithm given in Sect. 18.4.3.
5. Advance volume fractions f^k in time using (18.15).
6. Look for and conservatively redistribute any volume fraction undershoots ($f^k < 0$) or overshoots ($f^k > 1$).

In conclusion, we have presented operator-split and unsplit algorithms for the time integration of (18.12). These discretizations have assumed that the flow is incompressible.

18.4.5 Translation and Rotation Tests

Since translational and rotational flows do not induce topology change, the volume fractions associated with fluid bodies entrained in these flows are

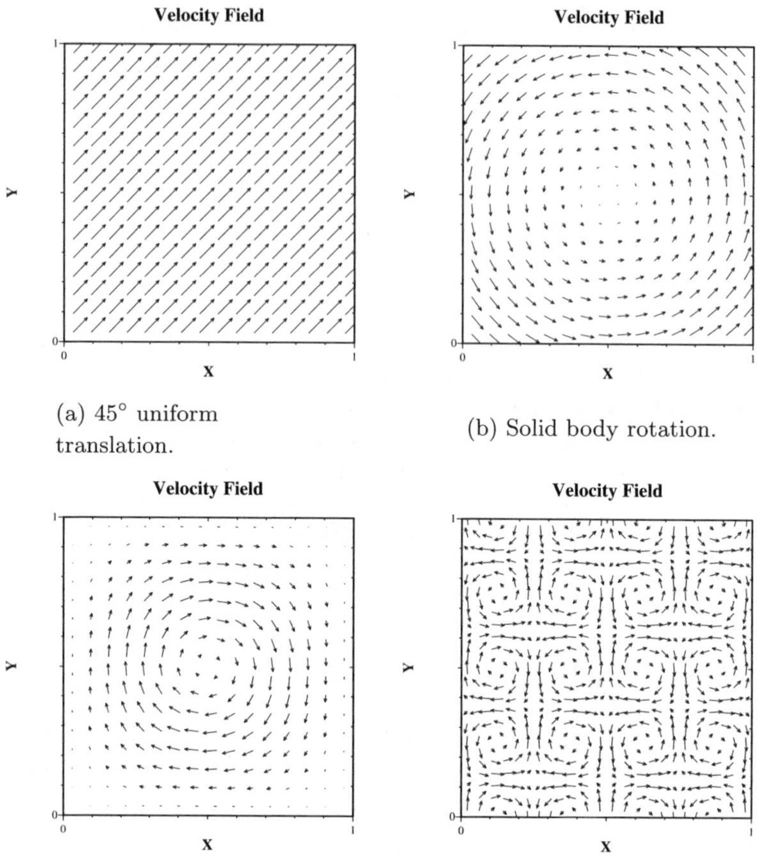

(a) 45° uniform translation.

(b) Solid body rotation.

(c) The single vortex.

(d) The deformation field.

Fig. 18.20. Velocity fields used to move the circular fluid body.

known exactly. In such cases, error norms can be defined based on some positive-definite measure of the differences observed between the computed and exact values of f. We choose to estimate computed errors in these problems with an L_1 norm, defined as

$$E^{L_1} = \sum_{grid} V_{i,j} \left| f_{i,j}^{\text{computed}} - f_{i,j}^{\text{exact}} \right| . \tag{18.16}$$

The E^{L_1} error defined above has units of area (or volume in 3-D), therefore its change with mesh size can be used to infer rates of convergence.

For both the translation and rotation test problems, a circular fluid body is placed in a unit square computational domain that is partitioned with either 32^2, 64^2, or 128^2 orthogonal, uniform cells. The circular body is represented with a scalar field that is unity and zero inside and outside the circle, respectively. For those cells containing the circular interface, the scalar field is set to a value between zero and one, in proportion to the cell volume truncated by the circle. This field represents a characteristic (or color) function, which for our purposes is the fluid volume fraction for a circular fluid body, i.e., the volume fraction is 100% inside the circle and 0% outside. All boundaries are periodic. For these tests, we use the unsplit time-integration scheme given by (18.15) and the least squares method for estimating the interface normal $\mathbf{n}$.

For the solid body translation problem, a uniform and constant velocity field having positive, equal components is imposed everywhere in the domain. This solenoidal field, shown in Fig. 18.20a, will cause fluid bodies to translate diagonally across the mesh at a 45° angle. The circular fluid body (radius 0.25), initially centered at $(0.50, 0.50)$, should return to its initial position after 1 time unit, allowing error measurement using (18.16). A CFL number of 1/2 is used. We note that the body should not change shape as a result of this movement.

We employ a piecewise constant method that represents a combination of the SLIC and VOF methods.[3] Computed piecewise constant L_1 errors are one to two orders of magnitude larger than the coarsest-grid (32^2) piecewise linear results, with the differences becoming even larger as the grid is refined. Convergence for this piecewise constant scheme is at best first order.

The piecewise linear scheme, on the other hand, preserves the circular shape after translation. The reconstructed interfaces and error contours are isotropic with respect to flow, i.e., exhibiting no bias toward flow direction. Mass conservation is exact. Convergence rates based on (18.16) are generally second-order, but they exhibit some dependence on the flow direction (relative to the grid), as shown in Table 18.4. The convergence rate is clearly second-order in most cases, with the highest errors generated by the 26.565° translation. These excellent translation results should be expected from a useful interface tracking method. Excellent translation performance is necessary, *but not sufficient*, for a method tasked to track interfaces in complex topology flows.

For the solid body rotation problem, a constant-vorticity velocity field is imposed at the center of the domain, as shown in Fig. 18.20b. This solenoidal field will cause all fluid bodies to rotate around this center. The circular fluid body (radius 0.15), initially centered at $(0.50, 0.75)$, should return to its initial position after π time units, allowing error measurement with (18.16). A CFL number of 1/2 is used, based on the maximum velocity in the domain.

[3] Like the VOF method, an interface normal $\mathbf{n}$ computed from a 3×3 stencil is used to determine interface orientation. Like the SLIC method, the interface is then reconstructed vertically or horizontally, depending upon the relative magnitudes of the $\mathbf{n}$ components.

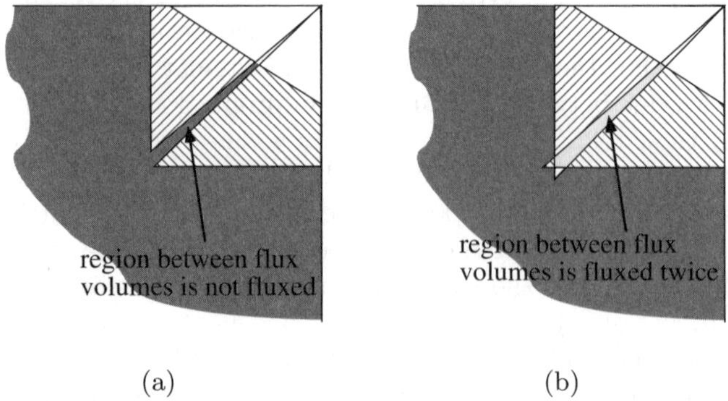

Fig. 18.21. Examples of two problems that can occur with the corner flux polygons in the presence of spatially-varying velocity fields. "Fluxed" refers to the movement of a quantity of volume from one cell to another. In (a) fluid is not fluxed and in (b) fluid is fluxed twice.

Table 18.4. L_1 error norms and convergence rates for a circular fluid body translated two domain diagonals at three different angles to the grid.

Grid	Error (0°)	Order (0°)	Error (26.565°)	Order (26.565°)	Error (45°)	Order (45°)
32^2	1.97×10^{-4}		1.99×10^{-3}		6.21×10^{-4}	
		2.67		2.16		1.33
64^2	3.09×10^{-5}		4.45×10^{-4}		2.47×10^{-4}	
		2.50		2.03		2.27
128^2	5.48×10^{-6}		1.09×10^{-4}		5.10×10^{-5}	

The body should not change shape as a result of this rotation. Solid body rotation results add little additional insight relative to the translation results, except that phase errors are now more apparent. Second-order accuracy is again exhibited, as shown in Table 18.5. Neither the translation or rotation problems pose serious problems for a well-designed interface tracking method. They will expose diffusion and dispersion problems typical of standard advection methods, but are not sufficient for methods designed specifically for interfaces. For both translation and rotation, the circular shape is preserved, hence we can declare victory *for now*, but will return to more stringent tests later.

Table 18.5. L_1 error norms and convergence rates for a circular fluid body rotated one revolution.

Grid	Error	Order
32^2	1.61×10^{-3}	
		2.19
64^2	3.54×10^{-4}	
		1.98
128^2	8.95×10^{-5}	

Problems can arise, however, when the velocity field possesses spatially-varying vorticity. This spatial variation can sometimes lead to fluid being multiply fluxed or not fluxed at all. These subtle fluxing errors, which we identify and correct, are shown in Fig. 18.21 for the two most common instances. We currently correct these situations with a local redistribution algorithm discussed later. The tendency for these problems to occur is greater for algorithms possessing multidimensional time-integration schemes that do *not* incorporate the corner flux corrections to the edge fluxes (i.e., the method shown in Fig. 18.18c).

As long as the CFL condition is met, one-dimensional (edge) fluxes are without systematic problems. For velocity fields that vary in space, however, multi-dimensional fluxes possess small inconsistencies. Small volumes of fluid can be fluxed twice or not at all. As a consequence, there is a propensity for methods based on multidimensional fluxes to produce small over/undershoots. This can be overcome through the use of a numerical velocity divergence definition,

$$\nabla \cdot \mathbf{u} = \sum_k \nabla \cdot \left(\mathbf{u} f^k\right),$$

which follows from the governing equation, and insures that the fractional volumes are volume filling.

18.5 Results For Vortical Flows

Flows that induce simple translation or solid rotation of fluid bodies do not adequately interrogate interface tracking methods for topology changes. Translation and rotation are useful debugging tests, but they are insufficient for definitive analysis, in-depth understanding, or final judgement. Difficult (yet simple) test problems having flows that bring about topology change elucidate algorithm strengths and weaknesses relevant to modeling interfacial flows. Also, with carefully controlled test problems, assessment of the interface tracking method is not obscured by subtleties of the algorithms in the flow solver.

Literature survey indicates that uniform translation is still the customary barometer for interface tracking methods. Solid-body rotation tests accompany translation tests in many analysis. An acceptable tracking method must translate and rotate fluid bodies without significant distortion or degradation of fluid interfaces. Mass should certainly be conserved rigorously in these cases. Translation and solid body rotation, however, enable only a *minimal* assessment of interface tracking algorithm integrity and capability because topology change is absent. Additional tests involving flows with nonuniform vorticity should be considered for a more complete assessment.

We therefore consider two 2-D test problems that more sufficiently challenge algorithm capabilities, provide meaningful metrics for measurement of algorithm performance, and are easy to implement. These problems, characterized by flows having non-constant vorticity, were introduced in [453] as proposed metrics for any method designed to track interfaces undergoing gross topology change. Beside inducing topology change, the test problems are representative of interfacial flows in real physical systems, e.g., instabilities such as the Rayleigh-Taylor, and Kelvin-Helmholtz instabilities, where sharp gradients in fluid properties and instabilities lead to vortical flow.

Our test problems possess vortical flows that stretch and potentially tear any interfaces carried within the flow. The first problem contains a single vortex that will spin fluid elements, stretching them into a filament that spirals toward the vortex center. The flow field is taken from the "vortex-in-a-box" problem introduced in [45, 166]. The second problem has a flow field characterized by sixteen vortices as introduced in [500]. This flow field causes fluid elements to undergo large topological change. In the converged limit, fluid elements will not tear, instead forming thin filaments. The flow field in both problems is solenoidal and is given cosinusoidal time-dependence following Leveque [340]. According to [340], the single vortex and deformation velocity fields can be multiplied by $\cos(\pi t/T)$, giving a flow that time-reverses (returns to its initial state) at $t = T$. In most of our tests we choose a period $T = 2$, hence the circular body will undergo large deformations until the first half period ($t = 1$), whereupon the flow will reverse, returning the circle to its initial undeformed state at the full period ($t = 2$). Error measurements are performed on the differences in data observed between $t = 0$ and $t = T$. These differences should ideally be zero, as the $t = 0$ and $t = T$ states should be identical.

All test problems have identical initial conditions: a circle (radius 0.15) is centered at $(0.50, 0.75)$ in a unit square computational domain. This setup is by no means necessary, other shapes including multiple shapes should be useful as well. The domain is partitioned with either 32^2, 64^2, or 128^2 orthogonal, uniform cells. All boundaries are periodic. A scalar field is initialized to unity and zero inside and outside the circle, respectively. For those cells containing the circular interface, the scalar field is set to a value between zero and one, in proportion to the cell volume truncated by the circle. This field represents a characteristic (or color) function, which for our purposes is the fluid volume fraction for a circular fluid body, i.e., the volume fraction is 100% inside the circle and 0% outside. For the following tests, unless otherwise stated, we employ a CFL number of one (based on the maximum velocity in the domain) and use Pilliod's method [425] for estimating the interface normal $\mathbf{n}$.

18.5.1 Single Vortex

A single vortex is imposed with a velocity field defined by the stream function [45],

$$\psi = \frac{1}{\pi} \sin^2(\pi x) \sin^2(\pi y), \tag{18.17}$$

where $u = -\partial\psi/\partial y$ and $v = \partial\psi/\partial x$. This solenoidal velocity field, which will deform bodies and promote topology changes, is shown in Fig. 18.20c. When the circular fluid body is placed in this field, it stretches and spirals inward toward the center of the domain, wrapping around the center approximately two and a half times by $t = 3$ (Fig. 18.22). The marker particles are initialized in a uniformly-spaced 4×4 array in each cell falling completely inside the circular fluid body. The 4×4 particle array is truncated in cells containing the circular interface and is absent in cells lying outside the circular interface. A more efficient approach to producing this solution would simply place particles on boundary of the shape [408].

The velocity field stretches and (non-ideally) eventually tears the initially circular fluid body as it becomes progressively entrained by the vortex. The entrainment is manifested as a long, thin fluid filament spiraling inward toward the vortex center. The under-resolved behavior of the PLIC method on the single-vortex problem is illustrated in Fig. 18.22. Here the solution becomes poor when interface topology is not resolved, i.e., as exhibited by single filament breaking into a series of fluid clumps that are each supportable by the reconstruction method. By $t = 3.0$ (Fig. 18.22d), the body has fragmented into numerous pieces. Despite the breakup, however, solution convergence does occur under grid refinement.

This behavior is reasonable and expected, given the assumptions inherent in the reconstruction, namely a piecewise linear interface approximation constrained by mass conservation. The breakup exhibited in Fig. 18.22 can be interpreted as an application of "numerical surface tension" along interfaces that are resolved inadequately. High curvature regions are those unresolvable regions having interfaces with a radii of curvature less than roughly a mesh spacing. As shown in [98] it may be advisable to adaptively stop using the

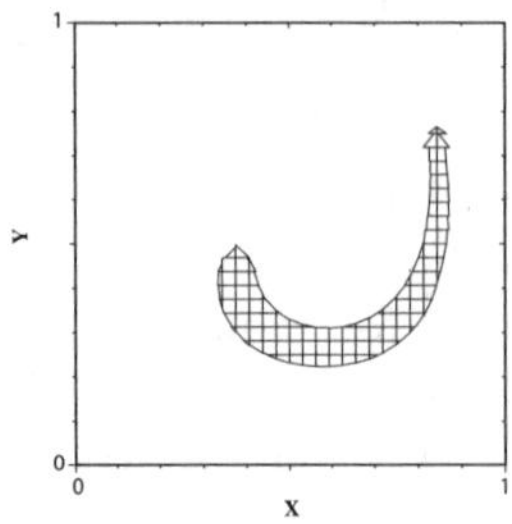

(a) $t = 0.75$.

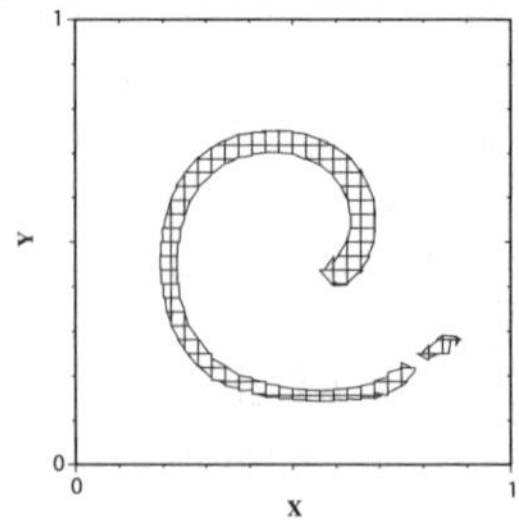

(b) $t = 1.50$.

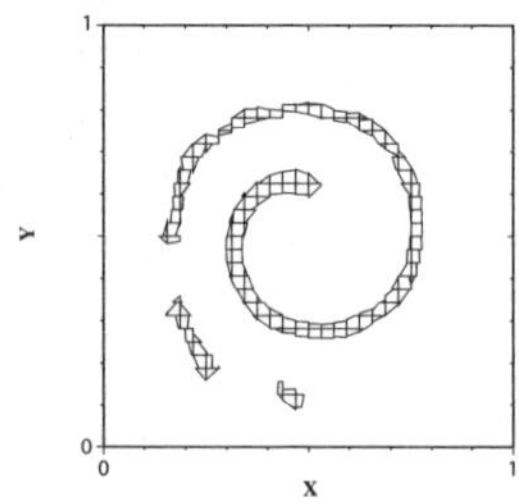

(c) $t = 2$.

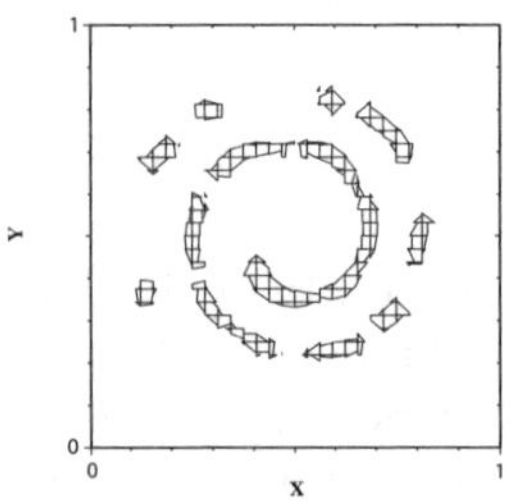

(d) $t = 3.00$.

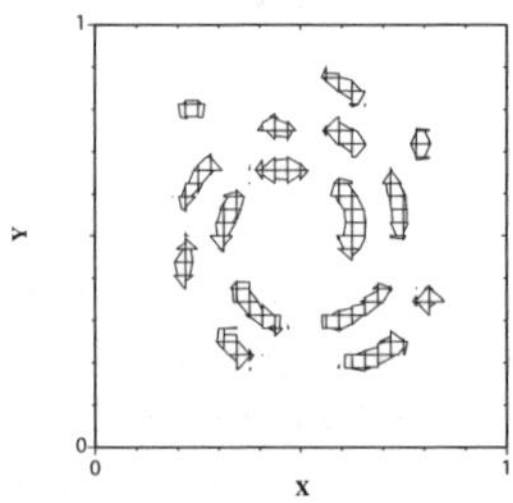

(e) $t = 3.75$.

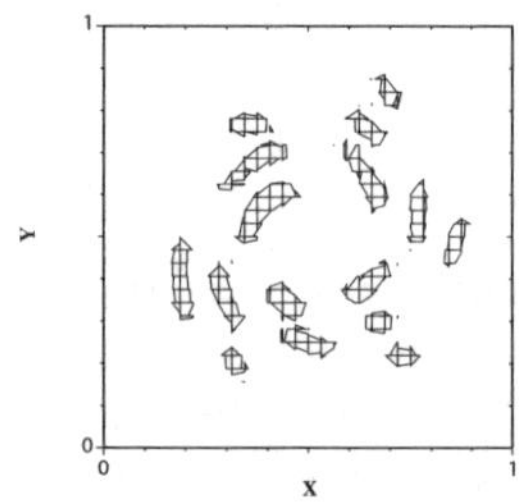

(f) $t = 4.50$.

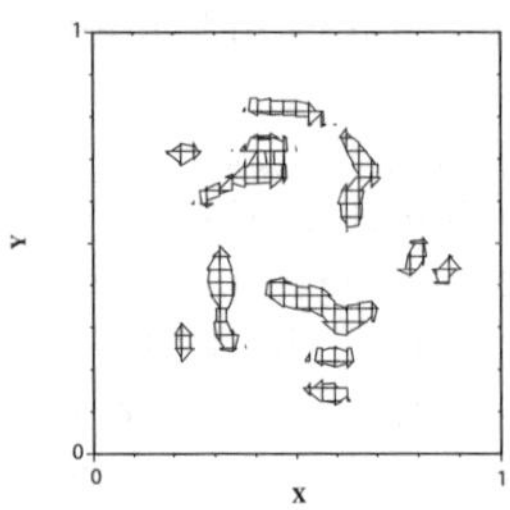

(g) $t = 5.25$.

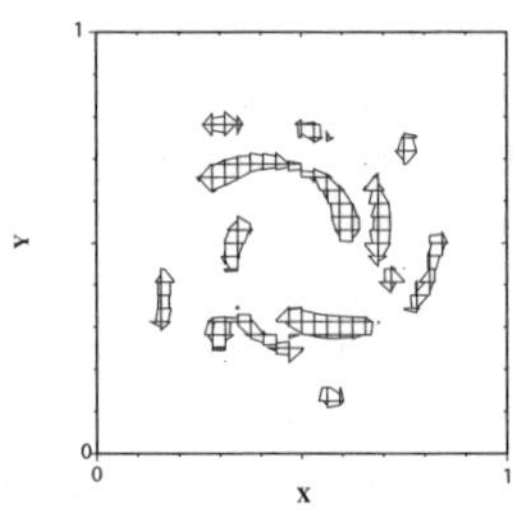

(h) $t = 6.00$.

Fig. 18.22. Results for long time integration of a circular fluid body placed in the single-vortex flow field on a 32^2 grid. The initial condition is a circle centered at $(0.5, 0.75)$ with a radius of 0.15.

compact interface description when the flow becomes under-resolved. This will diminish numerical surface tension significantly although diffusive effects take their place. The piecewise linear interface approximation immediately flattens these regions, effectively applying numerical surface tension. Thin filament regions can also be the recipients of numerical surface tension, because poor linear reconstructions occur in these regions from inaccurate interface normal estimations. Numerical surface tension in these high curvature and thin filament regions can be easily reduced (well below physical levels) with increased refinement.

By integrating to late times, it is possible to observe the behavior of a volume tracking method under rather extreme circumstances, whereby filaments become thinner than is supportable by the computational mesh. Convergence results are obtained by time-reversing the flow using Leveque's cosine term. As the reversal period T becomes longer, the fluid body evolves further away from its initial circular configuration, hence it must undergo increasingly complicated topological change to reassemble properly at $t = T$ (Fig. 18.23).

Convergence results for the single vortex velocity field indicate that the method is remarkably robust and resilient. The method exhibits second-order convergence, even for long periods ($T = 8$) after appreciable interface tearing and topological change has occurred. This is indicated by the L_1 error norms and convergence results shown in Table 18.6 for three different reversal periods. Fig. 18.23 illustrates that solution errors (roughly a measure of phase error) are more evident for longer reversal periods. The solution quality, however, increases remarkably as the grid is refined (shown here for a 32^2 grid). Convergence is aided by the regularity of the velocity field, but the improvement with grid refinement is quite profound.

Table 18.6. L_1 error norms and convergence rates for a circular fluid body placed in the time-reversed, single-vortex flow field.

Grid	Error ($T = 0.5$)	Order ($T = 0.5$)	Error ($T = 2.0$)	Order ($T = 2.0$)	Error ($T = 8.0$)	Order ($T = 8.0$)
32^2	7.29×10^{-4}		2.36×10^{-3}		4.78×10^{-2}	
		2.36		2.01		2.78
64^2	1.42×10^{-4}		5.85×10^{-4}		6.96×10^{-3}	
		1.86		2.16		2.27
128^2	3.90×10^{-5}		1.31×10^{-4}		1.44×10^{-3}	

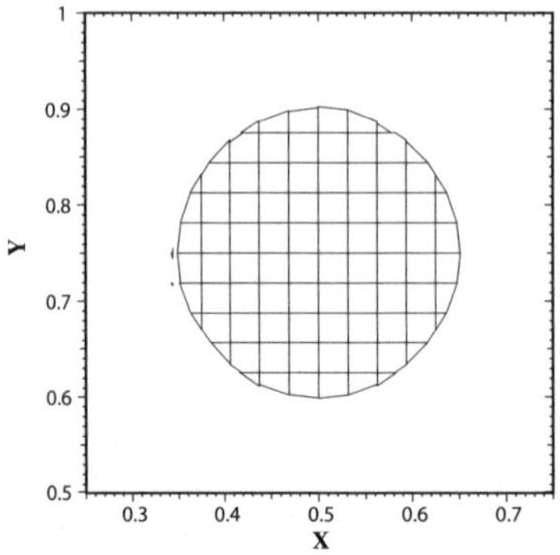

(a) Reconstructed interfaces at $t = T$ for $T = 0.5$.

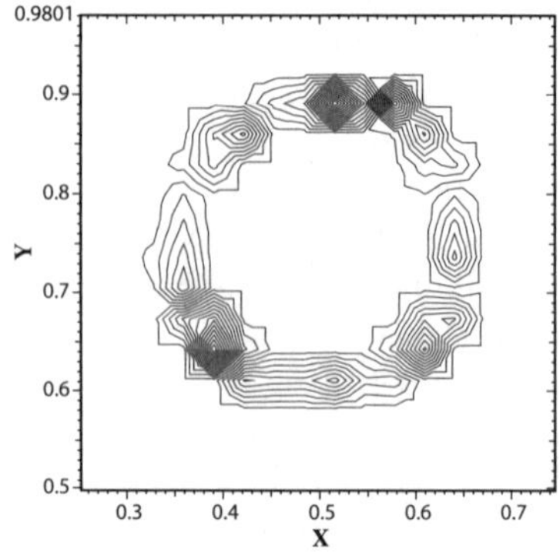

(b) L_1 error contours at $t = T$ for $T = 0.5$.

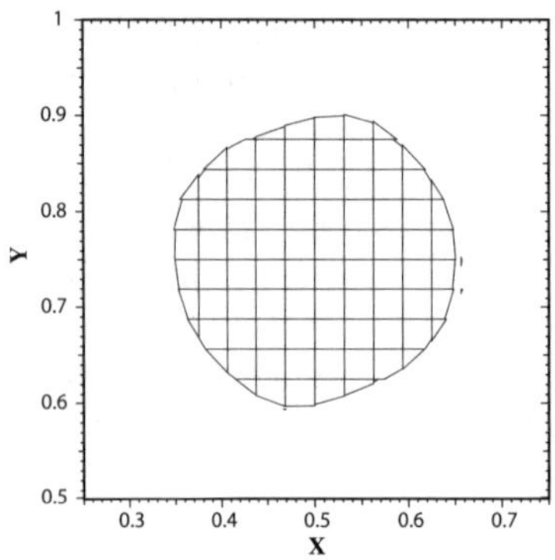

(c) Reconstructed interfaces at $t = T$ for $T = 2.0$.

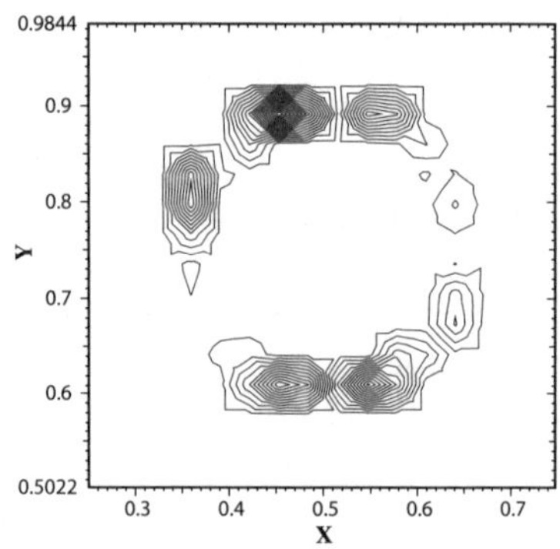

(d) L_1 error contours at $t = T$ for $T = 2.0$.

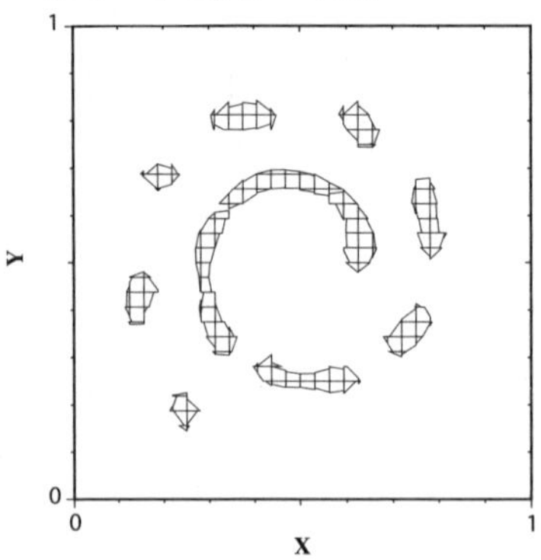

(e) Reconstructed interfaces at $t = T/2$ for $T = 8.0$.

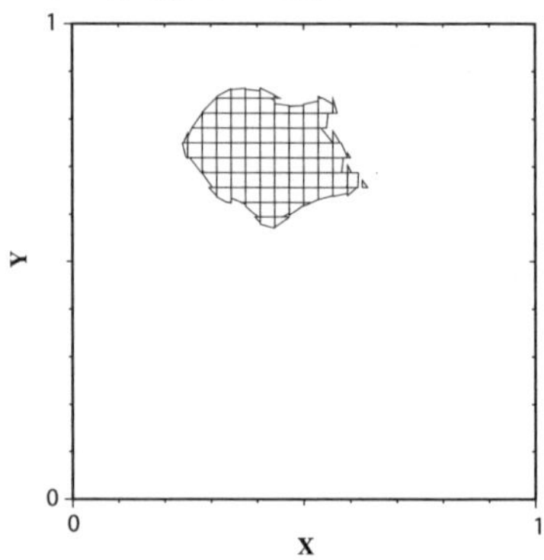

(f) Reconstructed interfaces at $t = T$ for $T = 8.0$.

Fig. 18.23. Results for a circular fluid body placed in the time-reversed, single-vortex flow field on a 32^2 grid.

Table 18.7. L_1 error norms and convergence rates for a circular fluid body placed in the time-reversed, deformation flow field.

Grid	Error ($T = 1.0$)	Order ($T = 1.0$)	Error ($T = 2.0$)	Order ($T = 2.0$)	Error ($T = 4.0$)	Order ($T = 4.0$)
32^2	5.20×10^{-3}		1.96×10^{-2}		4.68×10^{-2}	
		1.62		0.81		1.52
64^2	1.69×10^{-3}		1.12×10^{-2}		1.63×10^{-2}	
		1.95		0.91		0.84
128^2	4.36×10^{-4}		5.95×10^{-3}		9.08×10^{-2}	

18.5.2 Deformation Field

A complex velocity field given by the stream function [500]

$$\psi = \frac{1}{4\pi} \sin\left(4\pi\left(x + 1/2\right)\right) \cos\left(4\pi\left(y + 1/2\right)\right) , \tag{18.18}$$

induces even radical deformation and topology change of fluid bodies, providing a more stringent test than the field given by (18.17). The solenoidal velocity field corresponding to this stream function is shown in Fig. 18.20d.

Time reversal is again used to obtain quantitative results. The initial position of the circular body falls directly between two vortices, hence we find that the results fail to converge for long T. This is evident in Table 18.7, where second order convergence is realized only for $T = 1.0$. For longer periods, the results converge to only first order. This is indicative of the method being above the mesh spacing required for theoretical convergence for the given problem. The lack of qualitative similarity between the solutions obtained on the two finest grids at $T = 4$ (Fig. 18.24) is evidence for the lower convergence.

In Fig. 18.25 the deformation velocity field shows its ability to tear apart the circular body. Unfortunately this is not the true ideal solution, but rather a consequence of the lack of sufficient mesh resolution. Despite the severity of the interface deformation and topology change, mass conservation is maintained and the solution bears a qualitative resemblance to the true solution. Even with a coarse (32^2) resolution, the solution is not high quality, but its qualitative correctness exhibits the robustness one seeks in an interface tracking method.

Finally, we note that Enright and Fedkiw [181] have generalized each of these problems to three dimensions. Furthermore, they also added parti-

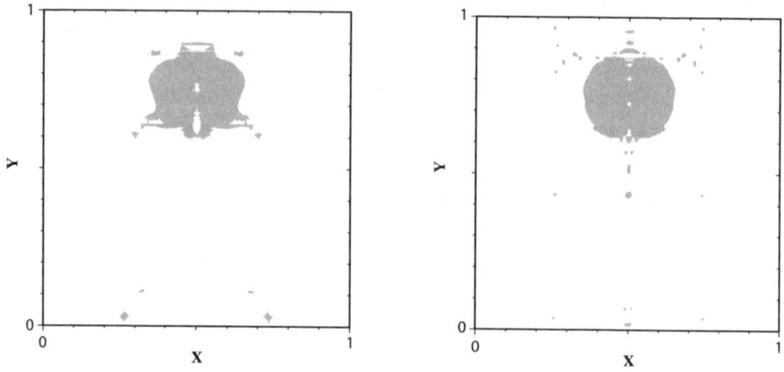

(a) 64^2 grid. (b) 128^2 grid.

Fig. 18.24. Results for a circular fluid body placed in the time-reversed, deformation flow field at $t = T$ for $T = 4$.

cles at the boundary of a level set algorithm to improve its accuracy, mass conservation and general ability to handle this class of flow.

We conclude our examples with a brief digression into operator-split, time-integration methods. Previously, results presented to this point are obtained with the unsplit time-integration scheme detailed in Section 18.4.4. When an operator-split, time-integration scheme is used for these tests, the observed convergence rates and error norms are similar to the presented unsplit results.

Operator-split, time-integration methods, however, are inferior for two important reasons: efficiency and symmetry preservation. Because volume tracking methods are dominated by the cost of reconstructing the interface, an operator-split method is roughly twice as expensive as an unsplit method (in 2-D and three times as expensive in 3-D) because one extra reconstruction is required. Operator-splitting also fails to maintain symmetry, even when sweep directions are alternated although results are often good enough for most applications. A worst case situation is evident in the results of Fig. 18.26, which illustrate that operator-split time-integration solutions are lower quality than the unsplit scheme (shown in Fig. 18.27). For these reasons we prefer methods based on unsplit time-integration schemes.

The results presented in this section are evidence for the topological changes a volume tracking method must manage while tracking a fluid body placed in the vortex and deformation flow fields. For performance of other tracking methods on these same problems, see the results in [453]. The methods tested in [453] are those based on particles, level sets [522], and standard upwind continuum advection schemes such as fourth-order PPM [120].

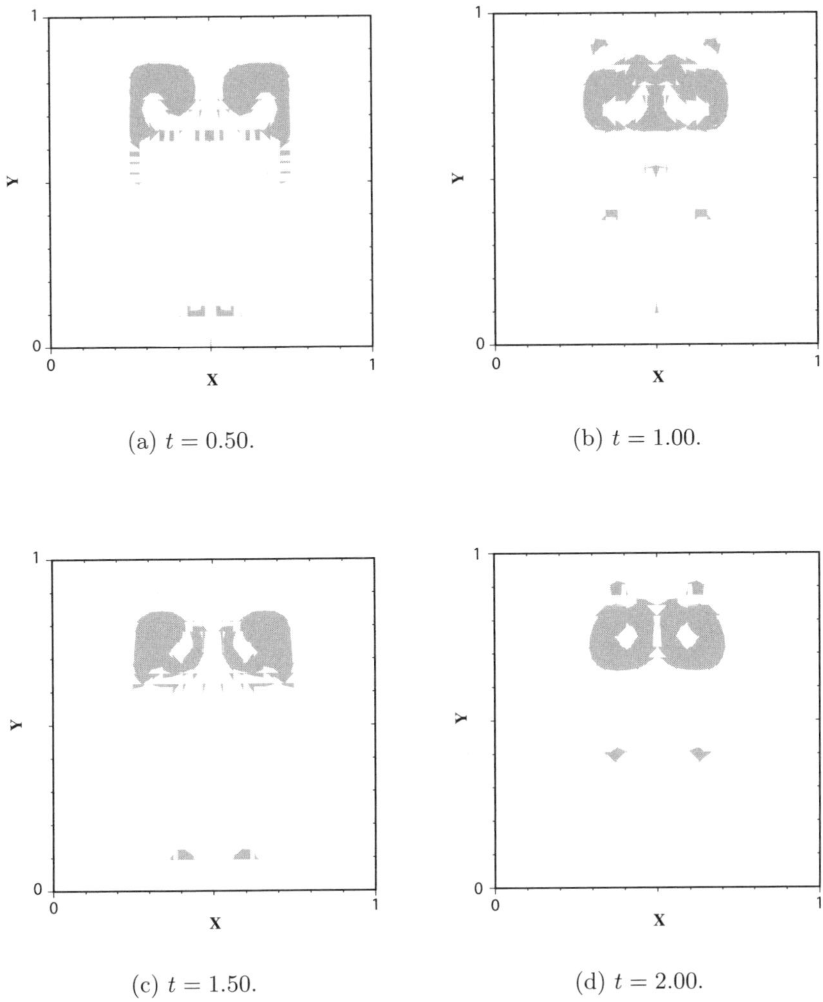

Fig. 18.25. Results for a circular fluid body placed in the deformation velocity field on a 32^2 grid.

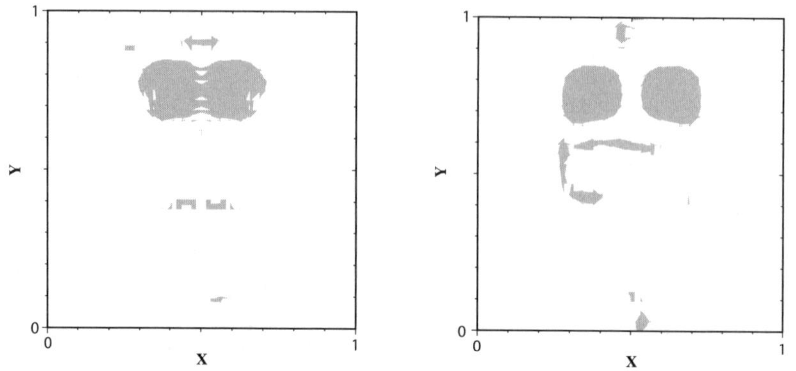

(a) Time-reversed flow. (b) Non-time-reversed flow.

Fig. 18.26. Results at $t = 4.0$ for a circular fluid body placed in the deformation flow field using an operator-split time-integration method on a 32^2 grid. In (a) the flow is time-reversed with $T = 4$. This should be compared with the unsplit results in Fig. 18.27.

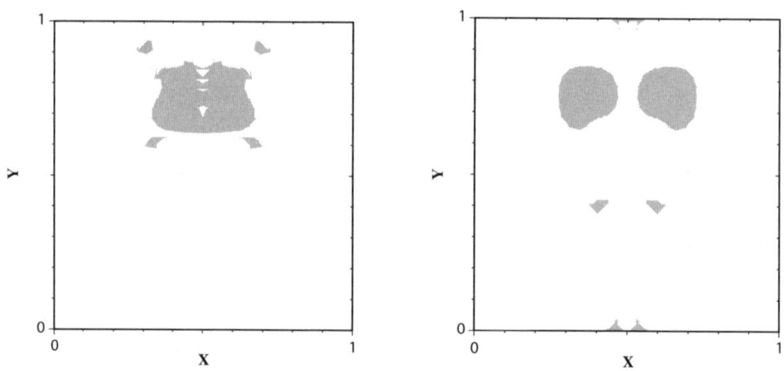

(a) Time-reversed flow. (b) Non-time-reversed flow.

Fig. 18.27. Results at $t = 4.0$ for a circular fluid body placed in the deformation flow field using an unsplit time-integration method on a 32^2 grid. In (a) the flow is time-reversed with $T = 4$. This should be compared with the operator split results in Fig. 18.26.

19. High-Resolution Methods and Turbulent Flow Computation

The development of highly accurate and efficient methods for the computation of turbulent flows is motivated by the broad spectrum of applications in science and engineering in which turbulence appears. We will briefly cover the spectrum of different approaches for modeling turbulent flows before focusing on the role of modern high-resolution methods in this enterprise. These methods are unique in their ability to integrate flows stably without appealing to strictly dissipative models for the sake of numerical stability. Because of this there is a necessary overlap between the classical modeling of turbulence and its computation through high-resolution methods.

In addition, high-resolution methods have the ability to compute flows that are extremely complex and difficult in practice with classical modelling approaches. In other words, they allow under-resolved flows (with respect to grid resolution) to be computed reliably with physically realizable results. High-resolution methods are used to simulate a broad variety of physical processes including unstable flows that are highly vortical leading to turbulence and the mixing of materials.

19.1 Physical Considerations

The nonlinear interactions induced by the hyperbolic part of the fluid dynamic equations (i.e., the nonlinear transport terms that are responsible for turbulence) naturally cause scale-to-scale transfer of energy, resulting ultimately in entropy production by viscosity at small length scales. Shock waves are the prototypical example of this. At large scales in the inertial range, the flow behaves (nearly) independently of viscosity. In addition to the viscous dissipation, important energy transfers known as *backscatter* move energy from small scales toward larger scales.

For example, rarefactions produce transfer of energy from small to large scales, purely through the hyperbolic terms in the governing equations. This process is not represented in those turbulence models that are purely dissipative, but found in models exhibiting self-similarity. Conversely, this effect is naturally present in the hyperbolic terms which display a natural scale invariance. Indeed, self-similarity is embedded in many high-resolution methods through the use of Riemann solvers (either exact or approximate in nature).

The Riemann solution, or equivalent dissipation mechanism, provides much of the dissipation used to stabilize the numerical method.

It is the dominance of the transport (hyperbolic) terms that leads to turbulence. As hyperbolic terms become more important, the problems become more sensitive to initial conditions in the presence of hydrodynamic instability. In compressible flows, the scale-changing phenomena cause wave steepening and shock waves. As an example of the nature of scale changing phenomena consider the inviscid Burgers' equation,

$$\frac{\partial U}{\partial t} + U\frac{\partial U}{\partial x} = 0 \rightarrow \frac{\partial U}{\partial t} + \frac{\partial}{\partial x}\left(\frac{1}{2}U^2\right) = 0 \,,$$

and derive an equation for the evolution of the gradient. Under the assumption of sufficient smoothness (indeed this equation is used to determine when smoothness breaks down) we obtain,

$$\frac{\partial}{\partial t}\left(\frac{\partial U}{\partial x}\right) + U\frac{\partial^2 U}{\partial x^2} + \left(\frac{\partial U}{\partial x}\right)^2 = 0 \,.$$

The last term controls the breakdown of smoothness when the velocity gradient is negative, i.e., $\partial U/\partial x < 0$, that it is compressive. In this case a shock will form, otherwise the flow expands in a rarefaction. Along characteristics, the solutions behave like a Ricatti ordinary differential equation. Analytically, this equation has solutions exhibiting a finite time singularity, thus indicating the blow-up of gradients in the solutions of the equations. The shock formation pushes information to small scales where dissipation asymptotically operates to destroy it thus creating an increase in entropy. In rarefactions the information moves to larger scales where smoothness is not threatened. Modern methods act appropriately in each circumstance providing dissipation where it is necessary.

High-resolution schemes combine the action of limiters to judiciously allow high accuracy while still relying upon a combination of conservation form [321] and entropy production to produce a unique (physically realizable) weak solution. As shown in [369], the property of conservation naturally produces the equations for the evolution of a control volume of fluid. It is the evolution equation for the finite volume of fluid that then is solved by the numerical method. Dissipation acts to regularize the flow, thereby allowing shock propagation to proceed physically even while it is unresolved on the computational mesh. The key assumption is that the desired goal is to propagate the discontinuity on a fixed number of discrete mesh cells (first suggested by Richtmyer in 1948 [445]).

In this chapter, we use the assumptions of large eddy simulation (LES) to explore the properties of high-resolution methods in relation to turbulence modeling. Some of the issues discussed here are similarly applicable in the context of unsteady RANS simulations. Here, in keeping with LES we assume that the numerical solution resolves the energy-containing range of the flow, i.e., that the grid scale is in the inertial range. In the following, we will develop a description of the effective model defined by high-resolution methods for hyperbolic PDEs. This will make use of the modified equations [592] describing the action of limiters in terms of nonlinear differential terms. As an archetypal example, the modified equation can be used to show that upwind differencing has a leading order error that is dissipative in nature, i.e., proportional to U_{xx}. For high-resolution schemes, this will result in a description of nonlinear eddy viscosity, solution adaptivity and scale-similarity arising naturally from a broad class of numerical methods.

Two issues need to be considered for the advancement of this concept: empirical evidence from the successful use of the idea and some sort of theoretical structure to build upon. Empirical evidence for this idea has been building for more than a decade as summarized in [150, 403]. There several issues are discussed: the character of an ideal subgrid model, and some fortunate circumstances for high-resolution schemes arising from the physics encoded into the numerical method. High-resolution schemes have a number of convenient properties that make them attractive as general tools for modeling turbulence. This of course assumes that there is some accuracy and fidelity in such a model. We will develop some structural evidence that strong connections exist with current practice in LES modeling. Next, we detail some theoretical scaling arguments linking conceptually the physics of shock waves (implicitly built into the numerical methods focused upon here) and turbulence (explicitly built into turbulence modeling).

As motivation for considering the utility of shock-capturing methods for turbulence, we consider the following similarity among several theoretical models. During the early 1940s, similar forms of dissipation were derived on both sides of the Atlantic. Kolmogorov [299] defined a dissipation of kinetic energy that was independent of the coefficient of viscosity in the limit of infinite Reynolds number; this theory was refined in [300]. In this form, the average time-rate-of-change of dissipation of kinetic energy, K, is given as

$$\langle K_t \rangle \, \mathcal{L} = \frac{5}{4} \left\langle (\Delta u)^3 \right\rangle . \tag{19.1}$$

In homogeneous, isotropic turbulence, this term is proportional to the average normal velocity difference at a length scale, $\mathcal{L}$, cubed. A length scale can be found using a velocity and an appropriate time interval. Note that this theory is analytic and independent of viscosity (although subtle arguments about viscous corrections at large, but finite Reynolds number continues to persist). Moreover, this theory provides a basis for the functional form of nonlinear eddy viscosity, i.e., [498]; this is discussed in more detail later.

In 1942, Bethe [282] derived the dissipation rate due to the passage of a shock wave (for a modern perspective on this relation see [383]). This rate depends on the curvature of the isentrope, $\mathcal{G}$, and on the cube of the jump in dependent variables across the shock as well as the sound speed, c:

$$T\Delta S = \frac{\mathcal{G}c^2}{6} \left(\frac{\Delta V}{V} \right)^3 . \tag{19.2}$$

Bethe defined this jump in terms of specific volume, V, but this can be restated in terms of velocity by applying the Rankine-Hugoniot conditions, $c_s \Delta V = -\Delta u$, where c_s is the shock speed. Both of these results are analytic. In each of these cases, the flow experiences an intrinsic asymmetry since the dissipative forces arise predominantly where velocity gradients are negative, i.e., compressive. As with Kolmogorov's relation, this only depends on the large scales in the flow and the small scale details of the dissipation are immaterial to the end result.

For Burgers' equation a similar result may be obtained [236],

$$\langle K_t \rangle \, \mathcal{L} = \frac{1}{12} \left\langle (\Delta u)^3 \right\rangle . \tag{19.3}$$

Again, this is an analytic result through the application of integration by parts, and the shock jump conditions. Next, we display the congruence of high-resolution numerical methods with this theory.

In a sense (19.3) is an entropy condition for "Burgers' turbulence" describing the minimum integral amount of inviscid dissipation for a physically meaningful solution. This dissipation is produced at the shocks and is a consequence of and proportional to the jump in dependent variables. Eyink [183] studied a conjecture by Kraichnan that the dissipation of kinetic energy as defined by the Kolmogorov similarity is both local as well as integral in nature (by definition, the shock dissipation is local).

These regularizations are the essence of the physical conditions that numerical methods must reproduce correctly. It is this idea, viz., the existence of a finite rate of dissipation independent of viscosity with an inherently local nature, that numerical methods are designed to reproduce. Modern high-resolution methods have an effective subgrid model that is inherently local. In addition, the algebraic form of the high-resolution methods has a great deal in common with scale-similarity forms of LES subgrid models coupled with a nonlinear eddy viscosity. This creates a coherent tie between the modern high-resolution, shock-capturing methods and LES subgrid models. We will build this deeper connection in the following pages.

One can show that control volume differencing can be viewed as a form of implicit spatial filtering. The consequence of control volume differencing is that the cell values are the cell average values for quantities, thus filtering the point values. Furthermore, control volume differencing naturally produces terms that are analogous to scale-similarity subgrid models. This analogy is predicated upon the structure of the modified equations for this class of methods [369].

In this chapter, we expand on this idea and show how high-resolution methods contain an implicit subgrid model that can be viewed as a dynamic mixed self-similarity model. We will show that methods can have much more in common with LES models than simply an elaborate nonlinear viscosity. It is the common (mis)perception that the implicit numerical viscosity is all that high-resolution methods have to offer. Indeed various elements of other types of subgrid models are found when looking at these methods through the lens of nonlinear truncation error. The dynamic aspect is associated with limiters whose effects vanish in resolved flows. The limiters are essential in modifying second-order dissipative terms into high-order nonlinear viscosity so commonly associated with this class of methods. Self-similarity comes from several sources, the control volume differencing and the slope limited interpolation (or similar concepts), which dynamically changes the nature of the high-order flux used locally.

19.2 Survey of Theory and Models

In CFD, there are three principal approaches which are used to compute turbulent flows. Ranked in order of computing expense, these are the Reynolds-Averaged Navier-Stokes equations (RANS), the Large Eddy Simulation (LES) and the Direct Numerical Simulation (DNS).

In the context of the RANS approach, the task of a turbulence model is to provide the RANS equations with closure relations for the Reynolds stresses $\overline{u_i' u_j'}$ $(i, j = 1, 2, 3)$ (see also Chap. 3). "Closure" identifies the process by which the stresses are related to known or determinable quantities: geometric parameters, flow scales and strains. The strains play an especially prominent role in the closure process, for they are the primary agency by which turbulence is generated and sustained. An indication of the importance attached to this linkage is provided by the Boussinesq stress-strain relationship

$$-\rho\overline{u_i' u_j'} = \mu_t\Big(\frac{\partial u_i}{\partial x_j} + \frac{\partial u_j}{\partial x_i} - \frac{2}{3}\frac{\partial u_k}{\partial x_k}\delta_{ij}\Big) - \frac{2}{3}\rho k \delta_{ij} , \tag{19.4}$$

which after we apply the incompressibility condition becomes

$$-\rho\overline{u_i' u_j'} = \mu_t\Big(\frac{\partial u_i}{\partial x_j} + \frac{\partial u_j}{\partial x_i}\Big) - \frac{2}{3}\rho k \delta_{ij} , \tag{19.5}$$

where μ_t is the eddy viscosity, k is the (kinematic) turbulence energy and $\delta_{ij} = 1$ for $i = j$ and 0 otherwise. The last term on the RHS of (19.5) is required to ensure that the normal stresses sum up to $2k$ in zero strain. On one hand, (19.5) expresses the basic fact that the level of turbulence mixing is strongly associated with straining. On the other hand, turbulence generally reacts slowly to external disturbances (including straining), a fact inconsistent with (19.5).

The interactions between stresses and strains can be illuminated by the Reynolds-stress equations (3.117) (as discussed in Chap. 3) that govern the evolution of the Reynolds stresses. If one looks at two simple turbulent flow situations such as the simple shear, in which the only major strain is $\partial u/\partial y$, and homogeneous compression, the following fundamental issues are revealed: In simple shear, the only normal stress generated by the shear strain is $\overline{(u')^2}$.

The other normal stresses are only finite because they receive a proportion of the turbulence energy contained in $\overline{(u')^2}$ via the redistributive pressure-strain process, which tends to steer turbulence toward isotropy, and does not contribute to the turbulence energy. This relationship is the reason that sheared turbulence is strongly anisotropic. Further, in the case of homogeneous compression the shear stress is generated by an interaction between the cross-flow normal stress and the shear strain. This highlights the need to determine accurately the turbulence anisotropy.

Because of the assumption that turbulence is isotropic in the smallest scales (thought to be, but not proven to be a good approximation at high Reynolds numbers, away from walls), there is no dissipative mechanism for the shear stress. However, the pressure-strain term provides a mechanism for that stress, which counteracts generation with the isotropization process. In other words, simple stress-balance considerations (analogous to Mohr's circle in Solid Mechanics) serve to show that the shear stress must decline and eventually vanish as the normal stresses approach a single value.

There are three principal families of models currently used in RANS computations (recent reviews can be found in [169, 334, 335]): linear eddy-viscosity models (LEVM); nonlinear eddy-viscosity models (NLEVM); and Reynolds-stress models (RSM). Some models do not fall clearly into any one of the above categories, straddling two categories or containing elements from more than one category. Within any one of the above major categories, there are dozens of variants, and the LEVM category, being the simplest, contains several sub-categories and is especially heavily populated with model variations, many differing from other forms by the inclusion of minor (though sometimes very influential) "correction terms", or different functional forms of model coefficient or even through slight differences in the numerical values of model constants.

To a considerable degree, this proliferation reflects a trend to adopt or adhere to simple (too simple) turbulence models for the modelling task at hand and then to add "patches" so as to "cure" specific ills for specific sets of conditions. Other not unimportant contributory factors are insufficiently careful and excessively narrow validation, yielding misleading statements on the predictive capabilities of existing models.

RANS computations are extensively used in practical engineering computations, especially for predicting steady-state solutions. Unsteadiness introduces a fundamental and profound uncertainty into the RANS approach: Reynolds-averaging, whether ensemble or time-based, assumes that the flow is statistically steady. At the very least, the time-scale associated with the organized unsteady motion must be substantially larger than the time scale of turbulent motion (see recent studies of turbulence models in unsteady aerodynamic flows [30, 31]). In other words, the two scales must be well separated. This condition may be satisfied in low-frequency unsteady flows, but the majority of turbulent flows does not fall into this category. Closure of the phase-averaged correlations is identical or very similar to that adopted for the conventional averaged correlations, and this inevitably leads to models which are formally identical to their steady counterparts. RANS-based modeling has an inherently empirical aspect as well as the issue of ensemble averaging built into many models.

In stark contrast DNS is an inherently numerical and intensively computational approach. The DNS approach has provided useful information in relation to the turbulent flow structure at relatively low Reynolds numbers and for simple geometries. The use of DNS for studying unsteady flows at high Reynolds numbers ($10^5 - 10^9$) is well beyond foreseeable computing power. For example, to compute the flow around an aircraft for one second of flight time, using a supercomputer of 10^{12} Flops, it is required several thousand years and 10^{16} grid points [291]. In addition to the computing power constraints, it is not certain that DNS provides sufficiently resolved results. To verify this, careful grid convergence studies would be required. This is not typically done in practice and one could argue that this is not feasible to be done due to inadequate computing power.

Large Eddy Simulation (LES) beckons in the distance as an alternative approach to RANS modeling, but poses substantial challenges in high-Re near-wall flows, especially in the presence of separation from gently curved surfaces, where resolution and thus computing-cost issues are critical. LES is more computationally challenging than RANS, but less than DNS. LES is certainly likely to be a useful (though expensive) approach in flows that are not strongly affected by viscous near-wall features. Various models for the subgrid scale stresses have been developed [209, 387, 499, 498, 345] and the quest for better representation of the SGS through an explicit model is an ongoing effort. Some additional remarks for the LES equations are made below.

The LES equations are derived (Sect. 3.8) on the basis of the assumption that filtering and differentiation commute [213], i.e., $\overline{\partial f/\partial x} = \partial \bar{f}/\partial x$, where the "overline" denotes the filtering operation. The above is satisfied if the filter width is constant but not otherwise [212]. By keeping constant the filter width, additional terms at the boundaries appear during the filtering procedure. The velocity terms vanish at solid boundaries due to the implementation of the no-slip boundary condition. However, the same does not occur for the pressure and viscous terms. To circumvent this difficulty one can use a variable filter width which separates the turbulent eddies into large-scale and small-scale eddies. The former are problem dependent whereas the latter may be modeled by the SGS model that represents the Kolmogorov cascade at small scales, which is assumed to be independent of the large scale field.

Such a separation of scales may be possible away from the solid boundaries, but cannot be applied close to the boundaries where turbulence manifests in the form of coherent structures which cannot be described by eddy-viscosity modeling. As a result, the eddies close to the boundary would still need to be resolved. Application of the variable filter width will eliminate the boundary terms appearing in the filtering operation of the derivatives but the LES equations are no longer valid [212]. The commutation errors can be removed if correction terms that account for these errors are introduced in the LES equations. This will, however, raise the order of the highest derivatives in the equations and at present there are no available methods to deal with this complexity. Relevant work to the above is the derivation of special filters that can eliminate the correction terms [578, 583]. However, these filters do not satisfy positivity for the turbulent kinetic energy [212].

The Kolmogorov spectrum [299, 300] describes how the energy density of turbulent structures decreases rapidly with increasing the wave number, where the Kolmogorov scale is the average scale at which the viscous dissipation dominates the inertial flow of the fluid. The downward transfer of energy from large to small scales is called the turbulent cascade process. The latter stops at the Kolmogorov scale, where an eddy is so small that it diffuses rapidly. Previous computations, experiments and theoretical analysis have shown that the physics of the turbulent cascade is controlled by the macroscopic scales of the flow and the process of dissipation of this energy due to molecular viscosity takes place primarily at scales considerably larger than the Kolmogorov scale. Another important issue is that the energy transfer is dominated by local interactions.

In other words, the energy does not skip from the large to the small scales, but the energy extraction from a given scale occurs as a result of interactions with eddies no more than an order of magnitude smaller. The above indicates that accurate simulation of turbulent flows can possibly be performed at scales much larger than the Kolmogorov scale.

19.3 Relation of High-Resolution Methods and Flow Physics

High-resolution methods are more than just numerical methods, the physics of flow is embedded at the core of the techniques. The nonlinearity that sets these methods apart from classical techniques provides guarantee of numerical stability and physical results through imposing physically realizable behavior on the results. It is this deep and abiding connection of numerical techniques to the physics that ultimately leads to their utility in computing a wide variety of flows. Below, we examine this connection of the numerical method to the physics important for turbulent flows. The mathematical description of weak solutions to the hydrodynamic equations forms a bridge between the physics and numerical techniques. This will lead to a careful study of the effective physical model that arises through the integration of the equations of hydrodynamics with high-resolution methods.

19.3.1 Numerical Considerations

For more than a decade there has been an increasing amount of evidence that high-resolution numerical methods for hyperbolic partial differential equations have an embedded (or "implicit") turbulence model [69, 148, 155, 201, 202, 349, 369, 371, 403, 429, 612, 613, 614]. We have discussed this class of methods in Chaps. 13-17. Here, we introduce this general class of methods and outline the basic structure of high-resolution methods as an effective turbulence model in the context of LES. This discussion is an extension of the MILES concept introduced by Boris, where monotone numerical algorithms are used for LES (MILES is an acronym for monotone integrated LES). We will discuss how the implicit modeling (henceforth labeled as implicit LES or ILES) includes elements of nonlinear eddy viscosity, scale-similarity and an effective dynamic model. In addition, we give examples of both success and failures with currently available methods and examine the effects of the embedded modeling in contrast to widely used explicit subgrid scale (SGS) models.

Philosophically, this approach differs greatly from the standard one because unresolved subgrid effects are not explicitly modeled. Classically, the effects of unresolved scales are modeled in their entirety. In that approach the numerical methods should be unobtrusive as possible. This means that approximation effects (i.e., truncation errors) should be as small as possible. Since dissipative effects are one of the preeminent effects from turbulent motion, requirements on the dissipative errors in the numerics are quite strict. Recent evidence shows that high-order upwinding with conservative differencing is acceptable with explicit SGS models [16, 149]. Thus, upwind methods, even high-order modern methods are greatly discouraged because of their intrinsically dissipative character. In particular for multimaterial or shock driven flows the classical approach tends to be quite limited because the numerical methods favored are quite fragile in the presence of discontinuities.

For MILES, Oran and Boris outline a set of essential aspects in an ideal subgrid model [403]:

- conserves mass, momentum and *total* energy;
- it smoothly connects different scales of the flowfield especially the large and subgrid scales;
- effectively dissipates energy at the grid scale;
- is flexible, allowing different physical models to be simulated;
- provides a model for transitional and laminar phenomena;
- is consistent with known scaling properties;
- is well matched to the numerical method;
- and is economical.

Oran and Boris then go on to describe four rather fortunate circumstances from physics that makes MILES possible:

1. the shape of a turbulent (Kolmogorov's) spectrum leads to a fortuitous rate of energy transfer with scale;
2. the tendency for energy to be transferred via local interactions;
3. the nature of the dynamics of large scale flow;
4. and the grid cutoff of monotone (nonoscillatory) methods.

19.3.2 Relation of High-Resolution Methods to Weak Solutions and Turbulence

It is useful to contrast certain aspects of turbulence modeling with numerical methods for hyperbolic PDEs. Hyperbolic PDEs are computed with two competing criteria in mind: a prescription of high accuracy coupled with guarantees against catastrophic failure due to nonlinear wave steepening or unresolved features. Nonlinear mechanisms (usually denoted as limiters) guard the method from such catastrophic failures by triggering entropy producing mechanisms that safeguard the calculation when the need is indicated by the structure of the solution.

Another important point to emphasize about the numerical methods for hyperbolic PDEs: the theory for both the numerical methods and more importantly for the physics of the flow is quite well developed in one dimension. The details of this physical theory have been well developed and are described by [383]. There the connection between a thermodynamically consistent equation of state and hyperbolic wave structure is elucidated. This follows the mathematical description due to Lax [318, 319] leading to the current numerical theory and analysis [339]. This combination has culminated with the availability of powerful numerical methods for several decades pervading many application areas in physics and engineering.

Nonetheless, open questions still exist in two or three dimensions, for example: are some multidimensional problems well-posed, i.e., stable [320]. In the case where the solution involves a vortex street, the solution to the two-dimensional Riemann problem shows a progression of greater and greater complexity as the mesh is refined. Could these sorts of solutions be related to turbulence [159, 488]?

The theory of weak solutions [317, 318, 319] is an extension to the theory of PDEs that allows us to study physically interesting solutions that are not continuous functions. Weak solutions use generalized functions, so that their convolution with test functions over any finite interval of space or time are measurable. PDEs with weak solutions arise as idealized models of physical processes in which asymptotically small effects are not explicitly modeled, but must ultimately be included in some manner. Of course, physically realizable solutions are continuous; however they may contain very steep transition regions (i.e., boundary layers), whose width is much smaller than the scales of interest to the modeler. This width is usually set by viscosity in fluid flows with high Re that weak solutions will be particularly useful.

A classic example of an interesting weak solution is the hydrodynamic shock, which frequently occurs in solutions of the inviscid, compressible Navier-Stokes equations. We emphasize that all real fluids have non-vanishing viscosity, with the consequence that all physical shocks have finite width. However, it is a remarkable property of shocks that the integral characteristics of the flow are independent of the actual value of viscosity. These integral characteristics include the shock speed, the jump conditions, the energy dissipation and entropy increase across the shock, all of which are determined by the large scales of motion.

While the magnitude of the viscosity is not important, it is crucial to maintain the inertial range effect of the viscous dissipation i.e., *the scaling in the weak solution*. Further, we note that the weak solution is not unique. Then the key to obtaining the physically relevant solution is to regularize the equations with dissipation. The principle of vanishing viscosity solutions is central to the design of high-resolution nonoscillatory solutions [339].

Although there are no shocks, it follows that for incompressible flows there are similar issues. Energy at large scales must be dissipated as the flow evolves, and again the mechanism is viscosity. In turbulent flows, eddy dynamics leads to a cascade of energy to ever smaller scales until viscous dissipation becomes sufficient. The expectation is that in high Re turbulent flows, these small scales need not be resolved so long as the energy dissipation has the appropriate form. Thus the strategy here will include the comparison of the energy dissipation in high-resolution methods with more standard LES closures.

To extend the discussion to numerical simulations, we provide several introductory observations. First, the minimal length scale in a simulation will be determined by the cell size. This reflects the usual distinction between DNS and LES. For numerical simulation, weak solutions become a matter of expediency rather than convenience. Second, the numerical solutions are not weak solutions; rather they are the convolution/approximations of the weak solutions with particular test functions, namely the heaviside functions on the cells. This identification is consistent with both the finite-volume nature of the high-resolution simulations and also with the spatial filtering practiced with LES modeling.

Next, we will review, compare and contrast the standard and implicit approaches to LES.

19.4 Large Eddy Simulation: Standard and Implicit

In the classical approach to simulating turbulent flows, either DNS or LES, one employs high-accuracy fluid solvers based on centered, compact or spectral schemes, where numerical dissipation can be minimized. In DNS, physical viscosity (ideally) provides all the dissipation necessary (in principle) to ensure numerical stability. In LES, one supplements the fluid solver with a subgrid model to represent the effects of the unresolved scales. We note that it is essential for the subgrid scale model to provide sufficient dissipation; otherwise energy in the smallest resolved scales will grow without bound, leading ultimately to numerical catastrophe. Beyond issues of stability, the combination of subgrid model and filtering should smooth the flow sufficiently to insure that the dissipation of the numerical algorithm is minimal.

The question of how to define the (numerical) transition between DNS and LES does not have a definitive answer. A typical definition for DNS is to require the cell size to be some (small) multiple of the Kolmogorov scale $\eta = \left(\nu^3/\dot{\epsilon}\right)^{1/4}$, where $\dot{\epsilon}$ is the turbulent dissipation rate. However, this is a global consideration and gradients with larger scale lengths will exist locally due to intermittency. One can conclude that the flowfield will contain unresolved regions with active length scales much smaller than η, or dissipation rates larger than $\dot{\epsilon} = \left\langle \nu \left(\nabla u\right)^2 \right\rangle$ [197, 428].[1] In addition, the latter implies unresolved local scales of time. The more general assumption is that the true smallest scale is proportional to the Kolmogorov scale. Furthermore, any statement of fully resolved flow should have explicit verification, for example, by convergence studies. After all, it is being used as a stand-in for reality and as such should be held to an especially high standard. It is not clear whether such verification occurs in practice, or is even feasible for many cases.

There are a number of types of LES models to consider. The crux of the modeling dilemma is that dissipative models are needed for numerical stability while nondissipative self-similar models are preferred based on theoretical and static (*a priori*) analysis [70, 209, 382]. One effective compromise is to use mixed models [382]. In the next subsection, we will demonstrate that high-resolution methods based on control volume differencing naturally produce the same differential terms as self-similar subgrid scale turbulence models while maintaining dissipativity where necessary, i.e., when the flow becomes under-resolved. The latter results from the intrinsically nonlinear nature of high-order nonoscillatory schemes, e.g., limiters.

An intermediate example between LES and high-resolution methods is the variational multiscale method of Hughes et al. [268]. In this approach, one performs an *a priori* scale separation into large and small scales. The small scales are regularized by an eddy viscosity while the large scales are solved with a high-order spectral method without subgrid dissipation. It is interesting that a spectral analysis of the dissipation in a high-resolution scheme shows qualitatively similar results regarding the presence of the sharp dissipation cutoff scale [142].

A direct relation exists between the original von Neumann-Richtmyer artificial viscosity [590] and the original Smagorinsky model [498], which has been unfortunately lost to the community consciousness. As recounted by

[1] The notation $\langle a \rangle$ is a spatial average.

Smagorinsky [499], in 1955 Phillip's original weather simulations suffered from "ringing" (oscillations) at late time. Jules Charney suggested that von Neumann's artificial viscosity be used to control this ringing and Smagorinsky was tasked with implementing a 3-D generalization of the (at that time) 1-D von Neumann-Richtmyer viscosity. This implementation became what is known as Smagorinsky's eddy viscosity. Since that time, the two approaches have evolved independently, but nonetheless still retain great similarity. While Smagorinsky's model led to LES modeling, von Neumann-Richtmyer viscosity was the forerunner of modern shock-capturing methods including nonoscillatory methods.

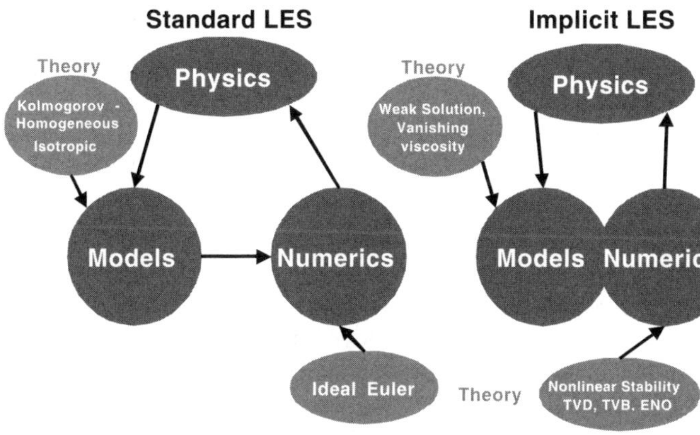

Fig. 19.1. This figure portrays the fundamental similarities in the modeling approaches described and analyzed here. In both cases some fundamental theoretical results are used to define models.

Fig. 19.1 shows a flowchart of how theory, modeling and numerical methods interact in LES and ILES. In standard LES, one applies the physical theories of homogeneous isotropic turbulence in the framework set forth by Kolmogorov to define subgrid models [428]. These models are coupled to the integration of the ideal (i.e., dissipation-free) Euler equations to provide the representation of reality. The ideal Euler equations are chosen so that the numerical method does not contaminate the solution with uncontrolled dissipation (i.e., what may be implicit modeling) [210].

With ILES, the model and numerics are necessarily merged with the modeling having theoretical foundations in vanishing viscosity used to select entropy-satisfying weak solutions. The numerical methods achieve high-resolution and nonlinear stability through the use of monotonicity, TVD, TVB, ENO or other physical/mathematical principles [315, 341]. Without these extensions allowing (at least) second-order accuracy, the vanishing viscosity approach produces first-order results that are not considered "high-resolution".

The key questions to answer are what leads to the similarity in the results found with these seemingly disparate approaches of LES and ILES, and what are the implications of this similarity as relates weak solutions to hyperbolic PDEs and turbulent flows? We will seek the answers to these questions, using modified equation analysis (MEA) to analyze LES models and ILES algorithms.

In closing the discussion connecting numerical analysis and subgrid models, it is important to consider the advice given by William of Occum, "It is vain to do with more what can be done with less".[2] It may be perceived as arguable as to whether explicit or implicit turbulence modeling is simpler. The weight of experience and familiarity favor the explicit models. We will offer two reasons to counterbalance this conclusion, one practical and one more philosophical.

From a practical point of view, high-resolution methods are already accepted as an accurate and efficient tool for simulating laminar flows. Thus their application to turbulent flows means that only one code is necessary, and further that the user needs not to determine *a priori* whether a particular flow is turbulent.

In addition, it is extremely difficult, if not impossible, to separate modeling errors from numerical errors in under-resolved flows. As we will demonstrate later in this chapter, the subgrid model are of the same order in their dependence on mesh quantities as the numerical truncation terms. Nevertheless, in classic LES, the development of subgrid models is approached independently of the fluid solver.

From a more philosophical viewpoint, we recognize that the PDEs themselves, like the numerical programs, are just models of physical reality. However, reality is understood through experiments, and experiments are necessarily carried out at discrete scales determined by the measuring device. Thus one might conclude that while the PDEs may describe nature, it is the discrete codes that better model experiments.

As first principle understanding of turbulence has proven to be a nearly intractable problem. Thus, the success of implicit turbulence modeling as a practical predictive tool in engineering and geophysics should not be ignored, despite its apparent lack of congruence with standard practice. The deeper

[2] Occum's Razor is a logical principle proposed by William of Occum, logician and philosopher, in the 14th century. The principle states, "Entities should not be multiplied unnecessarily". Over the years, mathematicians, philosophers, and scientists have adopted and revised this statement to mean: "When you have two competing theories which make the exact same predictions, the one that is simpler is the better" or "The simplest explanation for the same phenomenon is more likely to be accurate than the more complicated explanation".

issue to explore is the underlying reasons for this success and to understand what limitations may exist and what improvements may be possible.

The next section describes the first step in producing the necessary understanding of the models implicit in the methods via numerical analysis. This will provide a basic foundation from which the modeling associated with the methods can be more fully assessed and extended.

19.5 Numerical Analysis of Subgrid Models

In [149, 369] it was demonstrated that high-resolution approximations of Burgers' equation (see Chap. 5) provided some of the essential elements of an LES turbulence model, in particular as regards the dissipation of energy. As in [369], the methodology of modified equation analysis (MEA) is used, which derives the effective differential equation of a numerical algorithm as a means to analyze the algorithm's behavior [233, 259, 295, 592]. This technique was introduced and discussed in Chap. 6.

We will restrict the analysis to equations in one spatial dimension. There remain theoretical questions regarding the proper numerical regularization of the shear terms in multiple dimensions. However, we note that computational experiments [372] of 3-D turbulent flows governed by the incompressible Navier-Stokes equations verify the implicit turbulence modeling property by the high-resolution scheme MPDATA [503].

To demonstrate the further applicability of MEA, we will consider a simple example extending those available in Chap. 6. Consider the advection of a square wave by a constant velocity. After a short transient, both the leading and the trailing edges of the solution are well represented by hyperbolic tangents. We will compare two solvers, Fromm's method and van Leer's harmonic mean limiter. In Fig. 19.2, we compare the ratio of the errors as determined by MEA and by numerical experiments. Note that the analysis produces the qualitatively same basic structure as is seen experimentally. The numerical errors are less compact, due to higher-order dispersive errors that spread the structure. We note that these methods have the same leading order truncation error, but differ in the second term, which is of order h^3 (where h represents the grid-spacing) and dissipative. This term is:

$$\tau\left(U\right)=\frac{\left|E'\left(U\right)\right|\left(U_{xx}\right)^{2}}{8U_{x}}\,, \tag{19.6}$$

which is present in the van Leer's harmonic limiter; $E\left(U\right)$ is the advective flux. This term is the largest difference between the two solution methods, as is verified in the plots of the ratio of errors as predicted and as solved for the advection of a square wave.

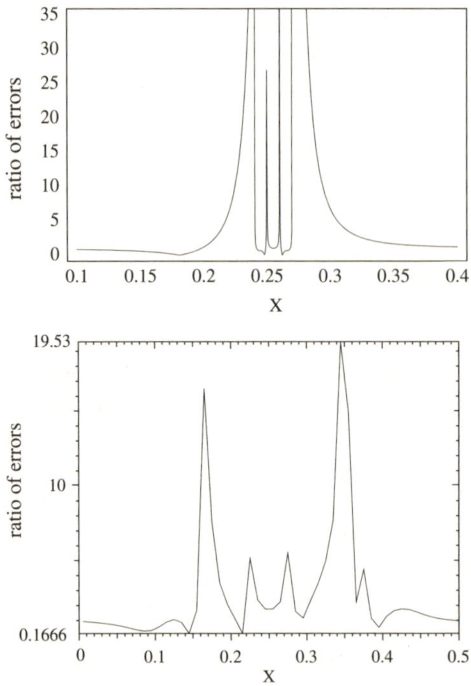

Fig. 19.2. The estimates of the relative error for the two advected profiles, the semi-analytic on the top and the numerically computed error on the bottom. The ratio of errors $E_{\text{vanLeer}}/E_{\text{Fromm}}$ is plotted for each.

19.6 ILES Analysis

19.6.1 Explicit Modeling

The modified equation analysis can be employed to examine several explicit LES models. These models are used in conjunction with the implicit assuption that numerical integration errors are small. The general form for a numerical/modeled solution is

$$U_t + \nabla \cdot E\,(U) = \nabla \cdot \tau\,(U)\,, \tag{19.7}$$

where the left-hand-side is the idealized inviscid equation and the right-hand-side is the subgrid model; the subscript t denotes the time derivative and $\tau(U)$ is the subgrid scale stress.

A starting point for explicit LES modeling is the Smagorinsky model [498] defined by

$$\tau\,(U) \sim C\Delta^2 \left\|\nabla U\right\| \nabla U\;, \tag{19.8}$$

where C is a constant and Δ is the cell size. Other dissipative models can be based on hyper-viscosity [70] in which the stress is proportional to odd order derivatives of degree higher than two, providing a sharper cutoff in spectral space. These have the general form:

$$\tau(U) \sim -\nabla^3 U; \nabla^5 U; -\nabla^7 U; \ldots \tag{19.9}$$

There are several models that are not strictly dissipative. As such these models do not assist the numerical stability of the overall solution methods used in LES. The self-similar model of Clark [382] has a simple differential form

$$\tau(U) \sim \Delta^2 \nabla U \nabla U \; . \tag{19.10}$$

Bardina [382] introduced another model based on filtering, which uses the difference in the subgrid term evaluated at two different filter sizes,

$$\tau(U) \sim \overline{E(\bar{U})} - \overline{\overline{E(U)}} \approx E''(U)\, \Delta^2 \nabla U \nabla U \; . \tag{19.11}$$

Finally, the dynamic Smagorinsky model of Germano et al. [209] compares the flow at two different filter scales to estimate a flow-dependent coefficient for the Smagorinsky model:

$$\tau(U) = C\Delta^4 \left\| \nabla U \right\| \nabla^3 U \; . \tag{19.12}$$

For simplicity, we will evaluate the differential form of these several models in one spatial dimension. For the basic Smagorinsky model, we use the form (19.8) directly

$$\tau(U) = C\Delta^2 \left| U_x \right| U_x, \tag{19.13}$$

Clark's model is simple as well and can be written directly

$$\tau(U) = C\Delta^2 \left(U_x \right)^2 . \tag{19.14}$$

For models based on filtering we need MEA to produce the differential forms. Evaluating Bardina's model using a box filter at 2Δ and 4Δ gives

$$\tau(U) = C\Delta^2 \left(U_x \right)^2 . \tag{19.15}$$

The dynamic Smagorinsky model can also be analyzed by this approach with the same width box filters, yielding

$$\tau(U) = C\Delta^4 \left| U_x \right| U_{xxx} \; . \tag{19.16}$$

Note that in the above expressions C is a constant, different for each model. To summarize, the Smagorinsky model (19.13) and the dynamic Smagorinsky model (19.16) are explicitly dissipative. The Clark's model (19.14) and the Bardina's model (19.15) are similar to the lowest order. However, all terms are not dissipative in either; i.e., the Bardina's and the Clark's models are unstable without additional dissipation. To provide such additional dissipation, mixed models are created where Smagorinsky's model is added to the self-similar model. We will show that ILES produces the same effect naturally including the same differential terms as the self-similar models, producing a mixed model through the nonlinear regularization associated with nonoscillatory differencing.

19.6.2 Implicit Modeling

The analysis of MPDATA (a high-resolution/high-order scheme that was discussed in Chap. 17) approximations of Burgers' equation in [369] shows that it is the finite volume form that mimics the explicit subgrid models of Clark and Bardina. Further, it is the nonoscillatory aspect of the approximation that yields the dissipative terms of the Smagorinsky models. Indeed, the MPDATA scheme (other schemes, as well) combines these two terms in an optimally dissipative fashion, i.e., the scheme is only dissipative in compressive regions for Burgers' equation. However, more generally we observe that the terms associated with explicit subgrid scale models arise in the high-resolution schemes as the combined result of flux-form (conservative) differencing and the fundamentally nonlinear nature of the nonoscillatory approximations, i.e., are proportional to $\partial^2 E/\partial U^2$. We recall that flux-form differencing is a direct consequence of averaging the inertial terms over a cell.

We will focus the analysis on spatial errors in one-dimensional high-resolution algorithms. We consider the following specific form

$$U_j^{n+1} = U_j^n - \frac{\Delta t}{\Delta x}\left[E\left(U_{j+1/2}\right) - E\left(U_{j-1/2}\right)\right] . \tag{19.17}$$

We use a high-resolution Godunov method based on reconstruction employing nonlinearly limited slopes (gradients), $S_j(U)$. With these slopes, we produce two edge values in each cell.

$$U_{j\pm1/2;L/R} = U_j \pm \frac{1}{2}S_j . \tag{19.18}$$

Thus, there are also two values at each node; these are resolved by a linearized Riemann solver:

$$\begin{aligned} E\left(U_{j+1/2}\right) &= \frac{1}{2}\left[E_{j+1/2;L} + E_{j+1/2;R}\right] \\ &\quad - \frac{|E'|}{2}\left(U_{j+1/2;R} - U_{j+1/2;L}\right) , \end{aligned} \tag{19.19}$$

where $E' \equiv \partial E/\partial U$. The use of a Godunov-type method is not essential to any result. However, the procedure outlined above provides a discrete algorithm that is amenable to computer analysis with tools like Mathematica (see App. A).

The general form of the modified equations from (19.17) implies an effective subgrid stress $\tilde{\tau}$ which, at order Δ^2 is

$$\tilde{\tau}_2(U) = c_1 E'(U)\, U_{xx} + c_2 E''(U)\left(U_x\right)^2 . \tag{19.20}$$

The subgrid stress τ_2 is a second-order approximation with constants c_1 and c_2 depending on the specific differencing scheme. In [369] it was shown that the second term, $E''(U)\left(U_x\right)^2$ is a consequence of the conservation form, and is not present if the differencing is not in conservation form. Furthermore, this term is identical to the leading order term for the self-similar model with standard LES. At the next order Δ^3,

$$\tilde{\tau}_3(U) = c_3 E'(U) U_{xxx} , \tag{19.21}$$

is a hyperviscosity term (19.9). Here τ_3 is a third-order term again with the constant c_3 depending on the specific numerical method.

19.6.3 Limiters

To evaluate these effective subgrid stresses, we must define the particular form of the limiter function for the slopes. The goal of all nonlinear limiters is similar – to produce dissipation wherever the flow becomes under-resolved. The underlying idea is to detect regions of the flow where linear high-order fluxes would produce oscillatory results and add dissipation locally by mixing in the low-order flux. Let us define $a = U_j - U_{j-1}$ and $b = U_{j+1} - U_j$. Then, the limiters we will consider can be defined compactly. For example, the van Leer harmonic mean limiter [570] is:

$$S_j = \frac{|b|\,a + |a|\,b}{|a| + |b|} , \tag{19.22}$$

and van Albada limiter [572] is:

$$S_j = \frac{b^2 a + a^2 b}{a^2 + b^2} , \tag{19.23}$$

Both of these limiters are analogous to data dependent weighted least square gradients. Applying MEA, we find that both limiters lead to the same form for the effective stress

$$\tilde{\tau}_3(U) = C\Delta^3 \left|E'(U)\right| \frac{(U_{xx})^2}{U_x} , \tag{19.24}$$

but with different constants, $C = 1/2$ for (19.23), and $C = 1/4$ for (19.22). These expressions can be derived by using the limiters to describe the reconstruction in a piecewise linear method. Upon defining the advective flux and then integrating the resulting equation in space to obtain the truncation error in divergence form, these terms describe the impact of the nonlinearly defined reconstruction in conservation form.

For the MPDATA scheme [503], whose limiter has the form

$$S_{j+1/2} = \left(1 - \left|\frac{b}{U_j + U_{j+1}}\right|\right) b , \tag{19.25}$$

and each cell edge is limited so that

$$U_{j+1/2,L} = U_j + \frac{1}{2} S_{j+1/2}, \quad U_{j+1/2,R} = U_{j+1} - \frac{1}{2} S_{j+1/2} , \tag{19.26}$$

the corresponding effective stress has the form of the Smagorinsky model and is of the order Δ^2,

$$\tilde{\tau}_2(U) = C\Delta^2 \left|E'(U)\right| |U_x|\, U_x . \tag{19.27}$$

The minmod function is another popular limiter. The usual form for such limiters, $\mathrm{minmod}(a,b) = \mathrm{sign}(a)\max[0, \min(|a|, \mathrm{sign}(a)\,b)]$, precludes analysis because of the min and max functions. Instead, we will write

$$\mathrm{minmod}(a,b) = \frac{1}{4}\left[\mathrm{sign}(a) + \mathrm{sign}(b)\right]\left[|a+b| - |a-b|\right], \tag{19.28}$$

where $\mathrm{sign}(a) = a/|a|$. The second-order ENO limiter [248], mineno, has a similar form:

$$\mathrm{mineno}(a,b) = \frac{1}{2}\mathrm{sign}(a+b)\left(|a+b| - |a-b|\right). \tag{19.29}$$

The analytic form of the minmod or mineno functions allow the use of symbolic algebra to perform the MEA. The minmod and mineno limiters have the same form for the effective subgrid stress

$$\tilde{\tau}(U) = C\Delta^2 \left|E'(U)\right| \frac{|U_x U_{xx}|}{U_x}. \tag{19.30}$$

The UNO scheme [251] uses three estimates of the slope for its limiter function, defined by

$$\mathrm{median}(a,b,c) = a + \mathrm{minmod}(b-a, c-a) . \tag{19.31}$$

Here, we will describe an elegant implementation [271]. The UNO scheme has the following steps:

1. Define first-order slopes,

   $$S_- = U_j - U_{j-1} \text{ and } S_+ = U_{j+1} - U_j ,$$

   and second-order slopes

   $$P_- = \frac{(3U_j - 4U_{j-1} + U_{j-2})}{2},$$

   $$P_0 = \frac{(U_{j+1} - U_{j-1})}{2},$$

   and

   $$P_+ = \frac{(-U_{j+2} + 4U_{j+1} - 4U_j)}{2}.$$

2. Conduct a slope selection based on accuracy and smoothness,

   $$Q_- = \mathrm{median}(S_-, P_-, P_0) \text{ and } Q_+ = \mathrm{median}(S_+, P_+, P_0).$$

3. Finally, choose the smoothest slope from these using a function,

   $$\mathrm{minmod}(Q_-, Q_+), \mathrm{mineno}(Q_-, Q_+) ,$$

   or Huynh's extended minmod,

   $$\mathrm{xm}(Q_-, Q_+) = \mathrm{median}(Q_-, Q_+, -Q_- - Q_+).$$

Its effective subgrid stress is

$$\tau(U) = C\Delta^3 \left|E'(U)\right| \frac{\left|U_x U_{xxx}\right|}{U_x} \,. \tag{19.32}$$

Next, we consider two higher-order schemes. We show only the effective subgrid stress, referring the reader to Chap. 17 for details of the schemes. A third-order WENO scheme [356] has a form like the van Leer (19.22) and van Albada (19.23) limiters,

$$\tilde{\tau}(U) = C\Delta^3 \left|E'(U)\right| \frac{\left(U_{xx}\right)^2}{U_x} \,. \tag{19.33}$$

Finally, a fifth-order WENO [280] has an effective subgrid stress

$$\tilde{\tau}(U) = C\Delta^5 \left|E'(U)\right| \frac{\left(U_{xxx}\right)^2}{U_x} \,. \tag{19.34}$$

We comment that WENO schemes are higher order in a point sense, not in a finite volume sense (recall the second example in Sect. 6.3 where the ostensibly higher-order method was shown to be less accurate than the lower-order scheme). In [369], it is argued that the physical dissipation should be of order Δ^2. Thus, one should question whether higher-order schemes are in any sense preferable.[3]

We close this section with a brief comment of the relation of these limiters to the multiscale method of Hughes mentioned in Sect. 19.4. In most cases we have discussed, the limiters are always "on" so their effect is felt even when the flow is fully resolved. Therefore, they do not form the basis of a variational multiscale method with its scale separation. However, the median limiter of the UNO scheme (by construction) uses the high-order method where the flow is resolved. The overall effect is to keep the strongly dissipative effects localized to the highest wave-numbers. This provides the same impact as the variational multiscale method in an implicit rather than explicit manner.

19.6.4 Energy Analysis

We begin the analysis of the impact of these models, LES and ILES, on the solution in terms of the evolution of energy. In flows described by (19.17) where there is no thermodynamics, the kinetic energy $\mathcal{E} \equiv 1/2U^2$ is also related to the negative of the entropy. Physically, energy at large scales cascades to smaller scales and finally is dissipated by viscosity. This dissipativity is the embodiment of the second law of thermodynamics; i.e., the entropy is an increasing function of time. For the numerical simulation, dissipativity guarantees the nonlinear stability of the algorithm. Of course these two properties are closely related, showing the intersection of physics and numerics. In fact,

[3] We always refer to nonlinear schemes.

the nonoscillatory property is a numerical constraint that enforces the second law on the simulated flow.

To derive the energy equation, we multiply the governing equation (19.17) by the dependent variable U. After some rearrangement, we derive

$$\mathcal{E}_t + \nabla \cdot U \left[E\left(U\right) - \tau\left(U\right)\right] = \left[\tau\left(U\right) - E\left(U\right)\right] \nabla U \,. \tag{19.35}$$

The term in divergence form on the LHS simply transports energy, but the right hand side will change the energy of the system. Eq. (19.35) was derived by explicitly arranging the equation to have these two types of terms. Ultimately, we seek forms for the subgrid term, τ that will provide for a guarantee of a dissipative result in a global sense. We note that the validity of this procedure depends to some degree on solution smoothness. Physically, such smoothness will only hold at small viscous scales and will generally break down in the inertial range. However, both explicit and implicit modeling will regularize the solution to the scale of the grid, allowing some confidence in the following analysis.

We now specialize the analysis to 1-D, where the τ term originating from the conservation form in (19.20) generates the term $E''\left(U\right)\left(U_x\right)^2$. Further, we will consider the specific form for the flux function $E = 1/2U^2$, appropriate for the Navier-Stokes equations. In fact, most physically interesting systems have a quadratic form. This alone accounts for the similarities between compressible and incompressible flows as well as Burgers' equation. Then the energy equation simplifies to

$$\mathcal{E}_t = \left[-\frac{1}{3}U^3\right]_x + U\tau_x = \left[-\frac{1}{3}U^3 + U\tau\right]_x - U_x\tau \,. \tag{19.36}$$

Now the term inside the gradient moves energy around, but (in a closed system) does not affect the global magnitude of energy

$$\bar{\mathcal{E}}(t) = \int \mathcal{E}(x',t)\,\mathrm{d}x' = -\int U_x(x',t)\tau(x',t)\,\mathrm{d}x' \,. \tag{19.37}$$

In a discrete simulation, the integral will be replaced by a sum over cells. Thus, if we can show that $\bar{\mathcal{E}}$ does not increase in time, then its value represents an upper bound on the energy in any cell guaranteeing the (energy) stability of the simulation. As the physical relations discussed at the beginning of this chapter imply, the amount of dissipation is not merely positive, but larger than a specific value determined by the large scale details of the flow. It is important for the subgrid model to reproduce these limits.

Next, we will describe the impact of some specific models. We begin by considering the 1-D Smagorinsky subgrid stress $\tau = C_s\left|U_x\right|U_x$, cf (19.8). Inserting into (19.37) and discarding the surface terms, we find

$$\bar{\mathcal{E}}(t) = -C\Delta^2 \int \left|U_x\right|^3 \,\mathrm{d}x' \le 0 \,. \tag{19.38}$$

That is, the Smagorinsky model is absolutely dissipative. The dynamic Smagorinsky model (19.16) has the global energy

$$\mathcal{E}(t) = -C\Delta^4 \int \left[|U_x| \, (U_{xx})^2 + U_x U_{xx} \, |U_{xx}| \right] \, \mathrm{d}x' \leq 0 \, . \tag{19.39}$$

So the dynamic Smagorinsky model is also dissipative.

This is not the case for the Clark's model (19.14) and the Bardina's model (19.15) whose global energy is

$$\bar{\mathcal{E}}(t) = -C\Delta^2 \int U^3(x',t) \, \mathrm{d}x' \, , \tag{19.40}$$

which may have either positive or negative sign unless the velocity field is asymmetric.

Now we shift our attention to the terms that come from the various monotonicity limiters. The van Leer's limiter global energy is

$$\bar{\mathcal{E}}(t) = -C\Delta^3 \int (U_{xx})^2 \mathrm{d}x' \leq 0 \, , \tag{19.41}$$

This term is inherently dissipative in nature. The same can be said for the fourth-order linear "hyper-viscosity" term U_{xxxx}, which produces a similar energy dissipation. Other limiters produce the same sort of term when analyzed.

To summarize, we find that all the high-resolution models that we have described are dissipative. However, all models are not equivalent. In terms of scaling with cell size Δ, we see the LES Smagorinsky model and the MPDATA ILES model (19.27) both scale with h^2, and in fact have equivalent forms for the subgrid stress τ and the global energy. The minmod and mineno ILES models (19.30) both have the same h^3 scaling and form for the global energy, but have different forms for τ. The ILES van Leer model (19.22), the van Albada model (19.23), and the third-order ENO schemes (19.33) all scale with Δ^3.

The results of this section show the equivalence of LES and ILES in many cases. Based on dimensional analysis, there are only limited choices for the form of the dissipation. In fact, in the absence of physical viscosity, the only available terms are the flow velocity and its derivatives, and the length scale h.

Not all the ILES models correspond to a known LES model; however their subgrid stresses, e.g., of the van Leer limiter (19.22) as well as others could be effective as LES models and warrant examination. It is important to note that the "control volume" term is always present at the leading order of the truncation error. Continuing this line of thought, no LES model has been found to have universal validity, and this argues that a more thorough exploration of the differences among ILES models would be useful. For example, some limiters (such as minmod, van Leer or ENO/WENO) are always present in the equations while other limiters (FCT and Fromm) are not active where the flow is deemed resolved. These latter ILES models are similar to the mixed LES models.

In the analysis of the dissipation, we have ignored the dispersive terms that integrate out in the global energy $\mathcal{E}$ balance, (19.35). These terms will change the distribution of the energy, but do not affect its global value.

19.7 Computational Examples

Below, we present examples from the implementation of high-resolution methods in turbulent flow simulations.

19.7.1 Burgers' Turbulence (Burgulence)

As a first example we examine the behavior of high-resolution methods with and without SGS models for simulations of Burgers' turbulence [303] (see also Chap. 5). The Burgers' equation can be considered as the one-dimensional analog of the Navier-Stokes equations. The problem of Burgers' turbulence is described by $\partial U/\partial t + U\ \partial U/\partial x = \nu\ \partial^2 U/\partial x^2$, subject to periodic boundary conditions $U(x,t) = U(x+l,t)$, $0 \leq x \leq l$, and a random initial condition for the velocity U (Fig. 19.3).

The random initial condition of Fig. 19.3 exhibits maximum value of the wave spectrum at $\log(k) = 1.283$ [149]. The velocity has become dimensionless by defining a characteristic length scale $L_o = 1/\log^{-1}(1.283)L = 0.052L$ (where L is an arbitrary unit of length; here $L = 1$), and a characteristic velocity u_o as the root mean square of the initial condition. The viscosity ν can then be defined by $\nu = (L_o u_o)/Re$, where Re is the Reynolds number. In [149] simulations were conducted for $Re = 6,000$ in a domain of length $l = 12L = 12$, using a fine grid (9,000 grid points) and a small time step ($\Delta t = 0.0001$). The obtained solution (henceforth labeled DNS) is grid and time-step independent (for finer grids and smaller time steps) and can thereby be considered as the "exact" solution.

In [149] coarsely-resolved simulations have also been carried out on a 700×100 space-time grid using different numerical schemes with and without different SGS models. The following numerical variants have been employed: (i) the characteristics-based (Godunov-CB) scheme of [148] without a SGS model (see Chap. 16 for details) ; (ii) the TVD-CB scheme of [149] without a SGS model (see Chap. 16 for details); (iii) the CB scheme in conjunction with the modified version of the dynamic SGS model [345] – the solution is labeled as "D-Model"; (iv) the CB scheme in conjunction with the structure-function SGS model [387] – the solution is labeled as "SF-Model". The results for the kurtosis distribution Fig. 19.4 reveal that: (i) modeling the unresolved scales through a SGS model does not always improve the results; for example, compare the Godunov-CB solutions with and without the dynamic model; (ii) high-resolution schemes designed to satisfy the total variation diminishing (TVD) condition can significantly improve the predictions without even using a SGS. Comparison of the average kurtosis value for the various computations

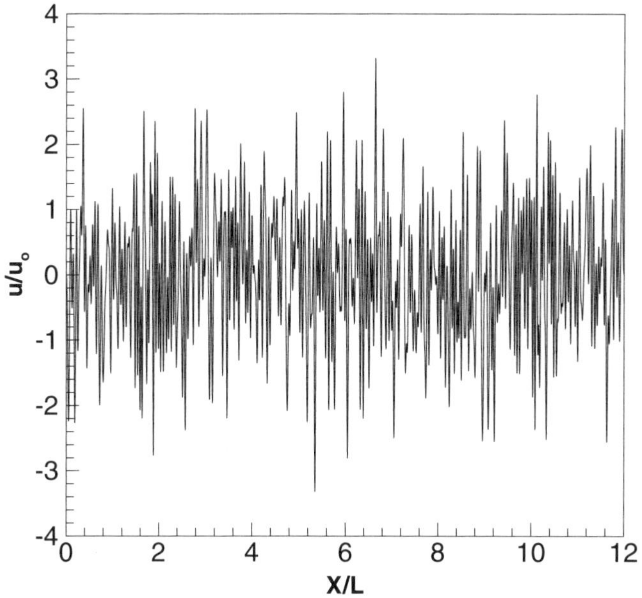

Fig. 19.3. Velocity profile considered as initial condition in the simulations of Burgers' turbulence.

including the case where the TVD-CB scheme is used in conjunction with the structure-function SGS model have shown that the computation based upon the TVD-CB scheme without a SGS model gives the closest agreement with the DNS solution [149].

19.7.2 Convective Planetary Boundary Layer

In previous studies, Smolarkiewicz and Prusa [504] as well as Margolin et al. [371], demonstrated that an atmospheric code based on MPDATA [503] can accurately reproduce (i.e., in close agreement with field/laboratory data and the existing benchmark computations) the structure of the convective planetary boundary layer. They carried out simulations with and without an explicit SGS turbulence model (see [504] for details of the SGS model). Their results showed that when an explicit turbulence model was implemented, MPDATA did not add any unnecessary diffusion. When no explicit turbulence scheme was employed, the high-resolution method itself appeared to include an effective SGS model. They also reported that using the explicit turbulence model with the eddy viscosity reduced by some factor, MPDATA

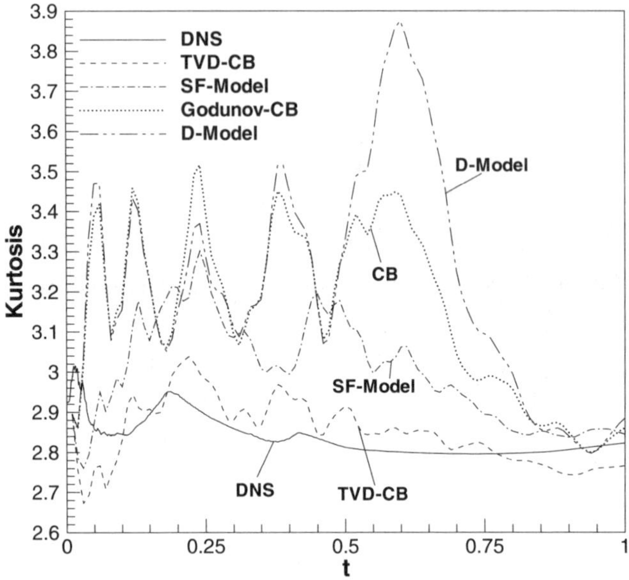

Fig. 19.4. Kurtosis distribution for simulation of Burgers' turbulence using characteristics-based (CB) and TVD schemes (TVD-CB) with and without SGS models.

added just enough dissipation. These numerical experiments demonstrate the self-adaptive character of the high-resolution method and suggest the physically realistic character of its truncation error (i.e., numerical dissipation).

Fig. 19.5 shows results for the resolved heat flux $\langle T'w' \rangle$ (normalized appropriately), where T and w are the temperature and vertical velocity, respectively, and primes denote deviation from the horizontal average $< \cdot\cdot >$ [371]. The three curves shown in the figure represent mean profiles from three different solutions: LES benchmark simulations of Schmidt and Schumann [484] using a centered numerical scheme both in space and in time; standard LES simulations with MPDATA in conjunction with a Smagorisnky-type LES model; simulations based on MPDATA with no explicit subgrid-scale model (i.e., ILES approach); circles represent field and laboratory data. The comparability of all the results with the data is excellent.

The most important result in Fig. 19.5 is the the accuracy of the high-resolution method in LES without need to resort to a SGS model. In contrast to linear methods, the success of high-resolution methods in turbulent flows is due to the self-adaptiveness of these schemes during the simulation. When the

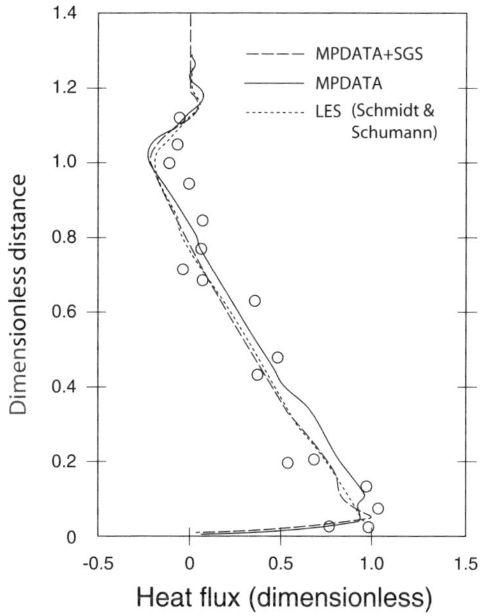

Fig. 19.5. Simulation results [504] for the convective atmospheric boundary layer using an explicit Smagorinsky-type SGS model (MPDATA+SGS), i.e., a standard LES approach; without an explicit SGS model (MPDATA curve), i.e., an ILES approach; and (standard) LES benchmark simulations of Schmidt and Schumann [484]. Reproduced with the kind permission of P. K. Smolarkiewicz and J. M. Prusa.

explicit SGS model is included, the resolved flow is sufficiently smooth and the part of the numerical algorithm that assures high-resolution properties is essentially switched off. When no explicit SGS model is used the high-resolution scheme adapts the numerics assuring solutions that are apparently as smooth as those generated with explicit SGS models. One should bear in mind that the dissipation of high-resolution methods cannot be universally quantified since the advective scheme can be effectively either non-dissipative or dissipative, depending upon the presence or absence, respectively, of an explicit SGS model [504].

A. MATHEMATICA Commands for Numerical Analysis

Below, the commands for conducting elementary numerical analysis using Mathematica are given. The first two sections describe the Fourier analysis of methods (first-order and second-order upwind) and the last section gives the modified equation analysis of nonlinear upwind differencing. These analyses are discussed in Chap. 6.

A.1 Fourier Analysis for First-Order Upwind Methods

```
(* Define the Fourier transform and grid relations *)

U[j_,t_]:= Cos[j t] + I Sin[j t]

u[j_]:= U[j,t]

(* edge value - constant *)

ue[j_]:= u[j]

(* cell update *)

u[0] - v(ue[0]-ue[-1])

Expand[%]

(* take apart into real and imaginary parts *)

realu = 1-v+v Cos[t];

imagu = Expand[-I(-\[ImaginaryI] v Sin[t])];

(* amplitude and phase errors *)

ampu = Sqrt[realu^2 + imagu^2];
```

```
phaseu = ArcTan[-imagu/realu]/(v t);

(* Taylor series expansion to get accuracy *)

Collect[Expand[Normal[Series[ampu,{t,0,4}]]],t]

Collect[Expand[Normal[Series[phaseu,{t,0,4}]]],t]

Plot3D[ampu,{t,0,Pi},{v,0,1},
  AxesLabel\[Rule]{StyleForm["t",FontSize\[Rule]18],
StyleForm["v",FontSize\[Rule]18], StyleForm["amp",FontSize\[Rule]18]}]

Plot3D[phaseu,{t,0.001,Pi},{v,0.001,1},
  AxesLabel\[Rule]{StyleForm["t",FontSize\[Rule]18],
StyleForm["v",FontSize\[Rule]18],StyleForm["phase",FontSize\[Rule]18]}]
```

A.2 Fourier Analysis for Second-Order Upwind Methods

```
(* Define the Fourier transform and grid relations *)

U[j_,t_]:= Cos[j t] + I Sin[j t]

u[j_]:= U[j,t]

(* edge value - constant + slope and time-centering *)

s[j_]:=1/2(u[j+1] - u[j-1])

ue[j_]:= u[j]+1/2(1-v) s[j]

(* cell update *)

u[0] - v(ue[0]-ue[-1])

Expand[%]

(* take apart into real and imaginary parts *)

(* amplitude and phase errors *)

ampf = Sqrt[realf^2 + imagf^2];

phasef = ArcTan[-imagf/realf]/(v t);
```

```
(* Taylor series expansion to get accuracy *)

Collect[Expand[Normal[Series[ampf,{t,0,4}]]],t]

Collect[Expand[Normal[Series[phasef,{t,0,4}]]],t]

Plot3D[ampf,{t,0,Pi},{v,0,1},
AxesLabel\[Rule]{StyleForm["t",FontSize\[Rule]18],
StyleForm["v",FontSize\[Rule]18],StyleForm["amp  ",FontSize\[Rule]18]}]

Plot3D[phasef,{t,0.001,Pi},{v,0.001,1},
AxesLabel\[Rule]{StyleForm["t",FontSize\[Rule]18],
StyleForm["v",FontSize\[Rule]18],StyleForm["phase ",FontSize\[Rule]18]}]
```

A.3 Modified Equation Analysis for First-Order Upwind

In order to make some of our points concrete we give a series of commands in Mathematica that will produce some of the analysis and figures shown in Chap. 6. The set of commands given below will produce the modified equation expressions given by (6.14) and the plots shown in Fig. 6.5.

```
(* Basic Setup *)
(* Grid Function defined in terms of h *)
U[m_]:= u[x + m h]

(* The interface average of the flux Jacobian *)
Df[m_]:= 1/2(D[f[U[m]], U[m]]+D[f[U[m+1]], U[m+1]])

(* Upwind flux expression *)
Fe[m_]:= 1/2(f[U[m]] + f[U[m+1]]) - 1/2 Abs[Df[m]] (U[m+1] - U[m])

(* defined flux function - Burgers' equation *)
f[x_]:= 1/2 x^2

(* flux difference *)
dxfe=(Fe[0] - Fe[-1])/h;

(* Expand the flux difference in a Taylor series and simplify *)
Collect[Simplify[Expand[Normal[Series[dxfe,{h,0,3}]]]],h]

(* Integrate this expression to get the expansion in terms of flux *)
Collect[Integrate[%,x],h]
```

```
(* Find this expression directly *)
Collect[Simplify[Expand[Normal[Series[Fe[j],{h,0,3}];]]],h];

(* Find the cell integral average in index space *)
Collect[Expand[Integrate[%,{j,-1,0}]],h]

(* set the mesh spacing *)
h=0.05

(* the true continuous function *)
v[x_]:= Tanh[10 x]

(* cell average value of v[x] *)
u[x_]:= (Log[Cosh[10 (x+h/2)] ]- Log[Cosh[10(x-h/2)]])/(10 h)

(* Order h truncation error *)
Oh[x_]:= 5 h Abs[Tanh[10 x]] Sech[10 x]^2

(* Order h^2 truncation error *)
Oh2[x_]:=-1/6 h^2 (100 Sech[10 x]^4 - 200 Sech[10 x]^2 Tanh[10 x]^2)

(* true continuous flux *)
dxf=D[f[v[x]],x];

(* create plots to compare the results *)
Plot[{dxf,dxfe},{x,-0.25,0.25},AxesLabel\[Rule]{x,"df/dx"},
  PlotStyle\[Rule]{{Thickness[0.005],Dashing[{0.02,0.02}]},}]

Plot[dxf-dxfe,{x,-0.25,0.25},AxesLabel\[Rule]{x, "Error in flux"}]

Plot[(Oh[x+h/2] -Oh[x-h/2] )/h,{x,-0.25,0.25},
  AxesLabel\[Rule]{x,"Order h terms"}]

Plot[(Oh2[x+h/2] -Oh2[x-h/2] )/h,{x,-0.25,0.25},
  AxesLabel\[Rule]{x,"Order h"^2 "terms"}]

Plot[(dxf-dxfe)-((Oh[x+h/2] + Oh2[x+h/2]- Oh[x-h/2] - Oh2[x-h/2])/h),
{x,-0.25,0.25},AxesLabel\[Rule]{x, "Error - Mod. Eqn."}]


(* define a small value *)
eps=0.00000001;
```

```
(* define the unperturbed error *)
h=h1;
error=Sqrt[(dxf -dxfe)^2];

(* define the perturbed error *)
h=h1+eps;
erroreps=Sqrt[(dxf -dxfe)^2];

(* define the convergence rate as a function of h1 *)
p=Log[erroreps/error]/Log[(h1+eps)/h1];

Plot[p,{h1,0.0000001,0.25}]
```

B. Example Computer Implementations

This appendix provides some brief examples of the implementation of several numerical methods. The purpose is to demonstrate how these methods might be implemented. The codes themselves do not constitute entire working codes, but they are parts of the codes used in the simulations. They have been written in Fortran 90.

B.1 Appendix: Fortran Subroutine for the Characteristics-Based Flux

The subroutine below presents the characteristics-based discretization [148, 156] of the advective flux **E** in curvilinear body-fitted coordinates.

```
!*****************************************************************
! Subroutine:  XiFlux
! Description: Calculation of advective flux using the
!              characteristics-based scheme.
! Arguments:
!            X,Y,Z       - Coordinates arrays
!            PS,US,VS,WS - Pressure, and velocity arrays
!            IE,JE,KE    - Array dimensions
!            BE          - Artificial compressibility parameter
!            LS          - .true. if left boundary is solid
!            RS          - .true. if right boundary is solid
!            IApp        - order of reconstruction
!            PH          - element of the continuity equation
!                          in the advective flux E
!            UH,VH,WH    - elements of the momentum equations
!                          in the advective flux E
!
! Note: This subroutine is provided as an example
!       of code structure only.
!*****************************************************************
 subroutine XiFlux(X,Y,Z,PS,US,VS,WS,PH,UH,VH,WH,IE,JE,KE,&
                  BE,LS,RS,IApp)
```

```
  implicit none
! Arguments
  real,    intent(in) :: BE
  logical, intent(in) :: LS,RS
  integer, intent(in) :: IE,JE,KE,IApp
  real, dimension (0:IE,0:JE,0:KE), intent(in)    :: X,Y,Z,PS
  real, dimension (0:IE,0:JE,0:KE), intent(inout) :: US,VS,WS
  real, dimension (IE,JE,KE), intent(inout)       :: PH,UH,VH,WH
! Temporary arrays for fluxes
  real, dimension (IE) :: XH,YH,ZH,RFL
! Array indices
  integer ::  I,J,K,I1,I2,J1,J2,K1,K2,IL,ILL,IR,IRR
! Temporary variables
  real :: HPL,HUL,HVL,HWL,HPLL,HULL,HVLL,HWLL,HPR,HUR,HVR,HWR, &
          HPRR,HURR,HVRR,HWRR,UN,VN,WN,PPL,UUL,VVL,WWL,PPR, &
          UUR,VVR,WWR,FFFA,FFFB,FFFC,FFFD,XET,YET,ZET, &
          XZE,YZE,ZZE,XIX,XIY,XIZ,XX,YY,ZZ, &
          B,EV0,EV1,EV2,S,GN0,GN1,GN2,U0,V0,W0,P1,U1,V1,W1,&
          P2,U2,V2,W2,P,U,V,W,R1
! IApp controls the order of accuracy for the
! high-order characteristic extrapolation.
! IApp=1,2,3 and 4 denotes the order of accuracy
  if(IApp.EQ.1) then
   FFFA=1.
   FFFB=0.
   FFFC=0.
   FFFD=0.
  else if(IApp.EQ.2) then
   FFFA=3./2.
   FFFB=1./2.
   FFFC=0.
   FFFD=0.
  else if(IApp.EQ.3) then
   FFFA=5./6.
   FFFB=1./6.
   FFFC=2./6.
   FFFD=0.
  else if(IApp.EQ.4) then
   FFFA=7./12.
   FFFB=1./12.
   FFFC=7./12.
   FFFD=1./12.
  end if
! Fluxes loop
```

```
 do K=2,KE-2
  K1=K-1
  K2=K+1
  do J=2,JE-2
   J1=J-1
   J2=J+1
! Extrapolate boundary conditions as inviscid for solid boundary
   if(LS) then
    US(1,J,K)=2.*US(2,J,K)-US(3,J,K)
    VS(1,J,K)=2.*VS(2,J,K)-VS(3,J,K)
    WS(1,J,K)=2.*WS(2,J,K)-WS(3,J,K)
    US(0,J,K)=2.*US(1,J,K)-US(2,J,K)
    VS(0,J,K)=2.*VS(1,J,K)-VS(2,J,K)
    WS(0,J,K)=2.*WS(1,J,K)-WS(2,J,K)
   end if
   if(RS) then
    US(IE-1,J,K)=2.*US(IE-2,J,K)-US(IE-3,J,K)
    VS(IE-1,J,K)=2.*VS(IE-2,J,K)-VS(IE-3,J,K)
    WS(IE-1,J,K)=2.*WS(IE-2,J,K)-WS(IE-3,J,K)
    US(IE,J,K)  =2.*US(IE-1,J,K)-US(IE-2,J,K)
    VS(IE,J,K)  =2.*VS(IE-1,J,K)-VS(IE-2,J,K)
    WS(IE,J,K)  =2.*WS(IE-1,J,K)-WS(IE-2,J,K)
   end if
   do IR=2,IE-1
    IL=IR-1
    ILL=IR-2
    IRR=IR+1
    HPL=PS(IL,J,K)
    HUL=US(IL,J,K)
    HVL=VS(IL,J,K)
    HWL=WS(IL,J,K)
    HPLL=PS(ILL,J,K)
    HULL=US(ILL,J,K)
    HVLL=VS(ILL,J,K)
    HWLL=WS(ILL,J,K)
    HPR=PS(IR,J,K)
    HUR=US(IR,J,K)
    HVR=VS(IR,J,K)
    HWR=WS(IR,J,K)
    HPRR=PS(IRR,J,K)
    HURR=US(IRR,J,K)
    HVRR=VS(IRR,J,K)
    HWRR=WS(IRR,J,K)
! High-order interpolation scheme (Section 16.4.5)
```

```
      PPL=FFFA*HPL-FFFB*HPLL+FFFC*HPR+FFFD*HPRR
      UUL=FFFA*HUL-FFFB*HULL+FFFC*HUR+FFFD*HURR
      VVL=FFFA*HVL-FFFB*HVLL+FFFC*HVR+FFFD*HVRR
      WWL=FFFA*HWL-FFFB*HWLL+FFFC*HWR+FFFD*HWRR
      PPR=FFFA*HPR-FFFB*HPRR+FFFC*HPL+FFFD*HPLL
      UUR=FFFA*HUR-FFFB*HURR+FFFC*HUL+FFFD*HULL
      VVR=FFFA*HVR-FFFB*HVRR+FFFC*HVL+FFFD*HVLL
      WWR=FFFA*HWR-FFFB*HWRR+FFFC*HWL+FFFD*HWLL
! Calculate metrics
      XET=0.5*(X(IR,J2,K)+X(IR,J2,K2)-X(IR,J,K)-X(IR,J,K2))
      YET=0.5*(Y(IR,J2,K)+Y(IR,J2,K2)-Y(IR,J,K)-Y(IR,J,K2))
      ZET=0.5*(Z(IR,J2,K)+Z(IR,J2,K2)-Z(IR,J,K)-Z(IR,J,K2))
      XZE=0.5*(X(IR,J,K2)+X(IR,J2,K2)-X(IR,J,K)-X(IR,J2,K))
      YZE=0.5*(Y(IR,J,K2)+Y(IR,J2,K2)-Y(IR,J,K)-Y(IR,J2,K))
      ZZE=0.5*(Z(IR,J,K2)+Z(IR,J2,K2)-Z(IR,J,K)-Z(IR,J2,K))
      XIX=YET*ZZE-ZET*YZE
      XIY=XZE*ZET-XET*ZZE
      XIZ=XET*YZE-XZE*YET
      B=SQRT(XIX**2+XIY**2+XIZ**2)
      XX=XIX/B
      YY=XIY/B
      ZZ=XIZ/B
! Middle velocity
      UN=0.5*(HUL+HUR)
      VN=0.5*(HVL+HVR)
      WN=0.5*(HWL+HWR)
! Zeroth, first and second eigenvalues (Section 16.4.3)
      EV0=UN*XX+VN*YY+WN*ZZ
      S=SQRT(EV0*EV0+BE)
      EV1=EV0+S
      EV2=EV0-S
! Upwinding along the zeroth characteristic (Section 16.4.5)
      GN0=SIGN(1.,EV0)
      U0=0.5*((1.+GN0)*UUL+(1.-GN0)*UUR)
      V0=0.5*((1.+GN0)*VVL+(1.-GN0)*VVR)
      W0=0.5*((1.+GN0)*WWL+(1.-GN0)*WWR)
! Upwinding along the characteristic corresponding to the
! eigenvalue EV1
      GN1=SIGN(1.,EV1)
      P1=0.5*((1.+GN1)*PPL+(1.-GN1)*PPR)
      U1=0.5*((1.+GN1)*UUL+(1.-GN1)*UUR)
      V1=0.5*((1.+GN1)*VVL+(1.-GN1)*VVR)
      W1=0.5*((1.+GN1)*WWL+(1.-GN1)*WWR)
! Upwinding along the characteristic corresponding to the
```

```
! eigenvalue EV2
      GN2=SIGN(1.,EV2)
      P2=0.5*((1.+GN2)*PPL+(1.-GN2)*PPR)
      U2=0.5*((1.+GN2)*UUL+(1.-GN2)*UUR)
      V2=0.5*((1.+GN2)*VVL+(1.-GN2)*VVR)
      W2=0.5*((1.+GN2)*WWL+(1.-GN2)*WWR)
! Characteristic-based calculation of the primitive variables
      R1=(0.5/S)*((P1-P2)+XX*(EV1*U1-EV2*U2)+YY*(EV1*V1-EV2*V2)+ &
                          ZZ*(EV1*W1-EV2*W2))
      U=XX*R1+U0*(YY*YY+ZZ*ZZ)-XX*(V0*YY+W0*ZZ)
      V=YY*R1+V0*(XX*XX+ZZ*ZZ)-YY*(U0*XX+W0*ZZ)
      W=ZZ*R1+W0*(XX*XX+YY*YY)-ZZ*(U0*XX+V0*YY)
      P=P1-EV1*(XX*(U-U1)+YY*(V-V1)+ZZ*(W-W1))
! Intercell flux (i-1/2) calculation (Section 16.4.3-16.4.6)
! Note that \Delta \xi = \Delta \eta =\Delta \zeta =1
! in the computational plane (see Section 4.1).
! mass conservation flux
      RFL(IR)=U*XIX+V*XIY+W*XIZ
! If the direction of discretization is normal to
! wall boundaries, then RFL(IR) below should be set
! equal to zero at wall boundaries!
      if((LS.and.IR.eq.2).or.(RS.and.IR.eq.IE-1)) RFL(IR)=0.
! x-momentum
      XH(IR)=U*RFL(IR)+P*XIX
! y-momentum
      YH(IR)=V*RFL(IR)+P*XIY
! z-momentum
      ZH(IR)=W*RFL(IR)+P*XIZ
     end do
! Discretization of the advective flux on the cell centers
     do I=2,IE-2
      I2=I+1
      PH(I,J,K)=RFL(I2)-RFL(I)
      UH(I,J,K)=XH(I2)-XH(I)
      VH(I,J,K)=YH(I2)-YH(I)
      WH(I,J,K)=ZH(I2)-ZH(I)
     end do
! Restore viscous boundary conditions for solid boundary
     if(LS) then
      US(1,J,K)=-US(2,J,K)
      VS(1,J,K)=-VS(2,J,K)
      WS(1,J,K)=-WS(2,J,K)
      US(0,J,K)=-US(3,J,K)
      VS(0,J,K)=-VS(3,J,K)
```

```
     WS(0,J,K)=-WS(3,J,K)
    end if
    if(RS) then
     US(IE-1,J,K)=-US(IE-2,J,K)
     VS(IE-1,J,K)=-VS(IE-2,J,K)
     WS(IE-1,J,K)=-WS(IE-2,J,K)
     US(IE,J,K)  =-US(IE-3,J,K)
     VS(IE,J,K)  =-VS(IE-3,J,K)
     WS(IE,J,K)  =-WS(IE-3,J,K)
    end if
   end do
  end do
  return
  end subroutine XiFlux
```

B.2 Fifth-Order Weighted ENO Method

Here we show the basic spatial interpolation routines. The example does not include the flux splitting used (which must be application-specific) or the recovery of the physical fluxes after their reconstruction.

B.2.1 Subroutine for Fifth-Order WENO

```
      Subroutine WENO_5 (u, area, vol, dfdx, f, src, nc)

c***********************************************************************
c
c Purpose:
c 5th order weighted ENO
c Jiang and Shu 1996
c
c***********************************************************************
c start of Subroutine WENO_5

      Implicit None

      Include "../header/param.h"
      Include "../header/problem.h"

c.... call list variables

      Integer nc                      ! number of cells
      Real u(0:nv,1-nbc:nc+nbc)! conserved variables
```

```
      Real f(0:nv,0:nc)          ! flux
      Real dfdx(0:nv,1:nc)       ! div(flux)
      Real src(0:nv,1:nc)        ! source term
      Real vol(0:nc+1)           ! cell volume
      Real area(0:nc)            ! cell edge area

c.... Local variables

      Integer i                          ! counter
      Integer k                          ! counter
      Real v(0:nv,-iw+1:iw)              ! local variables
      Real g(0:nv,-iw+1:iw)              ! local fluxes
      Real gm(0:nv,-iw+1:iw)             ! negative fluxes
      Real gp(0:nv,-iw+1:iw)             ! positive fluxes
      Real f3m(0:nv,0:2)                 ! 3rd order negative fluxes
      Real f3p(0:nv,0:2)                 ! 3rd order positive fluxes
      Real is(0:nv,0:2)                  ! smoothness detector
      Real w(0:nv,0:2)                   ! weights
      Real fm(0:nv)                      ! negative fluxes
      Real fp(0:nv)                      ! positive fluxes

      Real f5m(0:nv)                     ! 5th order negative fluxes
      Real f5p(0:nv)                     ! 5th order positive fluxes

c-----------------------------------------------------------------------

c.... Loop over edges and build stencil

      Do i = 0, nc

        v(0:nv,-iw+1:iw) = u(0:nv,i-iw+1:i+iw)

        Call FLUXES (v, g, -iw+1, iw)

        Call FLUX_SPLIT (v, g, gm, gp, -iw+1, iw)

c...... Do the stencil selection for f+

        Call FLUX_3RD (gp, f3p, 0, 1)

        Call WENO_5_SENSORS (gp, is, 0, 1)

        Call WENO_5_WEIGHTS (gp, is, w)
```

```
          fp(:) = w(:,0)*f3p(:,0) + w(:,1)*f3p(:,1) + w(:,2)*f3p(:,2)

c...... Do the stencil selection for f-

          Call FLUX_3RD (gm, f3m, 1, -1)

          Call WENO_5_SENSORS (gm, is, 1, -1)

          Call WENO_5_WEIGHTS (gm, is, w)

          fm(:) = w(:,0)*f3m(:,0) + w(:,1)*f3m(:,1) + w(:,2)*f3m(:,2)

          Call FLUX_RECOVER (v, fm, fp, -iw+1, iw)

          f(:,i) = fp(:) + fm(:)

        End Do

c.... Compute flux divergence and source

        Do k = 0, nv
          dfdx(k,1:nc) = (area(1:nc)*f(k,1:nc)
     &                  - area(0:nc-1)*f(k,0:nc-1)) / vol(1:nc)
        End Do

        Call GEO_SOURCE (u, f, area, vol, src, nc)

c-----------------------------------------------------------------------

        End Subroutine WENO_5

c end of Subroutine WENO_5
c><><><><><><><><><><><><><><><><><><><><><><><><><><><><><><><><><><><><>
```

B.2.2 Subroutine for Fifth-Order WENO's Third-Order Based Fluxes

```
        Subroutine FLUX_3RD (f, f3, ic, sgn)

c***********************************************************************
c
c Purpose:
c 3rd order accurate fluxes
c
```

```
c*********************************************************************
c start of Subroutine FLUX_3RD

      Implicit None

      Include "../header/param.h"

c.... call list variables

      Integer ic                    ! center zone
      Integer sgn                   ! -/+ 1 depending on wind
      Real f(0:nv,-2:3)             ! local variables
      Real f3(0:nv,0:2)             ! fluxes

c.... Local variables

      Integer a                     ! offset
      Integer b                     ! offset

c---------------------------------------------------------------------

      a = sgn
      b = 2*sgn

c.... Compute 3rd order fluxes

      f3(:,0) = sixth*(11.D0*f(:,ic) - 7.D0*f(:,ic-a) + 2.D0*f(:,ic-b))
      f3(:,1) = sixth*(2.D0*f(:,ic+a) + 5.D0*f(:,ic) - f(:,ic-a))
      f3(:,2) = sixth*(-f(:,ic+b) + 5.D0*f(:,ic+a) + 2.D0*f(:,ic))

c---------------------------------------------------------------------

      End Subroutine FLUX_3RD

c end of Subroutine FLUX_3RD
c><><><><><><><><><><><><><><><><><><><><><><><><><><><><><><><><><><>
```

B.2.3 Subroutine Fifth-Order WENO Smoothness Sensors

```
      Subroutine WENO_5_SENSORS (f, is, ic, sgn)

c*********************************************************************
c
c Purpose:
```

```
c COmpute WENO5 sensors
c
c*****************************************************************
c start of Subroutine WENO_5_SENSORS

      Implicit None

      Include "../header/param.h"

c.... call list variables

      Integer ic                    ! center zone
      Integer sgn                   ! -/+ 1 depending on wind
      Real f(0:nv,-2:3)             ! local variables
      Real is(0:nv,0:2)             ! fluxes

c.... Local variables
      Integer a                     ! offset
      Integer b                     ! offset

c-----------------------------------------------------------------
      a = sgn
      b = 2*sgn
c.... compute WENO5 smoothness sensors

      is(:,0) = 13.D0/12.D0 * (f(:,ic-b) - two*f(:,ic-a) + f(:,ic))**2
     &         + 0.25D0 * (f(:,ic-b)-four*f(:,ic-a)+three*f(:,ic))**2
      is(:,1)  = 13.D0/12.D0 * (f(:,ic-a) - two*f(:,ic) + f(:,ic+a))**2
     &         + 0.25D0 * (f(:,ic+a) -  f(:,ic-a))**2
      is(:,2)  = 13.D0/12.D0 * (f(:,ic) - two*f(:,ic+a) + f(:,ic+b))**2
     &         + 0.25D0 * (three*f(:,ic)-four*f(:,ic+a)+f(:,ic+b))**2

c-----------------------------------------------------------------

      End Subroutine WENO_5_SENSORS

c end of Subroutine WENO_5_SENSORS
c><><><><><><><><><><><><><><><><><><><><><><><><><><><><><><><><>
```

B.2.4 Subroutine Fifth-Order WENO Weights

```
      Subroutine WENO_5_WEIGHTS (f, is, w)

c*****************************************************************
c
```

```
c Purpose:
c WENO5 weights for fluxes
c
c*******************************************************************
c start of Subroutine WENO_5_WEIGHTS

      Implicit None

      Include "../header/param.h"

c.... call list variables

      Real f(0:nv,-2:3)          ! local variables
      Real is(0:nv,0:2)          ! smoothness sensors
      Real w(0:nv,0:2)           ! fluxes

c.... Local variable

      Real w0(0:nv)              ! sum of weights

      Real del(0:np)             ! small value

c-------------------------------------------------------------------
c.... Select weights to give 5th order accuracy

      del(:) = 1.0D-06 * Max(f(:,-2)**2, f(:,-1)**2,
     &                       f(:,0)**2, f(:,1)**2, f(:,2)**2,
     &                       f(:,3)**2) + 1.0D-15

      w(:,0) = 1.D0 / (is(:,0) + del(:))**2
      w(:,1) = 6.D0 / (is(:,1) + del(:))**2
      w(:,2) = 3.D0 / (is(:,2) + del(:))**2
      w0(:) = w(:,0) + w(:,1) + w(:,2)
      w(:,0) = w(:,0) / w0(:)
      w(:,1) = w(:,1) / w0(:)
      w(:,2) = w(:,2) / w0(:)

c-------------------------------------------------------------------

      End Subroutine WENO_5_WEIGHTS

c end of Subroutine WENO_5_WEIGHTS
c><><><><><><><><><><><><><><><><><><><><><><><><><><><><><><><><><><>
```

C. Acknowledgements: Illustrations Reproduced with Permission

Figures 10.2, 10.3, 16.7, 16.8, 16.9, 16.10, 16.11, 16.12 [157]: Reprinted from Journal of Computational Physics, 146, D. Drikakis, O. Iliev, D.P. Vassileva "A nonlinear full multigrid method for the three-dimensional incompressible Navier-Stokes equations," 301-321, Copyright (1998), with permission from Elsevier.

Figures 10.6, 10.7, 10.8 [158]: Reprinted from Journal of Computational Physics, 165, D. Drikakis, O. Iliev, D.P. Vassileva "Acceleration of multigrid flow computation through dynamic adaptation of the smoothing procedure," 566-591, Copyright (2000), with permission from Elsevier.

Figures 16.12 and 16.14 [366]: Reprinted from International Journal of Heat and Fluid Flow, 23, F. Mallinger and D. Drikakis, "Instability in three-dimensional, unsteady, stenotic flows," 657-663, Copyright (2002), with permission from Elsevier.

Figures 19.3 and 19.4 [149]: Reprinted from International Journal for Numerical Methods in Fluids "Embedded turbulence model in numerical methods for hyperbolic conservation laws," 39:763-781, D. Drikakis, 2002©, John Wiley & Sons Limited. Reproduced with permission.

Figures 16.13 [367]: Reprinted from Biorheology Journal "Laminar to turbulent transition in pulsatile flow through a stenosis," F. Mallinger and D. Drikakis, Biorheology Journal, 39, 437-441, 2002. IOS Press. Reproduced with permission.

Figures 11.1, 12.2, 12.3, 12.4, 12.5, 12.6, 12.10, 12.20, 12.23, 12.24, 12.25, 12.26, 12.27, 12.28 and 12.29 [450]: Reprinted from International Journal for Numerical Methods in Fluids, 28, W. J. Rider, "Filtering Non-Solenoidal Modes in Numerical Solutions of Incompressible Flows," 789-814, 1998©, John Wiley & Sons Limited. Reproduced with permission.

Figures 18.13, 18.14, 18.15, 18.16, 18.17, 18.18, 18.19, 18.21, 18.22, 18.23, 18.24, 18.25, 18.26 and 18.27 [454]: Reprinted from Journal of Computational Physics, 141, W. J. Rider and D. B. Kothe "Reconstructing Volume Tracking," 112-152, Copyright (1998), with permission from Elsevier.

References

1. L. M. Adams, R. J. Leveque, and D. M. Young. Analysis of the SOR iteration for the 9-point Laplacian. *SIAM Journal on Numerical Analysis*, 25:1156–1180, 1988.
2. F. L. Addessio, D. E. Carroll, J. K. Dukowicz, F. H. Harlow, J. N. Johnson, B. A. Kashiwa, M. E. Maltrud, and H. M. Ruppel. CAVEAT: A computer code for fluid dynamics problems with large distortion and internal slip. Technical Report LA–10613–MS, Los Alamos National Laboratory, 1986.
3. M. J. Aftosmis, M. J. Berger, and J. E. Melton. Robust and efficient Cartesian mesh generation for component-based geometry. *AIAA Journal*, 36:952–960, 1998.
4. R. E. Alcouffe, A. Brandt, , J. E. Dendy, Jr, and J. W. Painter. The multigrid method for the diffusion equation with strongly discontinuous ceofficients. *SIAM Journal on Scientific and Statistical Computing*, 2:430–454, 1981.
5. A. A. Aldama. *Filtering Techniques for Turbulent Flow Simulation.* Springer-Verlag, Berlin, 1990.
6. S. Allmaras. Analysis of semi-implicit preconditioners for multigrid solution of the 2-D compressible Navier-Stokes equations. *AIAA Paper 95-1651-CP*, 1995.
7. A. S. Almgren, J. B. Bell, P. Colella, and L. H. Howell. An adaptive projection method for the incompressible Euler equations. In J. L. Thomas, editor, *Proceedings of the AIAA Eleventh Computational Fluid Dynamics Conference*, pages 530–539, 1993. AIAA Paper 93–3345.
8. A. S. Almgren, J. B. Bell, P. Colella, and T. Marthaler. A Cartesian grid projection method for the incompressible Euler equations in complex geometries. *SIAM Journal on Scientific Computing*, 18(5):1289–1309, 1997.
9. A. S. Almgren, J. B. Bell, and W. Y. Crutchfield. Approximate projection methods: Part 1. inviscid analysis. *SIAM Journal on Scientific Computing*, 22:1139–1159, 2000.
10. A. S. Almgren, J. B. Bell, and W. G. Szymczak. A numerical method for the incompressible Navier-Stokes equations based on an approximate projection. *SIAM Journal of Scientific Computing*, 17:358–369, 1996.
11. A. Ålund, P. Lotstedt, and M. Silen. Parallel single and multiple grid solution of industrial compressible flow problems. Number L-0-1 R166 in SAAB Military Aircraft Report. AGARD, 1996.

12. R. W. Anderson and R. B. Pember. A dynamically adaptive arbitrary Lagrangian-Eulerian method for hydrodynamics. Technical Report UCRL-JC-150660, Lawrence Livermore National Laboratory, 2002.
13. R. W. Anderson and R. B. Pember. A dynamically adaptive arbitrary Lagrangian-Eulerian method for solution of the Euler equations. Technical Report UCRL-JC-151904, Lawrence Livermore National Laboratory, 2003.
14. E. Andserson, Z. Bai, C. Bischof, J. Demmel, J. Dongarra, J. DuCroz, A. Greenbaum, S. Hammarling, A. McKenney, S. Ostrouchov, and D. Sorensen. *LAPACK Users' Guide.* SIAM, Philidelphia, 1992.
15. W. Angelis, D. Drikakis, F. Durst, and W. Khier. Numerical and experimental study of the flow over a two-dimensional car model. *Journal of Wind Engineering and Industrial Aerodynamics*, 62:57–79, 1996.
16. A. Aprovitola and F. M. Denaro. On the application of congruent upwind discretizations for large eddy simulations. *Journal of Computational Physics*, 194:329–343, 2004.
17. A. Arakawa and V. R. Lamb. Computational design of the basic dynamical processes of the UCLA general circulation model. *Methods in Computational Physics*, 17:173–265, 1977.
18. C. Arakawa, Y. Qian, and M. Samejima. Turbulent flow simulation of Francis water runner with pseudo-compressibility. In J. B. Vos, editor, *Notes on Numerical Fluid Mechanics*, volume 35, pages 259–268, 1992.
19. U. M. Ascher and L. R. Petzold. *Computer Methods for Ordinary Differential Equations and Differential-Algebraic Equations.* SIAM, Philadelphia, PA, 1996.
20. N. Ashgriz and J. Y. Poo. FLAIR - flux line-segment model for advection and interface reconstruction. *Journal of Computational Physics*, 93:449–468, 1991.
21. G. Astarita and G. Marrucci. *Principles of Non-Newtonian Fluid Mechanics.* McGraw-Hill, 1974.
22. E. Aulisa, S. Manservisi, and R. Scardovelli. A mixed markers and volume-of-fluid method for the reconstruction and advection of interfaces in two-phase and free-boundary flows. *Journal of Computational Physics*, 188(2):611–639, 2003.
23. O. Axelsson and M. Neytcheva. Scalable parallel algorithms in CFD computations. In M. Hafez and K. Oshima, editors, *in Computational Fluid Dynamics Review*, pages 837–857, 1995.
24. O. Axelsson and M. Neytcheva. Some basic facts for efficient massively parallel computation. *CWI Quarterly*, 9:9–17, 1996.
25. K. Aziz and J. D. Hellums. Numerical solution of the three-dimensional equations of motion for laminar natural convection. *Physics of Fluids*, 10(2):314–324, 1967.
26. A. J. Baker. *Finite Element Computational Fluid Mechanics.* McGraw-Hill, New York, 1983.
27. N. S. Bakhvalov. On the convergence of a relaxation method with natural constraints on the elliptic operator. *USSR Comput. Math. Phys.*, 6:101–104, 1966.

28. D. Balsara and C.-W. Shu. Monotonicity preserving weighted essentially non-oscillatory schemes with increasingly high order of accuracy. *Journal of Computational Physics*, 160(2):405–452, 2000.
29. G. Barakos and D. Drikakis. Implicit-unfactored implementation of two-equation turbulence models in compressible Navier-Stokes methods. *International Journal for Numerical Methods in Fluids*, 28(1):73–94, 1998.
30. G. Barakos and D. Drikakis. Unsteady separated flows over manoeuvering lifting surfaces. *Philosophical Transactions: Mathematical, Physical and Engineering Sciences, Royal Society London, Series A*, 358:3279–3291, 2000.
31. G. Barakos and D. Drikakis. Computational study of unsteady flows around oscillating and ramping aerofoils. *International Journal for Numerical Methods in Fluids*, 42(2):163–186, 2003.
32. P. K. Barr and W. T. Ashurst. An interface scheme for turbulent flame propagation. Technical Report SAND82-8773, Sandia National Laboratories, 1984.
33. R. Barrett and et al. *Templates for the Solution of Linear Systems: Building Blocks for Iterative Methods*. SIAM, Philidelphia, 1994.
34. T. J. Barth. Recent developments in high order K-exact reconstruction on unstructured meshes. In *31st Aerospace Sciences Meeting and Exhibit*, Reno, Nevada, 1993. AIAA–93–0668.
35. T. J. Barth. Aspects of unstructured grids and finite-volume solvers for Euler and Navier-Stokes equations, 1995. VKI/NASA/AGARD Special Course on Unstructured Grid Methods for Advection Dominated Flows AGARD Publication R-787.
36. T. J. Barth. Parallel CFD algorithms on unstructured meshes, 1995. VKI/NASA/AGARD Lecture Series on Parallel Computing. AGARD Publication R-807.
37. T. J. Barth and P. O. Frederickson. Higher order solution of the Euler equations on unstructured grids using quadratic reconstruction. In *28th Aerospace Sciences Meeting and Exhibit*, Reno, Nevada, 1990. AIAA–90–0013.
38. T. J. Barth and D. C. Jesperson. The design and application of upwind schemes on unstructured meshes. In *27th Aerospace Sciences Meeting and Exhibit*, Reno, Nevada, 1989. AIAA–89–0366.
39. G. K. Batchelor. *The Theory of Homogeneous Turbulence*. Cambridge University Press, Cambridge, 1953.
40. H. Bateman. Some recent researches on the motion of fluids. *Monthly Weather Review*, 43:163–170, 1915.
41. P. Batten, C. Lambert, and D. M. Causon. Positively conservative high-resolution convection schemes for unstructured elements. *International Journal for Numerical Methods in Engineering*, 39:1821–1838, 1996.
42. R. M. Beam and R. F. Warming. An implicit factored scheme for compressible Navier-Stokes equations. *AIAA Journal*, 16:393–402, 1978.
43. G. S. Beavers, E. M. Sparrow, and R. A. Magnusson. Experiments on hydrodynamically developing flow in rectangular ducts. *International Journal of Heat Mass and Transfer*, 13:689–693, 1970.

44. J. Bell, M. Berger, J. S. Saltzman, and M. Welcome. Three dimensional adaptive mesh refinement for hyperbolic conservation laws. *SIAM Journal on Scientific Computing*, 15:127–138, 1994.
45. J. B. Bell, P. Colella, and H. M. Glaz. A second-order projection method of the incompressible Navier-Stokes equations. *Journal of Computational Physics*, 85:257–283, 1989.
46. J. B. Bell, P. Colella, and L. Howell. An efficient second-order projection method for viscous incompressible flow. In D. Kwak, editor, *Proceedings of the AIAA Tenth Computational Fluid Dynamics Conference*, pages 360–367, 1991. AIAA Paper 91–1560.
47. J. B. Bell, C. N. Dawson, and G. R. Shubin. An unsplit, higher order Godunov method for scalar conservation laws in multiple dimensions. *Journal of Computational Physics*, 74:1–24, 1988.
48. J. B. Bell, M. S. Day, A. S. Almgren, M. J. Lijewski, and C. A. Rendelman. A parallel adaptive projection method for low Mach number flows. *International Journal for Numerical Methods in Fluids*, 40(1–2):209–216, 2002.
49. J. B. Bell, M. S. Day, C. A. Rendleman, S. E. Woosley, and M. A. Zingale. Adaptive low Mach number simulations of nuclear flame microphysics. *Journal of Computational Physics*, 195(2):677–694, 2004.
50. J. B. Bell and D. L. Marcus. A second-order projection method variable-density flows. *Journal of Computational Physics*, 101:334–348, 1992.
51. M. Ben-Artzi and J. Falcovitz. A second order Godunov-type scheme for compressible fluid dynamics. *Journal of Computational Physics*, 55:1–32, 1984.
52. R. Benodekar, A. D. Gosman, R. I. Issa, and T. Marinaccio. Industrial thermofluids analysis with STAR-CD. *Proceedings of the International Conference of Science and Engineering on Supercomputers*, pages 181–188, 1990.
53. D. J. Benson. Computational methods in Lagrangian and Eulerian hydrocodes. *Computer Methods in Applied Mechanics and Engineering*, 99:235–394, 1992.
54. E. R. Benton and G. W. Platzman. A table of solutions of the one-dimensional Burgers equation. *Quarterly of Applied Mathematics*, pages 195–212, 1972.
55. M. J. Berger and P. Colella. Local adaptive mesh refinement for shock hydrodynamics. *Journal of Computational Physics*, 82:64–84, 1989.
56. B. Bernstein, E. A. Kearsley, and L. J. Zapas. On viscoelastic flows through abrupt contractions. *Transactions Society of Rheology*, 391:7, 1963.
57. F. Bianco, G. Puppo, and G. Russo. High-order central schemes for hyperbolic systems of conservation law. *SIAM Journal of Scientific Computing*, 21(1):294–322, 1999.
58. S. J. Billet and E. F. Toro. WAF-type schemes for multidimensional hyperbolic conservation laws. *Journal of Computational Physics*, 130:1–24, 1997.
59. S. J. Billett. *Numerical Aspects of convention diffusion problems.* MSc Thesis, Cranfield University, UK, 1991.
60. R. B. Bird, W. E. Stewart, and E. N. Lightfoot. *Transport Phenomena.* John Wiley & Sons, New York, 1960.

61. Å. Björk. *Numerical Methods for Least Squares Problems.* SIAM, Philidelphia, 1996.
62. D. L. Book, editor. *Finite-Difference Techniques for Vectorized Fluid Dynamic Calculations.* Springer-Verlag, 1981.
63. J. P. Boris. A fluid transport algorithm that works. In *Computing as a Language for Physics*, pages 171–189, Trieste, International Atomic Energy Commission, 1971.
64. J. P. Boris. On large eddy simulation using subgrid turbulence models. *Lecture Notes in Physics*, 357:344–351, 1990.
65. J. P. Boris and D. L. Book. Flux-corrected transport I. SHASTA, a fluid transport algorithm that works. *Journal of Computational Physics*, 11:38–69, 1973. Reprinted in Volume 135 Number 2, pp. 172–186, August 1997.
66. J. P. Boris and D. L. Book. Flux-corrected transport III. minimal-error FCT algorithms. *Journal of Computational Physics*, 20:397–431, 1976.
67. J. P. Boris and D. L. Book. *Solution of Continuity Equations by the Method of Flux- Corrected Transport*, volume 16, pages 85–129. Academic Press, 1976.
68. J. P. Boris, D. L. Book, and K Hain. Flux-corrected transport II: Generalizations of the method. *Journal of Computational Physics*, 18:248–283, 1975.
69. J. P. Boris, F. F. Grinstein, E. S. Oran, and R. J. Kolbe. New insights into large eddy simulation. *Fluid Dynamics Research*, 10:199–228, 1992.
70. V. Boure and S. A. Orszag. Local energy flux and subgrid-scale statistics in three-dimensional turbulence. *Journal of Fluid Mechanics*, 366:1–31, 1997.
71. R. L. Bowers. Personal Communication.
72. R. L. Bowers and J. R. Wilson. *Numerical Modeling in Applied Physics and Astrophysics.* Jones and Bartlett, 1991.
73. A. Brandt. A multilevel adaptive solutions of boundary value problems. *Mathematics and Computations*, 31:333–390, 1977.
74. A. Brandt. Multigrid Techniques: 1984 Guide with Applications to Fluid Dynamics. Technical report, The Weizmann Institute of Science, Rehovot, Israel, 1984.
75. A. Brandt. Rigorous local mode analysis of multigrid: I. constant coefficient two level cycle with L_2 norm. *SIAM Journal of Numerical Analysis*, 31:1695–1730, 1994.
76. M. Breuer and D. Hänel. A dual time-stepping method for 3D viscous incompressible vortex flows. *Computers and Fluids*, 22(4-5):467–484, 1993.
77. W. L. Briggs. *A Multigrid Tutorial.* SIAM, Philidelphia, 1987.
78. W. L. Briggs, V. E. Henson, and S. F. McCormick. *A Multigrid Tutorial.* SIAM, Philidelphia, 2000.
79. W. R. Briley and H. McDonald. Solution of the multidimensional compressible Navier-Stokes equations by a generalised implict method. *Journal of Computational Physics*, 24:372–397, 1977.
80. W. R. Briley and H. McDonald. The structure and use of linearized block implicit schemes. *Journal of Computational Physics*, 34:54–73, 1980.

81. W. R. Briley, S. S. Neerarambam, and D. L. Whitfield. Implicit lower-upper/approximate-factorization schemes for incompressible flows. *Journal of Computational Physics*, 128:32–42, 1996.
82. A. N. Brooks and T. J. R. Hughes. Streamline upwind/Petrov formulations for convection dominated flows with particular emphasis on the incompressible Navier-Stokes equations. *Computer Methods in Applied Mechanics and Engineering*, 32:199–259, 1982.
83. D. L. Brown, R. Cortez, and M. L. Minion. Accurate projection methods for the incompressible Navier-Stokes equations. *Journal of Computational Physics*, 168(2):464–499, 2001.
84. D. L. Brown, R. Cortez, and M. L. Minion. Accurate projection methods for the incompressible Navier-Stokes equations. *Journal of Computational Physics*, 168(2):464–499, 2001.
85. G. L. Brown and A. Roshko. On density effects and large structure in turbulent mixing layer. *Journal of Fluid Mechanics*, 64:775–816, 1974.
86. P.N. Brown and A.C. Hindmarsh. Matrix-free methods for stiff systems of ODE's. *SIAM Journal of Numerical Analysis*, 23:610–638, 1986.
87. G. Browning and H.-O. Kreiss. Problems with different time scales for non-linear partial differential equations. *SIAM Journal of Applied Mathematics*, 42:704–718, 1982.
88. P. E. O. Buelow, S. Venkateswaran, and C. L. Merkle. Effect of grid aspect ratio on convergence. *AIAA Journal*, 32(12):2401–2408, 1994.
89. P. E. O. Buelow, S. Venkateswaran, and C. L. Merkle. Stability and convergence analyis of implicit upwind schemes. *Computers and Fluids*, 30:961–988, 2001.
90. J. M. Burgers. Mathematical examples illustrating relations occurring in the theory of turbulent fluid motion. *Trans. Roy. Neth. Acad. Sci.*, 17:1–53, 1939.
91. J. M. Burgers. Statistical problems connected with asymptotic solution of one-dimensional nonlinear diffusion equation. In M. Rosenblatt and C. van Atta, editors, *In statistical models and turbulence*, pages 41–60. Springer, 1972.
92. J. M. Burgers. *The Nonlinear Diffusion Equation.* Reidel, Holland, 1974.
93. D. Calhoun and R. J. LeVeque. Solving the advection-diffusion equation in irregular geometries. *Journal of Computational Physics*, 156:1–38, 2000.
94. C. Canuto and A. Quarteroni. *Spectral and Higher Order Methods for Partial Differential Equations.* North Holland, 1990.
95. E. J. Caramana, D. E. Burton, M. J. Shashkov, and P. P. Whalen. The construction of compatible hydrodynamics algorithms utilizing conservation of total energy. *Journal of Computational Physics*, 146:227–262, 1998.
96. C. G. Caro, J. M. Fitzgerald, and R. C. Schroter. Atheroma and arterial wall shear observations, correlation and proposal of a shear dependent mass transfer mechanism for artherogenesis. *Proceedings: Mathematical, Physical and Engineering Sciences, Royal Society London, Series B*, 17(7):109–159, 1971.
97. F. Casier, H. Deconinck, and Ch. Hirsch. A class of central bidiagonal schemes with implicit boundary conditions for the solution of Euler's equations. *AIAA Journal*, 22:1556–1563, 1983.

98. G. Cerne, S. Petelin, and I Tiselj. Coupling fo the interface tracking and the two-fluid models for the simulation of incompressible two-phase flow. *Journal of Computational Physics*, 171(2):776–804, 2001.
99. S. R. Chakravarthy and S. Osher. A new class of high-accuracy TVD schemes for hyperbolic conservation laws. *AIAA Paper 85-0363*, 1985.
100. J. L. C. Chang and D.Kwak. On the method of pseudocompressibility for numerically solving incompressible flows. *AIAA Paper 84-0252*, 1984.
101. G. Chessire and W. D. Henshaw. Composite overlapping meshes for the solution of partial differential equations. *Journal of Computational Physics*, 90:1–65, 1990.
102. D. Choi and C. L. Merkle. Application of time-iterative schemes to incompressible flow. *AIAA Journal*, 23(10):1518–1524, 1985.
103. Y.-H. Choi and C. L. Merkle. The application of preconditioning to viscous flow. *AIAA Journal*, 23(10):1518–1524, 1993.
104. A. J. Chorin. A numerical method for solving incompressible viscous flow problems. *Journal of Computational Physics*, 2:12–26, 1967.
105. A. J. Chorin. Numerical solution of the Navier-Stokes equations. *Mathematics of Computation*, 22:745–762, 1968.
106. A. J. Chorin. On the convergence of discrete approximations to the Navier-Stokes equations. *Mathematics of Computation*, 23:341–353, 1969.
107. A. J. Chorin. Random choice solutions of hyperbolic systems. *Journal of Computational Physics*, 22:517–533, 1976.
108. A. J. Chorin. Flame advection and propagation algorithms. *Journal of Computational Physics*, 35:1–11, 1980.
109. A. J. Chorin and G. Marsden. *A Mathematical Introduction to Fluid Mechanics*. Springer-Verlag, 1993.
110. R. M. Christensen. *Theory of Viscoelasticity: An Introduction*. Academic Press, New York/London, 1971.
111. T. T. Clark. A numerical study of the statistics of a two-dimensional Rayleigh-Taylor mixing layer. *Physics of Fluids*, 15(8):2413–2423, 2003.
112. B. Cockburn and C.-W. Shu. TVB Runge-Kutta local projection discontinuous Galerkin finite-element method for conservation-laws 2. General framework. *Mathematics of Computation*, 52:411–435, 1989.
113. B. Cockburn and C.-W. Shu. The Runge-Kutta discontinuous Galerkin method for conservation laws V: Multidimensional systems. *Journal of Computational Physics*, 141:199–224, 1998.
114. W. J. Coirier and K. G. Powell. An accuracy assessment of Cartesian mesh approaches for the Euler equations. *Journal of Computational Physics*, 117:121–131, 1993.
115. P. Colella. Glimm's method for gas dynamics. *SIAM Journal on Scientific and Statistical Computing*, 3:76–110, 1982.
116. P. Colella. A direct Eulerian MUSCL scheme for gas dynamics. *SIAM Journal on Scientific and Statistical Computing*, 6:104–117, 1985.

117. P. Colella. Multidimensional upwind methods for hyperbolic conservation laws. *Journal of Computational Physics*, 87:171–200, 1990.
118. P. Colella, L. F. Henderson, and E. G. Puckett. A numerical study of shock wave refractions at a gas interface. In T. Pulliam, editor, *Proceedings of the AIAA Ninth Computational Fluid Dynamics Conference*, pages 426–439, 1989. AIAA Paper 89–1973–CP.
119. P. Colella and K. Pao. A projection method for low speed flows. *Journal of Computational Physics*, 149(2).
120. P. Colella and P. Woodward. The piecewise parabolic method (PPM) for gas-dynamical simulations. *Journal of Computational Physics*, 54:174–201, 1984.
121. L. Collatz. *The Numerical Treatment of Differential Equations.* Springer, New York, 1966.
122. R. Courant. *Calculus of Variations and Supplementary Notes and Exercise.* New York University, New York, 1956.
123. R. Courant, K. O. Friedrichs, and Lewy. On the partial difference equations of mathematical physics. *IBM Journal*, 11:215–235, 1967.
124. R. Courant and D. Hilbert. *Methoden der Mathematischen Physik.* Springer, Berlin, 1968.
125. L. D. Dailey. Evaluation of multigrid acceleration for preconditioned time-accurate Navier-Stokes algorithms. *Computers and Fluids*, 25:791–811, 1996.
126. Y. M. Dakhoul and K. W. Bedford. Improved averaging method for turbulent flow simulation. Part I: Theoretical development and application to Burgers transport. *International Journal for Numerical Methods in Fluids*, 6:49–64, 1986.
127. Y. M. Dakhoul and K. W. Bedford. Improved averaging method for turbulent flow simulation. Part II: Calculations and verification. *International Journal for Numerical Methods in Fluids*, 6:65–82, 1986.
128. S. B. Dalziel, P. F. Linden, and D. L. Youngs. Self-similarity and internal structure of turbulence induced by Rayleigh-Taylor instability. *Journal of Fluid Mechanics*, 399:1–48, 1999.
129. D. L. Darmofal and P. J. Schmid. The importance of eigenvectors for local preconditioners of the Euler equations. *Journal of Computational Physics*, 127:346–362, 1996.
130. D. L. Darmofal and K. Siu. A robust locally preconidtioned multigrid algorithm for the Euler equations. *AIAA Paper 98-2428*, 1998.
131. S. F. Davis. Simplified second-order Godunov-type methods. *SIAM Journal on Scientific and Statistical Computing*, 9:445–473, 1988.
132. P. M. de Zeeuw. *Acceleration of Iterative Methods by Coarse Grid Corrections.* PhD thesis, CWI Amsterdam, 1996.
133. R. DeBar. Fundamentals of the KRAKEN code. Technical Report UCIR-760, LLNL, 1974.
134. R. Dembo, S. C. Eisenstat, and T. Steihaug. Inexact newton methods. *SIAM Journal on Numerical Analysis*, 19:400–408, 1982.

135. I. Demirdžić and M. Perić. Finite volume for prediction of fluid flow in arbitraly shaped domains with moving boundaries. *International Journal for Numerical Methods in Fluids*, 10:771–790, 1990.
136. J. W. Demmel. *Applied Numerical Linear Algebra*. SIAM, Philidelphia, 1997.
137. J.E. Dendy. Black box multigrid. *Journal of Computational Physics*, 48:366, 1982.
138. J.E. Dendy. Black box multigrid for nonsymmetric problems. *Applied Mathematics and Computing*, 13:261, 1983.
139. B. Depres and F. Lagoutiere. Contact discontinuity capturing schemes for linear advection and compressible gas dynamics and its applications. *Journal of Scientific Computing*, 16(4):479–524, 2003.
140. C. R. DeVore. Flux-corrected transport techniques for multidimensional compressible magnetohydrodynamics. *Journal of Computational Physics*, 92:142–160, 1991.
141. G. Dimonte. Nonlinear evolution of Rayleigh-Taylor and Richtmyer-Meshkov instabilities. *Physics of Plasmas*, 6:209–215, 1999.
142. J. A. Domaradzki and K.-C. Loh. The subgrid-scale estimation model in the physical space representation. *Physics of Fluids*, 11:2330–2342, 1999.
143. J. Dongarra, J. Bunch, C. Moler, and G. Stewart. *LINPACK User's Guide*. SIAM, Philadelphia, 1979.
144. J. Douglas and J. E. Gunn. A general formulation of the alternating direction implicit method - Part I - Parabolic and hyperbolic problem. *Numerische Mathematik*, 82:428–453, 1964.
145. D. Drikakis. A parallel multiblock characteristics-based method for 3D incompressible flows. *Advances in Engineering Software*, 29:111–119, 1996.
146. D. Drikakis. Study of bifurcation flow phenomena in incompressible sudden-expansion flows. *Physics of Fluids*, 9(1):76–87, 1997.
147. D. Drikakis. Numerical issues in very large eddy simulation. CD-ROM Proceedings of ECCOMAS CFD 2001, Swansea, UK, September 2001.
148. D. Drikakis. Uniformly high order methods for unsteady incompressible flows. In E. F. Toro, editor, *Godunov Methods: Theory and Applications*, pages 263–283. Kluwer Academic Publishers, 2001.
149. D. Drikakis. Embedded turbulence model in numerical methods for hyperbolic conservation laws. *International Journal for Numerical Methods in Fluids*, 39:763–781, 2002.
150. D. Drikakis. Advances in turbulent flow computations using high-resolution methods. *Progress in Aerospace Sciences*, 39:405–424, 2003.
151. D. Drikakis. Development and investigation of VTVD schemes. *unpublished*, 2003.
152. D. Drikakis and F. Durst. Investigation of flux formulae in transonic shock wave/turbulent boundary layer interaction. *International Journal for Numerical Methods in Fluids*, 18(4):385–413, 1994.
153. D. Drikakis and F. Durst. Numerical simulation of three-dimensional incompressible flows by using high order schemes. In M. Feistauer, R. Rannacher, and

K. Kozel, editors, *Proceedings of the Second Summer Conference on Numerical Modelling in Continuum Mechanics: Theory, Algorithms, and Applications*, volume Vol. II, pages 79–88, Prague, 1994. Charles University.

154. D. Drikakis and F. Durst. Parallelization of inviscid and viscous flow solvers. *International Journal of Computational Fluid Dynamics*, 3:101–121, 1994.
155. D. Drikakis and B. Geurts. *Turbulent Flow Computation.* Kluwer-Academic Publishers, 2002.
156. D. Drikakis, P. A. Govatsos, and D. E. Papantonis. A characteristic-based method for incompressible flows. *International Journal for Numerical Methods in Fluids*, 19:667–685, 1994.
157. D. Drikakis, O. P. Iliev, and D. P. Vassileva. A nonlinear multigrid method for the three-dimensional incompressible Navier-Stokes equations. *Journal of Computational Physics*, 146(1):301–321, 1998.
158. D. Drikakis, O. P. Iliev, and D. P. Vassileva. Acceleration of multigird flow computations through dynamic adaptation of the smoothing procedure. *Journal of Computational Physics*, 165:566–591, 2000.
159. D. Drikakis, L. Margolin, and P. K. Smolarkiewicz. On "spurious" eddies. 40:313–332, 2002.
160. D. Drikakis and M. Schäfer. Comparison between a pressure correction method and an artificial compressibilty/characteristic based method in parallel incompressible fluid flow computations. In S. Wagner et al., editor, *Computational Fluid Dynamics' 94*, Proceedings of the second European Computational Fluid Dynamics Conference, Stuttgart, pages 619–626. John Wiley & Sons, 1994.
161. D. Drikakis, E. Schreck, and F. Durst. A comparative study of numerical methods for incompressible and compressible flows on different parallel machines. *AIAA Paper 94-0412*, January 1994.
162. D. Drikakis, E. Schreck, and F. Durst. Performance analysis of viscous flow computations on various parallel architectures. *Journal of Fluids Engineering*, 116(4):835–841, April 1994.
163. D. Drikakis and P. K. Smolarkiewicz. On spurious vortical structures. *Journal of Computational Physics*, 172:309–325, 2001.
164. D. Drikakis and S. Tsangaris. Local solution acceleration method for the Euler and Navier-Stokes equations. *AIAA Journal*, 30:340–348, 1992.
165. D. Drikakis and S. Tsangaris. Zonal-local solution method for the accelerated solution of the turbulent Navier-Stokes equations. *AIAA Journal*, 31(10):1759–1760, 1993.
166. J. K. Dukowicz. New methods for conservative rezoning (remapping) for general quadrilateral meshes *In Rezoning Workshop 1983.* Technical Report LA–10112–C, Los Alamos National Laboratory, 1984.
167. J. K. Dukowicz. Computational efficiency of the hybrid penalty-pseudocompressibility method for incompressible flow. *Computers and Fluids*, 23(2):479–486, 1994.

168. J. K. Dukowicz and J. W. Kodis. Accurate conservative remapping (rezoning) for arbitrary Lagrangian-Eulerian computations. *SIAM Journal on Scientific and Statistical Computing*, 8:305–321, 1987.
169. P. A. Durbin and B. A. Pettersson Reif. *Statistical Theory and Modeling for Turbulent Flows*. John Wiley & Sons, Chichester, England, 2001.
170. L. J. Durlofsky, B. Engquist, and S. Osher. Triangle based adaptive stencils for the solution of hyperbolic conservation laws. *Journal of Computational Physics*, 98:64–73, 1992.
171. D. R. Durran. *Numerical Methods for Wave Equations in Geophysical Fluid Dynamics*. Springer-Verlag, 1999.
172. A. S. Dvinsky and J. K. Dukowicz. Null-space-free methods for the incompressible Navier–Stokes equations on non-staggered curvilinear grids. *Computers and Fluids*, 22:685–696, 1993.
173. W. E and J. G. Liu. Projection method I. Convergence and numerical boundary-layers. *SIAM Journal on Numerical Analysis*, 32(4):1017–1057, 1995.
174. W. E and J. G. Liu. Projection method II. Godunov-Ryabenki analysis. *SIAM Journal on Numerical Analysis*, 33(4):1597–1621, 1996.
175. W. E and J. G. Liu. Projection method III. Spatial discretization on the staggered grid. *Mathematics of Computation*, 71:27–47, 2002.
176. W. E and C. W. Shu. A numerical resolution study of high order essentially non-oscillatory schemes applied to incompressible flow. *Journal of Computational Physics*, 110:39–48, 1994.
177. A. Eberle. Characteristic flux averaging approach to the solution of Euler's equations. In *VKI Lecture Series 1987-04*, Computational Fluid Dynamics, 1987.
178. A. Eberle, A. Rizzi, and E. H. Hirschel. *Numerical Solutions of the Euler Equations for Steady Flow Problems*, volume 34. Vieweg Verlag, Wiesbaden, 1992.
179. B. Einfeldt. On Godunov-type methods for gas dynamics. *SIAM Journal on Numerical Analysis*, 25:294–318, 1988.
180. B. Engquist and S. Osher. One-sided difference approximations for nonlinear conservation laws. *Mathematics of Computation*, 36:321–351, 1981.
181. D. Enright, R. Fedkiw, J. Ferziger, and I. Mitchell. A hybrid level set method for improved interface capturing. *Journal of Computational Physics*, 183(1):83–116, 2002.
182. J. L. Ericksen. Anisotropic fluids. *Arch. Rational Mech. Anal.*, 4:231, 1960.
183. G. L. Eyink. Local energy flux and the refined similarity hypothesis. *Journal of Statistical Physics*, 78:335–351, 1995.
184. J. Farmer, L. Martinelli, and A. Jameson. A fast multigrid method for solving incompressible hydrodynamic problems with free surfaces. *AIAA Paper*, 93-076, 1993.
185. R. Fedkiw and S. Osher. *Level Set Methods and Dynamic Implicit Surfaces*. Springer-Verlag, 2002.

186. R. P. Fedorenko. A relaxation method for solving elliptic difference equations. *USSR Comput. Math. Phys.*, 1:1092–1095, 1961.
187. J. H Ferziger. Large eddy simulation: An introduction and perspective. In O. Métais and J. Ferziger, editors, *New Tools in Turbulence Modelling*, pages 29–48. Les éditions de physique, Springer, 1997.
188. C. A. J. Fletcher. Burgers' equation: A model for all reasons. In J. Noye, editor, *Numerical Solutions of Partial Differential Equations*, pages 139–225. Noth-Holland Publishing Company, 1982.
189. C. A. J. Fletcher. A comparison of finite element and finite difference solutions of the one- and two-dimensional Burgers' equations. *Journal of Computational Physics*, 51:159–188, 1983.
190. C. A. J. Fletcher. *Computational Techniques for Fluid Dynamics: Volume I.* Springer-Verlag, 1988.
191. C. A. J. Fletcher. *Computational Techniques for Fluid Dynamics: Volume II.* Springer-Verlag, 1988.
192. D. J. E. Fletcher and D. F. Fletcher. A new volume of fluid advection algorithm: The STREAM scheme. *Journal of Computational Physics*, 162(1):1–32, 2000.
193. D. J. E. Fletcher and D. F. Fletcher. A new volume of fluid advection algorithm: the defined donating region method. *International Journal for Numerical Methods in Fluids*, 35(2):151–172, 2001.
194. R. Fletcher. Conjugate gradient methods for indefinite systems. In G. A. Watson, editor, *Proceedings of the Dundee Biennal Conference on Numerical Analysis 1974*, pages 73–89, 1975.
195. R.W. Freund. A transpose-free quasi-minimal residual algorithm for non-hermitian linear systems. *SIAM Journal on Scientific Computing*, 14:470–482, 1993.
196. R.W. Freund and N. M. Nachitgal. QMR: a quasi-minimal residual method for non-Hermitian linear systems. *Mathematics of Computation*, 60:315–339, 1991.
197. U. Frisch. *Turbulence: The Legacy of A. N. Kolmogorov.* Cambridge, 1995.
198. U. Frisch and S. A. Orszag. Turbulence: Challenges for theory and experiment. *Physics Today*, page 24, January 1990.
199. J. E. Fromm. A method for reducing dispersion in convective difference schemes. *Journal of Computational Physics*, 3:176–189, 1968.
200. B. Fryxell, K. Olson, P. Ricker, F. X. Timmes, M. Zingale, D. Q. Lamb, P. MacNeice, R. Rosner, J. W. Truran, and H. Tufo. Flash: An adaptive mesh hydrodynamics code for modeling astrophysical thermonuclear flashes. *Astrophysical Journal Supplementary Series*, 131:273–334, 2000.
201. C. Fureby and F. F. Grinstein. Monotonically integrated large eddy simulation of free shear flows. *AIAA Journal*, 37:544–557, 1999.
202. C. Fureby and F. F. Grinstein. Large eddy simulation of high Reynolds number free and wall bounded flows. *Journal of Computational Physics*, 181:68–97, 2002.

203. W. Gander and D. Gruntz. Derivation of numerical methods using computer algebra. *SIAM Review*, 41:577–593, 1999.
204. P. H. Gaskell and K. C. Lau. Curvature-compensated convective transport: SMART, a new boundedness-preserving transport algorithm. *International Journal for Numerical Methods in Fluids*, 8:617–641, 1988.
205. P. H. Gaskell and K. C. Lau. Unsteady flow: A curvature compensated approache. In C. Taylor, P. Gresho, R. L. Sani, and J Haüser, editors, *Numerical Methods in Laminar and Turbulent Flow: Volume 6*, pages 51–61, 1989.
206. I. R. Gatland. A weight-watcher's guide to least-squares fitting. *Computers in Physics*, 7(3):280–285, 1993.
207. C. W. Gear. *Numerical Initial Value Problems in Ordinary Differential Equations*. Prentice-Hall, Englewood Cliffs NJ, 1971.
208. M. Germano. A proposal for a redefinition of the turbulent stresses in the filtered Navier-Stokes equations. *Physics of Fluids*, 29(7):2323–2324, 1986.
209. M. Germano, U. Piomelli, P. Moin, and W. H. Cabot. A dynamic subgrid-scale eddy viscosity model. *Physics of Fluids*, 3:1760–1766, 1991.
210. S. Ghosal. An analysis of numerical errors in large-eddy simulations of turbulence. *Journal of Computational Physics*, 125:187–206, 1996.
211. S. Ghosal. Mathematical and physical constraints on LES. *AIAA Paper*, 98-2803, June 15-18 1998.
212. S. Ghosal. Mathematical and physical constraints on large-eddy simulation of turbulence. *AIAA Journal*, 37(4):425–433, 1999.
213. S. Ghosal and P. Moin. The basic equations for the large eddy simulation of turbulent flows in complex geometry. *Journal of Computational Physics*, 118:24–37, 1995.
214. J. Glimm. Solution in the large for nonlinear hyperbolic systems of equations. *Communications on Pure and Applied Mathematics*, 18:697–715, 1965.
215. S. K. Godunov. Finite difference method for numerical computation of discontinuous solutions of the equations of fluid dynamics. *Matematicheski Sbornik*, 47:271–306, 1959.
216. S. K. Godunov. Reminiscences about difference schemes. *Journal of Computational Physics*, 153(1):6–25, 1998.
217. S. K. Godunov, A. W. Zabrodyn, and G. P. Prokopov. A computational scheme for two-dimensional nonstationary problems of gas dynamics and calculation of the flow from a shock wave approaching steady-state. *USSR Journal of Computational Mathematics and Mathematical Physics*, 1:1187–1219, 1961.
218. R. J. Goldstein and D. K. Kreid. Measurement of laminar flow development in a square duct using a laser-doppler flowmeter. *Journal of Applied Mechanics, Series E*, 89:813–818, 1967.
219. G. H. Golub and C. F. Van Loan. *Matrix Computations*. The Johns Hopkins University Press, Baltimore, 1989.
220. S. Gottleib, C.-W. Shu, and E. Tadmor. Strongly stability preserving high-order time discretization methods. *SIAM Review*, 43:89–112, 2001.

221. J. J. Gottlieb and C. P. T. Groth. Assessment of Riemann solvers for unsteady one-dimensional inviscid flows of perfect gases. *Journal of Computational Physics*, 78:437–458, 1988.
222. P. A. Govatsos and D. E. Papantonis. A characteristic based method for the calculation of three-dimensional, incompressible and steady flows in hydraulic turbomachines and installations. *International Journal for Numerical Methods in Fluids*, 34:1–30, 2000.
223. A. E. Green and R. S. Rivlin. The mechanics of materials with memory. *Arch. Rational Mech. Anal.*, 1(Part 1):65, 1957.
224. A. Greenbaum. *Iterative Methods for Solving Linear Systems.* SIAM, Philidelphia, 1997.
225. J. A. Greenough, D. T. Graves, and P. Colella. Second-order method for interface reconstruction in orthogonal coordinate systems. Technical Report UCRL-JC-132902, Lawrence Livermore National Laboratory, 1998.
226. P. M. Gresho. Incompressible fluid dynamics: Some fundamental formulation issues. *Annual Review of Fluid Mechanics*, 23:413–453, 1991.
227. P. M. Gresho. Some current CFD issues relavent to the incompressible Navier–Stokes equations. *Computer Methods in Applied Mechanics and Engineering*, 87:201–252, 1991.
228. P. M. Gresho. Some interesting issues in incompressible fluid dynamics, bothin the continuum and in numerical simulation. *Advances in Applied Mechanics*, 28:45–140, 1992.
229. P. M. Gresho and R. L. Sani. On pressure boundary conditions for the incompressible Navier–Stokes equations. *International Journal for Numerical Methods in Fluids*, 7:1111–1145, 1987.
230. P. M. Gresho and R. L. Sani. *Incompressible Flow and the Finite Element Method: Advection-diffusion and isothermal laminear flow.* Wiley, 1998.
231. P. M. Gresho and R. L. Sani. *Incompressible Flow and the Finite Element Method.* John Wiley & Sons, 1999.
232. M. Griebel. Sparse grid methods, their parallelization and their application to cfd. In J. Haeuser, editor, *Parallel Computational Fluid Dynamics 1992*, pages 839–844. Elsevier, North-Holland, 1992.
233. D. F. Griffiths and J. M. Sanz-Serna. On the scope of the method of modified equations. *SIAM Journal on Scientific and Statistical Computing*, 7:994–1008, 1986.
234. H. Guillard and A. Angello. On the behavior of upwind schemes in the low Mach number limit: Ii. Godunov type schemes. *Computers and Fluids*, 33(4):655–675, 2004.
235. H. Guillard and C. Viozat. On the behaviour of upwind schemes in the low Mach number limit. *Computers and Fluids*, 28:63–86, 1999.
236. S. N. Gurbatov, S. I. Simdyankin, E. Aurell, U. Frisch, and G. Toth. On the decay of Burgers turbulence. *Journal of Fluid Mechanics*, 344:339–374, 1997.
237. B. Gustafson. Asymptotic expansions for hyperbolic problems with different time scales. *SIAM Journal of Numerical Analysis*, 17:623–634, 1980.

238. W. Hackbush. *Multigrid Methods and Applications.* Springer Verlag, Berlin, Heidelberg, New York, Tokyo, 1985.
239. J.M. Hammersley and D.C. Handscomble. *Monte Carlo Methods.* Chapman and Hall, 1964.
240. F. H. Harlow and A. A. Amsden. A simplified MAC technique for incompressible fluid flow calculations. *Journal of Computational Physics*, 6:322–325, 1970.
241. F. H. Harlow and J. P. Shannon. Distortion of a splashing liquid drop. *Science*, 157:547–550, 1967.
242. F. H. Harlow and J. P. Shannon. The splash of a liquid drop. *Journal of Applied Physics*, 38:3855–3866, 1967.
243. F. H. Harlow, J. P. Shannon, and J. E. Welch. Liquid waves by computer. *Science*, 149:1092–1093, 1965.
244. F. H. Harlow and J. E. Welch. Numerical calculation of time-dependent viscous incompressible flow of fluid with a free surface. *Physics of Fluids*, 8:2182–2189, 1965.
245. F. H. Harlow and J. E. Welch. Numerical study of large-amplitude free-surface motions. *Physics of Fluids*, 9:842–851, 1966.
246. A. Harten. The artificial compression method for computation of shocks and contact discontinuities. I. Single conservation laws. *Communications on Pure and Applied Mathematics*, 30:611–638, 1977.
247. A. Harten. High resolution schemes for hyperbolic conservation laws. *Journal of Computational Physics*, 49:357–393, 1983. Reprinted in Volume 135 Number 2, pp. 260–278, August 1997.
248. A. Harten, B. Engquist, S. Osher, and S. Chakravarthy. Uniformly high order accurate essentially non-oscillatory schemes, III. *Journal of Computational Physics*, 71:231–303, 1987.
249. A. Harten, J. M. Hyman, and P. D. Lax. On finite difference approximations and entropy conditions for shocks. *Communications on Pure and Applied Mathematics*, 29:297–322, 1976.
250. A. Harten, P. D. Lax, and B. van Leer. On upstream differencing and Godunov-type schemes for hyperbolic conservation laws. *SIAM Review*, 25:35–61, 1983.
251. A. Harten and S. Osher. Uniformly high-order accurate nonoscillatory schemes. I. *SIAM Journal on Numerical Analysis*, 24:279–309, 1987.
252. P.-M. Hartwich, C.-H. Hsu, and C. H Liu. Vectorizable implicit algorithms for the flux difference split, three-dimensional Navier-Stokes equations. *Journal of Fluids Engineering*, 110(1):297–305, 1988.
253. W. D. Henshaw. A fourth-order accurate method for the incompressible Navier-Stokes equations on overlapping grids. *Journal of Computational Physics*, 113:13–25, 1994.
254. J. G. Heywood, R. Rannacher, and S. Turek. Artificial boundaries and flux and pressure conditions for the incompressible Navier-Stokes equations. *International Journal for Numerical Methods in Fluids*, 22:325–332, 1996.

255. G. J. Hirasaki. *A General Formulation of the Boundary Conditions on the Vector Potential in Three-Dimensional Hydrodynamics.* PhD Thesis, Rice University, Houston, Texas, 1967.
256. C. Hirsch. *Numerical Computation of Internal and External Flows: Volume 1.* Wiley-Interscience, 1988.
257. C. Hirsch. *Numerical Computation of Internal and External Flows: Volume 2.* Wiley-Interscience, 1988.
258. C. Hirsch and N. Hakimi. Preconditioning methods for time-marching Navier-Stokes solvers. In *Solution Techniques for Large-Scale CFD Problems, CMAS*, pages 333–353. Jown Wiley & Sons, 1995.
259. C. W. Hirt. Heuristic stability theory for finite difference equations. *Journal of Computational Physics*, 2:339–355, 1968.
260. C. W. Hirt, A. A. Amsden, and J. L. Cook. An arbitray Lagrangian-Eulerian computing method for all flow speeds. *Journal of Computational Physics*, 14:227–253, 1974. Reprinted in Volume 135 Number 2, pp. 203–216, August 1997.
261. C. W. Hirt and F. H. Harlow. A general corrective procedure for the numerical solution of initial-value problems. *Journal of Computational Physics*, 2:114–119, 1967.
262. C. W. Hirt and B. D. Nichols. A computational method for free surface hydrodynamics. *Journal of Pressure Vessel Technology*, 103:136–140, 1981.
263. C. W. Hirt and B. D. Nichols. Volume of fluid (VOF) method for the dynamics of free boundaries. *Journal of Computational Physics*, 39:201–225, 1981.
264. K. S. Holian, S. J. Mosso, D. A. Mandell, and R. Henninger. MESA: A 3-D computer code for armor/anti-armor applications. Technical Report LA–UR–91-569, Los Alamos National Laboratory, 1991.
265. T. Y. Hou and B. T. R. Wetton. Second-order convergence of a projection scheme for the incompressible Navier-Stokes equations with boundaries. *SIAM Journal on Numerical Analysis*, 30(3):609–629, 1993.
266. L. H. Howell. A multilevel adaptive projection method for unsteady incompressible flow. In N. D. Melson, T. A. Manteuffel, and S. F. McCormick, editors, *Proceedings of the Sixth Copper Mountain Conference on Multigrid Methods*, pages 243–257, 1993.
267. T. J. R. Hughes, W. K. Liu, and A. Brooks. Finite element analysis of incompressible viscous flows by the penalty function formulation. *Journal of Computational Physics*, 30:1–60, 1979.
268. T. J. R. Hughes, L. Mazzei, and K. E. Jansen. Scale-invariand and turblence models for large-eddy simulation. *Computing and Visualization in Science*, 3:47–59, 2000.
269. A. C. Humphrey, A. M. K. Taylor, and J. H Whitelaw. Laminar flow in a square duct of strong curvature. *Journal of Fluid Mechanics*, 83:509–527, 1977.
270. W. Hundsdorfer, B. Koren, M. van Loon, and J. G. Verwer. A positive finite difference advection scheme. *Applied Mathemathics and Computation*, 117:35–46, 1995.

271. H. T. Huynh. Accurate upwind methods for the Euler equations. *SIAM Journal on Numerical Analysis*, 32:1565–1619, 1995.
272. R. I. Issa. Solution of the implicitly discretized fluid-flow equations by operator-splitting. *Journal of Computational Physics*, 62(1):40–65, 1986.
273. A. Jameson. Solution of the Euler equations for two dimensional transonic flow by a multigrid method. *Applied Mathemathics and Computation*, 13:327–356, 1983.
274. A. Jameson. A vertex based multigrid algorithm for three-dimensional flow calculations. *ASME Symposium on Numerical Methods for Compressible Flows*, December 1986.
275. A. Jameson. Computational transonics. *Communication Pure Applied Mathematics*, 41:507–549, 1988.
276. A. Jameson and T. J. Baker. Multigrid solution of the Euler equations for aircraft configurations. *AIAA Paper 84-0093*, 1984.
277. H. Jasak, H. G. Weller, and A. D. Gosman. High resolution nvd differencing scheme for arbitrarily unstructured meshes. *International Journal for Numerical Methods in Fluids*, 31(2):431–449, 1999.
278. D. Jespersen, T. Pulliam, and P. Bunning. Recent enhancements to OVERFLOW. *AIAA Paper 97-0644*, 1997.
279. G.-S. Jiang, D. Levy, C.-T. Lin, S. Osher, and E. Tadmor. High-resolution nonoscillatory central schemes with nonstaggered grids for hyperbolic conservation laws. *SIAM Journal of Numerical Analysis*, 35(6):2147–2168, 1998.
280. G.-S. Jiang and C.-W. Shu. Efficient implementation of weighted ENO schemes. *Journal of Computational Physics*, 126:202–228, 1996.
281. G.-S. Jiang and E. Tadmor. Nonoscillatory central schemes for multidimensional hyperbolic conservation laws. *SIAM Journal on Scientific and Statistical Computing*, 19(6):1892–1917, 1998.
282. J. N. Johnson. *Classic Papers in Shock Compression Science*. Springer-Verlag, 1998.
283. Y. Kallinderis and S. Ward. Prismatic grid generation for three-dimensional complex geometries. *AIAA Journal*, 31(10):1850–1856, 1993.
284. G. E. Karniadakis and S. J. Sherwin. *Spectral/hp Methods for CFD*. Oxford University Press, 1999.
285. B. A. Kashiwa, N. T. Padial, R. M. Rauenzahn, and W. B. VanderHeyden. A cell-centered ICE method for multiphase flow simulations. Technical Report LA–UR–93-3922, Los Alamos National Laboratory, 1993.
286. C.T. Kelly. *Iterative Methods for Linear and Nonlinear Equations*. SIAM Frontiers in Applied Mathematics, Philidelphia, 1995.
287. D. S. Kershaw. Incomplete Cholesky-conjugate gradient method for iterative solution of systems of linear equations. *Journal of Computational Physics*, 26:43–65, 1978.
288. R. Kettler. Analysis and comparison of relaxation schemes in robust multigrid and preconditioned conjugate gradient methods. In W. Hackbusch and U. Trottenberg, editors, *Multigrid Methods*, pages 502–534, 1982.

289. J. Kevorkian and J. D. Cole. *Perturbation Methods in Applied Mathematics.* Springer-Verlag, New York, 1981.

290. W.-W. Kim and S. Menon. An unsteady incompressible Navier-Stokes solver for large eddy simulation of turbulent flows. *International Journal for Numerical Methods in Fluids*, 31:983–1017, 1999.

291. W.-W. Kim and P. Moin. Tackling turbulence with supercomputers. *Scientific in American*, 276:62–68, 1997.

292. S. Klainerman and A. Majda. Compressible and incompressible fluids. *Communications on Pure and Applied Mathematics*, 35:629–653, 1982.

293. R. Klein. Asymptotic analyses for atmospheric flows and the construction of asymptotically adaptive numerical methods. *Zeitschrift fur Angewandte Mathematik und Mechanik*, 80(11–12):765–777, 2000.

294. R. Klein, R. Botta, T. Schneider, C. D. Munz, S. Roller, A. Meister, L. Hoffman, and T. Sonar. Asymptotic adaptive methods for multi-scale problems in fluid mechanics. *Journal of Engineering Mathematics*, 39(1–4):261–343, 2001.

295. D. A. Knoll, L. Chacon, L. G. Margolin, and V. M. Mousseau. Nonlinearly consistent approximations for multiple time scale systems. *Journal of Computational Physics*, Submitted, 2002. LA–UR–02–0927.

296. D. A. Knoll, P.R. McHugh, and V. A. Mousseau. Newton-Krylov-Schwarz methods applied to the tokamak edge plasma fluid equations. In D. E. Keyes, Y. A. Saad, and D. G. Truhlar, editors, *Domain-Based Parallelism and Problem Decomposition Methods in Computational Science and Engineering*, Minneapolis, Minnesota, April 1994. SIAM.

297. D. A. Knoll and W. J. Rider. A multigrid preconditioned Newton-Krylov method. *SIAM Journal of Scientific Computing*, 21(2):691–710, 1999. Also see LANL Report LA–UR–97–4013.

298. V. P. Kolgan. Application of the principle of minimum values of the derivative to the construction of finite-difference schemes for calculating discontinuous solutions of gas dynamics. *Scientific Notes of TsAGI*, 3:68–77, 1972.

299. A. N. Kolmogorov. Local structure of turbulence in an incompressible fluid at very high Reynolds numbers. *Doklady AN SSSR*, 30:299–303, 1941.

300. A. N. Kolmogorov. A refinement of previous hypothesis concerning the local structure of turbulence in viscous incompressible fluid at high Reynolds number. *Journal of Fluid Mechanics*, 13:82–85, 1962.

301. D. B. Kothe, J. R. Baumgardner, S. T. Bennion, J. H. Cerutti, B. J. Daly, K. S. Holian, E. M. Kober, S. J. Mosso, J. W. Painter, R. D. Smith, and M. D. Torrey. PAGOSA: A massively-parallel, multi-material hydro-dynamics model for three-dimensional high-speed flow and high-rate deformation. Technical Report LA–UR–92–4306, Los Alamos National Laboratory, 1992.

302. D. B. Kothe, W. J. Rider, S. J. Mosso, J. S. Brock, and J. I. Hochstein. Volume tracking of interfaces having surface tension in two and three dimensions. *AIAA 96–0859*, 1996.

303. R. H Kraichnan. Lagrangian-history statistical theory for Burgers' equation. *Physics of Fluids*, 11(2):265–277, 1968.

304. R. H Kraichnan. Models of intermittency in hydrodynamic turbulence. *Physical Review Letters*, 65(5):575–578, 1990.
305. H.-O. Kreiss. Problems with different time scales for partial differential equations. *Communications on Pure and Applied Mathematics*, 35:399–440, 1980.
306. D. N. Ku, D. P. Giddens, and J. M. Downing. Pulsatile flow and atherosclerosis in the humna carotid bifurcation. *Arteriosclerosis*, 5(3):293–302, 1985.
307. H. Kuerten and B. Geurts. Compressible turbulent flow simulation with a multigrid multiblock method. In Proceedings of Copper Mountain Multigrid Conference, pages 305–315, 1993.
308. R. Kupferman. A numerical study of the axisymmetric couette-taylor problem using a fast high-resolution second-order central scheme. *SIAM Journal on Scientific and Statistical Computing*, 20(3):858–877, 1998.
309. R. Kupferman and E. Tadmor. A fast high-resolution second-order central scheme for incompressible flows. *Proc. Nat. Acad. Sci. U.S.A*, 94:4848–4852, 1997.
310. D. Kwak, J. L. C. Chang, S. P. Shanks, and S. R. Chakravarthy. A three-dimensional incompressible Navier-Stokes flow solver using primitive variables. *AIAA Journal*, 24(3):390–396, 1986.
311. B. Lafaurie, C. Nardone, R. Scardovelli, S. Zaleski, and G. Zanetti. Modelling merging and fragmentation in multiphase flows with SURFER. *Journal of Computational Physics*, 113:134–147, 1994.
312. M. Lai, J. B. Bell, and P. Colella. A projection method for combustion in the zero Mach number limit. In J. L. Thomas, editor, *Proceedings of the AIAA Eleventh Computational Fluid Dynamics Conference*, pages 776–783, 1993. AIAA Paper 93–3369.
313. M. F. Lai. *A Projection Method for Reacting Flow in the Zero Mach Number Limit.* PhD thesis, University of California at Berkeley, 1993.
314. J. D. Lambert. *Computational Methods in Ordinary Differential Equations.* Wiley, New York, 1973.
315. C. Laney. *Computational Gas Dynamics.* Cambridge University Press, 1998.
316. B. E. Launder and D. B. Spalding. *Mathematical Models of Turbulence.* Academic Press, London, 1972.
317. P. D. Lax. Weak solutions of nonlinear hyperbolic equations and their numerical computation. *Communications on Pure and Applied Mathematics*, 7:159–193, 1954.
318. P. D. Lax. Shock waves and entropy. In E. H. Zarantonello, editor, *Contributions to Nonlinear Functional Analysis*, pages 603–634. Academic Press, 1971.
319. P. D. Lax. *Hyperbolic Systems of Conservation Laws and the Mathematical Theory of Shock Waves.* SIAM, Philidelphia, 1972.
320. P. D. Lax and Liu. Solution of two-dimensional Riemann problems of gas dynamics by positive schemes. *SIAM Journal of Scientific Computing*, 19(2):219–240, 1998.

321. P. D. Lax and B. Wendroff. Systems of conservation laws. *Communications on Pure and Applied Mathematics*, 13:217–237, 1960.
322. P.D. Lax. Weak solutions of nonlinear hyperbolic equations and their numerical computation. *Communications on Pure and Applied Mathematics*, 7:159–193, 1954.
323. S. K. Lele. Compact finite difference schemes with spectral-like resolution. *Journal of Computational Physics*, 103:16–42, 1992.
324. A. Leonard. Energy cascade for large eddy simulations of turbulent fluid flows. *Advances in Geophysics*, 18A:237–248, 1974.
325. B. P. Leonard. A stable and accurate convective modeling procedure based on quadratic upstream interpolation. *Computer Methods in Applied Mechanics and Engineering*, 19:59–98, 1979.
326. B. P. Leonard. Simple high-accuracy resolution program for convective modelling of discontinuities. *International Journal for Numerical Methods in Fluids*, 8:1291–1318, 1988.
327. B. P. Leonard. The ULTIMATE convective difference scheme applied to unsteady one-dimensional advection. *Computer Methods in Applied Mechanics and Engineering*, 88:17–74, 1991.
328. B. P. Leonard. Order of accuracy of QUICK and related convection-diffusion schemes. *Applied Mathematical Modelling*, 19(11):640–653, 1995.
329. B. P. Leonard, M. K. MacVean, and A. P. Lock. The flux integral method for multidimensional convection and diffusion. *Applied Mathematical Modelling*, 19(6):333–342, 1995.
330. B. P. Leonard and S. Mokhtari. Beyond first-order upwinding the ULTRA-SHARP alternative for non-oscillatory steady-state simulation of convection. *International Journal for Numerical Methods in Engineering*, 30:729–766, 1990.
331. B. P. Leonard and H. S. Niknafs. Sharp monotonic resolution of discontinuities without clipping of narrow extrema. *Computers and Fluids*, 19:141–154, 1991.
332. A. Lerat and R. Peyret. Non centered schemes and shock propagation problems. *Computers and Fluids*, 2:35–52, 1974.
333. A. Lerat and R. Peyret. The problem of spurious oscillations in the numerical solution of the equations of gas dynamics. In *Lecture Notes in Physics*, volume 35, pages 251–256, New York, 1986. Springer-Verlag.
334. M. A. Leschziner. Turbulence modelling for separated flows with anisotropy-resolving closures. *Philosophical Transactions, Series A*, (358):3244–3277, 2000.
335. M. A. Leschziner and D. Drikakis. Turbulence modelling and turbulent-flow computation in aeronautics. *The Aeronautical Journal*, 106:349–384, 2002.
336. M. Lesieur. *Turbulence in Fluids*. Kluwer, 1997.
337. M. Lesieur and O. Métais. New trends in large-eddy simulations of turbulence. *Annual Review of Fluid Mechanics*, 28:45–82, 1996.
338. M. Lesieur, C. Staquet, P. L. Roy, and P. Comte. The mixing layer and its coherence examined from the point of view of two-dimensional turbulence. *Journal of Fluid Mechanics*, 192:511–534, 1988.
339. R. J. Leveque. *Numerical Methods for Conservation Laws*. Birkhäuser, 1990.

340. R. J. Leveque. High-resolution conservative algorithms for advection in incompressible flow. *SIAM Journal on Numerical Analysis*, 33:627–665, 1996.
341. R. J. Leveque. *Finite Volume Methods for Hyperbolic Problems.* Cambridge University Press, 2002.
342. R. J. Leveque and O. Langseth. A wave propagation method for three-dimensional hyperbolic conservation laws. *Journal of Computational Physics*, 165(1):126–166, 2000.
343. D. Levy and E. Tadmor. Non-oscillatory central schemes for the incompressible 2D Euler equations. *Mathematical Research Letters*, 4:1–20, 1997.
344. F.-S. Lien and M. A. Leschziner. Upstream monotonic interpolation for scalar transport with application to complex turbulent flows. *International Journal for Numerical Methods in Fluids*, 19:527–548, 1994.
345. D. K. Lilly. A proposed modification of the Germano subgrid-scale closure method. *Physics of Fluids*, 4:633–635, 1992.
346. D. K. Lilly. The representation of small-scale turbulence in numerical simulation experiments. In *Proceedings of the IBM Scientific Computing Symposium on Environmental Sciences.* Yorktown Heights, USA, 1994.
347. F. B. Lin and F. Sotiropoulos. Strongly-coupled multigrid method for 3D incompressible flows using near-wall turbulence closures. *Journal of Fluids Engineering*, 119:314–324, 1997.
348. S.-Y. Lin and T.-M. Wu. An adaptive multigrid finite volume scheme for incompressible Navier-Stokes equations. *International Journal for Numerical Methods in Fluids*, 17:687–710, July 1993.
349. P. F. Linden, J. M. Redondo, and D. L. Youngs. Molecular mixing in Rayleigh-Taylor instability. *Journal of Fluid Mechanics*, 265:97–124, 1994.
350. P.-L. Lions. *Mathematical Topics in Fluid Mechanics. Volume 1: Incompressible Models.* Oxford Univ. Press, New York, 1996.
351. C. Liu, Z. Liu, and S. McCormick. An efficient multigrid scheme for elliptic equations with discontinuous coefficients. *Communications in Applied Numerical Methods*, 8:621–631, 1992.
352. F. Liu. Multigrid solution of the Navier-Stokes equations with a two-equation turbulence model. In D. A. Caughey and M. M. Hafez, editors, *Frontiers of Computational Fluid Dynamics*, pages 339–359. John Wiley and Sons, 1994.
353. F. Liu and J. Shanhong. Unsteady flow calculations with a multigrid Navier-Stokes method. *AIAA Journal*, 34(10):2047–2053, 1996.
354. J.-G. Liu and C.-W. Shu. A high order discontinuous Galerkin method for 2D incompressible flows. *Journal of Computational Physics*, 160(2):577–596, 2000.
355. X.-D. Liu. A maximum principle satisfying modification of triangle based adaptive stencils for the solution of scalar hyperbolic conservation laws. *SIAM Journal on Numerical Analysis*, 30:701–716, 1993.
356. X.-D. Liu, S. Osher, and T. Chan. Weighted essentially non-oscillatory schemes. *Journal of Computational Physics*, 115:200–212, 1994.
357. X.-D. Liu and E. Tadmor. Third order central scheme for hyperbolic conservation laws. *Numerical Mathematics*, 79:397–425, 1998.

358. B. P. Leonard A. P. Lock and M. K. MacVean. Order of accuracy of QUICK and related convection-diffusion schemes. *International Journal of Numerical Methods for Heat and Fluid Flow*, 5(4):341–377, 1995.
359. F. J. Lockett. Non-linear viscoelastic bodies. *Academic Press*, 1972.
360. R. Löhner, K. Morgan, J. Peraire, and M. Vahdati. Finite element flux-corrected transport (FEM-FCT) for the Euler and Navier-Stokes equations. *International Journal for Numerical Methods in Fluids*, 7:1093–1109, 1987.
361. P. Lotstedt. Improved convergence to the steady state of the Euler equations by enhanced wave propagation. *Journal of Computational Physics*, 114:34–44, 1994.
362. R. W. MacCormack. The effect of velocity in hypervelocity impact cratering. *AIAA Paper 69-354*, 1969.
363. P. W. MacDonald. The computation of transonic flow through two-dimensional gas turbine cascades. *ASME Paper*, 71-GT-89, 1989.
364. A. Majda and S. Osher. Propogation of error into regions of smoothness for accuracte difference approximations to hyperbolic equations. *Communications on Pure and Applied Mathematics*, 30:671–705, 1977.
365. A. Majda and J. A. Sethian. The derivation and numerical solution of the equations for zero Mach number combustion. *Combustion Science and Technology*, 42:185–205, 1985.
366. F. Mallinger and D. Drikakis. Instability in three dimensional unsteady stenotic flows. *International Journal of Heat and Fluid Flow*, 23:657–663, 2002.
367. F. Mallinger and D. Drikakis. Laminar to turbulent transition in pulsatile flows through a stenosis. *Biorheology Journal*, 39:437–441, 2002.
368. L. G. Margolin and J. J. Pyun. A method for treating hourglass patterns. In C. Taylor, W. G. Habashi, and M. M. Hafez, editors, *Numerical Methods in Laminar and Turbulent Flow*, pages 149–160, 1987.
369. L. G. Margolin and W. J. Rider. A rationale for implicit turbulence modeling. *International Journal for Numerical Methods in Fluids*, 39(9):821–841, 2002.
370. L. G. Margolin and P. K. Smolarkiewicz. Antidiffusive velocities for multipass donor cell advection. *SIAM Journal on Scientific Computing*, 20:907–929, 1998.
371. L. G. Margolin, P. K. Smolarkiewicz, and Z. Sorbjan. Large eddy simulations of convective boundary layers using nonoscillatory differencing. *Physica D*, 133:390–397, 1998.
372. L.G. Margolin, P.K. Smolarkiewicz, and A.A. Wyszogrodzki. Implicit turbulence modeling for high reynolds number flows. *Journal of Fluids Engineering*, 124:862–867, 2002.
373. R. S. Marshall, J. C. Heinrich, and O. C. Zienkiewicz. Natural convection in a square enclosure by a finite element penalty function method using primitive fluid variables. *Numerical Heat Transfer*, 1:315–330, 1978.
374. W. S. Martinson and A. I. Barton. A differentiation index for partial differential-algebraic equations. *SIAM Journal on Scientific and Statistical Computing*, 21:2295–2315, 2000.

375. Y. P. Marx. Time integration schemes for the unsteady incompressible Navier-Stokes equations. *Journal of Computational Physics*, 112(4):182–209, 1994.
376. D. Mateesc, M. P. Paidoussis, and F. Belanger. Experiments on an aerofoil at high angle of incidence in longitudinal oscillations. *Journal of Sound and Vibration*, 177(2):197–205, 1994.
377. D. Mavripilis. Unstructured grid techniques. *Annual Review of Fluid Mechanics*, 29:473–514, 1997.
378. S. McCormick. *Multilevel Adaptive Methods for Partial Differential Equations.* SIAM, Philadelphia, 1989.
379. B. E. McDonald. Flux-corrected pseudospectral method for scalar hyperbolic conservation laws. *Journal of Computational Physics*, 82:413–428, 1989.
380. G. R. McGuire and J. L. Morris. A class of second order accurate methods for the solution of systems of conservation laws. *Journal of Computational Physics*, 11:531–549, 1973.
381. P. R. McHugh and J. D. Ramshaw. Damped artificial compressibility iteration scheme for implicit calculations of unsteady incompressible flow. *International Journal for Numerical Methods in Fluids*, 21:141–153, 1995.
382. C. Meneveau and J. Katz. Scale-invariant and and turblence models for large-eddy simulation. *Annual Reviews in Fluid Mechanics*, 32:1–32, 2000.
383. R. Menikoff and B. J. Plohr. The Riemann problem for fluid flow of real materials. *Reviews of Modern Physics*, 61:75–130, 1989.
384. C. L. Merkle and M. Athavale. Time accurate unsteady incompressible flow algorithms based on artificial compressibility. *AIAA Paper 87-1137*, 1987.
385. B. Merriman. Understanding the Shu-Osher conservation finite difference form. Technical Report CAM 02-58, UCLA Department of Applied Mathematics, 2002.
386. L. Mesaros and P. Roe. Multidimensional fluctuation splitting schemes based on decomposition methods. In *Proceedings AIAA 12th Computational Fluid Dynamics Conference*, pages 582–591. AIAA, 1995.
387. O. Métais and M. Lesieur. Spectral large-eddy simulations of isotropic and stably-stratified turbulence. *Journal of Fluid Mechanics*, 239:157–194, 1992.
388. A. Michalke. On the inviscid instability of the hyperbolic tangent velocity profile. *Journal of Fluid Mechanics*, 19:543–556, 1964.
389. M. L. Minion. On the stability of Godunov-projection methods for incompressible flow. *Journal of Computational Physics*, 123(2):435–449, 1996.
390. J. E. Morel, T. A. Waring, and K. Smith. A linear-discontinuous spatial differencing scheme for Sn radiative transfer calculations. *Journal of Computational Physics*, 128:445–462, 1996.
391. K. Morgan, O. Hassan, J. Peraire, and N. P. Weatherill. Unstructured mesh methods for aerospace applications. In W. G. Habashi, editor, *In Solution Techniques for Large-Scale CFD Problems.* John Wiley and Sons, 1995.
392. S. J. Mosso, B. K. Swartz, D. B. Kothe, and R. C. Ferrell. A parallel, volume tracking algorithm for unstructured meshes. In P. Schiano, editor, *Parallel Computational Fluid Dynamics '96*, Italy, 1996.

393. W. A. Mulder. A new multigrid approach to convection problems. *Journal of Computational Physics*, pages 303–323, 1989.
394. C. D. Munz, S. Roller, R. Klein, and K. J. Geratz. The extension of incompressible flow solvers to the weakly compressible regime. *Computers and Fluids*, 32(2):173–196, 2003.
395. M. Muradoglou and D. A. Caughey. Implicit multigrid solution of the preconditioned multi-dimensional Euler equations. *AIAA Paper 98-0114*, 1998.
396. H. Nessyahu and E. Tadmor. Non-oscillatory central differencing for hyperbolic conservation laws. *Journal of Computational Physics*, 87(2):408–463, 1990.
397. B. D. Nichols and C. W. Hirt. Methods for calculating multi-dimensional, transient free surface flows past bodies. Technical Report LA–UR–75–1932, Los Alamos National Laboratory, 1975. Appeared in the First International Conference on Numerical Ship Hydrodynamics, Gaithersburg, MD 10/75.
398. W. F. Noh and P. R. Woodward. SLIC (simple line interface method). In A. I. van de Vooren and P. J. Zandbergen, editors, *Lecture Notes in Physics 59*, pages 330–340, 1976.
399. W. Noll. On the continuity of fluids and solid states. *J. Rational mech. Anal.*, 4:3, 1955.
400. M. L. Norman and K.-H. A Winkler. 2-D Eulerian hydrodynamics with fluid interfaces. In K.-H. A Winkler and M. L. Norman, editors, *Astrophysical Radiation Hydrodynamics*, pages 187–221, 1986.
401. J. G. Oldroyd. On the formulation of the rheological equations of state. *Proceedings: Mathematical, Physical and Engineering Sciences, Royal Society London, Series A*, 200(523), 1950.
402. J. G. Oldroyd. Non-newtonian effects in steady motion of some idealized elastico-viscous liquids. *Proceedings: Mathematical, Physical and Engineering Sciences, Royal Society London, Series A*, 245(278), 1958.
403. E. S. Oran and J. P. Boris. *Numerical Simulation of Reactive Flow.* Cambridge University Press, 2000.
404. E. S. Oran, J. P. Boris, and D. A. Jones. Reactive-flow computations on a connection machine. In K. W. Morton, editor, *Twelfth International Conference on Numerical Methods in Fluid Dynamics*, pages 318–322, 1990.
405. J. O'Rourke. *Computational Geometry in C.* Cambridge, 1993.
406. J. M. Ortega and W. C. Rheinboldt. *Iterative Solution of Nonlinear Equations in Several Variables.* Academic Press, New York, 1970.
407. S. Osher. Riemann solvers, the entropy condition and difference approximations. *SIAM Journal on Numerical Analysis*, 21:217–235, 1984.
408. S. Osher, 1996. e-mail communication.
409. S. Osher and S. Chakravarthy. Upwind schemes and boundary conditions with applications to Euler equations in general geometries. *Journal of Computational Physics*, 50:447–481, 1983.
410. S. Osher and F. Solomon. Upwind difference schemes for hyperbolic conservation laws. *Mathematics of Computation*, 38(158):339–374, 1982.

411. D. Pan and S. R. Chakravarthy. Unified formulation for incompressible flows. *AIAA Paper 89-0122*, 1989.
412. A. G. Panaras. The spatially non-uniform convergence of the numerical solution of flows. *Journal of Computational Physics*, 82(2):429–453, 1989.
413. G. Papadakis and G. Bergeles. A locally modified second order upwind scheme for convection terms discretization. *International Journal of Numerical Methods for Heat and Fluid Flow*, 5:49–62, 1995.
414. Th. Pappou and S. Tsangaris. Development of an artificial compressibility methodology using flux vector splitting. *International Journal for Numerical Methods in Fluids*, 25:523–545, 1997.
415. S. V. Patankar. *Numerical Heat Transfer and Fluid Flow.* Hemisphere, 1980.
416. S. A. Patel and D. Drikakis. Effects of preconditioning on the accuracy and efficiency of incompressible flows. Oxford, 2004.
417. R. B. Pember and R. W. Anderson. Comparison of staggered-mesh Lagrange plus remap and cell-centered direct Eulerian Godunov schemes for Eulerian shock hydrodynamics. Technical Report UCRL-JC-139820, Lawrence Livermore National Laboratory, 2000.
418. R. B. Pember, J. B. Bell, P. Coella, W. Y. Crutchfield, and M. L. Welcome. An adaptive Cartesian grid method for unsteady compressible flow in complex geometries. *Journal of Computational Physics*, 120:278–304, 1995.
419. R. B. Pember, L. H. Howell, J. B. Bell, P. Colella, W. Y. Cruchfield, W. A. Fiveland, and J. P. Jessee. An adaptive projection method for unsteady, low-mach number combustion. *Combustion Science and Technology*, 140(1–6):123–, 1998.
420. M. Perić, R. Kessler, and G. Scheuerer. Comparison of finite volume numerical methods with staggered and colocated grids. *Computers and Fluids*, 16:389–403, 1988.
421. M. Pernice and M. D. Tocci. A multigrid-preconditioned Newton-Krylov method for the incompressilbe Navier-Stokes equations. *SIAM Journal of Scientific Computing*, 23:398–418, 2001.
422. J. S. Perry, K. G. Budge, M. K. W. Wong, and T. G. Trucano. RHALE: A 3-D MMALE code for unstructured grids. In ASME, editor, *Advanced Computational Methods for Material Modeling, AMD-Vol. 180/PVP-Vol. 268*, pages 159–174, 1993.
423. R. Peyret and T. D. Taylor. *Computational Methods for Fluid Flow.* Springer, New York, 1983.
424. N. A. Pierce and M. B. Giles. Preconditioned multigrid methods for compressible flow calculations on stretched meshes. *Journal of Computational Physics*, 136:425–445, 1997.
425. J. E. Pilliod, Jr. An analysis of piecewise linear interface reconstruction algorithms for volume-of-fluid methods. Master's thesis, University of California at Davis, 1992.

426. J. E. Pilliod, Jr. and E. G. Puckett. Second-Order Volume-of-Fluid Algorithms for Tracking Material Interfaces, Submitted to the Journal of Computational Physics.
427. U. Piomelli. Large eddy simulation: achievements and challenges. *Progress in Aerospace Sciences*, 35:335–362, 1999.
428. S. B. Pope. *Turbulent Flows.* Cambridge University Press, 2000.
429. D. H. Porter, A. Pouquet, and P. R. Woodward. Kolmogorov-like spectra in decaying three-dimensional supersonic flows. *Physics of Fluids*, 6:2133–2142, 1994.
430. W. H. Press, S. A. Teukolsky, W. T. Vetterling, and B. P. Flannery. *Numerical Recipes in Fortran.* Cambridge, 1986.
431. E. G. Puckett. A volume of fluid interface tracking algorithm with applications to computing shock wave rarefraction. In *Proceedings of the 4th International Symposium on Computational Fluid Dynamics*, pages 933–938, 1991.
432. E. G. Puckett, A. S. Almgren, J. B. Bell, D. L. Marcus, and W. J. Rider. A second-order projection method for tracking fluid interfaces in variable density incompressible flows. *Journal of Computational Physics*, 130:269–282, 1997.
433. E. G. Puckett and J. S. Saltzman. A 3D adaptive mesh refinement algorithm for multimaterial gas dynamics. *Physica D*, 60:84–93, 1992.
434. J. Qiu and C.-W. Shu. Finite difference WENO schemes with Lax-Wendroff time discretization. *SIAM Journal on Scientific Computing*, 24(6):2185–2198, 2003.
435. J. J. Quirk. A contribution to the great Riemann solver debate. *International Journal for Numerical Methods in Fluids*, 18:555–574, 1992.
436. J. D. Ramshaw and G. L. Mesina. A hybrid penalty-pseudocompressibility method for transient incompressible fluid flow. *Computers and Fluids*, 20(2):165–175, 1991.
437. J. D. Ramshaw and V. A. Mousseau. Accelerated artificial compressibility method for steady-state incompressible flow calculations. *Computers and Fluids*, 18:361–367, 1990.
438. J. D. Ramshaw and V. A. Mousseau. Damped artificial compressibility method for steady-state low-speed flow calculations. *Computers and Fluids*, 20(2):177–186, 1991.
439. J. N. Reddy. On penalty function methods in the finite element analysis of flow problems. *International Journal for Numerical Methods in Fluids*, 2:151–171, 1982.
440. J. N. Reddy. Penalty finite element analysis of 3D Navier-Stokes equations. *Computer Methods in Applied Mechanics and Engineering*, 35:87–106, 1982.
441. W. H. Reed and T. R. Hill. Triangular mesh methods for the neutron transport equation. Technical Report LA–UR–73-479, Los Alamos National Laboratory, 1973.
442. M. Reiner. The coefficient of viscous traction. *Amer. J. Math.*, 67:350, 1946.
443. O. Reynolds. On the dynamical theory of incompressible viscous fluids and the determination of the criterion. *Philosophical Transactions: Mathematical,*

Physical and Engineering Sciences, Royal Society London, Series A, 186:123, 1895.

444. C. M. Rhie and W. L. Chow. A numerical study of the turbulent flow past an isolated airfoil with trailing edge separation. *AIAA Journal*, 21:1525–1532, 1983.
445. R. D. Richtmyer. Proposed numerical method for calculation of shocks. Technical Report LA–671, Los Alamos Scientific Laboratory, 1948.
446. R. D. Richtmyer. A survey of difference methods for non-steady gas dynamics. Technical Report NCAR 63–2, NCAR, 1963.
447. R. D. Richtmyer and K. W. Morton. *Difference Methods for Initial Value Problems*. Wiley-Interscience, 1967.
448. W. J. Rider. *The Design of High-Resolution Upwind Shock-Capturing Methods*. PhD thesis, University of New Mexico, 1992.
449. W. J. Rider. Filtering nonsolenoidal modes in numerical solutions of incompressible flows. Technical Report LA–UR–94–3014, Los Alamos National Laboratory, 1994.
450. W. J. Rider. Filtering non-solenoidal modes in numerical solutions of incompressible flows. *International Journal for Numerical Methods in Fluids*, 28:789–814, 1998.
451. W. J. Rider, J. A. Greenough, and J. R. Kamm. Extrema, accuracy and monotonicity preserving methods for compressible flows. *AIAA Paper 2003–4121*, 2003.
452. W. J. Rider, D. A. Knoll, and G. L. Olson. A multigrid Newton-Krylov method for multimaterial radiation diffusion. *Journal of Computational Physics*, 152(1):164–191, 1999.
453. W. J. Rider and D. B. Kothe. Stretching and tearing interface tracking methods. *AIAA Paper 95–1717*, 1995.
454. W. J. Rider and D. B. Kothe. Reconstructing volume tracking. *Journal of Computational Physics*, 141:112–152, 1998.
455. W. J. Rider and R. B. Lowrie. The use of classical Lax-Friedrichs and the discontinuous Galerkin method. *International Journal for Numerical Methods in Fluids*, 40(3-4):479–486, 2002.
456. W. J. Rider and L. G. Margolin. Simple modifications of monotonicity limiters. *Journal of Computational Physics*, 174:473–488, 2001.
457. B. Riemann. The propagation of planar air waves of finite amplitude. *Abhandlugen der Gessellschaft der Wissenschaften zu Goettingen*, 8:43–65, 1859. Translation reprinted in [282].
458. R. S. Rivlin. The hydrodynamics of non-Newtonian fluids. *Proceedings: Mathematical, Physical and Engineering Sciences, Royal Society London, Series A*, 193:260, 1948.
459. R. S. Rivlin and J. E. Eriksen. Stress-deformation relations for isotropic materials. *J. Rational Mech. Anal.*, 4:323, 1955.
460. P. J. Roache. *Computational Fluid Dynamics*. Hermosa, Albuquerque, New Mexico, 1988.

461. P. J. Roache. Quantification of uncertainty in computational fluid dynamics. *Annual Review of Fluid Mechanics*, 29:123–160, 1997.
462. P. J. Roache. *Verification and Validation in Computational Science and Engineering*. Hermosa Publishers, 1998.
463. P. L. Roe. Approximate Riemann solvers, parameter vectors, and difference schemes. *Journal of Computational Physics*, 43:357–372, 1981. Reprinted in Volume 135 Number 2, pp. 250–258, August 1997.
464. P. L. Roe. Numerical algorithms for the linear wave equation. *Royal Aircraft Establishment, UK*, (Technical Report 81047), 1981.
465. P. L. Roe. Some contributions to the modelling of discontinuous flows. In B. Engquist, editor, *Lectures in Applied Mathematics*, volume 22, pages 163–193, 1985.
466. P. L. Roe. Some contributions to the modelling of discontinuous flows. In B. Engquist, editor, *Lectures in Applied Mathematics*, volume 22, pages 163–193, 1985.
467. P. L. Roe and M. J. Baines. Asymptotic behaviour of some non-linear schemes for linear advection. In M. Pandolfi and R. Piva, editors, *Proceedings of the 5th GAMM Conference on Numerical Methods in Fluid Mechanics*, pages 283–290. Vieweg, 1992.
468. R. S. Rogallo and P. Moin. Numerical simulation of turbulent flows. *Annual Review of Fluid Mechanics*, 16:99–137, August 1984.
469. S. E. Rogers and D. Kwak. Upwind differencing scheme for the time-accurate incompressible Navier-Stokes equations. *AIAA Journal*, 28(2):253–262, 1990.
470. S. E. Rogers, D. Kwak, and C. Kiris. Steady and unsteady solutions of the incompressible Navier-Stokes equations. *AIAA Journal*, 29(4):603–610, 1991.
471. E. L. Rubin and S. Z. Burstein. Difference methods for the inviscid and viscous equations of compressible gas. *Journal of Computational Physics*, 2:178–196, 1967.
472. M. Rudman. Volume tracking methods for interfacial flow calculations. *International Journal for Numerical Methods in Fluids*, 24:671–691, 1997.
473. U. Ruede. On the multilevel adaptive iterative method. In T. Manteuffel, editor, *in Proceedings of the 2nd Copper Mountain Conference on Iterative Methods*, Denver, 1992. Univ. of Colorado.
474. U. Ruede. Fully adaptive multigrid methods. *SIAM Journal on Numerical Analysis*, 30:230–248, 1993.
475. Y. Saad. *Iterative Methods for Sparse Linear Systems*. PWS Publishing Company, Boston, 1996.
476. Y. Saad and M.H. Schultz. GMRES: A generalized minimal residual algorithm for solving non-symetric linear systems. *SIAM Journal on Scientific and Statistical Computing*, 7:856, 1986.
477. P. Sagaut. *Large Eddy Simulation for Incompressible Flows*. Springer, 2001.
478. J. S. Saltzman. Personal Communication.

479. J. S. Saltzman. Monotonic difference schemes for the linear advection equation in two and three dimensions. Technical Report LA–UR–87–2479, Los Alamos National Laboratory, 1987.
480. R. Scardovelli and S. Zaleski. Direct numerical simulation of free-surface and interfacial flow. *Annual Reviews of Fluid Mechanics*, 31:567–603, 1999.
481. R. Scardovelli and S. Zaleski. Analytical relations connecting linear interfaces and volume fractions in rectangular grids. *Journal of Computational Physics*, 164(1):228–237, 2000.
482. R. Scardovelli and S. Zaleski. Interface reconstruction with least-square fit and split Eulerian-Lagrangian advection. *International Journal for Numerical Methods in Fluids*, 41(3):251–274, 2003.
483. M. A. Schmatz, A. Brenneis, , and A. Eberle. Verification of an implicit relaxation method for steady and unsteady viscous flow problems. In *Validation of Computational Fluid Dynamics*, AGARD Conference Proceedings, AGARD CP-437, pages 15–1–15–33, 1988.
484. H. Schmidt and U. Schumann. Coherent structure of the convective boundary layer derived from large-eddy simulation. *Journal of Fluid Mechanics*, 200:511–562, 1989.
485. T. Schneider, T. Botta, K. J. Geratz, and R. Klein. Extension of finite volume compressible flow solvers to multi-dimensional, variable density zero Mach number flows. *Journal of Computational Physics*, 155(2):248–286, 1999.
486. E. Schreck and M. Perić. Computation of fluid flow with a parallel multigrid solver. *International Journal for Numerical Methods in Fluids*, 16:303–327, 1993.
487. E. Schreck and M. Schäfer. Numerical study of bifurcation in three-dimensional sudden channel expansions. *Computers and Fluids*, 29:583–593, 2000.
488. D. Serre. Systems of conservation laws: A challenge for the XXIst century. In *Mathematics Unlimited: 2001 and Beyond*, pages 1061–1080. Springer-Verlag, 2001.
489. J. A. Sethian. *Level Set Methods: Evolving Interfaces in Geometry, Fluid Mechanics, Computer Vision and Material Science.* Cambridge University Press, 1999.
490. K. Shahbazi, M. Paraschivoiu, and J. Mostaghimi. Second order accurate volume tracking based on remapping for triangular meshes. *Journal of Computational Physics*, 188(1):100–122, 2003.
491. Ch. Sheng, L. K. Taylor, and D. L. Whitfield. An efficient multigrid acceleration for solving the 3D incompressible Navier-Stokes equations in generalized curvilinear coordinates. *AIAA Paper 93-2335*, 1993.
492. B. R. Shin. Stable numerical method applying a total variation diminishing scheme for incompressible flow. *AIAA Journal*, 41(1):49–55, 2003.
493. C.-W. Shu. Total-variation-diminishing time discretizations. *SIAM Journal on Scientific and Statistical Computing*, 9:1073–1084, 1988.

494. C.-W. Shu. A survey of strong stability preserving high order time discretizations. In D. Estep and S. Tavener, editors, *Collected Lectures on the Preservation of Stability under Discretization*, pages 51–65, 2002.
495. C.-W. Shu and S. Osher. Efficient implementation of essentially non-oscillatory shock-capturing schemes. *Journal of Computational Physics*, 77:439–471, 1988.
496. C.-W. Shu and S. Osher. Efficient implementation of essentially non-oscillatory shock-capturing schemes II. *Journal of Computational Physics*, 83:32–78, 1989.
497. W. D. Shultz. *Two-dimensional Lagrangian hydrodynamic difference equations*, volume 4, pages 1–48. Academic Press, 1964.
498. J. Smagorinsky. General circulation experiments with the primitive equations. I. the basic experiment. *Monthly Weather Review*, 101:99–164, 1963.
499. J. Smagorinsky. The beginnings of numerical weather prediction and general circulation modeling: Early recollections. *Advances in Geophysics*, 25:3–37, 1983.
500. P. K. Smolarkiewicz. The multi-dimensional Crowley advection scheme. *Monthly Weather Review*, 110:1968–1983, 1982.
501. P. K. Smolarkiewicz. A fully multidimensional positive definite advection transport algorithm with small implicit diffusion. *Journal of Computational Physics*, 54:325–362, 1984.
502. P. K. Smolarkiewicz and L. G. Margolin. Variational elliptic solver for atmospheric applications. *Applied Mathematics and Computer Science*, pages 527–551, 1994.
503. P. K. Smolarkiewicz and L. G. Margolin. MPDATA: A finite-difference solver for geophysical flows. *Journal of Computational Physics*, 140:459–480, 1998.
504. P. K. Smolarkiewicz and J. M. Prusa. Forward-in-time differencing for fluids: Simulations of geophysical turbulence. In D. Drikakis and B. J. Geurts, editors, *Turbulent Flow Computation*, pages 279–312. Kluwer Academic Publishers, 2002.
505. W. Y. Soh and J. W. Goodrich. Unsteady solution of incompressible Navier-Stokes equations. *Journal of Computational Physics*, 79:113–134, 1988.
506. P. Sonneveld. CGS, a fast Lanczos-type solver for nonsymmetric linear systems. *SIAM Journal on Scientific and Statistical Computing*, 10:36–52, 1989.
507. R. Southwell. *Relaxation Methods in Engineering Science - A Treatise in Approximate Computation.* Oxford University Press, London, 1940.
508. C. G. Speziale. Galilean invariance of subgrid-scale stress models in the large eddy simulation of turbulence. *Journal of Fluid Mechanics*, 156:55–62, 1985.
509. R. J. Spiteri and S. J. Ruuth. A class of optimal high-order strong-stability-preserving time discretization methods. *SIAM Journal of Scientific Computing*, 40(2):469–491, 2002.
510. T. W. Spriggs, J. D. Huppler, and R. D. Bird. An experimental appraisal of viscoelastic models. *Trans. Soc. Rheol.*, 10:191, 1966.

511. J. Steelant, E. Dick, and S. Pattijn. Analysis of robust multigrid methods for steady viscous low Mach number flows. *Journal of Computational Physics*, 136:603–628, 1997.

512. J. L. Steger and P. Kutler. Implicit finite-difference procedures for the computation of vortex wakes. *AIAA Journal*, 15(4):581–261, 1977.

513. P. Steinle and R. Morrow. An implicit flux-corrected transport algorithm. *Journal of Computational Physics*, 80:61–71, 1989.

514. J. M. Straka, R. B. Wilhelmson, L. J. Wicker, J. R. Anderson, and K. K. Droegemeier. Numerical solutions of a non-linear density current: A benchmark solution and comparisons. *International Journal for Numerical Methods in Fluids*, 17:1–22, 1993.

515. G. Strang. On the construction and comparison of difference schemes. *SIAM Journal on Numerical Analysis*, 5:506–517, 1968.

516. J. C. Strikwerda. Finite difference methods for the Stokes and Navier–Stokes equations. *SIAM Journal on Scientific and Statistical Computing*, 5:56–68, 1984.

517. A. Suresh. Positivity-preserving schemes in multidimensions. *SIAM Journal on Scientific Computing*, 22(4):1184–1198, 2000.

518. A. Suresh and H. T. Huynh. Accurate monotonicity-preserving schemes with Runge-Kutta time stepping. *Journal of Computational Physics*, 136:83–99, 1997.

519. M. Sussman. A second order coupled level set and volume-of-fluid method for computing growth and collapse of vapor bubbles. *Journal of Computational Physics*, 187(1):110–136, 2003.

520. M. Sussman, A. S. Almgren, J. B. Bell, P. Colella, L. H. Howell, and M. L. Welcome. An adaptive level set approach for incompressible two-phase flows. *Journal of Computational Physics*, 148(1):81–124, 1999.

521. M. Sussman and E. G. Puckett. A coupled level set and volume-of-fluid method for computing 3D and axisymmetric incompressible two-phase flows. *Journal of Computational Physics*, 162(2):301–337, 2000.

522. M. Sussman, P. Smereka, and S. Osher. A level set approach for computing solutions to incompressible two-phase flow. *Journal of Computational Physics*, 114:146–159, 1994.

523. P. K. Sweby. High-resolution schemes using flux limiters for hyperbolic conservation laws. *SIAM Journal on Numerical Analysis*, 21:995–1011, 1984.

524. P. K. Sweby. Flux limiters. In F. Angrand, A. Dervieux, J. A. Desideri, and R. Glowinski, editors, *Numerical Methods for the Euler Equations of Fluid Dynamics*, pages 48–65, 1985.

525. P. K. Sweby. High resolution TVD schemes using flux limiters. In B. Engquist, editor, *Lectures in Applied Mathematics*, volume 22, pages 289–309, 1985.

526. P. Tamamidis, G. Zhang, and D. N. Assanis. Comparison of pressure-based and artificial compressibility methods for solving 3D steady incompressible viscous flows. *Journal of Computational Physics*, 124:1–13, 1996.

527. R. I. Tanner. *Engineering Rheology*. Oxford Engineering Science Series, New York, 2000.

528. E. Y. Tau. A second-order projection method for the incompressible Navier-Stokes equations in arbitary domains. *Journal of Computational Physics*, 115:147–152, 1994.
529. R. Temam. Une methode d'approximation de la solution des equations de Navier-Stokes. *Bull. Soc. Math. France*, pages 115–152, 1968.
530. R. Temam. *Navier-Stokes Equations*. North-Holland, Amsterdam, 1984.
531. P. D. Thomas and C. K. Lombard. Geometric conservation law and its applications to flow computations on moving grids. *AIAA Journal*, 17(10):1030–1037, 1979.
532. J. Thuburn. TVD schemes, positive schemes, and the universal limiter. *Monthly Weather Review*, 125:1990–1993, 1990.
533. J. Thuburn. Multidimensional flux-limited advection schemes. *Journal of Computational Physics*, 123:74–83, 1996.
534. V. A. Titarev. *Very High order ADER schemes for Nonlinear Conservation Laws*. Msc Thesis, Dept. of Computing and Mathematics, Manchester Metropolitan Univ., United Kingdom, 2001.
535. V. A. Titarev and E. F. Toro. ADER: Arbitrary High Order Godunov Approach. *Journal of Scientific Computing*, 17(1-4):609–618, 2002.
536. A. I. Tolstykh. *High Accuracy Non-Centered Compact Difference Schemes for Fluid Dynamic Applications*. World Scientific Publishing Co. Ltd., Singapore, 1994.
537. E. F. Toro. A weighted average flux method for hyperbolic conservation laws. *Proceedings: Mathematical, Physical and Engineering Sciences, Royal Society London, Series A*, A423:401–418, 1989.
538. E. F. Toro. Riemann problems and the WAF method for solving two-dimensional shallow water equations. *Proceedings: Mathematical, Physical and Engineering Sciences, Royal Society London, Series A*, A338:43–68, 1992.
539. E. F. Toro. Viscous flux limiters. In A. Rizzi J. B. Vos and I. L. Ryhming, editors, *Notes on Numerical Fluid Dynamics*, volume 35, pages 592–600. Vieweg, 1992.
540. E. F. Toro. The weighted average flux method applied to the time-dependent Euler equations. *Proceedings: Mathematical, Physical and Engineering Sciences, Royal Society London, Series A*, A341:499–530, 1992.
541. E. F. Toro. On Glimm-related schemes for conservation laws. In *Technical Report*, number MMU-9602, Manchester, UK, 1996. Department of Mathematics and Physics, Manchester Metropolitan University.
542. E. F. Toro. On two Glimm-related schemes for hyperbolic conservation laws. In *Proceedings of the Fifth Annual Conference of the CFD Society of Canada*, pages 3.49–3.54, Canada, 1997. University of Victoria.
543. E. F. Toro. *Riemann Solvers and Numerical Methods for Fluid Dynamics: A Practical Introduction*. Springer-Verlag, 1997.
544. E. F. Toro. *Riemann Solvers and Numerical Methods for Fluid Dynamics: A Practical Introduction*. Springer-Verlag, second edition, 1999.

545. E. F. Toro. Anomalies of conservative methods: Analysis, numerical evidence and possible cures. *International Journal of Computational Fluid Dynamics*, 11(2):128–143, 2002.
546. E. F. Toro, R. C. Millington, and L. A.M Nejad. Towards very high order godunov schemes. In E. F. Toro, editor, *Godunov Methods: Theory and Applications*, pages 905–937. Kluwer Academic/Plenum Publishers, 2001.
547. E. F. Toro, M. Spruce, and W. Speares. Restoration of the contact surface in the HLL-Riemann solver. *Shock Waves*, 4:25–34, 1994.
548. L. N. Trefethen. Group-velocity in finite-difference schemes. *SIAM Review*, 24(2):113–136, 1982.
549. L. N. Trefethen and D. Bau III. *Numerical Linear Algebra.* SIAM, Philidelphia, 1997.
550. U. Trottenberg, C. Oosterlee, and A. Schuller. *Multigrid.* Academic Press, 2001.
551. C. Truesdell and W. Noll. The non-linear field of mechanics. In *Handbuch der Physik*, volume III/3. Springer, 1965.
552. E. Turkel. Preconditioned methods for solving the incompressible and low speed compressible equations. *Journal of Computational Physics*, 72:277–298, 1987.
553. E. Turkel. A review of preconditioning methods for fluid dynamics. *Applied Numerical Mathematics*, 12:257–284, 1993.
554. E. Turkel. Preconditioning techniques in computational fluid dynamics. *Annual Review of Fluid Mechanics*, 31:385–416, 1999.
555. E. Turkel, A. Fiterman, and B. van Leer. Preconditioning and the limit to incompressible flow equations. In D. A. Caughey and M. M. Hafez, editors, *In Computing the Future: Frontiers of Computational Fluid Dynamics*, pages 215–234, New York, 1994. Wiley.
556. E. Turkel, R. Radespiel, and N. Kroll. Assessment of preconditioning methods for multidimensional aerodynamics. *Computers and Fluids*, 26(6):613–634, 1997.
557. D. L. Tweedt, R. V. Chima, and E. Turkel. Preconditioning for numerical simulation of low Mach number three-dimensional viscous turbomachinery flows. *AIAA Paper 97-1828*, 1997.
558. D. Unrau and D. W. Zingg. Viscous airfoil computations using local preconditioning. *AIAA Paper 97-2027*, 1997.
559. S. O. Unverdi and G. Tryggvason. A front-tracking method for viscous, incompressible, multi-fluid flows. *Journal of Computational Physics*, 100:25–37, 1992.
560. S. V. Utyuzhnikov, A. V. Konyukhov, D. V. Rudenko, S. A. Vasilenskii, A. F. Kolesnikov, and O. Chazot. A mathematically simple turbulence closure model for attached and separated turbulent boundary layers. *AIAA Paper 2003-3866*, 34th AIAA Plasmadynamics and Lasers Conference, 2003. accepted in AIAA Journal, 2004.

561. G. D. van Albada, B. van Leer, and W. W. Roberts. A comparative study of computational methods in cosmic gas dynamics. *Astronomy and Astrophysics*, 108:76–84, 1982.
562. H. van der Ven and J. J. W. van der Vegt. Space-time discontinuous Galerkin finite element method with dynamic grid motion for inviscid compressible flows: II. Efficient flux quadrature. *Computer Methods in Applied Mechanics and Engineering*, 191(41-42):4747–4780, 2002.
563. H. A. van der Vorst. Bi-CGSTAB: A fast and smoothly converging variant of Bi-CG for the solution of nonsymmetric linear systems. *SIAM Journal on Scientific and Statistical Computing*, 13:631–644, 1992.
564. J. P. van Doormaal and G. D. Raithby. Enhancements of the SIMPLE method for predicting incompressible fluid flows. *Numerical Heat Transfer*, 7:147–163, 1984.
565. W. van Hove and A. Arts. Comparison of several finite difference schemes for time marching methods as applied to one dimensional nozzle flow. In *Von Karman Institute Report*, number VKI-TN132, 1979.
566. B. van Leer. *The Choice of Differencing Schemes for Ideal Compressible Flow.* PhD thesis, University of Leiden, 1970.
567. B. van Leer. Towards the ultimate conservative differencing scheme I: The quest of monotonicity. In *Lecture Notes in Physics*, pages 163–168, Berkeley, California, 1973.
568. B. van Leer. Towards the ultimate conservative difference scheme. II. Monotonicity and conservation combined in a second-order scheme. *Journal of Computational Physics*, 14:361–370, 1974.
569. B. van Leer. Towards the ultimate conservative difference scheme. III. Upstream-centered finite-difference schemes for ideal compressible flow. *Journal of Computational Physics*, 23:263–275, 1977.
570. B. van Leer. Towards the ultimate conservative difference scheme. IV. A new approach to numerical convection. *Journal of Computational Physics*, 23:276–299, 1977.
571. B. van Leer. Towards the ultimate conservative difference scheme. V. A second-order sequel to Godunov's method. *Journal of Computational Physics*, 32:101–136, 1979. Reprinted in Volume 135 Number 2, pp. 229–248, August 1997.
572. B. van Leer. Flux-vector splitting for the Euler equations. *Lecture Notes in Physics*, 170:507–512, 1981.
573. B. van Leer. On the relation between the upwind-differencing schemes of Godunov, Engquist-Osher and Roe. *SIAM Journal on Scientific and Statistical Computing*, 5:1–20, 1984.
574. B. van Leer. Upwind-difference methods for aerodynamic problems governed by the Euler equations. In B. Engquist, editor, *Lectures in Applied Mathematics*, volume 22, pages 327–336, 1985.
575. B. van Leer. Upwind and high-resolution methods for compressible flow: From donor cell to residual-distribution schemes. *AIAA Paper 2003–3559*, 2003.

576. B. van Leer, W. T. Lee, and P. L. Roe. Characteristic time-stepping or local preconditioning of the Euler equations. *AIAA Paper 91-1552*, 1991.
577. B. van Leer, L. Mesaros, C. H Tai, and E. Turkel. Local preconditioning in a stagnation point. In *Proceedings AIAA 12th Computational Fluid Dynamics Conference*, pages 88–101, 1995.
578. H. Vanderven. A family of large eddy simulation filters with nonuniform filter widths. *Physics of Fluids*, 7(5):1171–1172, 1995.
579. S. P. Vanka. Block implicit solution of the Navier-Stokes equations in primitive variables. *Journal of Computational Physics*, 65:138–158, 1986.
580. J. Vankan. A 2nd-order accurate pressure-correction scheme for viscous incompressible flow. *SIAM Journal on Scientific and Statistical Computing*, 7(3):870–891, 1986.
581. A. A. Varonos and G. C. Bergeles. Development and assessment of a variable-order non-oscillatory scheme for convection term discretization. *International Journal for Numerical Methods in Fluids*, 26(1):1–16, 1998.
582. A. A. Varonos and G. C. Bergeles. A multigrid method with. higher-order discretization schemes. *International Journal for Numerical Methods in Fluids*, 35(4):395–420, 2001.
583. O. V. Vasilyev, T. S. Lund, and P. Moin. A general class of commutative filters for les in complex geometries. *Journal of Computational Physics*, 146(1):82–104, 1998.
584. P. S. Vassilevski. Preconditioning nonsymmetric and indefinite finite element matrices. *Journal Numerical Linear Algebra with Applications*, 1:59–76, 1992.
585. A. E. P. Veldman. "Missing" boundary conditions? Discretize first, substitute next, and combine later. *SIAM Journal on Scientific and Statistical Computing*, 11:82–91, 1990.
586. V. Venkatakrishnan. Perspective on unstructured grid flow solvers. *AIAA Journal*, 34:533–547, 1996.
587. J. Vierendeels, K. Riemslagh, and E. Dick. Multigrid semi-implicit line-methods for viscous incompressible low mach number flows on high aspect ratio grids. *Journal of Computational Physics*, 154:310–341, 1999.
588. M. Vinokur. Conservation equations of gas dynamics in curvilinear coordinate systems. *Journal of Computational Physics*, 14:105–125, 1974.
589. H. Viviand. Conservative forms of gas dynamic equations. *Rech. Aerosp.*, 1971-1:65–68, 1974.
590. J. von Neumann and R. D. Richtmyer. A method for the numerical calculation of hydrodynamic shocks. *Journal of Applied Physics*, 21:232–237, 1950.
591. B. Vreman, B Geurts, and H. Kuerten. Discretization error dominance over subgrid terms in large eddy simulation of compressible shear layers in 2D. *International Journal for Numerical Methods in Engineering*, 10:785–790, 1994.
592. R. F. Warming and B. J. Hyett. The modified equation approach to the stability and accuracy analysis of finite-difference methods. *Journal of Computational Physics*, 14:159–179, 1974.

593. R.F. Warming, P. Kutler, and H. Lomax. Second and third order non centered difference schemes for non-linear hyperbolic equations. *AIAA Journal*, 11:189–195, 1973.
594. J. M. Weiss and W. A. Smith. Preconditioning applied to variable and constant density flows. *AIAA Journal*, 33:2050–2057, 1995.
595. P. Wesseling. *An Introduction to Multigrid Methods.* Wiley, 1992.
596. P. Wesseling. *Principles of Computational Fluid Dynamics.* Springer-Verlag, Berlin Heidelberg, 2000.
597. P. Wesseling and C. W. Oosterlee. Geometric multigrid with applications to computational fluid dynamics. *Journal of Computational and Applied Mathematics*, 128:311–334, 2001.
598. B. T. R. Wetton. Error analysis for Chorin's original fully discrete projection method and regularizations in space and time. *SIAM Journal on Numerical Analysis*, 34(5):1683–1697, 1997.
599. B. T. R. Wetton. Error analysis of pressure increment schemes. *SIAM Journal on Numerical Analysis*, 38(1):160–169, 2000.
600. J. L. White and A. B. Mentzer. Development of constitutive equations of polymeric mets and solutions. *Journal of Applied Polymer Science*, 7:1867, 1963.
601. G. B. Whitham. *Linear and Nonlinear Waves.* Wiley, 1974.
602. D. C. Wilcox. *Turbulence Modeling for CFD.* DCW Industries, 1993.
603. S. Wolfram. *The Mathematica Book: Fourth Edition.* Cambridge University Press, 1999.
604. P. R. Woodward. Piecewise-parabolic methods for astrophysical fluid dynamics. In K.-H. A Winkler and M. L. Norman, editors, *Astrophysical Radiation Hydrodynamics*, pages 245–326, 1986.
605. J. Xu. New class of iterative methods for non-self-adjoint or indefinite problems. *SIAM Journal of Numerical Analysis*, 29:303–319, 1992.
606. J. Xu. A novel two–grid method for semilinear elliptic equations. *SIAM Journal on Scientific and Statistical Computing*, 15:231–237, 1994.
607. H. Yang. An artificial compression method for ENO schemes: The slope modification method. *Journal of Computational Physics*, 89:125–160, 1990.
608. J.-Y. Yang, S.-C. Yang, Y.-N. Chen, and C.-A. Hsu. Implicit weighted ENO schemes for the three-dimensional incompressibble Navier-Stokes equations. *Journal of Computational Physics*, 146:464–487, 1998.
609. H. C. Yee. Upwind and symmetric shock capturing schemes. In *NASA Technical Memorandum 89464*, volume 31, Moffett Field, CA, 1987. NASA Ames Research Centre.
610. D. L. Youngs. Time-dependent multi-material flow with large fluid distortion. In K. W. Morton and M. J. Baines, editors, *Numerical Methods for Fluid Dynamics*, pages 273–285, 1982.
611. D. L. Youngs. An interface tracking method for a 3D Eulerian hydrodynamics code. Technical Report 44/92/35, AWRE, 1984.

612. D. L. Youngs. Three-dimensional numerical simulation of turbulent mixing by Rayleigh-Taylor instability. *Physics of Fluids A*, 3:1312–1320, 1991.
613. D. L. Youngs. Numerical simulation of mixing by Rayleigh-Taylor and Richtmyer-Meshkov instabilities. *Lasers and Particle Beams*, 12:725–750, 1994.
614. D. L. Youngs. Application of MILES to Rayleigh-Taylor and Richtmyer-Meshkov mixing. *AIAA Paper 2003-4102*, 2003.
615. S. T. Zalesak. Fully multidimensional flux-corrected transport algorithms for fluids. *Journal of Computational Physics*, 31:335–362, 1979.
616. C. Zenger. Sparse grids. In W. Hackbusch, editor, *Notes on Numerical Fluid Mechanics*, volume 31 of *Parallel Algorithms for Partial Differential Equations*, Braunschweig, 1991. Vieweg Verlag.
617. Y. Zhao and C. H. Tai. High-order characteristics-based method for incompressible flow computation on unstructured grids. *AIAA Journal*, 39(7):1280–1287, 2001.
618. X. Zheng, C. Liao, C. Liu, C. H Sung, and T. T. Huang. Multigrid computation of incompressible flows using two-equation turbulence models: Part 1 - numerical method. *Journal of Fluids Engineering.*, 119:893–899, 1997.
619. J. Zhu and M. A. Leschziner. A local oscillation-damping algorithm for higher-order convection schemes. *Computer Methods in Applied Mechanics and Engineering*, 67:355–366, 1988.
620. J. Zhu and W. Rodi. Zonal finite-volume computations for incompressible flows. *Computers and Fluids*, 20:411–420, 1991.
621. O. C. Zienkiewicz. Constrained variational principles and penalty function methods in finite element analysis. In G.A. Watson, editor, *Lecture Notes in Mathematics*, Conf. on the Numerical Solution of Differential Equations, pages 207–214, Berlin, 1974. Springer.
622. O. C. Zienkiewicz and P. N. Godbole. Viscous, incompressible flow with special reference to non-newtonian (plastic) fluids. In R.H. Gallagher et al., editor, *Finite Elements in Fluids*, volume 1, pages 25–55, London, 1975. Wiley.
623. G. Zwas. On two step lax-wendroff methods in several dimensions. *Numerische Mathematik*, 20:350–355, 1973.

Index

accuracy, 79
ACM, 447, 465
Adams-Bashforth, 316
ADER scheme, 448
adiabatic coefficient, 16
advection-diffusion equation, 75
ALE, 156, 498
amplification error, 83
anti-symmetric tensor, 8
antidiffusion, 472
arbitrary Lagrangian-Eulerian, 156, 498
artificial compressibility, 3
– artificial compressibility parameter, 180
– basic formulation, 173
– boundary conditions, 190
– convergence to incompressible limit, 174
– eigenstructure, 177
– explicit solvers, 183
– implicit solvers, 184
– local time step, 191
– preconditioning, 176
– unsteady flows, 188
artificial compression method, 447, 465

backward differentiation formula, 119
backward Euler, 116
Barth, T. J., 327
BDF, 119
Bell, Colella and Glaz, 209, 301, 309
Bell, J., 209, 309
Book, D. L., 472
Boris, J. P., 295, 298, 472
boundary conditions, 190
boundary conditions, high-order interpolation, 396
Boussinesq approximation, 23
Brandt, A., 192, 196
Briley, W. R., 184, 185
bulk pressure equation, 24
Burgers' equation, 76, 311, 532, 546
– entropy condition, 532
Burgers' turbulence, 532

cell Reynolds number, 271
centered schemes, 305, 347
CFL
– condition, 311
– condition, multidimensional, 310
– limiter, 458
– number, 85, 311, 343, 472
Chakravarthy's and Osher's TVD scheme, 414
Chakravarthy, S. R., 414
Chang, J. L. C., 174
characteristic, 153
characteristic polynomial, 82
characteristics, 18
characteristics-based
– TVD-CB scheme, 422
– TVD-SBE scheme, 422
characteristics-based scheme, 376, 384, 469, 552
– advective flux calculation, 396, 441
– flux limiter, 381
– high-order interpolation, 393
– results, 397, 421
– three-dimensional reconstruction, 389

- TVD flux, 404
- two-dimensional reconstruction, 392
- unstructured grid, 404
Choi, D., 160
Chorin, A., 71, 160, 173, 209, 495, 497
Clark's model, 544
Clark, T., 488
Cockburn, B., 467
Colella, P., 147, 209, 300, 301, 309, 310, 318
collocated divergence, 239
collocated gradient, 239
compressible Euler equations, 16
compressible flow, 3, 14, 16
compressible solvers, 147
computational geometry, 499, 500
conjugate gradient method, 126, 127
conservation form, 16, 530
conservation laws, 100
consistency, 79, 81
continuity equation, 10, 11
control volume, 10
convergence, 79
convergence rate, 323
corner transport upwind, 310
- stability, 345
Courant
- condition, 311
- number, 343
Courant, R., 72, 390
Crank-Nicholson method, 116
curvilinear coordinates, 51

dense linear algebra, 122, 330
differential-algebraic equations, 100
diffusion, 83, 90, 315
dimensional splitting, 386
direct Eulerian, 153
direct numerical simulation, 533
discontinuous Galerkin method, 467
discrete divergence, 238
discrete gradient, 238, 331
dispersion, 83, 90
dissipation, 17
dissipation independent of viscosity, 532
divergence
- cell-centered, 214
- MAC staggered, 219
- third-order cell-centered, 223
- vertex-centered, 217
divergence-free condition, 21, 22, 336
DNS, 533
double mixing layer problem, 351
Drikakis, D., 192, 201, 301, 384

eigenvalues, 18
eigenvectors, 18
Einfeldt, B., 421
energy analysis, 549
energy equation, 14
- enthalpy, 15
- kinetic, 14
- pressure, 15, 20
- temperature, 15, 23, 24
- total, 14
ENO, 209, 297, 300, 301
ENO schemes, 305, 429, 433, 448, 463, 465
- ACM, 447
- using fluxes, 436
entropy, 14, 17
entropy condition, 530
equation of motion, 20, 22, 23
equation of state, 14, 24
essentially nonoscillatory schemes, 433
Euler equations, 16
Eulerian, 9, 16, 19
explicit solvers, 183

FCT scheme, 295, 472
filter, 47, 274
- box, 48
- Gaussian, 48
- halo, 279
- projection, 256, 276
- top-hat, 48
finite element, 467
first law of thermodynamics, 14

first-order, 148
first-order upwind, 85, 90
flotsam and jetsam, 497
flux form, 16
flux limiter, 373, 404
– "Viscous" TVD limiters, 424
– approach, 373
– characteristics-based/Lax-Friedrichs scheme, 376
– construction, 374
– Godunov/Lax-Wendroff TVD scheme, 375
flux reconstruction, 156
flux-corrected transport, 295, 298, 472
flux-splitting, 157, 297
FORCE scheme, 358, 361, 374
– variants, 363
forward Euler, 80
Fourier analysis, 79, 83, 239, 343
Fourier number, 344
Fourier series, 83
Fourier stability analysis, 343
fourth-order differencing, 90
Fromm's scheme, 316, 543
fundamental derivative, 17

Gauss-Seidel iteration, 123, 124, 132, 187
– red-black, 132
Generalized Minimum Residual Algorithm, 128
generalized Riemann problem, 448
genuine nonlinearity, 18
geometric conservation law, 63
Glaz, H., 209, 309
Global limiters, 333
Godunov's method, 148, 296, 374, 467
– first-order, 309
– high-resolution, 316
– second-order, 298
Godunov's theorem, 2, 429, 538
Godunov, S. K., 2, 147, 296, 472, 538
Godunov-type methods, 3
gradient
– cell-centered, 214
– MAC staggered, 219
– third-order cell-centered, 223
– vertex-centered, 217
Green-Gauss vortices, 323
grid
– A, B, C, 210
– body-fitted, 51
– C-type, 53
– calculation of metrics, 55
– geometric conservation law, 65
– Jacobian, 56
– Jacobian for a 3-D grid, 60
– Jacobian for moving grid, 65
– MAC, 210
– moving grid, 63
– O-type, 53
– staggered, 210
– structured, 51
– unstructured, 51, 404
GRP, 448

Hancock's method, 102, 152, 310, 488
Hancock, S., 152
Harlow, F. H., 209, 219
Harten, A., 295, 301, 304, 416, 447
heat conduction, 14
Helmholz decomposition, 71, 212
heuristic, 90
high-order edges, 462
high-order interpolation, 393
high-order schemes, 429
– interpolation, 393
high-resolution methods, 1, 295
– characteristics-based scheme, 376
– circumventing Godunov's theorem, 2
– flow physics, 536
– flux limiter approach, 373
– for projection methods, 309
– properties, 301
– strict conservation form, 373
Hilbert, D., 390
Hill, T., 467
Hirt, C. W., 90, 156, 495
HLL scheme, 416
– wave speed, 420

HLLC scheme, 419
– wave speed, 420
HLLE scheme, 421
Hodge decomposition, 71, 212
Huynh, H. T., 455, 458
hybrid method, 472
hyperviscosity, 544

ideal gas, 16
idempotent, 71, 212, 239, 242
ILES, 543, 546
implicit large eddy simulation, 543, 546
implicit methods, 251
implicit solver, 184
– approximate factorization, 185
– implicit unfactored, 186
– time-linearized Euler, 184
incompressible fluid flow equations, 67
inertial, 1
interface normal, 505
interface reconstruction, 502
internal energy, 14

Jacobi iteration, 123, 133
Jacobian
-D grid, 60
– approximate, 142
– implicit approximate factorization method, 185
– inviscid flux, artificial compressibility method, 178
– Jacobian-free algorithm, 141
– Krylov iteration, 142
– Newton's and Newton-Krylov methods, 139
– Newton's methods, 129
– of the coordinates transformation, 56
Jameson, A., 192

kinetic energy, 14
kinetic energy dissipation, 531
Kolmogorov, A. N., 531
Krylov, 491
Krylov subspace methods, 123, 126, 255
Kwak, D., 174, 184

Lagrange-remap, 155, 300
Lagrangian, 9, 16
Lagrangian equations, 155
laminar flow, 1
large eddy simulation, 47, 533, 539
– subgrid model, 537
Lax equivalence theorem, 79, 80
Lax, P. D., 19, 79, 147, 416
Lax-Friedrichs, 313
Lax-Friedrichs flux, 441, 444, 467, 469
Lax-Friedrichs scheme, 348, 374
Lax-Wendroff, 158, 448, 476
Lax-Wendroff method, 100
Lax-Wendroff scheme, 353, 374, 475
– family of schemes, 357
– Richtmyer's variant, 355
– Zwas's variant, 355
Lax-Wendroff theorem, 147
least squares, 329, 506–509, 517
LeBlanc, J., 496
Legendre polynomial, 467
LES, 47, 533, 539
level sets, 526
limiter, 373, 455, 467, 538, 547
– accuracy and monotonicity-preserving, 455
– characteristics-based scheme, 376
– edge limiter, 461
– extended minmod, 456
– fourth-order, 318
– Fromm's, 317, 336
– geometric, 329
– median, 456, 547
– minbar, 456
– minbee, 409
– mineno, 319, 547
– minmod, 319, 336, 416, 428, 455, 456, 547
– monotone, 298
– slope, 456
– slope limiter, 461
– superbee, 319, 336, 404
– TVD, 336
– UNO, 319, 547

– van Albada, 319, 409, 547
– van Leer, 319, 409, 547
– viscous TVD, 424
line intersection, 500
linear multi-step methods, 113
– Adams-Bashforth, 113
– Adams-Moulton, 116
– SSP, 114
linear multistep methods, 81
linearly degenerate, 18
Liu-Tadmor third-order centered scheme, 369
local time step, 191
Los Alamos, 83
low-Mach number
– asymptotics, 20
– derivation of the incompressible equations, 20
– scaling, 20

MAC method, 209
MAC projection, 315
MacCormack's scheme, 354
Margolin, L., 475, 543, 546
marker-and-cell, 252
mass conservation, 10
mass conservation equation, 20, 23
material derivative, 9
Mathematica, 83, 90
McDonald, H., 184
McHugh, P. R., 182
MEA, 90
mean-preserving interpolation, 297
median function, 455
Merkle, C., 160, 162
method-of-lines, 158, 209
method-of-lines approach, 103
metrics, 60
MILES, 472
mixing layer, 421
model equations, 75
modified equation analysis, 90, 475, 476, 546
momentum equation, 11
monotone, 472, 538
monotone limiter, 333
Monotone schemes, 305
monotonicity, 303, 333, 455, 465
monotonicity-preserving, 455, 458
Morel, J., 467
MPDATA, 475
– third-order, 476
MPWENO schemes, 458
multigrid, 130, 192, 255, 491
– adaptive multigrid, 201
– adaptivity criterion, 202
– artificial compressibility, 192
– examples, 205
– full approximation storage, 196
– full-multigrid, full approximation storage, 193
– post-relaxation, 197
– pre-relaxation, 197
– preconditioner, 138, 255
– short-multigrid, 192
– three-grid approach, 192
– transfer operators, 198
MUSCL scheme, 396

Navier-Stokes equations
– advective form, 38
– artificial-compressibility formulation, 71
– compressible, 16
– constant density fluid, 31
– divergence form, 38
– hybrid formulation, 73
– LES form, 47
– penalty formulation, 72
– pressure-Poisson formulation, 70
– projection formulation, 71
– quadratically conserving form, 39
– Reynolds-Averaged Navier-Stokes form, 43
– rotational form, 38
– skew symmetric form, 38
– vorticity-velocity formulation, 70
– vorticity/stream-function formulation, 67

– vorticity/vector-potential formulation, 69
Nessyahu-Tadmor's second-order scheme, 364
neutron transport, 467
Newton iterations, 187
Newton's Method, 139
Newton-Krylov, 139, 140
Nichols, B. D., 495
Noh, W., 495, 497
non-Newtonian constitutive equations, 33
nondimensionalization, 39
nonlinear stability, 455
nonoscillatory, 296, 301, 547
nonoscillatory methods, 545
normalized value diagram schemes, NVD, 255
number of extrema diminishing property, 369
numerical analysis, 79
numerical linear algebra, 121
– exact cell-centered projection, 217
– exact vertex projection, 218
– order of operations, 121–124, 126, 131
numerical stability, 311

ODE, 79
operator splitting, 494–498, 510, 513–515, 525
order of accuracy, 297, 377, 393
order of operations, 121
ordinary differential equation, 79
Osborne Reynolds, 43
Osher's method, 412
Osher, S., 109, 412, 414

Patankar, S. V., 252
Peclet number, 76
penalty methods, 3
phase error, 83
piecewise linear interface calculation, 502
piecewise linear method, 147
Piecewise Parabolic Method, 147, 300, 320, 526
PLIC, 502
PLIC, Piecewise Linear Interface Calculation, 499, 500, 521
PLM schemes, 460
point location, 500
polygon operations, 500
positive schemes, 382
PPM, 462
preconditioned-compressible solvers, 147
preconditioner
– multigrid, 491
preconditioning, 160
– differential form, 169
– for compressible equations, 161
– of numerical dissipation, 167
predictor-corrector, 92, 118, 152
pressure
– loss of accuracy, 100
– thermodynamic, 12
pressure correction method, 3
pressure Poisson equation, 70, 212, 251
pressure Poisson method, 3
pressure scaling, 21
primitive variables, 19
projection, 3
– approximate, 209, 237
– cell-centered, 214
– continuous, 209
– discrete, 213
– exact, 209
– MAC, 209, 219
– marker and cell, 209
– null space, 216, 217
– stability, 212, 217
– Strikwerda, 209, 223
– third-order cell-centered, 223
– truncation error, 216
– variable density, 213, 479
– vertex-centered, 217
Puckett, E. G., 495, 498, 506

QR decomposition, 123, 330

QUICK, 254

Ramshaw, J. D., 73, 182, 183
Random Choice Method, 359
Rankine-Hugoniot, 17
Rankine-Hugoniot conditions, 532
rarefaction, 530
Rayleigh-Taylor instability, 488
reconstruction, 147
Reed, W., 467
regularization, Tikhonov, 330
remap, 155
residual smoothing, 171
Reynolds number, 1, 42, 325
– infinite limit, 531
Reynolds-Averaged Navier-Stokes, 43, 533
Rhie and Chow, 252
Richtmyer-Morton scheme, 354
Rider, W. J., 274
Riemann problem, 298
Riemann solver, 3, 148, 311, 373, 384, 406, 409, 412, 416, 419, 421, 467
– exact, 312
– Harten-Lax-van Leer, 313
– Lax-Friedrichs, 313
– Local Lax-Friedrichs, 313
– Roe, 313, 314
Roe flux, 444, 469
Roe's method, 409
Roe, P., 166, 409
Rogers, S. E., 184
Rudman, M., 497
Runge-Kutta, 316
Runge-Kutta method, 81, 92, 103, 183, 467
– classical, 107
– Heun's method, 104
– modified Euler, 104
– SSP, 105, 106
– TVD, 104, 106, 467

second law of thermodynamics, 14, 17
second-order upwind, 86
self-similar, 529
– solution, 16
self-similar solution, 16
self-similarity, 544
SHARP, 254
shock formation, 17
shock wave, 14, 17
Shu, C.-W., 109, 209, 444, 467
sign-preserving, 455
sign-preserving limiters, 333
SIMPLE, 251, 252
SIMPLER, 252, 253
singular value decomposition, SVD, 123, 330
SLIC, Simple Line Interface Calculation, 496, 497, 505, 517
Smagorinsky model, 544, 546
SMART, 254
Smolarkiewicz, P. K., 363, 475, 543, 553
sound speed, 15
sparse linear algebra, 122
specific heat, 15
stability, 79, 85
– -stability, 79, 81
– A-stability, 116
– Fourier, 216
– time integrators, 82
staggered grid, 210, 252
steepened transport method, 465
steepeners, 465
steepening, 530
Stokes equations, 32
Strang splitting, 310
stress tensor, 12, 13
– Newtonian fluid, 27
– Reynolds stress tensor, 47
Strikwerda, J., 209
strongly stability preserving, 99
subgrid models, 544
successive over-relaxation, SOR, 125
Suresh, A., 455, 458
SVD, 330
symbol, 132, 216, 240, 270
symbolic algebra, 83, 90
symmetric tensor, 8

symmetry, 312

Tadmor, E., 364, 369
Taylor series, 79, 80, 83, 90, 139
Temam, R., 72, 174
thermal conductivity, 14
Tikhonov-type method, 331
Toro, E. F., 358, 361, 406, 408, 419, 424, 448
total variation, 301
total variation diminishing, 99, 295, 304
total variation non-increasing (TVNI), 304
transformation of the equations, 57
transition, 402
truncation error, 82, 90, 216, 240, 537
turbulence, 1, 402, 531–533
turbulent flow, 42, 529
– closure, 43
– ILES computational examples, 552
– physical considerations, 529
Turkel, E., 160, 162, 163, 176
TVB, 300
TVD, 3, 99, 300, 304, 305, 373
TVD method, 295, 373
TVD Runge-Kutta, 183
TVD-CB scheme, 404, 422, 552
TVD-SBE scheme, 422

UHO scheme, 469
under-resolved, 296, 529, 540, 547
uniformly high-order scheme, 469
uniformly nonoscillatory, 547
universal limiter, 384
unsplit, 491, 498, 499, 510, 512, 513, 515, 517, 526
upstream differencing, 90
upwind schemes, 305
upwinding, 312

van Albada limiter, 396
van Leer's method, 298, 543
van Leer, B., 147, 160, 166, 295, 298, 416, 467
very high-order schemes, 429
viscous stress, 12
viscous terms, 60
– curvilinear coordinates, 62
– discretization, 62
VOF, volume-of-fluid, 488, 490, 497
volume tracking, 490
von Neumann stability analysis, 83, 343
von Neumann, J., 79, 83, 544
von Neumann-Richtmyer viscosity, 298
vortex-in-a-box, 323

weak form, 467
weak solution, 530, 538
weighted average flux method, 406
weighted average flux method, TVD version, 408
weighted least squares, 329, 338
Wendroff, B., 147
WENO schemes, 300, 439, 458, 463, 465
– ACM, 447
– fifth-order, 444
– fourth-order, 442
– third-order, 441
Wesseling, P., 192
Woodward, P. R., 147, 300, 495, 497

Yang, H., 447
Youngs' method, 495, 496, 498, 506, 508
Youngs, D. L., 495, 496, 498, 506, 508

Zalesak, S. T., 472
zero Mach number equations, 24
Zienkiewicz, C., 72

Printing: Mercedes-Druck, Berlin
Binding: Stein + Lehmann, Berlin